Encyclopedic Handbook of

INTEGRATED OPTICS

Encyclopedic Handbook of

INTEGRATED OPTICS

Edited by
KENICHI IGA *and* YASUO KOKUBUN

Taylor & Francis
Taylor & Francis Group

Boca Raton London New York

A CRC title, part of the Taylor & Francis imprint, a member of the
Taylor & Francis Group, the academic division of T&F Informa plc.

Reprint, 2018

Published in 2006 by
CRC Press
Taylor & Francis Group
6000 Broken Sound Parkway NW, Suite 300
Boca Raton, FL 33487-2742

Printed and bound in India by Bhavish Print Solutions Pvt.Ltd.,

International Standard Book Number-10: 0-8247-2425-9 (Hardcover)
International Standard Book Number-13: 978-0-8247-2425-2 (Hardcover)
Library of Congress Card Number 2005047022

Library of Congress Cataloging-in-Publication Data

Encyclopedic handbook of integrated optics / Kenuchi [i.e. Kenichi] Iga and Yasuo Kokaqbun [i.e. Kobubun], editors.
    p. cm.
  ISBN 0-8247-2425-9 (alk. paper)
  1. Integrated optics--Handbooks, manuals, etc. I. Iga, Ken'ichi, 1940- II. Kokubun, Y.

TA1660.E53 2005
621.36'93--dc 22   2005047022

Taylor & Francis Group
is the Academic Division of Informa plc.

Visit the Taylor & Francis Web site at
http://www.taylorandfrancis.com

and the CRC Press Web site at
http://www.crcpress.com

For sale in India, Pakistan, Nepal, Bhutan, Bangladesh and Sri Lanka only.

Foreword

In the past two decades optical fiber communications has totally changed the way the world communicates and transports information. It is a technological revolution that has fundamentally transformed the core of telecommunications, its basic science, and its industry. Meeting the explosive global demand for new information services including data, Internet, and broadband services, strong technical growth in this technology is now providing terabit/second capacities per optical fiber and has led to the deployment of more than a half terameter of fiber around the world. As unit costs keep coming down, fiber applications are expanding from the early long-distance undersea and terrestrial applications to shorter distance metropolitan and local applications such as fiber to the home and office. At the same time the complexity of optical systems is steadily increasing due to innovations like wavelength-division-multiplexing (WDM) and the transition from simple point-to point transmission to WDM networking.

Integrated optics was conceived in analogy to electronic integrated circuits to handle increased systems complexity and to reduce the cost of packaging and of subsystems. Its early successes included integrated guided-wave wavelength filters and WDM multiplexers, WDM laser sources such as distributed-feedback (DFB) lasers providing spectral control, and chips integrating lasers with high-speed modulators. Continuing success has matched increasing system complexity with higher levels of integration in integrated optics.

This encyclopedic handbook is about integrated optics. It is designed to serve both as a handbook for speedy look-up as well as a textbook providing deeper information. The volume takes a broad view of the subject and includes information on tunable laser sources, VCSELS, single-photon sources, micro-electromechanical systems (MEMS), and nanophotonics. I expect it will serve as a highly useful reference for students, teachers, managers, scientists and engineers, as well as the interested public.

Herwig Kogelnik

Contents

Preface

The field of integrated optics has progressed dramatically since the early stages of optical fiber communication research. Two books on integrated optics have been published by Marcel Dekker, Inc.; one was edited by Lynn Hutcheson in 1987 and the other was edited by Edmond J. Murphy in 1999. Murphy said in the preface of his book that

> Integrated optic (IO) devices have moved from an exciting, fast moving research and development phase to a more exciting, faster moving commercial deployment phase. This book focuses on technical developments while capturing a flavor of commercial success by incorporating contributions from leaders of commercially successful enterprises. Despite the maturation of the technology in some areas, there is much fertile ground for future research and development work.

The present editors have a similar feeling and have encountered more advanced concepts and technology in the last several years due to the explosive development of the Internet followed by wavelength division multiplexing (WDM) systems. Most advanced integrated optics technology supported the infrastructure of this kind of optical fiber networks. We would say that the fiber and laser turned into "fibest and lasest." There are two important issues in integrated optics: (1) a basic unchanged concept such as how the optical wave can be confined in a small volume and how the laser works, and (2) the rapidly evolving frontier of device and integration technology.

Therefore, the editors decided to include as many of the most important issues concerning integrated optics as possible so that we could cover all related fields. To accomplish this, we asked leading writers from around the world to describe the principles and advanced technologies.

This book presents the works of fifty-three contributors. The reader is thus provided with basic knowledge as well as the most advanced technology currently obtainable. The references at the end of each entry and the index are included to provide a means of cross-referencing related subjects. To our knowledge, no similar handbook dealing with integrated optics has ever been published.

We wish to sincerely thank the contributing authors for their fine work. We also wish to thank the staff at Marcel Dekker, Inc., for its professional assistance and efforts to keep the book on schedule.

Finally, we want to thank all the contributors to this book. We wish to express our deepest appreciation to Dr. Yasuharu Suematsu, Prof. Emeritus of Tokyo Institute of Technology; Dr. Herwig Kogelni; and Dr. P. K. Tien, Lucent Technology for providing us with a golden opportunity to study integrated optics.

Kenichi Iga and Yasuo Kokubun
Editors

How to Read this Handbook

In this handbook, there are both very brief articles that offer only a definition and the principle of operation, and longer articles that provide more detailed discussion of technical terms in integrated optics. These longer articles cover the basic principle, mathematical equations, the principle of operation, device fabrication and testing, and system applications.

This book can be used either as a handbook similar to a dictionary or as a text for students. Readers using this book as a handbook will find referring to the index very convenient. When using this handbook as a text, the reader will find useful the guidance contained below.

CONTENTS

Polarization Control
Stacked Planar Optics

4. Lasers and Amplifiers for Integrated Optics
Distributed Bragg Reflector (DBR) Laser
Distributed Feedback (DFB) Laser
Erbium-Doped Fiber Amplifier (EDFA)
Integrated Twin-Guide (ITG) Laser
Optical Parametric Amplifier (OPA)
Raman Amplifier
Semiconductor Optical Amplifier (SOA)
Single Photon Source
Tunable Semiconductor Lasers
Vertical Cavity Surface Emitting Laser (VCSEL)

5. Modulators and Switches
Frequency Chirping
Lithium Niobate (LN) Modulator
Modulation Limit of Semiconductor Lasers
Optical Switch
Thermo-Optic Devices
Traveling-Wave Electroabsorption Modulators

6. Applied Integrated Optics in Photonics
Micro-Electro-Mechanical Systems (MEMS)
Optical Disk Pickup
Optical Interconnect
Optical Parallel Processors
Optoelectronic Integrated Circuit (OEIC)
Planar Lightwave Circuit (PLC)
RF Spectrum Analyzer
Three R Circuit
Transmitter/Receiver
Wavelength Multiplexer/Demultiplexer (MUX/DEMUX)

Introduction

Kenichi Iga and Yasuo Kokubun

Since the laser appeared in 1960, a new field of optoelectronics has been born and the necessity for combined optical devices in order to realize various optical functions has emerged. The hybrid formation of small optical devices was considered first and this idea was called "microoptics" [1]. It is still used in actual optoelectronic systems such as optical fiber communications, optical disks, etc. Also appearing was the use of a planar optical dielectric waveguide for forming a function of optical circuitry [2]. The technical idea of integrated optics was proposed by generalizing this concept [3, 4].

Following the achievement of low loss silica fibers and the continuous operation of semiconductor lasers around 1970, optical fiber communication became a reality in the late 1970s. Various optical components based upon microoptics and guided optics supported those systems. Then, in the mid-1980s, the digital optical disk (also called a compact disc) was developed and another optoelectronic market appeared.

In 1980 and successive years, semiconductor lasers emitting 1.3 and 1.5 μm wavelengths began to be produced to match ultra-low loss silica fibers; these enabled long haul and undersea optical fiber cables to cover most of the world's basic communication infrastructures. It is further noted that the introduction of optical amplifiers, including erbium-doped optical fiber amplifiers (EDFA) and semiconductor optical amplifiers (SOA), made a drastic change in the lightwave transmission system. In the mid-1990s, an undersea cable using optical amplifiers without electronic repeaters was able to connect more than 9000 km of cable length.

Since 1999, the demand for enhanced information capacities has increased markedly due to the widespread acceptance and use of the Internet. An economically viable solution for solving this problem utilizing technology was to introduce a wavelength division multiplexing (WDM) system. In the WDM system, various kinds of optical components have been employed based upon microoptics and integrated optics technology.

In the 21st century, the necessity for large scale optical systems will increase in order to meet the expansion of the requirements for information handling. The importance of integrated optics will also increase.

This book was compiled for the purpose of covering the vast field of integrated optics, presenting its principles, history of development, design rules, representative performances, and important references. The editors expect that our readers will be scientists, engineers, managers of technology, teachers, and graduate students. The *Encyclopedic Handbook of Integrated Optics* has been organized to serve as a source of answers to the reader's questions and a source the reader may consult for information concerning the future development of integrated optics.

REFERENCES

1. T. Uchida, M. Furukawa, I. Kitano, K. Koizumi, and H. Matsumura, A light focusing fiber guide, *IEEE J. Quantum Electronics*, vol. QE-5, no. 6, pp. 331, June 1969.

2.   A. Yariv and R. C. Leite, Dielectric waveguide mode in light propagation in p-n junction, *Appl. Phys. Lett.*, 2, 25, 1963.
3.   R. Shubert and J. H. Harris, Optical surface wave on thin films and their application to integrated data processors, *IEEE. Trans. Microwave Theory & Tech.,* vol. MTT-16, p. 1048, Dec. 1968.
4.   S. E. Miller, Integrated optics : An introduction, Bell Syst. *Tech. J.,* 48(7), 2059, September 1969.

ACOUSTO-OPTICAL DEVICES

Harald Herrmann

INTRODUCTION

The diffraction of light by a sound wave is a well-known phenomenon, which was first predicted by Brillouin (1922) and experimentally discovered by Debye and Sears (1932). At present, devices based on this diffraction have found a variety of applications in bulk optics. Bragg type modulators, frequency shifters, and deflectors are commercially available and find widespread usage.

A sound wave propagating in an optically transparent solid material induces a periodic density modulation and, hence, a modulation of the refractive index for the optical wave. This gives rise to the photoelastic coupling between the optical field and the sound wave, which in practical devices is usually an ultrasonic wave in the frequency range of several 10 to several 100 MHz.

With the growing interest in integrated optical devices starting in the 1970s, intense research and development activities were carried out to realize integrated acousto-optical components, and to integrate them with other optical components to optical circuits of high functionality.

Integrated acousto-optical devices have been developed in several materials. However, it is pointed out that for most devices lithium niobate ($LiNbO_3$) is the best choice. $LiNbO_3$ offers many advantages. At present, high quality wafers are available with up to 5 in. diameter. With Ti-indiffusion and proton exchange, two well-established mature technologies exist to fabricate low-loss optical waveguides. The photoelastic properties allow an efficient acousto-optical interaction and a direct excitation of surface acoustic waves is possible in the piezoelectric substrate.

In the early years of guided-wave acousto-optics, most of the developed devices utilized Bragg-type modulators in analogy to bulk optical components. Integrated circuits have been demonstrated, for instance, for RF-signal (radio frequency-signal) processing and as space switch modules [1,2]. More recently, most R&D activities are focusing on the development of wavelength-selective devices mainly for wavelength division multiplexing (WDM) transmission systems [3].

This chapter provides an overview of guided-wave acousto-optical devices, which is organized in the following way. In the next section a very brief introduction of surface acoustic waves is given. These types of acoustical waves are used for most of the integrated acousto-optical devices. Subsequently a short review on Bragg-type acousto-optical devices in planar waveguides and some of their applications are given. An extensive part of this chapter covers acousto-optical devices based on collinear polarization conversion and some of their applications in WDM systems (last section).

SURFACE ACOUSTIC WAVES

Guided optical waves in waveguides (either planar or strip guides) are tightly confined to a region close to the substrate surface. To obtain an efficient interaction between an acoustic wave and the guided optical waves, the acoustic field must also be localized in this region; however, this does not apply to bulk acoustic waves. They extend over the whole substrate volume and only a very small part of the wave may interact with the optical fields.

To overcome this problem, surface acoustic waves (SAW) are used for integrated acousto-optical devices. These SAWs are special types of acoustical waves that are bounded close to the surface of the crystal. Their penetration depth into the material is, typically, in the range of one or a few acoustical wavelengths.

Surface acoustical waves have been predicted as a solution of the acoustical wave equation by Lord Rayleigh in 1888 [4] (therefore, SAWs are also called "Rayleigh waves"). Nowadays, SAWs find a wide range of applications, in particular, in RF-electronics, where SAW devices are used for signal processing and frequency filtering. It is beyond the scope of this chapter to discuss SAW properties in detail. A wide range of textbooks and tutorial articles exist on this topic (see e.g., References 5 and 6).

In most materials, SAWs have a complex structure. They are neither truly longitudinal nor truly transversal waves. The amplitude and the direction of the mechanical displacement of a point in a crystal are strongly dependent on the distance from the substrate surface. Moreover, in piezoelectric materials (such as $LiNbO_3$, ZnO, etc.) the relative mechanical displacements induce local electric fields resulting in a coupling between the mechanical motion and an electromagnetic field. In a quasi-static approximation, SAWs are usually described by a vectorial field with four components: three belonging to the mechanical displacements u_i along the coordinate axes and the fourth is the electrostatic potential Φ.

An example of a SAW is shown in Figure 1. The left side of the figure qualitatively illustrates the structure of the SAW propagating in X-cut $LiNbO_3$ along the Y-direction. The elongation amplitudes u_i are strongly enhanced for a better visualization. The diagram on the right side of the Figure 1 quantitatively illustrates the amplitudes of the elongation and the potential as function of the depth coordinate.

Excitation of SAWs is usually accomplished by applying a sinusoidal RF-signal to inter-digital transducer electrodes. In piezoelectric substrates, such electrodes are directly deposited on the crystal surface, whereas for nonpiezoelectric substrates a piezoelectric interfacial layer is required. The simplest transducer structure consists of a double comb-like structure with interleaving finger pairs as shown in Figure 2. The period must match the SAW wavelength Λ. The efficiency and the bandwidth for the excitation are mainly determined by the transducer design. The optimum structure strongly depends on the specific application. Sophisticated structures such as unidirectional transducers, chirped and curved transducers have been developed; some of these examples are discussed in Reference 7.

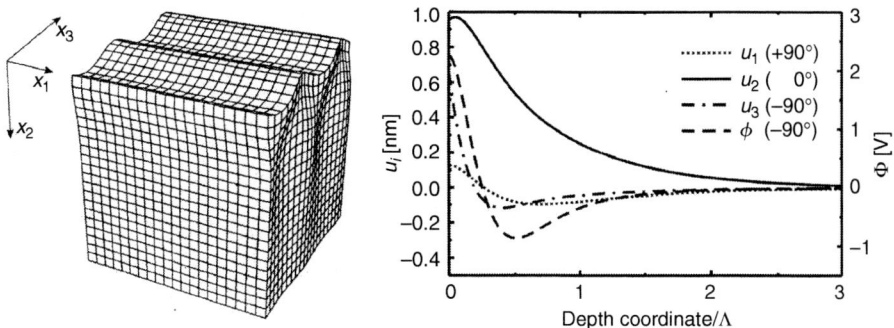

Figure 1 SAW on X-cut LiNbO$_3$ substrate propagating along the Y-direction. On the left side the SAW is shown with strongly enhanced amplitudes to illustrate the structure. In the right diagram the amplitudes are plotted as a function of the normalized depth coordinate (power density 100 W/m). The numbers in brackets in the legend give the phase-shift relative to the u_2 component

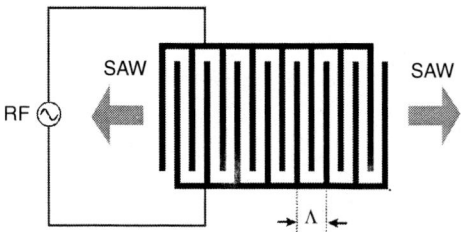

Figure 2 Interdigital transducer electrodes for the excitation of SAWs. The electrodes are directly deposited on the piezoelectric substrate

PLANAR WAVEGUIDE TYPE ACOUSTO-OPTICAL DEFLECTORS

As previously discussed, Bragg-type acousto-optical deflectors and modulators are well-established bulk optic components with a variety of applications. Therefore, it was obvious to develop equivalent integrated optical circuits. In Figure 3 the basic structure of such a device is shown. It consists of a planar optical waveguide that confines the guided optical waves in a thin layer below the substrate surface. A SAW is excited on the substrate surface providing a good overlap with the optical fields for an efficient interaction.

The SAW induces a (traveling) periodic modulation of the refractive index for the optical wave. This modulation acts like an optical phase grating and, hence, results in diffraction. Let Q be defined as:

$$Q = \frac{2\pi\lambda L}{n\Lambda^2}$$

with λ and n being the wavelength and the effective refractive index of the optical wave, respectively. Λ is the wavelength and L is the aperture of the SAW. For an interaction with a narrow aperture SAW ($Q \ll 1$) (Raman–Nath regime) a number of side orders symmetrically displaced from the transmitted (undiffracted) beam occur. The frequency of the kth sidelobe is shifted by k times the acoustic frequency.

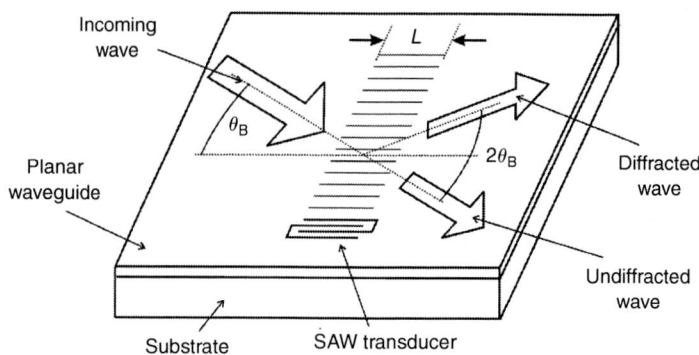

Figure 3   Basic structure of an integrated acousto-optical Bragg deflector in a planar waveguide

In practice, devices with larger SAW aperture are more important (Bragg regime), that is, $Q \gg 1$. In this case, Bragg diffraction can be obtained under the right condition. If the incidence angle Θ_B (see Figure 3) is adjusted to satisfy the Bragg-condition

$$\sin \Theta_B = \frac{\lambda}{2n\Lambda},$$

the diffraction results only in one side order propagating at $2\Theta_B$. The frequency of the diffracted light is either upshifted or downshifted from that of the incident light by the frequency of the SAW. The upshift or downshift corresponds, respectively, to the case in which the incident wave vector is at an angle larger or smaller than 90° from the SAW wave vector. The diffraction efficiency is proportional to $\sqrt{P_a}$, where P_a is the power of the acoustic wave.

Based on such integrated acousto-optical Bragg deflectors a variety of devices have been developed for several different application areas. A detailed overview can be found in Reference 2. For instance, RF-signal processing devices have been demonstrated for spectrum analyzes, convolution, pulse chirping and compression, etc. Furthermore, for applications in optical computing acousto-optical modules have been developed, for instance, for vector–matrix or matrix–matrix multiplication.

Another group of devices are mainly developed for applications in optical communications. In Figure 4 an example of such a circuit is shown. The device acts as a 4 × 4 space switch [8]. The four input signals $I_1, \dots, I_4$ enter channel waveguides. The interface between the channel waveguides and the planar waveguides coincides with the front focal plane of the collimating lens. As the channel waveguides are positioned off the lens axis, the resulting expanded and collimated beams propagate under a certain tilt angle with respect to the lens axis. Without acousto-optical deflection, the beams are refocused by the rear-focusing lens and directed to the monitor output ports $M_1, \dots, M_4$, respectively. Four transducers $T_1, \dots, T_4$ are appropriately tilted to excite SAWs, which can interact with the corresponding beams $I_1, \dots, I_4$, respectively. Via the frequency of the SAW the deflection angle is adjusted to route the beams into one of the output ports. The device has been realized in Y-cut LiNbO$_3$. Channel and planar waveguides are fabricated using the Ti-indiffusion technology. The lenses are formed by an additional proton exchange resulting in an increased refractive index in the exchanged areas.

As discussed earlier, a great variety of integrated acousto-optical devices based on Bragg-diffraction in planar waveguides have been studied. Successful applications of these devices have been presented. There is not yet a commercial exploitation of these results; however, in the past decade there were only little research activities in this area.

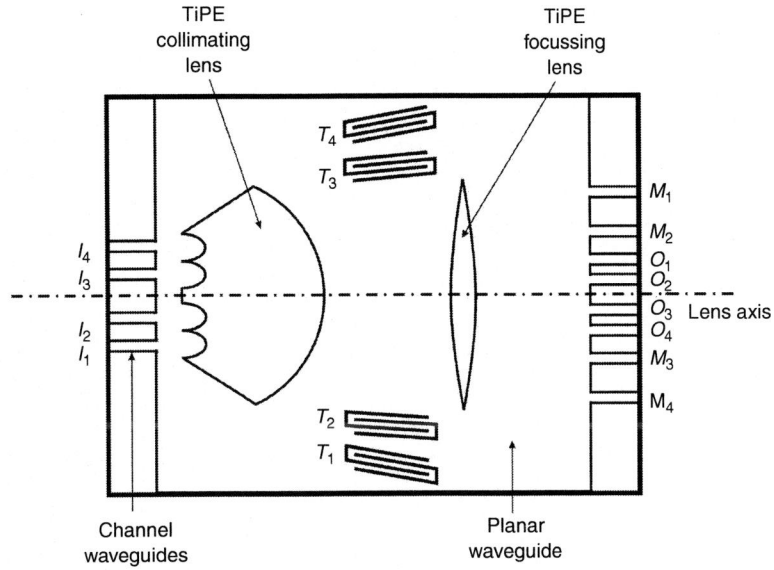

Figure 4 Integrated circuit of a 4 × 4 acousto-optical space switch module [8]

DEVICES BASED ON COLLINEAR ACOUSTO-OPTICAL POLARIZATION CONVERSION

Since about 15 years, another type of acousto-optical device has received remarkable attention [3,9]. These devices are based on a *collinear* acousto-optical interaction in optical strip waveguides resulting in a wavelength-selective polarization conversion. The main driving force for research and development activities in this area was the objective to develop tunable wavelength-selective devices, which are required in WDM communication systems.

The basic building blocks of these devices are acousto-optical polarization converters and polarization splitters. Details of their operation characteristics will be discussed in the following sections. However, to understand the device principles, it is sufficient to know that in the acousto-optical converter a *wavelength-selective polarization conversion* is induced by a surface acoustical wave. Its frequency determines the optical wavelength at which the polarization conversion occurs. The polarization splitters/combiners separate/combine the transverse electric (TE) and transverse magnetic (TM) polarized guided optical waves.

By combining acousto-optical polarization converters with polarization splitters (or polarizers) a whole family of integrated wavelength-selective devices can be obtained (Figure 5). Most important are wavelength filters, wavelength-selective switches, and add-drop multiplexers, which are schematically sketched in Figure 5. An acousto-optical converter between two polarization splitters forms a single-stage bandpass filter. Polarization independent operation is achieved by applying the principle of polarization diversity, that is, the two polarization components are converted separately and recombined by the rear polarization splitter. To improve the performance characteristics one can cascade two filters forming a double-stage tunable wavelength filter. A wavelength-selective 2 × 2 switch is realized using two converters and two splitters. The device allows the routing of each incoming wavelength channel independent of the switching state of the

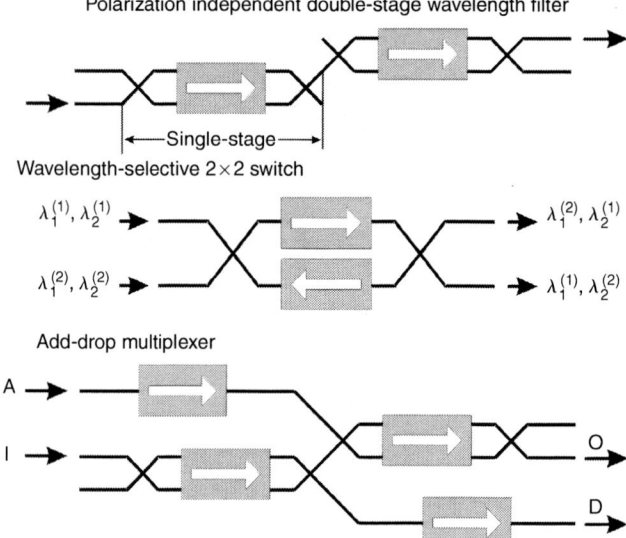

Figure 5 Schematic drawing of the most important integrated acousto-optical devices. The basic structure of a wavelength filter, a wavelength-selective switch and an add-drop multiplexer is shown. All devices can be realized by combining the basic building blocks acousto-optical polarization converter and polarization splitter

other channels. The add-drop multiplexer consists of four converters and four splitters. Wavelength channels can be inserted and extracted from the transmission line.

Such integrated acousto-optical devices are of particular interest for WDM systems as they offer some unique features, such as broad tuning range with electronic control, fast tuning speed, and especially simultaneous multiwavelength operation. In the following, the basic building blocks of these devices are briefly discussed. Subsequently, a description of the state-of-the-art of integrated acousto-optical devices for WDM systems as well as some applications are given.

Acousto-Optical Polarization Converters

The central building block of integrated acousto-optical devices is the acousto-optical polarization converter. Due to the interaction of a SAW with optical waves guided in Ti-indiffused stripe waveguides (fabricated in X-cut, Y-propagating LiNbO$_3$), a wavelength-selective polarization conversion, that is, TE $\rightarrow$ TM or TM $\rightarrow$ TE, is performed.

A propagating SAW induces a periodic perturbation of the dielectric tensor, which results in a coupling of orthogonally polarized optical modes. To achieve an efficient polarization conversion the interaction process must be phase-matched: The difference between the wave numbers of the optical modes must be compensated by the wave number of the SAW, that is, the phase-matching condition

$$\frac{\left| n_{\text{eff}}^{\text{TE}} - n_{\text{eff}}^{\text{TM}} \right|}{\lambda} = \frac{f_{\text{SAW}}}{c_{\text{SAW}}}$$

must be fulfilled. f_{SAW} and c_{SAW} are the frequency and the velocity of the SAW, respectively; $n_{\text{eff}}^{\text{TE}}$ and $n_{\text{eff}}^{\text{TM}}$ are the effective indices of the TE and TM polarized optical modes. The phase-matching condition makes the conversion process wavelength selective. The optical wavelength λ of the modes to be phase-matched can be adjusted via the frequency of the SAW. For wavelengths

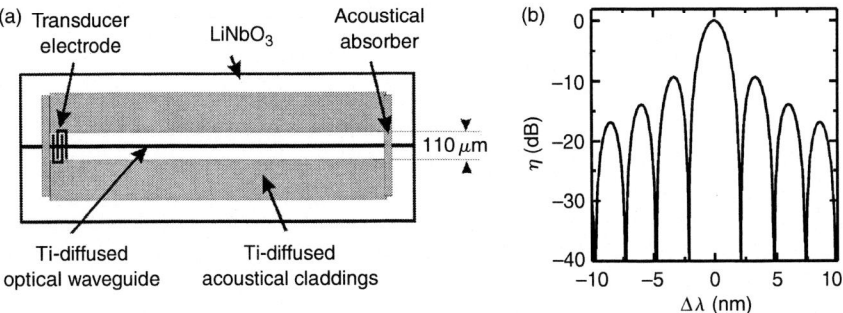

Figure 6   Acousto-optical polarization converter with straight acoustical waveguide (a) and the corresponding calculated conversion characteristic assuming a 12 mm long acousto-optical interaction length (b)

in the third communication window, around $\lambda = 1.55\ \mu$m, the SAW frequency for phase-matching is around 170 MHz with a tuning slope of about 8 nm/MHz; these properties are determined by the birefringence of $LiNbO_3$.

Nowadays, most of the acousto-optical devices take advantage of integrated acoustical waveguides to confine the SAWs into localized regions, yielding large acoustical power densities even at low or moderate overall acoustic power levels (Figure 6). Such acoustical waveguides can be fabricated by a Ti-indiffusion into the cladding region of the guide, which stiffens the material and, hence, increases the acoustic velocity [10,11]. The SAW is guided in the undoped region between the Ti-diffused claddings. Other types of acoustical guides are film-loaded strip or slot type waveguides [12–14]. Due to a film stripe deposited on top of the substrate, the SAW propagation velocity is locally changed and guiding can be obtained.

Moreover, besides straight acoustical waveguides even more complex guiding structures, for example acoustical directional couplers, can be obtained. Such directional couplers have been used to improve the spectral conversion characteristic as discussed in the following paragraphs.

In a simple (unweighted) acousto-optical converter an optical waveguide is embedded in a straight acoustical guide (Figure 6) [10,11]. The SAW propagates co- or contradirectional to the optical waves. The theoretical conversion characteristic, that is, the converted power as function of the optical wavelength, is a sinc²-function as shown in Figure 6(b). The spectral half-width of the curve is proportional to $1/L$ with L being the interaction length. Severe disadvantages of such devices are the high sidelobes of about -10 dB.

The spectral conversion characteristic is approximately given by the Fourier transform of the interaction strength. Therefore, to suppress the sidelobes one can apply a weighted coupling scheme (apodization). Instead of an abrupt change of the interaction strength a soft onset and a soft cutoff is required. This can be achieved using an acoustical directional coupler [15–17]. The optical waveguide is embedded in one arm of the acoustical directional coupler (Figure 7). The SAW is excited in the other arm and couples into the adjacent guide and back again. Therefore, sidelobes of the conversion characteristic are strongly suppressed (>20 dB) as shown in Figure 7(b). Alternatively, some authors used a tilted film-loaded acoustical waveguide in order to achieve the apodization [12,13], that is, the direction of the acoustical guide is slightly tilted with respect to the direction of the optical guide.

At perfect phase-matching, the conversion efficiency η, that is the ratio of the converted optical power to the input power, is given by

$$\eta = \sin^2(\gamma\sqrt{P_a}L),$$

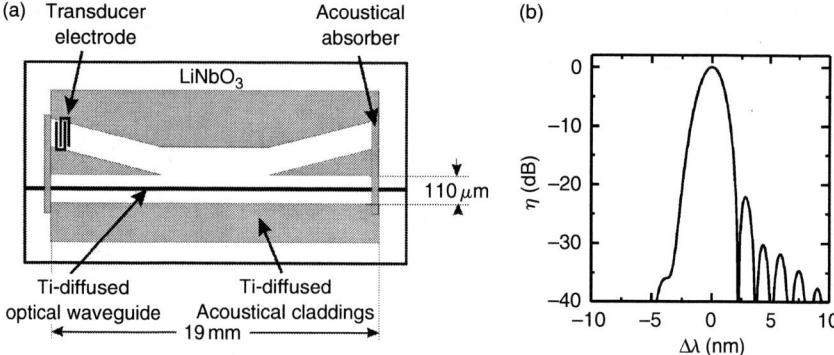

Figure 7 Acousto-optical polarization converter with acoustical directional coupler for weighted coupling (a) and the corresponding calculated conversion characteristic (b)

where γ is a constant determined by the overlap integral between the normalized optical mode fields of both polarization and the acoustical mode, L the interaction length and P_a the power of the acoustical wave. By adjusting the acoustical power the conversion, efficiency can be controlled.

The acousto-optical polarization conversion is accompanied by a frequency shift. The frequency of the converted optical wave is shifted by the SAW frequency. The direction of the shift depends on the direction of conversion, that is, TE→TM or TM→TE, and on the propagation direction of the SAW relative to the propagation direction of the optical waves.

A unique feature of the acousto-optical mode converters is their multiwavelength capability. By simultaneously exciting several acoustical waves at different frequencies in the polarization converter, a simultaneous conversion at different optical wavelengths can occur [18]. However, by multiwavelength operation additional crosstalk can be induced [19–21].

Polarization Splitters and Polarizers

Polarization splitters are applied to separate the TE and TM components of an incoming wave and route them to different optical waveguides. Several concepts have been used to realize such polarization splitters for integrated acousto-optical devices. It is beyond the scope of this chapter to discuss them in detail. Currently, most devices use a passive directional coupler structure fabricated by solely applying the Ti-indiffusion technique [22,23]. Taking advantage of the polarization dependent refractive index profiles, the couplers have been designed to route TE-polarized waves to the cross-state output and TM-polarized waves to the bar-state output of the structure. With an optimized design of such a structure splitting ratios exceeding 20 dB can be obtained.

Alternatively, for some devices polarizers instead of polarization splitters might be suitable as well. Although several types of integrated polarizers have been developed, their fabrication requires further technological processes making the fabrication of the whole device more complex. Therefore, polarization splitters are used more often instead of polarizers.

Tunable Wavelength Filters

The first polarization dependent filters consisted of an integrated acousto-optical polarization converter between crossed polarizers (either external or integrated) [10,24]. Polarization insensitivity has been obtained by applying the principle of polarization diversity, that is, by a separate processing of the polarization components [25–27]. Improved filters use cascaded

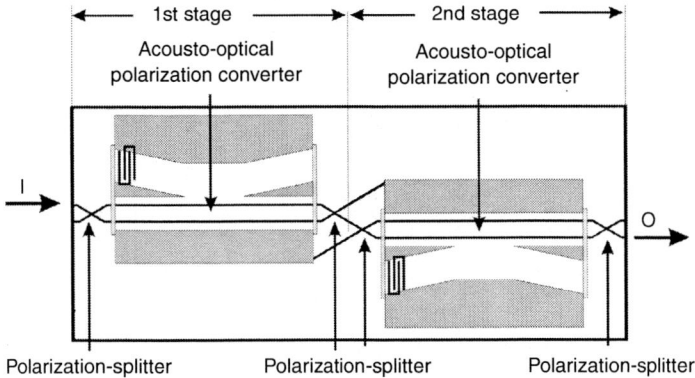

Figure 8 Integrated optical, acoustically tunable double-stage wavelength filter realized by cascading two single-stage filters on a common substrate

structures to provide a double-stage filtering [22,28] or even multistage filtering [12] with the advantage of strongly suppressed baseline levels and reduced sidelobes.

As an example, a polarization-independent double-stage wavelength filter with weighted coupling in each stage is shown in Figure 8 [28]. The incoming wave is split into its polarization components in the first polarization splitter. They are routed to separate optical waveguides that are embedded in a common acoustical waveguide which is one branch of an acoustical directional coupler. After passing this polarization converter, the signals are recombined by the second polarization splitter. As the state of polarization of the phase-matched waves has been changed, they are separated from the unconverted ones. The converted waves are routed to the second filter stage whereas the unconverted ones are fed into a waveguide, which is terminated on the substrate outside the interaction area.

Such double-stage filters offer several advantages. First, the filter characteristics of the overall device is the product of the filter characteristics of the individual stages. This yields a strong suppression of the sidelobes and narrowing of the spectral filter response. Even if one stage has a nonideal performance, for example, large sidelobes or bad splitting ratios of the polarization splitters, cascading with the other stage still results in good overall device performance. Second, due to the double-stage design there is no net frequency shift imposed on the waveguide modes. The opposite frequency shifts of the polarization components in the first stage are compensated by reverse frequency shifts in the second stage.

In Figure 9 the transmission of a pigtailed and packaged wavelength filter is shown. Two curves are drawn corresponding to an input polarization with minimum and maximum insertion loss at the peak transmission, respectively. (The state of polarization at the input of the device cannot be determined after pigtailing. Therefore, the minimum and maximum insertion loss has been used as criteria to adjust the input polarization.) The bandwidth (full-width at half maximum) is 1.6 nm. There are no pronounced sidelobes in the filter characteristics and the baseline, that is, the residual transmission at a wavelength far away from the filter peak, is about 35 dB below the transmission maximum. The polarization dependence is quite small: Only a small shift of about 0.07 nm occurs for the peaks of maximum transmission. The overall insertion loss (fiber-to-fiber) is <4.2 dB with polarization dependence smaller than 0.1 dB. The tuning range of the filter exceeds the spectral range for typical WDM applications. This filter can be tuned from 1530 to 1570 nm without readjusting the drive power of about 100 mW for both stages together.

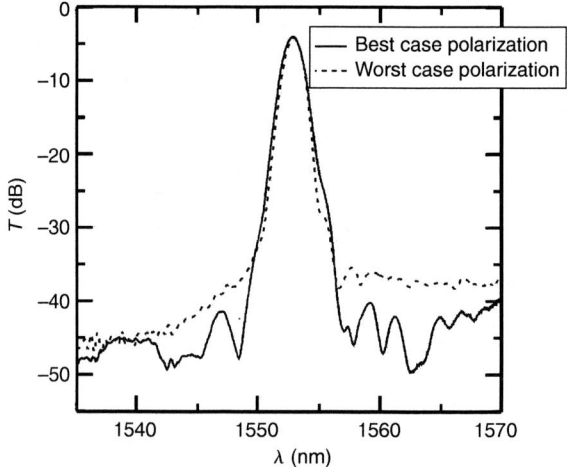

Figure 9 Measured spectral bandpass characteristic of the double-stage wavelength filter. The two curves correspond to an input state of polarization with minimum and maximum insertion loss at the peak wavelength, respectively

A further narrowing of the spectral width is required for most WDM applications, as, nowadays, the channel spacing in dense WDM systems is only 100 GHz ($\approx$0.8 nm) or 50 GHz. An increase in the interaction length is required to reduce the spectral width of the filter. However, this is limited by the maximum available device length. To overcome this restriction a folded geometry has been developed [12]. It utilizes waveguide reflectors to fold the waveguides. Such waveguide reflectors consist of a directional coupler structure, which is cut at the center of the intersecting region. The end faces are coated with a metallic mirror. In this way, a three-stage wavelength filter has been demonstrated. The spectral 3 dB-bandwidth could be reduced to 0.37 nm and the sidelobe suppression is better than 27 dB.

Wavelength-Selective Switches and Add-Drop Multiplexers

A wavelength-selective switch is a 2×2 switch matrix allowing the individual routing of the wavelength channels of a WDM-transmission line to the cross- and bar-state outputs of the device. In Figure 10 the design of a single-stage device is shown [23]; it consists of two acousto-optical polarization converters and two polarization splitters. The light entering, for instance, through input port i1 is divided into its TE and TM polarized components by the first splitter, with the second splitter acting as combiner. The optical power is routed to output port o1 ("bar-state") if no mode conversion is performed. For converted waves the state of polarization changes and, therefore, these signals are routed to output port o2 ("cross-state"). The SAWs in the mode converters propagate into opposite directions to achieve identical frequency shift for TE and TM components. This avoids beating effects that would seriously disturb the performance of the switch.

Typical spectral switching characteristics are shown in Figure 11. The spectral half-width is about 2 nm. Fiber-to-fiber insertion losses are typically in the range of 3 to 5 dB. The sidelobe suppression is about -15 to -20 dB. The extinction of the notch curve is, typically, <25 dB if the input light is exactly TE or TM polarized. However, as the notch curves for the two polarizations do not exactly coincide, extinction of the notch curve for unpolarized input (as shown in the diagram) is typically limited to -15 to -20 dB. The tuning range of such devices

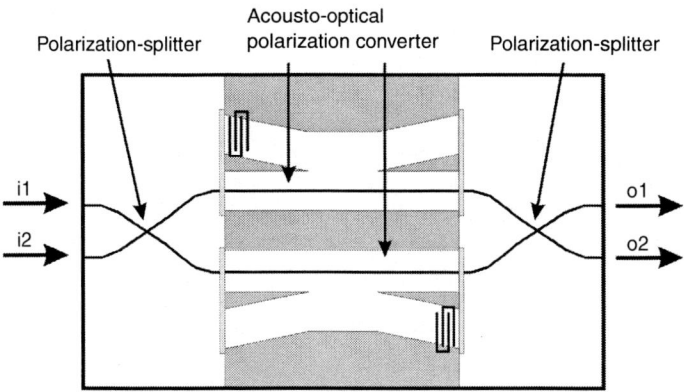

Figure 10   Single-stage integrated acousto-optical wavelength-selective 2 × 2 switch

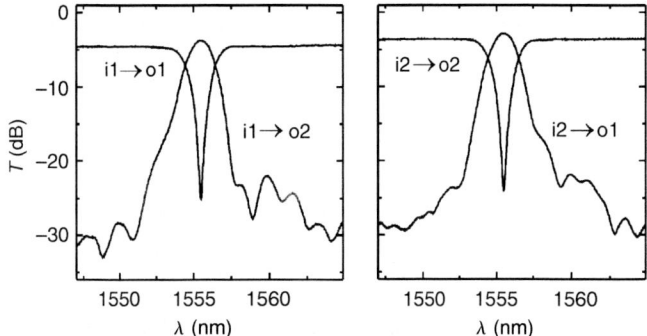

Figure 11   Measured switching characteristics of a single-stage 2 × 2 switch. In the diagrams the transmission characteristics for unpolarized input waves are shown. The switch has been operated with an acoustical wave to route optical signals at λ = 1556 nm to the cross-state

exceeds 70 nm. The switching time is determined by the build-up time of the acoustical wave in the converter structure. For 20 mm long converters it is about 5 μsec.

Crosstalk suppression is the most stringent requirement for components applied in WDM transmission systems. With single-stage switches, the necessary crosstalk suppression cannot be achieved. Therefore, a double-stage design must be used leading to dilated switches [21,29,30]. Combining four single-stage switches yields a device with crosstalk suppression in the order of ε^2, if ε is the crosstalk of a single switch. Till recently, no monolithically integrated dilated acousto-optical switch has been realized.

A first step toward a monolithically integrated dilated switch is the double-stage add-drop multiplexer shown in Figure 12 [31]. Add-drop multiplexers do not require the full functionality of a 2 × 2 switch as no direct routing from the add- to the drop-port is required. Therefore, a partially dilated switch can be used to form an improved add-drop multiplexer. The circuit consists of two switches in series and two frequency shifters. In the first switch the drop-function is performed and in the second switch the add-function. Due to this local separation, that is, spatial dilation of the add- and drop-functions, crosstalk between add- and drop-ports is strongly suppressed (<50 dB). Moreover, the device acts as double-stage notch filter for the transmission from I to O resulting in a suppression of crosstalk due to incomplete dropping of about 26 dB.

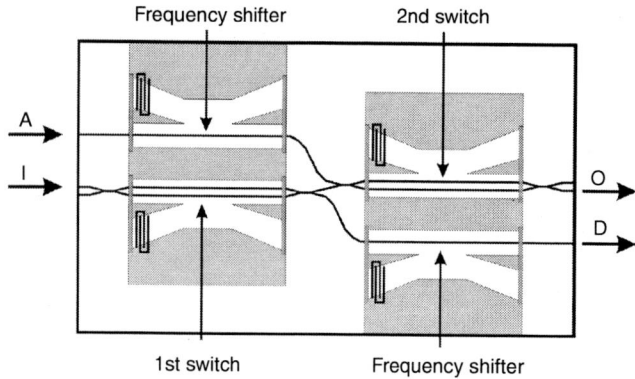

Figure 12   Double-stage add-drop multiplexer

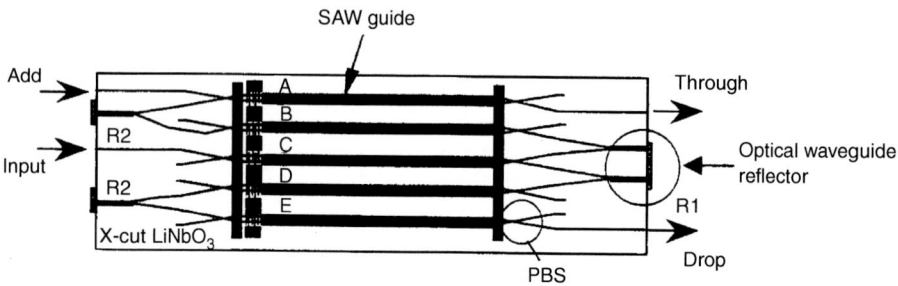

Figure 13   Folded multistage add-drop multiplexer with waveguide reflectors (From T. Nakazawa: *Proceedings of 10th European Conference on Integrated Optics (ECIO'01)*, Paderborn, [2001]. With permission.)

The frequency shifters in the add- and drop-arms are required to compensate the frequency shift induced by the acousto-optical polarization conversion in the switches.

Another add-drop multiplexer in a folded arrangement is shown in Figure 13 [32]. It again uses a folded geometry with waveguide reflectors to increase the effective interaction length. With a three-stage filtering in the drop-arm as well as three-stage notch filter for the transmission from input- to through-port, a good crosstalk suppression is obtained. In Figure 14 results of a multichannel operation of this device are shown. The device was operated to drop simultaneously eight channels separated by 1.6 nm. A sidelobe suppression better than 35 dB in the drop-port and an isolation of >40 dB in the through-port has been achieved. The overall fiber-to-fiber insertion losses are about 9 dB.

Some Applications in WDM Systems

It is beyond the scope of this chapter to discuss details of WDM systems using acousto-optical devices. However, a brief overview demonstrating the potential is given.

The simplest application of acousto-optical wavelength filters is the usage as channel selector in front of a receiver. Another filter application using specific acousto-optical properties is the power equalization at different wavelength channels [33]. Taking advantage of the multi-wavelength capabilities of such a device, the transmission of each channel can be adjusted by applying a set of RF-signals with appropriate frequencies and power levels.

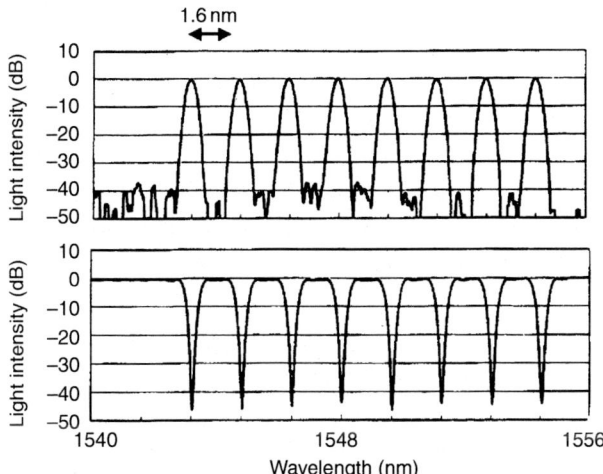

Figure 14   Simultaneous multiwavelength performance of the add-drop multiplexer. Top. Normalized transmission at the drop-port, bottom: normalized transmission at the through-port (From T. Nakazawa: *Proceedings of 10th European Conference on Integrated Optics (ECIO'01)*, Paderborn, [2001]. With permission.)

Acousto-optical filters may also be used as a wavelength-selective element inside a laser cavity. In this way a rapid tuning of the laser wavelength via the SAW frequency is accomplished. This application of the acousto-optical filter has been demonstrated in an erbium-doped fiber ring laser [34] and with a semiconductor optical amplifier [35]. Moreover, even an intrinsic doping of the filter itself by Er-indiffusion has been demonstrated resulting in a tunable wavelength filter with gain [36] and in a tunable Er:Ti:LiNbO$_3$ waveguide laser [37].

Acousto-optical switches and add-drop multiplexers are used in dynamically reconfigurable WDM networks. The multiplexers can be used in network nodes to insert or extract information at certain wavelengths to/from the transmission line. Due to the tunability of the devices the configuration is flexible; the devices can for example, be applied in systems with dynamic wavelength reuse. Within several microseconds — that is, the build-up time of the acoustic wave in a polarization converter — the wavelength for polarization conversion can be switched. Several prototype networks using integrated acousto-optical add-drop multiplexers have been successfully demonstrated (see, e.g., References 38 to 40).

The 2 × 2 wavelength-selective switches form the basic building blocks of more complex switching nodes as required for optical cross-connects. By combining several such 2 × 2 matrices, $n \times n$-switches can be realized. The architecture is strongly simplified in comparison with WDM nodes without wavelength selective switches. As a first example, a 4 × 4 acousto-optical switching node consisting of five 2 × 2 switches has been demonstrated [3,41]. With this node the applicability of these devices in dynamically reconfigurable WDM-systems could successfully be demonstrated. Besides improved crosstalk suppression and narrower channel spacing, a higher state of integration in order to obtain more complex switching nodes with minimum optical devices is one of the challenges for future activities.

REFERENCES

1.   C.S. Tsai (Ed.): "*Guided-Wave Acousto-Optics*," Springer Series in Electronics and Photonics 23, Springer-Verlag, Berlin (1990).

2. C.S. Tsai: "Integrated acoustooptic circuits and applications," *IEEE Trans. Ultrasonics, Ferroelectrics Frequency Control*, 39, 529–554 (1992).

3. M.K. Smit, T. Koonen, H. Herrmann, and W. Sohler: "Wavelength-selective devices," in N. Grote and H. Venghaus (eds.): *Fibre Optic Communication Devices*, Springer-Verlag, Berlin (2001).

4. Lord Rayleigh: "On waves propagated along the plane surface of an elastic solid," *Proc. London Math. Soc.*, 17, 4–11 (1888).

5. M. Feldmann and J. Hénaff: *Surface Acoustic Waves for Signal Processing*, Artech House, Boston (1989).

6. C. Campbell: *Surface Acoustic Wave Devices for Mobile and Wireless Communication*, Elsevier, Amsterdam (1998).

7. T.M Reeder: "Excitation of surface acoustic waves by use of interdigital electrode transducers," Chap. 4 in: C.S. Tsai (Ed.): *Guided-Wave Acousto-Optics*, Springer Series in Electronics and Photonics 23, Springer-Verlag, Berlin (1990).

8. A. Kar-Roy and C.S. Tsai: "8×8 symdmetric nonblocking integrated acoustooptic space switch module on $LiNbO_3$," *IEEE Photon. Technol. Lett.*, 4, 731–734 (1992).

9. D.A. Smith, R.S. Chakravarthy, Z. Bao, J.E. Baran, J.L. Jackel, A. d'Alessandro, D.J. Fritz, S.H. Huang, X.Y. Zou, S.M. Hwang, A.E. Willner, and K.D. Li: "Evolution of the acousto-optic wavelength routing switch," *J. Lightwave Technol.*, 14, 1005–1019 (1996).

10. J. Frangen, H. Herrmann, R. Ricken, H. Seibert, W. Sohler, and E. Strake: "Integrated optical, acoustically tunable wavelength filter," *Electron. Lett.*, 25, 1583–1584 (1989).

11. D.A. Smith and J.J. Johnson: "Low drive power integrated acousto-optic filter on X-cut, Y-propagating $LiNbO_3$," *IEEE Phot. Technol. Lett.*, 3, 923–925 (1991).

12. T. Nakazawa, S. Taniguchi, and M. Seino: "$Ti:LiNbO_3$ acousto-optical tunable filter (ATOF)," *Fujitsu Sci. Tech. J.*, 35, 107–112 (1999).

13. H. Mendis, A. Mitchell, I. Belski, M. Austin, and O.A. Peverini: "Design, realisation and analysis of an apodized film-loaded acousto-optic tunable filter," *Appl. Phys.* B, 73, 489–493 (2001).

14. O.A. Peverini, H. Herrmann, and R. Orta: "Film-loaded SAW waveguides for integrated acousto-optical polarization converters," *IEEE Trans. Ultrasonics, Ferroelectrics and Frequency Control*, 51(10), 1298–1307 (2004).

15. H. Herrmann and St. Schmid: "Integrated acousto-optical mode convertors with weighted coupling using surface acoustic wave directional couplers," *Electron. Lett.*, 28, 979–980 (1992).

16. D.A. Smith and J.J. Johnson: "Sidelobe suppression in an acousto-optic filter with a raised cosine interaction strength," *Appl. Phys. Lett.*, 61, 1025–1027 (1992).

17. H. Herrmann, U. Rust, and K. Schäfer: "Tapered acoustical directional couplers for integrated acoustooptical mode converters with weighted coupling," *J. Lightwave Technol.*, 13, 364–374 (1995).

18. K.W. Cheung, S.C. Liew, C.N. Lo, D.A. Smith, J.E. Baran, and J.J. Johnson: "Simultaneous five-wavelength filtering at 2.2 nm wavelength separation using integrated optic acousto-optic tunable filter with subcarrier detection," *Electron. Lett.*, 25, 636–637 (1989).

19. F. Tian and H. Herrmann: "Interchannel interference in multiwavelength operation of integrated acousto-optical filters and switches," *J. Lightwave Technol.*, 13, 1146–1154.

20. M. Fukutoku and K. Oda: "Optical beat-induced crosstalk of an acousto-optic tunable filter for WDM network application," *J. Lightwave Technol.*, 13, 2224–2235 (1995).

21. J.L. Jackel, M.S. Goodmann, J.E. Baran, W.J. Tomlinson, G.K. Chang, M.Z. Iqbal, G.H. Song, K. Bala, C.A. Brackett, D.A. Smith, R.S. Chakravarthy, R.H. Hobbs, D.J. Fritz, R.W. Ade, and K.M. Kissa: "Acousto-optic tunable filters (AOTFs) for multi-wavelength optical cross-connects: crosstalk considerations," *J. Lightwave Technol.*, 14, 1056–1066 (1996).

22. F. Tian, Ch. Harizi, H. Herrmann, V. Reimann, R. Ricken, U. Rust, W. Sohler, F. Wehrmann, and S. Westenhöfer: "Polarization independent integrated optical, acoustically tunable double stage wavelength filter in $LiNbO_3$," *J. Lightwave Technol.*, 12, 1192–1197 (1994).

23. F. Wehrmann, Ch. Harizi, H. Herrmann, U. Rust, W. Sohler, and S. Westenhöfer: "Integrated optical, wavelength selective, acoustically tunable 2×2 switches (add-drop multiplexers) in $LiNbO_3$," *IEEE J. Selected Top. Quantum Electron.*, 2, 263–269 (1996).

24. B.I. Heffner, D.A. Smith, J.E. Baran, A. Yi-Yan, and K.W. Cheung: "Integrated-optic, acoustically tunable infrared optical filter," *Electron. Lett*, 24, 1562–1563 (1988).

25. D.A. Smith, J.E. Baran, K.W. Cheung, and J.J. Johnson: "Polarization-independent acoustically tunable optical filter," *Appl. Phys. Lett.*, 56, 209–211 (1990).

26. K.W. Cheung, D.A. Smith, J.E. Baran, and J.J. Johnson: "1 Gb/s system performance of an integrated, polarization-independent, acoustically tunable optical filter," *IEEE Phot. Technol. Lett.*, 2, 271–273 (1990).

27. T. Pohlmann, A. Neyer, and E. Voges: "Polarization independent Ti:LiNbO$_3$ switches and filters," *IEEE J. Quantum Electron.*, 27, 602–607 (1991).

28. H. Herrmann, K. Schäfer, and Ch. Schmidt: "Low-loss tunable integrated acousto-optical wavelength filter in LiNbO$_3$ with strong sidelobe suppression," *IEEE Photon. Technol. Lett.*, 10, 120–123 (1998).

29. D.A. Smith, A. d'Alessandro, J.E. Baran, D.J. Fritz, and R.H. Hobbs: "Reduction of crosstalk in an acousto-optic switch by means of dilation," *Opt. Lett.*, 19, 99–101 (1994).

30. J.L. Jackel, M.S. Goodmann, J. Gamelin, W.J. Tomlinson, J.E. Baran, C.A. Brackett, D.J. Fritz, R. Hobbs, K. Kissa, R. Ade, and D.A. Smith: "Simultanoues and independent switching of 8 wavelength channels with 2 nm spacing using a wavelength-dilated acousto-optic switch," *IEEE Photon. Technol. Lett.*, 8, 1531–1533 (1996).

31. H. Herrmann, A. Modlich, Th. Müller, and W. Sohler: "Double-stage, Integrated, Acousto-optical Add-drop Multiplexers with Improved Crosstalk Performance," IEE Conference publication No. 448, Vol. 3 (ECOC '97), pp. 10–13 (1997).

32. T. Nakazawa: "Integrated acousto-optic tunable filter for optical add/drop multiplexers," in: *Proceedings of the 10th European Conference on Integrated Optics (ECIO'01)*, Paderborn, 2001.

33. S.F. Su, R. Olshansky, G. Joyce, D.A. Smith, and J.E: Baran: "Gain equalization in multiwavelength lightwave systems using acousto-optic tunable filters," *IEEE Photon. Technol. Lett.*, 4, 269–271 (1992).

34. D.A. Smith, M.W. Maeda, J.J. Johnson, J.S. Patel, M.A. Saifi, and A. Von Lehman: "Acoustically tuned erbium-doped fiber ring laser," *Opt. Lett.*, 16, 387–389 (1991).

35. K. Morito, K. Takabayashi, K. Takada, N. Hashimoto, M. Doi, S. Tomabechi, T. Takeuchi, G. Nakagawa, H. Miyata, and T. Nakazawa: "Widely wavelength tunable lasers (>100 nm) using SOA and AOTF," *Ninth Optoelectronics and Communications Conference/Third International Conference on Optical Internet (OECC/COIN2004)*, Technical digest, pp. 700–701, Pacifico Yokohama (2004).

36. R. Brinkmann, M. Dinand, I. Baumann, Ch. Leifeld, W. Sohler, and H. Suche: "Acoustically tunable wavelength filter with gain," *IEEE Photon. Technol. Lett.*, 6, 519–521 (1994).

37. K. Schäfer, I. Baumann, W. Sohler, H. Suche, and S. Westenhöfer: "Diode-pumped and packaged, acoustooptically tunable Ti:Er:LiNbO$_3$ waveguide laser of wide tuning range," *IEEE J. Quantum Electron.*, 33, 1636–1641 (1997).

38. W.I. Way, D.A. Smith, J.J. Johnson, and H. Izadpanah: "A self-routing WDM high capacity SONET ring network," *IEEE Photon. Technol. Lett.*, 4, 402–405 (1992).

39. F. Arecco, F. Casella, E. Iannone, A. Mariconda, S. Merli, F. Pozzi, and F. Veghini: "A transparent, all-optical metropolitan network experiment in a field environment: The 'PROMETEO' self-healing ring," *J. Lightwave. Technol.*, 15, 2206–2213 (1997).

40. T. Chikama, H. Onaka, and S. Kuroyanagi: "Photonic networking using optical add-drop multiplexers and optical cross-connects," *Fujitsu Sci. Technol. J.*, 35, 46–55 (1999).

41. H. Herrmann, A. Modlich, Th. Müller, W. Sohler, and F. Wehrmann: "Advanced integrated, acousto-optical switches, add-drop multiplexers and WDM cross-connects in: LiNbO$_3$," in: *Proceedings of the 8th European Conference on Integrated Optics (ECIO '97)*, 1997, pp. 578–581.

ADD/DROP FILTER

Yasuo Kokubun

Add/Drop filter is an optical wavelength filter that can extract one or a few wavelengths from busline waveguide and can also merge the same wavelength to busline waveguide. Basically, a 2×2 port device configuration is needed, that is, input port, drop port, through port, and add port. The wavelength intended to be extracted is transmitted to the drop port and other wavelengths pass straight through to the through port. In addition, the same wavelength can be incident on the add port and is transmitted to the through port, i.e. the wavelength is added to other wavelengths. Microring resonator filter and acousto-optic filter are categorized as Add/Drop filter, which should be distinguished from wavelength multi/demultiplexers, such as arrayed waveguide grating (AWG) demultiplexer. A few 1×1 port devices such as the fiber Bragg grating and multilayer thin film filter, can also be used as an Add/Drop filter by attaching circulators to the input and output ports, or by incorporating the grating into two arms of Mach–Zehnder interferometer.

In general, wavelength demultiplers, such as AWG, have the function that all multiplexed wavelengths are simultaneously directed into different output ports according to the wavelength, and have been used to expand the transmission capacity of trunk lines. On the other hand, an Add/Drop filter is used in the future photonic network, in which wavelength is used as the label for the routing of signals.

ARRAYED WAVEGUIDE GRATING (AWG)

Hiroshi Takahashi

INTRODUCTION

Since the mid-1990s, wavelength division multiplexing (WDM) transmission systems are needed to meet the huge demand for data communication resulting from the worldwide spread of the Internet. In these systems, signals at different wavelengths are mixed and transmitted through a single optical fiber, and this technology provides us with a high per-fiber transmission capacity and low communication costs. A wavelength multi/demultiplexer is a key device in such WDM systems. As described in the chapter entitled *Wavelength multi/demultiplexer for WDM*, an *arrayed-waveguide grating* (AWG) is a key optical component for realizing a multi/demultiplexer, especially in dense WDM systems with 16 or more wavelength channels that are arranged with a spacing of 100 GHz (approximately 0.8 nm in the 1.55 μm band) or less.

The AWG (also called a PHASAR) is a kind of transmission grating that consists of hundreds of channel waveguides of different lengths. These waveguides are fabricated on a substrate by using planar lightwave circuit (PLC) technology that includes glass film deposition, photolithography, and dry etching. As shown in Figure 1, the AWG is integrated with input and output waveguides and two slab waveguides with a focusing function, and together they act as a wavelength multi/demultiplexer in the same way as a conventional diffraction grating-based spectrometer. The features of AWG multi/demultiplexers are their compact size, stable operation in the presence of mechanical vibration, high long-term reliability, and mass producibility.

The concept of the AWG, in which circular arc waveguides are concentrically arrayed in a planar waveguide substrate, was first reported by Smit in 1988 [1]. He showed that the input light beam was focused and the arrayed waveguide acted as a phased array antenna, as is well known in microwave remote sensing. He also suggested that this approach can be applied to a wavelength multi/demultiplexer. H. Takahashi described the first demultiplexing function in 1990 [2]. He demonstrated that densely packed wavelength channels with a 1 nm spacing could be demultiplexed by using a high diffraction order obtained by employing a large length difference between arrayed waveguides. Dragone [3] showed that the device could be used as an N-input and N-output wavelength selective star coupler. Then, in 1991, Vellekoop and Smit [4], Takahashi et al. [5], and Dragone et al. [6] independently reported one-chip AWG wavelength multi/demultiplexers that included input and output waveguides and two focusing slab waveguides. An important issue to overcome before this device could be used as a wavelength multi/demultiplexer in practical WDM systems was the device's polarization dependence. The polarization dependence results from waveguide birefringence induced by thermal stress caused by the difference between the thermal expansion coefficients of the waveguide layer and the substrate. Even if the birefringence is as small as 1×10^{-4}, it results in a 0.1 nm wavelength shift, which is not permissible in 0.8 nm spacing WDM systems. A polarization cancellation technique that uses a half wavelength plate [7,8] is commonly used because it works even when the birefringence is not reduced. Insertion loss is another issue. This is reduced by using a silica-based waveguide [9]. The propagation loss is <0.03 dB/cm and the single-mode fiber coupling loss is 0.4 dB. At present, almost all commercial AWG multi/demultiplexers are made of silica waveguides. InP-based waveguides are also used for the monolithic integration of AWG multi/demultiplexers with laser diodes and photo diodes [10]. This technology has the potential to provide us with one-chip WDM transmitters and receivers [11].

FUNDAMENTALS OF AWG

Figure 1 shows the waveguide layout of the AWG multi/demultiplexer, which is the same as that of a conventional spectrometer. The concave alignment of the waveguide ends of the AWG acts

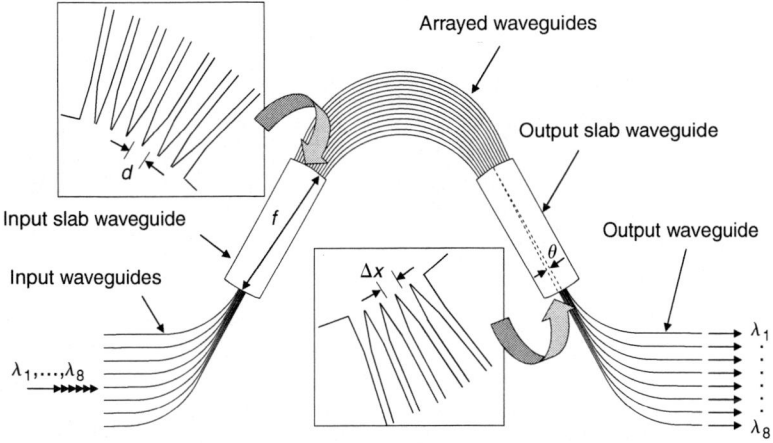

Figure 1   Waveguide layout of AWG wavelength multi/demultiplexer

as a lens and the AWG acts as a diffraction grating. Accordingly, lights with different wave-lengths are focused at corresponding ports when a wavelength multiplexed light is launched into the input port. The diffraction angle θ and wavelength λ satisfy the following grating equation:

$$n_c \cdot \Delta L + n_s \cdot d \cdot \sin\theta = m \cdot \lambda, \tag{1}$$

where

n_c: effective refractive index of channel waveguides in AWG

n_s: effective refractive index of slab waveguide

d : grating pitch (distance between waveguide ends of AWG)

ΔL: length difference between channel waveguides in AWG

m: diffraction order (natural number)

$\theta = 0$ is a direction aiming at the center output waveguide. The wavelength of the light that trav-els in this direction is called the center wavelength. As the wavelength changes from the center wavelength, the focal point moves at a speed of $dx/d\lambda$, which is called the linear dispersion of the grating where x is a coordinate along the focal line. The linear dispersion is obtained from Eq. (1) as follows:

$$\frac{dx}{d\lambda} = \frac{f \cdot m}{n_s \cdot d} \cdot \frac{n_g}{n_c}, \tag{2}$$

where

$$n_g = n_c - \lambda \cdot \frac{dn_c}{d\lambda} \tag{3}$$

is the group refractive index. f is the focal length (radius) of the slab waveguide. In order to obtain a high wavelength resolution, we need a large linear dispersion. This is realized with a long focal length f and a small pitch d, as well as a conventional grating. In an AWG, we can obtain an extremely large linear dispersion with a high diffraction order m simply by designing the waveguide length. This is the most important characteristic feature of the AWG and a wave-length channel spacing of 0.8 nm or less is easily obtained in AWG multi/demultiplexers. Figure 2 illustrates the interference in the output slab waveguide. Figure 2(a) depicts the situation when the input light wavelength is the center wavelength and Figure 2(b) shows the same for a slightly (e.g., 0.8 nm) longer wavelength. The diffracted light is focused at the center output waveguide as shown in Figure 2(a). Since the length difference ΔL in the AWG is very large, even a small 0.8 nm wavelength shift results in a clear phase difference at the arrayed waveguide ends in Figure 2(b). As a result, the focal point is shifted to the output waveguide end adjacent to the center.

SPECTRAL CHARACTERISTICS

Figure 3 shows a typical insertion loss spectrum for all the output channels. The bold line is the spectrum of channel 1 and the thin lines are for channels 2 to N. Because the AWG has a peri-odic spectrum, channel 1 has plural transmission wavelengths including $\lambda 1$. The spacing between them is called the free spectral range (FSR). Therefore, the wavelength range available

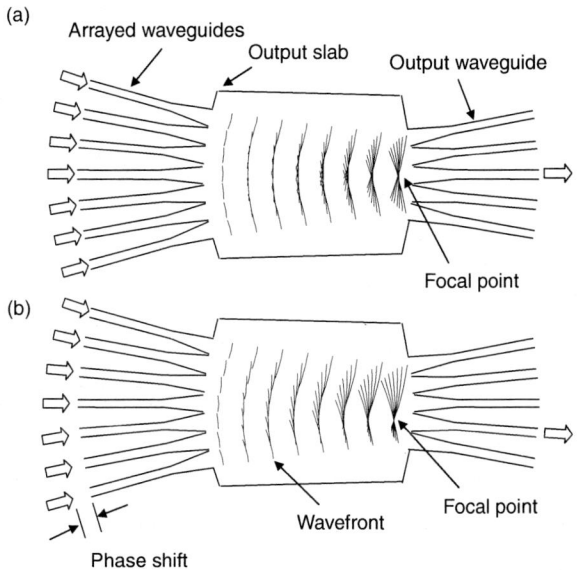

Figure 2 Interference in output slab waveguide (a) when λ = center wavelength (b) when λ = center wavelength + $\Delta\lambda$

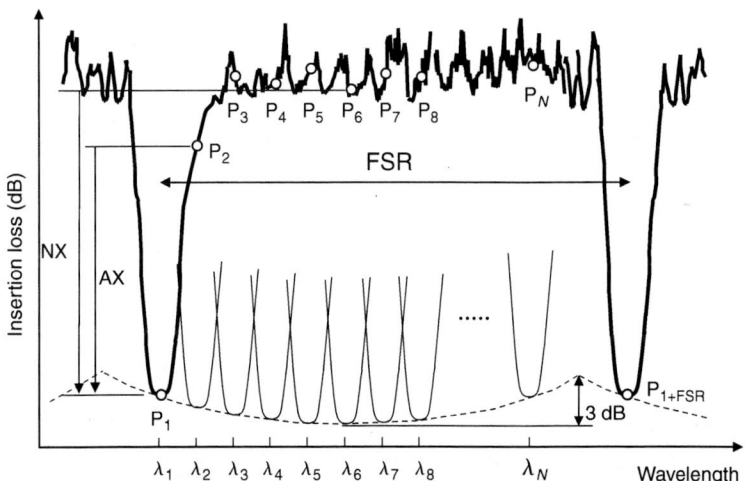

Figure 3 Insertion loss spectra of AWG multi/demultiplexer and definitions

for WDM is limited to the FSR. The FSR in terms of frequency is given by

$$FSR = \frac{c}{n_g \cdot \Delta L} \quad (Hz),\tag{4}$$

where c is the light velocity in a vacuum. As described in fundamentals of AWG section, we can obtain high wavelength resolution with a large ΔL but this reduces the FSR. So proper FSR selection is very important with regard to AWG multi/demultiplexer design. The insertion loss dependence among the channels is drawn as a dashed envelope line and the loss is lower in the central region than in the outer region. The FSR is usually set at or above 1.5 times the WDM bandwidth (channel spacing × channel number).

In Figure 3 AX indicates the ratio between the losses at the selected channel and adjacent channels. This is called the adjacent channel crosstalk and is usually −30 dB. The ratio to the other channels is called the nonadjacent channel crosstalk (NX) and is usually −35 to−50 dB. Figure 4 is a typical measured insertion loss spectrum for a 0.8 nm (100 GHz) spacing, 48-channel AWG multi/demultiplexer made of silica-based waveguides [9]. The insertion loss is 3.2 to 3.9 dB. There is very little variation in the insertion loss values or the shapes of the spectral curves.

The spectral curve, near the transmission wavelength, is given by the convolution integral of the electrical field distributions at the input and the output waveguide apertures because the transmittance is given by the coupling coefficient of the output waveguide and the focused beam electrical field, which is a projection of the input waveguide electrical field. Figure 5 shows the shape of the input waveguide end and the obtained spectral curve. When the input waveguide end is

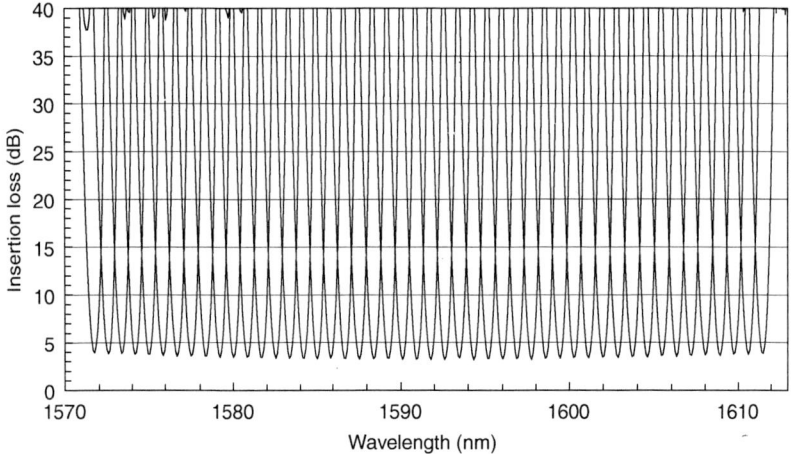

Figure 4   Typical insertion loss spectrum of AWG multi/demultiplexer (100 GHz spacing, 48 channel, Gaussian)

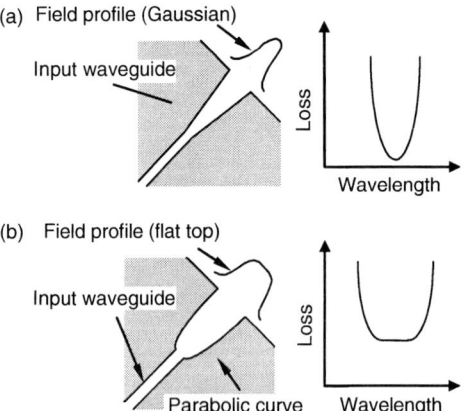

Figure 5   Relation between input waveguide shape and spectrum

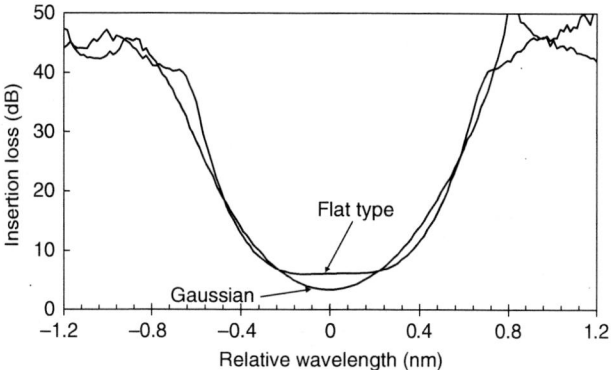

Figure 6   Spectral curve of Gaussian and flat type AWG

widened with a linear taper (Figure 5[a]), the electrical field is approximately Gaussian, as is the spectral curve [12]. In contrast, when the parabolic curve is used, the electrical field is a mixture of the fundamental and 2nd order modes (Figure 5[b]). This results in a flat spectral curve [13]. Figure 6 shows the typical spectra of Gaussian type and flat type AWGs made of silica-based waveguides. The 1 dB bandwidths for the Gaussian and flat type, AWGs are 0.22 and 0.4 nm, respectively, and their 3 dB bandwidths are 0.4 and 0.6 nm, respectively. With the flat type AWG, the 1 and 3 dB bandwidths are wider than with the Gaussian type but the insertion loss at the channel center is 2.5 dB larger. This is due to a mismatch between the electrical fields at the input and the output waveguide ends. The flat type is commonly used for commercial WDM systems regardless of the large insertion loss because this is easily compensated for by using an erbium-doped fiber amplifier. A medium spectrum between the flat and Gaussian spectra is easily obtained by changing the input waveguide end shape, and this design is used for WDM systems that take equal account of insertion loss and bandwidth.

CHANNEL COUNT, INSERTION LOSS, AND TEMPERATURE DEPENDENCE

The channel count of commercial AWG multi/demultiplexers is 8 to 64 because no more than 64 channels are currently required in dense WDM systems. Meanwhile, increasing the channel count is very important with regard to research since AWGs are expected to be employed in noncommunication fields, for example, as a one-chip spectrometer. From Eq. (2), the maximum channel count M is given by

$$M = \frac{\text{FSR}}{\Delta\lambda} = \frac{\lambda \cdot f}{n_s \cdot d \cdot \Delta x},$$ (5)

where the FSR in terms of wavelength is, approximately, given by

$$\text{FSR} = \frac{\lambda}{m} \cdot \frac{n_c}{n_g} \quad (\text{m})$$ (6)

and $\Delta\lambda$ and Δx are the wavelength channel spacing and the output waveguide pitch at the junction with the slab waveguide, respectively. Equation (5) shows that a large focal length f is required for a higher channel count. Accordingly, we need a large number of arrayed waveguides to receive all the diffracted light from the input waveguide. Figure 7 is a sketch of a 25 GHz spacing, 400 channel AWG multi/demultiplexer composed of silica-based waveguides with a refractive index

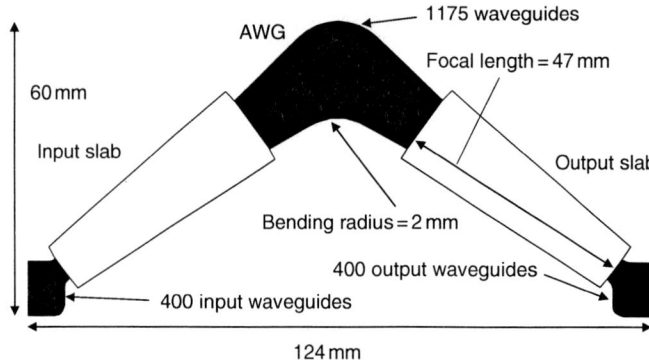

Figure 7   Waveguide pattern of 25 GHz spacing, 400 channel AWG

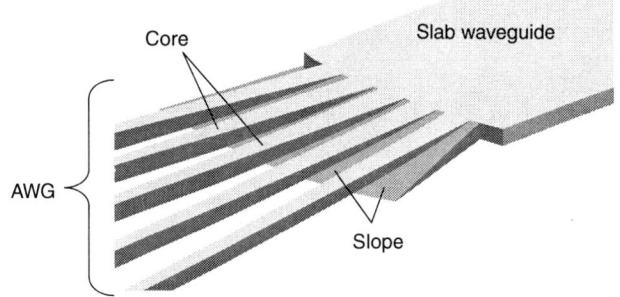

Figure 8   Slope structure at slab/channel waveguide junction for loss reduction

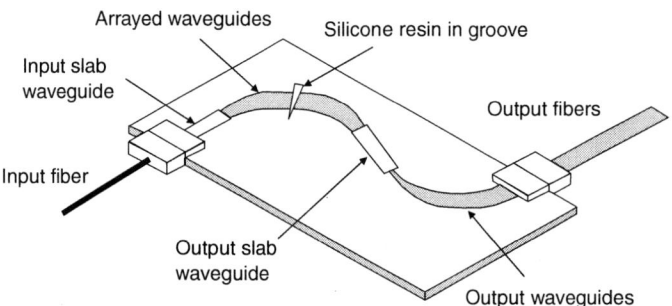

Figure 9   Athermal AWG multi/demultiplexer with silicone resin inserted

difference of 1.5% [14]. The focal length is 47 mm, the number of arrayed waveguides is 1175 and the chip size is 124 × 60 mm. The uniformity of the refractive index and waveguide core size (thickness and width) is improved so that it becomes very high over this large area because poor uniformity would deform the spectral curve. This makes the fabrication process very important with regard to obtaining a high channel count AWG.

Insertion loss reduction is also an important issue. In a silica-based AWG, the waveguide propagation loss and fiber coupling loss are negligible and the main cause of the insertion loss is mode conversion loss at the slab-channel waveguide junction. As shown in Figure 2, the waveguide ends of the arrayed waveguide are widened but there is actually a 1 to 2 μm gap between the waveguide cores due to imperfect etching resolution. This gap disturbs the smooth adiabatic transition from the plane wave to the multichannel waveguide propagation mode. This loss is reduced by the slope between the tapered waveguides as shown in Figure 8. The slopes are formed because the etching rate is low in the narrow gap during the core patterning process [15]. A loss reduction of 1 dB has been confirmed experimentally.

The center wavelength is given by $n_c \Delta L/m$ from Eq. (1) and depends on the temperature because n_c depends on the temperature. For example, the sensitivity is 0.01 nm/ °C in a silica-based AWG. The value is approximately ten times greater in InP-based and polymer-based AWGs. Even the silica-based AWG requires temperature control when used in 0.8 nm spacing WDM systems. The temperature dependence can be eliminated by inserting silicone resin with a negative temperature coefficient as shown in Figure 9 [16].

REFERENCES

1. M.K. Smit, "New focusing and dispersive planar component based on an optical phased array," *Electron. Lett.*, 24, 385–386, 1988.
2. H. Takahashi, S. Suzuki, K. Katoh, and I. Nishi, "Arrayed-waveguide grating for wavelength division multi/demultiplexer with nanometer resolution," *Electron. Lett.*, 26, 87–88, 1990.
3. C. Dragone, "Optimum design of a planar array of tapered waveguides," *J. Optical Soc. Am.* A, 7, 2081–2093, 1990.
4. A.R. Vellekoop, and M.K. Smit, "Four-channel integrated-optic wavelength demultiplexer with week polarization dependence," *J. Lightwave Technol.*, 9, 310–314, 1991.
5. H. Takahashi, S. Suzuki, and I. Nishi, "Multi/demultiplexer for nanometer-spacing WDM using arrayed-waveguide grating," Integrated Photonics Research, OSA, paper PD-1, 1991.
6. C. Dragone, C.A. Edwards, and R.C. Kistler, "Integrated optics N × N multiplexer on silicon," *IEEE Photon. Technol. Lett.*, 3, 896–899, 1991.
7. H. Takahashi, Y. Hibino, and I. Nishi, "Polarization-insensitive arrayed-waveguide grating wavelength multiplexer on silicon," *Opt. Lett.*, 17, 499–501, 1992.
8. Y. Inoue, Y. Ohmori, M. Kawachi, S. Ando, T. Sawada, and H. Takahashi, "Polarization mode converter with polyimide half waveplate in silica-based planar lightwave circuits," *IEEE Photonics Technol. Lett.*, 6, 626–628, 1994.
9. A. Himeno, K. Kato, and T. Miya., "Silica-based planar lightwave circuits," *IEEE J. Selected Top. Quantum Electron.*, 4, 913–924, 1998.
10. M.K. Smit and C. van Dam, "PHASAR-based WDM-devices: principles, design and applications," *IEEE J. Selected Top. Quantum Electron.*, 2, 236–250, 1996.
11. Y. Yoshikuni, "Semiconductor arrayed waveguide gratings for photonic integrated devices," *IEEE J. Selected Top. Quantum Electron.*, 8, 1102–1114, 2002.
12. H. Takahashi, K. Oda, H. Toba, and Y. Inoue, "Transmission characteristics of arrayed waveguide N × N wavelength multiplexer," *J. Lightwave Technol.*, 13, 447–455, 1995.
13. K. Okamoto and A. Sugita, "Flat spectral response arrayed-waveguide grating multiplexer with parabolic waveguide horns," *Electron. Lett.*, 32, 1661–1662, 1996.
14. Y. Hida, Y. Hibino, T. Kitoh, Y. Inoue, M. Itoh, T. Shibata, A. Sugita, and A. Himeno, "400-channel arrayed-waveguide grating with 25 GHz spacing using 1.5%-Δ waveguides on 6-inch Si wafer," *Electron. Lett.*, 37, 576–577, 2001.
15. A. Sugita, A. Kaneko, K. Okamoto, M. Itoh, A. Himeno, and Y. Ohmori, "Very low insertion loss arrayed-waveguide grating with vertically tapered waveguides," *IEEE Photon. Technol. Lett.*, 12, 1180–1182, 2000.
16. Y. Inoue, A. Kaneko, F. Hanawa, H. Takahashi, K. Hattori, and S. Sumida, "Athermal silica-based arrayed-waveguide grating multiplexer," *Electron. Lett.*, 33, 1945–1946, 1997.

ATHERMAL COMPONENTS

Yasuo Kokubun

INTRODUCTION

The temperature dependence of lightwave devices is a serious and outstanding problem. This characteristic is mostly caused by the temperature dependence of optical path length.

For example, the filtering characteristics of lightwave devices such as Fabry–Perot and ring resonators, Mach–Zehnder type narrow-band filters, gratings, etc., vary with temperature. This dependence is expressed by

$$\frac{d\lambda_0}{dT} = \left(\frac{1}{L}\frac{dS}{dT}\right)\frac{\lambda_0}{2n_{eq}}, \tag{1}$$

where λ_0 is the central wavelength of the filter, L is the length of the device, n_{eq} is the equivalent index of the waveguide, and S is the optical path length which will be discussed later. Since the value of $(1/L)(dS/dT)$ is in the order of 10^{-5} [1/K], the temperature dependence of the central wavelength of narrow-band filters is about 0.1 nm/K (corresponding to 1 GHz/K in frequency in the 1.55 μm wavelength range). The lasing Wavelength of LDs also depends on the temperature with the coefficient of about 0.1 nm/K, owing to the temperature dependence of optical path length of the laser cavity. Thus, the dense WDM systems with a very close channel spacing, for example, several tens of GHz, require a precise temperature control unit to avoid the crosstalk due to this temperature dependence of central wavelength. This result in the deterioration of reliability and the device is no more suitable as a compact component.

To avoid the use of a precise and sophisticated temperature control unit, a technique to eliminate the temperature dependence of optical path length is required.

RELATIVE AND ABSOLUTE ATHERMAL CONDITIONS

Relative Athermal Condition

Here, we consider a transparent bulk material of which the refractive index is n and the length is L, between two points A and B that are located at a distance L' in air as shown in Figure 1. The optical path length S' between A and B is given by

$$S' = nL + (L' - L)$$
$$= (n-1)L + L'. \tag{2}$$

Now let us assume that the length L' in the air is independent of temperature. Then the temperature dependence of the optical path length $(1/L)(dS'/dT)$ is obtained by differentiating Eq. (2) with T and normalized by L as

$$\frac{1}{L}\frac{dS'}{dT} = \frac{dn}{dT} + (n-1)\frac{1}{L}\frac{dL}{dT},$$
$$= \frac{dn}{dT} + (n-1)\alpha, \tag{3}$$

where α is the thermal expansion coefficient $(\alpha = (1/L)(dL/dT))$.

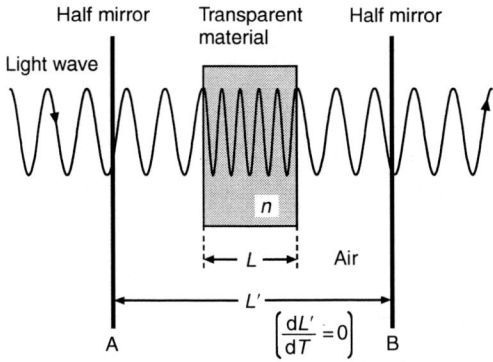

Light wave

$\left[\dfrac{dL'}{dT} = 0\right]$

Figure 1   Definition of relative optical path length

Then the athermal condition is derived from Eq.(3) as

$$\frac{dn}{dT} + (n-1)\alpha = 0. \tag{4}$$

This Eq. (4) is called the relative athermal condition. This case corresponds to solid state lasers of which cavity is held by some special materials with low thermal expansion coefficient and only the gain material is heated. The athermal condition of the focal length of convex lenses is also given by Eq. (4).

Absolute Athermal Condition

Next, we consider the case where the light propagates in a uniform material $(L = L')$. In this case, the optical path length S is given by

$$S = nL. \tag{5}$$

Then, the temperature dependence of the optical path length $(1/L)(dS/dT)$ is obtained by

$$\frac{1}{L}\frac{dS}{dT} = \frac{dn}{dT} + n\frac{1}{L}\frac{dL}{dT},$$

$$= \frac{dn}{dT} + n\alpha, \tag{6}$$

and the athermal condition is derived from Eq. (6) as

$$\frac{1}{n}\frac{dn}{dT} + \alpha = 0. \tag{7}$$

This Eq. (7) is called the absolute athermal condition (Figure 2). This case corresponds to the filters and resonators made of uniform material or optical waveguide.

PRINCIPLES OF AN ATHERMAL WAVEGUIDE

As described in the preceding subsection, absolute athermal condition of homogeneous bulk matertial is given by Eq.(7). In the remainder of this chapter, the absolute athermal condition

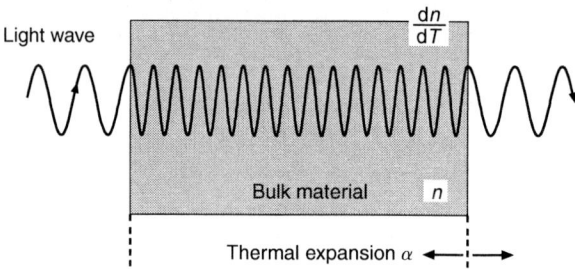

Figure 2   Definition of absolute optical path length

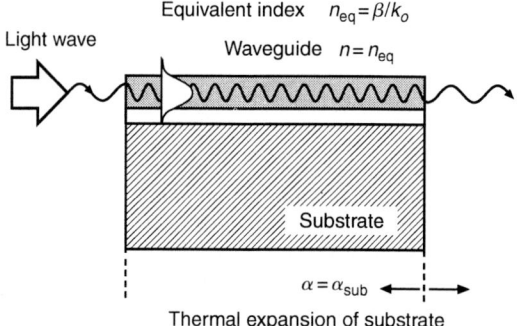

Figure 3   Definition of optical path length in waveguide

will be discussed. Since the parameters α, dn/dT, and n belong to the same material and these values are usually positive, that is, both the length of material and the refractive index increase with temperature, it is difficult to find an athermal material satisfying Eq. (7).

On the other hand, the temperature dependence of the optical path length of an optical waveguide formed on a substrate is given by

$$\frac{1}{L}\frac{dS}{dT} = \frac{dn_{eq}}{dT} + n_{eq}\alpha_{sub} \tag{8}$$

as shown in Figure 3, where n_{eq} is the equivalent index of the waveguide defined by $n_{eq} = \beta/k_0$ and α_{sub} is the thermal expansion coefficient, not of the waveguide material but of the substrate. Since the thickness of substrate is usually more than the waveguide layer, the thermal expansion of the waveguide is determined by that of the substrate. From Eq. (8), the athermal condition for a waveguide is derived as

$$\frac{1}{n_{eq}}\frac{dn_{eq}}{dT} + \alpha_{sub} = 0. \tag{9}$$

There are two methods to satisfy Eq. (9); one is to use a special substrate with a negative value of α_{sub} as used in the optical fiber [1,2] and the other is to use a special waveguide satisfying Eq. (9) by choosing a suitable combination of α_{sub}, refractive indices of core and cladding (either of which has to have a negative value of dn/dT), and the dimensions of the core and the

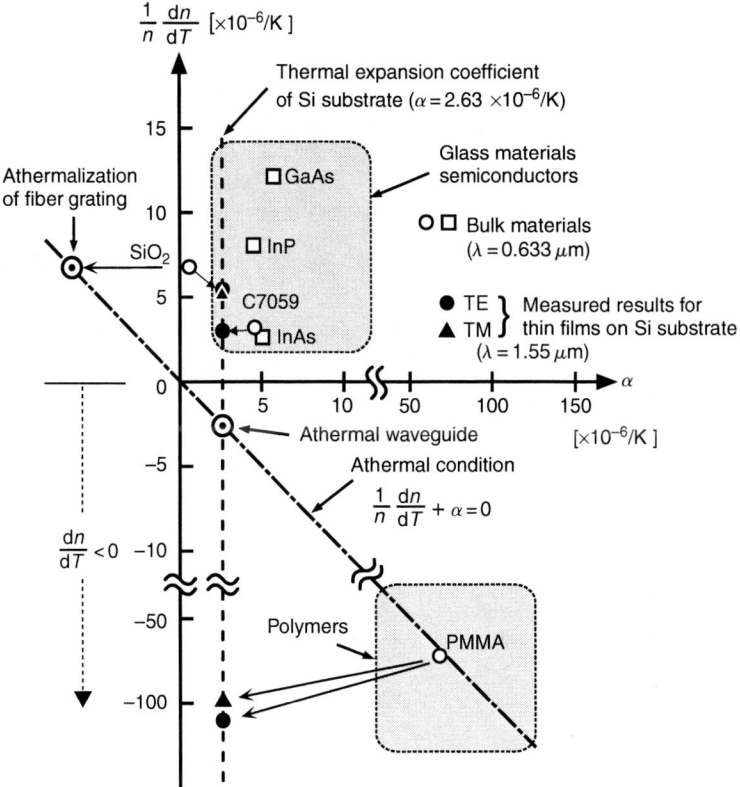

Figure 4   Plots of $(1/n)(dn/dT)$ and α

overcladding layers [3–8]. An optical waveguide designed to satisfy Eq. (9) using the latter method is called an *athermal waveguide*.

The key point of an athermal waveguide is that we can choose the values of the two parameters, $(1/n_{eq})(dn_{eq}/dT)$ and α_{sub}, independently in contrast with bulk materials, because the first term of Eq. (9) is determined by the core and the cladding materials while the second term is determined by the substrate material.

Since most of the optical materials have positive values of α_{sub} and dn/dT, the athermal waveguide requires a special material, which has a negative value of dn/dT to realize $dn_{eq}/dT < 0$. Figure 4 shows the plots of some optical materials for $(1/n)(dn/dT)$ as the ordinate and the thermal expansion coefficient α as the abscissa. It can be seen that it is difficult to be satisfied by a single bulk material as indicated by $\bigcirc$ (glass materials) and $\square$ (semiconductors). On the other hand, the athermal condition given by Eq. (9) is illustrated by the dash-dotted line. Since optical polymers usually have large negative value of dn/dT, we can design an athermal waveguide on a Si substrate by utilizing these materials having negative value of dn/dT. Although the plotted point for PMMA (Poly-Methyl-Methacrylate) looks like it is just on the line indicating the athermal condition in this figure, this point is not actually on the line.

Table 1 summarizes the refractive index n, temperature coefficient of refractive index dn/dT, and the thermal expansion coefficient α of some optical materials.

Table 1

Refractive Index and its Temperature Coefficient of Optical Waveguide Materials

Material	Wave-length	$\lambda = 0.633\ \mu m$		$\lambda = 1.3\ \mu m$		$\lambda = 1.55\ \mu m$		α (Bulk) $[10^{-6}/]$	α_v (Film) $[10^{-6}/]$
		n	dn/dT (σ) $[10^{-6}/]$	n	dn/dT (σ) $[10^{-6}/]$	n	dn/dT (σ) $[10^{-6}/]$		
C7059 Glass	TE	1.540	6.47 (—)	1.543	4.63 (0.183)	1.541	4.22 (0.248)	4.50	8.24
	TM	1.540	5.48 (—)	1.543	5.13 (0.265)	1.541	5.19 (0.238)		
SiO$_2$	TE	1.465	7.58 (—)	1.455	7.75 (0.604)	1.451	9.77 (0.334)	0.55	−3.61
	TM	1.465	7.69 (—)	1.455	7.45 (0.363)	1.451	9.92 (0.640)		
TiO$_2$/SiO$_2$ TiO$_2$:Wt.5%	TE	—	— (—)	—	— (—)	1.466	9.77 (—)	−0.11	−5.60
	TM	—	— (—)	—	— (—)	1.466	9.92 (—)		
TiO$_2$/SiO$_2$ TiO$_2$:Wt.8.5%	TE	—	— (—)	1.487	7.87 (0.228)	1.484	7.75 (0.361)	−0.21	−4.03
	TM	—	— (—)	1.488	7.67 (0.373)	1.484	7.67 (0.332)		
PMMA	TE	1.49	−167 (—)	1.479	−176 (16.5)	1.478	−163 (19.5)	67.9	198
	TM	1.49	−142 (—)	1.479	−167 (29.7)	1.478	−145 (10.3)		
PMMA-TFMA TFMA:Wt. 49%	TE	—	— (—)	—	— (—)	1.445	−144 (—)	67.9	198
	TM	—	— (—)	—	— (—)	1.445	−105 (—)		
BCB	TE	—	— (—)	1.546	— (—)	1.537	−70.4 (5.58)	52	151
	TM	—	— (—)	—	— (—)	—	— (—)		
Ta$_2$O$_5$/SiO$_2$ Ta$_2$O$_5$:8 mol%	TE	—	— (—)	—	— (—)	1.558	10.7 (0.773)	0.85	−2.72
	TM	—	— (—)	—	— (—)	1.559	9.73 (0.503)		
Ta$_2$O$_5$/SiO$_2$ Ta$_2$O$_5$:17 mol%	TE	—	— (—)	—	— (—)	1.653	11.1 (1.22)	3.98	6.67
	TM	—	— (—)	—	— (—)	1.655	9.58 (1.13)		

EXAMPLE OF AN ATHERMAL WAVEGUIDE

Let us design a three-dimensional athermal waveguide using a polymer as the material with negative value of dn/dT. For example, if the rectangular waveguide shown in Figure 5 is adopted, the temperature coefficient of equivalent refractive index dn_{eq}/dT can be expanded in terms of

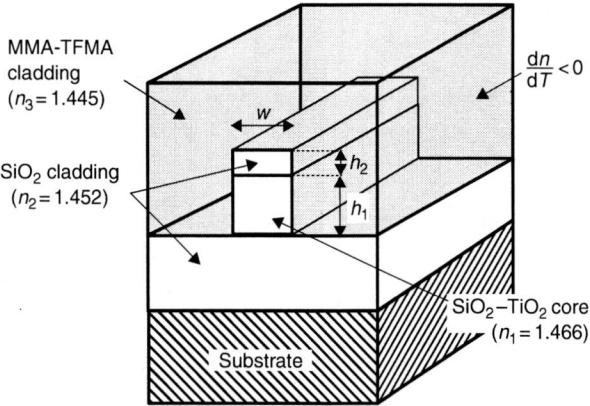

Figure 5 Example of an athermal waveguide structure

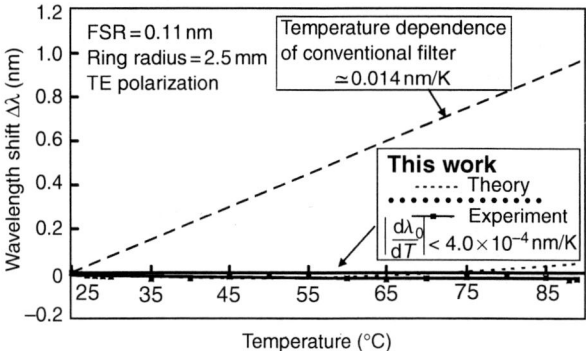

Figure 6 Center wavelength shift of ring resonator versus temperature

elemental parameters as

$$\frac{dn_{eq}}{dT} = \frac{\partial n_{eq}}{\partial n_1}\frac{dn_1}{dT} + \frac{\partial n_{eq}}{\partial n_2}\frac{dn_2}{dT} + \frac{\partial n_{eq}}{\partial n_3}\frac{dn_3}{dT}$$

$$+ w\frac{\partial n_{eq}}{\partial w}\alpha_{sub} + h_1\frac{\partial n_{eq}}{\partial h_1}\alpha_1^v + h_2\frac{\partial n_{eq}}{\partial h_2}\alpha_2^v, \qquad (10)$$

where h_1 and W are the thickness and width of the core, respectively, and h_2 is the thickness of the SiO_2 upper-loaded cladding, as shown in Figure 5. α_2^v and α_2^v are the vertical thermal expansion coefficients of the core and the SiO_2 upper cladding layer ($\alpha_1^v = -5.60 \times 10^{-6}/K$, $\alpha_2^v = -4.03 \times 10^{-6}/K$). The vertical thermal expansion coefficient was calculated by $\alpha_i^v = 3\alpha_i - 2\alpha_{sub}Z$ ($i = 1$ or 2) to make the volume expansion coefficient equal to that to bulk. The partial derivative terms of the right-hand side in Eq. (10) can be calculated by some numerical mode solver like the Finite Element Method.

Since the confinement factor of polymer overcladding layer decreases with the increase of temperature due to its negative value of dn/dT. The optimum value of temperature coefficient of optical path length $(-1/L)(dS/dT)$ is not zero but a small negative value at room temperature [8].

Figure 6 shows the measured center wavelength shift of a ring resonator made of an athermal waveguide. In this device, the waveguide structure shown in Figure 5 was adopted and a copolymer of PMMA and TFMA (Tri-Fluoroethyl-Methacrylate) was used as the overcladding with a negative value of dn/dT. The copolymer of PMMA and TFMA (TFMA 49 wt.%) was 1.445 at $\lambda = 1.55\ \mu$m. The measured values are plotted by closed square in Figure 6. The maximum wavelength shift was <0.03 nm in a wide temperature range of 25 to 89°C. In this device the optimum value of $(1/L)(dS/dT)$ was designed to be $-2.24 \times 10^{-6}[1/K]$.

ATHERMALIZATION OF MULTIPATH INTERFEROMETERS

Since the central wavelength of multipath interferometers such as a Mach-Zehnder interferometer and an Arrayed Waveguide Grating (AWG) filters is determined by the optical path difference, the central wavelength can be made independent of temperature by equating the temperature coefficient of the optical path length difference. For example, let us suppose the Mach-Zehnder interferometer as shown in Figure 7. If the equivalent refractive index n_{eq} is uniform in the whole waveguide region, the optical path length difference ΔS between arm #1 and #2 is determined in terms of the path length and the equivalent index n_{ex} by

$$\Delta S = n_{ex}(L_1 - L_2) = n_{ex}\Delta L. \tag{11}$$

However, if the arm #2 is composed of two different regions with different equivalent indices n_{eq}^a and n_{eq}^b and their temperature coefficients as shown in Figure 7, the optical path length difference ΔS is given by

$$\Delta S = n_{ex}^a L_1 - (n_{eq}^a L_a + n_{eq}^b L_b) = n_{ex}^a (L_1 - L_a) - n_{eq}^b L_b. \tag{12}$$

In this case, assuming the thermal expansion coefficient α is uniform in the whole substrate region, the temperature coefficient of optical path length difference is expressed by

$$\frac{\partial \Delta S}{\partial T} = \left(\frac{\partial n_{eq}^a}{\partial T} + \alpha n_{eq}^a\right)(L_1 - L_a) - \left(\frac{\partial n_{eq}^b}{\partial T} + \alpha n_{eq}^b\right)L_b. \tag{13}$$

Equating $\partial \Delta S/\partial T$ to be zero, the following athermal condition is obtained [9].

$$\frac{\left((\partial n_{eq}^a / \partial T) + \alpha n_{eq}^a\right)}{\left((\partial n_{eq}^b / \partial T) + \alpha n_{eq}^b\right)} = \frac{L_1 - L_a}{L_b} \tag{14}$$

To satisfy Eq. (14), the relation is $(\partial n_{eq}^a/\partial T) < (\partial n_{eq}^b/\partial T)$ required.

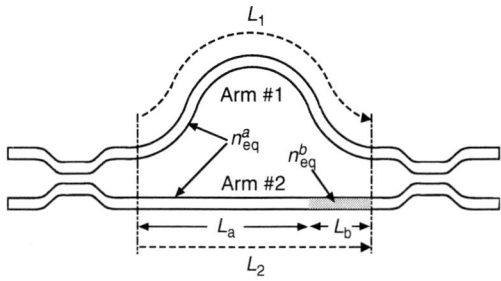

Figure 7 Athermalization of Mach-Zehnder interferometer

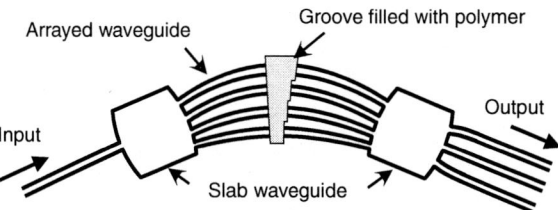

Figure 8   Athermalization of AWG filter

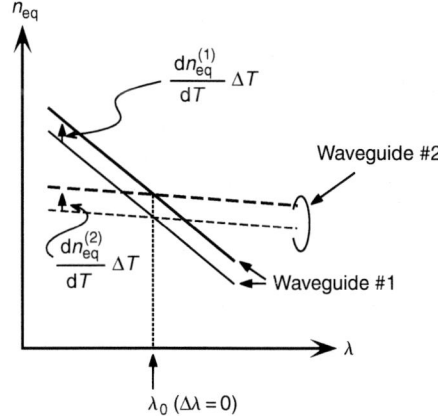

Figure 9   Athermalization of asymmetric directional coupler add/drop filter

The same technique can be applied to an AWG filter [10]. Since this condition does not require a special material with negative value $\partial n/\partial T$, this has been used for a semiconductor AWG filter [10].

Another athermalization method of AWG is the use of polymer material filled in a groove as shown in Figure 8 [11]. This method was applied to silica AWG filters.

ATHERMALIZATION OF ASYMMETRIC COUPLER FILTERS

The asymmetric directional coupler add/drop filter requires another method. The operation principle of the asymmetric directional coupler filter is that the light propagating in one of asymmetric coupled waveguides #1 is transmitted to another waveguide #2 at the wavelength at which the equivalent index of waveguide #1 $n_{eq}^{(1)}$ is equal to that of waveguide #2 $n_{eq}^{(2)}$. Since the wavelength dependence (dispersion) of an equivalent index of two waveguides are different in the asymmetric directional coupler, the coupling occurs at only one specific wavelength as shown in Figure 9. If the temperature dependence of equivalent index of waveguide #1 ($\partial n_{eq}^{(1)}/\partial T$), is different from that of waveguide #2 ($\partial n_{eq}^{(2)}/\partial T$) the crossing point of dispersion curves shifts in the wavelength axis, and so the center wavelength has the temperature dependence. However, if $\partial n_{eq}^{(1)}/\partial T$ is equal to $\partial n_{eq}^{(2)}/\partial T$ the center wavelength is independent of temperature as shown in Figure 9. This method was applied to the vertically coupled ARROW-type asymmetric directional coupler filter [12].

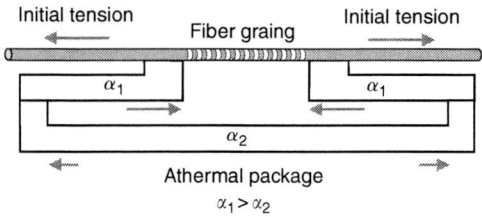

Figure 10   Athermalization of fiber Bragg grating

Table 2
Comparison of Athermalization Methods of Waveguide Filters

Method Filter	Athermal waveguide [3]-[8]	Combination of two different $\dfrac{1}{L}\dfrac{dS}{dT}$ [9]–[11]	Independent control of n_{eq} and $\dfrac{dn_{eq}}{dT}$ [12]
Mach-Zehnder type	Applicable	Applicable	Not applicable
AWG	Applicable	Applicable	Not applicable
DFB and DBR types	Applicable	Not applicable	Not applicable
Resonator (FP, Ring)	Applicable	Not applicable	Not applicable
Asymmetric directional coupler	Not applicable	Not applicable	Applicable

ATHERMALIZATION OF FIBER BRAGG GRATING

In optical fibers, the athermalization of optical path length can be achieved by using some special methods to reduce the thermal expansion as shown in Figure 4. One of methods to reduce the thermal expansion is the use of special polymer jacket with a negative value of α [1,13]. Another method is the use of athermal package, which is composed of two different materials with large and small thermal expansion α_1 and α_2 as shown in Figure 10.

SUMMARY

In the beam components such as multilayer thin film filter and Virtually imaged phased array (VIPA), a method that utilizes the difference of α between the film and the substrate has been proposed [2]. Table 2 summarizes the athermalization methods of waveguide type optical filters.

REFERENCES

1. T. Shirasaki, "Temperature-independent narrow-band filter," in Proceedings of the *ECOC'96*, Oslo, WeD.1.6 (1996).
2. R. Kashyap, S. Hornung, M. H. Reeve, and S. A. Cassidy, "Temperature desensitisation of delay in optical fibres for sensor applications," *Electron. Lett.*, 19, 1039–1040 (1983).
3. Y. Kokubun, F. Funato and M. Takizawa, "Athermal waveguides for temperature-independent lightwave devices," *IEEE Photon. Technol. Lett.*, 5, 1297–1300 (1993).

4. Y. Kokubun, M. Takizawa, and S. Taga, "Three-dimensional athermal waveguides for temperature inde-pendent lightwave devices," *Electron. Lett.*, 30, 1223–1224 (1994).
5. S. Taga, H. Tanaka, and Y. Kokubun, "Three-dimensional athermal waveguide at 1.3 μm wavelength for temperature independent lightwave devices," *Opt. Rev.*, 3, 478–480 (1996).
6. Y. Kokubun, S. Yoneda, and H. Tanaka, "Temperature-independent narrow-band optical filter at 1.3 μm wavelength by an athermal waveguide," *Electron. Lett.*, 32, 1998–2000 (1996).
7. Y. Kokubun, S. Yoneda, and H. Tanaka, "Temperature-independent narrow-band optical filter by an ather-mal waveguide," *IEICE Trans. Electron.*, E80-C, 632–639 (1997).
8. Y. Kokubun, S. Yoneda, and S. Matsuura, "Athermal narrow-band optical filter at 1.55 μm wavelength by silica-based athermal waveguide," *IEICE Trans. Electron.*, E-81C, 1187–1194 (1998).
9. H. Tanobe, Y. Kondo, Y. Kadota, H. Yasaka, and Y. Yoshikuni, "A temperature insensitive InGaAsP-InP optical filter," *IEEE Photon. Technol. Lett.*, 8, 1489–1491 (1996).
10. H. Tanobe, Y. Kondo, Y. Kadota, K. Okamoto, and Y. Yoshikuni, "Temperature insensitive arrayed waveguide gratings on InP substrates," *IEEE Photonics Technol. Lett.*, 10, 235–237 (1998).
11. Y. Inoue, A. Kaneko, F. Hanawa, H. Takahashi, K. Hattori, and S. Sumida, "Athermal silica-based arrayed-waveguide grating multiplexer," *Electron. Lett.*, 33, 1945–1947 (1997).
12. S. Suzuki, S. Endo, S. Sato, W. Pan, S. T. Chu, and Y. Kokubun, "Design of temperature independent Add/Drop filter using vertical coupled ARROW filter," *Jpn. J. Appl. Phys.*, 39, 1497–1502 (2000).
13. T. Kakuta, and S. Tanaka, "LCP coated optical fiber with zero thermal coefficient of transimission delay time," in *Proceedings of the Thirty-Sixth International Wire and Cable Symposium*, pp. 234–240 (1987).
14. H. Takahashi, K. Oda, H. Toba, and Y. Inoue, "Transmission characteristics of arrayed waveguide $N \times N$ wavelength multiplexer," *J. Lightwave Technol.* 13, 447–455 (1995).
15. K. Tada, Y. Nakano, and A. Ushirokawa, "Temperature compensated coupled cavity diode lasers," *Opt. Quantum Electron.*, 16, 463–469 (1984).

ATTENUATOR

Yasuo Kokubun

In the optical communication system, the power of the optical signal should be greater than the minimum detectable power of the receiver (receiver sensitivity) and should also be smaller than the maximum allowable power of the receiver to assure the allowable bit error rate. Attenuator is used to adjust the optical power to be in the allowable power range of receiver. Also in front of the optical amplifier, the optical power should be adjusted in a specific range, typically between –30 and –10 dBm, and an attenuator is sometimes required. As fixed attenuators, optical fibers with absorptive dopants or thin metal films are used.

In the wavelength division multiplexing (WDM) system, the multiplexed signals are simulta-neously amplified. Even when the power levels of all wavelength channels are equalized, the power profile of wavelength channels is not uniform after passing through the optical amplifier owing to the gain spectrum. When the wavelength multiplexed signal passes several amplifiers in a long distance optical fiber transmission system, the nonuniformity of power profile of wave-length channels is accumulated and emphasized. As a result, some of wavelength channels cannot satisfy the allowable power range of optical amplifier. To solve this problem, variable optical attenuator (VOA) is used to adjust the uniformity of wavelength channels in WDM systems. MEMS type and waveguide type VOA have been reported.

D

DIRECTIONAL COUPLER

Yasuo Kokubun

A 2×2 port device for which the transfer matrix is expressed as:

$$\begin{bmatrix} B_1 \\ B_2 \end{bmatrix} = \begin{bmatrix} t_{11} & t_{12} \\ t_{21} & t_{22} \end{bmatrix} \cdot \begin{bmatrix} A_1 \\ A_2 \end{bmatrix} \tag{1}$$

The transfer matrix has a nonzero and equal off-diagonal components t_{12} and t_{21} ($t_{12} = t_{21}$) and also equal diagonal components t_{11} and t_{22} ($t_{11} = t_{22}$), where the port labels are defined as shown in Figure 1.

When there is no loss in the device, the transfer matrix is unitary and Eq. (1) can be rewritten as:

$$\begin{bmatrix} B_1 \\ B_2 \end{bmatrix} = \begin{bmatrix} \cos\theta & j\sin\theta \\ j\sin\theta & \cos\theta \end{bmatrix} \cdot \begin{bmatrix} A_1 \\ A_2 \end{bmatrix} \tag{2}$$

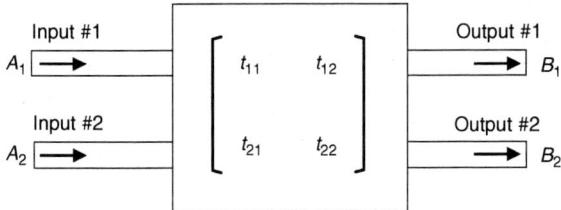

Figure 1 Definition of field amplitudes of input and output optical signals and transfer matrix

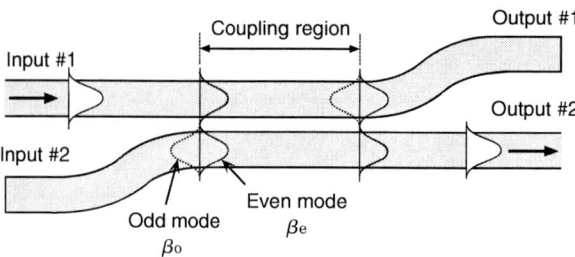

Figure 2 Waveguide directional coupler

When the parameter θ is $\pi/4$, the optical power is equally divided into output ports, and this device is called 3 dB coupler.

In the actual device, beam optics such as thin film or half mirror with microlenses and waveguide are used. In the optical waveguide, a special phenomenon called waveguide coupling is utilized to form the directional coupler as shown in Figure 2 (See Optical Coupling in Waveguides).

DISTRIBUTED BRAGG REFLECTOR LASER

Shigehisa Arai

A laser consisting of a distributed-Bragg-reflector (DBR) having a strong wavelength selection mechanism in its reflectivity due to periodicity in the refractive index of the waveguide or the resonator (also called grating), is called as a DBR laser. A laser consisting of a periodic structure on an active region is called a distributed feedback (DFB) laser, in which periodicities in both refractive index and optical gain (can be regarded as an imaginary part of the refractive index) are the areas of concern. In 1971, Kogelnik and Shank [1] demonstrated the first DFB laser using dye active medium, after which the first DBR laser was demonstrated by Kaminow and Weber [2].

The transmission bandwidth of ultimately low-loss optical fibers (whose minimum loss is about 0.2 dB/km at around 1.55 μm wavelength) is limited by the inverse of product of the chromatic dispersion of the optical fiber and the effective spectral width of the optical signal. Also, the lasing spectrum of ordinary Fabry–Perot cavity lasers showed randomness in their time-resolved property. To counter these problems, a dynamic-single-mode (DSM) laser that operates in single longitudinal mode under rapid direct modulation was required for wideband optical communications [3]. Semiconductor-based DBR lasers as well as DFB lasers have been investigated especially for light sources in wideband optical communications because of their superior single-mode operation characteristics.

Simple DBR laser with relatively long cavity is not in practical use now; however the concept of DBR laser is widely used in vertical cavity surface emitting lasers (VCSELs) and wavelength tunable lasers.

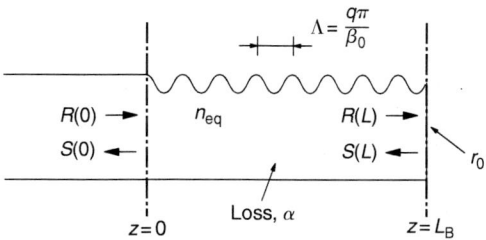

Figure 1 Periodic structure with reflection facet on one side. $R(0)$ and $S(0)$ denote amplitudes of incident and reflected waves, respectively

WAVELENGTH SELECTIVITY OF DBR

In a waveguide with a periodic index structure as shown in Figure 1, the reflectivity r (for electric field) of a DBR, in which the refractive index sinusoidally changes with the period $\Lambda = q\pi/\beta_0 = q\lambda_B/2$ (q is a positive integer representing the order of the grating, λ_B is the Bragg wavelength), can be expressed by the ratio $R(z)/S(z)$ of amplitudes of incident and reflected waves at position $z = 0$ [4].

$$r = S(0)/R(0)$$
$$= \frac{\gamma r_0 \exp\{-j(2\beta_0 L_B)\}\cosh(\gamma L_B) - \left[((\alpha/2) + j\delta)r_0 \exp\{-j(2\beta_0 L_B)\} + j\kappa\right]\sinh(\gamma L_B)}{\gamma\cosh(\gamma L_B) + \left[((\alpha/2) + j\delta) + j\kappa r_0 \exp\{-j(2\beta_0 L_B)\}\right]\sinh(\gamma L_B)}, \quad (1)$$

where r_0, $\beta_0 (= q\pi/\Lambda)$, α, and κ are the amplitude of reflectivity on the right-hand side facet, the propagation constant at the Bragg wavelength, an absorption coefficient for optical power (not for electric field), and the coupling coefficient (for electric field) between forward and backward propagating waves, respectively. δ is the wave number detuning from the Bragg wavelength and is given by $\delta = \beta - \beta_0 = 2n_{eq}\pi/\lambda - q\pi/\Lambda$, where n_{eq} is an equivalent refractive index of the waveguide, γ satisfies the dispersion relation

$$\gamma^2 = \left(\frac{\alpha}{2} + j\delta\right)^2 + \kappa^2. \quad (2)$$

When the end facet reflectivity is negligibly small, the fundamental reflection characteristics of DBR can be expressed by

$$r = \frac{-j\kappa\tanh(\gamma L_B)}{\gamma + ((\alpha/2) + j\delta)\tanh(\gamma L_B)} \equiv |r|\exp(-j\phi) \quad (3)$$

Figure 2 shows the wavelength dependences of the power reflectivity ($R = |r|^2$) and the phase delay (ϕ) of the reflected wave. The maximum attainable reflectivity decreases with an increase in the ratio $\alpha/(2\kappa)$, a low-loss waveguide is essential for high reflectivity DBRs. As can be seen, the phase delay at the Bragg condition ($\delta = 0$) is $\pi/2$ (not 0), the active region length should be properly selected to set the resonant mode at the Bragg wavelength. In other words, a laser with double-sided DBRs, the active region length (the distance between the index maximum points in each DBR) should be odd times of 1/4 wavelength (λ/n_{eq}) hence the round trip phase shift becomes odd times of π. This design policy has been adopted in not only special types of

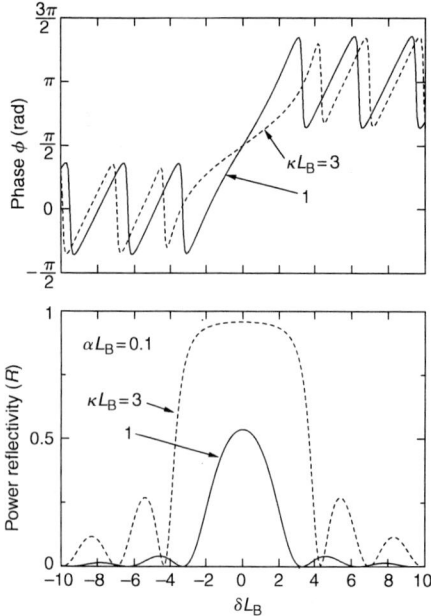

Figure 2   Power reflectivity (lower) of DBRs and phase delay of reflected lightwaves as a function of detuning parameter δL_B

index-coupled DFB lasers the so called $\pi/2$- (or quarter wavelength) phase-shifted and phase-adjusted DFB lasers, but also in VCSELs with DBRs on both sides.

EFFECTIVE LENGTH OF DBR

The optical power confined in a DBR region exponentially changes, as shown in Figure 3(a), where forward (or backward) propagating optical power decreases (or increases) with propagation. In a DBR of a low-loss waveguide without facet reflection, the total optical power confined in the DBR can be expressed as a product of the incident optical power and the effective length L_{eff} of the DBR which is approximately given by [5].

$$L_{eff} \equiv \frac{1}{2}\frac{\partial \phi}{\partial \delta}. \tag{4}$$

When κL_B is large and $\alpha/(2\kappa)$ is negligibly small, L_{eff} approaches to $1/(2\kappa)$. Figure 3(b) shows calculated L_{eff} as a function of the DBR length for various coupling strength κL_B.

LASING CONDITIONS OF DBR LASER

Figure 4 shows a schematic diagram of a DBR laser consisting of active and DBR regions on both sides coupled to each other via a certain waveguide coupling method. The DBR structure is similar to a conventional Fabry-Perot type laser; except that it has DBR regions at both ends of the active waveguide instead of cleaved facets. The DBR regions consist of low-loss optical waveguides with corrugations (grating) on or beneath the waveguides, which are made of compound crystals with an energy gap wider than that of the active layer. By taking the power

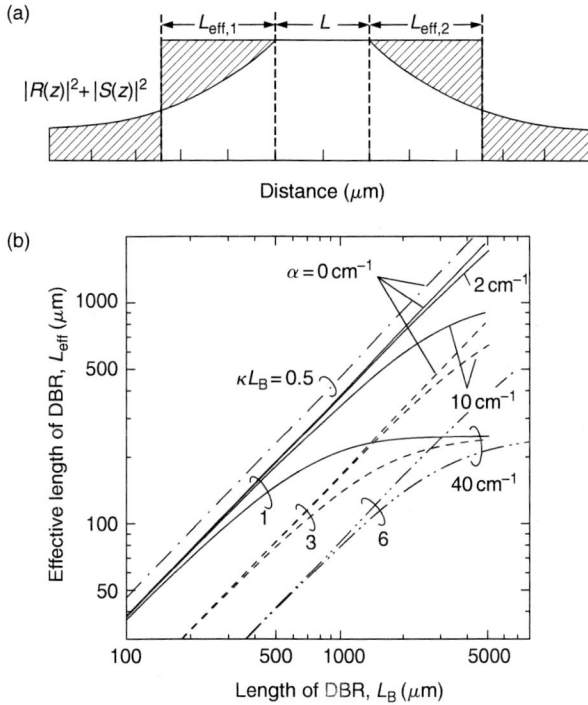

Figure 3 (a) An illustration explaining the concept of the effective length of DBR, and (b) calculated effective length of DBR as a function of actual length [5]

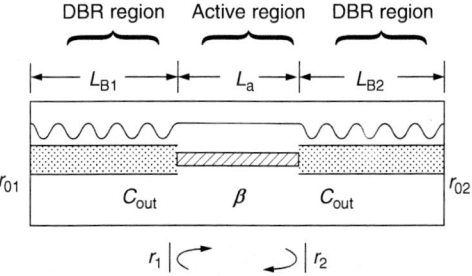

Figure 4 Analysis model of a DBR laser consisting of DBRs on both sides of the active region

coupling efficiency, C_{out}, between the active and DBR regions, the lasing condition for a DBR laser is

$$C_{out}^2 |r_1||r_2| \exp\left\{(\xi g_{th} - \alpha) L_a\right\} \exp\left\{-j(\phi_1 + \phi_2 + 2\beta L_a)\right\} = 1, \qquad (5)$$

where ξ and g_{th} are the optical confinement factor and the threshold gain of the active layer, respectively, α is the absorption coefficient of the waveguide in the active region, which are defined for optical power. The propagation constant of the active waveguide is β.

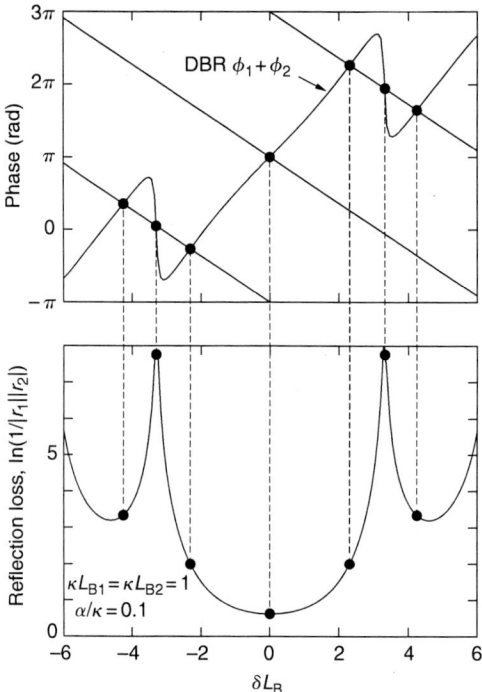

Figure 5   Reflection loss (lower) and phase delay (upper) of a function of detuning parameter δL_B of a DBR laser as shown in Figure 4

From the real part of Eq. (5), threshold gain of the DBR laser is

$$\xi g_{th} = \alpha + \frac{1}{L_a} \ln\left(\frac{1}{C_{out}^2 |r_1||r_2|} \right). \tag{6}$$

Since C_{out} directly affects both reflectivity and transmittivity of DBR, therefore, increasing the threshold current and decreasing the differential quantum efficiency, it is essential for high performance DBR lasers to adopt a certain waveguide coupling structure with very high C_{out}. Various structures of waveguide coupling have been developed for this purpose, such as, a large optical cavity (LOC), a taper coupling (TPC), a butt-joint built-in (BJB), and a bundle-integrated-guide (BIG) structures. About 95 to 99% of C_{out} can be obtained in a BIG structure with relatively large fabrication tolerance [6].

From Eq. (4), the phase delay due to reflection from the DBR ϕ_i can be approximately expressed by (Figure 5)

$$\phi_i = {\pi}\!/\!{2} + 2\delta L_{eff,i} \qquad (i = 1, 2) \tag{7}$$

By substituting Eq. (7) into the imaginary part of the Eq. (5), we can obtain

$$\phi_1 + \phi_2 + 2\beta L_a = 2q\pi,$$
$$2\delta\left(L_{eff,1} + L_{eff,2} + \beta L_a \right) = \left(2q - 1 \right)\pi. \qquad (q : \text{integer}) \tag{8}$$

The resonant mode spacing $\Delta\lambda_{sp}$ between neighboring longitudinal modes of a DBR laser is then given by

$$\Delta\lambda_{sp} = \frac{\lambda^2}{2\left\{n_{eff,a}L_a + n_{eff,b}\left(L_{eff,1} + L_{eff,2}\right)\right\}},$$ (9)

where $n_{eff,a}$ and $n_{eff,b}$ are effective refractive indices of waveguides in the active and DBR regions taking into account the chromatic dispersion. This approximation fits well when the active region length is similar or larger than the effective length of DBR and number of resonant modes exist in the stobband of the DBR.

REFERENCES

1. H. Kogelnik and C. V. Shank, "Stimulated emission in a periodic structure," *Appl. Phys. Lett.*, 18, 152–154 (1971).
2. I. P. Kaminow and H. P. Weber, "Poly (methylemethacrylate) dye laser with internal diffraction gating resonator," *Appl. Phys. Lett.*, 18, 497–499 (1971).
3. Y. Suematsu, S. Arai, and K. Kishino, "Dynamic-single-mode semiconductor lasers with a distributed reflector, *J. Lightwave Technol.*, LT-1, 161–176 (1983).
4. S. Wang, "Principles of distributed feedback and distributed Bragg-reflector lasers," *IEEE J. Quantum Electron.*, QE-10, 413–427 (1974).
5. F. Koyama, Y. Suematsu, S. Arai, and T. Tanbun-Ek, "1.5–1.6 μm GaInAsP/InP dynamic-single-mode (DSM) lasers with distributed Bragg reflector," *IEEE J. Quantum Electron.*, QE-19, 1042–1051 (1983).
6. Y. Tohmori, K. Komori, S. Arai, and Y. Suematsu, "1.5–1.6 μm GaInAsP/InP bundle-integrated-guide (BIG) distributed-Bragg-reflector (DBR) lasers," *Trans. IEICE Jpn.*, E70, 494–503 (1987).

DISTRIBUTED FEEDBACK LASER

Shigeyuki Akiba

THEORETICAL ANALYSIS OF DISTRIBUTED FEEDBACK STRUCTURE

Brief History

Distributed feedback (DFB) laser was proposed and demonstrated by Kogelnik and Shank in a dye laser [1,2]. Semiconductor DFB laser was first realized for AlGaAs/GaAs [3,4], and then for InGaAsP/InP material for 1.55 μm fiber transmission [5]. These early stage DFB lasers were made on uniform corrugation grating for distributed reflection. However, it has been known [1] that the uniform grating manifests two lowest-threshold modes and is not convenient for stable single mode operation.

Haus and Shank [6] has shown that the introduction of phase-shift into grating should lead to stable single mode operation at the exact Bragg wavelength, where the distributed reflection is strongest and therefore the lowest-threshold current can be obtained. Utaka et al. [7,8] demonstrated that such a prediction was right.

Derivation of the Coupled-Wave Equation in a Generalized form with Phase-Shifted Grating

Figure 1 shows the theoretical model of a Bragg waveguide with a phase-shift, and the structural parameters denoted here are of a general case [9]. Two gratings located in the left- (region 1) and

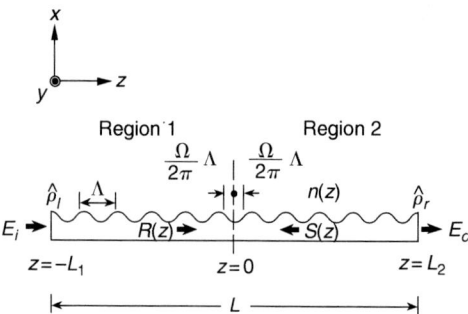

Figure 1   Theoretical model of a Bragg waveguide with a phase shift

right- (region 2) hand sides are phase-shifted by 2Ω Λ is a corrugation grating pitch, κ is the coupling coefficient defined later. $L\ (=L_1 + L_2)$ is the length of the Bragg waveguide, $\hat{\rho}_1$ and $\hat{\rho}_r$ are the field reflection coefficients at left and right ends, respectively. A general Bragg waveguide has a spatial modulation of the refractive index $n(z)$ and the gain constant $\alpha(z)$ of the form

$$n(z) = n_0 + \Delta n \cos(2\beta_0 z \pm \Omega) \quad \left(z \gtrless 0 \right),$$

$$\alpha(z) = \alpha_0 + \Delta\alpha \cos(2\beta_0 z \pm \Omega) \quad \left(z \gtrless 0 \right),$$

(1)

where n_0 and α_0 are the mean values and Δn and $\Delta\alpha$ are the amplitudes of the variation. Note that the signs of Ω in Eq. (1) is inverted in the right (region 2) and the left (region 1) halves. β_0 is given by

$$\beta_0 = \pi / \Lambda = n_0 \omega_0 / c,$$

(2)

where ω_0 is Bragg angular frequency. As far as a DFB laser with moderate corrugation depth is concerned, $\Delta n/n_0$ is much larger than $\Delta\alpha/\alpha_0$, and hence only an index coupling ($\Delta n \neq 0$, $\Delta\alpha = 0$ or $\alpha = \alpha_0$) is assumed in the following, although gain coupling condition ($\Delta\alpha \neq 0$) can also be obtained when a depth of corrugation is fairly large (>0.1 μm) or the active layer thickness is modulated. This model is based on the wave equation for the electric field $E(z)$ and is in the form of

$$\left(\frac{\partial^2}{\partial z^2} + k^2 \right) E(z) = 0.$$

(3)

Wave number k for the electric field of propagating light is expressed by

$$k^2 = \beta^2 + j2\,\alpha\beta + 4\kappa\beta \cos\{2\beta_0 z \pm \Omega\},$$

(4)

assuming $\alpha \ll \beta_0$ and $\Delta n \ll n_0$. α and β are a gain constant and a propagation constant, respectively. Considering the wavelength range near the Bragg wavelength, a right-going wave $R(z)\exp(-j\beta_0 z)$ and a left-going wave $S(z)\exp(j\beta_0 z)$ couple with each other by the periodic refractive index change, and thus the electric field is written as the sum

$$E(z) = R(z)\exp(-j\beta_0 z) + S(z)\exp(j\beta_0 z).$$

(5)

The following two equations, that is, the so-called coupled-wave equations [2], are obtained by substituting Eq. (5) into Eq. (3) and neglecting second derivatives of the gradually varying amplitudes $R(z)$ and $S(z)$:

$$-\frac{\partial R}{\partial z}+(\alpha - j\delta\beta)\, R = j\kappa\, \exp(\pm j\Omega)\, S, \qquad \left(z \gtrless 0\right),$$

$$\frac{\partial S}{\partial z}+(\alpha - j\delta\beta)\, S = j\kappa\, \exp(\mp j\Omega)\, R, \qquad \left(z \gtrless 0\right), \tag{6}$$

where κ is the coupling coefficient and it is given by

$$\kappa = \Delta n\, \pi\, /\, \lambda_0 = k_0\, \Delta n\, /\, 2, \tag{7}$$

where λ_0 is the Bragg wavelength. $\delta\beta$ is the deviation of the propagation constant from the Bragg wave frequency, which is given by

$$\delta\beta = \frac{\left(\beta^2 - \beta_0^{\,2}\right)}{2\beta_0} \approx \beta - \beta_0 = \frac{n_0\left(\omega - \omega_0\right)}{c}. \tag{8}$$

General solutions of the coupled-wave Eqs. (6) are in the form of

$$R(z) = r_1\, \exp(\gamma z) + r_2\, \exp(-\gamma z),$$

$$S(z) = s_1\, \exp(\gamma z) + s_2\, \exp(-\gamma z),$$

$$\gamma^2 = (\alpha - j\delta\beta)^2 + \kappa^2. \tag{9}$$

From these forms, we obtain the following eigenvalue equation,

$$\frac{j\left(\hat{\Gamma}^2 + \kappa^2 e^{2\gamma L_1}\right)\rho_1 + \kappa\hat{\Gamma}\left(1 - e^{2\gamma L_1}\right)}{j\kappa\hat{\Gamma}\left(1 - e^{2\gamma L_1}\right)\rho_1 + \kappa^2 + \hat{\Gamma}^2\, e^{2\gamma L_1}} \cdot \frac{j\left(\hat{\Gamma}^2 + \kappa^2 e^{2\gamma L_2}\right)\rho_r + \kappa\hat{\Gamma}\left(1 - e^{2\gamma L_2}\right)}{j\kappa\hat{\Gamma}\left(1 - e^{2\gamma L_2}\right)\rho_r + \kappa^2 + \hat{\Gamma}^2\, e^{2\gamma L_2}} = -e^{j2\Omega},$$

where

$$\rho_1 = \hat{\rho}_1\, \exp\left\{-j\left(\beta_0 L - \Omega\right)\right\}, \qquad \rho_r = \hat{\rho}_r\, \exp\left\{-j\left(\beta_0 L + \Omega\right)\right\},$$

$$\Gamma = \gamma + \alpha - j\delta\beta, \qquad \hat{\Gamma} = -\gamma + \alpha - j\delta\beta, \tag{10}$$

The boundary conditions at the Ω-shift position and at both ends determine the eigenvalues, which are the gain constant α and the propagation constant $\delta\beta$.

In the following, some results of numerical calculations of the eigenvalue equation are described. For the simplicity, the position of the phase-shift is located at the center of the DFB waveguide, that is, $L_1 = L_2$.

Resonant Modes (DFB modes)

Figure 2 shows the threshold gain α_{th} and the propagation constant in the form of $\delta\beta$, both normalized by the DFB length L, of resonant modes. Reflectivity at both ends is assumed to be zero. Parameter is the phase-shift Ω. The normalized coupling coefficient κL is set to be two. $\Omega = 0$ corresponds to a uniform corrugation and two identical modes (•) with the same lowest-threshold gain $\alpha_{th}L$ exist at symmetric deviations from the Bragg wavelength. The Bragg waveguide exhibits high reflectivity for $\delta\beta L$ between those two lowest modes, a so-called stop-band. Introduction of a finite phase-shift Ω breaks the symmetric modes and the mode with positive $\delta\beta L$ side becomes the lowest. With increasing Ω from 0 to $\pi/2$, the lowest order mode moves toward smaller $\delta\beta L$, that is, approaches the Bragg wavelength and the $\alpha_{th}L$ is decreased. When $\Omega = \pi/2$, $\alpha_{th}L$ of the lowest order mode becomes minimum at the exact Bragg wavelength, which is at the center in the stopband, and symmetric spectrum appears for higher order modes. Considering the first-order corrugation, the phase-shift Ω of $\pi/2$ corresponds to a quarter wavelength of the light. Therefore, this type of Bragg waveguide is called $\lambda/4$ shifted corrugation grating.

Threshold Gain Difference and Single Mode Condition

The power intensity ratio of a main mode to a side mode S_0/S_1, which is called a side mode suppression ratio (SMSR), may be a direct index of single longitudinal mode (SLM) operation. However, SMSR is not easily derived from the parameters of the DFB waveguide structure, since it depends on conditions of modulation and nonlinear effects such as gain saturation. Instead of SMSR, it may be convenient to use a threshold gain difference $\Delta\alpha_{th}$ between the lowest-order mode and the second lowest-order mode defined by Eq. (11) as a parameter of SLM condition

$$\Delta\alpha_{th} = \alpha_{th}\left(\lambda_1\right) - \alpha_{th}\left(\lambda_0\right). \tag{11}$$

The relation between SMSR and $\Delta\alpha_{th}$ can be estimated from a multimode rate equation. In most situations, $\Delta\alpha_{th}$ of 10 cm^{-1} is sufficient to maintain SMSR over 35 dB even under 100% modulation at the resonant frequency, the hardest modulation condition.

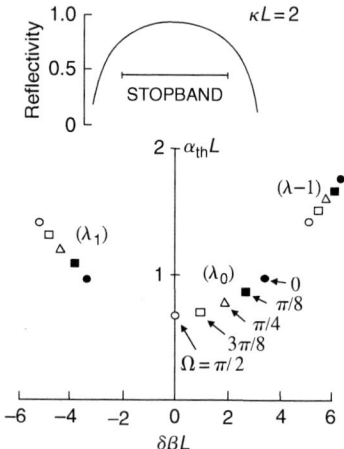

Figure 2   Threshold gain α_{th} and propagation constant in the form of $\delta\beta$, both normalized by the DFB length L, of resonant modes

Spectra below Threshold

Behavior of longitudinal mode below threshold can be analyzed by calculation of transmitting spectra using the equations developed earlier. The power transmitting ratio T for the model depicted in Figure 1 can be calculated by setting $E_i = R(-L/2)\exp(j\beta_0 L/2)$ and $E_o = R(L/2)\exp(-j\beta_0 L/2)$ and is given by

$$T = \left|\frac{E_o}{E_i}\right|^2 = \left|\frac{4e^{-j\Omega}\gamma^2}{\left(\Gamma e^{-\gamma L/2} - \hat{\Gamma}e^{\gamma L/2}\right)^2 - \kappa^2 e^{-2j\Omega}\left(e^{-\gamma L/2} - e^{\gamma L/2}\right)^2}\right|, \tag{12}$$

where nonreflective ends are assumed, that is, $\rho_l = \rho_r = 0$. Figure 3 shows the power transmitting ratio T calculated by Eq. (12) for gratings with (a) $\Omega = 0$ and (b) $\pi/2$. With increasing α, two symmetric modes outside the stopband, which is defined by $-\kappa L < \delta\beta L < \kappa L$, increases for $\Omega = 0$, whereas the mode at the Bragg wavelength increases for $\Omega = \pi/2$. Finally T attains infinity at a certain gain, which corresponds to the threshold gain α_{th}. Threshold condition is given by setting the denominator of Eq. (12) to be zero in this case. In DFB lasers, the ratio of the main to side mode and the main mode position in the stopband below threshold, for example, $I = 0.9I_{th}$, can be good measures to evaluate the SLM property for actual devices.

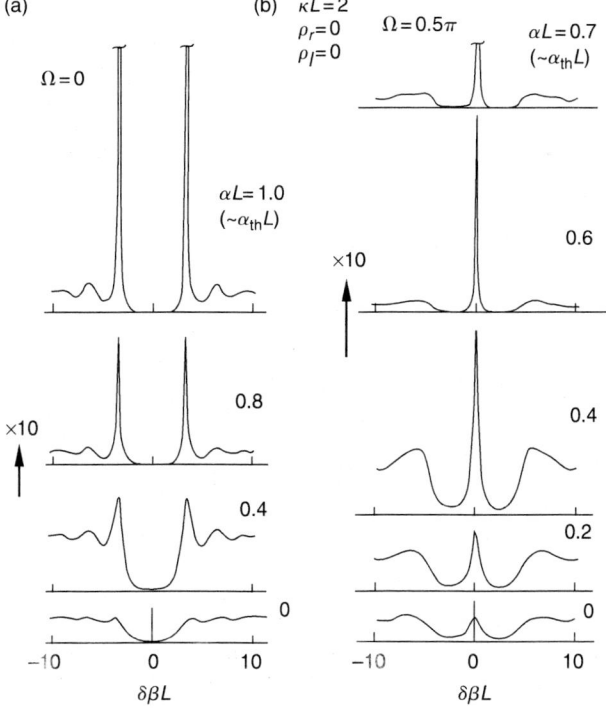

Figure 3   Power transmitting ratio T as a function of $\delta\beta$ for gratings with (a) $\Omega = 0$ and (b) $\pi/2$

Threshold Current and Differential Quantum Efficiency

In the following, we consider an actual DFB laser as depicted in Figure 4. The So-called double-channel buried-hetero-structure (DC-PBH) is adopted for current confinement and stabilizing the lateral mode. A typical stripe width is 1.5 to 2.0 μm. The phase-shifted DFB structure as shown in Figure 4 requires suppression of the end reflectivity for a stable single mode operation. Effective reflectivity at the laser facet is usually reduced by so-called AR-coating but very pre-cise control in the coating process is required. On the other hand, nonreflective DFB laser end can be easily obtained by the window structure, where the active stripe is buried by InP in axial direction in addition to a lateral direction [10]. In such a structure, the effective reflectivity < 0.1% is obtained with the window region length l_{w1} and l_{w2} over 20 μm. A combination of window structure and AR-coating at the facet may be an ideal DFB end structure.

The threshold current and the differential quantum efficiency are important factors in the designing phase. Since the threshold gain for a conventional FP type laser corresponds to the sum of a mirror loss and a waveguide loss, the same concepts can be applied to the threshold gain α_{th} for a DFB laser. Note that α_{th} as discussed previously is the threshold gain for the elec-tric field and therefore when discussing gains and losses in terms of power, $2\alpha_{th}$ should be used. The threshold current density of a DFB laser is expressed by [5]

$$J_{th} = \frac{eB_{eff}}{A_0^2} d \left(\frac{2\alpha_{th}}{\xi} + \alpha_{in} + \alpha_{ac} + \frac{1-\xi}{\xi}\alpha_{ex} \right)^2, \tag{13}$$

where e is the electronic charge, B_{eff} is the effective recombination constant, A_0 is the slope of the gain versus carrier density curve. α_{in} is the extrapolated loss with no current injection. d is the thickness of active layer, ξ is the optical confinement factor, α_{ac} and α_{ex} are absorption losses in the active layer and the cladding layer, respectively. In the actual calculations, values described in [5] are used in the following.

On the other hand the external differential quantum efficiency is derived from the method for FP type lasers as follows:

$$\eta_d = \eta_i \, r_c \frac{2\alpha_{th}}{2\alpha_{th} + \xi\alpha_{ac} + (1-\xi)\alpha_{ex}}, \tag{14}$$

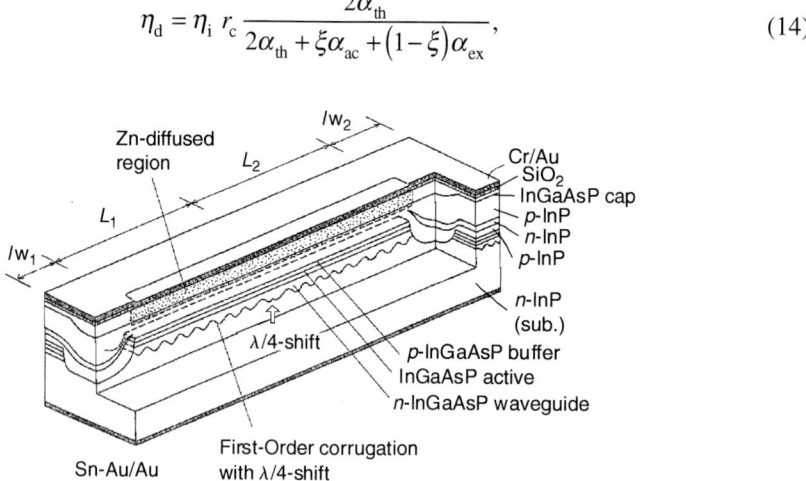

Figure 4   Schematic structure of $\lambda/4$ shifted DFB laser

where η_i is the internal quantum efficiency and r_c is the ratio of the current injected into the active layer to the total current. η_d described here is the sum of the output from both facets. As the grating coupling strength κL increases over three, the differential quantum efficiency becomes as small as 10 to 15% per facet or less, and the longitudinal mode selectivity also decreases. Therefore, such a large κL, although reduces the threshold current, is not desirable for the overall performance of the lasers. In order to expect the differential quantum efficiency more than 20% per facet, the optimum grating coupling strength κL is ~2.0 or smaller, and the resultant κ should be ~50 cm^{-1} when the DFB region length of $L = 300 - 400$ μm is considered.

The earlier-discussed designing may be changed, if asymmetric $\lambda/4$ shifted DFB structures, that is, $L_1 \bullet L_2$, for higher quantum efficiency are considered [11].

Polarization Characteristics

It is known that the transverse electric (TE) mode is favored in a FP type laser diode because of the reflectivity difference at cleaved facets between TE and (transverse magnetic) TM modes [12]. In order to stabilize the main longitudinal mode at the Bragg wavelength in $\lambda/4$ shifted DFB lasers, nonreflective ends are desirable in terms of a large threshold gain difference and reproducibility, as mentioned earlier. In this case, the coupling coefficient κ and the optical confinement factor ξ in the active layer play important roles in polarization behavior.

Figure 5 shows the computed coupling coefficient κ [13,14] and the threshold gain α_{th} of the $\lambda/4$ shifted DFB laser, in which the waveguide layer ($h = 0.1$ μm), the active layer ($d = 0.1$ μm), and the buffer layer ($t = 0.1$ μm) for 1.3, 1.55, and 1.3 μm wavelengths, respectively, and these three InGaAsP are sandwiched by InP as shown in the inset (see Figure 4 as an experimental model). Fundamental transverse mode, the lowest longitudinal mode at the Bragg wavelength, and zero reflectivity at both ends are assumed. κ is proportional to the corrugation depth a with the rectangular and the sinusoidal shapes as long as shallow corrugations are concerned. The difference of κ between the TE and TM modes is ~10% and the resultant threshold gain difference $\{\alpha_{th}$ (TM) $- \alpha_{th}$ (TE)$\}$, corresponding to the mirror loss difference in an FP laser, is ~5 to 6 cm^{-1} for the DFB region length $L = 350$ μm. The estimated threshold gain difference due to different κ values is not large enough to stabilize the TE mode. Therefore, it is true that the TM mode suppression condition in DFB lasers may be strict compared with FP lasers.

The optical confinement factor ξ, which is also important in an FP laser with a weakly-guiding double heterostructure, and the threshold current density J_{th} of the TE and TM modes

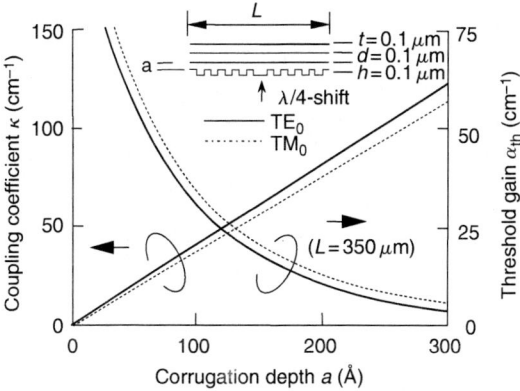

Figure 5 Computed coupling coefficient κ and the threshold gain α_{th} of the TE and TM modes as a function of the corrugation depth a

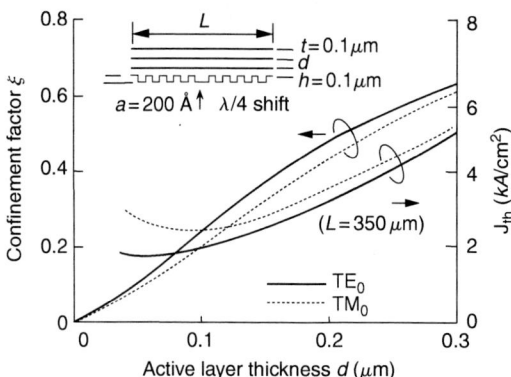

Figure 6 Optical confinement factor ξ and threshold current density J_{th} of the TE and TM modes as a function of the active layer thickness d

are shown in Figure 6 as a function of the active layer thickness d. The corrugated waveguide similar to that in Figure 5 is assumed as shown in the inset but the depth a of the rectangular shape corrugation is fixed to be 200 Å, which induces $\kappa^{TE} = 82$ cm^{-1} and $\kappa^{TM} = 75$ cm^{-1} for the given layer thickness when $d = 0.1$ μm. J_{th} includes both effects from κ and ξ. Now we assess the TE mode stability from the ratio J_{th} (TM)/J_{th} (TE). The ratio of ~1.3 at $d = 0.1$ μm corresponds to the mirror loss difference $2\Delta\alpha_{th}$ of 15 to 20 cm^{-1}. However, note that J_{th} (TM)/J_{th} (TE) is strongly dependent on the active layer thickness. Therefore, the active layer thickness is a key parameter and should be controlled as thin as 0.1 μm to keep the ratio J_{th} (TM)/J_{th} (TE) large.

ACTUAL DFB LASERS AND CHARACTERISTICS

Fabrication of λ/4 Shifted Corrugated Grating—Nega-Posi Photoresist Method

First-order corrugation pitch Λ is ~2400 Å for 1.5 μm range InGaAsP/InP DFB laser with effective refractive index of ~3.5. It is rather difficult to utilize a conventional photolithography technique using ultraviolet light and a photo-mask for fabricating corrugation because the resolution is the order of the grating pitch or even poor. There are several alternative fabrication methods for phase-shifted corrugation grating. Electron beam lithography [15] and x-ray lithography [16] are applied to make corrugation with a phase shift; however, the production throughput may not be high enough. A holographic exposure technique utilizing the interference between two coherent beams is highly convenient to make corrugations because of its simplicity, precision, and high throughput. However, the key issue in the holographic exposure method is how to introduce a phase-shift. In the phase-shift mask method [17] it may be rather difficult to control the amount of the phase shift exactly.

As an example, Figure 7 depicts a negative and a positive (nega-posi) photoresist method, which is a simple and promising method for fabricating of λ/4 shifted corrugation [7,10]. The process is as follows:

1. Nonexposed positive photoresist stripe of typically 400 Å thickness and 500 μm width is formed by a conventional photolithography.

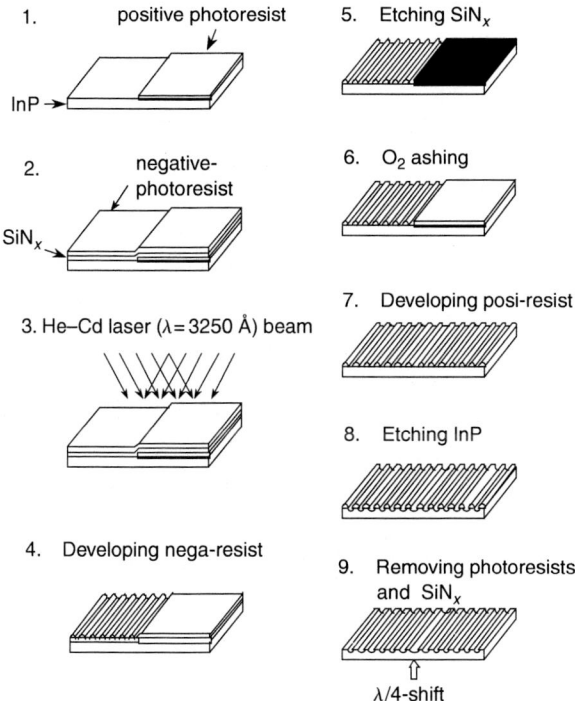

Figure 7 Process flow to fabricate the quarter-wave-shifted corrugation grating by the negative and positive photo-resist method

2. 70 Å thick SiN$_x$ film is deposited by ECR plasma deposition technique at room temperature as an intermediate layer which prevents from the mixing of the negative and the positive photoresists. Then the negative photoresist of 300 Å thickness is spin coated.
3. Two beam holographic exposure is carried out using He–Cd laser ($\lambda = 3250$ Å).
4. The negative photoresist is developed.
5. The SiN$_x$ film is etched by buffered HF.
6. Surface damaging layer of the positive photoresist is removed by O$_2$ plasma ashing.
7. The positive photoresist is developed.
8. InP is etched by an HBr + HNO$_3$ solution.

The phase of grating in the positive photoresist region and the negative photoresist region are shifted exactly by a half pitch because of the inverse nature of the development in positive and negative photoresists.

Lasing Characteristics

DFB lasers as shown in Figure 4 are grown on the substrate with corrugated surface as described earlier by either a liquid phase or vapor phase epitaxial technique.

Figure 8 is spectra below the threshold obtained from DFB lasers as shown in Figure 4, where (a) is for the case of $\Omega = 0$, that is, uniform corrugation without a phase shift and (b) $\Omega = \pi/2$, that is, quarter-wave-shifted corrugation. The spectra are well compared with the theoretical prediction as shown in Figure 3(a) and 3(b).

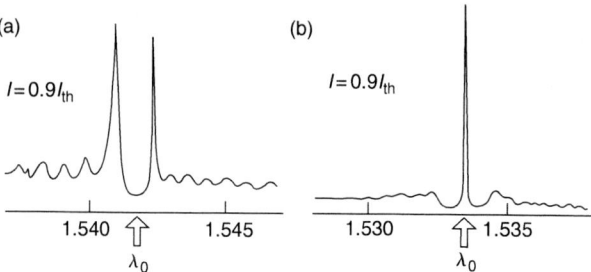

Figure 8   (a) Experimental spectra of uniform DFB laser and (b) quarter-wave-shifted DFB laser below the threshold current, which clearly manifest the theoretical prediction in Figure 3. λ_0 is the Bragg wavelength

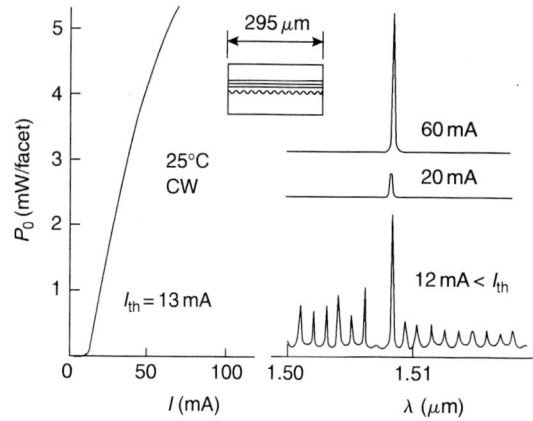

Figure 9   Output versus current curve and spectra of DFB laser with uniform corrugation with cleaved facet with the reflectivity of 33% at both ends

Figure 9 is light output versus current characteristics with corresponding spectra for the uniform corrugation DFB lasers, where the end of the DFB structure is a cleaved facet with reflectivity of about 33%. It should be noted that the finite reflectivity at the DFB ends will produce asymmetry for the two lowest-threshold modes and may lead to single mode operation as shown in Figure 9 [18,19]. In Figure 10 the light output versus current characteristics as well as spectra is shown, where the quarter-wave-shifted DFB structure with nonreflective end is adopted exactly the same as in Figure 4. The two spectra above the threshold current show that the SMSR is as high as 30 dB or more and it is sufficient for stable single mode operation.

IMPORTANT NOTES

Spatial Hole Burning

The analyses described earlier assume that the average refractive index, that is, n_0 in Eq. (1) is constant along the laser length and unchanged. In an actual DFB lasers, however, the refractive index n_0 is a function of the location z and the current density. This fact causes the variation of

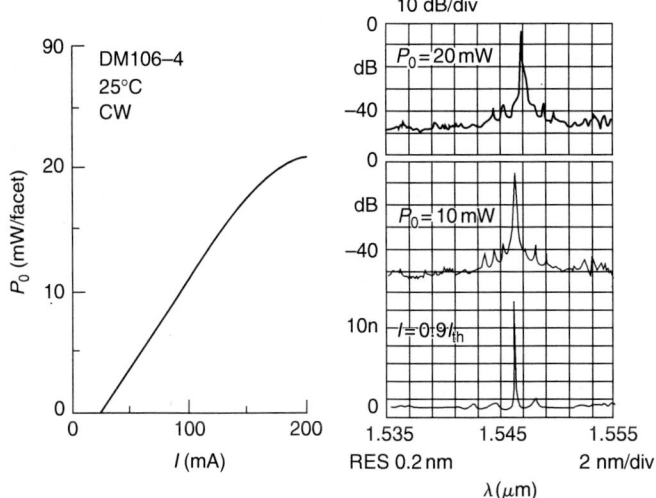

Figure 10 Output versus current curve and spectra of DFB laser with quarter-wave-shifted corrugation exactly the same as shown in Figure 4

the distribution of the light intensity along the laser cavity and eventually causes nonlinear light output characteristics as well as unfavorable spectral behaviors [20]. This effect is sometimes called spatial hole burning and becomes prominent when κL is two or larger. In an actual application, therefore, κL must be controlled to 1 to 1.5, although a unique method to avoid the effect has been studied [21].

Gain Coupling DFB Laser

The DFB lasers with periodic modulation in reflective index have been described and such DFB lasers are based on refractive index coupling, that is, Δn is finite and $\Delta \alpha$ is zero in Eq. (1). As mentioned earlier, such refraction coupling usually yields two lowest-threshold modes and the phase-shift structure is strongly recommended for stable single mode operation. But if $\Delta \alpha$ is finite, which is now gain-coupling mechanism, uniform grating gives us the lowest-threshold mode at the Bragg wavelength and should lead to stable single mode operation [1]. However, it is rather difficult to realize such gain coupling structure [22] and in most applications refractive index coupling DFB lasers are widely used.

REFERENCES

1. H. Kogelnik and C. V. Shank, "Stimulated emission in a periodic structure," *Appl. Phys. Lett.*, 18, 152–154, 1971.
2. H. Kogelnik and C. V. Shank, "Coupled-wave theory of distributed feedback lasers," *J. Appl. Phys.*, 43, 2327–2335, 1972.
3. M. Nakamura, A. Yariv, H. W. Yeng, S. Smokeh, and H. L. Garvin, "Optically pumped GaAs surface laser with corrugation feedback," *Appl. Phys. Lett.*, 24, 224–225, 1973.
4. D. R. Scifres, R. D. Burnham, and W. Streifer, "Distributed-feedback single heterojunction GaAs diode laser," *Appl. Phys. Lett.*, 25, 203–206, 1974.
5. K. Sakai, K. Utaka, S. Akiba, and Y. Matsushima, "1.5 μm range InGaAsP/InP distributed-feedback lasers," *IEEE J. Quantum Electron.*, QE-18, 1272–1278, 1982.

6. H. A. Haus and C. V. Shank, "Antisymmetric taper of distributed feedback lasers," *IEEE J. Quantum Electron.*, QE-12, 532–539, 1976.

7. K. Utaka, S. Akiba, K. Sakai, and Y. Matsushima, "$\lambda/4$-shifted InGaAsP/InP DFB lasers by simultaneous holographic exposure of positive and negative photoresists," *Electron. Lett.*, 20, 1008–1010, 1984.

8. K. Utaka, S. Akiba, K. Sakai, and Y. Matsushima, "Longitudinal-mode behavior of $\lambda/4$-shifted InGaAsP/InP DFB lasers," *Electron. Lett.*, 21, 367–369, 1985.

9. S. Akiba, M. Usami, and K. Utaka, "1.5 μm $\lambda/4$-shifted InGaAsP/InP DFB lasers," *IEEE J. Lightwave Technol.*, LT-5, 1564–1573, 1987.

10. S. Akiba, K. Utaka, K. Sakai, and Y. Matsushima, "Distributed feedback InGaAsP/InP lasers with window region emitting at 1.5 μm range," *IEEE J. Quantum Electron.*, QE-19, 1052–1056, 1983.

11. M. Usami, S. Akiba, and K. Utaka, "Asymmetric $\lambda/4$-shifted InGaAsP/InP DFB lasers," *IEEE J. Quantum Electron.*, QE-23, 815–821, 1987.

12. T. Ikegami, "Reflectivity of mode at facet and oscillation mode on double-heterostructure injection lasers," *IEEE J. Quantum Electron.*, QE-8, 470–476, 1972.

13. W. Streifer, D. R. Scifres, and R. D. Burnham, "Coupling coefficients for distributed feedback single- and double-heterostructure diode lasers," *IEEE J. Quantum Electron.*, QE-11, 867–873, 1975.

14. W. Streifer, D. R. Scifres, and R. D. Burnham, "TM-mode coupling coefficients in guided-wave distributed feedback lasers," *IEEE J. Quantum Electron.*, QE-12, 74–78, 1976.

15. K. Sekartedjo, N. Eda, K. Furuya, Y. Suematsu, F. Koyama, and T. Tanbun-ek, "1.5-μm phase-shifted DFB lasers for single-mode operation," *Electron. Lett.*, 20, 80–81, 1984.

16. M. Nakao, K. Sato, T. Nishida, T. Tamamura, A. Ozawa, Y. Saito, I. Okada, and H. Yoshihara, "1.55 μm DFB laser array with $\lambda/4$-shifted first-order gratings fabricated by x-ray lithography," *Electron. Lett.*, 25, 148–149, 1989.

17. M. Okai, S. Tsuji, N. Chinone, and T. Harada, "Novel method to fabricate corrugation for a $\lambda/4$-shifted distributed feedback laser using a grating photomask," *Appl. Phys. Lett.*, 55, 415–417, 1989.

18. W. Streifer, R. D. Burnham, and D. R. Scifres, "Effect of external reflections on longitudinal modes of distributed feedback lasers," *IEEE J. Quantum Electron.*, QE-11, 154–161, 1975.

19. K. Utaka, S. Akiba, K. Sakai, and Y. Matsushima, "Effect of mirror facet on lasing characteristics of distributed feedback InGaAsP/InP laser diodes at 1.5 μm range," *IEEE J. Quantum Electron.*, QE-20, 230–245, 1984.

20. H. Soda, Y. Kotaki, H. Sudo, H. Ishikawa, S. Yamakoshi, and H. Imai, "Stability in single longitudinal mode operation in GaInAsP/InP phase-adjusted DFB lasers," *IEEE J. Quantum Electron.*, QE-23, 804–814, 1987.

21. M. Usami and S. Akiba, "Suppression of longitudinal spatial hole-burning effect in $\lambda/4$-shifted DFB lasers by nonuniform current distribution," *IEEE J. Quantum Electron.*, QE-25, 1245–1253, 1989.

22. Y. Nakano, Y. Luo, and K. Tada, "Facet reflection independent, single longitudinal mode oscillation in a GaAlAs/GaAs distributed feedback laser equipped with a gain-coupling mechanism," *Appl. Phys. Lett.*, 55, 1606–1608, 1989.

E

ERBIUM-DOPED FIBER AMPLIFIER

Masataka Nakazawa

PRINCIPLE OF ERBIUM-DOPED FIBER AMPLIFIER

An erbium-doped fiber amplifier (EDFA) has five main optical components: an erbium-doped fiber, a laser diode (LD) to pump the fiber, a wavelength-dependent fiber coupler, a polarization-independent optical isolator, and an optical bandpass filter. Figure 1 shows the configuration of an EDFA operating in the 1.5 μm band that utilizes the stimulated emission of Er^{3+} ions from $^4I_{13/2}$ to $^4I_{15/2}$ [1–4]. The LDs (EDFA pump sources) operate at wavelengths which include 1.48 [5], 0.98 [6], and 0.82 μm [7]. Fabry–Perot multimode LDs, namely InGaAsP and InGaAs type utilizing a strained superlattice [6], and GaAlAs type, operate at the above mentioned wavelengths, respectively. The wavelength-dependent fiber coupler (or a dielectric mirror) is used to couple the signal and pump beams into the erbium-doped fiber. Conventionally, a 1.48 μm/1.53 to 1.55 μm coupler with a coupling efficiency exceeding 90% is used.

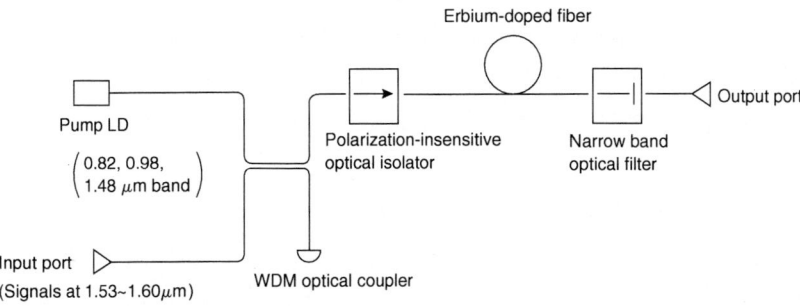

Figure 1 Configuration of EDFA. An EDFA is composed of five main optical components: a semiconductor laser, a fiber coupler, a polarization-independent optical isolator, erbium-doped fiber, and a narrow-band optical filter

Advantages of EDFA

    High net gain (>35 dB)
    Low coupling losses
    Good noise figure (3~5 dB)
    Polarization-insensitive gain characteristics
    Low interchannel crosstalk (>1 MHz)
    Wide bandwidth (>3700 GHz)
    Large energy storage

Figure 2   Advantages of EDFA

The polarization-insensitive optical isolator plays a very important role in suppressing laser oscillation and amplified spontaneous emission (ASE). Since the Er^{3+} ions in optical fibers are fused in the glass, its fluorescence and gain characteristics are intrinsically polarization independent. The advantage of using a polarization-insensitive isolator is that uniform gain can be obtained for an arbitrary input polarization (whereby ordinary and extraordinary lights are spatially separated and propagated in the same yttrium iron garnet (YIG), and the output signal is obtained by adding the two polarized beams at the output end). To obtain a gain higher than 40 dB [8], an isolator should also be installed at the output stage between the erbium-doped fiber and the optical fiber.

It is also important to insert a narrow bandpass optical filter with a bandwidth of 1 to 3 nm, in the EDFA after the erbium-doped fiber. This eliminates ASE and improves the signal to noise ratio. By inserting such a narrowband filter, even when the EDFA is cascaded, ASE growth can be suppressed, which means that only the signal is allowed to pass. As shown in Figure 1, the signal is led to an erbium-doped fiber from the input port through a wavelength-dependent, wavelength division multiplexed (WDM) coupler and an optical isolator. Thus, the amplified signal is obtained from the output port through a narrow band optical filter.

As regards the optimization of the Er^{3+} ion doping concentration, it has been established that an erbium fiber with an extremely high index difference Δ between the core and the cladding has high energy conversion efficiency. By using a high pump power density unfavorable pump absorption is avoided, and highly efficient pumping is realized at relatively low concentrations between 100 and 1000 ppm. However, it is important to note that a high Δ makes it possible to realize a high gain coefficient in an EDFA, whereas the difference between the transmission and the erbium fiber spot sizes leads to increased splice loss.

With these optical components, an EDFA can amplify optical signals in the 1.5 μm band. The advantages of the EDFA are summarized in Figure 2.

ABSORPTION CHARACTERISTICS

Figure 3 shows the wavelength dependence of the absorption coefficient of a 1 m long erbium-doped fiber whose doping concentration is 320 ppm. Figure 4 shows the corresponding energy diagram. These figures show the absorption peaks at 1.48 ($^4I_{15/2}-^4I_{13/2}$), 0.98 ($^4I_{15/2}-^4I_{11/2}$), and 0.8 μm ($^4I_{15/2}-^4I_{9/2}$), and the absorption coefficient around 1.48 μm is 1/2–1/3 of that at 0.98 μm. The absorption at 0.82 μm, which is approximately 0.3 dB/m, is much smaller than the other two. The absorption coefficient at 0.8 μm is as high as 0.75 dB/km, but this wavelength suffers from excited state absorption (ESA) and hence the efficiency is low [9,10]. It is also characteristic in that the absorption is accompanied by the green luminescence of erbium-doped fiber because of the direct spontaneous emission from the upper state to the ground state. However,

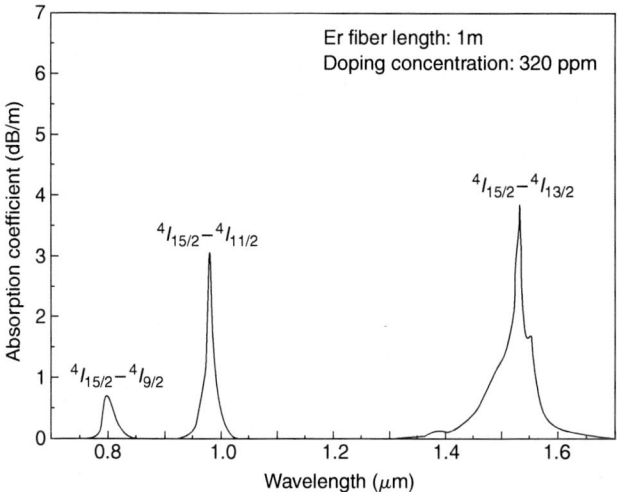

Figure 3 Wavelength dependence of absorption coefficient of erbium-doped fiber. There are absorption peaks at 1.48 μm ($^4I_{15/2} - ^4I_{13/2}$), 0.98 μm ($^4I_{15/2} - ^4I_{11/2}$), and 0.8 μm ($^4I_{15/2} - ^4I_{9/2}$)

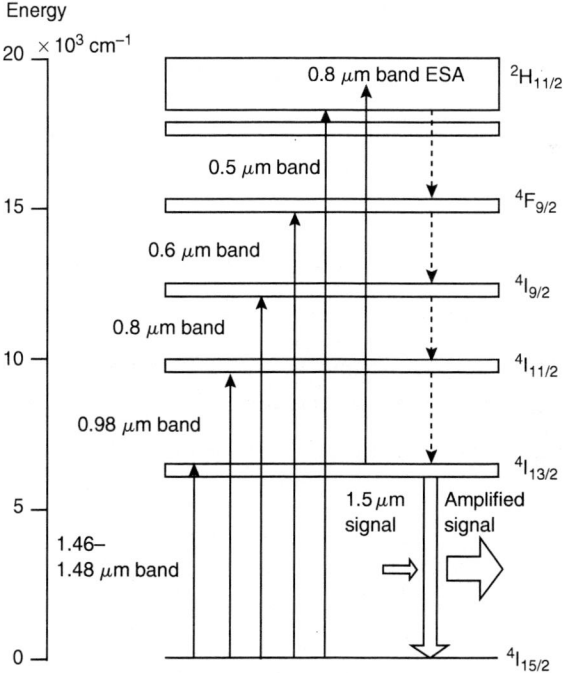

Figure 4 Energy diagram of erbium ion pumping. Signal amplification occurs between $^4I_{15/2}$ and $^4I_{13/2}$. There are many pumping schemes of up to 0.5 μm

it has been reported that pumping at 0.82 μm, where the ESA is negligible, produces a high gain by increasing the fiber length to compensate for the low absorption [11].

The absorption characteristics in the 1.48 μm band can change greatly depending on the glass composition and the codopants. For example, with GeO_2/SiO_2, there is a sharp absorption peak at 1.535 μm, whereas with Al_2O_3/SiO_2, the absorption peak is shifted to slightly shorter wavelengths near 1.530 μm, and the full width at half maximum of the absorption is broadened to 42 nm (twice of that with GeO_2/SiO_2, which is 23 nm). In general, by codoping with Al_2O_3, the Er^{3+} ion configuration becomes different from that of GeO_2/SiO_2, resulting in the filling of fluorescence defects around 1.54 μm as seen with GeO_2/SiO_2, and the broadening of the fluorescence intensity distribution to longer and shorter wavelengths [12]. Therefore, erbium-doped fiber codoped with Al_2O_3 is widely used for the simultaneous amplification of multichannel (WDM) signals and ultrashort pulse amplification.

OPTICAL AMPLIFICATION CHARACTERISTICS

An EDFA operates as a three-level optical amplifier and, therefore, has different gain saturation characteristics from those of a four-level amplifier. Here we describe a simple rate equation analysis to characterize the EDFA, and several experimental results related to gain saturation power, signal power, and their dependence on the pump wavelengths, for comparison with the analysis.

Analysis of EDFA Amplification Characteristics

In general, it can be assumed that the population inversion in optical amplifiers is formed uniformly in the optical signal propagation direction. However, with the EDFA the amount of population inversion varies with pump absorption, since the formation of the population inversion is dependent on the propagation of the pump beam. This feature is responsible for the change in noise growth depending on whether the pumping direction is forward or backward, and the change in the saturation parameter along the propagation axis.

As shown in Figure 5, let the number of Er^{3+} ions per volume of electron distribution present on energy levels E_3 (pump level), E_2 (upper level), and E_1 (ground level) be N_3, N_2, and N_1, respectively. The rate equation for the number of Er ions can be expressed as follows:

$$\frac{dN_3}{dt} = \left(W_p B_{13}\right) N_1 - \left(W_p B_{31}\right) N_3 - A_{32} N_3 - A_{31} N_3, \tag{1a}$$

$$\frac{dN_2}{dt} = -\left(W_s B_{21}\right) N_2 + \left(W_s B_{12}\right) N_1 + A_{32} N_3 - A_{21} N_2, \tag{1b}$$

$$\frac{dN_1}{dt} = \left(W_s B_{21}\right) N_2 - \left(W_s B_{12}\right) N_1 - \left(W_p B_{13}\right) N_1 + \left(W_p B_{31}\right) N_3$$
$$+ A_{21} N_2 + A_{31} N_3, \tag{1c}$$

where W_p and W_s are the pump and the signal energy densities, respectively, A_{ij} is the coefficient of spontaneous emission from the i to j level, and B_{ij} is the coefficient of spontaneous emission

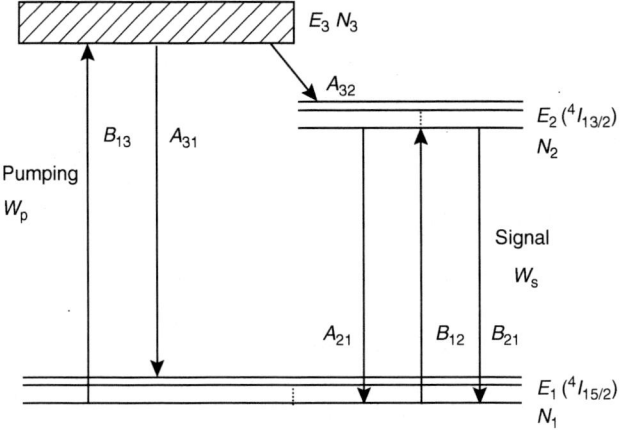

Figure 5   Energy diagram of EDFA for rate-equation analysis. E_3 is the pump level, E_2 ($^4I_{13/2}$) is the upper level of the stimulated emission, and E_1 ($^4I_{15/2}$) is the lower level. On each energy level, N_3, N_2, and N_1, Er^{3+} ions per volume exist

(absorption). By using the fact that N_3 is smaller than N_1 and N_2, the number of Er^{3+} ions per volume, ρ, is given by

$$\rho = N_1 + N_2 + N_3 \cong N_1 + N_2. \tag{2}$$

Since $dN_1/dt = 0$, $dN_2/dt = 0$, and $dN_3/dt = 0$ in the steady state, from Eq. (1a) we have

$$N_3 = \frac{W_p B_{31}}{A_{31} + A_{32} + W_p B_{31}} N_1. \tag{3}$$

Here, by noting $A_{31} \ll A_{32}$ because most of the transition from E_3 goes to E_2, and $W_p B_{31} \ll A_{32}$ because the relaxation of $N_3(t)$ is sufficiently fast, we substitute Eq. (1b) with $dN_2/dt = 0$ into Eq. (3) and obtain

$$\frac{N_2}{N_1} = \frac{W_p B_{13} + W_s B_{12}}{A_{21} + W_s B_{21}} \tag{4}$$

and

$$N_1 + N_2 = \frac{W_p B_{13} + W_s B_{12} + W_s B_{21} + A_{21}}{W_s B_{21} + A_{21}} N_1,$$

$$\cong \rho. \tag{5}$$

Let $\sigma_s^a(v)$ and $\sigma_s^e(v)$ be the absorption and emission cross-sections at the signal wavelengths, respectively, $\sigma_p^a(v)$ be the absorption cross-section of the pump beam, and $A_{21} = 1/\tau$. From

Eqs. (4) and (5), we obtain

$$N_1 = \frac{I_s\left(\sigma_s^e \tau / hv_s\right) + 1}{I_p\left(\sigma_p^a \tau / hv_p\right) + I_s\left(\sigma_s^e + \sigma_s^a\right)(\tau / hv_s) + 1}\, \rho, \tag{6a}$$

$$N_2 = \frac{I_s\left(\sigma_s^a \tau / hv_s\right) + I_p\left(\sigma_p^a \tau / hv_p\right)}{I_p\left(\sigma_p^a \tau / hv_p\right) + I_s\left(\sigma_s^e + \sigma_s^a\right)(\tau / hv_s) + 1}\, \rho, \tag{6b}$$

where I_p and I_s are the pump intensity (power density) and the signal intensity, respectively.

As it is well known, the changes in the signal and the pump intensities along the z (propagation) direction in the gain medium are described by

$$\frac{dI_s}{dz} = \left(\sigma_s^e N_2 - \sigma_s^a N_1\right) I_s, \tag{7a}$$

$$\frac{dI_p}{dz} = -\sigma_p^a N_1 I_p. \tag{7b}$$

With $I_p^{th} = hv_p / \sigma_p^a \tau$, $\bar{I}_p = I_p / I_p^{th}$, $\bar{I}_s = \gamma I_s / I_p^{th}$, $\gamma = \sigma_s^a v_p / \sigma_p^a v_s$, and $k_s = \sigma_s^e / \sigma_s^a$ in Eqs. (6) and (7), we have

$$\frac{d\bar{I}_s}{dz} = \frac{k_s \bar{I}_p - 1}{\bar{I}_p + (1 + k_s)\bar{I}_s + 1}\, \rho \sigma_s^a \bar{I}_s, \tag{8a}$$

$$\frac{d\bar{I}_p}{dz} = \frac{k_s \bar{I}_s + 1}{\bar{I}_p + (1 + k_s)\bar{I}_s + 1}\, \rho \sigma_p^a \bar{I}_p. \tag{8b}$$

If E_1 and E_2 do not contain fine structures, $k_s = \sigma_s^e / \sigma_s^a = 1$. However, because the $^4I_{13/2}$ and $^4I_{15/2}$ levels contain several fine Stark structures, σ_s^e is not equal to σ_s^e and becomes a function of wavelength [13]. Nevertheless, in a rough calculation, we can assume $k_s = 1$.

Equations (8a) and (8b) are important equations that describe the propagation of the signal and the pump light. The gain coefficient $g(\bar{I}_s)$, which includes saturation as a function of signal $\bar{I}_s$, satisfies

$$\frac{1}{\bar{I}_s}\frac{d\bar{I}_s}{dz} = g(\bar{I}_s)$$

and from Eq. (8a), this is given by

$$g(\overline{I}_s) = \left[1 + \frac{\overline{I}_s}{(\overline{I}_p+1)/(1+k_s)}\right]^{-1}\left(\frac{k_s\overline{I}_p-1}{\overline{I}_p+1}\right)\rho\sigma_s^a.\tag{9}$$

This can be rewritten as the following familiar form:

$$g(\overline{I}_s) = \frac{g_0}{(1+\overline{I}_s)/\overline{I}_{sat}},\tag{10}$$

where

$$\overline{I}_{sat} = \frac{\gamma I_{sat}}{I_p^{th}} = \frac{\overline{I}_p+1}{1+k_s}.\tag{11}$$

The unsaturated gain g_0 is given by

$$g_0 = \frac{k_s\overline{I}_p-1}{\overline{I}_p+1}\sigma_s^a\rho.\tag{12}$$

The pump and the signal power, P_p and P_s, are given by $P_p = I_p A_p$, $P_s = I_s A_s$, respectively, where A_p and A_s are their effective cross-sections. From Eq. (11), by setting $k_s \cong N_1$ we obtain

$$I_{sat} \cong \frac{1}{1+k_s}\frac{h\nu_s}{\sigma_s^a\tau}(I_p+1).\tag{13}$$

This equation indicates that the saturation intensity is a function of pump power. We can represent Eq. (13) in terms of the saturation power P_{sat}:

$$P_{sat} \cong \frac{1}{1+k_s}\left[\frac{1}{\gamma}P_p + \left(\frac{h\nu_s}{\sigma_s^a\tau}\right)A_s\right].\tag{14}$$

When we use $k_s = 0.71$, $1/\gamma = 0.29$, which are parameters that provide a good fit with the experimental results, we can obtain a saturation power of approximately 18 mW when, for example, $P_p = 100$ mW. Here, we used the parameters of the present erbium-doped fiber: $\sigma_s^a = 3.3 \times 10^{-25}$ m^2, $\tau = 14$ msec, and $A_s = 4 \times 10^{-11}$ m^2. The values of k_s and γ depend on the codopant material, and σ_s^a also depends on the erbium doping concentration.

EDFA Gain Characteristics

Figure 6(a) shows the gain $G = P_s^{out}/P_s^{in}$ obtained by solving Eqs. (10) and (11), where the horizontal axis is the pump power. Here the equations are solved using the following parameters corresponding to the experimental condition: the erbium fiber is 100 m long, the doping concentration is 92 ppm, the signal wavelength is 1.545 μm, the input signal level is −45 dBm, and the optical losses at the pump and the signal wavelengths are 0.13 and 0.42 dB/m, respectively. The circles indicate the experimental result and the solid curve is the numerical result. This figure shows that the two results are in good agreement and the pump threshold exists at around $P_p = 6.7$ mW, where the net gain is zero, and a higher gain is obtained by increasing the pump

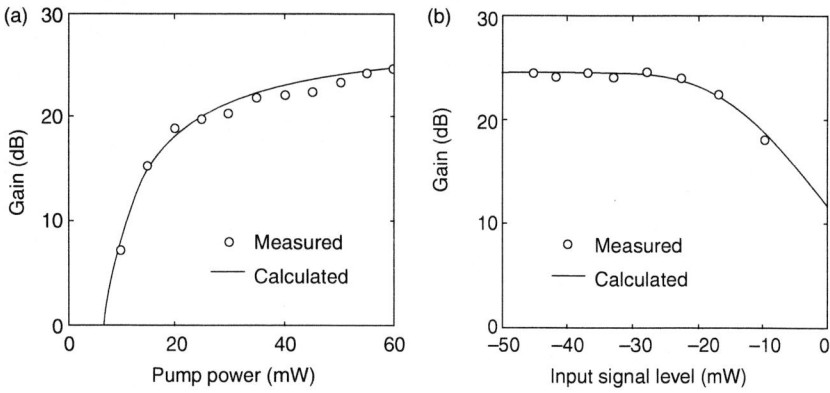

Figure 6 Dependence of erbium-doped fiber gain on (a) pump power and (b) input signal power level. In (a), there is a clear pump threshold. The gain increases as the pump power increases, and the gain is saturated for a pump power of approximately 20 mW. In (b), the gain is constant at small input power levels, but saturation starts to occur at an input power of about −20 dBm

power. When the pump power is increased beyond 20 mW, significant gain saturation is observed.

Figure 6(b) shows the gain dependence on the input signal power level for a pump power of 59 mW. As seen earlier, the experimental results denoted with circles are in good agreement with the numerical results denoted with the solid curve. The gain is constant for a small input power, and the small signal gain G_0 is 24.5 dB. An increase in the signal power level leads to gain saturation, and the input intensity for saturation defined as −3 dB down is −5.2 dBm.

Figure 7 shows the high gain characteristics obtained with 1.48 μm pumping. The 100 m long erbium-doped fiber had a doping concentration of 100 ppm, and was composed of GeO$_2$/SiO$_2$. The pump threshold was $P_p = 28$ mW and a gain as high as 46.5 dB was obtained with $P_p = 130$ mW and . The inset shows the spectrum of the pumping LDs at the two wavelengths used. P_{sat} is +11 dBm and the maximum output reached 30 mW. These results show that the EDFA operates as a power amplifier. Recently, an EDFA output power exceeding 1 W was possible by increasing the pump power.

Figure 8(a) and 8(b) show the gain dependence on pump wavelength in the 1.48 and 0.98 μm bands, respectively. In Figure 8(a), the gain decreases monotonically when the pump wavelength is set at longer wavelengths of 1.50 to 1.51 μm. This is because the population inversion approaches the resonance pumping of a two-level system as the pump wavelength approaches the spontaneous emission wavelength and thus $N_2 \cong N_1$.

By contrast in Figure 3, when the pump wavelength is set at shorter wavelengths of below 1.45 μm the absorption at $^4I_{13/2}$ becomes very small, and so the gain approaches zero. Therefore, the optimum-pumping wavelength is between the two wavelengths, and a high gain can be realized over a wide pumping wavelength region of 1.450 to 1.490 μm. This means that an InGaAsP LD with Fabry–Perot type and multimode oscillation (spectral width ~20 nm) can be suitably employed as an efficient pump source for erbium-doped fiber [5].

There is a distinct difference between Figure 8(b) and 8(a) in terms of the gain dependence on pump wavelength. In the former the effective pumping wavelength band for 0.98 μm is 10 nm, ranging from 0.975 to 0.985 μm. The effective bandwidth is narrower than that of the 1.48 μm band, and the gain varies greatly at both longer and shorter wavelengths. This change is a direct consequence of the sharp change in the absorption coefficient at 0.98 μm as shown in Figure 3. Hence, the choice of pumping LD is limited. Recently, high power 0.98 μm LDs with a long lifetime have been developed, and low-noise EDFA technology has been established.

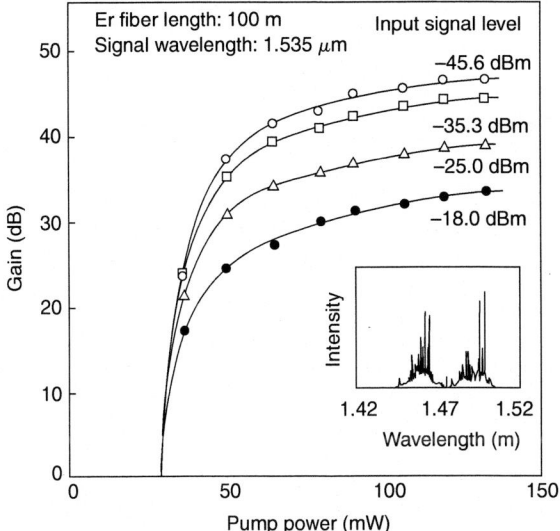

Figure 7 High gain characteristics of EDFA. Gain as high as 46.5 dB can be realized for a pump power of about 130 mW and an input signal power level of −45.6 dB. The inset shows the spectrum of the pumping LD at two wavelengths. The erbium fiber was 100 m long and the doping concentration was 100 ppm

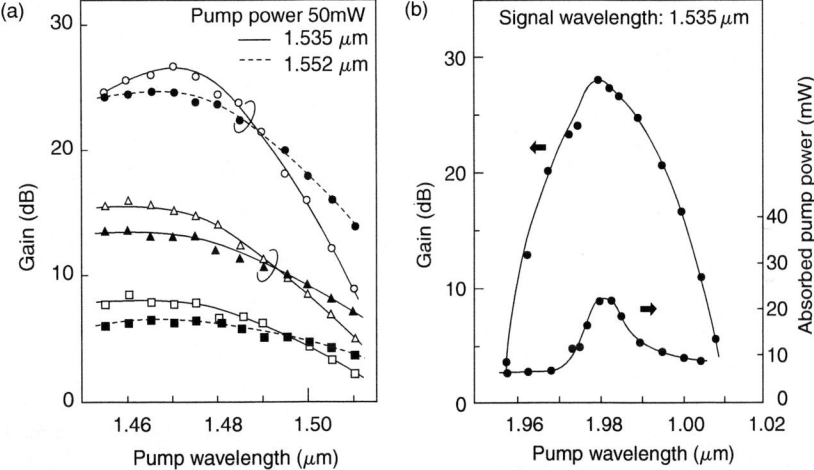

Figure 8 Gain dependence on pump wavelength. (a) is the result for pumping in the 1.48 μm region. There is a wide pumping wavelength band (as wide as 35 nm) that ranges from 1.450 to 1.485 μm. (b) The effective pumping region in the 0.98 μm band is 10 nm near 0.98 μm, which is relatively narrower than that with 1.48 μm pumping (2001)

REFERENCES

1.   E. Desurvire, *Erbium-Doped Fiber Amplifiers*, John Wiley & Sons, New York (1994).
2.   D. Bayart, "Erbium-doped and Raman fiber amplifiers," *C. R. Physique*, **4**, 65–74 (2003).
3.   J. A. Buck, "Optical fiber amplifiers," in *Handbook of Optics IV*, Optical Society of America, Chap. 5, pp. 5.1–5.8 (2001).
4.   G. P. Agrawal, *Fiber-Optic Communication Systems*, 2nd ed., pp. 391–420 (1997).

5. M. Nakazawa, K. Suzuki, and Y. Kimura, "Efficient Er^{3+}-doped optical fiber amplifier pumped by a 1.48 μm InGaAsP laser diode," *Appl. Phys. Lett.*, **54**, 295–297 (1989).

6. S. Uehara, M. Horiguchi, T. Takeshita, M. Okayasu, M. Yamada, M. Shimizu, O. Kogure, and K. Oe, "0.98-μm InGaAs strained quantum well lasers for erbium-doped fiber optical amplifiers," in *Tech. Dig. International Conference on Integrated Optics and Optical Fiber Communication '89, Kobe*, 20PDB-11 (1989).

7. K. Suzuki, Y. Kimura, and M. Nakazawa, "High gain Er^{3+}-doped fiber amplifier pumped by 820 nm GaAlAs laser diodes," *Electron. Lett.*, **26**, 948–949 (1990).

8. Y. Kimura, K. Suzuki, and M. Nakazawa, "46.5 dB gain in Er^{3+}-doped fiber amplifier pumped by 1.48 μm GaInAsP laser diodes," *Electron. Lett.*, **25**, 1656–1657 (1989).

9. R. I. Laming, S. B. Poole, and E. J. Tarbox, "Pump excited-state absorption in erbium-doped fibers," *Opt. Lett.*, **13**, 1084–1086 (1988).

10. P. R. Morkel and R. I. Laming, "Theoretical modeling of erbium-doped fiber amplifiers with excited-state absorption," *Opt. Lett.*, **14**, pp. 1062–1064 (1989).

11. M. Nakazawa, Y. Kimura, and K. Suzuki, "High gain erbium fiber amplifier pumped by 800 nm band," *Electron. Lett.*, **26**, 548–549 (1990).

12. S. B. Poole, "Fabrication of Al_2O_3 co-doped optical fibers by a solution-doping technique," *Tech. Dig. European Conference on Optical Communications '88*, Brighton, U.K. pp. 433–435 (1988).

13. E. Desurvire and J. R. Simpson, "Amplification of spontaneous emission in erbium-doped single-mode fibers," *J. Lightwave Technol.*, **LT-7**, 835–845 (1989).

FIBER BRAGG GRATING (FBG)

Kazaro Kikuchi

INTRODUCTION

The fiber Bragg grating (FBG) is one of the key elements in the established and emerging fields of optical communication systems. Their applications also spread into the area of optical fiber sensing. In this item, we review the FBG technology including fabrication techniques, advanced gratings structures, and packaging considerations. In addition, we describe various applications of the FBG in optical transmission systems.

HISTORY OF FBG DEVELOPMENT

Hill and co-workers [1,2] at the Communications Research Centre in Canada discovered the grating formation in optical fibers in 1978, for the first time. In one of their experiments to investigate the nonlinear property of silica–glass optical fibers, they found that launching of an Argon-ion laser beam at 488 nm (blue/green) on an optical fiber developed reflection of the incident beam over time. This phenomenon is attributed to the Bragg reflection as follows: The standing wave is set up in the core of the fiber by a 4% Fresnel reflection of the beam from the cleaved end of the fiber. The standing wave gradually induces the refractive index variation along the fiber according to the spatial power variation, forming the Bragg grating. The refractive index of the fiber is changed by photon absorption, which is called the photosensitivity of the glass material.

Subsequent detailed research on the photosensitivity by Lam and Garside [3] showed that the grating strength increased as a function of the square of the light intensity. This suggested that a two-photon absorption mechanism contributed to the grating formation. Thereby, instead of using the 488-nm light to write the grating, a single photon with the doubled photon energy is found to be more effective for the grating formation process. From this viewpoint, lights, whose wavelengths are in the ultraviolet (UV) region between 244 to 260 nm, are employed for inscribing the grating. However, the self-organized grating only allowed inscription of the grating whose Bragg wavelength was at the writing wavelength. This fact prevented FBG-based devices from operating in the telecommunication window at 1550 nm. In 1989, Meltz et al. [4] demonstrated the transverse exposure of the UV-radiation onto the fiber core for grating inscription. This fabrication method opened up the possibility of the fiber grating operating at an arbitrary wavelength, because the Bragg wavelength was no longer restricted by the writing wavelength. In 1993, Hill introduced a robust fabrication technique based on the phase-mask method. This method greatly relaxed the stringent requirements on making FBGs. From that time, a lot of interesting researches on FBGs have been carried out, and various applications of FBGs have been found in optical communication systems.

PHOTOSENSITIVITY AND PHOTOSENSITIZATION FOR GRATING FORMATION

Using the photosensitivity of silica–glass optical fibers, we can produce a permanent refractive-index modulation, when the fiber-core is exposed with an intense UV-interference pattern. The underlying mechanism of the change in the refractive index is still a controversial issue. However, it is believed that defects in the silica–glass are the origin of the photosensitivity. Defects that are created during the fiber drawing process cause the oxygen deficient matrix in the glass. They have UV-absorption band in the range of 240 to 250 nm with the peak absorption at 242 nm.

To explain the mechanisms of the photosensitivity, several models were proposed. Basically, they can be categorized into two groups that are the color-center model and the structural model. The color-center model, which was proposed by Hand and Russell [5], assumed that incident UV-radiation bleached the germanium oxygen-deficient center and modified the absorption property of the medium [6,7]. The photo-induced change in the absorption bands was associated with the change in the refractive index through the Kramers–Kronig relation. On the other hand, the structural model proposed that the intense UV-radiation might bring the structural rearrangement of the glass matrix, possibly densification, which will induce the refractive-index change [8–10]. Generally speaking, the combination of the above-mentioned mechanisms takes place during the actual Bragg gratings formation.

In order to fasten the FBG inscription speed, it is desirable to enhance the photosensitivity of standard silica fibers. One of the methods is to co-dope GeO_2 and B_2O_3 in the core. To further enhance the photosensitivity, the photosensitive fiber should also be pre-processed with techniques such as "hydrogen loading" [11] and "flame brushing" [12]. These methods can induce a large magnitude of the refractive index change when pre-processed fibers are exposed to the UV-light. The refractive-index change as high as the order of 10^{-2} is attainable by using the hydrogen-loaded photosensitive fiber. Advantages of hydrogenation include the higher refractive-index change, shorter FBGs writing time, lower birefringence, and freedom to control the photosensitivity level of different kinds of fibers. The enhanced photosensitivity, however, is lost if the hydrogen diffuses out. The flame brushing technique has a major benefit over the hydrogen loading technique, that is the photosensitization is permanent. However, the flame-brushed fiber becomes weak due to high temperature involved during the photosensitization treatment.

PROPERTIES OF FBGs

We first define the Bragg wavelength λ_B given by the following relation:

$$\lambda_B = 2n_{eff}\Lambda,$$

(1)

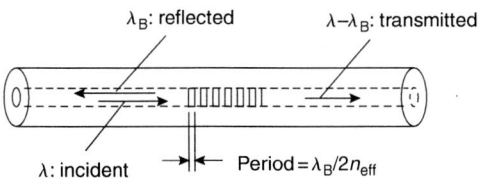

Figure 1   Basic structure of fiber Bragg gratings

where n_{eff} is the modal refractive index and Λ represents the index modulation period of the grating. Assume that the incident light fulfils the Bragg condition given by Eq. (1). The incident light will be Fresnel-reflected at one of the grating plane. Under the Bragg condition, the reflected light is exactly in-phase with reflected lights from other grating planes. Since all of these reflected lights will be constructively interfered at the incident edge of the FBG, the incident light is totally reflected as shown in Figure 1. On the other hand, other wavelength components will pass through the FBG. Thus, the FBG acts as a band-pass filter at the reflection port and a band-rejection filter at the transmission port. Such interferometric effect usually exhibits very narrow bandwidth characteristics of filters.

The refractive-index profile of an FBG can be expressed as:

$$n(z) = n_0 + m(z) \cdot \Delta n \cdot \cos\left(\frac{2\pi z}{\Lambda} + \phi(z)\right),$$ (2)

where n_0 is the averaged refractive index, Δn is the induced modulation index typically ranging between 10^{-5} and 10^{-3}, and the grating period is symbolized by Λ. The symbol $m(z)$ is the taper function, and $\phi(z)$ accounts for the grating chirp. Figure 2 summarizes different types of gratings and their optical characteristics.

The uniform FBG has a rectangular index envelope $m(z) = 1$ while $\phi = 0$. This grating is simple to fabricate and has an almost flat top reflection spectrum. However, serious side lobes appear on both edges of the stop band. Such side lobes are undesirable in a DWDM system, because they may cause crosstalk between adjacent WDM channels. In order to suppress the side lobes, a taper function is imposed on the envelope of the grating index profile. In this case, m is no longer a constant but is a function of z. The method of tapering the index changes is referred to apodization. The apodized FBG shows significant improvement on the spectral characteristic with the side lobe suppression >30 to 40 dB.

Another type of FBG is the chirped FBG, which can compensate for dispersion of the standard single-mode fiber. This requires that the period of the refractive index vary linearly toward the end of the grating. The change in the period with the position broadens the spectrum, and induces the time delay between different wavelength components; thus, the properly designed chirped FBG can compensate for fiber dispersion within its bandwidth.

Apart from the aforementioned grating types, phase-shift gratings and sampled gratings have been demonstrated. They have special optical characteristics, which are useful for specific

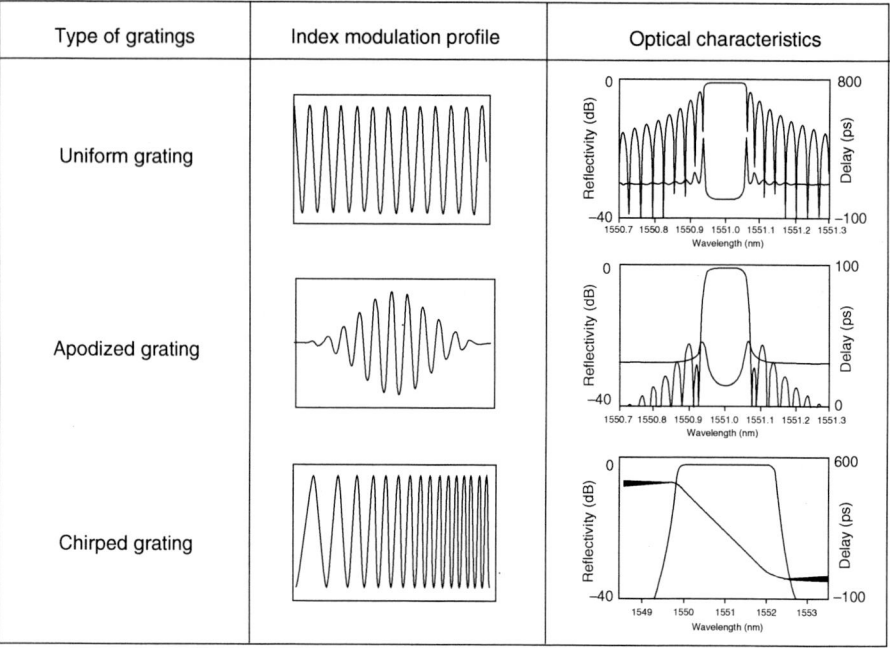

Figure 2   Types of gratings with their index modulation profiles and optical characteristics

applications in optical communication systems. More details on these kinds of gratings can be found in References [13,14].

Since the Bragg wavelength is dictated by the modal refractive index and grating period, strain and temperature change given to an FBG will alter its Bragg wavelength, which enables the FBG to become an adaptive device. Such tunability of the Bragg wavelength can be understood from the following formula [15]:

$$\Delta\lambda_{\mathrm{B}} = \lambda_{\mathrm{B}}\left[\left(1-\rho_e\right)\varepsilon_z + \left(\alpha_{\Lambda} + \alpha_{\mathrm{n}}\right)\Delta T\right], \tag{3}$$

where λ_{B} is the Bragg wavelength in the idle condition. The effective strain-optic constant is represented by $\rho_e \cong 0.22$ while ε_z is the amount of strain applied on the grating. The symbol $\alpha_{\Lambda} = (1/\Lambda)(\partial\Lambda/\partial T)$ is the thermal expansion coefficient of the fiber ($\sim$0.55 × 10^{-6} for silica) and $\alpha_{\mathrm{n}} = (1/n_{\mathrm{eff}})(\partial n_{\mathrm{eff}}/\partial T)$ represents the thermo-optic coefficient ($\sim$8.6 × 10^{-6} for the Ge-doped silica core fiber). It is found that α_{n} is an order of magnitude larger than α_{Λ}. Therefore, when temperature is changed, the main shift in the Bragg wavelength is dominated by the thermo-optic effect.

For an FBG with the center wavelength at 1550 nm, the strain sensitivity is about 1.209 pm/$\mu\varepsilon$, whereas the temperature sensitivity is approximately 11 pm/°C. Such dependence of the Bragg wavelength on strain and temperature can be used to realize tunable optical devices for optical communications. On the other hand, FBGs also can be exploited to use as strain or temperature sensors through interrogation of the wavelength shift due to these two factors.

FABRICATION TECHNIQUES

Exposing the fiber core to an intense interference pattern of the UV-radiation, we can induce the refractive-index modulation in the fiber core through the photosensitivity effect. Among various FGB writing methods, the transverse interference-fringe method is the most widely employed. This method can further be classified to the two-beam interference method and the phase-mask method.

Two Beam Interference Method

This method is also termed as the transversely holographic writing technique, which was first demonstrated by Meltz et al. [4]. The UV-laser light is split equally into two beams and then recombined back on the photosensitivity fiber at a mutual angle of 2θ by means of two UV-mirrors as depicted in Figure 3.

The advantage of this technique is that an FBG can be formed with an arbitrary spatial period Λ simply by varying the angle θ. The relation between the angle θ and the spatial period Λ is given as

$$\Lambda = \frac{\lambda_{uv}}{n_{uv} \sin\theta},\tag{4}$$

where λ_{uv} is the wavelength of the UV-light, and n_{uv} is the fiber refractive index at λ_{uv}. The angle 2θ between the two beams determines the Bragg wavelength according to Eqs. (1) and (2). The flexibility brought by this holographic technique enables fabrication of FBGs with Bragg wavelengths around 1550 nm for the use in optical communication systems.

However, this technique requires a rather complex writing setup and is susceptible to vibration. The UV-laser must have a high degree of coherent in order to produce stable interference fringes after recombination of the two divided beams. Qualified candidates are a frequency-doubled Argon ion laser, an excimer laser using KFr or ArF, and a frequency-doubled copper-vapor laser. In view of these stringent requirements, another technique based on the phase mask was proposed and has gained popularity among researchers and manufacturers for making FBGs.

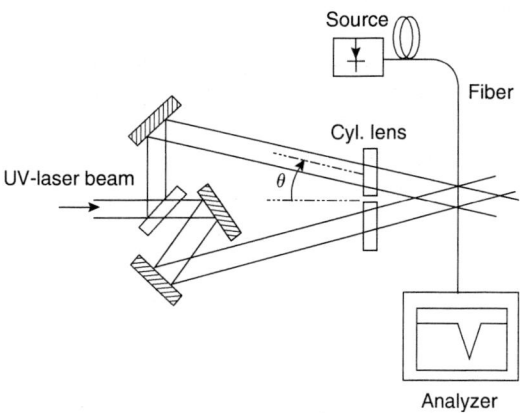

Figure 3   Holographic interference beams writing setup

Phase Mask Technique

This is a much simpler arrangement for fabricating FBGs [16,17]. The spatially modulation of light intensity, which is essential for FBG inscription, is set up by using a phase mask. The phase mask is a silica plate transparent to UV-radiation, and one of its surfaces is etched with a relief grating. The phase mask diffracts the incident UV-beam into different angles. By an appropriate design on the depth of the grating pitch, the zeroth order diffraction can be suppressed as low as 5% of the incident beam intensity, and then the UV-radiation is equally diffracted into +1 and −1 order. Both of the diffracted beams will interfere each other and produce a periodic intensity variation in the adjacent area of the relief grating. Such intensity variation of the UV-light will be imprinted on the photosensitive fiber placed in the interference region as shown in Figure 4. Scanning the UV-beam along the grating of the phase mask, we can fabricate an FBG, whose length is equal to the phase mask length. It should be pointed out that the interference pattern created by the phase mask has a period half of the phase mask periodicity.

The phase mask technique has relaxed the stringent requirement for highly coherent UV-sources. The fabrication setup has better tolerance to vibration than the two-beam interference method, and one can use less expensive optical mounts. With this simpler setup, high performance FBGs can be fabricated. However, one drawback in the phase mask technology is that the grating pitch of the phase mask determines the Bragg wavelength of the inscribed FBG. Nevertheless, we can achieve a small amount of wavelength tuning by straining the fiber before writing the FBG. After the strain is released, a wavelength shift toward the shorter wavelength side will be observed, and the amount of the wavelength shift depends on the pre-applied strain. This simple method allows wavelength tuning within ~2 nm.

Advanced structures such as apodized FBGs and chirped FBGs can also be fabricated by using the phase mask technique. Apodization can be realized by the piezo-electric dithering method [18]. The phase mask is set on a stage with a piezo-electric transducer (PZT), and it position can be dithered by the PZT. The voltage is applied to the PZT in the following way to realize the apodazation function: When the UV-beam is scanned at the edges of the grating, a larger amplitude of the dithering voltage is applied to the PZT in order to reduce the visibility of the refractive index modulation. In the middle of the grating, on the other hand, the phase mask is almost not dithered, which allows us to form the refractive index modulation with high visibility.

Another apodization process involves the dual exposures process [19]. The first exposure, where the beam scanning velocity is varied on the phase mask, makes the apodized grating. In such a case, however, the average of the modulated refractive-index profile is not zero, causing side lobes in the filtering spectrum. To make the average to be zero, the second exposure is

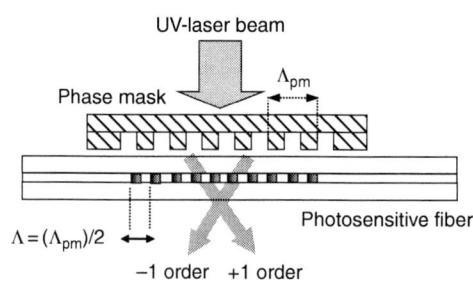

Figure 4   Phase-mask method where the incident UV-light is diffracted to +1 and −1 order to form an interference pattern

carried out without a phase mask. The beam is scanned with an inverse velocity profile of the first exposure or by using an amplitude mask with the inverse apodization function.

Complex gratings such as chirped FBGs and phase-shifted gratings could also be fabricated by the phase mask method. Usually, the easiest way is to employ a chirped or phase-shifted phase mask. Apart from linearly chirped FBGs, the phase mask can also be incorporated with nonlinear chirp. Nonlinearly chirped FBGs mainly target for dispersion-slope compensation applications.

SYSTEM APPLICATIONS OF FBGs

Sharp filtering characteristics, direct connectivity to fibers, and ease of manufacturing are very attractive features of FBGs when they are applied to wavelength-division multiplexed (WDM) systems. In addition to band-pass filters and band-rejection filters, FBGs have many other applications such as wavelength lockers for stabilizing laser spectra, gain equalizers for EDFAs, optical add-drop multiplexers, and dispersion compensators. This section deals with these applications of FBGs,

Wavelength-Stabilized Lasers

When an FBG is coupled to a semiconductor optical amplifier, the oscillation wavelength of the laser is defined by the Bragg wavelength of the FBG. Since the temperature sensitivity of the Bragg wavelength is small, the oscillation wavelength can be stabilized precisely to the ITU grid [20]. The FBG is also applied to the wavelength stabilization of 980-nm semiconductor lasers for EDFA pump sources as shown by Figure 5. The FBG acts as wavelength locker to produce a narrow spectrum near the Bragg wavelength. The narrow and stabilized spectrum of 980-nm lasers is of importance for pumping Erbium–Ytterbium fiber amplifiers.

The grating can be directly written inside the rare-earth-doped fiber-laser cavity with a $\pi/2$-phase shift in the middle of the cavity length. Such phase-shifted FBG enables the fiber DFB laser to oscillate at the Bragg wavelength. The fiber distributed feedback (DFB) laser has shown superior low noise, high output power, and narrow linewidth [21,22]. Another advantage of this type of laser is that its oscillation wavelength can be tuned without mode hopping just by changing the Bragg wavelength. Set et al. [23] have demonstrated 30-nm wavelength tuning of the fiber DFB laser by using the beam bending technique.

Gain Equalizers for EDFAs

In wideband WDM systems, the non-flat gain of Erbium-doped fiber amplifiers impairs the system performance. Therefore, gain equalization is necessary in such systems and various devices have already been developed. Among them, the gain-flattening filter (GFF) using the

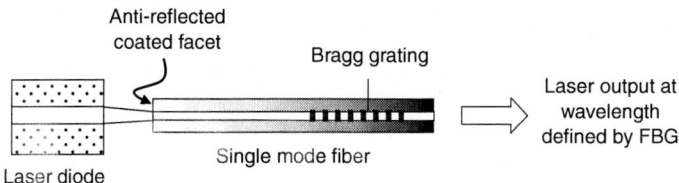

Figure 5 Wavelength locker for semiconductor laser

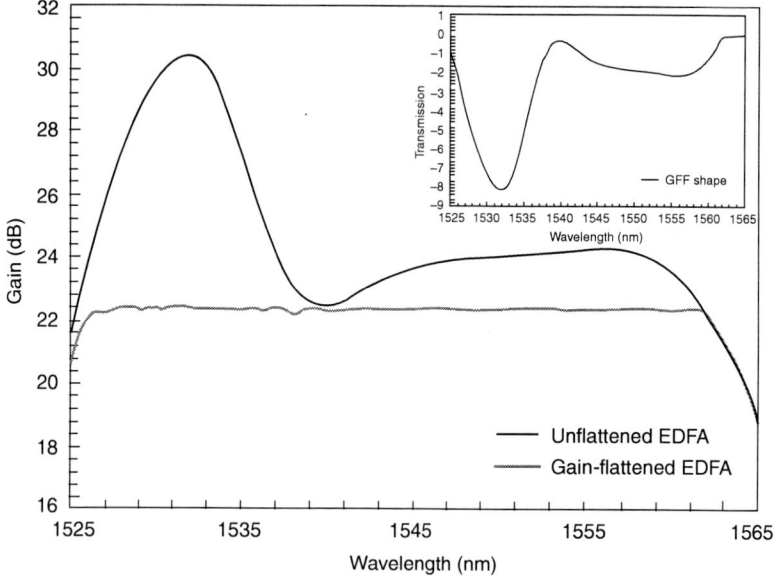

Figure 6   Gain spectrum of a gain-flattened EDFA by using a FBG-based gain equalizer

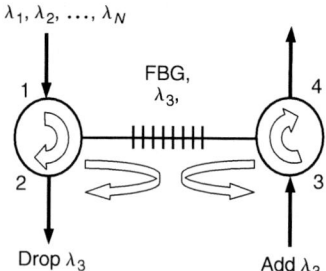

Figure 7   FBG-based optical add-drop multiplexer

chirped FBG meets requirements for practical gain equalizers [24]. The GFF is fabricated by changing the grating period so that the refractive-index modulation realizes the required transmission loss spectrum, which should have an inverse shape of the EDFA gain spectrum (see Figure 6). This device can cover a large optical bandwidth over >35 nm.

Optical Add-Drop Multiplexers

The FBG can be used to construct an optical add-drop multiplexer (OADM) in WDM networks. The configuration of the OADM is shown in Figure 7, where two circulators are connected with an FBG. Wavelength channels $\lambda_1, \lambda_2, \ldots, \lambda_N$ are coming from the network and incident on the OADM from port 1. Only the wavelength that coincides with the Bragg wavelength λ_k of the FBG is dropped from port 2. Other nonresonant wavelengths pass through the FBG, and come

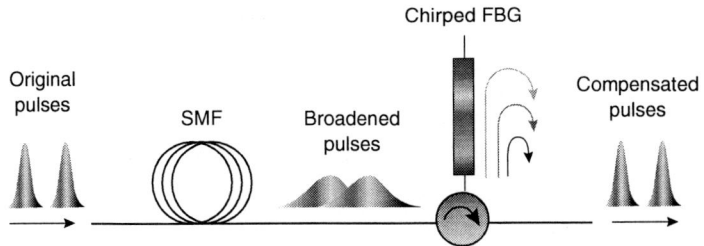

Figure 8   Principle of dispersion compensation by chirped fiber Bragg grating

back to the network from port 4. On the other hand, the wavelength λ_k can be added to the network from port 3.

Chirped FBGs for Dispersion Compensation

One of the most useful applications of FBGs is the dispersion compensator that can compensate for group-velocity dispersion (GVD) of fibers for transmission [25–27]. The standard single-mode fiber (SMF) has a zero-GVD wavelength at 1.3 μm; therefore, it has an anomalous GVD as large as 16 ps/nm/km at 1.5 μm. Chirped FBGs can be designed to have GVD that compensates for fiber dispersion precisely.

Chirped FBGs usually work in the reflection mode as shown in Figure 8. Due to anomalous GVD of SMF, the longer wavelength component of the signal travels through the fiber slower than the shorter wavelength component, resulting in the pulse-width broadening. The grating period of the chirped FBG is linearly chirped, and becomes shorter from the input end toward the far end. Figure 8 illustrates the working principle of the chirped FBG as a dispersion compensator. The longer wavelength component is reflected at the input end of the FBG, while the shorter one at the far end of the grating; thereby, a delay is induced between these wavelength components; the broadened pulse is restored to its original width after passing through the chirped grating.

For higher bit rate systems >40 Gbit/s, the high-order dispersion effect needs to be compensated. Nonlinearly chirped FBGs, with their periods varying nonlinearly along their length, can be used to compensate for the dispersion-slope effect [28,29].

DEVICE PACKAGING

FBGs are subjected to change their characteristics with changes in external parameters such as strain and temperature. This is because the Bragg wavelength shifts due to variations of these parameters. In actual optical networks, temperature variation ranges from –5 to 70°C. As the temperature sensitivity of FBGs is ~11 pm/°C, FBGs will suffer approximately a 1-nm wavelength shift, which is unacceptable for DWDM systems with 100- or 50-GHz channel spacing.

Hammon et al. [30] have shown that the temperature sensitivity of a packaged FBG is successfully reduced to as low as 0.4 pm/°C. In their technique depicted in Figure 9, the FBG is pre-tensioned before it is glued on a hybrid substrate that is composed of two materials with different thermal expansion coefficients. The temperature change causes the length of the substrate, either releasing or increasing the tension on the FBG; thus, the temperature dependent refractive index is cancelled by such strain effect. However, this packaging requires rather complex mechanical arrangement.

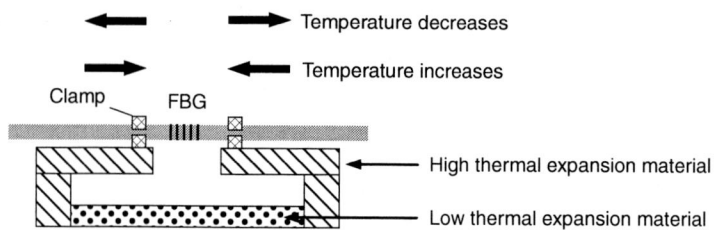

Figure 9   Athermal package for FBG

Another much simpler method has been developed, which is based on a negative-expansion ceramic substrate (NECS) [31]. The NECS has a negative expansion coefficient of $-8.2 \times 10^{-6}/°C$. In the athermalized FBG package, the pre-tensioned FBG is mounted and glued on the groove on the substrate. Any temperature elevation causes contraction of the substrate, which relieves the pre-stretched grating accompanying with a Bragg wavelength shift to the blue side of the spectrum. This shift cancels out the red shift of the Bragg wavelength by the thermo-optic effect; therefore, the Bragg wavelength is unchanged even by the temperature increase. However, this substrate shows small hysteresis < 0.03 nm for temperature cycle tests between −40 and 85°C. In addition, the bonding material used for gluing FBGs is also an important aspect. Good adhesive should be chosen to faithfully hold the strained FBG over the device lifetime.

CONCLUSION

It is anticipated that FBGs will continue to play significant roles in DWDM systems: The sharp filtering response of FBGs is especially important for DWDM systems with extremely narrow channel spacing. For high-speed transmission, dispersion compensators are the most attractive application of FBGs. In addition, it should be noted that tunability of the Bragg wavelength as well as the GVD value will extend potential application areas of FBGs in future dynamically reconfigurable networks.

REFERENCES

1. K.O. Hill et al., "Photosensitivity in optical fiber waveguides: Application to reflection filter fabrication," *Appl. Phys. Lett.*, 32, 1978, 647–649.
2. B.S. Kawasaki et al., "Narrow-band Bragg reflectors in optical fibers," *Opt. Lett.*, 3, 1978, 66–68.
3. D.K. Lam and B.K. Garside, "Characterization of single-mode optical fiber filters," *Appl. Opt.*, 20, 1981, 440–445.
4. G. Meltz, W.W. Morey, and W.H. Glenn, "Formation of Bragg gratings in optical fibers by a transverse holographic method," *Opt. Lett.*, 14, 1989, 823–825.
5. D.P. Hand and P. St. J. Russell, "Photoinduced refractive-index changes germanosilicate fibers," *Opt. Lett.*, 15, 1990, 102–104.
6. W.W. Morey et al., *OSA Opt. Photon News*, 1, 1990, 14–17.
7. R. Kashyap, *Fiber Bragg Gratings*, Academic Press, San Diego, CA, 1999.
8. B. Poumellec et al. "UV induced densification during Bragg grating inscription in Ge:SiO$_2$ preforms: Interferometric microscopy investigations," *Optical Mater.*, 4, 1995, 404–409.
9. M. Douay et al. "Densification involved in the UV-based photosensitivity of silica glasses and optical fibers," *IEEE J. Lightwave*, 15, 1997, 1329–1342.

10. H.G. Limberger et al. "Compcation- and photoelastic-induced index changes in fiber Bragg gratings," *Appl. Phys. Lett.*, 68, 1996, 3069–3071.

11. P.J. Lemaire, R.M. Atkins, V. Mizrahi, and W.A. Reed, "High pressure H_2 loading as a technique for achieving ultrahigh UV photosensitivity and thermal sensitivity in GeO_2 doped optical fibres," *Electron. Lett.*, 29, 1993, 1191–1193.

12. F. Bilodeau, B. Malo, J. Albert, D.C. Johnson, and K.O. Hill, "Photosensitization of optical fiber and silica-on-silicon/silica waveguides," *Opt. Lett.*, 18, 1993, 953–955.

13. R. Kashyap, P.F. McKee, and D. Armes, "UV written reflection grating structure in photosensitive optical fibres using phase-shifted phase masks," *Electron. Lett.*, 30, 1994, 1977–1978.

14. B.J. Eggleton, P.A. Krug, L. Poladian, and F. Ouellette, "Long period superstructure Bragg gratings in optical fibres," *Electron. Lett.*, 30, 1994, 1620–1622.

15. G. Meltz, "Overview of fiber grating-based sensors," in *Proceedings of SPIE, Distributed and Multiplexed Fibre Optics Sensors* VI, Denver, CO, 1996, Vol. 2838, 1–21.

16. K.O. Hill, B. Malo, F. Bilodeau, D.C. Johnson, and J. Albert, "Bragg gratings fabricated in monomode photosensitive optical fiber by UV exposure through a phase mask," *Appl. Phys. Lett.*, 62, 1993, 1035–1037.

17. D.Z. Anderson, V. Mizrahi, T. Erdogan, and A.E. White, "Production of in-fibre gratings using a diffractive optical element", *Electron. Lett.*, 29, 1993, 566–568.

18. M.J. Cole, W.H. Loh, R.I. Laming, M.N. Zervas, and S. Barcelos, "Moving fibre/phase mask-scanning beam technique for enhanced flexibility in producing fibre gratings with a uniform phase mask," *Electron. Lett.*, 31, 1995, 92–94.

19. B. Malo, S. Theriault, D.C. Johnson, F. Bilodeau, J. Albert, and K.O. Hill, "Apodised in-fibre Bragg grating reflectors photoimprinted using a phase mask," *Electron. Lett.*, 31, 1995, 223–225.

20. R.J. Cambell et al. "Wavelength stable uncooled fibre grating semiconductor laser for use in an all optical WDM access network," *Electron. Lett.*, 32, 1996, 119–120.

21. W.H. Loh and R.I. Laming, "1.55 μm phase-shifted distributed feedback fibre laser," *Electron. Lett.*, 31, 1995, 1440–1442.

22. W.H. Loh, B.N. Samson, L. Dong, G.J. Cowle, and K. Hsu, "High Performance Single Frequency Fiber Grating-based Erbium:Ytterbium-codoped Fiber Lasers," *J. Lightwave Technol.*, 16, 1998, 114–118.

23. S.Y. Set, M. Ibsen, C.S. Goh, and K. Kikuchi, "Simple broadrange tuning of fibre-DFB lasers," in *Proceedings of the ECOC'01*, paper Tu.F.3.4, Vol. 2, 2001, 200–201.

24. M. Rochette, M. Guy, S. LaRochelle, J. Lauzon, and F. Trepanier, "Gain Equalization of EDFA's with Bragg Gratings," *IEEE Photon. Technol. Lett.*, 11, 1999, 536–538.

25. K.O. Hill, S. Theriault, B. Malo, F. Bilodeau, T. Kitagawa, D.C. Johnson, J. Albert, K. Kataoka, and K. Hagimoto, "Chirped in-fibre Bragg grating dispersion compensators: Linearization of the dispersion characteristics and demonstration of dispersion compensation in a 100 km, 10 Gbit/s optical fiber link," *Electron. Lett.*, 30, 1994, 1755–1756.

26. J.A.R. William, I. Bennion, K. Sugden, and N.J. Doran, "Fiber dispersion compensation using a chirped in-fibre Bragg grating," *Electron. Lett.*, 30, 1994, 985–987.

27. R. Kashyap, S.V. Chernikov, P.F. McKee, and J.R. Taylor, "30 ps chromatic dispersion compensation of 400 fs pulses at 100 Gbit/s in optical fibers using an all fiber photoinduced chirped reflection grating," *Electron. Lett.*, 30, 1994, 1078–1080.

28. M. Durkin, M. Ibsen, M.J. Cole, and R.I. Laming, "1m long continuously-written fibre Bragg gratings combined second- and third-order dispersion compensation," *Electron. Lett.*, 33, 1997, 1891–1893.

29. J.A.R. Williams, L.A. Everall, I. Bennion, and N.J. Doran, "Fiber Bragg grating fabrication for dispersion slope compensation," *IEEE Photon. Technol. Lett.*, 8, 1996, 1187–1189.

30. T.E. Hammon, J. Bulman, F. Ouellette, and S.B. Poole, "A Temperature compensated optical fibre Bragg grating band rejection filter and wavelength reference," *OECC'96 Technical Digest*, paper 18C1-2, 1996, 350–351.

31. S. Yoshihara, T. Matano, H. Ooshima, and A. Sakamoto, "Reliability of Athermal FBG component with negative thermal expansion ceramic substarte," *IEICE 2003*, paper C-3-73, 2003, 206.

FOUR WAVE MIXING

Kazuo Kuroda

Four wave mixing, a third-order optical parametric process, is the nonlinear optical process in which four waves interact with each other through the third-order optical nonlinearity [1]. In this process, three waves form a nonlinear polarization at the frequency of the fourth wave. Let the frequencies and the wave vectors of these four waves be (ω_j, k_j), $(j = 1, ..., 4)$. The wave function is then expressed as $E_j(r, t) = A_j(r) \exp[i(k_j \cdot r - \omega_j t)]$, where $A_j(r) = |A_j(r)| \exp[i\phi_j(r)]$ is a complex amplitude. There are two possible relations of the frequencies that satisfy the conservation of photon energies: (i) $\omega_4 = \omega_1 + \omega_2 + \omega_3$, and (ii) $\omega_3 + \omega_4 = \omega_1 + \omega_2$. The phase matching condition or the conservation of photon momenta is associated with each case: (i) $k_4 = k_1 + k_2 + k_3$, and (ii) $k_3 + k_4 = k_1 + k_2$.

The case (i) involves the third harmonic generation and the third-order sum frequency generation. The case (ii) is much more interesting because the conservation law of photon energies allows the degenerate case where all the interacting waves have the same frequency $(\omega_1 = \omega_2 = \omega_3 = \omega_4 = \omega)$. In this case, the nonlinear polarization is given by

$$P_{NL}(\omega_4) = 3\varepsilon_0 \chi^{(3)} A_1 A_2 A_3^* e^{i(k_1 + k_2 - k_3)r - i\omega_4 t},$$

where $\chi^{(3)}$ is the third-order nonlinear susceptibility. Since this nonlinear polarization generates the fourth wave, $A_4 \propto A_1 A_2 A_3^*$.

An important application of degenerate four wave mixing is the generation of phase conjugate wave or simply the phase conjugation. The phase conjugate wave has the same wavefront as the incident wave; however, it propagates backward. If the amplitude of incident wave is expressed as $E(r, t) = A(r) \exp i(k \cdot r - \omega t)$, then its phase conjugation is given by $E_{pc}(r, t) = A^*(r) \exp i(-k \cdot r - \omega t)$. Spatial part of the phase conjugate wave function is just the complex conjugate of the incident wave. A device that generates phase conjugate wave is called a phase conjugate mirror.

The phase conjugation is equivalent to the time reversal. Straightforward application of the time reversal is the correction of phase aberration. Suppose that, the incident wave propagates in the nonuniform medium, such as a laser amplifier, where the temperature is fluctuated by intense pumping (Figure 1). Then, the incident wave suffers severe phase aberration during the propagation and the transmitted wave is then reflected by the phase conjugate mirror and goes back into the same nonuniform medium. When the phase conjugate wave returns on the initial plane, the phase aberration is completely corrected by virtue of the time reversal operation. This technique is useful for high-power laser amplifier systems.

There are two common configurations for the phase conjugation by four wave mixing. The degenerate four wave mixing is shown in Figure 2. The pump waves E_1 and E_2 are counter-propagating and uniform plane waves, that is, relation $k_1 + k_2 = 0$ is satisfied, and $A_1 A_2$ is almost constant. Then the phase conjugate wave is proportional to the complex conjugate of the signal wave $(A_4 \propto A_3^*)$, and it propagates backward $(k_4 = -k_3)$. The generated wave E_4 is the phase conjugation of the signal wave E_3. In this configuration, the phase matching condition is always fulfilled. Reflectivity of phase conjugate wave is given by

$$R = \left(\frac{3\pi\chi^{(3)}L}{\varepsilon_0 cn^2 \lambda} \right)^2 I_1 I_2,$$

where L is the interaction length, n is the refractive index, λ is the wavelength, and I_1 and I_2 are intensities of pump waves.

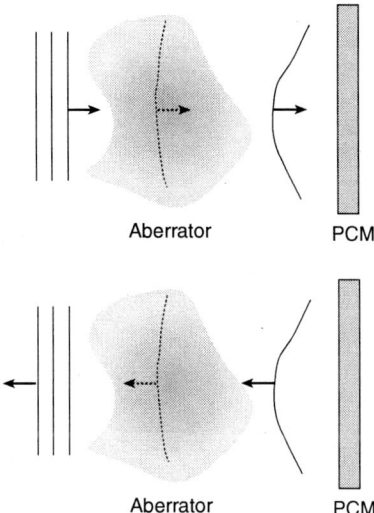

Figure 1 Correction of phase distortion by the phase conjugation. PCM: phase conjugate mirror

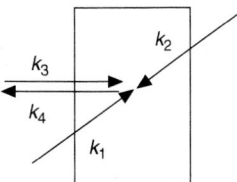

Figure 2 Degenerate four wave mixing

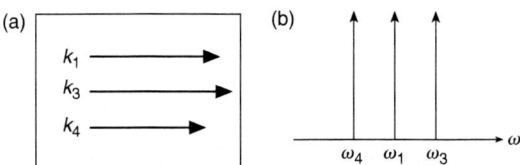

Figure 3 Nearly degnerate three wave mixing. (a) wavevectors (b) frequencies in spectral region

The other configuration is nearly degenerate three wave mixing. As shown in Figure 3(a), all the waves propagate in the same direction. In this configuration, $\omega_1 = \omega_2$ and $k_1 = k_2$, that is, only one wave is used for pumping. Thus, actually three waves participate in this process. The signal frequency ω_3 should be slightly shifted from the pump frequency ω_1, otherwise it is difficult to distinguish the signal from the pump. The frequency of phase conjugate wave is given by $\omega_4 = 2\omega_1 - \omega_3$, which is located symmetrically with respect to the pump frequency ω_1, as shown in Figure 3(b). In this configuration, there remains a phase mismatch; however, it is small if the frequency difference $\Delta\omega = \omega_3 - \omega_1$ is small. This configuration is used not only for the phase conjugation but also for the frequency conversion and the optical parametric amplification in optical fibers [2].

REFERENCES

1. R. W. Boyd, *Nonlinear Optics*, 2nd ed. (Academic Press, New York, 2003).
2. G. P. Agrawal, *Nonlinear Fiber Optics*, 3rd ed., Chap. 10 (Academic Press, New York, 2001).

FREQUENCY CHIRPING

Fumio Koyama

INTRODUCTION

Dynamic single-mode semiconductor lasers that operate at a fixed single mode under rapid direct modulation [1,2], such as distributed Bragg reflector (DBR) and distributed feedback (DPB) lasers, have been developed for high capacity and long-haul single-mode fiber communication systems at the wavelength region of 1.5 to 1.6 μm [3,4]. In this wavelength region, finite spectral width of a light source causes pulse broadening of the transmitted signal due to the effect of the chromatic dispersion of conventional single-mode fibers. An important question arose on the single-mode lasers concerned with the maximum transmission bandwidth of the single-mode fiber: how large is the dynamic spectral width of a directly modulated laser? The wavelength of the lasing mode swings around its central wavelength due to the

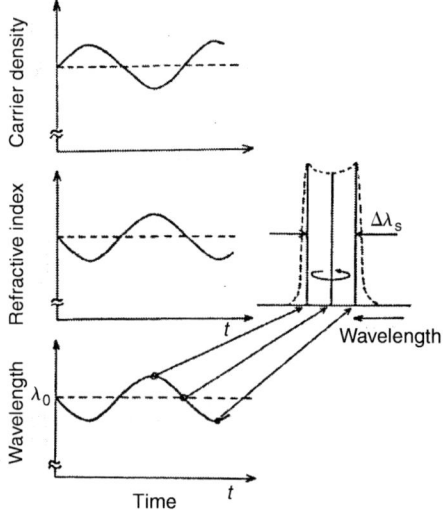

Figure 1 Frequency (wavelength) chirping caused by carrier effect

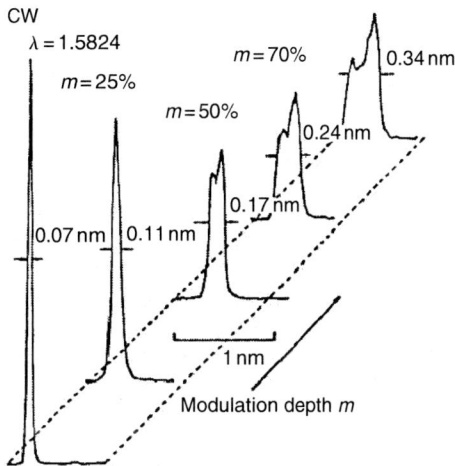

CW
λ = 1.5824
m = 25%
m = 70%   0.34 nm
m = 50%
0.24 nm
0.17 nm
0.07 nm   0.11 nm
1 nm
Modulation depth m

Figure 2   Measured lasing spectra of directly modulated DBR laser with different modulation depth (From F. Koyama, S. Arai, Y. Suematsu, and K. Kishino, *Electron. Lett.*, 17, 938–940, 1981. With permission.)

variation of refractive index of the active layer under direct modulation as shown in Figure 1 [5–7]. This phenomenon is called frequency chirping or wavelength chirping [8]. The dynamic spectral width of a directly modulated single-mode laser was observed to be a few angstrom [9]. Figure 2 shows the measured lasing spectra of a directly modulated single-mode laser [9]. The nature of the frequency chirp were made clear [10,11]. These works gave rise to an estimation of the transmission bandwidth at the wavelength region of 1.55 μm [11]. In addition, the transmission properties of a chirped pulse through a single-mode fiber have been studied theoretically and experimentally [12] and it was pointed out that the frequency chirping caused a power penalty.

On the other hand, an external modulation is useful for eliminating this problem. Some types of external intensity modulators such as directional-coupler type modulator, electro-absorption modulator, and Mach-Zehnder interferometer type modulator were developed. Some transmission experiments employing an external modulation technique have been reported for the purpose of eliminating the frequency chirping. Also, high-speed external modulators using a quantum-well structure and the monolithic integration of a modulator and a laser have been intensively developed. It was pointed out that some frequency chirping is caused by phase modulation due to a refractive index change in a loss modulator [13]. It is important to make clear the nature of the frequency chirping in external intensity modulators because it determines the ultimate transmission bandwidth of single-mode fiber systems employing an external modulation.

FORMULA OF FREQUENCY CHIRPING

Frequency Chirping under Direct Modulation

The variation of the lasing wavelength of directly modulated semiconductor lasers is caused by the temperature change and the carrier density modulation in the active layer. Figure 3 shows the measured spectral width of a directly modulated semiconductor laser as a function of modulation frequency. The spectral width is maximized at a resonance-like frequency. The thermal effect is negligible under high-frequency modulation of more than several hundreds megahertz,

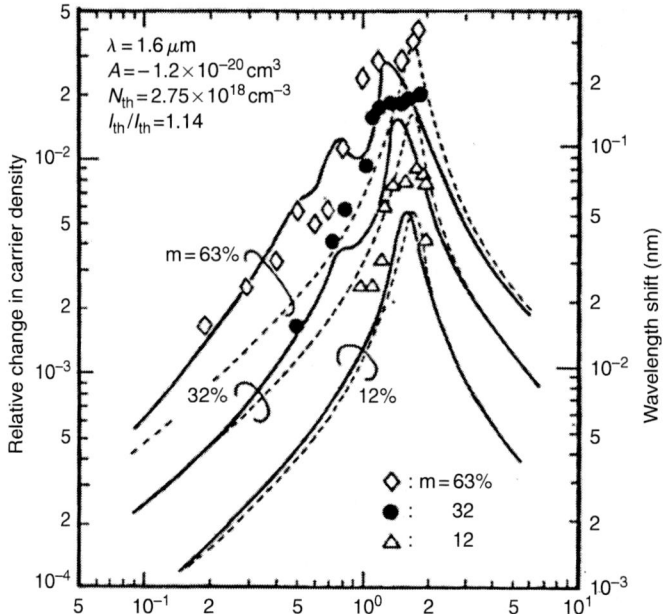

Figure 3   Amount of frequency chirping as a function of modulation frequency (From K. Kishino, S. Aoki, Y. Suematsu, *IEEE J. Quantum Electron.*, QE-18, 582–595. With permission.)

which is the case in optical communication systems. Therefore, only the carrier effect is considered. The carrier density modulation that causes the frequency chirp can be calculated by using the rate equations. In the calculation, we assume the shape of the modulated light intensity. Henry derived a relationship between the phase and the light intensity in the laser cavity for the analysis of the static linewidth of lasers [14]. The temporal angular frequency $\omega(t)$ of a directly modulated laser is given in Reference 11 by

$$\omega(t) = \omega_0 + \frac{\alpha}{2} \cdot \frac{1}{S} \frac{dS}{dt}, \tag{1}$$

where ω_0 is the angular frequency under CW operation, S is the photon density and α is the linewidth enhancement factor, which is defined by Henry as the ratio between the changes of the real part and imaginary part of the refractive index of the active layer [14]. α has been measured to be from 3 to 7 in GaAs and GaInAsP lasers. Harder et al. indicated that the frequency modulation is coupled to the intensity modulation through the susceptibility of the gain medium and that the coupling constant is the linewidth enhancement factor under small-signal sinusoidal modulation [15].

Equation (1) indicates that the frequency chirp even under large signal modulation can be obtained analytically if the instantaneous light intensity is given. Thus the electric field of the instantaneous output light intensity of a directly modulated laser can be expressed as follows:

$$E \propto \sqrt{S} \exp[j \int \omega(t) dt]. \tag{2}$$

The frequency chirp of a laser under pulse modulation is derived in the following. We assume that the instantaneous photon density, which is proportional to the output light intensity, is a Gaussian-shaped pulse with a temporal full-width $2T$ at the $1/e$ points as shown in the following equation.

$$S = S_b + S_m \exp\left[-\left(\frac{t}{T}\right)^2\right],$$

(3)

where the modulation amplitude and the bias level of the photon density are S_m and S_b, respectively. By substituting (3) into (1), the temporal angular frequency under deep-pulse modulation is approximately given by

$$\omega(t) = \omega_0 - \alpha \frac{t}{T^2}.$$

(4)

The frequency chirp depends on the optical pulse width and the α-parameter. Equation (4) shows that the frequency chirp is unidirectional and linear as a function of time, and the direction is determined by the sign of the α-parameter.

Calculating the Fourier transform of the electric field E, we obtain the spectral width under direct modulation as follows:

$$\Delta\omega = \frac{2\sqrt{1+\alpha^2}}{T}.$$

(5)

The dynamic spectral width under direct modulation is determined by the linewidth enhancement factor α and modulated optical pulse-width T and is independent of the modulation amplitude. Equation (5) shows that the amount of the frequency chirp increases by $\sqrt{1+\alpha^2}$ [11].

Nonlinear Effect

Some nonlinear effects such as spectral hole burning and spatial hole burning cause additional term in frequency chirping. These nonlinear effects give us nonlinear gain, which is proportional to the output intensity of a semiconductor laser. Thus, the carrier density of an active layer is changed to be proportional to the output. According to this effect, the formula can be expressed by the following equation [10]:

$$\Delta\omega(t) = \frac{\alpha}{2}\left(\frac{1}{S}\frac{dS}{dt} + \varepsilon S\right),$$

(6)

where ε is the nonlinear gain coefficient, which is originated from the spectral hole burning as well as spatial hole burning. When the output of a laser is increased, this term cannot be neglected.

Frequency Chirping in External Modulators

Electro-absorption Modulator

The frequency chirping is caused by the refractive index change of an active layer due to carrier density modulation in the direct modulation of semiconductor lasers. To eliminate the chirping, external modulators have been used for long-haul fiber transmission. However, chirp-like spectral broadening also must be investigated. One possible cause is the wavelength shift of the semiconductor laser due to an external reflection outside of the external modulator. This is because the coupled power of the externally reflected light to the laser cavity is varied with

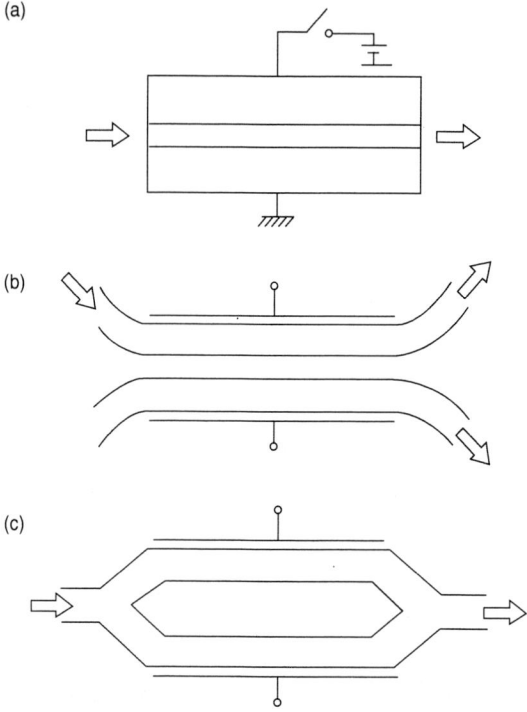

Figure 4   Schematics of various external modulators (a) Electro-absorption modulator, (b) Directional coupler type, and (c) Mach-Zehnder interferometer type

intensity modulation, which changes the laser wavelength [16]. This sort of chirping can be removed by adopting an optical isolator and reducing the reflection by means of AR coating to the end-facet of the modulator. Another possible cause is the phase modulation due to the refractive index change of medium inside an external modulator. The relation between the frequency chirping due to phase modulation and intensity modulation is derived [13].

Here we describe the frequency chirping for three types of external modulators as shown in Figure 4. One of the external modulators using semiconductor materials is an electro-absorption modulator with a variable loss to the laser light passing through the modulator [17]. This type of external modulators can be made with the same process as semiconductor lasers, which enables the monolithic integration of lasers and modulators [18,19]. Intensity modulation can be obtained by changing the loss of a modulator. In order to change the loss in a modulator, the electro-absorption effect in quantum-well as well as waveguide or thin film structure bulk crystal has been utilized.

Here we obtain a formula for the chirping in a loss modulator. In general, if we vary the loss that is due to the imaginary part n'' of the refractive index, the real part n' will suffer some of the modulation according to the Kramers–Kronig relation. Therefore, this causes the phase variation of transmitted light through an external modulator together with intensity modulation. Assuming that the input light into the modulator has a constant amplitude E_1 and constant angular frequency ω_0, the amplitude E_2 and the relative phase change ϕ of the output are given by

$$\left|E_2\right| = \left|E_1\right| \cdot \exp\left(-k_0 n'' L\right), \tag{7}$$

$$\phi = -k_0 n' L, \tag{8}$$

where k_0 is the propagation constant in free space and L is the length of an external modulator. Combining Eqs. (7) and (8), we obtain the following equation, which relates the instantaneous intensity and the phase ϕ of the output light:

$$\frac{d\phi}{dt} = \omega_c t \frac{\alpha}{2} \cdot \frac{l_c}{S_c} \tag{9}$$

where $\alpha = \Delta n' / \Delta n''$ is the relative change of the real part $\Delta n'$ and the imaginary part $\Delta n''$ of the refractive index. The derivative of the phase in the left side of Eq. (9) corresponds to the instantaneous angular frequency shift. It is noteworthy that this relation is identical to that in the direct modulation of a semiconductor laser given by Eq. (1).

When a loss is changed by a carrier injection, α is equal to the linewidth enhancement factor derived by Henry for semiconductor lasers. This value ranges from 2 to 7 for GaAs and GaInAsP conventional lasers and from 1.5 to 1.7 for a quantum-well laser. Therefore, the external modulator using the same process as lasers gives almost the same frequency chirping as the direct modulation, whereas ringing of carriers due to relaxation oscillation in external modulation is not noticeable as in lasers, and the chirping can be reduced by choosing some suitable waveforms of modulation signals. Suzuki et al. obtained the α-parameter for the electro-absorption type bulk modulator, which was 1 to 2 for a discrete modulator [20]. Also, the α-parameter for a loss modulator using a quantum-well structure was experimentally estimated to be 0.9 [21]. It was shown that the α-parameter in a quantum well structure strongly depends on the operating wavelength and applied electric field. Zero chirp or negative chirp can be possible by choosing appropriate operating conditions.

Directional Coupler Type Modulator

Optical switching devices or modulators built by electrically switching a directional coupler from the crossover state to the straight-through state have been widely studied. Their switching characteristics can be analyzed by using the coupled mode theory. Only one waveguide is assumed to be initially excited by the light with a constant amplitude and constant frequency. The output intensity of both waveguides is modulated by varying the propagation constant of each waveguide. We found that the intensity of light, which has been coupled from the initially-excited waveguide can be modulated without phase modulation, when the propagation constant of the two waveguides changes by the same amount $\Delta\beta$ but in the opposite sign [13]. This is due to the compensation of the phase shift of the coupled light. On the other hand, the phase of light straight through the waveguide initially excited changes together with the intensity modulation. Thus, the frequency chirping in the cross-port can be compensated to zero and that of the straight-port can be designed. This result can be used for the pre-chirp technique in fiber transmission systems.

Mach-Zehnder interferometer type modulator

The frequency chirping of a Mach-Zehnder interferometer type modulator with the device length of L is also considered. We find that the phase modulation can be completely compensated if the propagation constants of two waveguides are changing by the same amount $\Delta\beta$ with an opposite sign [13]. Thus, the α-parameter defined in Eq. (9) in this type of modulator is equal to zero.

The frequency chirping in three types of the external intensity modulators treated here can be summarized by the same Eq. (9), which simply uses the α-parameter. The α-parameter for a loss modulator employing the carrier injection may be almost same as direct laser modulation, which depends on operating wavelength and is ranging from 2 to 7. On the other hand, that of a loss modulator using the electro-absorption in a bulk and quantum-well experimentally exhibited smaller value. It depends on operating wavelength and the applied electric field, and therefore the optimization of device structure and operating condition enables the reduction of the frequency chirping [22]. In principle, the frequency chirping of directional-coupler type and Mach-Zehnder interferometer type modulators can be eliminated.

EFFECT ON FIBER TRANSMISSION BANDWIDTH

The frequency chirping affects the transmission properties of a single-mode fiber. We found that the frequency chirping in an external modulation is expressed by the same equation as in a direct modulation. By using this result, we can obtain the transmission bandwidth of a single-mode fiber including the effect of frequency chirping in light sources. The following dispersion formula is used to evaluate the bandwidth [12]:

$$E_{out}(t) = \int_{-\infty}^{\infty} \varepsilon(\omega) \exp[j\{\omega t - \beta L - \beta'(\omega - \omega_0) - \frac{\beta''}{2}(\omega - \omega_0)^2\}]d\omega, \tag{10}$$

where $E_{out}(t)$ is the electric field at the fiber end, β, β' and β'' are the propagation constant, its first and second derivatives of the fiber with respect to the angular frequency, and $\varepsilon(\omega)$ is the spectrum of a light source. Now we consider the pulse response of a single-mode fiber, assuming the input pulse has a Gaussian shape. We obtain the maximum corresponding bit rate B as follows [13], which is defined by the reciprocal of the attainable minimum output pulse width:

$$B = \frac{1}{2\sqrt{2(\sqrt{\alpha^2 + 1} + \alpha) \cdot |\beta'' L|}}. \tag{11}$$

Figure 5 shows the calculated product of the maximum bit rate B and the square root of the fiber length L as a function of α-parameter, where a wavelength and a fiber dispersion are assumed to be 1.55 μm and 20 ps/km/nm, respectively. In the range of the α-parameter from −4 to 0, the transmission bandwidth exceeds the value that is limited by the sideband of the modulation signal, which is due to the pulse compression caused by "blue shift chirp". This technique is called the pre-chirp for expanding the transmission bandwidth [23].

The modulation waveform is also important for the transmission characteristics. The change of the intensity causes the chirping according to Eq. (9). Therefore, we can see that the reduction

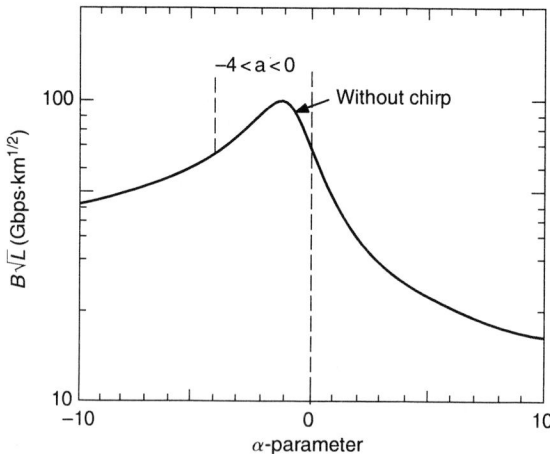

Figure 5 Calculated transmission bandwidth of a single-mode fiber versus α-parameter (From F. Koyama and K. Iga, *IEEE J. Lightwave Technol.*, 6, 87–93, 1988. With permission.)

of the transient time in a pulse eliminates the chirping. Thus, wideband external modulation with optimized pulse shape can be effective to reduce the effect of the chirping.

The frequency chirp is particularly important for long-haul fiber transmission systems such as submarine cable systems, since chirpless transmitters are essentially needed for this purpose. Therefore, Mach-Zehnder interferometer type external modulators have been widely used for current long-haul transmission systems.

REFERENCES

1. K. Utaka. K. Kobayashi, K. Kishino, and Y. Suematsu,"1.5–1.6 μm GaInAsP/InP integrated twin-guide lasers with first order distributed Bragg reflector," *Electron. Lett.*, 16, 455–456, 1980.
2. Y. Suematsu, S. Arai, and K. Kishino, "Dynamic single-mode semiconductor lasers with a distributed reflector," *J. Lightwave Technol.*, LT-1, 161–176, 1983.
3. K. Utaka, S. Akiba, K. Sakai, and M. Matsushima, "Room temperature CW operation of distributed-feedback buried-hetero-structure InGaAsP/InP lasers emitting at 1.57 μm," *Electron Lett.*, 17, 961–963, 1981.
4. T. Matsuoka, H. Nagai, Y. Itaya, Y. Noguchi, U. Suzuki, and T. Ikegami, "CW operation of DFB-BH GaInAsP/InP lasers in 1.5° μm wavelength region," *Electron. Lett.*, 18, 27–28, 1982.
5. J. M. Osterwalder and B. J. Rickett, "Frequency modulation of GaAlAs injection lasers at microwave frequency rate," *IEEE J. Quantum Electron.*, QE-16, 250–252, 1980.
6. K. Kishino, S. Aoki, and Y. Suematsu, "Wavelength variation of 1.6 μm wavelength buried heterostructure GaInAsP/InP lasers due to direct modulation," *IEEE J. Quantum Electron.*, QE-18, 343–351, 1982.
7. S. Kobayashi, Y. Yamamoto, M. Ito, and T. Kimura, "Direct frequency modulation in AlGaAs semiconductor lasers," *IEEE J. Quantum Electron.*, QE-18, 582–595, 1982.
8. C. Lin, "Picosecond frequency chirping and dynamic line broadening in GaInAsP/InP injection lasers under fast excitation," *Appl. Phys. Lett.*, 42, 141–143, 1983.
9. F. Koyama, S. Arai, Y. Suematsu, and K. Kishino, "Dynamic spectral width of rapidly modulated 1.58 μm GaJnAsP/JnP buried heterostructure distributed- Bragg-reflector integrated-twin-guide lasers," *Electron. Lett.*, 17, 938–940, 1981.
10. T. L. Koch and J. E. Bowers, "Nature of wavelength chirping in directly modulated semiconductor lasers," *Electron. Lett.*, 20, 1038–1040, 1984.
11. F. Koyama and Y. Suematsu, "Analysis of dynamic spectral width of dynamic single mode lasers and related transmission bandwidth of single-mode fibers," *IEEE J. Quantum Electron.*, QE-21, 292–297, 1985.
12. D. Marcuse, "Pulse distortion in single-mode fibers, 3: Chirped pulses," *Appl. Opt.*, 20, 3573–3579, 1981.
13. F. Koyama and K. Iga, "Frequency chirping in external modulators," *IEEE J. Lightwave Technol.*, 6, 87–93, 1988.
14. C. H. Henry, "Theory of the linewidth of semiconductor lasers," *IEEE J. Quantum. Electron.*, QE-18, 259–264, 1982.
15. C. Harder, K. Vahala, and A. Yariv, "Measurement of the linewidth enhancement factor of semiconductor lasers," *Appl. Phys. Lett.*, 42, 328–330, 1983.
16. K. Matsumoto, "Study on Integrated External Modulators," Tokyo Institute of Technology, Tokyo, Japan, Masters thesis, 1982.
17. T. H. Wood, C. A. Burrus, D. A. B. Miller, D. S. Chemla, T. C. Damen, A. C. Cossard, and W. Wiegmann, "131-ps optical modulation in semiconductor multiple quantum wells (MQW's)," *IEEE J. Quantum Electron.*, QE-21, 117–118, 1985.
18. Y. Kawamun. K. Wakita, Y. Yoshikuni, Y. Itaya, and H. Asahi, "Monolithic integration of InGaAsP/InP DPB lasers and InGaAs/InAIAs MQW optical modulators," *Electron. Lett.*, 22, 242–243, 1986.
19. M. Suzuki and Y. Noda, "Monolithic Integration of InGaAsP/InP Distributed-Feedback Laser and Electroabsorption Modulator by Vapor-Phase Epitaxy," paper presented at OFC/IOOC'87, Reno. paper TRF4, 1987.

20. M. Suzuki, Y. Noda, and Y. Kushiro, "Characterization of dynamic spectral width of an InGaAsP/InP electroabsorption light modulator," *IECE Jpn,* E69, 395–398, 1986.

21. T. H. Wood, R. W. Tkach, and A. R. Chraplyvy, "Observation of Low Chirp in GaAs Multi-Quantum-Well Intensity Modulation," paper presented at OFC/IOOC '87, Reno. paper WO5, 1981.

22. F. Koyama, K. Y. Liou, A. G. Dentai, G. Raybon, C. A. Burrus, "Measurement of chirp parameter in GaInAs/GaInAsP quantum well electroabsorption modulators by using intracavity modulation," *IEEE Photonics Technol. Lett.,* 5, 1389–1393, 1993.

23. T. Saito, N. Henmi, S. Fujita, M. Yamaguchi, and M. Shikada, "Prechirp technique for dispersion compensation for a high-speed long-span transmission," *IEEE Photonics Technol. Lett.,* 3, 74–76, 1991.

I

INTEGRATED TWIN-GUIDE LASER

Katsumi Kishino and Yasuharu Suematsu

INTRODUCTION

Integrated twin-guide (ITG) lasers were proposed and developed as one of the integrated lasers [1–4] in 1974. Integrated lasers are kind of lasers, which enable a monolithic connection from the laser section to a low-loss output waveguide to be realized. On extension of the output waveguide, other lightwave elements could be connected. Therefore it could open the way to make photonic integrated circuits, in which monitors, amplifiers, switches, couplers, and so on would be monolithically integrated. Because of the potentiality, this concept of lasers was enthusiastically investigated in the mid-1970s. Various types of integrated laser were demonstrated employing different coupling schemes, that is, phase coupling [1–4], direct coupling [5], evanescent coupling [6,7], taper coupling [8], butt-jointed built-in structure [9] and so on.

Historically speaking, integrated lasers were demonstrated first on the base of phase coupling scheme, which was AlGaAs/GaAs ITG lasers [1–4]. AlGaAs/GaAs ITG lasers were operated under the optical pumping [1,3] and then the current injection [2] both in 1974. Figure 1 shows the basic structure of ITG lasers, which consists of coupled active and output waveguides (waveguide 1 and 2, respectively). The lasing light generated in the active waveguide transfers into the output waveguide by phase coupling, as is the case with directional couplers. An important aspect of ITG lasers was a single longitudinal mode operation and a concept of the dynamic single mode laser was inspired by these studies [10]. In those days, the standard Fabri–Perot lasers operated in multi-longitudinal modes because the laser technology was at an early preliminary stage. Among them, a fine lasing spectrum of ITG lasers came as a surprise to researchers. The reason why it was in ITG is that the lasing mode is selected by the double cavity effect as clarified theoretically [4,11]. The detail is described later in this section.

In early 1970s, the subject in integrated optics was to discover how to fabricate laser mirrors without cleavage of crystal. In regular semiconductor lasers, cleaved facets are used as

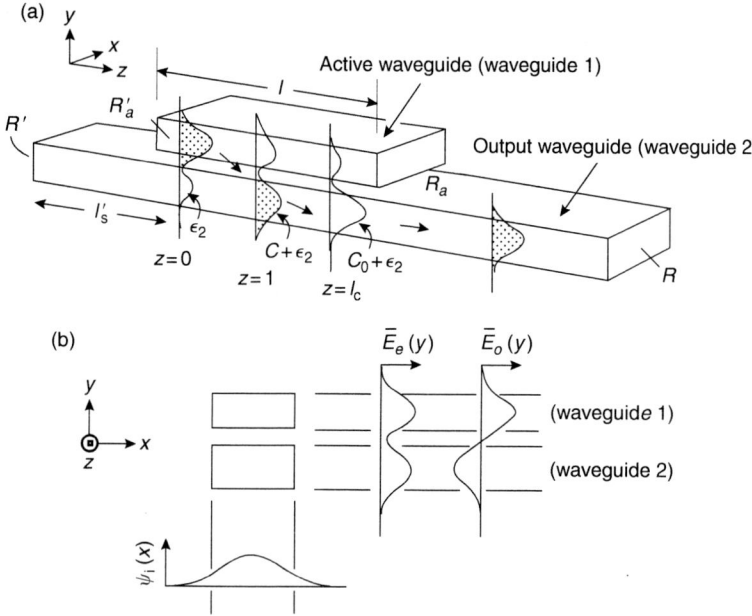

Figure 1 Schematic diagram of integrated twin-guide lasers, which consist of coupled active waveguide (waveguide 1) and output waveguide (waveguide 2). The maximum coupling coefficient C_0 indicates fractional power transfer at the coupling length l_c and ε_1 and ε_2, uncoupled factors in waveguide 1 and 2, respectively (After Figure 11.11 in Y. Suematsu and A. R. Adams (Eds.), *Handbook of Semiconductor Laser and Photonic Integrated Circuits*, Chapman & Hall, London (1994). With permission.)

reflectors, but it is difficult to integrate monolithically other lightwave elements with a semiconductor laser, because the laser crystal is interrupted at the cleaved facet. In distributed feedback (DFB) lasers, the periodic corrugation on the active layer supplied a new reflection scheme [12,13], but in which low-loss waveguides were not integrated.

In ITG structure, the mirrors were fabricated first by RF back-sputtering [1–3] and then by wet chemical etching [14]. The schematic diagram of this type of ITG lasers is shown in Figure 2(a). The etching was stopped at the middle place of two waveguides to fabricate end mirrors at the active waveguides. Nowadays the same thing can be easily realized by a well-developed dry etching. It was, however, difficult to make smooth mirror facets in perpendicular by a chemical etching. Wet chemical etching produces in many cases crystallographical facets, giving rise to oblique facets, though an isotropic etching by special combinations of chemicals [14] could produce a perpendicular face at the bottom, but without enough reproducibility. This type of AlGaAs ITG lasers showed a single axial-mode operation due to double cavity effect as shown in Figure 2(b).

Meanwhile ITG lasers with distributed Bragg reflector (DBR), which is shown in Figure 3(a), opened a new stage in ITG-research [15,16]. In the structure the corrugation gratings (DBRs) are prepared on both wings of output waveguide, which are stuck outside the coupled active waveguide region. The light transferred into the output waveguide is reflected at DBRs, and is again coupled back to the active waveguide, making the round-trip oscillation loop. As the reflection at ends of active waveguide is not always preferable, oblique crystallographical etched-facets are employed. The wavelength selectivity in reflection of DBR can supply the dynamic single mode operation of DBR–ITG lasers [17–24], which is an important

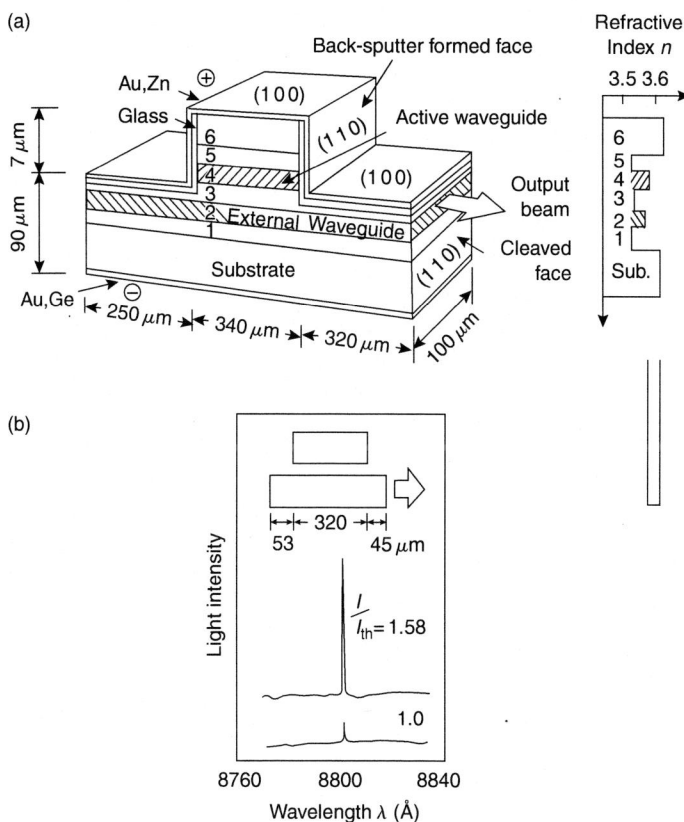

Figure 2 Schematic diagram of (a) AlGaAs/GaAs ITG lasers (From Y. Suematsu, K. Kishino, and K. Hayashi, *Trans. IECE Jpn.*, 58-C, 654–660 (1975). With permission.) and (b) and the lasing spectra

property for the application to optical communications. One of the examples of single mode spectra for this type of AlGaAs DBR–ITG lasers is shown in Figure 3(b) [15]. In addition the output light is extracted outside supporting in the low-loss waveguide, so that this structure functions as an integration source with other lightwave elements.

To construct efficiently coupled waveguides, the key issue is to realize the degenerated coupling system, so that the phase velocities of two modes supported in each waveguides are equal when two waveguides are well separated. If the waveguide parameters are fabricated precisely to the degenerated system, the efficient light coupling from one waveguide to another is realized. On the contrary, when the phase velocity is mismatched, the light coupling drops at a fast clip, so that fabrication tolerance of the coupling system is very severe. ITG laser crystals were grown using liquid phase epitaxy (LPE), which is able to control the thickness within margins of error around 10%. Obtaining efficient coupling, therefore, was frequently difficult. In order to avoid the difficulty, the coupled waveguide in ITG was designed to be strong coupling when two waveguides were placed very close to each other. In a strong coupling system, the uncoupled factor ε_1 increases, which lowers the maximum coupling coefficient C_0 (see Figure 1). The perfect power transfer from one waveguide to another does not occur even for the degenerated case. However, abrupt reduction in coupling caused by the fabrication error is appropriately relaxed on the sacrifice of reduced C_0 [25]. As a result an efficient coupling in

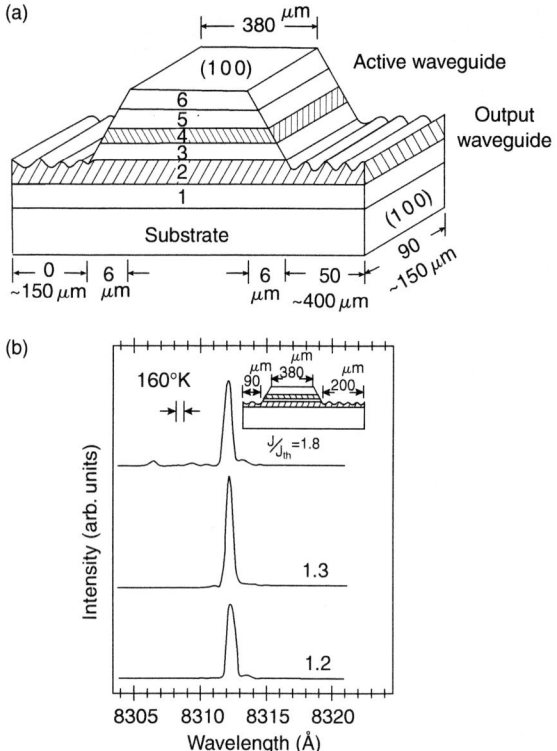

Figure 3   Schematic diagram of AlGaAs/GaAs ITG lasers with distributed Bragg reflectors (DBRs) (a) and the lasing spectra of the DBR–ITG laser (b) (After K. Kawanishi, Y. Suematsu, and K. Kishino, *IEEE J. Quantum Electron.*, QE-13, 64–65 (1997). With permission.)

ITG was realized using the LPE technology [15–24]. Such a design principle of strongly coupled ITG-waveguides is described later.

Meanwhile the molecular beam epitaxy (MBE) and metal-organic chemical vapor deposition (MOCVD) developed as of late, and the layer thickness can be controlled in tolerance <1%. So the ITG structure can enjoy its inherent ability of a 100% coupling, for stepping up to the higher stage. As described earlier, by making good use of recent technologies, integrated twin-guide lasers may give God's gift in a future integrated optics. In this section, therefore the basic concepts of ITG structure are described.

WAVEGUIDE DESIGN FOR THE ITG COUPLING SYSTEMS

The strongly coupled waveguide system like ITG can be analyzed on a base set of two guided modes (an even mode $E_e(x, y)$ and odd mode $E_o(x, y)$) (see Figure 1[b]) which are given by an exact solution of the electromagnetic wave equation of the system. When high-order modes are absent, the electromagnetic field can be written as

$$E(x,y,z) = A_e(x,y)\,\exp(-j\beta_e z) + A_o E_o(x,y)\,\exp(-j\beta_o z), \tag{1}$$

where A_e and A_o are amplitudes, and the guided modes $E_i(x,y)$ are given by

$$E_i(x,y) = \phi_i(x)\bar{E}_i(y) \quad i = e,\,o \tag{2}$$

and normalized by

$$\iint_{-\infty}^{\infty} E_i(x,y)H_i^*(x,y)\,dx\,dy = 1. \tag{3}$$

Examples of field distributions of even and odd modes in the degenerated coupling-system are shown in Figure 4. At a glance it can be understood that the weak coupling theory built on the mode-coupled equation breaks down in the strong-coupling case. According to the mode-coupled equation, two complex-modes (even and odd modes) are expressed by the combination of guided modes supporting in each waveguides. When two waveguides are relatively far apart, the combinations can approximate field profiles of the coupled waveguide system very well, like Figure 4(a), but for close waveguides, as shown in Figure 4(b) and 4(c), it brings about a serious discrepancy.

The transfer of power from one waveguide to another in a strong-coupling system can be calculated just as power-distribution change with propagation in the corresponding waveguide part of the system. Note that the even and odd modes have symmetrical and anti-symmetrical field profiles, respectively. When the even and odd modes interfere in the phase of Figure 4, the field distributes preferentially around waveguide 1 (active waveguide), because the fields cancel each other around waveguide 2. These two modes travel with slightly different phase velocities, and the phase difference between the modes increases gradually along the waveguide

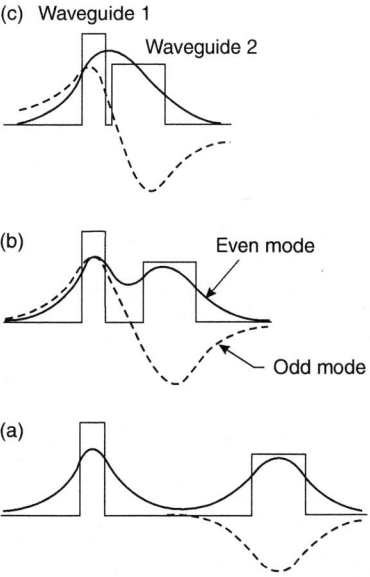

Figure 4 Field profiles of even and odd modes supported in degenerated coupling system: (a) $T_d = 1.44\pi$; (b) 0.36π, and (c) 0.036π (after Figure 11.13 in Y. Suematsu and A. R. Adams (Eds.), *Handbook of Semiconductor Laser and Photonic Integrated Circuits*, Chapman & Hall, London (1994). With permission.)

axis, so that the two modes interfere constructively and destructively, alternately, along one waveguide. If the field profiles of the even and odd modes coincided perfectly apart from the phase, the two modes would cancel out in active waveguide-part at a certain phase (when $(\beta_e - \beta_o)l = (2m + 1)\pi (m = 0, 1, 2, ...)$), so a complete exchange of light power between two waveguide-parts, which occurs periodically, would be expected. Here the coupling length l_c is defined as $\pi/(\beta_e - \beta_o)$. In principle, the profiles of two modes are not the same in the optimum condition of phase-matching (degenerated system), though the difference becomes infinitely small by increasing the waveguide separation. Thus complete light transfer does not occur and a part of the light power remains uncoupled. The proportion of the power which is uncoupled is called the uncoupled factor [25], and denoted ε_1 and ε_2 for the two waveguides. The uncoupled factors increase with decreasing the waveguide separation as understood from Figure 4.

The electromagnetic analysis deduces the uncoupled factor as follows [25]:

$$\varepsilon_2 = 1 - \frac{1}{2}\left[\xi_e + \xi_o + \left\{(\xi_e - \xi_o)^2 + (\xi_{eo} + \xi_{oe})^2\right\}^{1/2}\right],$$
(4)

where ξ_i and ξ_{ij} are given by

$$\xi_i = \int_{-\infty}^{\infty} \psi_i(x)\psi_{ih}^*(x)\,dx \int_0^{\infty} \overline{E}_i(y)\overline{H}_i^*(y)\,dy \quad (i = e, o)$$

$$\xi_{ij} = \int_{-\infty}^{\infty} \psi_j(x)\psi_{ih}^*(x)\,dx \int_0^{\infty} \overline{E}_j(y)\overline{H}_i^*(y)\,dy \qquad (i \neq j, \; j = e,$$
(5)

where $\psi_{ih}(x)$ denotes the magnetic component of $\psi_i(x)$ and the origin of y-axis is the middle place of two waveguides.

The coupling coefficient C is defined as the ratio of the light transferred from one waveguide to another over a waveguide of length l to the total propagating light power, which is [25]

$$C = C_0 \sin^2\left(\pi l / (2l_c)\right),$$
(6)

where

$$C_0 = 1 - \varepsilon_1 - \varepsilon_2 = \frac{\left(\xi_{eo} + \xi_{oe}\right)^2}{\left\{(\xi_e - \xi_o)^2 + (\xi_{eo} + \xi_{oe})^2\right\}^{1/2}}$$
(7)

$$l_c = \pi/(\beta_e - \beta_o).$$
(8)

The coupling coefficient C changes with the length of a coupled waveguide system. When the length is $(2s + 1)\,l_c$, by means of the coupling length l_c, the power transfer form one waveguide to the other is maximum, that is, $C = C_0$. The maximum coupling coefficient C_0 gives a measure for the evaluation of coupling ability, and in principle, it is always less than unity as deduced from Eq. (8). As ordinarily $\psi_e(x) \approx \psi_o(x)$, it is shown from Eq. (5) that ξ_i and ξ_{ij} approximate those of a slab waveguide with a structure uniform along the x-direction. From Eq. (7) therefore C_0 is approximately given by the value calculated for a slab waveguide [25]. Figure 5 shows the relationship between C_0 and the dimensions of the waveguides.

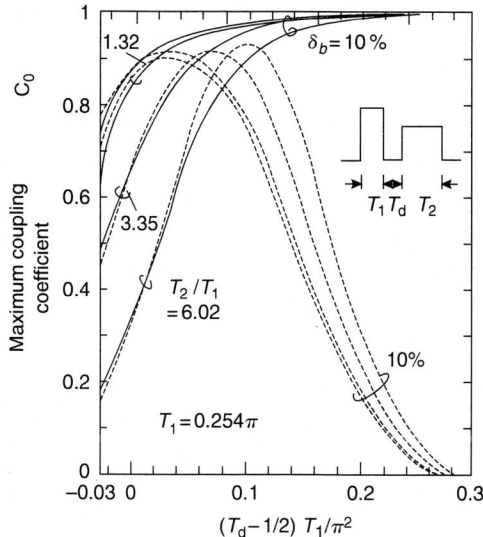

Figure 5 Maximum coupling coefficient C_0 as a function of waveguide parameter $(T_d - 1/2)T_1/\pi^2$. δ_b denotes the thickness deviation of waveguide 2 from the phase-matching condition (After Y. Suematsu and K. Kishino, *Radio Sci.*, 12, 587–592 (1977). With permission.)

In the ITG-waveguides, the bandgap energy of core materials of the active waveguide, where laser gain produces, is smaller than that of the low-loss output waveguide, so that the cores of two waveguides possess a different refractive index. To build up, therefore the phase matching between two waveguides, the core thicknesses are designed different. Modal property of waveguide could be treated universally using the normalized thickness of waveguide. T_i in Figure 5 is the normalized thickness of waveguide i, and is given by $\sqrt{(n_i^2 - n_3^2)}\pi d_i / \lambda$ $(i = 1, 2)$, where d_i and n_i are the core thickness and refractive index, and n_3 are the clad refractive indices, respectively, and λ is the wavelength. The waveguide separation d_3 is normalized similarly as $T_d = \sqrt{(n_i^2 - n_3^2)}\pi d_3 / \lambda$. Numerically shown, it is, that the coupled waveguide systems with the same values of T_1, T_2, and T_d have very similar values of C_0.

Figure 5 shows C_0 calculated as a function of $(T_d - 1/2)T_1/\pi^2$. The parameter is T_2/T_1, which is changed from 1, 1.32, 3.35, and to 6.02, at $T_1 = 0.254\pi$. This figure gives a universal diagram with respect to T_1, as long as T_2/T_1 is not so large and T_d is not small. The solid lines correspond to the degenerated case where the phase velocities of waveguides 1 and 2 are equal to each other (phase-matching condition) and the broken lines are for a thickness of waveguide 2 deviating +10% from the phase-matching thickness b_0. That is, $\delta_b = 10\%$ where $\delta_b = (b - b_0)/b_0$. If the coupled waveguide system is perfectly fabricated with $\delta_b = 0\%$, a complete power transfer is expected for a large waveguide separation T_d, that is, for weak coupling. However, when the thickness deviates from the degenerated system, the power transfer vanishes abruptly as shown in Figure 5. An extremely high fabrication tolerance of dimension is necessary for realizing a high coupling.

As decreasing the waveguide separation, the coupling length l_c becomes small so that light coupling is produced with a shorter waveguide length. But in such strong coupling, C_0 decreases below unity even for the phase-matching condition (degenerated system), and therefore no complete power transfer from one waveguide to the other produces for a waveguide length of $(2m + 1)l_c$. This phenomenon could not be explained by a weak-coupling theory based on the mode-coupled equation. An important thing is that the thickness deviation does not bring about a serious decrease in C_0, when the two waveguides are so close. It is, therefore, introduced from Figure 5 that a design guideline for coupled waveguides with a higher coupling is $(T_d - 1/2)T_1/\pi^2 \approx 0.05$, where the coupling efficiencies C_0 as high as 90% can be obtained under the phase-mismatching $(\delta_b = 10\%)$. Rewriting the equation

$$\frac{d_3}{\lambda} \approx \frac{1}{2\pi\sqrt{n_1^2 - n_3^2}} + \frac{0.05}{(n_1^2 - n_3^2)d_1 / \lambda},\tag{9}$$

which gives the guideline for designing the waveguide dimensions of ITG waveguides.

GaAs/AlGaAs ITG lasers consisted of a GaAs active and an $Al_{0.27}Ga_{0.73}As$ separation layers as reported in [10,14–16]. The corresponding refractive indices of these layers and the lasing wavelength are $n_1 = 3.6$, $n_3 = 3.49$, and $\lambda = 0.88$ μm, respectively and a typical thickness of active layer was 0.5 μm. By substituting these into Eq. (9), d_3 is calculated to be 0.26 μm. The ITG lasers were fabricated with 0.3 μm-thick separation layers.

THEORETICAL ANALYSIS OF INTEGRATED TWIN-GUIDE LASERS

Matrix Method for ITG Analysis

The oscillation condition of the ITG lasers shown in Figure 6 can be analyzed by use of the matrix method [4,11]. The length of coupled twin-guide region is l, and those of the right and left arms of output waveguide, l_s and l_s', respectively. The end mirrors of active waveguide have the reflectivity of R_a and R_a', respectively, and those of output waveguide, R and R'. The field E, which propagates in the twin-guide region, is expressed by the combination of even and odd modes as follows:

$$E = \left[A_e E_e(x,y)\exp(-j\beta_e z) + A_o E_o(x,y)\exp(-j\beta_o z)\right] \cdot \exp\left[G(z) - a(z)\right],\tag{10}$$

where A_e and A_o are the amplitudes of the even and odd modes, and $G(z)$ and $a(z)$ are the gain and loss functions, respectively. The guided mode supported in the arm region of output waveguide is termed s-mode and the field is expressed with the absorption coefficient α_s by

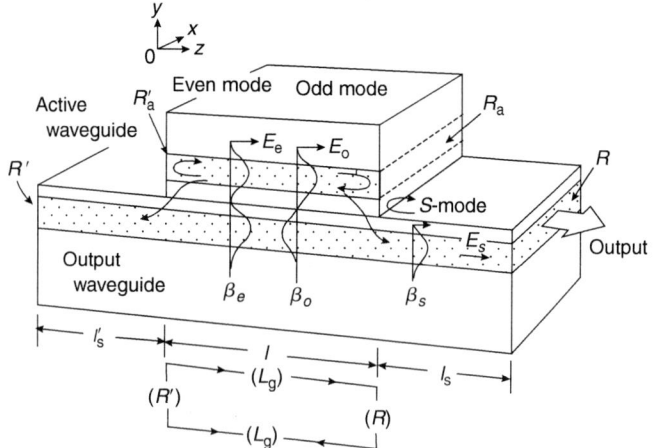

Figure 6   Analytical model of basic ITG structures (After Y. Suematsu, K. Kishino, and T. Kambayashi, *IEEE J. Quantum Electron.*, QE-13, 619–622 (1977). With permission.)

$$E = A_s E_s(x, y)\exp(-j\beta_s z) \cdot \exp(-\alpha_s z). \tag{11}$$

The medium gain g exists only in the active layer, and the medium loss coefficient for each ITG waveguide layer is α_q. The gain and loss functions $G(z)$ and $a(z)$, that is, the gain and loss received by the coupled wave E, can be calculated by the integration weighing the lateral distributions of medium gain and loss coefficients with the fractional power profile of coupled wave against the total propagation power. As discussed in the last section, the coupled wave propagates transferring the light power back and forth between the two waveguides for every coupling length l_c, so that the field profile of coupled wave distributes alternately between two waveguide regions. $G(z)$ and $a(z)$, therefore become complicated functions of z, reflecting the field profile change, and depend on the propagation direction. For the wave propagating toward $(\pm)$-direction of z-axis, they are expressed as follows:

$$G^{\pm}(z) = g \cdot \left[\xi_4\right]^{\pm}_{\text{coup}} \cdot z,$$

$$a^{\pm}(z) = \left(\sum_{q=1}^{5} \alpha_q \cdot \left[\xi_q\right]^{\pm}_{\text{coup}}\right) \cdot z, \tag{12}$$

$$\left[\xi_q\right]^{\pm}_{\text{coup}} = \frac{\left[\xi_{eq} + \left|A_o^{(\pm)} / A_e^{(\pm)}\right|^2 \xi_{oq} + 4\left|A_o^{(\pm)} / A_e^{(\pm)}\right|\xi_{eoq} \cdot \cos(\Delta\beta z / 2 - \varphi) \cdot \sin(\Delta\beta z / 2) / \Delta\beta z\right]}{\left(1 + \left|A_o^{(\pm)} / A_e^{(\pm)}\right|^2\right)}, \tag{13}$$

where $[\xi_q]^{\pm}_{\text{coup}}$ indicates the fraction of power in the qth layer for the wave traveling toward $(\pm)$ direction of z-axis, respectively. The amplitudes $A_e^{(+)}$ and $A_o^{(+)}$ are those of even and odd modes incident to the right end place of twin-guide at the right side, respectively $A_e^{(-)}$ and $A_o^{(-)}$ and , those to the left end. The value $A_o^{(\pm)} / A_e^{(\pm)}$ of is determined from the oscillation condition of ITG lasers as given in Eq. (22), where φ is the phase difference between $A_o^{(\pm)}$ and $A_e^{(\pm)}$. The fourth layer is active layer and $[\xi_4]_{\text{coup}}$ indicates the fraction of power in the active layer.

At the coupling length l_c, $\Delta\beta l_c = \pi$ is satisfied. When the length of coupling region l is much longer than the coupling length l_c and the light power is transferring many times between two waveguides, the second term of the numerator in Eq. (13) becomes negligibly small and $[\xi_q]_{\text{coup}}$ is expressed by a simple form. In the degenerated case, as $\xi_{e4} = \xi_{o4} = \xi$ is satisfied, $[\xi_q]_{\text{coup}} = \xi_q$. Thus the gain and loss functions are given independently on the propagation direction by,

$$G(z) - a(z) = g\xi_4 z - \sum_{q=1}^{5} \alpha_q \xi_q z. \tag{14}$$

At the end place of the coupling region (i.e., twin-guide), the active waveguide is terminated at the reflection mirror, while the output waveguide continues without light reflection (see Figure 6). When the coupled wave is incident on the end place, the upper-half fields of even and odd modes are reflected at the active waveguide end, and the field components are divided into three parts; that is, the reflection of each mode, the coupling from one mode to another, and the coupling to s-mode. The s-mode travels along the arm region and is reflected at the output waveguide end.

The reflection coefficient $r_i (i = e,o)$ of mode i and the coupling coefficient $r_{ij} (i \neq j, i, j = e,o)$ from mode j to mode i at the end place of twin-guide coupling region are given by

$$r_i = \sqrt{R_a} \xi_i \exp(j\theta_a), \quad r_{ij} = \sqrt{R_a} \xi_{ij} \exp(j\theta_a), \quad r_s = \sqrt{R} \exp(j\theta), \qquad (15)$$

where $\sqrt{R_a} \exp(j\theta_a)$ and $\sqrt{R_a} \exp(j\theta_a)$ show the reflective coefficients of electric field at the active and output waveguide ends, respectively, and ξ_i, ξ_{ij} are given in Eq. (5). The coupling coefficient from mode i to s-mode t_i is expressed by

$$t_i = \int_{-\infty}^{\infty} \psi_i(x)\psi_{sh}^*(x)\,dx \int_{-\infty}^{0} \bar{E}_i(y)\bar{H}_s^*(y)\,dy \ (i = e,o). \qquad (16)$$

The incident amplitudes of even and odd modes to the end place of twin-guide at the right side are $A_e^{(+)}$ and $A_o^{(+)}$, respectively. From the discussion above, it is clear that the amplitudes of reflected waves at the active waveguide end, $A_e^{(+)r}$ and $A_o^{(+)r}$ are expressed by matrix

$$\begin{pmatrix} A_e^{(+)r} \\ A_o^{(+)r} \end{pmatrix} = \begin{pmatrix} r_e & r_{eo} \\ r_{oe} & r_o \end{pmatrix} \begin{pmatrix} A_e^{(+)} \\ A_o^{(+)} \end{pmatrix} = (R_g) \begin{pmatrix} A_e^{(+)} \\ A_o^{(+)} \end{pmatrix}, \qquad (17)$$

where (R_g) is called reflection matrix.

Similarly, various kinds of matrix are defined in Table 1; that is, the propagation matrix of even and odd modes along the twin-guide region $(L_p^{\pm})$ that of s-mode along the arm region (L_s^{+}), the transmission matrix from mode i $(i = e,o)$ to s (T_{out}) and that from mode s to i (T_{in}), and the reflection matrix of s-mode (R_s). Using these matrices, the electromagnetic response of various ITG-type devices can be conveniently analyzed as discussed in References 4 and 11.

An ITG structure of Figure 1, if $R = R' = 0$, functions as a light amplifier. The light coming from the left arm (amplitude A_s^{in}) experiences a round trip traveling inside ITG region and passes through that into the right arm region, making the amplified light output with the amplitude A_s^{out}. The amplitude can be easily obtained in relation to A_s^{in} by the matrix method as follows.

$$\begin{pmatrix} A_s^{out} \\ 0 \end{pmatrix} = \frac{(T_{out})(L_g^+)(T_{in})}{(E) - (R_g)(L_g^-)(R_g)(L_g^+)} \begin{pmatrix} A_s^{in} \\ 0 \end{pmatrix}, \qquad (18)$$

where $R_a = R_a'$ and (E) is unit matrix. From Eq. (18) the amplification gain is given by A_s^{out} / A_s^{i}.

In this section, the oscillation condition of ITG lasers is analyzed. At the beginning, the equivalent reflection matrix is defined at the end place of twin guide, taking the light reflection from both active and output waveguide ends into account. Thus the equivalent reflection matrix (R_T) is given by

$$(R_T) = \begin{pmatrix} R_e & R_{eo} \\ R_{oe} & R_o \end{pmatrix} = (R_g) + (T_{in})(L_s)(R_s)(T_{out}). \qquad (19)$$

By use of (R_T), the analysis is performed with a simplified equation. The amplitudes of incident waves into the right end place of the twin-guide region, $A_e^{(+)}$ and $A_o^{(+)}$, are amplified after one round trip (see Figure 6), as

Table 1

Expression of reflection, propagation and transmission matrices for ITG structure

Reflection Matrix (R)	Propagation Matrix (L)	Transmission Matrix (T)
$\begin{pmatrix} A_e^{(2)} \\ A_o^{(2)} \end{pmatrix} = (R_g) \begin{pmatrix} A_e^{(1)} \\ A_o^{(1)} \end{pmatrix}$	$\begin{pmatrix} A_e^{(2)} \\ A_o^{(2)} \end{pmatrix} = (L_g^{\pm}) \begin{pmatrix} A_e^{(1)} \\ A_o^{(1)} \end{pmatrix}$	$\begin{pmatrix} A_s \\ 0 \end{pmatrix} = (T_{\text{out}}) \begin{pmatrix} A_e \\ A_o \end{pmatrix}$
$(R_g) = \begin{pmatrix} r_e & r_{eo} \\ r_{oe} & r_o \end{pmatrix}$	$(L_g^{\pm})^a = \begin{pmatrix} L_{ge}^{\pm} & 0 \\ 0 & L_{go}^{\pm} \end{pmatrix}$	$(T_{\text{out}}) = \begin{pmatrix} t_e & t_o \\ 0 & 0 \end{pmatrix}$
$\begin{pmatrix} A_s^{(2)} \\ 0 \end{pmatrix} = (R_s) \begin{pmatrix} A_s^{(1)} \\ 0 \end{pmatrix}$	$\begin{pmatrix} A_s^{(2)} \\ 0 \end{pmatrix} = (L_s) \begin{pmatrix} A_s^{(1)} \\ 0 \end{pmatrix}$	$\begin{pmatrix} A_e \\ A_o \end{pmatrix} = (T_{\text{in}}) \begin{pmatrix} A_s \\ 0 \end{pmatrix}$
$(R_s) = \begin{pmatrix} r_s & 0 \\ 0 & 0 \end{pmatrix}$	$(L_s)^b = \begin{pmatrix} L_s & 0 \\ 0 & 0 \end{pmatrix}$	$(T_{\text{in}}) = \begin{pmatrix} t_e & 0 \\ t_o & 0 \end{pmatrix}$

a $\quad L_{gi}^{\pm} = \exp\{(G^{\pm}(z - z_o) - a^{\pm}(z - z_o))/2 - j\beta_i(z - z_o)\} \qquad (i = e, o)$

b $\quad L_s = \exp\{-a_s(z - z_0)/2 - j\beta_s(z - z_o)\}$

Source: After Y. Suematsu, K. Kishino, and K. Hayashi, *Trans. IECE Jpn.,* 58-C, 654–660 (1975). With permission.

$$\begin{pmatrix} A_e'^{(+)} \\ A_o'^{(+)} \end{pmatrix} = (L_g^{+})(R_T')(L_g^{-})(R) \begin{pmatrix} A_e^{(+)} \\ A_o^{(+)} \end{pmatrix} \tag{20}$$

The oscillation condition for the laser is obtained by equating the matrix of one round trip to unit matrix (E) as follows:

$$\det\left\{(L_g^{+})(R_T')(L_g^{-})(R_T) - (E)\right\} = 0, \tag{21}$$

where (R_T') is the equivalent reflection matrix for the left side.

The ratio of $A_o^{(+)}/A_e^{(+)}$, when the oscillation condition is satisfied, is simply given from Eqs. (20) and (21)

$$A_o^{(+)}/A_e^{(+)} = (1 - R_e'R_eL_{ge}^{+}L_{ge}^{-} - R_{eo}'R_{oe}L_{ge}^{+}L_{go}^{-})/(R_{eo}'R_oL_{ge}^{+}L_{go}^{-} + R_e'R_{eo}L_{ge}^{+}L_{ge}^{-}). \tag{22}$$

Threshold Gain and Differential Quantum Efficiency of ITG Lasers

The characteristics of ITG lasers could not be intuitively recognizable, because they have complicated cavities as against a regular Fabry–Perot laser. Equivalent reflection and transmission coefficient, R_{eq} and T_{eq} are defined, which enables us to treat the ITG lasers as an ordinary single-guide laser. R_{eq} is the overall power reflectivity at the end plane of the twin-guide, and T_{eq} is the-fractional radiated power from the mirror of the arm. When the incident wave is $A_e^{(+)}E_e + A_o^{(+)}E_o$, by use of Eq. (19), the reflection wave is expressed as $(R_e A_e^{(+)} + R_{eo} A_o^{(+)})E_e + (R_{oe} A_e^{(+)} + R_o A_o^{(+)})E_o$. Thus R_{eq} is given as

$$R_{eq} = \left[\left| R_e + R_{eo}(A_o^{(+)} / A_e^{(+)}) \right|^2 + \left| R_{oe} + R_o \left(A_o^{(+)} / A_e^{(+)} \right) \right|^2 \right] \Big/ \left(1 + \left| A_o^{(+)} / A_e^{(+)} \right|^2 \right). \qquad (23)$$

Using (L_s) and (T_{out}) in Table 1, T_{eq} is,

$$T_{eq} = (1-R)|L_s|^2 \left[|t_e|^2 + \left| t_o A_o^{(+)} / A_e^{(+)} \right|^2 + 2 Real\left(t_e^* t_o A_o^{(+)*} / A_e^{(+)} \right) \right] \Big/ \left(1 + \left| A_o^{(+)} / A_e^{(+)} \right|^2 \right), \qquad (24)$$

where the asterisks denote the complex-conjugate quantities. The $A_o^{(+)} / A_e^{(+)}$ in the above equations is obtained from the oscillation condition and given in Eq. (22).

Using the equivalent parameters, the threshold gain can be written simply by

$$G(l) - a(l) = \frac{1}{2} \ln \left[R_{eq} \cdot R'_{eq} \right] \qquad (25a)$$

and

$$G(l) = (G^+ (l) + G^- (l))/2, \qquad a(l) = (a^+ (l) + a^- (l))/2, \qquad (25b)$$

where $G^{\pm} (l)$, $a^{\pm} (l)$ are the gain and loss function for the traveling waves toward $(\pm)$-direction of z-axis, respectively, and R'_{eq} is the equivalent reflection coefficient at the other side.

The external differential quantum efficiency for the output light from the R_{eq} side, η_d is expressed as follows:

$$\eta_d = \frac{T_{eq} \cdot \eta_i}{\left(1 + \sqrt{R_{eq} / R'_{eq}} \right)\left(e^{\alpha l} - \sqrt{R_{eq} R'_{eq}} \right)}, \qquad (26)$$

where η_i is the internal quantum efficiency.

In the weakly-coupled degenerated waveguides, the power profiles of even and odd modes are fitted to each other as shown in Figure 4(a) so that $\xi_e = \xi_o = \xi_{eo} = 1/2$, $\xi_{iq} = \xi_{eoq}$ $(i = e,o)$ and $t_e = -t_o = 1/\sqrt{2}$. In this case, the above equations for ITG lasers are simplified and the coupling coefficient C is given by $C = \sin^2 [(\beta_e - \beta_o)l / 2]$ $(C_0 = 1)$. When $l > l_c$, the gain and loss functions are approximately given by $G(l) - a(l) = (g - \alpha)\xi_4 l$, where the absorption other than active layer is neglected.

Measurement of Coupling Coefficient and Coupling Length of ITG Coupling System

Investigating the dependence of threshold current density of ITG lasers on the length of twin-guide region, the coupling coefficient and coupling length of twin-guide structure were estimated [26]. Very high coupling efficiencies between active and output waveguides for

AlGaAs/GaAs ITG structure grown by liquid phase epitaxy were shown experimentally. The highest C_0 value was evaluated to be as high as 90%, and the coupling length was about 250 μm for a typical ITG waveguide structure.

As discussed in detail in Reference 26, the threshold current density of ITG lasers given in Figure 7 depend sensitively on the length of twin-guide region l, because the coupling coefficient C of Eq. (6) is a function. The threshold gain of the ITG laser can be derived from Eq. (25a)

$$(g-\alpha)\xi_4 l = \ln\frac{1}{R|1-2C|} \Bigg/ \left(1 - \frac{\sin(\pi l/l_c)}{\pi l/l_c}\right). \tag{27}$$

The gain is a strong function of the twin-guide length, and it becomes infinity periodically, at which the light is incident into the zero-reflectance end mirror of active waveguide. On the other hand, when the ITG structure is cleaved at the twin-guide ends and the active and output

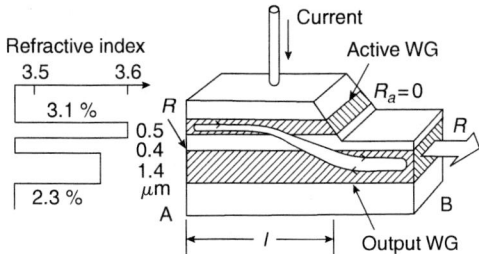

Figure 7 Schematic diagram of AlGaAs/GaAs ITG-type lasers (After K. Utaka, Y. Suematsu, K. Kishino, and H. Kawanishi, *Trans. IECE Jpn.* E62, 319–323 (1979). With permission.)

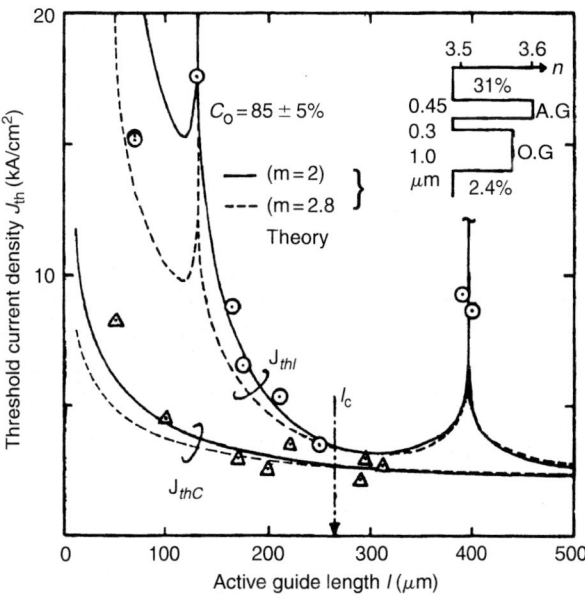

Figure 8 Threshold current densities J_{thl} of ITG-type lasers (circle points) and J_{th} of cleaved ITG lasers (triangles) versus the active waveguide length l, where $l = 8890$ Å, $C_0 = 85 \pm 5\%$, and $l_c = 265$ μm (After K. Utaka, Y. Suematsu, K. Kishino, and H. Kawanishi, *Trans. IECE Jpn.* E62, 319–323 (1979). With permission.)

waveguides are terminated with common facets, the threshold gain does not depend on the coupling property. By comparing the threshold current densities of an ITG-type laser and a cleaved ITG laser, J_{thl} and J_{theC}, respectively, the maximum coupling coefficient C_0 can be estimated. Figure 8 shows one of the experimental results for estimating the coupling property of AlGaAs ITG structures [27]. The vertical axis is the threshold current densities of ITG-type and cleaved ITG lasers and the horizontal one is the active waveguide length l. The theoretical calculation was fitted to the experimental values for estimating the maximum coupling coefficient and the coupling length. In this case, C_0 and l_c were estimated to be $85 \pm 5\%$ and $265 \ \mu m$, respectively.

Axial-Mode Selectivity of Various ITG Lasers

Various types of ITG lasers are devised by combination of the basic ITG lasers and the DBRs, as schematically summarized in Figure 9. The double-resonator type of ITG lasers, Type 1, is the basic ITG laser structure, which has four mirrors at both ends of the active and output waveguides, and axial-mode selectivity is expected due to the double-resonance effect. For the DBR-type (Type 2), distributed Bragg reflectors (DBRs) are prepared on the arms of output waveguide. The reflectivity of DBRs is peaked at the Bragg wavelength, and an axial mode around the Bragg wavelength is selected. In the tandem-connection type (Type 3), two active waveguide regions are connected by a common output waveguide.

   To simplify the analysis, the lasing characteristics of ITG lasers are analyzed assuming the simplified coupling condition above (weak coupled degenerate system). The minimum threshold gains for Type 1–3 are approximately given by

$$G(l)_{min} \cong a(l) - \ln\left[(1-C)\sqrt{R_a R_a'} + C\left(\sqrt{R_a R'} + \sqrt{R_a' R}\right)/2 \right] \quad (R_a R_a' \ge RR')$$

$$G(l)_{min} \cong a(l) - \ln\left[(1-C)\sqrt{RR'} + C\left(\sqrt{R_a R'} + \sqrt{R_a' R}\right)/2 \right] \quad (R_a R_a' \le RR') \tag{28}$$

   R for Type 2 is the power reflectivity of the DBR at the Bragg wavelength and shown by $R = \tanh^2(\kappa L)$, where κ and L are the coupling coefficient and length of the DBR, respectively. R of Type 3 indicated in Figure 9 is the effective power reflectivity, which is the fractional power returned from the right-hand side of the lasing section. As the DBR-type ITG lasers do not need the reflection at the active waveguide ends ($R_a = R_a' = 0$), $G(l)_{min}$ is expressed by

$$G(l)_{min} \cong a(l) - \ln\left[(1-C)\sqrt{RR'} \right]. \tag{29}$$

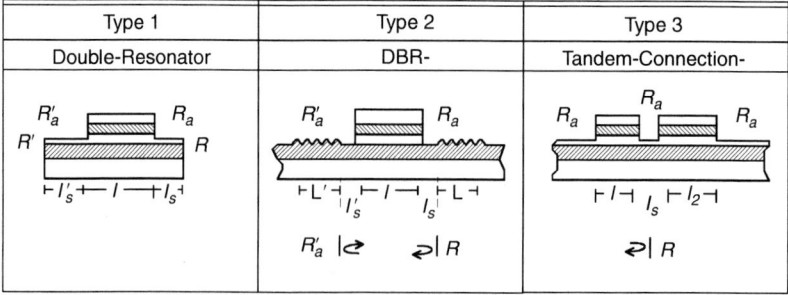

Figure 9   Schematic diagram of various ITG lasers (After Y. Suematsu, K. Kishino, and T. Kambayashi, *IEEE J. Quantum Electron.*, QE-13, 619–622, (1977) With permission.)

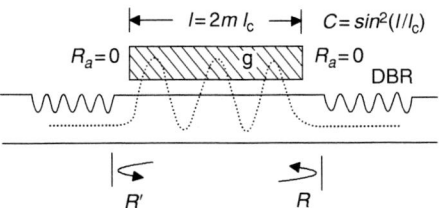

Figure 10   Lasing operation of DBR–ITG lasers

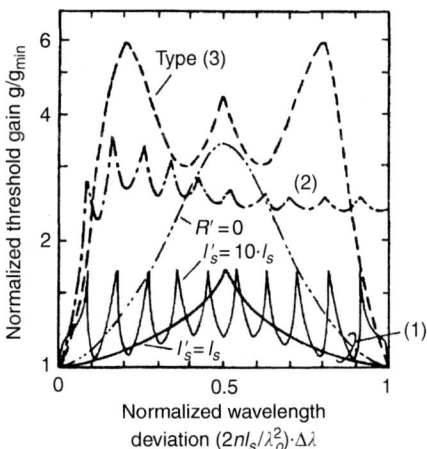

Figure 11   Dependencies of normalized threshold gain for various types of ITG lasers on normalized wavelength deviation (After K. Utaka, Y. Suematsu, K. Kishino, and H. Kawanishi, *Trans. IECE Jpn.* E62, 319–323 (1979). With permission.)

When $C = 0$, that is, $l = 2m \cdot l_c$ (m: integer) in Eq. (29), the light coupled into active waveguide returns completely to the output waveguide as schematically shown in Figure 10. The lasing oscillation occurs between two DBR mirrors and the threshold gain is given by R and R'. Dependences of threshold gains for various ITG lasers on wavelength are calculated from Eq. (25) as shown in Figure 11 [11]. Laser parameters are assumed to be $R_a = R_a' = 0.8$, $C_0 = 1$, $C = 0.5$, $a(l) = 0.15$ commonly, and in the case of Type 1, $R = R' = 0.3$ for solid curve and $R = 0.3$, $R' = 0$ for two-dot chain curve. The gain curve for Type 2 given by a dotted-dashed line is calculated assuming $\kappa L = 1$, $L = L' = 10 l_s'$, $l_s = l_s'$. For Type 3 (dashed curve), the parameters of right-hand side laser cavity are $l_2/l = 1.2$, $G_2(l) = 0.4756$, and $C_2 = 0.6545$. The vertical axis is threshold-gain normalized by the minimum value given by Eq. (28) and the horizontal the wavelength deviation $\Delta\lambda$ from the wavelength $\lambda_0 = 4nl_s / (2m_0 + 1)$ at which threshold gain becomes minimum. Where m_0, n, and l_s are integer, the refractive index of the active waveguide, and the arm length, respectively. The threshold gains for Type 1 and 3 shown in Figure 11 are periodic with respect to the normalized wavelength deviation $(2nl_s/\lambda_0^2)\Delta\lambda$. For these types of lasers, the minimum threshold gains appear alternately on the normalized wavelength deviation axis with the periodicity of unity. The threshold gain for Type 1 with $l_s' = 10 l_s$ becomes maximum periodically at the interval determined by the sum of both arm lengths; that is, $\lambda_0^2 /(2n(l_s + l_s'))$. The threshold gain of Type 2 increases sharply with $\Delta\lambda$, as shown by dotted-dashed curve in Figure 5, where the wavelength dependency of the reflectivity of the DBR is taken into account. For Type 3, the threshold gain shows a sharp increase with the wavelength

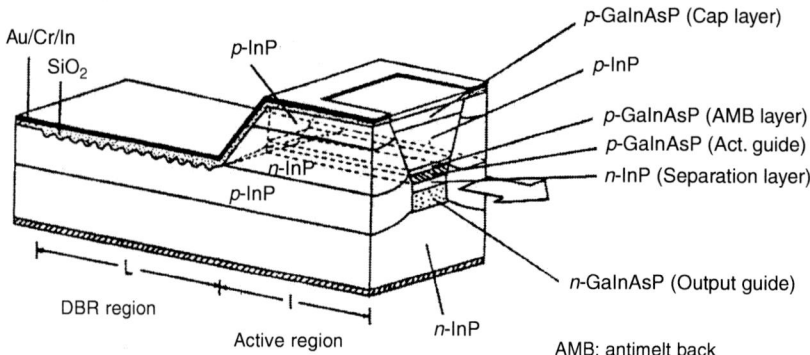

Figure 12   The 1.5–1.6 μm wavelength GaInAsP/InP dynamic single mode lasers with BH–DBR–ITG structure (After T. Tanbun-EK, S. Arai, F. Koyama, K. Kishino, S. Yoshizawa, T. Watanabe, and Y. Suematsu, *Electron. Lett.*, 17, 967–968 (1981). With permission.)

deviation and goes down again. If the l_s value is selected to be small, the next minimum point is taken away toward low gain wavelengths, so that a high axial selectivity is expected. In the experiment, AlGaAs-ITG lasers of Type 1 and 2 operated under the single longitudinal mode based on the axial mode selectivity predicted in this section, as shown in Figure 2(b) and Figure 3(b).

The DBR–ITG structures were successfully applied for the demonstration of GaInAsP/InP dynamic single mode (DSM) lasers [21] with the lasing wavelengths of 1.5 to 1.6 μm [22–25], as discussed in the chapter of Distributed Bragg Reflector. In this section, the result is briefly introduced. Figure 12 shows a schematic structure of GaInAsP/InP buried heterostructure distributed Bragg reflector integrated twin-guide (BH–DBR–ITG) lasers [24,25]. The narrow striped GaInAsP-based active and output waveguide layers, which were clad along the vertical direction by InP layers, were embedded in lateral, into *pn* reverse-junction InP current blocking layers. The waveguide width was 3 μm and the thickness of active and output waveguide layers were 0.2 and 1.4 μm, respectively. The bandgap wavelength of the active waveguide was designed to be 1.6 μm and that of the output waveguide, 1.4 μm. The first order corrugation grating with the period of 236 nm was formed on the arm region of output waveguide. Reflecting the axial mode selectivity of the DBR–ITG cavity, the single longitudinal mode operation around 1.55 μm in wavelength was obtained at the injection current no less than 1.54 times the threshold current. The lasing spectra of a rapidly modulated 1.58 μm BH–DBR–ITG laser with modulation frequency from 0.5 to 3 GHz were shown in Figure 13 [24]. Direct bias current.was fixed at 1.2 times the threshold current and the modulation current with a modulation depth of 100% was applied. In the device, a stable single-mode operation was realized at modulation frequencies from 0.25 to 3 GHz, and the dynamic spectral broadening usually seen in Fabry-Perot-type lasers was not observed at any frequency. The line broadening of the lasing mode under modulation was observed in Figure 13. This phenomenon is called by as the dynamic wavelength shift, or the wavelength chirping, which is caused by the refractive-index variation of the active layer under direct modulation [28]. The wavelength shift is maximized at the resonance-like frequency of modulation. In this experiment, it was 0.27 nm at 1.8 GHz.

These dynamic single mode operations of DBR–ITG lasers had contributed to the dramatic improvement of transmission bandwidth for optical communication systems based on conventional single-mode fibers with wavelength dispersion.

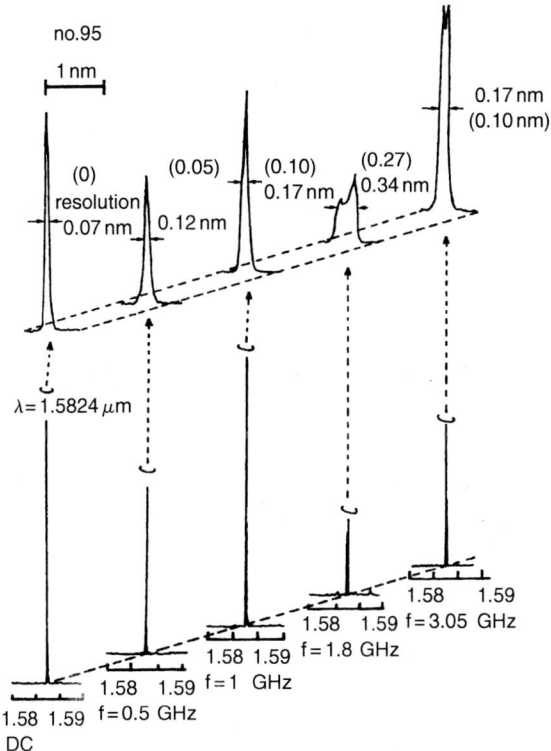

Figure 13   Lasing spectra of 1.58 μm BH–DBR–ITG laser under stationary condition and high-speed modulation with frequency from 0.5 to 3 GHz with modulation depth of 100%. Dynamic single mode operation of the laser was demonstrated (After F. Koyama, S. Arai, Y. Suematsu, and K. Kishino, *Electron. Lett.*, 17, 938–940 (1981).)

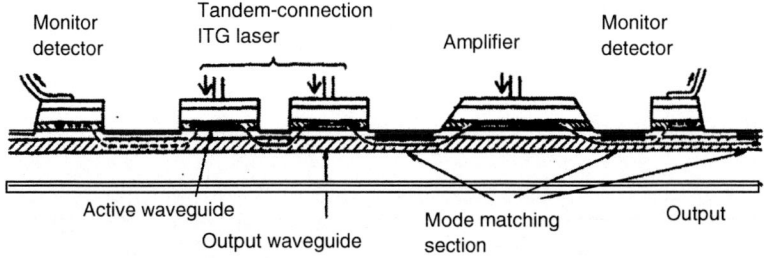

Figure 14   An example of photonic integrated circuits based on ITG structure (After Y. Suematsu, in *Proceedings of the International Conference of Integrated Optics and Optical Communications*, B1-1, Tokyo (July 1977). With permission.)

INTEGRATION OF OPTICAL ELEMENTS BASED ON ITG STRUCTURES

By epitaxial growth of the semiconductor multi-layers, the integrated twin-guide structures, which consist of the active and output waveguide layers, are prepared. As shown in Figure 13, if the active waveguide layer is etched off except for the parts assigned for lasers, amplifiers, modulators, and monitors, each active element is integrated monolithically on the low loss output waveguide, and a photonic integrated circuit can be fabricated [29–31]. The output

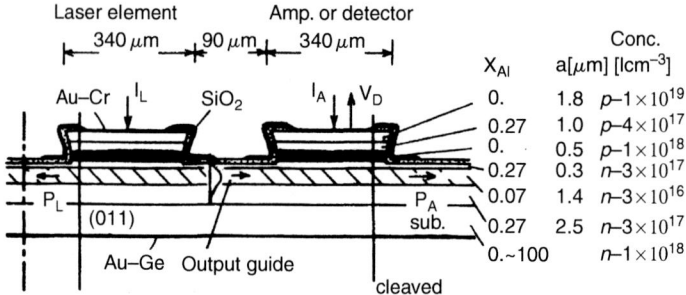

Figure 15 Schematic diagram for monolithic integration of laser and amplifier or detector based on AlGaAs–ITG structure (After K. Kishino, Y. Suematsu, K. Utaka, and H. Kawanishi, *Jpn. J. Appl. Phys.* 17, 589–590 (1978). With permission.)

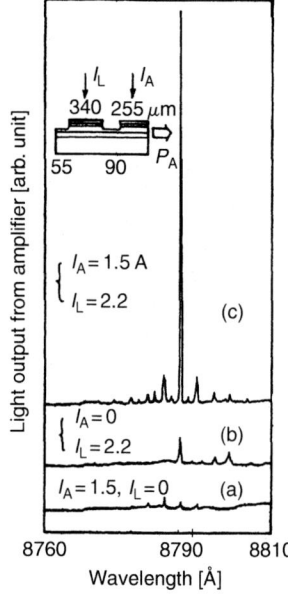

Figure 16 Light spectrum (a) for amplifier element ($I_L = 0$), (b) for laser element ($I_A = 0$), and (c) for amplified spectrum (After K. Kishino, Y. Suematsu, K. Utaka, and H. Kawanishi, *Jpn. J. Appl. Phys.* 17, 589–590 (1978). With permission.)

waveguide is etched further to integrate DBRs, filters, and other passive elements. In Figure 14, the tandem-connection type ITG laser consisting of two laser cavities is fabricated by making etched mirrors at both ends of two active waveguides.

The first success of monolithic integration of a laser and a laser amplifier or a laser and a detector was obtained based on AlGaAs twin-guide structures [31]. The ITG integrated device is schematically shown in Figure 15. Six layers of $Al_xGa_{1-x}As$ were grown on the n-type (1 0 0) GaAs substrate by the conventional liquid-phase epitaxy. The two active elements of mesa structures were fabricated by a wet chemical etching, and the end mirror surfaces of active elements were covered with SiO_2 film and then with Au film. The integrated device was evaluated under the pulse current injection. As the laser amplifier in the structure contained a resonator, it showed the characteristics of an injection locked amplifier with an amplification factor of more than ten, and a maximum output power of about 130 mW. When no electrical bias was applied to the laser amplifier, it functioned as a detector. The voltage detected with a 50 Ω load increased linearly as a function of the relative power measured at the laser side. Figure 16 shows an example of the lasing spectrum of a fabricated device; (a) spectrum of the amplifier element when the current was not applied to the laser element, (b) that of the laser element when the current was not applied to the amplifier element, and (c) that of the amplified output when the currents were applied to both the laser and amplifier elements. In (a) and (b), the mode separation of each element was the same as those calculated from the resonator length of each element, which showed that each element operated independently. However, when both currents were applied simultaneously, only one mode, at which both elements had the same wavelength, could be amplified strongly. The amplification factor estimated at the output end of the amplifier was about 15.

REFERENCES

1. Y. Suematsu, M. Yamada, and K. Hayashi, "A multi-hetero-AlGaAs laser with integrated twin-guide," *Proc IEEE*, 63, 208–209 (1975).
2. M. Yamada, K. Hayashi, Y. Suematsu, and K. Kishino, "Integrated twin-guide AlGaAs lasers with current injection pumping," *Trans. IECE of Jpn.* 58-C, 162–163 (1975).
3. Y. Suematsu, M. Yamada, and K. Hayashi, "Integrated twin-guide AlGaAs laser with multi-hetero structure," *IEEE J. Quantum Electron.*, QE-11, 457–463 (1975).
4. Y. Suematsu, K. Kishino, and K. Hayashi, "Theory of integrated twin guide lasers," *Trans. IECE Jpn.*, 58-C, 654–660 (1975).
5. C. E. Hurwitz, J. A. Rossi, J. J. Hsieh, and C. M. Wolf, "Integrated GaAs-AlGaAs double-heterostructure lasers," *Appl. Phys. Lett.*, 27, 241–243 (1975).
6. J. C. Campbell and D. W. Bellavance, "Monolithic laser/waveguide coupling by evanescent fields," *IEEE J. Quantum Electron.*, QE-13, 253–255 (1977).
7. J. L. Mertz and R. A. Logan: "Integrated GaAs–Al$_x$Ga$_{1-x}$As injection lasers and detectors with etched reflectors," *Appl. Phys. Lett.*, 30, 530–533 (1977).
8. F. K. Reinhart and K. Kishino, "GaAs–AlGaAs double heterostructure lasers with taper coupled passive waveguide," *Appl. Phys. Lett.*, 26, 516–518 (1975).
9. Y. Abe, K. Kishino, T. Tanbun-ek et al., "Room-temperature CW operation of 1.6 mm GaInAsP/InP buried-heterostructure integrated laser with butt-jointed built-in distributed-Bragg-reflection waveguide," *Electron. Lett.*, 18, 410–411 (1982).
10. M. Yamada, H. Nishizawa, and Y. Suematsu, "Mode selectivity in integrated twin-guide lasers," *Trans. IECE of Jpn.*, E59, 9–10 (1976).
11. Y. Suematsu, K. Kishino, and T. Kambayashi, "Axial mode selectivities for various types of integrated twin-guide lasers," *IEEE J. Quantum Electron.*, QE-13, 619–622 (1977).
12. H. Kogelnik and C. V. Shank, "Stimulated emission in a periodic structure," *Appl. Phys. Lett.*, 18, 152–154 (1971).
13. M. Nakamura, K. Aiki, J. Umeda, and A. Yariv, "CW operation of distributed feedback GaAs-GaAlAs diode lasers at temperature up to 300K," *Appl. Phys. Lett.*, 27, 403–405 (1975).
14. T. Kambayashi and Y. Suematsu, "Microfabrication for semiconductor integrated optical circuit by chemical etching," in *Proceedings of the International Conference of Integrated Optics and Optical Communications*, O1-3, Tokyo (July 1977).
15. K. Kawanishi, Y. Suematsu, and K. Kishino, "GaAs–AlGaAs integated twin-guide lasers with distributed bragg reflectors," *IEEE J. Quantum Electron.*, QE-13, 64–65 (1977).
16. H. Kawanishi and Y. Suematsu, "Temperature charactersitics of GaAs–AlGaAs integrated twin-guide laser with distributed bragg reflectors," *Jpn. J. Appl. Phys.*, 17, 1599–1603 (1978).
17. H. Kawanishi, Y. Suematsu, K. Utaka, Y. Itaya, and S. Arai, "GaInAsP/InP injection laser partially loaded with first order distributed bragg reflector," *IEEE J. Quantum Electron.*, QE-15, 701–706 (1979).
18. K. Utaka, Y. Suematsu, K. Kobayashi, and H. Kawanishi, "GaInAsP/InP integrated twin-guide lasers with first-order distributed bragg reflectors at 1.3 μm wavelength," *Jpn. J. Appl. Phys.*, 19, L137–L140 (1980).

19. K. Utaka, K. Kobayashi, K. Kishino, and Y. Suematsu, "1.5–1.6 μm GaInAsP/InP integrated twin-guide lasers with first-order distributed bragg reflectors," *Electron. Lett.*, 16, 455–456 (1980).

20. Y. Sakakibara, K. Furuya, K. Utaka, and Y. Suematsu, "Single mode oscillation under high-speed direct modulation in GaInAsP/InP integrated twin-guide lasers with distributed Bragg reflectors," *Electron. Lett.*, 16, 456–458 (1980).

21. Y. Suematsu, S. Arai, and K. Kishino, "Dynamic-single-mode semiconductor lasers with distributed reflector," (Invited), *IEEE J. Lightwave Technol.*, LT-1, 161–176 (1983).

22. K. Kobayashi, K. Utaka, Y. Abe, and Y. Suematsu: "CW operation of 1.5–1.6 μm wavelength GaInAsP/InP BH integrated twin guide lasers with distributed bragg reflector," *Electron. Lett.*, 17, 366–368 (1981).

23. K. Utaka, K. Kobayahsi, F. Koyama, Y. Abe, and Y. Suematsu, "Single wavelength operation of 1.53 μm GaInAsP/InP BH integrated twin guide lasers with distributed Bragg reflector under direct modulation up to 1 GHz," *Electron. Lett.*, 17, 368–369 (1981).

24. F. Koyama, S. Arai, Y. Suematsu, and K. Kishino, "Dynamic spectral width of rapidly modulated 1.58 μm GaInAsP/InP buried-heterostructure distributed-Bragg-reflector integrated-twin-guide lasers," *Electron. Lett.*, 17, 938–940 (1981).

25. T. Tanbun-Ek, S. Arai, F. Koyama, K. Kishino, S. Yoshizawa, T. Watanabe, and Y. Suematsu, "Low threshold current CW operation of GaInAsP/InP buried heterostructure distributed Bragg-reflector integrated-twin-guide laser emitting at 1.5–1.6 μm," *Electron. Lett.*, 17, 967–968 (1981).

26. Y. Suematsu and K. Kishino, "Coupling coefficient in strongly coupled dielectric waveguides," *Radio Sci.*, 12, 587–592 (1977).

27. K. Utaka, Y. Suematsu, K. Kishino, and H. Kawanishi, "Measurement of coupling coefficient and coupling length of GaAs/AlGaAs integrated twin-guide injection lasers prepared by liquid-epitaxy," *Trans. IECE of Jpn.*, E62, 319–323 (1979).

28. K. Kishino, S. Aoki, and Y. Suematsu, "Wavelegnth variation of 1.6 μm wavelegnth buried heterostructure GaInAsP/InP lasers due to direct modulation," *IEEE J. Quantum Electron.*, QE-18, 343–351 (1982).

29. Y. Suematsu, "Monolithic integration of optical circuits and related twin-guide lasers" (Invited), in *Proceedings of the International Conference of Integrated Optics and Optical Communications*, B1-1, Tokyo, (July 1977).

30. Y. Suematsu and A. R. Adams (Eds.), *Handbook of Semiconductor Lasers and Photonic Integrated Circuits*, Chapman & Hall, London (1994).

31. K. Kishino, Y. Suematsu, K. Utaka, and H. Kawanishi, "Monolithic Integration of Laser and Amplifier/Detector by Twin-Guide Structure," *Jpn. J. Appl. Phys.*, 17, 589–590 (1978).

ISOLATOR AND CIRCULATOR

Tetsuya Mizumoto

Isolators and circulators are nonreciprocal devices, whose transmission characteristics change depending on the propagation direction. An ideal isolator transmits incident lightwave with no attenuation in a forward direction, while it completely attenuates backward reflections. The optical isolator is usually used to protect optical active devices such as laser diodes and optical amplifiers from unwanted reflections and to stabilize their operations. The optical circulator is used to construct optical functional devices and/or subsystems. For example, an add-drop multiplexer is constructed with the optical circulator in combination with a fiber Bragg grating.

RARE EARTH IRON GARNET

A magneto-optic effect is indispensable to realize optical nonreciprocal devices such as isolators and circulators. In optical fiber communication wavelength bands, a rare earth iron garnet is known to be the best candidate to provide a magneto-optic effect because of its large figure of merit, which is defined by the ratio of Faraday rotation coefficient to optical loss per unit propagation distance.

The group of rare earth iron garnet has the unit chemical composition of $R_3Fe_5O_{12}$, where R represents a rare earth element. $Y_3Fe_5O_{12}$, abbreviated as YIG, is the most familiar material in this group. The rare earth iron garnet has a low optical absorption window in a wavelength range between 1.1 and 5 μm [1], and, hence, is useful as a constituent magneto-optic material at the optical fiber communication wavelength bands of 1.3 and 1.55 μm. The rare earth iron garnet is grown by liquid phase epitaxy (LPE), sputter epitaxy, and flux method.

The LPE of rare earth iron garnet was initially developed for growing a thin crystal layer. It has been modified to grow a crystal of some hundreds-μm thickness, which is used as a bulk Faraday rotator [2–4]. In the LPE, chemical elements consisting of garnet crystal are supplied as oxides like R_2O_3 and Fe_2O_3. They are melted in a flux typically composed of PbO and B_2O_3 at the temperature ranging from 700 to 1000°C. When a garnet substrate is dipped in the melt of super cooling state, a garnet crystal is grown on the substrate. The substrate is usually made of nonmagnetic garnet such as $Gd_3Ga_5O_{12}$ (GGG) and $(GdCa)_3(GaMgZr)_5O_{12}$ (NOG). The composition of grown crystal is determined by the melt composition. The lattice constant changes depending on the rare earth element included in the garnet crystal [5]. When growing the iron garnet in the LPE, much attention has to be paid not to include a divalent ion such as Pb^{2+} and tetravalent ions such as Pb^{4+} and Pt^{4+}, since these ions change Fe^{3+} to Fe^{4+} and Fe^{2+} for charge compensation and cause large absorption in a near-infrared wavelength region.

High figure of merit is desirable for the iron garnet used in the isolator and circulator. It is known that the Faraday effect is remarkably enhanced by substituting rare earth elements with Bi [6] and Ce. Ce substituted YIG [7, 8], which is usually grown by the sputter epitaxy process, exhibits the largest Faraday rotation at 1300 and 1550 nm wavelength bands. Figure 1 shows

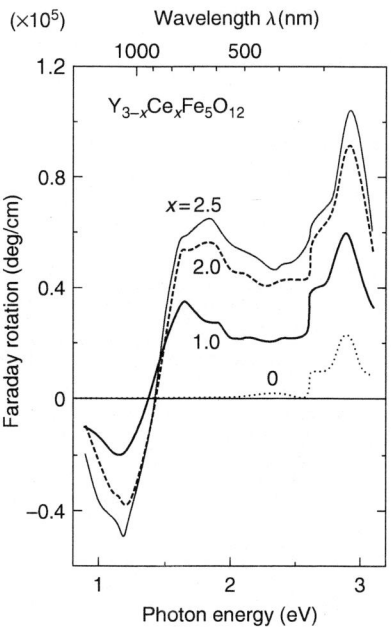

Figure 1   Wavelength dependence of Faraday rotation in $(Y_{3-x}Ce_x)Fe_5O_{12}$ (From M. Gomi, H. Furuyama, and M. Abe, *J.Appl. phys.*, 70, 7065–7067 [1991]. With permission.)

the wavelength dependence of Faraday rotation in a $(Y_{3-x}Ce_x)Fe_5O_{12}$ layer grown by sputter epitaxy [7].

It is requested to saturate the magnetization of Faraday rotator along the light propagation direction for making full use of the Faraday rotation and for reducing the diffraction loss due to magnetic domains. It is desirable to minimize the strength of magnetic field required to saturate the magnetization for simplifying an external magnetic circuit.

In the bulk isolator, the birefringence of Faraday rotator should be minimized to realize high isolation. When we consider applications to waveguide devices, care must be taken to reduce scattering loss as well as optical absorption in the garnet layer. In general, the liquid phase and the sputtering epitaxy are used to grow the garnet layer. The refractive index is adjusted by controlling the garnet composition. In case of LPE, controlling the growth temperature and/or the rotation speed of substrate during the crystal growth enables to control the index of grown layer [9].

PRINCIPLE OF OPERATION AND PERFORMANCE

Isolator

Bulk Type Optical Isolator

To understand the operation principle of a waveguide optical isolator, it is useful to review the working principle of a bulk type optical isolator. The basic configuration of the bulk isolator is shown in Figure 2. Lightwave incident to a left polarizer P_1 passes through a Faraday rotator, in which the polarization plane is rotated by 45° with respect to an incident polarization. The rotation angle is adjusted by the length of Faraday rotator. The polarization plane of right polarizer P_2 is matched with the rotated polarization so as to pass the lightwave with no attenuation. In the reverse propagation, reflected unpolarized lightwave becomes linearly polarized after passing through the polarizer P_2 and is propagated in the Faraday rotator. In the Faraday rotator, the polarization plane is rotated by another 45°, which, in turn, results in a cross-polarized sate with respect to the polarization direction that is allowed to pass through P_1. Therefore, the reflected lightwave cannot pass through the isolator. When the angle of Faraday rotation differs from the

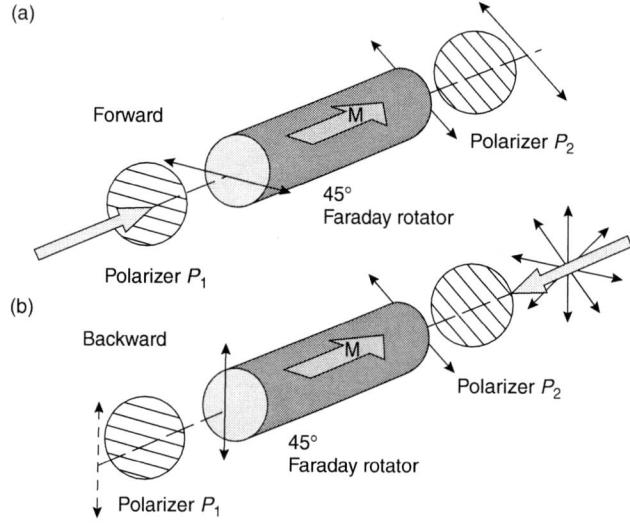

Figure 2   Basic configuration of a bulk isolator

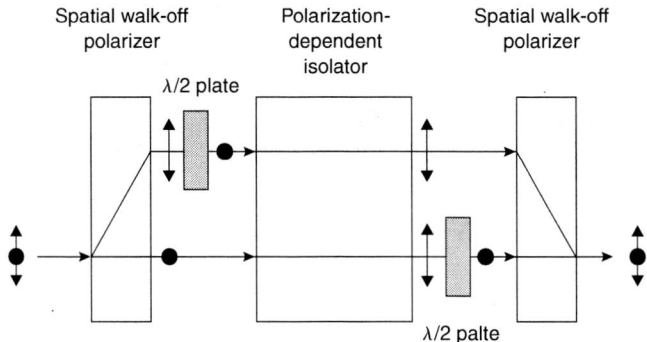

Figure 3 Polarization-independent optical isolator (From K. Shiraishi, *Electron. Lett.*, 27, 302–303 [1991]. With permission.)

designed value of 45°, the insertion loss is increased in the forward propagation. In the backward propagation, the isolation that is defined by the loss of backward propagating lightwave is degraded.

In the configuration shown in Figure 2, the lightwave having a polarization state that is mismatched with the transmitting polarization direction of polarizer P_1 suffers an insertion loss. That is, the performance is dependent on the polarization state of lightwave. A polarization-independent optical isolator does work as an isolator for any incident polarization. Figure 3 shows an example of configuration [10]. The incident lightwave is split into two orthogonally polarized components by a polarization splitter. After one of the two polarizations is rotated by 90° through a $\lambda/2$ plate, two polarization components are incident on the polarization dependent isolator. Before exiting the isolator, the polarization is re-rotated by 90° through a $\lambda/2$ plate, which results in a polarization state identical to the incident wave.

Fiber-in-line Type Optical Isolator

A fiber-in-line type optical isolator has high connectivity with other fiber optic components. In the fiber-in-line isolator, the components needed for isolator operation are micro-optic devices and installed in-between input and output fibers. Since a lightwave transmitted through an optical fiber is usually in a random polarized state, it is needed for the isolator to work as a polarization-independent device.

The typical structure of the fiber in-line optical isolator is shown in Figure 4 [11]. Basically, its configuration and principle of operation are similar to those of bulk type isolator. The coupling efficiency of lightwave into a fiber is strongly dependent on the beam position [11–14]. That is, the forward traveling lightwave is transmitted through an output fiber with low loss, while the backward light beam is shifted transversely and is not coupled into the fiber. This mechanism is employed to reject the backward light instead of a polarizer.

Because the lightwave propagates in a micro-optic components that has no waveguide mechanism, care must be taken not to increase the insertion loss due to the diffraction of light beam [15]. This is achieved either by collimating the beam with a set of lens or by employing a spot-size converter fiber, which is fabricated by tapering a core diameter through an appropriate thermal process [16]. In a lens-less configuration, the propagation length of light beam should be minimized to reduce the diffraction loss. Polarization splitters with large splitting angle as well as a Faraday rotator with a large Faraday rotation coefficient are needed to achieve this. As a polarization splitter, rutile and calcite crystals are commonly used. To obtain a larger splitting

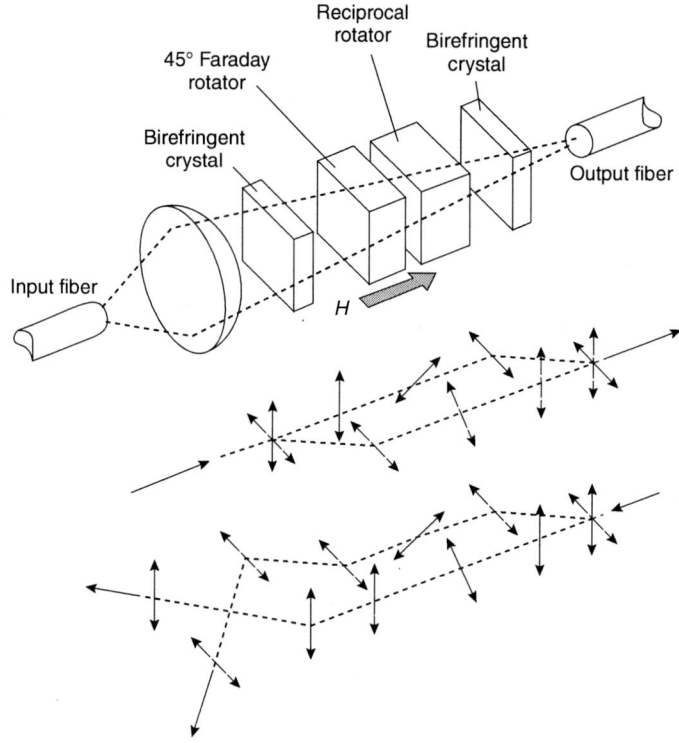

Figure 4   Structure of a fiber in-line optical isolator (From T. Matsumato, Trans. IECE, J62-C, 505–512 [1979, in Japanese]. With permission.)

angle, an artificial device with a laminated structure of high-index and low-index dielectrics [14] is used.

Waveguide Isolator

In the waveguide isolator, the propagation of eigen mode, which is classified into quasi TE and TM modes in a channel waveguide, is strongly dependent on the waveguide structure. This should be taken into account in designing the device.

The waveguide isolators investigated so far are classified into mode conversion type and phase shifter type. The former makes use of the conversion between TE and TM modes, while the latter nonreciprocal phase shift. Since the phase shift isolator operates in a single polarization, it has the advantage of not needing phase matching between two orthogonal polarizations.

Mode Conversion Type

The mode conversion in a waveguide isolator is similar to the rotation of polarization in a bulk isolator. In the waveguide device, the output lightwave is usually needed to be the identical polarization to the input one. Therefore, the unidirectional mode converter is to be realized that provides no mode conversion in a forward propagation and a complete TE–TM mode conversion in a backward propagation [17–19].

The unidirectional mode converter is constructed by combining nonreciprocal and reciprocal mode conversion. In the forward direction, two mode conversions cancel each other, while,

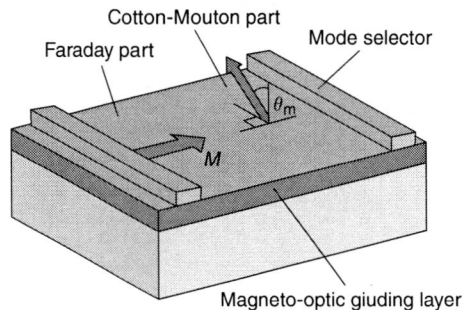

Cotton-Mouton part

Faraday part

Mode selector

θ_m

M

Magneto-optic giuding layer

Figure 5 Bi-sectional type waveguide optical isolator composed of Faraday and Cotton-Mouton mode converters

in the backward direction, they add to provide 100% conversion. When only magneto-optic effects are used for these functions, Faraday and Cotton-Mouton effects are used for nonreciprocal and reciprocal mode conversions, respectively [20–22]. Faraday effect is produced by a magnetization component parallel to the propagation direction of lightwave, while Cotton-Mouton effect is given by a magnetization component perpendicular to the lightwave propagation. The device in which the two mode converters are connected in tandem as shown in Figure 5, is called a bi-sectional type.

The phase matching between TE and TM modes is a critical issue in a mode conversion isolator. In a bulk Faraday rotator, a linearly polarized wave rotates its polarization plane maintaining the polarization state. The isolator works properly, when two orthogonal polarizations propagate in the same velocity, that is, birefringence-free in the rotator. Similarly, in the waveguide isolator, only when TE and TM modes have an identical propagation constant, that is, in a phase-matched state, a complete mode conversion is realizable. It should be noted that, in a mode conversion isolator based on Faraday and Cotton-Mouton effects, slight phase mismatch between TE and TM modes in a Faraday section can be compensated for by choosing proper conditions in a Cotton-Mouton section for realizing a unidirectional mode converter that outputs a linearly polarized wave [23]. In this particular case, the magnetization direction is to be adjusted properly in the Cotton-Mouton section depending on the phase mismatch in the Faraday section. The magnetization direction in the Cotton-Mouton section corresponds to the optical axis in a birefringent crystal.

The propagation constant of the mode guided in a waveguide varies depending on the waveguide structure like the thickness and width of waveguide as well as the refractive indices of materials constructing the waveguide. Therefore, for achieving the phase matching, it is needed to control these waveguide parameters properly. Applying external stress [24] and controlling the temperature [25] have been investigated as ways of adjusting the phase mismatch as a post fabrication tuning technique. When a waveguide is composed of a rare earth iron garnet grown on a garnet substrate, the lattice mismatch between the grown layer and the substrate brings about stress-induced [26] and growth-induced [27] birefringence in a guiding layer.

In the bi-sectional structure, the magnetization directions in respective sections are perpendicular to each other. The following approach has been proposed to realize such a magnetic structure. First, a garnet guiding is grown on an angled substrate whose crystal orientation is slightly offset from [1 1 1] direction. This yields a garnet layer having a magnetization tilted from the substrate normal with an appropriate angel suitable for Cotton-Mouton effect [28]. Then, by annealing the layer locally with a laser beam, the in-plane magnetization region is produced [29]. The magnetization in this region is aligned suitable for Faraday effect by applying

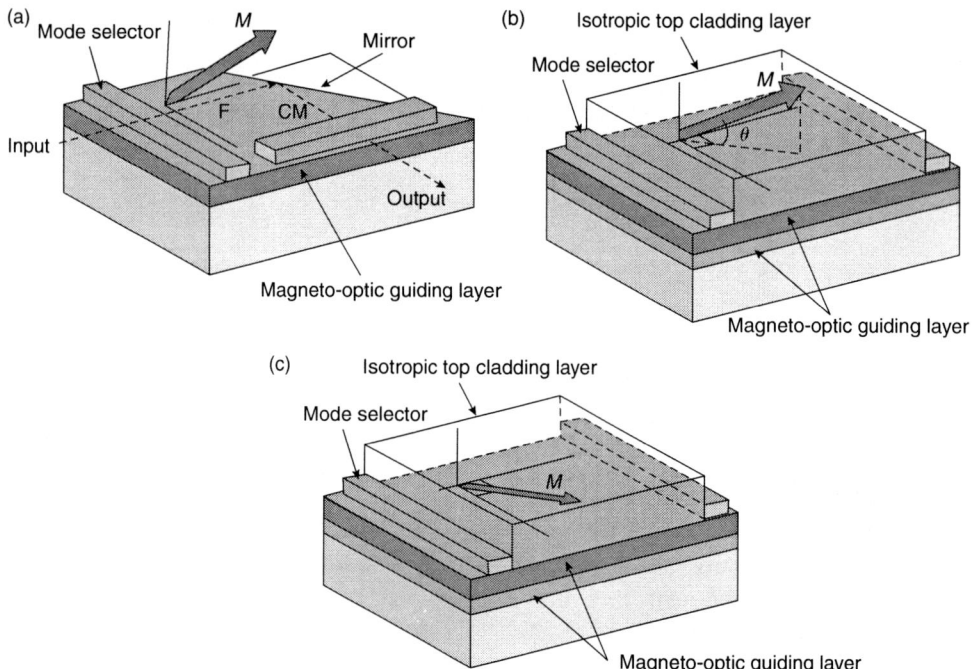

Figure 6 Mono-sectional type waveguide optical isolators. (a) F and CM stand for a Faraday and a Cotton-Mouton section, respectively. (b) Magnetization M is aligned in one direction with an elevation angle θ with respect to a layer plane. (c) Magnetization is aligned on a layer plane

a weak external magnetic field along the light propagation direction. Using a GGG substrate with 8° offset angle from [1 1 1] orientation, $(BiGdLu)_3(FeGa)_5O_{12}$ is grown by LPE. The fabricated isolator exhibited the isolation of 12.5 dB [22].

In order to circumvent the complexity of aligning the magnetization in two perpendicular directions in the bi-sectional configuration, a mono-sectional isolator has also been investigated, in which the magnetization is uniformly aligned in a certain direction along the entire device length. Figure 6 shows several types of proposed mono-sectional isolators. Figure 6(a) shows the mirror-type, in which the optical path is reflected by 90° with a corner mirror to provide a Faraday rotation and a reciprocal mode conversion [30,31]. In Figure 6(b), the magnetization is aligned in a plane transverse to the propagation direction with a certain angle with respect to the film normal [32]. Reciprocal and nonreciprocal mode conversions are brought about by respective magnetization components. The isolation of 13 dB has been reported in a fabricated device. The device shown in Figure 6(c) uses a mode conversion and nonreciprocal phase shift experienced by TM modes [33]. This configuration has the advantage of relatively easy magnetization control, since it is aligned on a film plane.

As another type of mono-sectional mode conversion isolator, a semi-leaky isolator shown in Figure 7 has been investigated [34]. This device utilizes a reciprocal TE–TM mode conversion provided by an off-angled birefringent crystal like $LiNbO_3$ together with a nonreciprocal conversion provided by Faraday effect. In the forward propagation, two mode conversions are cancelled with a proper choice of the offset angle of birefringent crystal. In the backward propagation, they add up because of the nonreciprocal nature of Faraday effect. Therefore, an incident TE mode is converted into a TM mode. Since the converted TM mode is radiated into the birefringent superstrate and does not appear in the incident port, the device works as an isolator.

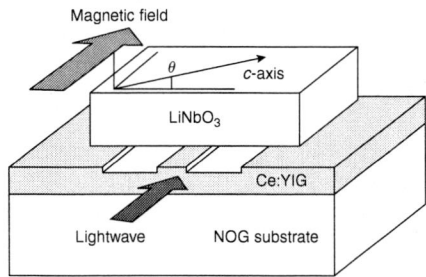

Figure 7   Semi-leaky optical isolator

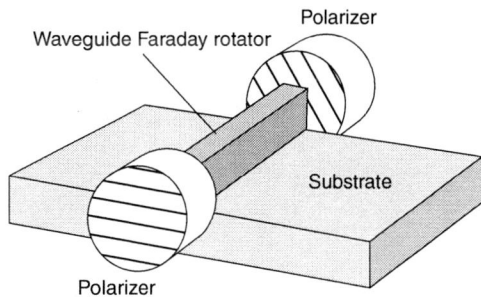

Figure 8   Hybrid configuration composed of a waveguide Faraday rotator and micro-optic polarizers

Some preliminary experiment has been reported in literatures [35,36]. The optical contact between the birefringent crystal used as a superstrate and the magneto-optic guiding layer is a critical issue for achieving sufficient performance.

A hybrid configuration has also been studied, where a waveguide Faraday rotator is sandwiched with micro-optic polarizers as is shown in Figure 8 [37–39]. An isolation of 30 dB has been reported in a device fabricated with a buried waveguide YIG Faraday rotator [40].

Nonreciprocal Phase Shift Type

When TM modes propagate in a waveguide including the magneto-optic material whose magnetization is aligned transversely to the light propagation direction on its film plane as shown in Figure 9, they experience nonreciprocal phase shift [41]. That is, their propagation constants vary, when the propagation direction is reversed.

The nonreciprocal phase shift per unit propagation distance is calculated as a function of guiding layer thickness at a wavelength of 1.55 μm and is shown in Figure 10. Figure 10(a) corresponds to the case where the guiding layer is composed of a magneto-optic material, while Figure 10(b) to the case where the magneto-optic material is used in a cladding layer. In both cases, the nonreciprocal phase shift takes a maximum value, when the waveguide is close to cut-off. In Figure 10(a), two layered structures air/ $(CeY)_3Fe_5O_{12}$/ NOG and air/ $(LuNdBi)_3Fe_5O_{12}$/ GGG are assumed. The Faraday rotation of $(CeY)_3Fe_5O_{12}$ and $(LuNdBi)_3Fe_5O_{12}$ are $-4500°$/cm and $-500°$/cm, respectively, at this wavelength. The refractive index of NOG and GGG is 1.95. In Figure 10(b), the guiding layer is assumed to be a GaInAsP layer, which is lattice-matched to InP and has bandgap wavelength of 1.25 μm and refractive index of 3.36. $(CeY)_3Fe_5O_{12}$ is placed in an upper cladding layer. Nonreciprocal phase shift is proportional to Faraday effect as

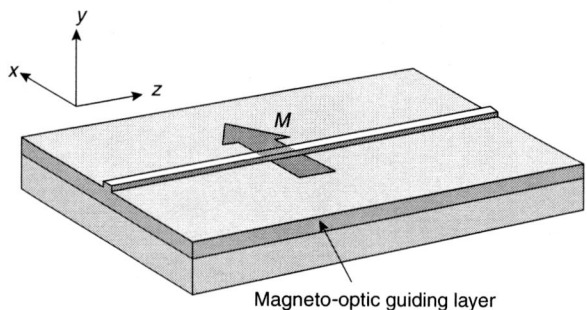

Figure 9 Configuration of optical nonreciprocal phase shifter, where a magnetization M is aligned transversely to the light propagation direction z on a magneto-optic guiding layer

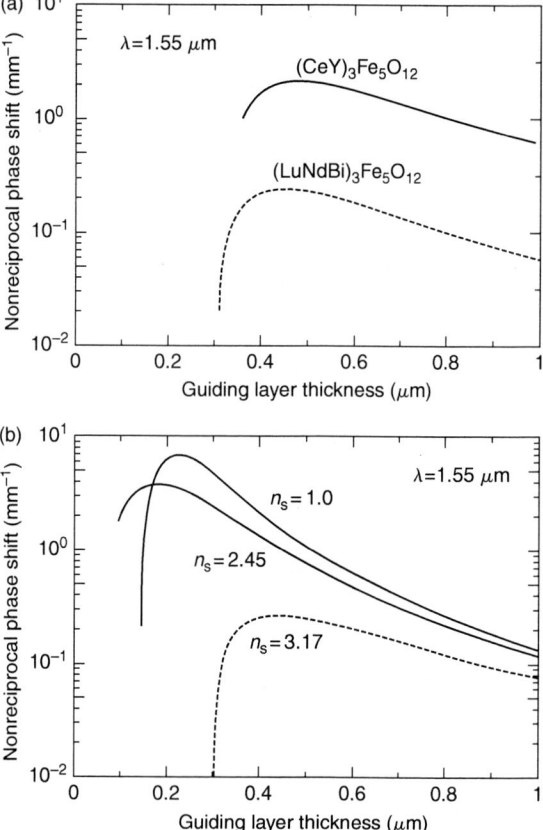

Figure 10 Calculated optical nonreciprocal phase shift in layered structures of air/ $(CeY)_3Fe_5O_{12}$/ NOG and air/ $(LuNdBi)_3Fe_5O_{12}$/ GGG in (a) and $(CeY)_3Fe_5O_{12}$/ GaInAsP ($\lambda_g = 1.25$ μm)/ substrate in (b), where the refractive index of substrate n_s is taken as a parameter

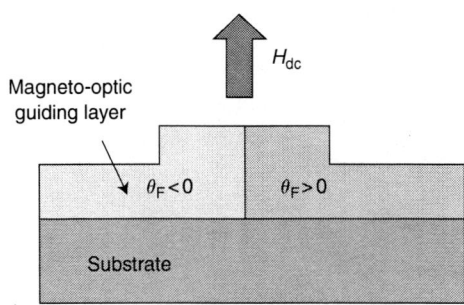

Figure 11 Optical nonreciprocal phase shifter for TE modes

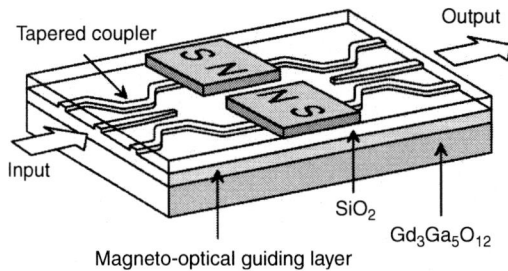

Figure 12 Interferometric waveguide isolator

is shown in Figure 10(a), because it is brought about by the 1st order magneto-optic effect like a Faraday rotation. In the case that a magneto-optic material is used in an upper cladding layer, nonreciprocal phase shift is enhanced when the refractive index of substrate is reduced.

According to the perturbation theory, the nonreciprocal phase shift is provided by the magneto-optic coupling between transverse and longitudinal electric field components of TM modes [42]. This, in turn, gives the conclusion that the phase shift is proportional to the difference in the squared amplitude of the TM mode's transverse magnetic field component, when the magneto-optic material is used as a guiding layer. Therefore, for obtaining the nonreciprocal phase shift, it is required to use an asymmetric index distribution. The larger the difference in the refractive index between upper and under cladding layers, the more the phase shift. Also, it can be understood that the phase shift becomes large where the mode is close to cutoff.

To enhance the nonreciprocal phase shift, it is effective to tailor the direction of magnetization locally in anti-parallel manner [43]. In an ordinary configuration shown in Figure 9, only TM modes experience the nonreciprocal phase shift. The magnetic structure should be modified as is shown in Figure 11 so as to work for TE modes [44]. Polarization independent configuration has also been proposed [45,46].

An interferometric waveguide isolator is constructed by using Y branches or 3-guide symmetric tapered couplers as shown in Figure 12. In the forward propagation, input lightwave is split into two waves with equal amplitude and phase. Two waves experience nonreciprocal phase shift with opposite sign in respective arms, because the magnetization is aligned in an anti-parallel direction. The phase difference between two waves provided by the nonreciprocal magneto-optic effect is set to be 90°. This phase difference is canceled by an additional reciprocal phase shift,

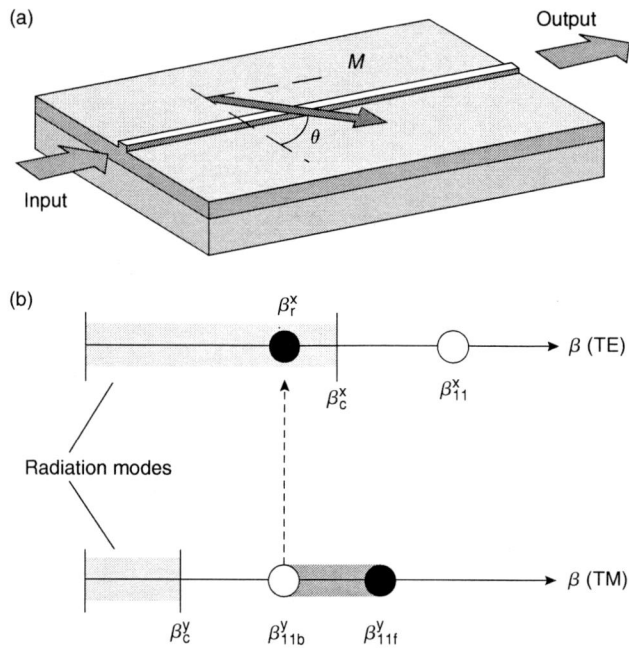

Figure 13 Waveguide isolator employing a nonreciprocal conversion to radiation modes. (a) Structure and (b) diagram of propagation constants

which is provided by an optical path difference between two arms of interferometer. Thus, the forward waves become in phase and are coupled into the output port of coupler. In the reverse direction, the nonreciprocal phase shift changes its sign. Therefore, two waves propagating in two arms of interferometer become out-of-phase and are not coupled out from the input port. In case of Y branch, the out-of-phase component is radiated by virtue of the single mode nature of input waveguide [47]. A maximum isolation of 19 dB is reported in Reference [48]. Also, the waveguide isolator having a semiconductor guiding layer, which is highly compatible for integrating with semiconductor optical active devices, was investigated in this configuration [49]. An isolation of 5 dB has been demonstrated.

The cutoff condition is dependent on the propagation direction due to nonreciprocal phase shift. That is, lightwave is confined in the waveguide core in the case of forward propagation, while it is leaky in the backward direction. This can be utilized to construct an isolator [50]. In the channel waveguide, this is applicable to the nonreciprocal mode conversion to a transverse radiation mode (Figure 13) [51].

Loss Compensation Type

As a third category of optical isolator, loss compensation type optical isolator has been investigated [52,53]. The schematic structure is illustrated in Figure 14. The magnetic cladding material provides a direction dependent loss in addition to a direction dependent phase shift. That is, the loss provided to the forward propagating wave is set to be smaller than that of backward wave. Thus, optical isolator can be constructed by compensating for the forward loss by virtue of the gain of semiconductor amplifier.

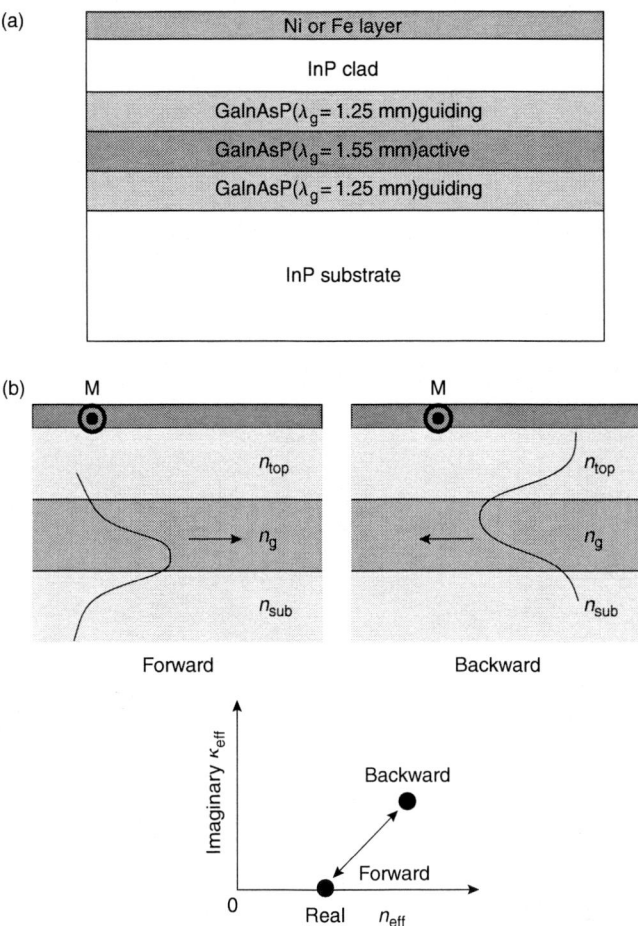

Figure 14   Loss compensation isolator (a) structure and (b) principle of operation

Circulator

Bulk Type Optical Circulator

The circulator utilizes nonreciprocal rotation of polarization and is composed of a Faraday rotator and polarization beam splitters [54–57]. The principle of operation is illustrated in Figure 15. The light beam incident on port-1 is split into two cross-polarized components with polarization beam splitter. While passing through the Faraday rotator and reciprocal rotator, the polarization rotates by 90° with respect to the incident polarization. Two polarization components are combined in polarization splitter, and exit from port-2. Since in the reverse direction, the rotation of polarization in the reciprocal rotator is canceled in the Faraday rotator, the light beam incident on port-2 does not experience rotation of polarization. As a result, the beam is output from port-3. Similarly, the input from port-3 is output from port-4, and that from port-4 exits from port-1. The isolation can be improved by using the combination of a polarization beam splitter and a birefringent crystal to suppress the leakage from isolated ports [58,59].

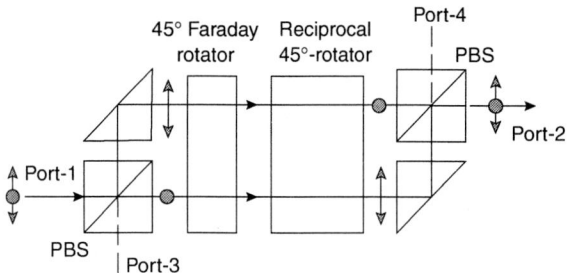

Figure 15   Bulk optical circulator. PBS stands for a polarization beam splitter (From H. Imamura, H. Iwasaki, K. Kubodera, Y. Torii, and J. Noda, *Electron. Lett.*, 15, 830–831 [1979]. With permission.)

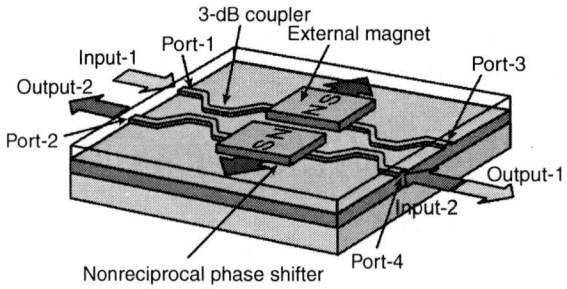

Figure 16   Waveguide optical circulator employing nonreciprocal phase shift

Waveguide Optical Circulator

A waveguide circulator is constructed using the nonreciprocal phase shift in combination with the interferometric structure [60]. When we construct a Mach-Zehnder interferometer so as to give 0° and 180° phase difference between two arms for forward and backward propagation, respectively, output is obtained in a cross state or a bar state with respect to the input port. Such phase differences can be achieved using nonreciprocal and reciprocal phase sifters as are mentioned in the previous section. The circulator operation is shown in Figure 16. The lightwave incident on port-1 is output from port-4, because total phase difference between two arms becomes zero. For the lightwave incident on port-4, the phase difference becomes 180 degree due to the nonreciprocal phase shift. The output appears in port-2. Similarly, the input from port-2 is output from port-3, and that from port-3 returns to port-1. The waveguide circulator based on this principle is reported in Reference 60.

Another type of circulator is investigated using the nonreciprocal phase shift in a coupled waveguide [61]. The concept of this configuration is that the coupled waveguide becomes symmetric or asymmetric depending on the propagation direction because of the direction-dependent nature of waveguide effective index [62]. The coupled waveguide is designed so as to be a complete coupled waveguide for the symmetric configuration, and the input lightwave is coupled out from the cross port. In the reverse direction, the coupled waveguide becomes asymmetric, and can be designed so that the input lightwave is output from the bar port. Adjusting the coupling coefficient and coupling length of the coupler can achieve this.

In addition, combining polarization rotations of a 45° Faraday rotation and a half-wave plate in a Mach-Zehnder interferometer, polarization-independent circulator operation is demonstrated

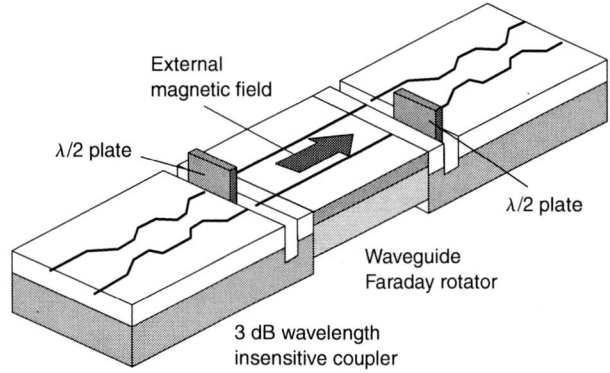

External
magnetic field

λ/2 plate

λ/2 plate

Waveguide
Faraday rotator

3 dB wavelength
insensitive coupler

Figure 17 Waveguide polarization independent optical circulator in a hybrid configuration (From N. Sugimoto, T. Shintaka, A. Tata, H. Terai, M. Shimokozono, E. Kubota, M. Ishii, and Y. Inoue, *IEEE Photonics Technol. Lett.*, 11, 355–357 [1999]. With permission.)

in a hybrid waveguide form (Figure 17) [63]. In this configuration, the polarization independent operation has been achieved. The LPE-grown La- and Ga-substituted YIG are used as a Faraday rotator. The polyimide waveplate and 3 dB directional coupler composed of silica based PLC are used to construct the device. An isolation of 17.1 to 27.0 dB has been obtained together with 3.0 to 3.3 dB insertion loss.

Temperature Dependence

For practical applications, temperature independent operation of optical devices is required. In the case of optical isolators and circulators, the temperature dependence of performance is predominated by the temperature dependence of Faraday rotation angle. Tailoring the composition of rare earth iron garnet can minimize the temperature dependence of Faraday rotator. When one substitutes iron with non-magnetic elements to compensate lattice mismatch between a magneto-optic garnet crystal and a substrate, the temperature dependence increases due to the decreased Curie temperature of the magneto-optic garnet. Contrary to this, substitution of a rare earth element with Bi contributes to an increase in the Curie temperature and decreases the temperature dependence at room temperature. Low temperature dependence of Faraday rotation $d\theta/dT = -0.04$ deg/°C has been achieved at a wavelength of 1.31 μm in $(YbTbBi)_3Fe_5O_{12}$[64]. Also, cascading two garnet crystals having a temperature coefficient of opposite sign can compensate the temperature dependence of Faraday rotation.

In addition to reducing the temperature dependence of Faraday rotator, cascading the isolators that exhibit different temperature dependence is applicable to cancel and flatten the temperature dependence of entire performance [65]. This scheme is also applicable to reduce the wavelength dependence of isolators.

REFERENCES

1. D. L. Wood and J. P. Remeika, "Effect of impurities on the optical properties of yttrium iron garnet," *J. Appl. Phys.*, 38, 1038–1045 (1967).
2. T. Aoyama, T. Hibiya, and Y. Ohta, "A new Faraday rotator using a thick Gd:YIG film grown by liquid-phase epitaxy and its applications to an optical isolator and optical switch," *IEEE/OSA J. Lightwave Technol.*, LT-1, 280–285 (1983).

3.   T. Hibiya, T. Ishikawa, and Y. Ohta, "Growth and characterization of 300-μm thick Bi-substituted gadolinium iron garnet films for an optical isolator," *IEEE Trans. Mag.*, MAG-22, 11–13 (1986).

4.   K. Nakajima, Y. Nomi, H. Ishikawa, and K. Machida, "A new improved $(YbTbBi)_3Fe_5O_{12}$ epitaxial thick film," *IEEE Trans. Mag.*, 24, 2565–2567 (1988).

5.   P. Hansen, W. Tolksdorf, and K. Witter, "Recent advances of bismuth garnet materials research for bubble and magneto-optical applications," *IEEE Trans. Mag.*, MAG-20, 1099–1104 (1984).

6.   H. Takeuchi, "The Faraday effect of bismuth substituted rare-earth iron garnet," *Jpn. J. Appl. Phys.*, 14, 1903–1910 (1975).

7.   M. Gomi, H. Furuyama, and M. Abe, "Strong magneto-optical enhancement in highly Ce-substituted iron garnet films prepared by sputtering," *J. Appl. Phys.*, 70, 7065–7067 (1991).

8.   T. Shintaku and T. Uno, "Preparation of Ce-substituted yttrium iron garnet films for magneto-optic waveguide devices," *Jpn. J. Appl. Phys.*, 35, 4689–4691 (1996).

9.   W. Tolksdorf, H. Dammann, E. Pross, B. Strocka, H. J. Tolle, and P. Willich, "Growth of yttrium iron garnet multi-layers by liquid phase epitaxy for single mode magneto-optic waveguides," *J. Crystal Growth*, 83, 15–22 (1987).

10. K. Shiraishi, "New configuration of polarization-independent isolator using a polarization-dependent one," *Electron. Lett.*, 27, 302–303 (1991).

11. T. Matsumoto, "Polarization-independent isolators for fiber optics," *Trans. IECE*, J62-C, 505–512 (1979, in Japanese).

12. M. Shirasaki and K. Asama, "Compact optical isolator for fibers using birefringent wedges," *App. Opt.* 21, 4296–4299 (1982).

13. K. W. Chang and W. V. Sorin, "Polarization independent isolator using spatial walkoff polarizer," *IEEE Photonics Technol. Lett.*, 1, 68–70 (1989).

14. K. Shiraishi and S. Kawakami, "Spatical walk-off polarizer utilizing artificial anisotropic dielectrics," *Opt. Lett.*, 15, 516–518 (1990).

15   K. Shiraishi, T. Chuzenji, and S. Kawakami, "Polarization-independent in-line optical isolator with lens-free configuration," *J. Lightwave Technol.*, 10, 1839–1842 (1992).

16. K. Shiraishi, T. Yanagi, Y. Aizawa, and S. Kawakami, "Fiber-embedded in-line isolator," *J. Lightwave Technol.*, 9, 430–435 (1991).

17. S. Wang, M. Shah, and J. D. Crow, "Studies of the use of gyrotropic materials for mode conversion in thin-film optical-waveguide applications," *J. Appl. Phys.*, 43, 1861–1875 (1972).

18. S. Yamamoto and T. Makimoto, "Circuit theory for a class of anisotropic and gyrotropic thin-film optical waveguides and design of nonreciprocal devices for integrated optics," *J. Appl. Phys.*, 45, 882–888 (1974).

19. J. Warner, "Nonreciprocal magneto-optic waveguides," *IEEE Trans. MTT*, MTT-23, 70–78 (1975).

20. J. P. Castera and G. Hepner, "Isolator in integrated optics using the Faraday and Cotton-Mouton effects," *IEEE Trans. Mag.*, MAG-13, 1583–1585 (1977).

21. Y. Miyazaki, "Optical propagation and mode conversions in magnetic garnet film waveguide," *J. Mag. Soc. Jpn.*, 6, 254–259 (1982, in Japanese).

22. K. Ando, T. Okoshi, and N. Koshizuka, "Waveguide magneto-optic isolator fabricated by laser annealing," *Appl. Phys. Lett.*, 53, 4–6 (1988).

23. T. Mizumoto, Y. Kawaoka, and Y. Naito, "Waveguide-type optical isolator using the Faraday and Cotton-Mouton effects," *Trans. IECE Jpn.*, E69, 968–972 (1986).

24. H. Dammann, E. Pross, G. Rabe, W. Tolksdorf, and M. Zinke, "Phase matching in symmetrical single-mode magneto-optic waveguides by application of stress," *Appl. Phys. Lett.*, 49, 1755–1757 (1986).

25. J. P. Castera, P. L. Meunier, J. M. Dupont, and A. Carenco, "Phase matching in magneto-optic YIG films by waveguide temperature control," *Electron. Lett.*, 25, 297–298 (1989).

26. G. Hepner, J. P. Castera, and B. Desormiere, "Studies of magnetooptical effects in garnet thin film waveguides," *AIP Conf. Proc.*, 29, 658–659 (1975).

27. K. Ando, N. Koshizuka, T. Okuda, and Y. Yokoyama, "Growth-induced optical birefringence in LPE-grown Bi-substituted iron garnet films," *Jpn. J. Appl. Phys.*, 22, L618–L620 (1983).

28. K. Ando, N. Takeda, T. Okuda, and N. koshizuka, "Waveguide mode conversion by magnetic linear birefringence of Bi-substituted iron garnet films titled from (111)," *J. Appl. Phys.*, 57, 718–722 (1985).

29. K. Ando, Y. Yokoyama, T. Okuda, and N. koshizuka, "Localized modification of magnetic anisotropy in LPE iron garnet films by laser annealing," *J. Magnetism Magn. Mat.*, 35, 350–352 (1983).

30. H. Hemme, H. Dotsch, and H. -P. Menzler, "Optical isolator based on mode conversion in magnetic garnet films," *Appl. Opt.*, 26, 3811–3817 (1987).
31. K. Ando, "Waveguide optical isolator: a new design," *Appl. Opt.*, 30, 1080–1084 (1991).
32. Y. Miyazaki and K. Taki, "An analysis of isolation properties of mono-sectional optical isolators using Bi:YIG thin-films," *Trans. IEICE Jpn.*, E72, 742–750 (1989).
33. T. Mizumoto and Y. Naito, "Waveguide-type optical isolator with in-plane magnetization structure," in *Proceedings of the Third European Conference on Optics*, The Hague, SPIE, Vol. 1274, pp. 220–228 (March 12–13, 1990).
34. S. Yamamoto, Y. Okamura, and T. Makimoto, "Analysis and design of semileaky-type thin-film optical waveguide isolator," *IEEE J. Quantum Electron.*, QE-12, 764–770 (1976).
35. S. T. Kirsh, W. A. Biolsi, S. L. Blank, P. K. Tien, R. J. Martin, P. M. Bridenbaugh, and P. Grabbe, "Semi-leaky thin-film optical isolator," *J. Appl. Phys.*, 52, 3190–3199 (1981).
36. T. Mizumoto, H. Yokoi, M. Shimizu, and S. Narikawa, "Semi-leaky optical isolator fabricated by wafer bonding of magneto-optic garnet and $LiNbO_3$," in *Proceedings of the IEEE LEOS Annual Meeting 2001*, La Jolla, WW3, Vol.2, pp. 580–581 (November, 2001).
37. R. Wolfe, J. Hegarty, J. F. Dillon, Jr., L. C. Luther, G. K. Celler, L. E. Trimble, and C. S. Dorsey, "Thin-film waveguide magneto-optic isolator," *Appl. Phys. Lett.*, 46, 817–819 (1985).
38. P. Friez, J. Machui, and P-L. Meunier, "An approach towards a single mode garnet rib waveguide calculations and experiments," in *Advances in Magneto-Optics, Proceedings of the International Symposium on Magneto-Optics, J. Magn. Soc. Jpn.*, 11, 385–388 (1987).
39. E. Pross, W. Tolksdorf, and H. Dammann, "Yttrium iron garnet single-mode buried channel waveguides for waveguide isolators," *Appl. Phys. Lett.*, 52, 682–684 (1988).
40. H. Dammann, E. Pross, G. Rabe, and W. Tolksdorf, "The 45deg-waveguide-isolator," in *proceedings of the Seventh International Conference on Integrated Optics and Optical Fiber Communications*, Kobe, 20B2-7, Vol.3, pp. 102–103 (1989).
41. F. Auracher and H. H. Witte, "A new design for an integrated optical isolator," *Opt. Commun.*, 13, 435–438 (1975).
42. M. Shamonin and P. Hertel, "Analysis of nonreciprocal mode propagation in magneto-optic rib-waveguide structures with the spectral-index method," *Appl. Opt.*, 33, 6415–6421 (1994).
43. M. Wallenhorst, M. Niemoller, H. Dotsch, P. Hertel, R. Gerhardt, and B. Gather, "Enhancement of the nonreciprocal magneto-optic effect of TM modes using iron garnet double layers with opposite Faraday rotation," *J. Appl. Phys.*, 77, 2902–2905 (1995).
44. A. F. Popkov, M. Fehndrich, M. Lohmeyer, and H. Dotsch, "Nonreciprocal TE-mode phase shift by domain walls in magnetooptic rib waveguides," *Appl. Phys. Lett.*, 72, 2508–2510 (1998).
45. J. Fujita, M. Levy, R. M. Osgood, Jr., L. Wilkens, and H. Dotsch, "Polarization-independent waveguide optial isolator based on nonreciprocal phase shift," *IEEE Photonics Technol. Lett.*, 12, 1510–1512 (2000).
46. O. Zhuromskyy, H. Dotsch, M. Lohmeyer, L. Wilkens, and P. Hertel, "Magnetooptical waveguides with polarization-independent nonreciprocal phase shift," *Jo Lightwave Technol.*, 19, 214–221 (2001).
47. Y. Okamura, H. Inuzuka, T. Kikuchi, and S. Yamamoto, "Nonreciprocal propagation in magnetooptic YIG rib waveguides," *J. Lightwave Technol.*, LT-4, 711–714 (1986).
48. J. Fujita, M. Levy, R. M. Osgood, Jr., L. Wilkens, and H. Dotsch, "Waveguide optical isolator based on Mach-Zehnder interferometer," *Appl. Phys. Lett.*, 76, 2158–2160 (2000).
49. H. Yokoi, T. Mizumoto, N. Shinjo, N. Futakuchi, and Y. Nakano, "Demonstration of an optical isolator with a semiconductor guiding layer that was obtained by use of a nonreciprocal phase shift," *Appl. Opt.*, 39, 6158–6164 (2000).
50. H. Hemme, H. Dotsch, and P. Hertel, "Integrated optical isolator based on nonreciprocal-mode cut-off," *Appl. Opt.*, 29, 2741–2744 (1990).
51. T. Shintaku and T. Uno, "Optical waveguide isolator based on nonreciprocal radiation," *J. Appl. Phys.*, 76, 8155–8159 (1994).
52. M. Takenaka and Y. Nakano, "Proposal of a novel semiconductor optical waveguide isolator," in *proceedings of the 11th International Conference on Indium Phosphide and Related Materials*, Davos, Switzerland, TuA3-5, pp. 289–292 (1999).
53. W. Zaets and K. Ando, "Optical waveguide isolator based on nonreciprocal loss/gain of amplifier covered by ferromagnetic layer," *IEEE Photonics Technol. Lett.*, 11, 1012–1014 (1999).

54. H. Imamura, H. Iwasaki, K. Kubodera, Y. Torii, and J. Noda, "Simple polarization-independent optical circulator for optical transmission systems," *Electron. Lett.*, 15, 830–831 (1979).

55. T. Matsumoto, "Polarization-independent optical circulator coupled with multimode fibers," *Electron. Lett.*, 16, 8–9 (1980).

56. M. Shirasaki, H. Kuwahara, and T. Obokata, "Compact polarization-independent optical circulator," *Appl. Opt.*, 20, 2683–2687 (1981).

57. I. Yokohama, K. Okamoto, and J. Noda, "Polarization-independent optial circulator consisting of two fiber-optic polarising beam splitters and two YIG spherical lenses," *Electron. Lett.*, 22, 370–371 (1986).

58. Y. Fujii, "High-isolation polarization-independent optical circulator coupled with single-mode fibers," *J. Lightwave Technol.*, 9, 456–460 (1991).

59. Y. Fujii, "Polarization-independent optical circulator having high isolation over a wide wavelength range," *IEEE Photonics Technol. Lett.*, 4, 154–156 (1992).

60. T. Mizumoto, H. Chihara, N. Tokui, and Y. Naito, "Verification of waveguide-type optical circulator operation," *Electron. Lett.*, 26, 199–200 (1990).

61. T. Mizumoto, K. Ohchi, and Y. Naito, "Measurement of optical nonreciprocal phase shift in a Bi-substituted $Gd_3Fe_5O_{12}$ film and application to waveguide-type optical circulator," *J. Lightwave Technol.*, LT-4, 347–352 (1986).

62. T. Shintaku and T. Uno, "Directional coupler type optical circulator," *Trans. IEICE Jpn.*, E73, 474–476 (1990).

63. N. Sugimoto, T. Shintaku, A. Tate, H. Terui, M. Shimokozono, E. Kubota, M. Ishii, and Y. Inoue, "Waveguide polarization-independent optical circulator," *IEEE Photonics Technol. Lett.*, 11, 355–357 (1999).

64. K. Nakajima, Y. Asahara, K. Machida, and Y. Fujii, "Improvement of the temperature coefficient of Faraday rotation in Bi-substituted garnet film," *Trans. IEICE*, J70-C, 120–121 (1987, in Japanese).

65. K. Shiraishi and S. Kawakami, "Cascaded optical isolator configuration having high-isolation characteristics over a wide temperature and wavelength range," *Opt. Lett.*, 12, 462–464 (1987).

LAMBDA PLATE

Yasuo Kokubun

A bulk component made of birefringent material of which phase retardation between ordinary and extraordinary axes is π or $\pi/2$. Quartz, rutile, and some polymer thin films such as polyimide and liquid-crystalline polymer are used as birefringent materials. The plate with the retardation of π is called half-wave plate and with the retardation of $\pi/2$ is called quarter-wave plate. The combination of half- and quarter-wave plates can convert any polarization state to linear or other desired polarization state by adjusting their primary axes.

LIGHT

Kenichi Iga

The frequency of the electromagnetic wave is 10^{14} to 10^{15} Hz and the velocity of light in vacuum is $c = 299792485$ m/sec which was determined in 1983. The origin of their generation can be categorized as follows:

1. Black body radiation (high temperature object)
2. Transition of energy states (LED, laser, fluorescence, etc.)
3. Bramstrahlung (decelerated electrons)
4. Synchrotron radiation (electrons of circular motion with high speed)
5. Cherenkov radiation (moving charged particles with velocity larger than c/n, n: refractive index of medium, etc.)
6. γ (Gamma) decay

The understanding of these effects can be attributed to the change of velocity in moving electrons. The quantum of light is a boson with polarization having no mass, momentum $\hbar k$, and energy $h\omega$ ($\hbar = h/2\pi$, $k = 2\pi/\lambda$, $\omega = 2\pi \times$ frequency, $\lambda = c/\omega$).

LITHIUM NIOBATE MODULATOR

Masayuki Izutsu

Lithium Niobate (LN, $LiNbO_3$) is widely used to build various types of light modulators of integrated optic structure. Electro-optic light modulators play important roles in optical fiber communication systems as external modulators with a fast response and small signal distortion/chirp parameters, as well as in microwave photonics to mix up optical signals with microwave signals to produce modulated optical outputs, where generated optical sidebands consist of sideband components with a frequency separation between adjacent components that is equal to the modulating microwave frequency. The performances of these devices has been dramatically improved through the introduction of the concept of integrated optics, together with optical waveguide structures [1]. Light modulators with a frequency range of over 40 or 60 GHz are now commercially available with several hundred mW of driving power and optical insertion losses less than several dB. Packaging techniques have also been improved substantially, and recent small-mount modulator elements can be packed into standard rack-mounts with semiconductor devices. In addition, intricate functional devices have become available through the integration of different elements on a single waveguide substrate.

Because of its excellent piezoelectric, pyroelectric, optoelectric, and nonlinear optic properties, LN is a well known artificially synthesized ferroelectric crystal and has been extensively studied for various device applications. It is chemically stable and insoluble in water and organic solvents, and is a colorless or light yellow crystal with a density of 4.64 g/cm^3 and a hardness 6 on Moh's hardness scale.

The crystal structure of LN is very near to perovskite, composed of lightly deformed oxygen octahedra containing a cation, stacking along *c*-axis (threefold axis) and being connected with the neighboring octahedra through an oxygen bonding. The cation arrangement along *c*-axis follows the sequence (Nb, vacancy, Li), (Nb, vacancy, Li), ..., Nb and Ti, respectively, slightly below and substantially above the center of oxygen octahedra. This causes strong spontaneous polarization at room temperature and its crystal symmetry belongs to the point group 3 m in the trigonal ferroelectric phase [2].

The LN has an optical birefringence and is a negative uniaxial crystal with respective ordinary and extraordinary refractive indices, n_o and n_e, of around 2.28 and 2.20 in visible wavelength range. The electro-optic coefficient is a physical value given by a third rank tensor and is, in many cases, represented by a six by three matrix form. From the crystallographic symmetrity, the coefficient for LN, a point group 3 m crystal, has the nonvanishing components: $r_{13} = r_{23} \sim 8.6$, $r_{33} \sim 30.8$, $r_{22} = -r_{12} = -r_{61} \sim 3.4$, and $r_{42} = r_{51} \sim 28$ in pm/V, so that the coefficients are in the form [3–5].

$$r = \begin{pmatrix} 0 & r_{12}\,(=-r_{22}) & r_{13} \\ 0 & r_{22} & r_{23}\,(=r_{13}) \\ 0 & 0 & r_{33} \\ 0 & r_{42} & 0 \\ r_{51}\,(=r_{42}) & 0 & 0 \\ r_{61}\,(=-r_{22}) & 0 & 0 \end{pmatrix} \tag{1}$$

The electro-optic effect is described as a change in the index ellipsoid, or indicatrix, given by

$$x^2/n_x^2 + y^2/n_y^2 + z^2/n_z^2 = 1, \tag{2}$$

when the crystal axes are parallel to the x-, y-, and z-direction. The index ellipsoid is proportional to the surface of constant energy density in terms of D space of the crystal where D is the electric flux density vector, and the change in the energy density due to the applied external electric field causes a small change in the ellipsoid surface. When an external electric field of E_k ($k = x, y, z$ or 1, 2, 3) is applied, the index ellipsoid is distorted as,

$$x^2(1/n_x^2 + r_{1k}E_k) + y^2(1/n_y^2 + r_{2k}E_k) + z^2(1/n_z^2 + r_{3k}E_k)$$
$$+ 2yzr_{4k}E_k + 2zxr_{5k}E_k 2xyr_{6k}E_k = 1. \qquad (3)$$

In the case of LN, the above equation becomes, for applied field E_z,

$$x^2(1/n_o^2 + r_{13}E_z) + y^2(1/n_o^2 + r_{13}E_z) + z^2(1/n_e^2 + r_{33}E_z) = 1 \qquad (4)$$

with the help of Eqs. (1) and (3). Since the effect is sufficiently small, the perturbed principal indices can be written as

$$n_x = n_o - \left(\frac{1}{2}\right)n_o^3 r_{13} E_z,$$

$$n_y = n_o - \left(\frac{1}{2}\right)n_o^3 r_{13} E_z, \qquad (5)$$

$$n_z = n_e - \left(\frac{1}{2}\right)n_o^3 r_{33} E_z.$$

The last term of the each equation above indicates the index change due to the electro-optic effect.

If we apply the external field in y-direction, E_y, to 3 m crystals, the deformed index ellipsoid is given by

$$x^2(1/n_o^2 - r_{22}E_y) + y^2(1/n_o^2 + r_{22}E_y) + z^2/n_e^2 + 2yzr_{42}E_y = 1. \qquad (6)$$

According to the last term on the left, a small rotation of the principal axes occurs in the yz-plane in addition to the change in ordinary refractive index, $(\frac{1}{2}) n_o^3 r_{22}Ey$ in the x- and y-directions. When the applied field is in x-direction, Eq. (3) is written as,

$$x^2/n_o^2 + y^2/n_o^2 + z^2/n_e^2 + 2zxr_{42}E_x - 2xyr_{22}E_x = 1. \qquad (7)$$

Rotational change of principal axes occur only in the zx- and xy-planes.

It is not difficult to fabricate optical waveguides with acceptable performance levels on the LN crystal surface, and it is one of the most popular materials as the substrate for various functional devices of integrated optic structure. While there are several different methods for waveguide fabrication, the most important and widely used is the thermal in-diffusion of Ti metals. Oxidized transition metals were initially tested, but now Ti metal in-diffusion is the current accepted standard method of waveguide fabrication.

First, a negative pattern of the desired optical waveguide layout is fabricated on the LN substrate through photoresist, and the Ti metal is deposited over the patterned resist film through electron-beam vacuum evaporation to a thickness of several tens of nanometers. By using the lift-off technique, the waveguide pattern of the Ti metal is placed directly on the substrate, and then is thermally in-diffused into LN substrate. To suppress the unwanted out-diffusion of substances contained in the crystal during the process, thermal diffusion is carried out under wet-argon or wet-oxygen gas atmosphere with a temperature slightly obove 1000°C for approximately 10 h [6].

Optical waveguides obtained through Ti-metal thermal in-diffusion are called graded-index channel waveguides with a maximum index change of around 2%. Index profiles are said to be Gaussian or complementary error functions in the depth direction depending on the condition of fabrication, and the lateral direction features a more complicated profile depending on the width of the initial Ti metal pattern. Diffusion constants of Ti metals are different for crystalline z- and x- or y-directions, and the cross section of the resulting waveguide strongly depends on the crystal axis of the substrate. It is not difficult to have a Ti diffused waveguide with a propagation loss of <0.5 or 0.3 dB/cm in 1.3 or 1.5 μm range, and loss increase due to the so-called optical damage will occur in the visible range with an incident optical power of more than several milliwatts. In any case, the parameters given are only typical examples, because fabrication conditions and obtained waveguide properties depend on the individual performances of LN crystal.

In an electro-optic light modulator, a refractive index of the guide is changed according to the applied modulating voltage to give the transmitting light wave a phase retardation that is proportional to the voltage, so that the electro-optic phase modulator is rather straight forward, while for light intensity modulation or switching, Mach–Zehnder (MZ) type interferometric structures have largely been used for guided-wave structure.

The response of electro-optic effect is so fast that it adjoins the nonlinear optic effect by which the index changes with the optical field amplitude. There is hence no limitation in principle in outlining the device as having an ultrafast response time or an operation frequency up to or beyond the sub-millimeter range, except in certain absorption bands of the electro-optic material (from around 6 to 60 THz and above 600 THz for LN). Meanwhile, from the practical point of view, the design of the electrical circuit of the device—that is, the structure of modulator electrodes and the feeding method of modulating signals—is especially important in order to realize ultrafast light modulator and other functional devices.

A simple way to apply the modulating electric field is to treat the modulator electrode as a lumped capacitor. If the modulating frequency is low enough to assume the voltage on the electrode is uniform over the electrode length L, and is constant during the transit time nL/c for the light wave to pass through under the modulator electrode, we can regard the modulator as a lumped capacitance. For an LN waveguide modulator of an electrode length of 1 cm, a 3 dB bandwidth with 50 Ω parallel termination becomes 2 or 3 GHz. When bandwidth is not important, the use of longer electrode is advantageous since the required drive voltage is proportional to the inverse of electrode length L, together with the bandwidth.

An effective way to bring up the modulating frequency to the microwave and millimeter-wave region is the use of traveling-wave mode of operation. Modulating mw/mmw are made to propagate in the same direction as the lightwave. The bandwidth is limited by an accumulating phase difference between light and modulating waves during their propagation through the modulator, and is inversely proportional to the product of a velocity mismatch between two waves and the interaction length (or an electrode length). Selection of the modulator electrode structure as a part of wide-band mw/mmw circuits for the modulating signal becomes another important issue for smooth frequency responses. To avoid unwanted reflections at discontinuities in the circuit, an asymmetric coplanar strip line and a three-electrode coplanar waveguide structures were successfully adopted.

With reduced interaction length, faster operation will be achieved, although higher drive power is needed to decrease the modulation efficiency since required drive power is inversely proportional to the squared interaction length. A method to simultaneously achieve efficient and fast or wideband operation is to reduce the velocity mismatch between the light and modulating waves of the traveling-wave modulator. In the case of LiNbO₃ waveguides with a coplanar electrode structure, the lightwave travels twice as fast as the modulating microwave, so that a decrease in lightwave velocity or increase in modulating wave velocity is required for the reduction of the mismatch. Several attempts have already been made to realize extremely high modulation efficiency for 10 or 20 GHz light modulators. Among them, the use of extra thick (5 to 7 μm) electrode is an accepted design practice that decreases the velocity mismatch by increasing the portion of modulating electric field traveling outside the crystalline substrate of high dielectric constant. The use of a groove between parallel coplanar electrodes and extremely thin substrate is also applied to reduce the velocity mismatch, together with the impedance matching with feeder lines.

As a different way to achieve light modulation in higher frequency range with lower driving power, the use of band operation has been proposed. So far as conventional lumped or traveling wave structures are employed to built modulators, wider bandwidth will be needed to modulate lightwaves at higher frequencies because the lower end of their operation bands are rooted at zero frequency. To double the bandwidth, the electrode length should be halved, and with fixed drive power, the modulation depth decreases to a quarter of the initial value. The band modulation scheme reduces the drive power dramatically at the expense of narrowing the bandwidth. The concept of band-limited modulation is believed to play an important role in future photonic systems, and intensive research and development is now underway.

For light intensity modulation or switching, MZ type and directional-coupler type structures have largely been used. The acousto-optic effect is also available for constructing wideband light control devices. Using the Bragg diffraction of the guided light by a surface acoustic wave, light deflectors/modulators of several GHz bandwidth have already been experimentally demonstrated. Directional coupler type switches were investigated intensively at the first stage of the development of guided-wave optical switches, although only MZ type intensity modulators have survived as commercial models for above gigahertz operation frequency.

The integrated optic MZ interferometer consists of a waveguide Y junction to divide the input light wave into two parallel waveguides, and two waves from parallel waveguides are recombined again at the other Y junction at the output end. The operation of the waveguide Y junction is identical to that of a beam splitter (or combiner) of conventional bulk optics, and also to the 180 degree hybrid in microwave range if radiation from the junction point is taken into account as the hidden forth port. Figure 1 shows a typical schematic of a MZ interferometric light intensity modulator using the LN waveguide.

One of the most important assets of guided-wave devices is the possibility of combining different devices or elements on a single substrate to realize optical integrated circuits. Integration of ultrafast guided-wave components onto a substrate provides compact and stable optical circuits that enable us to construct novel optoelectronic functional devices difficult to be realized without the integrated optics technologies.

Switch matrices composed of multiple optical switch elements have been attracting attention for use in space-division switching in optical communication systems, and a good deal of work has been reported. Even though each switch element of a matrix does not make a high-speed response, it can treat lightwave signals modulated by microwave or millimeter waves. For the integration of high-speed electro-optic devices, however, not many examples have been reported because of difficulties in integration. Examples are single sideband (SSB) modulators/frequency shifters, signal sampler/multipliers, a time multi/demultiplexers, analog to digital converters, and microwave phase shifters especially for antenna applications (Figure 2) [7].

Among them, especially interesting is the SSB modulator of parallel MZ structure with additional phase and intensity modulation electrodes to give a signal modulation on a generated optical single sideband, which, now commercially available, is operated up to 20 GHz baseband frequency range at 1.5 μm wavelength. With the travelling-wave electrode, the device has a

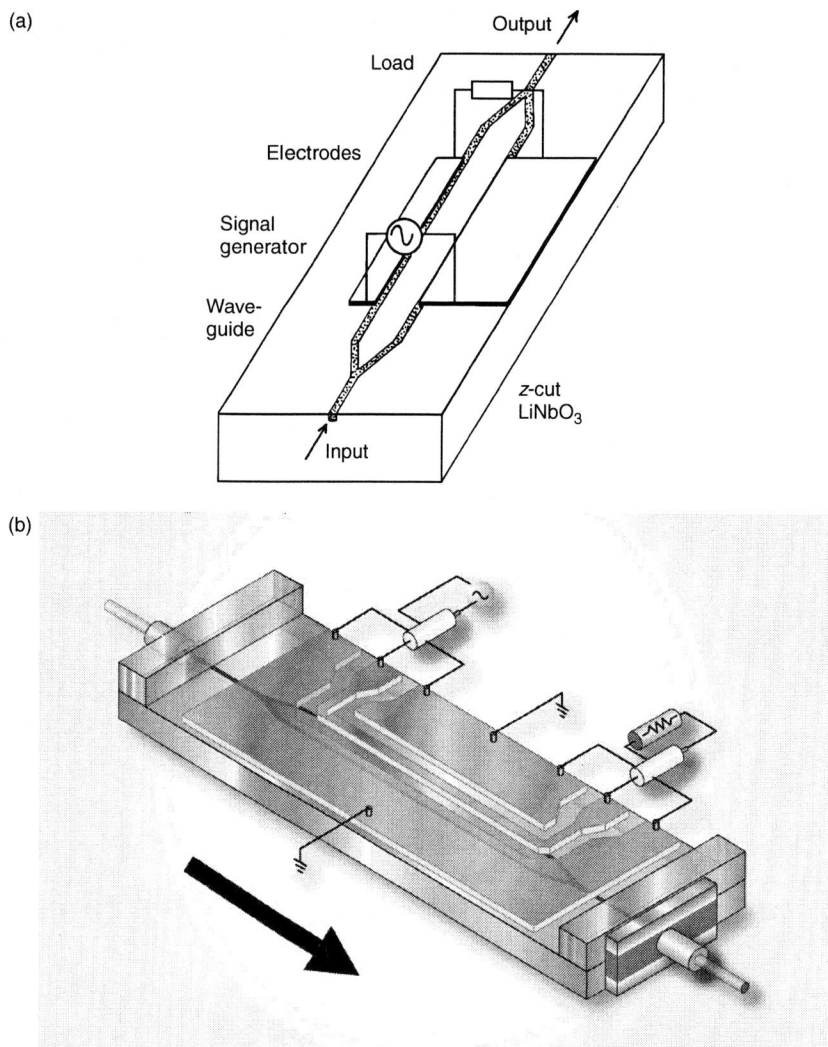

Figure 1   (a) and (b) MZ type light intensity modulator using z-cut LN substrate

wide bandwidth and is packed into a sealed metal case with fibre pig tails so that it is applicable to practical photonic systems immediately.

Other interesting feats are the use of periodically polarization reversed LN (PPLN) substrates for novel light modulators to achieve higher efficiency in the interaction between light and microwave signals. Examples include SSB modulators using PPLN and terahertz mixers. It is important to introduce optical gains in the LN substrate to realize integrated optic circuits of advanced generation, so that complicated optical circuits can be packed on a single substrate by compensating unavoidable optical transmission losses. The combination of electro-optic modulators with other elements on a single substrate is extremely important for the next stage of the research on LN integrated optic devices.

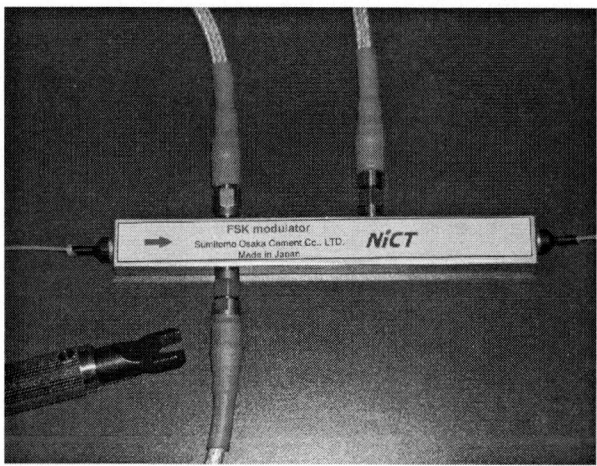

Figure 2   FSK modulator based on a SSB modulator structure

REFERENCES

1.   Theodor Tamir (Ed.), *Guided-wave Optoelectronics*, Springer-Verlag, Heidelberg, 1988.
2.   Yuhum Xu, *Ferroelectric Materials and Their Applications*, North-Holland, Amsterdam, 1991, Chap. 5, pp. 217/245.
3.   J. F. Nye, *Physical Properties of Crystals*, Oxford University Press, Oxford, 1985.
4.   I. P. Kaminow and A. E. Siegman (Eds.), *Laser Devices and Applications*, IEEE Press, 1973, pp. 296/312.
5.   A. Yariv and P. Yeh, *Optical Waves in Crystals*, John Wiley & Sons, New York, 1984.
6.   H. Nishihara, M. Haruna, and T. Suhara, *Optical Integrated Circuits*, McGraw-Hill, New York, 1989.
7.   T. Sueta and T. Okoshi (Eds.), *Ultrafast and Ultra-Parallel Optoelectronics*, John Wiley & Sons, New York, 1995, p. 349.

M

MICRO-ELECTRO-MECHANICAL SYSTEMS

Hiroyuki Fujita

INTRODUCTION

The key concept of micro electro-mechanical systems (MEMS) technologies is to extend the VLSI fabrication capability to realize three-dimensional (3D) microsystems, which are composed of electrical, mechanical, chemical, and optical elements. Using VLSI fabrication processes such as photolithography, film deposition, and etching, it is possible to obtain submicrometer-precision structures in a large quantity with excellent alignment between each other.

The research and development of MEMS have made a remarkable progress since 1988 when an electrostatic micromotor, the size of a human hair, was operated successfully. We can build 3D microstructures on a silicon substrate and operate micromotors/actuators without any difficulty [1]. Commercial products such as inkjet printers, integrated accelerometers, and projection displays using micromirrors are successful in the market. Other promising applications of MEMS for the near future will be in optics [2,3], data storage devices, fluidics, biotechnologies, and scanning probe microscopes.

The optical application is one of the most prospective applications, because the MEMS technology has superior features for the optical field. Movable structures such as micromirrors can be made by micromachining processes. Precise V-grooves and alignment structures can be defined by dry and wet etching. Most of the optical components such as lasers, photodetectors, mirrors, and optical waveguides are fabricated by semiconductor process, which is compatible with micromachining processes in many cases. In addition, it provides key functional devices for optical communication networks, a market that is expanding rapidly.

This section describes in brief, the summary of current MEMS technology and its application to optics. It also deals with typical examples among those activities including optical switches, scanners, sensors, integrated optical encoders, tunable VCSELs/filters, and spatial light modulators.

MEMS TECHNOLOGY

The MEMS research [4] has its root in silicon sensor research. Micromachining technologies based on semiconductor processes have been used to make microsensors, their packages, and microstructural devices [5]. Sensor research first evolved toward MEMS research at the International Conference on Solid State Sensors and Actuators (Transducers-87), held in Tokyo in June 1987 [6]. At the conference, the presentations on surface micromachining of gears and sliding stages were among the most remarkable ones. An integrated servo system for mass flow control and some active devices such as an electrostatic actuator were also reported. The first IEEE MEMS workshop, called the Micro Robots and Teleoperators Workshop, was held in November, 1987 [7]. Scientists and engineers have been investigating materials, fabrication processes, device and system design, and applications of MEMS since then. In Japan, this research field is commonly referred to as micromachine.

In particular, MEMS involves two major features:

- Many structures can be obtained simultaneously by preassembly and batch processes.
- Electronic circuits and sensors can be integrated to obtain smart microsystems.

With these features, high-performance and complex systems that include many actuators with corresponding sensors and controllers can be mass produced in a cost-effective manner.

Micromachining

Table 1 summarizes micromachining processes that are in common use today. The crystallographic dependence of etching speed of a single-crystal silicon in such etchants as KOH and TMAH (tetramethyl ammonium hydroxide) is utilized for wet anisotropic etching. The etching speed of (111) plane is much slower than other crystallographic orientations. Well-defined microstructures surrounded by (111) planes can be fabricated.

Table 1
Micromachining processes, features, and applications

Micromachining process	Features	Applications
Crystallographic wet etching (single-crystal silicon)	Precise 3D structures defined by crystal planes	Sensitive membrane for pressure sensors, V-grooves for optical fiber alignment, mirror-flat surfaces
Anisotropic dry etching (silicon)	3D structures of various shapes determined by a mask process	Microactuators generating large forces, freely-shaped microstructures
Surface micromachining (polysilicon thin film, other thin films)	Ultra-fine structures, good compatibility with CMOS circuit process	Integrated sensor array Actuators with control circuits (e.g., DMD[a] display)
Hinged 3D structure (polysilicon film)	Microstructures folded up from the substrate to form 3D shapes	Micro-optical devices on a silicon chip
Replica processes: LIGA[b], molding, hexil[c] (metals, polymers, polysilicon, glass)	Many replicas of a 3D master mold are obtained, various materials are available	Microfluidic chip by injection molding, glass chip by hot embossing

[a] DMD stands for digital micromirror device and consists of a million of movable micromirrors on CMOS circuits [20]. It is a commercial product of TI, Inc.
[b] LIGA is a 3D micromachining method combining x-ray deep lithography, electroforming, and injection molding.
[c] hexil is a 3D micromachining method to obtain 10 to 50 μm thick polysilicon structures by depositing a CVD polysilicon layer on a silicon substrate with deep-RIE trenches.

A new technique called surface micromachining emerged in the 1980s. For example, thin films of polysilicon and metals are patterned into the shape of a gear. Patterned thin films are released from the substrate by etching away an easily resoluble material placed between the structural film and the substrate. Researchers were able to make rotating micromotors and linear actuators by this process.

Microstructures fabricated by surface micromachining [4] are planar in nature and have thickness of up to 10 μm in most cases. Some applications require thicker or 3D-complex structures. Modifications of surface micromachining have been attempted. One technique is to fold up micromachined plates from the substrate to construct a 3D structure [8].

In addition to thin-film 3D processes, deep dry etching of silicon has become very popular over the last five years. Thick microstructures from 50 to 500 μm, with width/thickness of 1/20, can be fabricated. Electroplating through thick resist patterns creates metallic 3D microstructures. These structures have aspect ratios (the height divided by the width) in the range of 20 to 50 nm. Precision is typically 10 to 100 nm, but reaches a few nanometer in certain specific processes.

The variety of materials have also widened. We can micromachine compound semiconductors (e.g., GaAs), polymers, metals, ceramics and biorelated materials, as well as silicon.

Microactuators

Microactuators are the key devices for MEMS to perform physical functions. Many types of microactuators have been successfully operated. They are categorized in two types: one, based on driving forces and, the other, based on mechanisms. Force can be generated in the space between stationary and moving parts using electric, magnetic, and flow fields. Materials such as piezoelectric, magnetostrictive and photostrictive have intrinsic actuation capabilities. Thermal expansion [9] and phase transformations such as the shape-memory effect and bubble formation cause shape or volume changes. With regard to mechanism, coping with friction, which dominates over inertia forces in microworld, makes a major difference. One solution is to suspend moving parts by flextures, although displacement is generally limited by the support. Reducing friction between stationary and freely moving parts is a challenging problem if no support is used. On the other hand, friction-driven mechanisms are widely adopted [10,11]. An ultimate solution is levitation that removes the friction completely; the repulsive force between a permanent magnet and a superconducting material [12] or controlled air flow from micronozzles [13] has been used successfully to levitate micro-objects.

OPTICAL MEMS

When MEMS is applied to optical devices and systems, it is necessary to incorporate many other technologies. Optical MEMS is based on semiconductor, optical integration, planar light waveguide circuits, and optical packaging technologies as well as MEMS technology (Figure 1). Some of these technologies are selected and merged together in order to realize application specific devices or subsystems.

Table 2 summarizes optical MEMS devices and applications together with basic micromachined elements used in the device. Basic elements include linearly or rotationally movable mirrors, shutters, movable lenses, micro apertures, which are precise structures for alignment. These elements are sometimes used in arrayed configuration. Interferometers, switches, and scanners are built on the elements for particular applications.

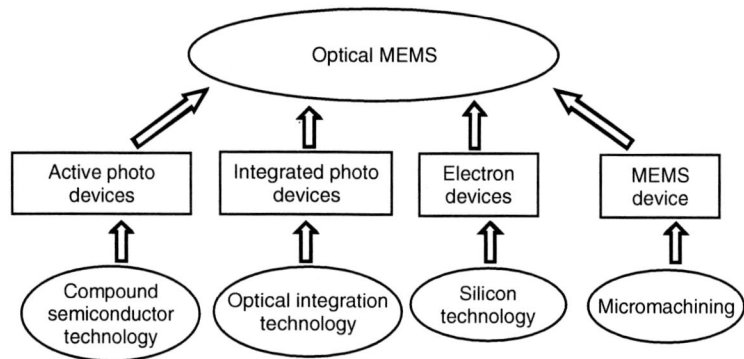

Figure 1   Basic technologies of optical MEMS

Table 2
MEMS optical devices

Devices	Basic elements	Applications
Interferometer	Linear motion mirror Optical waveguide Tunable grating	Wavelength tunable laser/detector/filters Spectrometer Sensors (displacement, pressure, chemical) Modulator Attenuator Display (if arrayed)
Scanner	Torsional motion mirror Rotating mirror Movable lens/prism	Printer Barcode reader Projection display Laser range finder Imager Size measurement Data storage
Switch/attenuator	Movable mirror/shutter Microactuated fiber/waveguide	Optical communication Optical interconnection Optical measurement apparatus
Chopper	Linear motion shutter Rotating shutter	Pyroelectric sensor Modulator for active optical sensor
Spacial light modulator	Arrayed movable mirror Deformable mirror	Display Compensator for optical distortion
Micro lens	Refraction lens Diffraction lens	Autofocusing Collimator Display (if arrayed)
Evanescent light devices	Nanometer aperture Planar waveguide	Near-field scanning optical microscope Ultra-high density optical data storage Particle manipulation
3D micro-optical devices	Hinged 3D structure 3D assembled structure LIGA structure	Free-space optical system Pig-tailed system
Optical devices integrated with micro fluidic system	Laser/detector/lens/fiber with micro channel/chamber	Fluorescent detection Integrated laser tweezer

Optical MEMS are categorized in three classes with respect to working principles. The first class depends on simple geometrical optics. Scanners with movable mirrors and lenses, optical switch using mirrors, arrayed mirror displays, and micromechanical shutters belong to the first class. The second class utilizes the wave nature of light. Typical examples are: wavelength-tunable lasers and photo detectors based on tunable Fabry–Perot interferometers, gratings with movable elements and spatial light modulators. Other types of tunable interferometers are also included. The third class depends on evanescent light such as scanning near-field optical microscopes (SNOM) using micromachined nanoapertures, optical sensors using evanescent light coupled with environment to measure, and high-density optical data storages based on near-field coupling. Future devices combining MEMS and photonic band-gap crystals are also included in this class.

IMPACT OF MEMS TO OPTICS

Integration of Devices

Both monolithic and hybrid integration are possible. Optical systems composed of many devices can be realized with or without minimum assembly process. A highly functional optical system can be packaged in a small volume. A wavelength-tunable laser [14] was obtained by integrating a movable Fabry–Perot cavity with VCSEL (vertical cavity surface emitting laser). A V-shaped laser diode with etched mirrors, optical waveguides made of polyimide, and two photodiodes are monolithically integrated to form an optical microencoder [15] and a displacement sensor (Figure 2). A laser diode was precisely fixed by microstructures on a silicon substrate and aligned to a microlens; the collimated beam was coupled to other micro-optical elements. Thus, a free-space micro-optical system [16] can be constructed on the silicon substrate (see Figure 3 and Figure 4).

Accurate Pre/Passive Alignment

The alignment between elements in optical MEMS is either determined in the monolithic integration process or ensured by structures for passive alignment in the hybrid integration case and for optical fiber pig-tailing. Note that the alignment accuracy depends on the lithography and etching steps, which can be controlled within submicrometers. Bulk micromachining technique based on deep dry etching was utilized to make mechanical connectors, such as pins and receptacle holes, directly on chips [17]. These chips, which contained MEMS devices, V-grooves, or electrical contacts, were assembled precisely into 3D optical MEMS.

Motion and Feature Size Comparable to Wavelength

The motion produced by microactuators is typically 1 to 10 μm in full stroke with precision of <1%. The feature sizes of microstructures are in the same range. Because these are in the same order as the wavelength of light, MEMS devices are very effective for interacting with light. Though microactuators have improved a lot and may produce forces up to a few Newtons, they are more suitable to driving small objects such as micromirrors. A tunable IR filter was developed based on a microactuated adjustable grating [18]. Fabry–Perot, Michelson, and Mach–Zehnder interferometers incorporated with microactuated structures have been demonstrated. Micromirrors driven by microactuators are most commonly used components for scanners, switches, displays, and optical modulators [19].

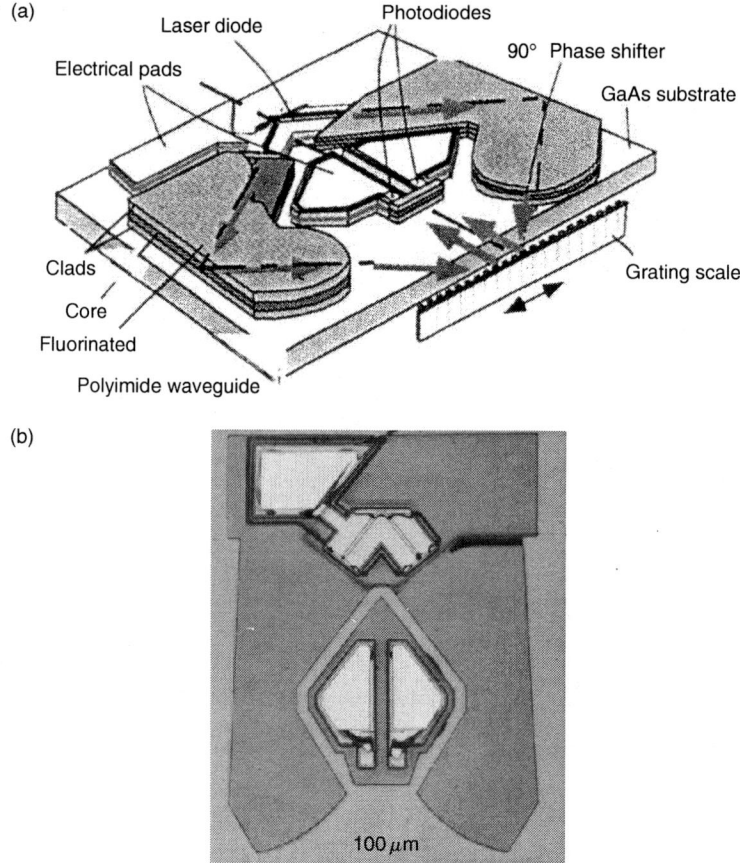

Figure 2   (a) Schematic representation of integrated optical encoder and (b) Photograph of a chip. A V-shaped laser, polymer waveguides, and two photodiodes are integrated. The waveguide provides two interferometric paths with phase difference of one quarter of the wavelength, because one end of the guide has a step structure to make the optical path length different. Both ends are shaped into a cylindrical lens structure. As the reference scale moves, two intereferometric outputs are detected by the photodiodes. The outputs are sinusoidal signals with a 90° phase difference; this phase difference gives the directional information of motion (From R. Sawada, in *Proceedings of the Eighth International Conference on Solid-State Sensors & Actuators [TRANSDUCERS '95]*, Vol. 1, Stockholm, Sweden, June 25–29 [1995], pp. 281–284. With permission.)

Arrayed and Repetitive Structures

The micromachining process enables easy fabrication of many devices in an array or structures with repetitive small features. Integration with electrical circuits enables the individual control of each element in a large array. Millions of tortionally movable mirrors are integrated in an array on video RAM [20]; this device (DMD) is the key component for a projection display that may replace liquid crystal-based system. Similar arrays of mirrors that move linearly and tortionally serve as a spatial phase modulator [21]; they are useful for compensating fluctuations of propagating media or aberration of optical components. A MEMS display based on movable gratings in an array was also demonstrated [22].

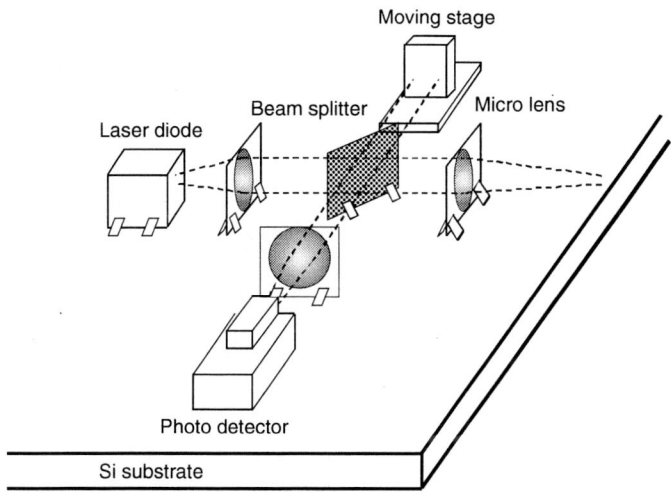

Figure 3 The concept of a free-space micro-optical bench (MOB) proposed by Wu et al. [16], which is the integration of an optical system such as microlenses, mirrors, gratings, lasers, and detectors onto a single chip. They used polysilicon hinged structures to build optical components such as mirrors and gratings located perpendicular to the wafer surface (From M. C. Wu, L. Y. Lin, S. S. Lee, and K. S. J. Pister, *Sensors Actuators*, A 50, 127–134 [1995]. With permission.)

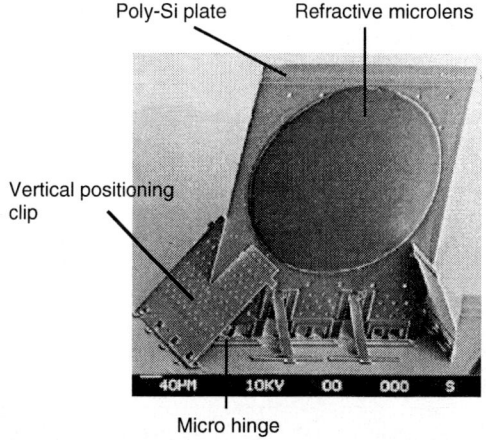

Figure 4 SEM micrograph of a 3D microlens with precision lens mount (From M. C. Wu, L. Y. Lin, S. S. Lee, and K. S. J. Pister, *Sensors Actuators*, A 50, 127–134 [1995]. With permission.)

Wavelength Independence

MEMS devices, especially those based on micromirrors, are far less sensitive to wavelength than solid-state nonlinear optical devices. In addition, they are thermally stable. Therefore, MEMS devices are suitable to wavelength division multiplexing (WDM) optical networks [23]. Also, wavelength tunable devices based on MEMS have wide tunable ranges [12].

Free-Space Optics with Short Propagation Path

Because MEMS devices are so small and can be densely placed together, free-space optical systems with propagation paths of 1 to 10 mm are possible. Light beams can be reflected in arbitrary angles and multiple beams do not interfere with each other in free space. With these features, large scale matrix switches based on free-space optics have been investigated [24].

High Sensitivity and Fast Response

The heat capacity decreases in proportion to the cube sizes while the heat transfer depends on the surface area that is in proportion to the square sizes. Therefore, the scaling law states that thermal sensitivity increases and thermal time constant decreases when things become smaller.

Opto-thermal sensors and actuators are profited from these features. Uncooled IR cameras composed of arrayed microbolometers have been demonstrated [25]. The bolometer is 50 μm in size and supported above the substrate by thermal insulation legs. Typical values for heat capacity, thermal time constant, and thermal noise are 1 nJ/K, 1 msec, and 40 mK respectively. Electronic circuits were integrated in microbolometer arrays. High frequency resonator having the resonant frequency over 1 MHz could be actuated by modulated laser light through the opto-thermal effect [26].

Near-Field Optical Elements

Micro electro-mechanical systems structures provide convenient means to handle evanescent light; openings with nanometer apertures and exposed planar waveguides are often used to general evanescent light field. Near-field scanning optical microscopes utilize micromachined cantilevers with sharp tips and nanometer apertures on top of them [27]. Figure 4 shows the tip with nanometer aperture. Very small particles are aligned and transported by the evanescent light generated above a planar waveguide core [28].

Optical Tweezers

Small particles, typically a few micrometers in size, can be manipulated by strongly focused light [29]. Though optical tweezers based on MEMS are only emerging, the author believes that an integrated MEMS optical tweezer will be very useful for nanotechnologies; it will be possible to handle small particles or even molecules in microchemical systems.

Local Servo Feedback

Sensors and control circuits as well as microactuators can be integrated by micromachining. Therefore, it is possible to detect the motion of the actuator by embedded sensors [30] and to precisely adjust the motion by the local controller; this enables quick and easy servo feedback. Such servo system can be applied to tracking in optical disks, autofocusing, active alignment by a microactuator, stabilizing both optical and mechanical resonance, and compensating thermal drifts.

Hermetic/Vacuum Packaging

Micro electro-mechanical systems devices are small enough to be easily encapsulated in a hermetic or vacuum package. Such packages ensure stable operation and high reliability for scanners and reflective light modulators, for example, the DMD display [20]. The performance of resonant scanners is improved a lot by vacuum packaging [31]. Vacuum packaging is also effective for room-temperature IR cameras in increasing the sensitivity of microbolo-meters [25].

Issues against Future Growth

As discussed earlier, there are many advantages of optical MEMS. However, there are some issues, to hinder wide application of optical MEMS devices. First of all, small mirrors and lenses suffered with diffraction; the diameter of reflected optical beam tends to increase with propagation. Some micromirrors are so thin that it may deform because of internal stress, thermal expansion mismatch between layers, and dynamic loading during high frequency operation. Thick mirrors are more stable but limit operational bandwidth and require more driving force. Therefore, careful optimization is necessary to satisfy application requirements. CAD programs that can handle combined analysis of mechanical, electrical, and optical problems should be developed.

Packaging should be improved for easy interconnection of optical MEMS devices with optical fibers and electrical circuits, and for long life and high reliability. Evaluation methodology of life-time and long-term reliability should also be developed especially for arrayed devices. MEMS foundry service for low cost, flexible, and small quantity manufacturing must be established.

APPLICATIONS OF OPTICAL MEMS

The most promising applications with large impact are devices for optical communication networks [32] such as micromechanical switches, programmable equalizers, attenuators with fast response, wavelength tunable lasers, and filters (Table 3). In addition, applications in information apparatus, namely displays, optical date storage and scanners for printers, barcode readers and sensors, are also of high prospect.

MEMS Optical Attenuator

A fast response programmable attenuator is a useful device in optical communication networks for suppressing changes in the network, equalizing power of WDM optical signals, and compensate attenuation between fiber links with different transmission lengths. MEMS optical attenuators have been developed with various attenuation elements; namely, tunable

Table 3
Opportunities of MEMS in optical communication networks

Optical MEMS	Application in the Network
Large-scale optical matrix switch	Optical cross connect for network reconfiguration
Small-scale optical matrix switch	Optical add-drop multiplexer, back-up module, testing device for networks, LAN router
Programmable equalizer [a]	Equalizer for fiber optical amplifier
Attenuator with fast response [b], [c]	Protection against transient-networks
Tunable laser	Back-up for WDM lasers, wavelength convertor
Tunable filter	Optical add-dropmultiplexer, optical network unit

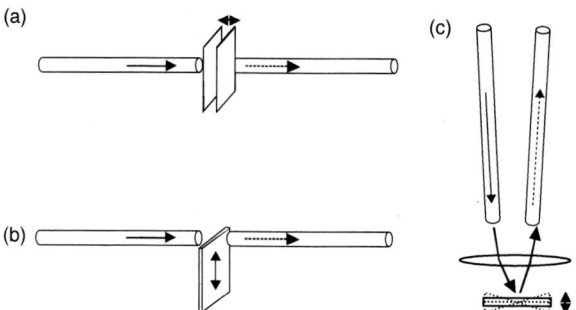

Figure 5 MEMS variable optical attenuators based on (a) Fabry–Perot interferometer, (b) movable shutter, and (c) tilting mirror

Fabry–Perot interferometers [32], movable shutters [33], and tilting/translating mirrors [34]. Schematic representations of these devices are shown in Figure 5.

A 2D Optical Scanner

A 2D optical scanner is a fundamental element for wide varieties of applications such as a laser radar for automobile collision avoidance, a laser range finder, a scanning laser display, a 3D analog beam steering switch [24], and a barcode reader. Mechanical structures to enable two-degree-of-freedom motion include integration of one-degree-of-freedom mirrors [35], multimodal, for example, bending and torsion, vibration of a flexible support [36], and gimbals [37]. Static operation over large angles is the most difficult task. High-frequency resonating vibration of large mirrors is the next difficult one. Many actuation principles are used to generate enough force to achieve high speed operation and large angle of motion. Electrostatic force [35,37], electromagnetic force, piezoelectric force, and magnetostrictive force [36] have been employed.

Fast and precise control of tilting angles are essential for arrayed 2D scanners in a 3D analog beam steering switch. In order to complete the switching operation within tens of milliseconds and to achieve the accuracy, there are three possible strategies for beam position control. Let us discuss the case of electrostatically driven mirrors.

1. Open-loop control, in which predetermined voltage corresponding to the target position are applied to driving electrodes, is fast and easy, but it requires calibration of control voltage for each channel [37], and is susceptible to drift caused by mechanical, electrical, and thermal effects.

2. Close-loop control based on mirror angle, in which tilted angles of the mirror are detected by integrated sensors and are adjusted to desired values, is one of two close-loop strategies. Piezoresistive strain sensors or capacitive distance sensors can detect the angle. The control scheme is efficient to improve the positioning accuracy and to compensate causes of long-term drift. The advantage is that the feedback loop is closed locally; this means signals from integrated sensors are processed by circuits placed close to the mirror and control outputs can be fed to electrodes directly. One of the major features of MEMS, that is, cointegration of sensors, control ciruits, and actuators, is beautifully adopted. It does not, however, overcome problems due to relative position drift between fibers, lenses, and mirrors.

3. Close-loop control based on optical signal at the output, in which the optical signal intensity (or position, if possible) in the output fiber is measured, and mirrors are controlled to maximize it, is the second close-loop strategy. The control variable itself, which is the optical signal intensity in the output fiber, is the same as the final performance criteria.

Therefore, it is the ultimate strategy. In practice, however, it is a very challenging task, because one needs to adjust two-degree-of-freedom in a pair of mirrors based on just one variable. Further, keeping one value at its maximum/minimum is generally a difficult control problem as one also needs to extract some light from the output. The feedback signal must be returned from the output port to the array of mirrors.

REFERENCES

1. Special issue on "Integrated sensors, microactuators, and microsystems (MEMS)," *Proc. IEEE*, 86, 1531–1787 (1998).
2. H. Fujita and H. Toshiyoshi, "Micro-optical devices," in *SPIE Handbook of Microlithography, Micromachining and Microfabrication*, Vol. 2, P. Rai-Choudhury, Ed. (SPIE, Washington, 1997).
3. K. E. Petersen, "Micromechanical light modulator array fabricated on silicon," *Appl. Phys. Lett.* 31, 521 (1977).
4. R. T. Howe, R. S. Muller, K. J. Gabriel, and W. S. N. Trimmer, "Silicon micromechanics: sensors and actuators on a chip," *IEEE Spectrum* 27, 29–35 (1990).
5. K. Petersen, "Silicon as a mechanical material," *Proc. IEEE*, 70, 420–457 (1982).
6. *Proceedings of the 4th International Conference on Solid-State Sensors & Actuators*, Tokyo, Japan, June 2–5, 1987.
7. *Proceedings of the IEEE Micro Robots and Teleoperators Workshop*, Hyannis, MA, November 9–11, 1987.
8. D. S. Gunawan, L.-Y. Lin, and K. S. J. Pister, "Micromachined corner cube reflectors as a communication link," *Sensors Actuators*, A 46–A 47, 580–583 (1995).
9. F. C. M. Van De Pol, D. G. J. Wonnink, M. Elwenspoek, and J. H. J. Fluitman, "A thermo-pneumatic actuation principle for a microminiature pump and other micromechanical devices," *Sensors Actuators*, 17, 139–143 (1989).
10. M. Ataka, A. Omodaka, N. Takeshima, and H. Fujita, "Polyimide bimorph actuators for a ciliary motion system," *J. Microelectromech. Syst.*, 2, 146–150 (1993).
11. T. Higuchi, M. Watanabe, and K. Kudoh, "Precise positioner utilizing rapid deformations of a piezo electric element," *J. Jpn Soc. Precision Eng.* 54, 75–80 (1988).
12. Y.-K. Kim, M. Katsurai, and H. Fujita, "A levitation-type linear synchronous microactuator using the Meissner effect of high-Tc superconductors," *Sensors Actuators*, 29, 143–150 (1991).
13. S. Konishi and H. Fujita, "A conveyance system using air flow based on the conccept of distributed micro motion systems," *J. Microelectromech. Syst.*, 3, 54–58 (1994).
14. M. C. Larson and J. S. Harris, Jr., "Wide and continuous wavelength tuning in a vertical-cavity surface-emitting laser using a micromachined deformable-membrane mirror," *Appl. Phys. Lett.*, 68, 891–893 (1996).
15. R. Sawada, "Integrated optical encoder," in *Proceedings of the Eighth International Conference on Solid-State Sensors & Actuators (TRANSDUCERS '95)*, Vol. 1, Stockholm, Sweden, June 25–29 (1995), pp. 281–284.
16. M. C. Wu, L. Y. Lin, S. S. Lee, and K. S. J. Pister, "Micromachined free-space integrated micro-optics," *Sensors Actuators*, 50, 127–134 (1995).
17. A. Tixier, Y. Mita, S. Oshima, J.-P. Gouy, and H. Fujita, "3D-microsystem packaging for interconnecting electrical, optical, and mechanical microdevices to the external world," in *Proceedings of the IEEE International Conference on Micro Electro Mechanical Systems* (MEMS 2000), Miyazaki, Japan, January 23–27 (2000) pp. 698–703.
18. H. Guckel, "Micro electro-mechanical positioning systems for optical instruments and switches," in *Technical Digest of International Conference on Optical MEMS and Their Applications* (MOEMS 97), Nara, Japan (1997), pp. 229–232.
19. H. Fujita and H. Toshiyoshi, "Micro-optical devices," in *SPIE Handbook of Microlithography, Micromachining and Microfabrication*, Vol. 2, P. Rai-Choudhury, Ed. (SPIE, Washington, 1997).

20. P. F. van Kessel, L. J. Hornbeck, R. Meier, and M. R. Douglass, "A MEMS-based projection display," *Proc. IEEE*, 86, 1687–1704 (1998) (also see http://www.ti.com/dlp/resources/library/configurations.shtml).

21. J. G. D. Su, J. Duparre, L. Fan, P. K. C. Wang, and M. C. Wu, "Micromachined tip-tilt micromirror arrays with large strokes," in *Proceedings of the Third International Conference on Micro Opto Electro Mechanical Systems* (MOEMS 99), Mainz, Germany (1999), p. 39.

22. R. B. Apte, F. S. C. Sandejas, W. C. Banyai, and D. M. Bloom, "Deformable grating light valves for high resolution displays," in *Proceedings of the Solid-State Sensor and Actuator Workshop*, South Carolina, U.S.A., June 13–16 (1994), pp. 1–6.

23. J. E. Ford, V. A. Aksyuk, D. J. Bishop, and J. A. Walker, "Wavelength add-drop switching using tilting micromirrors," *IEEE J. Light. Technol. Lett.*, 17, 904–911 (1999).

24. L. Y. Lin, E. L. Goldstein, and R. W. Tkach, "Free-space micromachined optical switches for optical networking," *IEEE J. Select. Top. Quantum Electron.*, 5, 4–9 (1999).

25. B. E. Cole, R. E. Higashi, and R. A. Wood, "Monolithic two-dimensional arrays of micromachined microstructures for infrared applications," *Proc. IEEE*, 86, 1679–1686 (1998).

26. H. Ukita, Y. Uenishi, and H. Tanaka, "A photomicrodynamic system with a mechanical resonator monolithically integrated with laser diode on gallium arsenide," *Science*, 260, 786–789 (1993).

27. P. N. Minh, T. Ono, and M. Esashi, "A novel fabrication method of the tiny aperture tip on silicon cantilever for near field scanning optical microscopy," in *Technical Digest of IEEE Workshop on Micro Electro Mechanical Systems*, Orland, FL, January (1999), pp. 360–365.

28. S. Kawata and T. Tani, "Optically driven mie particles in an evanescent field along a channeled waveguide," *Opt. Lett.*, 21, 1768–1770 (1996).

29. E. Higurashi, H. Ukita, H. Tanaka, and O. Ohguchi, "Optically induced rotation of anisotropic micro-objects fabricated by surface micromachining," *Appl. Phys. Lett.*, 64, 2209–2210 (1994).

30. M. Ikeda, H. Goto, M. Sakata, and S. Wakabayashi, "Two-dimensional silicon micromachined optical scanner integrated with photo detector and piezoresistor," in *Proceedings of the Eighth International Conference on Solid-State Sensors and Actuators* (TRANSDUCERS 95), Vol. 1, Stockholm, Sweden, June 25–29 (1995), pp. 293–296.

31. A. Garnier, T. Bourouina, H. Fujita, E. Orsier, T. Masuzawa, T. Hiramoto, and J. C. Peuzin, "A fast, robust and simple 2D micro-optical scanner based on contactless magnetostrictive actuation," in *Proceedings of the IEEE International Conference on Micro Electro Mechanical Systems* (MEMS 2000), Miyazaki, Japan, January 23–27 (2000), pp. 715–720.

32. L. Y. Lin and E. L. Goldstein, "Opportunities and challenges for MEMS in lightwave communications," *IEEE J. Select. Top. Quantum Electron.*, 8, 163–172 (2002).

33. C. R. Giles, V. Aksyuk, B. Barber, R. Ruel, L. Sultz, and D. Bishop, "A silicon MEMS optical switch attenuator and its use in lightwave subsystems," *IEEE J. Select. Top. Quantum Electron.*, 5, 18–25 (1999).

34. H. Toshiyoshi, K. Isamoto, A. Morosawa, M. Tei, and H. Fujita, "A 5-volt operated MEMS variable optical attenuator," in *Proceedings of the 12th International Conference on Solid-State Sensors, Actuators and Microsystems* (TRANSDUCERS 03), Boston, MA, June 8–12 (2003), pp. 1768–1775.

35. P. M. Hagelin and O. Solgaard, "Optical raster-scanning display based on surface-micromachined polysilicon mirrors," *IEEE J. Select. Top. Quantum Electron.*, 5, 67–74 (1999).

36. T. Bourouina, G. Reyne, A. Debray, H. Fujita, A. Ludwig, E. Quandt, H. Muro, T. Oki, and A. Asaoka, "Integration of two degree-of-freedom magnetostrictive actuation and piezoresistive detection: application to a two-dimensional optical scanner," *J. Microelectromech. Syst.*, 11, 355–362 (2002).

37. H. Toshiyoshi, W. Piyawattanametha, C.-T. Chan, and M. C. Wu, "Linearization of electrostatically actuated surface micromachined 2D optical scanner," *J. Microelectromech. Syst.*, 10, 205–214 (2001).

MICROLENS

Kenichi Iga

A microlens is defined as a lens with relatively small dimensions ranging from a few microns to several millimeters [1] for its use in microoptics [2]. Wide applications of microlens arrays for arrayed device coupling and parallel optical image processing have been developed [3]. The arrayed microoptics technology would be very helpful for advanced ultra-parallel optical systems. This concept has become important in order to avoid complexities in optical coupling between various optical components. A key component is a microlens and several methods to realize this has been developed.

A graded-index microlens, as shown in Figure 1, was developed along with the production of graded-index multimode fibers by using an ion exchange technique, so-called selfoc lens [2]. The refraction of the rod is prepared to have a distribution that is described as:

$$n^2(r) = n^2(0)[1 - (gr)^2], \tag{1}$$

where $n(0)$ is the index of the center of the rod, and g is the index gradient along the radial direction. It is assumed that the index has an axial symmetry. The principle of lens action is based on the light ray deflection by a gradient of refractive index of the medium as shown in Figure 1. The pitch of sinusoidal ray trajectory L_p is described by the equation:

$$L_p = 2\pi/g. \tag{2}$$

Real and virtual images can be made inside the lens medium.

A distributed-index *planar microlens* array as shown in Figure 2 which is made with a glass substrate by a selective ion exchange technique, was developed since 1980 [3]. The 10×10 cm squared array samples are commercially available.

One of the applications of planar microlens arrays is a self-aligned put-in microconnector as shown in Figure 3 [4]. It consists of a number of microconnectors utilizing the coupling

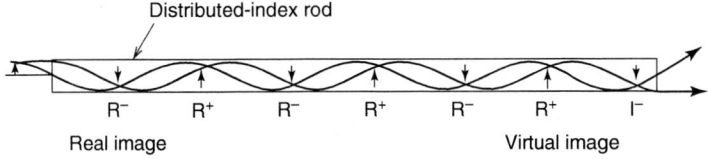

Figure 1 A graded-index microlens and ray trajectory

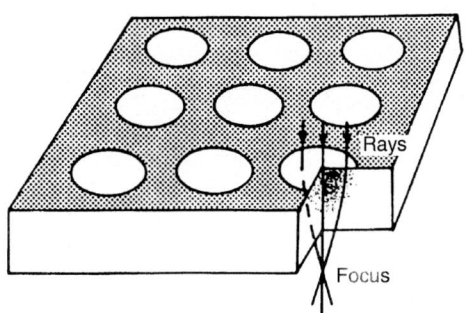

Figure 2 A planar microlens

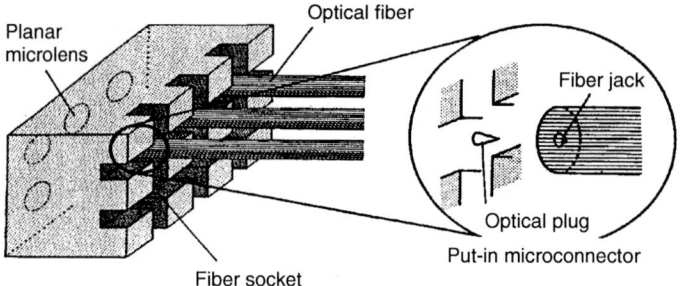

Figure 3   A put-in microconnector

between the optical plugs on the microlens substrate and the fiber jacks made on the core of fibers. The light coupling characteristic of fabricated devices has been examined. The coupling loss of 3 dB has been achieved without any precise optical alignment. The excess loss and deviation will be reduced by the reduction of angle errors in the inserted fibers.

In addition, a variety of fabrication methods for microlens arrays have been developed; for example, etching method, molding, ink-jet, reforming by surface tension, and so on.

REFERENCES

1.   K. Iga, M. Oikawa, S. Misawa, J. Banno, and Y. Kokubun, "Stacked planar optics: an application of the planar microlens," *Appl. Opt.*, 21, 3456, 1982.
2.   T. Uchida, M. Furukawa, I. Kitano, K. Koizumi, and H. Matsumura, "A light-focusing fiberguide," Paper presented at the IEEE Conference on Laser Engineering and App., Washington, D.C., May 1969; *IEEE J. Quantum Electron.*, QE-5, 6, 331, 1969.
3.   M. Oikawa, K. Iga, and T. Sanada, "A distributed index planar microlens made of plastics," *Jpn. J. Appl. Phys.*, 20, L51–L54, 1981.
4.   A. Sasaki, T. Baba, and K. Iga, "Put-in microconnectors for alignment-free coupling of optical fiber arrays," *IEEE Photon. Technol. Lett.*, 4, 908–910, 1992.

MICRO-RING RESONATOR CIRCUIT

Yasuo Kokubun

Ring resonator is a resonator that has two input and output ports as shown in Figure 1. When a ring resonator is coupled to only one busline waveguide with input and output ports, the circuit is called "all-path filter." Microring resonator is a waveguide type ring resonator circuit that has a very small radius in the order of several tens to hundreds micron of magnitude.

Let us suppose, a ring resonator with radius R is coupled to busline and add/drop waveguides as shown in Figure 1. When the propagation constants in the ring resonator and the busline and add/drop waveguides are all the same, the coupling strength of optical field in the coupling region can be expressed by $\cos(\kappa l)$ to the through port and $j \sin(\kappa l)$ to the drop port. Then the transfer function of optical power to the drop port is expressed by

$$T(\lambda) = \frac{|Y|^2}{(1-X)^2 + 4X \sin^2(\phi(\lambda)/2)},\tag{1}$$

where the parameters X, Y, and $\phi(\lambda)$ are given by

$$X = \cos^2(\kappa l)\exp\left(-\frac{\alpha L}{2}\right),\tag{2}$$

$$Y = \sin^2(\kappa l)\exp\left(-\frac{\alpha L}{4}\right),\tag{3}$$

$$\phi(\lambda) = \beta L = \frac{2\pi n_{eq}L}{\lambda}.\tag{4}$$

Here L $(=2\pi R)$ is the path length in the ring, n_{eq} is the equivalent refractive index in the ring and busline waveguides $(=\beta/k_0, k_0$ is the propagation constant in a vacuum), and α is the propagation loss coefficient.

The theoretical drop port response calculated by Eqs. (1)–(4) is shown in Figure 2. The response shape is expressed by so-called Airy function, and is sometimes approximated by Lorentzian function. In this calculation, the propagation loss in the ring waveguide was assumed

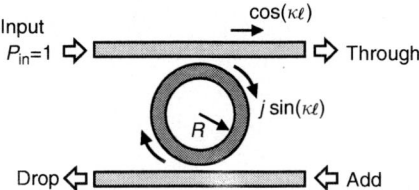

Figure 1 Basic structure of ring resonator circuit

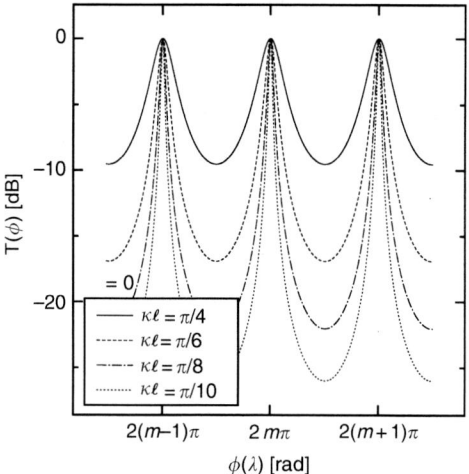

Figure 2 Theoretical drop port spectrum response

to be zero ($\alpha = 0$) and the parameter is the product of coupling coefficient and path length in the coupling region. The peak wavelength is given by the wavelength satisfying $\phi(\lambda) = 2\,m\pi$, where m is the resonance order, and is expressed by

$$\lambda_m = \frac{n_{eq} L}{m}.$$

(5)

On the other hand, the spacing between peak wavelengths, which is called free spectral range (FSR), is given by

$$\mathrm{FSR} = \Delta\lambda = -\frac{\lambda^2}{2n_{eff}\pi R},$$

(6)

where n_{eff} is the effective equivalent index defined by

$$n_{eff} = n_{eq}\left(1 - \frac{\lambda}{n_{eq}} \cdot \frac{dn_{eq}}{d\lambda}\right).$$

(7)

It is seen from Figure 2, the bandwidth (full width at half maximum, FWHM) depends on the coupling strength of pass band and the rejection (cross talk) in the rejection band depend on the coupling strength between the busline waveguide and the ring resonator. In other words, the finesse that is defined by the ratio of FSR to bandwidth (FWHM).

As an example, let us design the FSR to be 20 nm and the bandwidth to be 0.4 nm at the wavelength of 1550 nm, which correspond to the finesse of 50. From Eq. (6), the product $n_{eq}R$ is calculated to be 19.12 μm. When the effective equivalent index is 1.7, using a high index contrast waveguide, the required ring radius is obtained to be 11.2 μm. Therefore, to achieve a wide FSR (>10 nm), a small ring radius (<20 μm) is required. On the other hand, the coupling strength required to achieve the finess of 50 is obtained to be 6% ($\sin^2(\kappa l) = 0.06$). In the vertically coupled microring resonator (VCMRR) shown in Figure 3, the propagation constants in

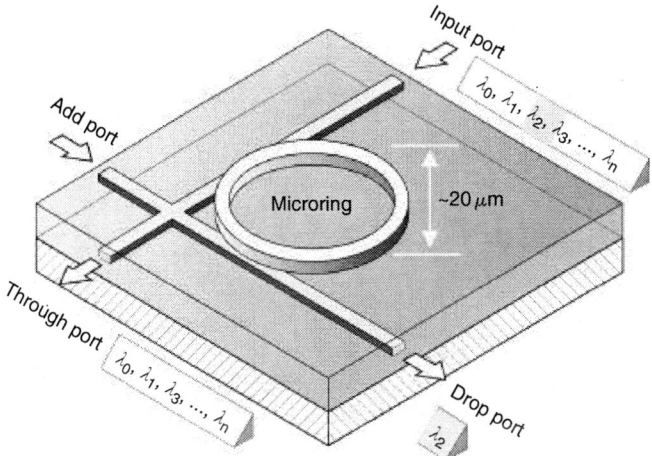

Figure 3   Vertically coupled microring resonator

the busline waveguide and the microring are not exactly equal to each other. Consequently, the coupling strength cannot reach 100%. However, although the coupling is a kind of asymmetric coupling, its strength that ranges from several to 15% can be easily obtained by controlling the thickness of buffer layer, and so the achievable finess is larger than 50.

Since the first proposal of optical waveguide ring resonator [1], it has been utilized as the cavity of ring laser and some sensor applications [2]. However, the ring radius was in the order of millimeter, and so the FSR was too small to be used as the wavelength filter for dense WDM systems. In 1997, an MIT group proposed the microring resonator for an add/drop filter, of which ring radius is in the order of several tens of micron [3]. To realize this tiny ring radius, a high index contrast waveguide of which index difference is larger than 20% is needed, and in late 1997 and 1998, microring resonator filter was demonstrated using a semiconductor waveguide with air cladding [4–6]. However, since the structure of this device is planar, where the ring and the busline waveguides were formed in the same layer, a very precise fabrication technique was needed to form a very fine gap between the busline and the microring.

On the other hand, to solve this problem, a VCMRR [7], as shown in Figure 3, was proposed and demonstrated in 1999. Also, in the same year the VCMRR with cross grid busline topology was demonstrated [8]. This filter element has the following advantages:

1. Owing to the stacked configuration, the upper and the lower waveguides play different roles, that is, the lower buried channel waveguides serve as input/output busline waveguides, while the ring functions as the frequency selective element.
2. Owing to the high index contrast in the upper layer, a ultra-compact ring resonator with very small radius of 10–20 μm is possible, which exhibit the bandwidth of 0.1–1.0 nm and the FSR of 10–25 nm. This narrow bandwidth leads to the demonstration of high Q (1500–15,000) micro-ring resonator filters.
3. Owing to the ultra-compact element and the cross grid configuration, a dense integration up to 10^4 to 10^5 devices/cm^2 is possible.
4. The coupling strength between the ring and the busline waveguides can be controlled more precisely than the lateral coupling, because the vertical separation is obtained by the well-controlled deposition, rather than etching of fine gaps.
5. In the vertical coupling, the fabrication tolerance of the lateral misalignment between upper microring and lower busline waveguide is relaxed, because the overlap of field profile is

the maximum at the point of zero offset, and is much less sensitive to the offset than in lateral coupling.

6. The cross grid topology makes it possible to adapt the same configuration to many coupled ring topologies, such as the series coupled, parallel coupled, and cascaded topologies.

REFERENCES

1. E. A. J. Marcatili, "Bends in optical dielectric guides," *Bell Syst. Technol. J.*, 48, 2103–2132, 1969.
2. S. Suzuki, K. Shuto, and Y. Hibino, "Integrated optic ring resonators with two stacked layers of silica waveguides on Si," *IEEE Photon. Technol. Lett.*, 4, 1256–1258, 1992.
3. B. E. Little, S. T. Chu, H. A. Haus, J. Foresi, and J.-P. Laine, "Microring resonator channel dropping filters," *J. Lightwave Technol.*, 15, 998–1005, 1997.
4. S. C. Hagness, D. Rafizadeh, S. T. Ho, and A. Taflove, "FDTD microcavity simulations: design and experimental realization of waveguide-coupled single-mode ring and whispering-gallery-mode disk resonators," *J. Lightwave Technol.*, 15, 2154–2165, 1997.
5. D. Rafizadeh, J. P. Zhang, S. C. Hagness, A. Taflove, K. A. Stair, S. T. Ho, and R. C. Tiverio, "Waveguide-coupled AlGaAs/GaAs microcavity ring and disk resonators with high finesse and 21.6 nm free spectral range," *Opt. Lett.*, 22, 1244–1246, 1997.
6. B. E. Little, J. S. Foresi, G. Steinmeyer, E. R. Thoen, S. T. Chu, H. A. Haus, E. P. Ippen, L. C. Kimerling, and W. Greene, "Ultra-compact Si/SiO$_2$ microring resonator optical channel dropping filters," *IEEE Photon. Technol. Lett.*, 10, 549–551, 1998.
7. B. E. Little, S. T. Chu, W. Pan, D. Ripin, T. Kaneko, Y. Kokubun, and E. Ippen, "Vertically coupled glass microring resonator channel dropping filters," *IEEE Photon. Technol. Lett.*, 11, 215–217, 1999.
8. S. T. Chu, B. E. Little, W. Pan, T. Kaneko, S. Sato, and Y. Kokubun, "An eight-channel add-drop filter using vertically coupled microring resonators over a cross grid," *IEEE Photon. Technol. Lett.*, 11, 691–693, 1999.

MODE SCRAMBLER

Yasuo Kokubun

The branch ratio of multimode waveguide branching depends strongly on the mode excitation distribution. As a result, when the multimode waveguide branching device is connected to optical fibers, the branching ratio fluctuates depending on the temporal change of modal distribution in the fiber. In addition, it is difficult to design the branching ratio of cascaded branching waveguides. Mode scrambler is used to obtain a fixed modal distribution to stabilize the branching ratio of multimode waveguide branching devices. A zigzag waveguide pattern is used for step index multimode waveguide and a cascaded tapered waveguide is used for distributed index multimode waveguide.

MODULATION LIMIT OF SEMICONDUCTOR LASERS

Fumio Koyama

The rapid increase in network traffics has made high capacity optical communication systems including short reach systems crucial. For this purpose, high speed semiconductor lasers have been developed. Direct modulation of semiconductor lasers has been used for simple configuration of transmitter modules, while the frequency chirp is caused with intensity modulation. The light output of semiconductor lasers is modulated with the change of injection current as shown in Figure 1. In order to avoid turn-on delay time, bias injection current is typically set either above the threshold or just below the threshold. The direct modulation characteristics of semiconductor lasers can be examined by using rate equations. If we assume small signal modulation, we are able to obtain an analytic expression of amplitude modulation responses. Ikegami and Suematsu reported the existence of resonance-like phenomenon in directly modulated semiconductor lasers [1]. Under small-signal analysis of the rate equations, we obtain the following expression of the relaxation oscillation frequency.

$$f_r = \frac{1}{2\pi}\sqrt{\frac{\xi g S}{\tau_p}} = \frac{1}{2\pi}\sqrt{\frac{g P_{\text{out}}}{\eta_d h\omega V_m}}. \tag{1}$$

Here, we define as follows: S: photon density, g: differential gain, ξ: optical confinement factor, τ_p: photon lifetime, P_{out}: output power, η_d: differential quantum efficiency, V_m: mode volume, and $h\omega$: photon energy.

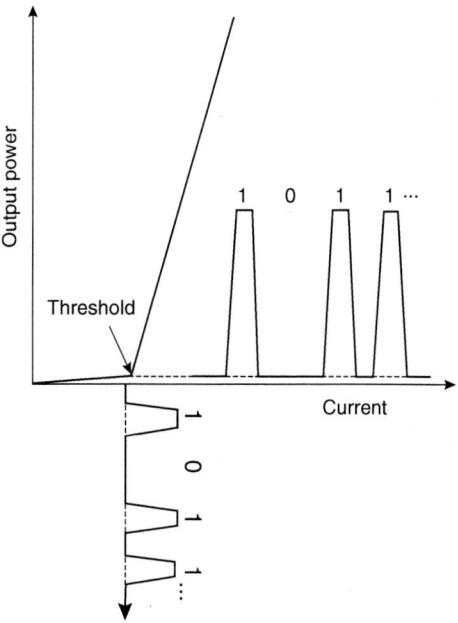

Figure 1   Schematic of direct modulation of semiconductor lasers

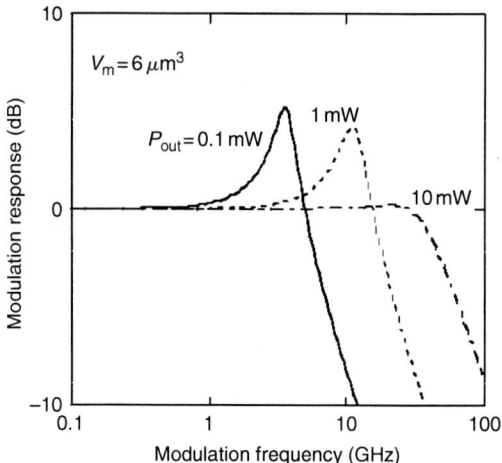

Figure 2   Calculated small-signal modulation response

Figure 2 shows the calculated small signal modulation response for surface emitting lasers with a small mode volume. By increasing the output power, the relaxation oscillation frequency is increased in proportion to the square of the output power divided by the volume of a laser cavity. Thus, the modulation speed can be determined by the power density stored in a laser cavity. The 3 dB-modulation bandwidth under small signal modulation can be given by

$$f_{3dB} = \sqrt{1+\sqrt{2}}\, f_r \cong 1.55 f_r. \tag{2}$$

In order to increase the relaxation oscillation frequency, it is effective to increase the differential gain and to reduce photon lifetime. For this purpose, short cavity lasers and modulation doping quantum-well lasers have been studied [2] and a modulation bandwidth of over 30 GHz in multiple quantum-well lasers has been demonstrated [3,4]. Large signal modulation up to 40 Gbps was also demonstrated [5]. The advance in laser bandwidth makes direct modulation a possible scheme for high-speed applications. There are certain limitations for the modulation bandwidth with increasing the output power of semiconductor lasers. The following issues are such limitations of practical semiconductor lasers:

1. Thermal heating
2. Damping due to nonlinear gain
3. Parasitic capacitance

As the output power increases, the relaxation oscillation frequency also increases and suppresses the resonance-peak simultaneously. This is due to the damping effect [6]. The damping factor Γ is expressed by

$$\Gamma = \Gamma_0 + K \cdot f_r^2, \tag{3}$$

where K is called the K-factor [6], which includes the nonlinear gain coefficient. Several nonlinear effects such as spatial hole burning, spectral hole burning, and carrier heating may contribute to the K-factor. Thus, the maximum modulation bandwidth is limited by the K-factor; however, is not practically achievable due to the constrains on the applied bias current level.

Here, we discuss on how to exceed the limit of a relaxation oscillation frequency. The use of an optical external feedback has been studied for increasing the modulation bandwidth [7,8]. In addition, the extension of modulation bandwidth in an external cavity DBR laser beyond a relaxation frequency limit has been predicted [9]. The injection-locking technique offers another possibility to increase the modulation bandwidth for direct modulation [10]. Recently, it was estimated that the modulation bandwidth can be extended to reach 50 GHz under strong injection-locking scheme.

A monolithic external cavity surface emitting laser (VCSEL) was proposed recently, for increasing the modulation bandwidth [11], in which the backside of a substrate can be used for external optical feedback. A multilayer-coated concave mirror can be formed on the backside surface of a substrate. Part of the optical output from the VCSEL is injected to the laser cavity after one round trip of the external cavity. The single-mode condition of external cavity lasers can be kept, which is dependent on an external mirror reflectivity and cavity length. The phase of external feedback light is an important parameter for modulation characteristics. The small-signal modulation response with an external cavity length of about 300 μm and an external mirror reflectivity of 0.2 was carried out. The calculated 3 dB modulation-bandwidth could be increased from 20 to 32 GHz with out-of-phase coupling of external light. That is attributed to an effective loss modulation induced by optical feedback, which is a different phenomenon from Reference 9. Figure 3 shows the calculated 3 dB-modulation bandwidth versus the relaxation frequency for different nonlinear gain parameters. The modulation bandwidth can be enhanced by 50% even for high relaxation frequencies and this leads us to exceed the limit of a damping effect as shown in Figure 3. If the relaxation oscillation frequency is 20 GHz without optical feedback, the modulation bandwidth can be enhanced to be over 40 GHz.

Further, the simulation of NRZ quasi-random large signal modulation is shown, where the bit rate is 40 Gbpsec. The calculated results of eye patterns of directly modulated VCSELs with and without optical feedback are illustrated in Figure 4(a) and 4(b), respectively. The extinction ratio is 10 dB and the pattern-effect can be avoided with optical feedback. Clear eye opening can be expected with the help of optical feedback.

Iga investigated the limit of modulation speed of semiconductor lasers for various modulation of parameters such as the laser gain coefficient, recombination lifetime of carriers, and the

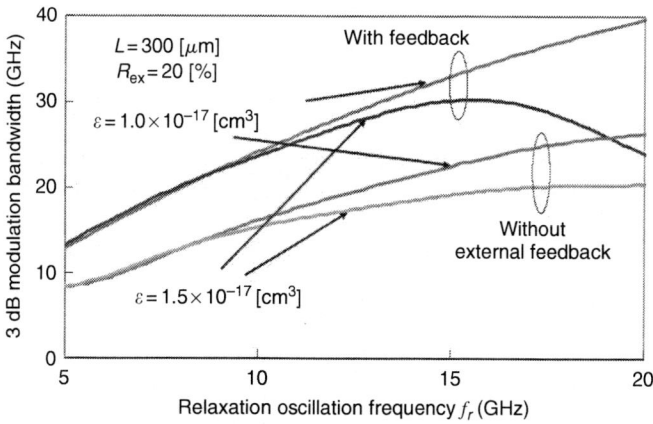

Figure 3   Calculated 3 dB-modulation bandwidth vs. relaxation oscillation frequency with optical feedback (After T. Ota, T. Uchida, M. Arai, T. Kondo, and F. Koyama, in *Proceedings of the Ninth OptoElectronics and Communications Conference*, OECC/COIN2004, Kanagawa, Japan, 15E1-4, July 12–16, 2004. With permission.)

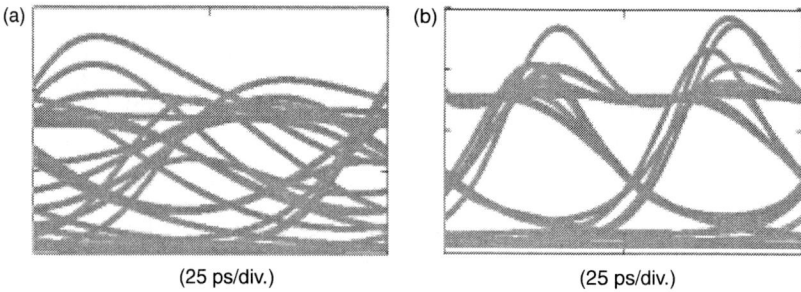

(25 ps/div.)   (25 ps/div.)

Figure 4   Calculated 40 Gbps NRZ eye patterns (a) with and (b) without external optical feedback (After T. Ota, T. Uchida, M. Arai, T. Kondo, and F. Koyama, in *Proceedings of the Ninth OptoElectronics and Communications Conference*, OECC/COIN2004, Kanagawa, Japan, 15E1-4, July 12–16, 2004. With permission.)

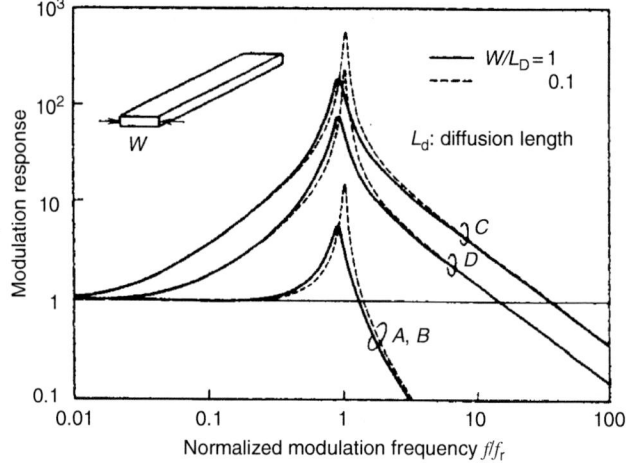

Figure 5   Calculated modulation response for various parametric modulation schemes. A: Injection modulation, B: Carrier lifetime modulation, C: Gain modulation, and D: Photon lifetime modulation (After K. Iga, *IEICE Jpn.*, 68, 471–420, 1985. With permission.)

cavity Q values in comparison with the conventional direct modulation scheme. Figure 5 shows the calculated small-signal modulation response versus the modulation frequency for various modulation schemes. The maximum modulation frequency for the gain and Q value modulation is enhanced to be ten times larger than the relaxation oscillation frequency.

REFERENCES

1.   T. Ikegami and Y. Suematsu, "Resonance-like characteristics of the direct modulation of a junction laser," *Proc. IEEE*, 55, 22–24, 1967.
2.   K. Uomi, "Modulation-doped multi-quantum well lasers," *J. Appl. Phys.*, 29, 81–87, 1990.
3.   J. D. Ralston, S. Weisser, K. Eisele, R. E. Sah, E. C. Larkins, J. Rosenzweig, J. Fleissner, and K. Bender, "Low-bias-current direct modulation up to 33 GHz in InGaAs/GaAs/AlGaAs pseudomorphic MQW ridge-waveguide lasers," *IEEE Photon. Technol. Lett.*, 6, 1076–1079, 1994.

4. Y. Matsui, H. Murai, S. Arahira, S. Kutsuzawa, and Y. Ogawa, "30-GHz bandwidth 1.55-μm strained-compensated InGaAlAs–InGaAsP MQW laser," *IEEE Photon. Technol. Lett.*, 9, 25–27, 1997.

5. K. Sato, "Semiconductor light sources for 40-Gb/s transmission system," *J. Lightwave Technol.*, 20, 2035–2043, 2002.

6. R. Olshansky, P. Hill, V. Lanzisera, and W. Powazinik, "Frequency response of 1.3 μm InGaAsP high speed semiconductor lasers," *IEEE J. Quantum Electron.*, 23, 1410–1418, 1987.

7. K. Y. Lau and A. Yariv, "Ultra-high speed semiconductor lasers," *IEEE J. Quantum Electron.*, 21, 121–138, 1985.

8. K. Vahala, J. Pasiaski, and A. Yariv, "Observation of modulation speed enhancement, frequency modulation suppression and phase noise reduction by detuned loading in a coupled-cavity semiconductor laser," *Appl. Phys. Lett.*, 46, 1025–1027, 1985.

9. O. Kjebon, R. Schatz, S. Lourdudoss, S. Nilsson, B. Stalnacke, and L. Backbom, "30 GHz direct modulation bandwidth in detuned loaded InGaAsP DBR lasers at 1.55 μm wavelength," *Electron. Lett.*, 33, 488–489, 1997.

10. L. Chrostowski, C. Chih-Hao, and C. J. Chang-Hasnain, "Enhancement of dynamic range in 1.55 μm VCSELs using injection locking," *IEEE Photon. Technol. Lett.*, 15, 490–492, 2003.

11. T. Ota, T. Uchida, M. Arai, T. Kondo, and F. Koyama, "Enhanced modulation bandwidth of surface-emitting laser with external optical feedback," in *Proceedings of the Ninth OptoElectronics and Communications Conference,* OECC/COIN2004, Kanagawa, Japan, 15E1-4, July 12–16, 2004.

12. K. Iga, "Modulation limit of semiconductor lasers by some parametric modulation schemes," *IEICE Jpn.*, 68, 417–420, 1985.

MULTI-MODE INTERFERENCE DEVICES

Katsuyuki Utaka

INTRODUCTION

So far, most of the optical devices are believed to exhibit high performances by adopting single-mode waveguide configurations; thanks for stable operation without unwanted mode couplings and interferences, large bandwidth and so on. On the other hand, multi-mode waveguide devices have shown their advantages for only large fabrication tolerance, easy optical coupling and, resultantly, low-cost devices. These understandings are mostly true, but it is noteworthy that positive utilizations of mode interference in a multi-mode waveguide were also proposed to manipulate image formation by Bryngdahl and Ulrich [1,2]. And multi-mode interference (MMI) waveguides have been spotlighted as compact functional optical devices, and various optical devices such as couplers, combiners, splitters and optical switches have been proposed and demonstrated for higher performances [3–9]. The theoretical configurations have also been established [10,11]. The main advantage of MMI waveguide devices compared to other similar waveguide devices is rather short coupling region as is not with single-mode waveguide couplers, which need multi-staged configuration in some cases and adiabatic mode-adjusting regions, as for the Y-junctions. Due to the efficiency of image transformation, the MMI waveguide can shorten the device length.

In this section, the operational principle, theoretical treatment [10] and fundamental characteristics of mode interference in the multi-mode interference (MMI) waveguides are described. Device applications of the MMI waveguides are also introduced.

FUNDAMENTAL THEORY

Beat Length

The main feature of the MMI waveguide is that it supports plural eigen modes, and they interfere along the propagation to show peculiar lateral field distribution variation depending on the propagated distances. As a result, the total field distribution is self-imaged at periodic distances, which are determined by the smallest common multiples of the propagation constants of eigen modes. These field variation behaviors in the MMI waveguide can be analyzed by several methods, such as eigen mode and beam propagation analyses. Due to their superior utility, beam propagation analysis is widely used. Wide-angle beam propagation method (WA-BPM) is quite effective for the application of rather wide waveguide analysis such as MMI waveguides, in which the Fresnel approximation cannot be assumed anymore, as long as reflection at the ends of the MMI waveguide can be neglected [12]. If the existence of end reflection has to be taken into account, rigorous simulation by finite-difference time-domain (FD-TD) analysis is necessary, which needs tremendous CPU time and memory.

The theoretical model of an MMI waveguide is shown in Figure 1. Here, we define the parameters used, as follows. Effective refractive indices of the MMI waveguide region and its cladding region are n_1 and n_2, respectively; widths of the MMI waveguide and access waveguides are W and w, respectively; and an offset between the center of the access waveguide and the edge of the MMI waveguide is s. In order to evaluate the propagation and output characteristics of the MMI waveguide, a beat length of the two lowest-order guided modes L_π is a very important parameter, and its derivation will be explained first [10].

The dispersion relation is expressed as

$$k_{yv}^2 + \beta_v^2 = k_0^2 n_1^2, \tag{1}$$

where

$$k_0 = 2\pi / \lambda_0, \tag{2}$$

$$k_{yv} = (v+1)\pi / W_{ev}. \tag{3}$$

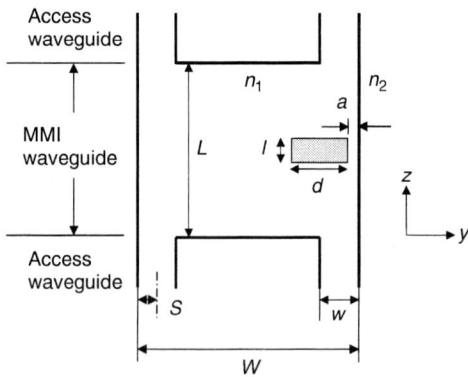

Figure 1 Theoretical model of MMI waveguide and MIPS-P

Here, W_{ev} is an effective MMI waveguide width for νth mode which an evanescent field out of the MMI waveguide is taken into account. In general, it depends on the lateral mode number, but it is assumed to be almost the same as that of the fundamental mode, as $W_{ev} \sim W_{e0} = W_e$. From Eqs. (1) to (3) and assuming $k_{yv}^2 \ll k_0^2 n_1^2$,

$$\beta_\nu \sim k_0 n_1 - (\nu + 1)^2 \pi \lambda_0 / 4 n_1 W_e^2. \tag{4}$$

Here, L_π is defined as a beat length of the two lowest-order modes, and is derived as

$$L_\pi = \pi / (\beta_0 - \beta_1) \sim 4 n_1 W_e^2 / 3 \lambda_0. \tag{5}$$

Optimal lengths of the MMI waveguides for various applications such as a connector, a switch, a splitter and a combiner are designed by taking the beat length into consideration, and that is why the beat length is one of the very important factors for characterization of MMI waveguides.

Fields in the MMI Waveguide

Fields in the MMI waveguide are originally excited by an input field, and propagate with each propagation constant given by Eq. (4). The field distribution at an input end of the MMI waveguide is given by

$$\psi(y,0) = \Sigma c_\nu \phi_\nu(y), \tag{6}$$

where c_ν is a field excitation coefficient, and at a distance z

$$\psi(y,z) = \Sigma c_\nu \phi_\nu(y) \exp\{j(\omega t - \beta_\nu z)\}.$$

It can be rewritten by taking relative phases of high-order modes to that of the fundamental mode as

$$\psi(y,z) = \Sigma c_\nu \phi_\nu(y) \exp\{j(\beta_0 - \beta_\nu)z\}. \tag{7}$$

From Eq. (4), the phase term is rewritten for the MMI length of L, as

$$\psi(y,z) = \Sigma c_\nu \phi_\nu(y) \exp\{j\nu(\nu+2)L/3L_\pi\}. \tag{8}$$

The field distribution at an output end of the MMI waveguide can be reconstructed as single self-images of the input field distribution depending on whether the phase terms are all $2m\pi$ or alternately $2m\pi$ and $(2m-1)\pi$. An integer number of the term $\nu(\nu+2)$ is very important for specifying the interference characteristics of the MMI waveguide, and two kinds of the interference behaviors are classified depending on which eigen modes are excited in the MMI waveguide, that is, general interference and restricted interference.

General interference

The case that there is no restriction to excited eigen modes gives "general interference," in which, in general, all of the eigen modes may propagate in the MMI waveguide. From Eq. (8)

and its subsequent discussion, the reconstructed field distribution is the same as that of the input one for the phase term of $2m\pi$, that is, a direct single-image, and the inverted one for the phase term of $(2m - 1)\pi$, that is, a mirrored single-image. Therefore, these single self-images appear at the positions of every $3L_\pi$, that is,

$$L = p(3L_\pi), \tag{9}$$

where p is an integer, and an even and an odd p give a direct and a mirrored single-image, respectively.

It is noted that at the positions given by

$$L = 3L_\pi / N \tag{10}$$

with $N = 2, 3, \ldots$, multi-fold images are formed, where the image number is N. Therefore, if the length is chosen to be $L = 3L_\pi/2$, that is, $N = 2$, the output light distribution gives two peaks, and such an MMI waveguide can be used as a 3 dB power splitter with a 2×2 MMI structure if access output waveguides are formed to couple each peak into them. Figure 2 shows the calculated light propagation behavior and output distribution of the MMI waveguide whose width is 12 μm. The calculation was done by the wide-angle finite difference beam propagation method (FD-BPM) [12]. Throughout the calculations described hereafter, the width of the access waveguide and an offset of the access waveguide center are $\omega = 2$ μm and $s = 1$ μm, respectively. The length is depicted for the case of $L = 3L_\pi$, and the result clearly indicates that the output is extracted from the cross port, as shown in the figure, following the result of Eq. (9). It is also evident that two peaks appear at the position of $L = 3L_\pi/2$, that is, for the case of $N = 2$ in Eq. (10).

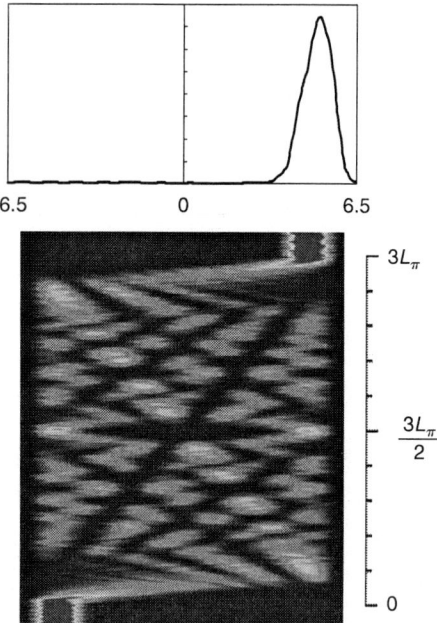

Figure 2   Light propagation behavior and output distribution of general interference for an MMI length corresponding to $3L_\pi$

Restricted interference

In the case that excited modes in the MMI waveguide by the input light are restricted, its interference behavior is different from the one for the general interference to exhibit shorter periodicity of self-imaging. This is called "restricted interference."

From Eq. (8), single-imaging is attained at $L = p(L_\pi)$, where p is 0, 1, 2, ..., if some vth modes are not excited, that is, $c_v = 0$, for $v = 2, 5, 8, ...,$ since $v(v + 2)$ becomes three times multiple. This means that the periodicity for the single-imaging renders one-third of the general interference and more compact devices can be realized. This excited-mode restriction can be attained by forming the input access waveguide at the position such that $s = W/3$. The light propagation behavior of this restricted interference is drawn in Figure 3. Three periods during the length of $L = 3L_\pi$ are evident.

When an access waveguide is formed at the center of the MMI waveguide width, only symmetric modes can be excited, which is called "symmetric interference." This makes $c_v = 0$ for odd v, and consequently the single-imaging periodicity is reduced to $L = p(3L_\pi/4)$, where p is 0, 1, 2, ..., four times shorter than that of the general interference. Figure 4 shows the light propagation behavior for the center input of the MMI waveguide to show four periods of self-imaging.

STRUCTURAL DEPENDENCE OF THE CHARACTERISTICS

MMI Length and Width

Since MMI waveguide devices utilize mode interference among plural transverse modes, it is very important to clearly define the modes that are involved in such mode interference. From

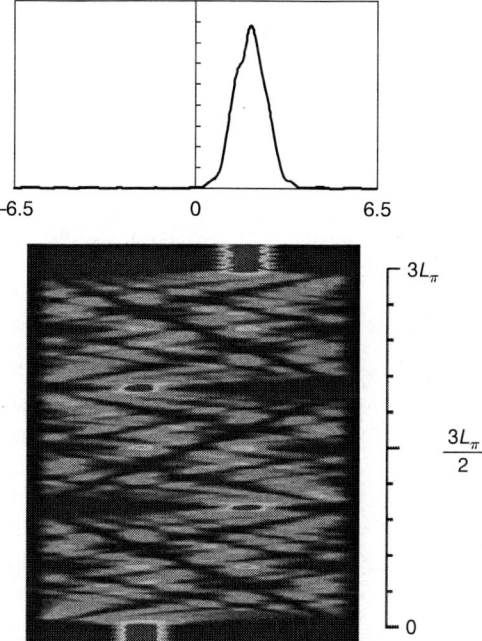

Figure 3   Light propagation behavior and output distribution of restricted interference for an MMI length corresponding to $3L_\pi$

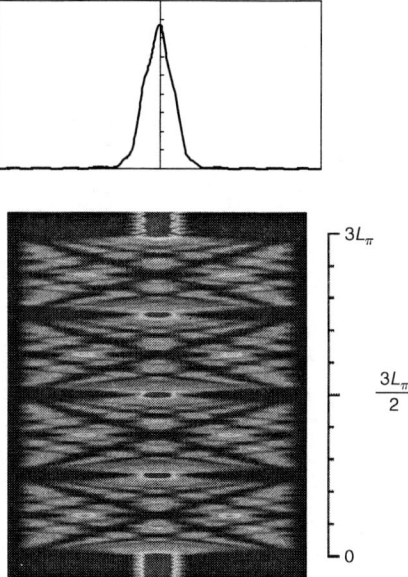

Figure 4 Light propagation behavior and output distribution of symmetric interference for an MMI length corresponding to $3L_\pi$

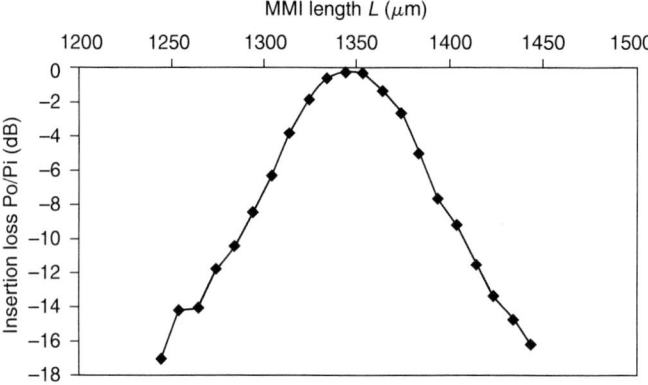

Figure 5 Insertion loss as a function of MMI length

this point of view, dependence of the output characteristics of the MMI waveguide on its structural parameters is discussed. The device output characteristics can be evaluated by the beat length as a parameter, and therefore, the most important equation we need to think about first is Eq. (5), $L_\pi = 4n_1 W_e^2/3\lambda_0$. For the application of a 2 × 2 coupler or an optical switch, they need an MMI length of $3L_\pi$. The calculated insertion loss is shown in Figure 5 as a function of an MMI length with an MMI width of 12 μm. An optimum MMI length is known to be 1344 μm, and tolerance in length for 1 dB loss is about 30 μm, which is long enough to control in fabrication process.

On the other hand, the insertion loss as a function of an MMI width is shown in Figure 6 with an MMI length of 1344 μm. In order to keep a low insertion loss below 1 dB, very small tolerance in MM width (within 0.2 μm) is allowed, as long as the MMI length is fixed. These results indicate the necessity of critical control of an MMI width.

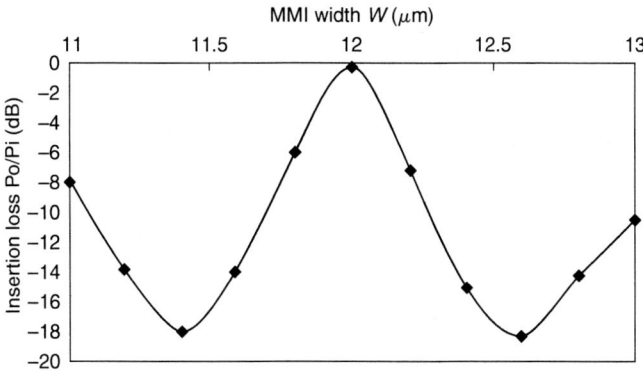

Figure 6   Insertion loss as a function of MMI width

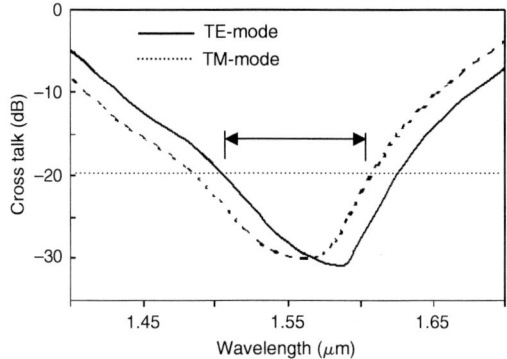

Figure 7   Cross talks as a function of wavelength for TE and TM modes

Wavelength and Polarization

Figure 7 shows the wavelength dependence of cross talks of an MMI waveguide with a length of $3L_\pi$ for TE and TM modes. The wavelength dependences come from that the beat length is dependent on a wavelength, as given by Eq. (5). Shorter-wavelength shift that is characteristic for the TM mode is attributed to a smaller equivalent refractive index of the MMI waveguide. As a result, an effective polarization-independent bandwidth of the MMI waveguide under a criterion of the cross talk below -20 dB is wide of about 70 nm, which can be increased by improving the waveguide structure for less polarization-independence.

DEVICE APPLICATIONS

1 × N Power Splitter

The most useful application of the MMI waveguide is $1 \times N$ power splitters. As was explained, in the cases of an MMI waveguide length of $L = 3L_\pi/N$ for general interference, $L = L_\pi/N$ forrestricted interference, and moreover, $L = (L_\pi/4)/N$ for symmetric interference, multi-spotted light distribution, thus almost equally-divided N outputs are obtained if output access waveguides are formed at appropriate positions. Especially, by utilizing symmetric interference,

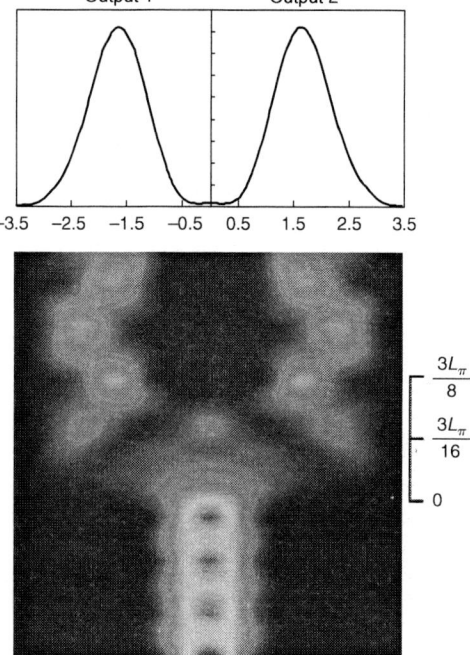

Figure 8   Characteristics of 1 × 2 power splitter

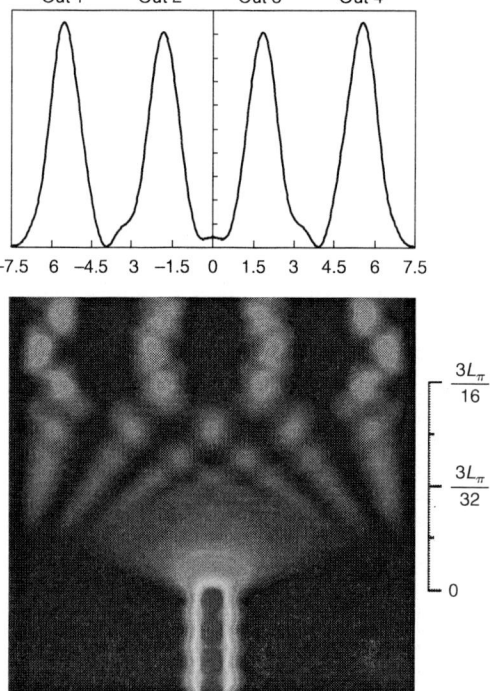

Figure 9   Characteristics of 1 × 4 power splitter

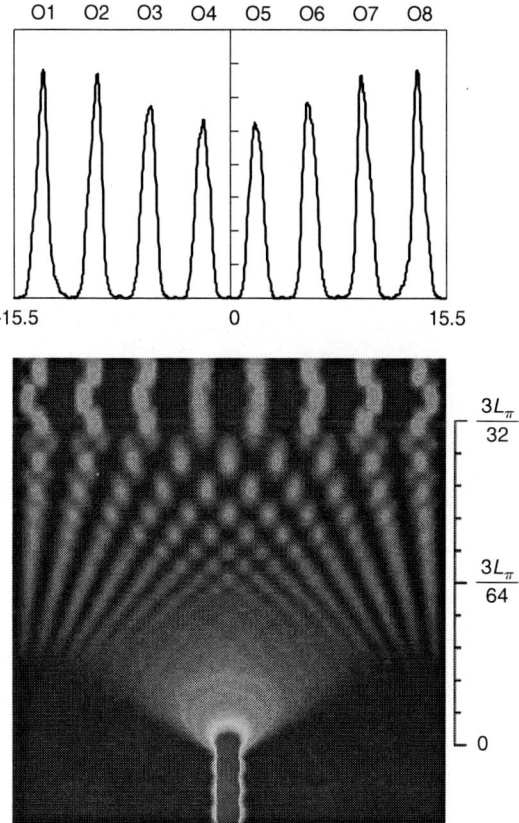

Figure 10 Characteristics of 1 × 8 power splitter

very compact and high performance power splitters can be realized due to high symmetry of the light field distribution in the MMI waveguide. Characteristics of 1 × 2, 1 × 4, and 1 × 8 power splitters are shown in Figures 8 to 10, respectively. Respective device lengths are $L = 3L_\pi/8, 3L_\pi/16$, and $3L_\pi/32$, and due to the necessity of a wider MMI width for larger N, actual MMI lengths are 46, 113, and 248 μm with MMI widths of 6, 14, and 30 μm for $N = 2$, 4, and 8, respectively. Probably due to the effect of evanescent field of the MMI waveguide, there is a tendency that output intensities for outer spots are a little larger than those of inner ones, but clear divided-spots appear at the output end with small excess losses. These results exhibit the effectiveness of symmetric-interference MMI waveguide as power splitters.

$N \times 1$ **Power Combiner**

Power combiners are important devices for light collection in arrayed waveguide devices and integrated light sources. The symmetric interference is again effective as power combiners as the opposite phenomena of power splitters. The calculated results are shown in Figures 11 to 13 for $N = 2$, 4, and 8, respectively. It is liable that the inputs into outer access waveguides show worse output intensity distributions, which lead to relatively larger losses, because of high-order eigen-mode excitation, but by designing an appropriately wider MMI waveguide width the uniformity of the intensity distribution can be much improved. The length of an 8 × 1 MMI power

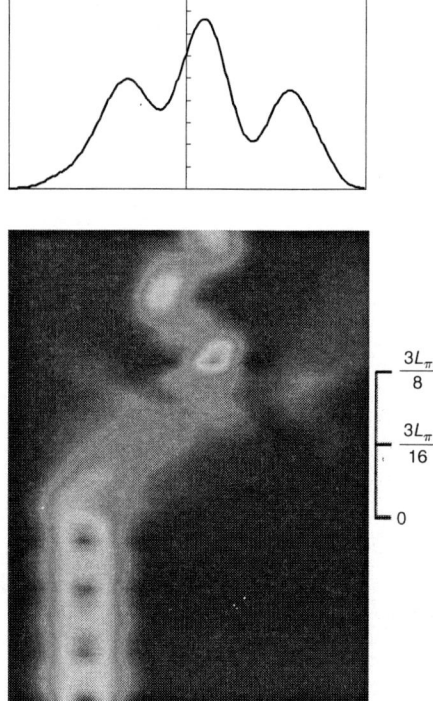

Figure 11   Characteristics of 2 × 1 power combiner

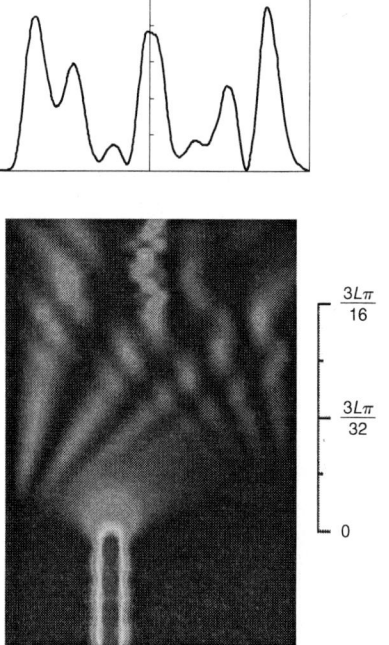

Figure 12   Characteristics of 4 × 1 power combiner

combiner is as small as 248 μm, and very compact devices can be provided using MMI waveguides.

$N \times N$ Coupler

$N \times N$ structures can be applied to a coupler or a switch by introducing a modulation region. The light propagation in the MMI waveguide for an $N \times N$ structure obeys general interference, as was explained above. Therefore, the length for a single imaging is given by $3L_\pi$. If it is applied to a power splitter and a power combiner, the device length may be chosen to be $3L_\pi/N$, where the input power is divided into $1/N$, as is expressed by Eq. (10). Among some applications, a 2×2 structure may be the most useful one, such as a power splitter and a combiner in a Mach–Zehnder interferometer.

2×2 Multimode Interferometer Optical Switch

Heretofore, passive MMI waveguides have been explained, and by introducing the mechanism of refractive-index modulation an optical switching device can be realized. Multi-mode interference photonic switch with partial index-modulation region (MIPS-P) [13] is such a device, and the structural model is shown in Figure 1. A compact 2×2 optical switch can be realized with MMI width and length of 12 and 1400 μm, respectively, whose relation corresponds to $3L_\pi$ for general interference. The refractive index modulation region with a length and a width of l and d, respectively, is formed at one side of the middle position of the MMI waveguide.

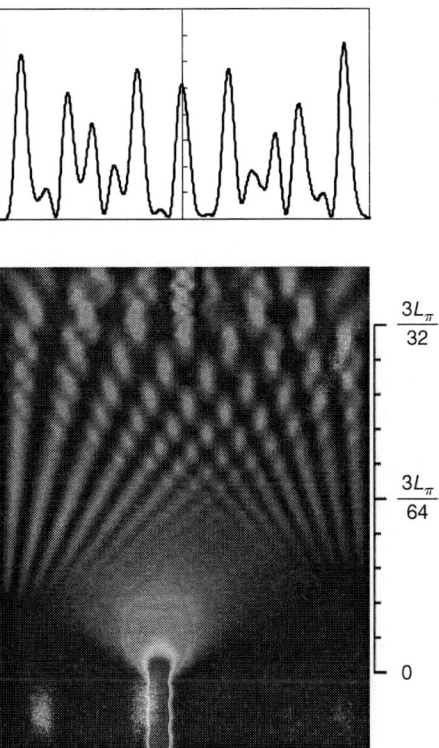

Figure 13   Characteristics of 8 × 1 power combiner

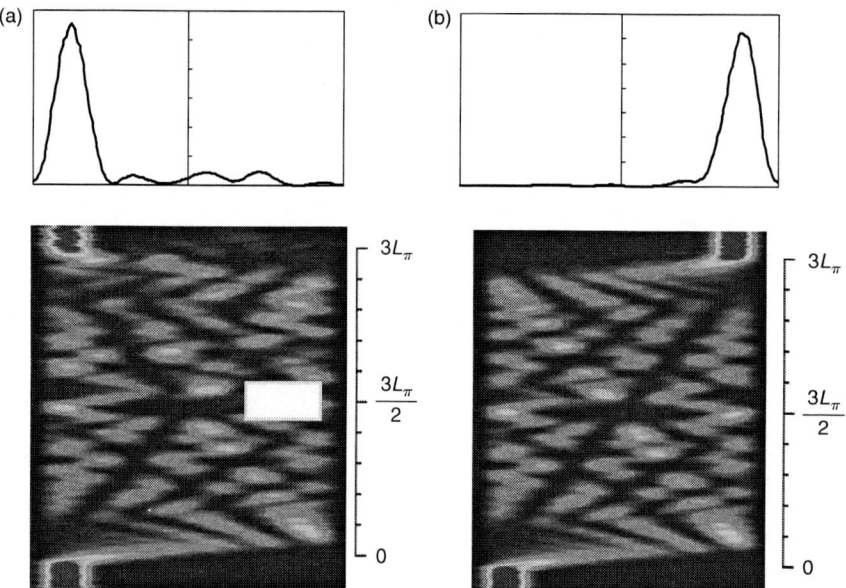

Figure 14   Characteristics of optical switch, MIPS-P; (a) with partial-index modulation and (b) without modulation corresponding to π-phase shift

Refractive index change is presumed to be realized by, for example, current injection, which is suitable for high speed switching operation. Characteristics with and without modulations are drawn in Figure 14(a) and 14(b), respectively. Here, the refractive index change of the partial region is about −0.24%, and a required length of the modulation region for π-phase shift is 150 μm. Relatively low cross talk of about −20 dB or less can be expected by optimizing the MMI structure, as discussed previously.

CONCLUSION

Fundamental characteristics of MMI waveguides are described. Theoretical backgrounds to evaluate the characteristics are explained to stress the importance of a beat length of the MMI waveguide, and two kinds of mode interferences, that is, general and restricted ones, are explained. Parameter dependences of the characteristics, such as MMI width and length, wavelength and polarization, are discussed. Some applications of the MMI waveguides as functional optical devices are introduced.

REFERENCES

1.   O. Bryngdahl, *J. Opt. Soc. Am.*, 63, 416–418 (1973).
2.   R. Ulrich, *Opt. Commun.*, 13, 259–264 (1975).
3.   E. C. M. Pennings, R. J. Deri, A. Scherer, R. Bhat, T. R. Hayes, N. C. Andreadakis, M. K. Smit, L. B. Soldano, and R. J. Hawkins, *Appl. Phys. Lett.*, 59, 1926–1928 (1991).
4.   J. M. Heaton, R. M. Jenkins, D. R. Wight, J. T. Parker, J. C. H. Birbeck, and K. P. Hilton, *Appl. Phys. Lett.*, 61, 1754–1756 (1992).
5.   P. A. Besse, E. Gini, M. Bachmann, and H. Melchior, *J. Lightwave Technol.*, 14, 2286–2293 (1996).

6. G. A. Fish, L. A. Coldren, and S. P. Denbaars, *IEEE Photon. Technol. Lett.*, 10, 230–232 (1998).

7. H. Okayama, H. Yaegashi, and M. Kawahara, in *Proceedings of the Fifth Optoelectronics Conference.* (OEC'94), 15B3-6 (1994).

8. J. C. Campbell and T. Li, *J. Appl. Phys.*, 10, 6149–6154 (1979).

9. S. Nagai, G. Morishima, M. Yagi, and K. Utaka, *Jpn. J. Appl. Phys.*, 38, 1269–1272 (1999).

10. L. B. Soldano and E. C. M. Pennings, *J. Lightwave Technol.*, 13, 615–627 (1995).

11. P. A. Besse, M. Bachmann, H. Melchior, L. B. Soldano, and M. K. Smit, *J. Lightwave Technol.*, 12, 1004–1009 (1994).

12. G. R. Hadley, *Opt. Lett.*, 17, 1743–1745 (1992).

13. M. Yagi, S. Nagai, H. Inayoshi, and K. Utaka, *Electron. Lett.*, 36, 533–534 (2000).

N

NANOPHOTONICS

Satoshi Kawata, Yasushi Inouye, and Hong-Bo Sun

INTRODUCTION

If photons can enter the nano-world, a lot of exciting new technologies and applications are expected. These include, for example (i) control of atoms with photons and nano-fabrication with controlled atoms; (ii) an optical microscope that can observe and analyze individual molecules; (iii) reading and controlling the sequence of DNA bases by photons; and (iv) direct control of protein molecules inside cells *in vivo*. Immediate industrial applications contain quantitative performance control of quantum wires and quantum dots, ultra-high density optical memory, and nano integrated circuits. It is also interesting to see how photonics is linked to physics of quantum effect, mesoscopic effect, and size effect that appear in the nano-world.

Photon energy is low compared to other quanta so that it gives less damage to biological tissues, living cells, and various organic materials. However, the low quanta energy means a long wavelength. The light spot that is tightly focused by a lens spreads to the size of about a wavelength, that is, several hundreds of nanometers, because of the diffractive nature of light. Therefore, we cannot observe or fabricate structures finer than the wavelength. This is the limit of classical optics.

The photon energy of the visible light is about several electron volts (eV). Wavelength of a few nanometer comes from photons of several kilo electron volts (keV), which lie in the x-ray wave range. In order to detect, control, and manipulate nano-structures with the visible light, photons of a few eVs have to overcome a "mountain of nano" (or a "barrier of nano") with a height of keV energy (Figure 1[a]). If the photon in the visible or infrared range could overcome the barrier and exceed the limitation of the wavelength, it becomes possible to observe single molecules and distinguish molecular structures as well as nanostructures with nanometric resolution. For example, the frequency of the visible and the infrared light match that of molecular vibration, it is possible to excite desired molecular vibration to distinguish different molecular bonds.

Three methods have been suggested to overcome the mountain of nano. First, we may "dig a tunnel" at the bottom of a high mountain. If the width of the mountain is narrow, the photon

165

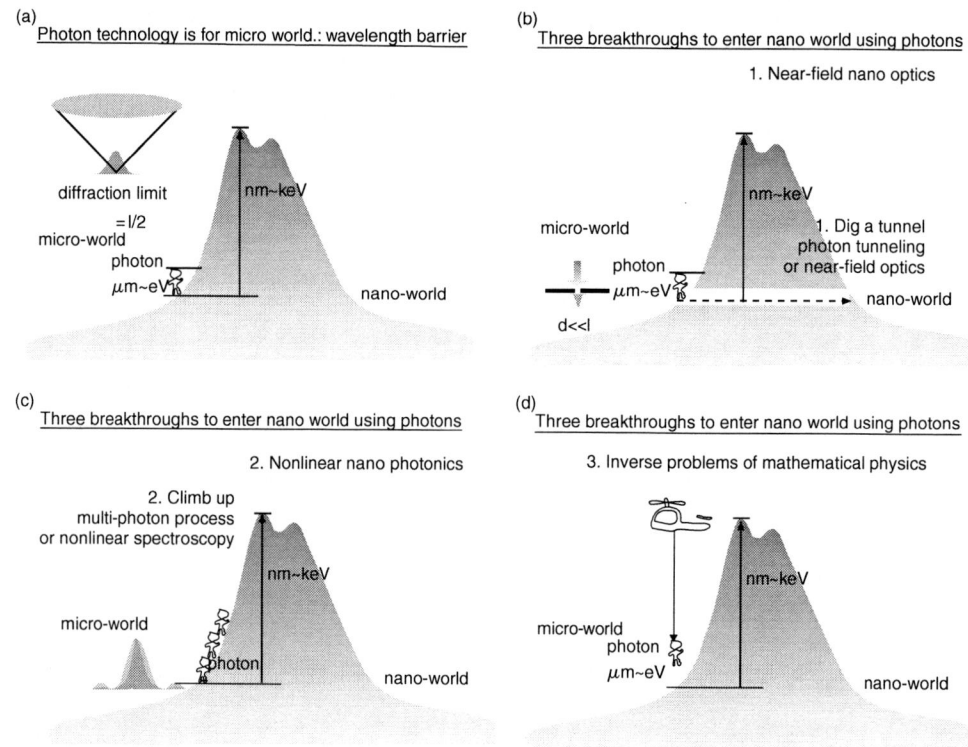

Figure 1 The wavelength barrier between the micro- and nano-worlds and three technologies used to overcome the limit existing in classical optics. (a) Barrier of the wavelength, (b) photon tunneling: near field optics, (c) multiple photons: nonlinear optics, and (d) super resolution achieved by solving the math-physical inverse problem

can reach the nano-world from the micro-world by means of photon tunneling (Figure 1[b]). Near-field optics works on this principle [1,2]. Second, the height barrier of the nano-world may be transcended by joint effort of multiple photons (Figure 1[c]). Mutli-photon absorption and nonlinear spectroscopy are generally based on this mechanism. Third, a photon can be lifted by a "special equipment." The helicopter in Figure 1(d) could be a math-physical technique. The super resolution is achieved by solving an inverse problem for a particular measurement system with restricted physical conditions as the boundary condition, for example, light intensity or density should not have a negative sign and the size of the actual object is finite.

In this chapter, we will discuss how photons attempt to enter the nano-world by the first two approaches, near-field optics and nonlinear optics.

PHOTON TUNNELING

When the boundary surface of a prism where light is reflected due to total internal reflection is brought near another prism within 100 nm, a part of the reflected light passes through it (Figure 2). This is the tunneling phenomenon of photon, as first described in Optik by Newton [3]. When the gap between the two prisms is small, that is, less than several tens of nanometers, the probability of the photon tunneling arises and ultimately reaches 100% when the two prisms are

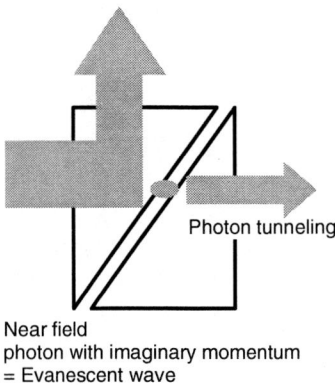

Photon tunneling

Near field
photon with imaginary momentum
= Evanescent wave

Figure 2   Photon tunneling through the total internal reflection

in contact with each other. When the gap is >100 nm, the tunneling does not occur. The range of <100 nm to 1 nm is the region of interest in near-field optics. The decay of transmittance versus distance does not obey exponential law when two boundary surfaces are in near-field [4] because of the electromagnetic interactions of dipoles on the two surfaces of the prisms.

Why a photon can be tunneled from a high refractive index medium (e.g., a prism) to a low index one (e.g., the air gap)? This is because photons with imaginary momentum are created in the medium of low refractive index at the total internal reflection condition. The momentum in the surface normal direction is an imaginary quantity. The imaginary momentum is converted to a real quantity by another prism and, as a result, the light propagates through the gap. The light whose momentum is an imaginary quantity in one direction is called evanescent wave. Its amplitude decreases with the distance from the boundary surface, and the probability of tunneling increases as the gap decreases.

NEAR-FIELD OPTICS

Evanescent wave is generated when light is irradiated to an aperture smaller than the wavelength in a fashion similar to the prism case. The evanescent wave exists only at the vicinity of the aperture and does not propagate through it (Figure 3[a]). However, when another aperture is drawn near, the photons start propagating in the same way as the two-prism case (Figure 3[b]).

If one of the aperture is assumed to be the sample structure and the other as a near-filed probe, intensity distribution corresponding to the sample surface is obtainable by detecting the light from the probe while scanning the probe in near field. This is the principle of Near-Field Scanning Optical Microscope (NSOM). If the two apertures are brought in contact with each other, they become one and photon tunneling does not occur any more (Figure 3[c]). This means that the probe is not only a device to read the evanescent field but also an output coupler in terms that it generates multiple interaction between two apertures. This principle is much different from the imaging theory of optical microscopy in classical optics (Abbe's diffraction theory [5]). The image contrast depends on the distance between the sample and the probe. The amplitude and polarization state of output light differ by shapes and materials (complex dielectric constant) of samples and the probe, the polarization of incident light and the incident angle [4,6,7]. These interesting features have been derived with Finite-Difference Time-Domain method (FDTD) calculation of Maxwell's equations [6,8,9].

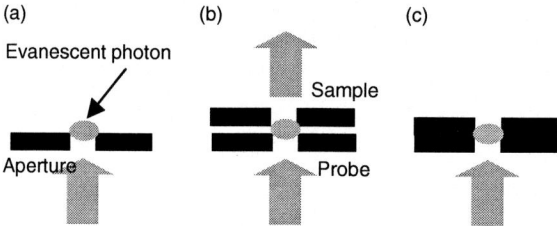

Figure 3   Photon tunneling on small apertures

Although Synge had suggested the principle of NSOM in 1928 [10], it was until 1972 that Ash and Nicholls demonstrated NSOM experimentally for the first time. They succeeded in resolving 0.5-mm structures using 3-cm wavelength microwave [11]. The resolution of the wavelength is $\lambda/60$. Figure 4(a) shows the prototype of NSOM with a slit probe. NSOM with visible light was reported for the first time in 1984 by a group in IBM Zurich [12]. It is said that this work was stimulated and technically supported by their neighbored group, which succeeded in developing scanning tunneling microscope (STM). The probe is a sharpened quartz coated with a metal, and the aperture is fabricated on the tip (Figure 4[b]). Since Betzig in Bell laboratory, AT&T utilized an aperture probe made from an optical fiber by sharpening and coating the fiber tip end in 1990 [13], the aperture probe started to appear in market.

APERTURELESS PROBE

The evanescent field could be created not only at the aperture probe, but also possibly on a structure smaller than the wavelength when the light is incident on it. We suggested the use of a scatterer for generating the evanescent wave rather than the aperture (Figure 4[c]) [14,15]. The small scatterer itself is an electric dipole, which induces the electric field when the light is incident on it. On the other hand, in case of aperture probe, a virtual magnetic dipole is generated, which produces the evanescent wave on the other side of the screen [16,17]. The scatterer, therefore, generates a more intense electric field. The evanescent wave around the scatterer illuminates the localized area of the sample and induces the interaction when the light is irradiated to the scatter at the vicinity of the sample. Imaging is achieved by scanning the scatterer as the near-field probe in the same way as atomic force microscopic (AFM) or STM.

Laser trapping technique can control and manipulate a particle three dimensionally [20]. While dielectric-particle trapping is well known [18], the metallic particle is generally considered as impossible to be trapped because the scattering force, which repels metallic particles, is dominant compared with the gradient force. However, we demonstrated that the 40-nm gold particle is trapped in the highest intensity region in the focus spot because the gradient force becomes dominant over the scattering force for the gold particle of this particular small size [15]. For infrared trapping beam, the focus spot size is about 1 μm. The trapped particle fluctuates inside the focus spot due to Brownian motion. We succeeded in suppressing the fluctuation within 2 nm with a feedback system, by which the particle position is monitored and the trapping force is increased when the particle displacement from the focus center increases to pull back the particle. Figure 5 shows the NSOM image of DNA labeled with fluorescent dyes by the trapped-probe scanning [19].

The near-field imaging with a trapped gold particle as a probe is limited to observation of the biological sample in solution, because the electrostatic force is much stronger than the laser

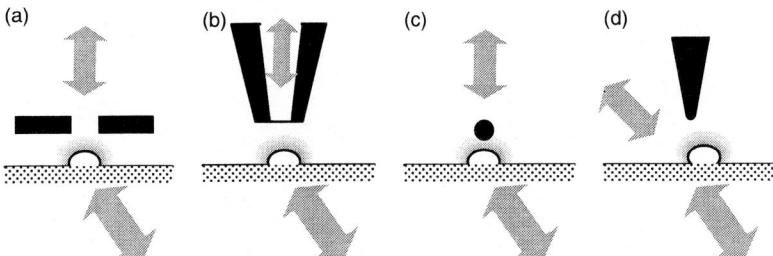

Figure 4 Probes of near-field optical microscope: (a) micro-slit probe; (b) optical-fiber typed aperture probe; (c) small particle probe; and (d) metallic needle probe

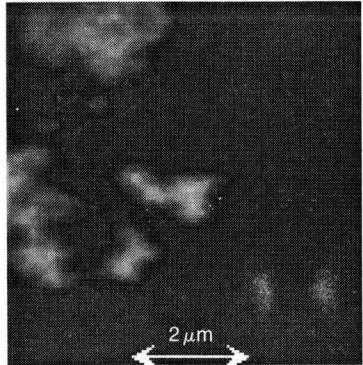

Figure 5 Near-field image of DNA stained with fluorescent dyes. A trapped gold nano-particle was utilized as the detection probe

trapping force in air or vacuum. A metallic needle with sharpened tip is more appropriate for the near-field probe in air or vacuum as shown in Figure 4(d). When the metallic probe is inserted to an existing near-field, the light is scattered at the tip. Silver or gold is particularly effective for scattering visible light because surface plasmon polariton resonates with the incident light and the enhancement of electromagnetic field is significant [20]. We proposed and demonstrated this idea in 1994, and Boccara in France and Wickramasinghe in America gave similar reports independently soon after [21,22].

The probe consisting of only metal does not need any aperture, which is used to tunnel the evanescent field constructed on the sample surface. Compared to the dielectric probe with the small aperture at the tip end, this new type of probe is called apertureless probe. The apertureless probe has several advantages: no artifact reflecting the tip structure is superposed on the near-field image because of the simplicity and sharpness of the tip structure; there is no attenuation of the scattered light as encountered in propagation in the optical fiber, where the transmittance is of the order of 10^{-5}; finally the photon coupling efficiency is high because of the electromagnetic field enhancement by the local-mode surface-plasmon resonance [1].

Optical image of the sample surface can be observed with nano-metric resolution by detecting the scattered light while scanning the metallic probe on the sample surface [20]. Figure 6(a) shows the near-field image of a silicon grating with the groove width of 60 nm and the periodicity of 240 nm [23]. Compared to the STM image obtained simultaneously, the NSOM image has

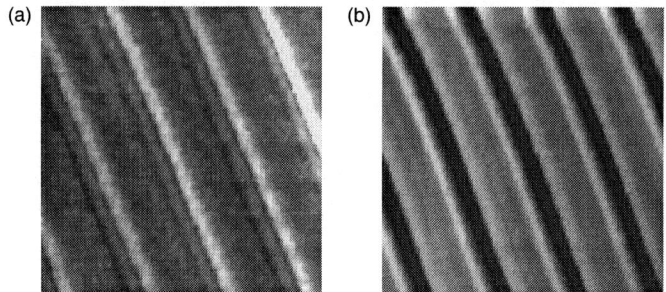

Figure 6   (a) Near-field image of a silicon grating with 240-nm periodicity observed with the metallic needle probe. (b) The STM image

the appearance of solidity. The contrast of the optical image changes in accordance with the direction of illumination. In this observation the light is illuminated in an oblique incidence. From the edge response, we recognize that the resolution is 6 nm, which is beyond the diffraction limit of light.

TIP-ENHANCED NEAR-FIELD RAMAN SPECTROSCOPY AND IMAGING

Near-field optical technology has been developed for Raman spectroscopy, enabling molecular vibrations to be measured with nanometric spatial resolution. However, it is difficult to detect near-field Raman signals because of the small cross-section of Raman scattering ($\sim 10^{-30}$ cm^2), which is much smaller than that of infrared absorption ($\sim 10^{-20}$ cm^2). It is quite time consuming to measure the near-field Raman signal using an apertured fiber probe. One of the methods to overcome the difficulty is the use of the field enhancement effect locally induced by the metallic tip where localized surface plasmon polaritons are excited due to photons coupling with free electrons. Shown in Figure 7(b) is the intensity of light field scattered at the tip obtained with a numerical analysis [24]. Figure 7(a) represents the calculation model. A silver metallic tip with a radius of 20 nm is placed in contact with a glass substrate (refractive index: 1.5). The silver tip is illuminated at 45° incidence by a plane wave traveling from the substrate. The wavelength of the incident field is 488 nm and its polarization is the transverse magnetic (TM) mode (i.e., p-polarization). The finite-differential time-domain method was employed in the calculation [25]. A localized and enhanced field spot is observed around the tip in the figure. The size of the small spot is about 30 nm, which approximately corresponds to the radius of the tip. The peak intensity of the small spot is enhanced by a factor of, ~80 compared with intensity of the incident field. The localized photons are employed for nano-light source of Raman spectroscopy.

Figure 8 shows a NSOM system using a metallic tip for local enhancement of Raman scattering [14–16]. A light field from an Ar$^+$ laser enters the epi-illumination optics after being expanded and collimated. A part of the illumination light, which corresponds to the numerical aperture (NA) < 1.0, is rejected by inserting a mask in front of a beam splitter. The annular illumination is then focused onto the sample surface by using an oil immersion objective lens (60×, NA ~ 1.4). The focused light spot consists only of an evanescent field, because only the component of NA > 1.0 is transmitted through the mask. By approaching the metallic tip toward this evanescent focused spot, a part of the evanescent light is scattered and converted to propagating light. A small and intense light spot with the size corresponding to the radius of the cantilever tip, is generated at the tip. The scattered light is collected by the same objective used to focus the laser, and is detected by a polychromator with a liquid-nitrogen-cooled CCD camera.

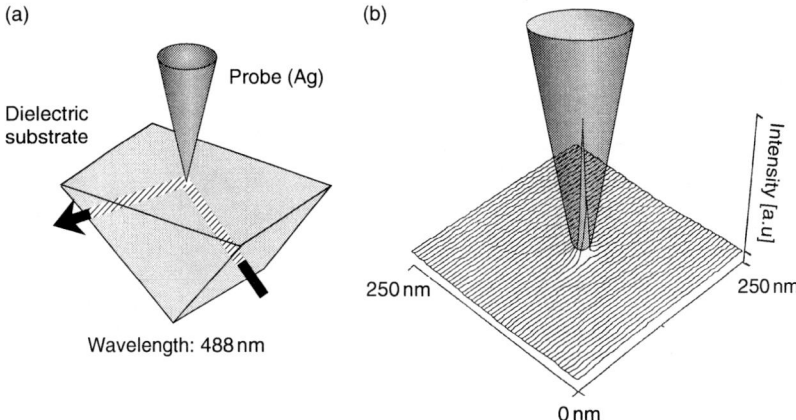

Figure 7   Numerical analysis of light–field intensity at the silver near-field tip. (a) The model for simulation and (b) the light intensity distribution

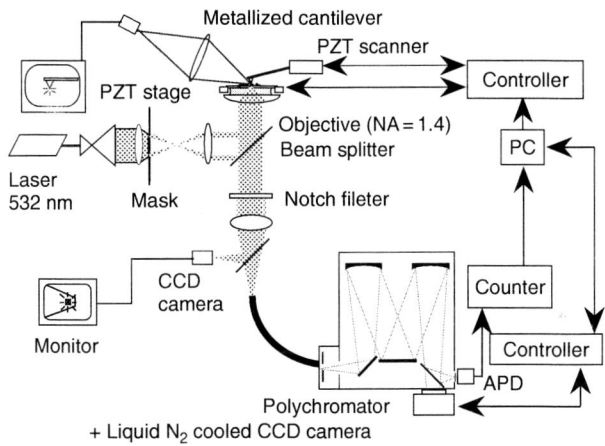

Figure 8   Illustration of the NSOM system that uses a metallic tip

Excitation light or Rayleigh scattering is rejected completely by a notch filter. The sample was scanned with piezoelectric transducers in the x–y plane while the motion of the tip in the z-direction was controlled via a commercial AFM feedback system. A silicon cantilever, which was coated with a silver film of 40-nm thickness, was used as the metallic tip.

Figure 9 shows a tip-enhanced near-field Raman spectrum of a single nanocrystal of adenine, which was detected with our developed tip-enhanced near-field Raman microscope with a silver tip. Adenine nanocrystals dissolved in ethanol were cast and spread out on a coverslip. The crystal size is laterally 7 nm × 20 nm and vertically 15 nm. The far-field Raman spectrum of the same sample without a metallic probe tip is also shown for comparison. In the tip-enhanced Raman spectrum, several characteristic Raman peaks of adenine molecules are enhanced and become visible such as those at 736 and 1330 cm^{-1}, while no Raman peak at these positions is observed when the probe is withdrawn from the near-field

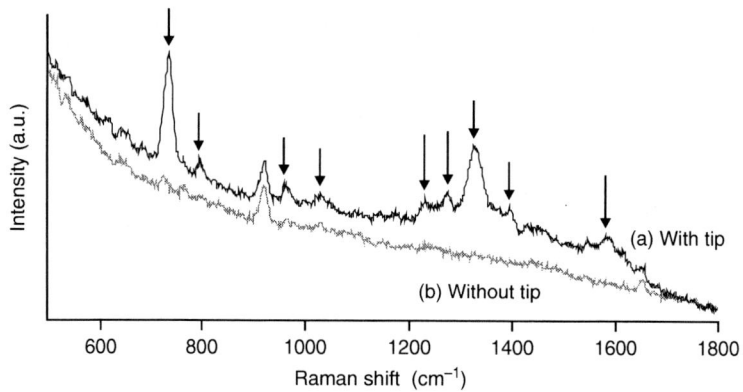

Figure 9   Near-field and far-field Raman spectra of a single nanocrystal of adenine. (a) A metallic tip is drawn near to the near-field of the sample and (b) far-field

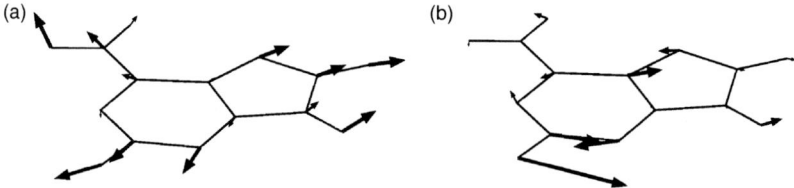

Figure 10   Molecular vibration modes. (a) A whole molecule and (b) a diazole

region. The experimental condition was exactly the same for both experiments except for the tip-sample distance. The Raman scattering cross-section of the molecules just beneath the enhanced electric field is strongly increased. The strong Raman peaks at 736 and 1330 cm^{-1} are assigned to ring breathing mode of a whole molecule and ring breathing mode of a diazole, respectively as shown in Figure 10(a). Other Raman peaks are also assigned to the normal modes of adenine molecules on the basis of density functional (B3LYP) calculation in combination with the 6–311+G** basis set (Figure 10[b]) [15–17]. The observed tip-enhanced near-field Raman peaks, the black line in Figure 9(b), are slightly shifted from the far-field ones obtained from a bulk sample, the gray line in Figure 9(b). For example, the ring breathing mode at 720 cm^{-1} in far-field Raman spectrum of bulk sample is shifted to 736 cm^{-1} in the tip-enhanced near-field Raman spectrum while the ring breathing mode of a diazole is not shifted. These phenomenon of spectral shift is in good agreement with surface enhanced Raman scattering (SERS) spectra of adenine molecules [18], and ensure that the metallic probe tip works as a surface enhancer for SERS effect.

NANOPHOTONICS ON THE BASIS OF OPTICAL NONLINEARITY

Photons enter the nano-world in near-field optics by means of tunneling of evanescent electromagnetic field. The shallow penetration nature of the evanescent wave restrains the near-field optical microscopy a surface probing approach. For three-dimensional (3D)/far-field detecting and recording, we have to resort to an alternative mechanism, that is, nonlinear laser-matter interaction. For optically inducing any change in material performance, the photon energy must be first absorbed by the material. If the photon energy, $h\nu$, is equal or larger than the energy gap between the ground state and the excited state, or in other words, between the valence band and the conduction band, E_g, the absorption is a linear process or one-photon absorption. In our everyday realm of experience, most optical phenomena are linear, that is, optical properties are independent of the intensity of the incident light. However, electrons could be excited by simultaneous absorption of two or more photons with identical energy of $h\nu < E_g$. This process is known as two- or multi-photon absorption, a typical kind of nonlinear optical effect [26]. The nonlinearity in light–matter interactions contributes to the breakthrough of diffraction limit. For example, for two-photon absorption, the absorption probability is proportional to the square of light intensity, instead of the light intensity itself. This narrows the size of light excitation spot by a factor of 1.4 due to the reduced half-width of light response curve. Moreover, if a threshold exists in material response to light excitation, the photochemical/physical reactions could be further confined to a volume much smaller than that defined by the diffraction limit.

Take two-photon photopolymerization as an example to explain how the sub-diffraction-limited spatial resolution has been achieved. Experimentally observed for the first time in 1965 [27], two-photon photopolymerization has been utilized for micro-nanofabrication since 1997 for the first time by some of the current authors [28]. The basic idea of pinpoint two-photon photopolymerization is scanning a tightly focused red or NIR laser beam according to a preprogrammed pattern. Since the chemical reaction of polymerization was confined to occur solely at the close vicinity of the focal spot, pinpoint material solidification and pattern drawing can be achieved. The process of two-photon photopolymerization could be roughly described as follows:

$$I \rightarrow 2R', \tag{1}$$

$$R' + M \rightarrow R\text{-}M', \tag{2}$$

$$R\text{-}M' + M \rightarrow R\text{-}M'_2, \tag{3}$$

$$R\text{-}M'_n + M \rightarrow R\text{-}M'_n, \tag{4}$$

$$R\text{-}M_m + R\text{-}M_n \rightarrow R\text{-}M_{m+n} - R, \tag{5}$$

where the meaning of symbols are: I, photoinitiator, R', radicals, M, monomer, and R-M′, monomer radical. At the beginning of the reaction, the photoinitiator was dissociated into radicals by simultaneous absorption of two photons. A radical molecule reacts with a monomer molecule (small molecules in liquid status), forming a monomer radical. Eqs. (1) and (2) describe the initiating process. The monomer radical expands by continuously absorbing new monomer molecules, which is a chain propagation reaction as indicated by Eqs. (3) and (4). With the reduction of monomer concentration, two monomer radical molecules may meet with each other and are combined into a macromolecule, of which the aggregation takes solid form. This is the reaction termination process [29].

Since the two-photon absorption cross-section is extremely small, a high photon flux density is required for the excitation. In time domain, efforts were devoted to use femtosecond laser pulses, where photon energy was accumulated into series of sharp peaks. A typical laser of 100-fsec pulsewidth, 780-nm wavelength, and 80-MHz repetition rate has an average output of >1 Ws. In order to achieve high resolution in the spatial domain, a high NA objective lens is utilized to focus a femtosecond laser beam. The diffraction limit is

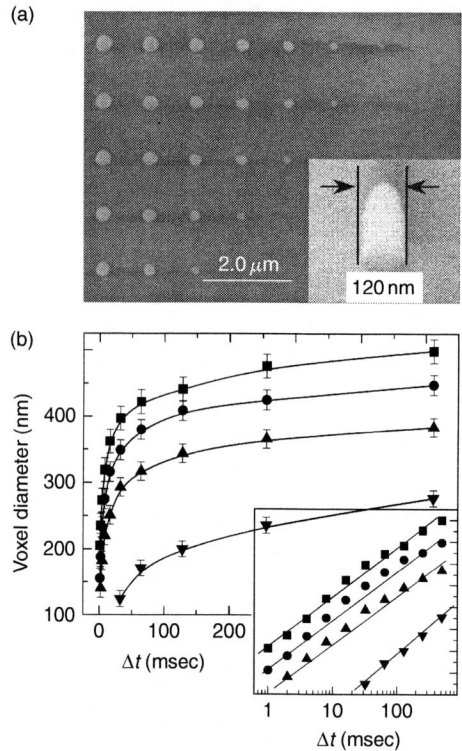

Figure 11 Sub-diffraction limit spatial resolution. (a) SEM image of voxels that are two-photon polymerized under different conditions; (b) exposure time-dependent lateral voxel size under laser pulse energy of 163, 137, 111, and 70 pJ, respectively

defined as $1.22\lambda/\text{NA}$. When $\text{NA} = 1.4$, $\lambda = 780$ nm, and the refractive index of the resin is 1.52, the diffraction-limited spot size is 460 nm. Two strategies as mentioned earlier, the quadratic dependence of photopolymerization rate on the laser light intensity and the threshold effect, were employed to exceed the diffraction limit. The mechanism of threshold is radical quenching by oxygen molecules. The oxygen molecules are naturally dissolved and exist in photopolymerizable resins [30]. If the light intensity is weak, the concentration of photo-generated radicals is so low that they may be completely scavenged. By carefully controlling the laser light intensity, it is possible to control that photopolymerization occurs only at the central portion of the laser focal spot with volume smaller than that defined by the diffraction limit [31]. Figure 11(a) is the scanning electron microscopic (SEM) image of volume elements (voxels) of photopolymerization, which are solidified under various light intensity and exposure duration. Figure 11(b) is exposure time-dependent lateral voxel size under laser pulse energy of 163, 137, 111, and 70 pJ (from the top to the bottom), respectively. The size of the smallest isolated voxels as those shown in Figure 11 is 120 nm. 50-nm resolution has been realized in practical fabrication, for example, the ends of the bull horn as shown in Figure 12. This is approximately one-tenth of the diffraction limit. For this reason, we call this technology sub-diffraction-limited laser fabrication.

THREE-DIMENSIONAL LASER FABRICATION FOR MICRO-NANOMACHINES

The basic idea of three-dimensional (3D) laser fabrication is quite simple [28,32]. A femtosecond near-infrared laser is tightly focused into a photopolymerizable resin, and then the focal spot is scanned relative to the substrate according to a preprogrammed 3D CAD pattern. Since photopolymerization reaction is confined to occur only at close vicinity of the focal spot, solidified structures are created following the scanning locus. After the pinpoint two-photon writing, the entire structure emerges into an organic solvent like acetone, by which unsolidified liquid resin is dissolved and removed, while the solid skeleton remains. Photopolymerized structures have real physical shape, contrasting with those image-like structures recorded in solid matrixes. Hence, not only optical components, but also micromechanical devices as well as microelectromechanical system (MEMS) could be produced. With the sub-diffraction-limit fabrication accuracy, it is possible to fabricate devices of nanoscale size or with nano-features. Figure 12 is the SEM image of a micro-bull sculpture, which consists of smooth and tough surfaces, curvatures, and sharp horns. It is a good proof of the feasibility of depicting sub-diffraction-limit features by two-photon photopolymerization. The 10-μm long and 7-μm high bulls are the smallest animals ever made artificially, and are as small as the red blood cell. Their volume allows us to send them, actually micromachines of this size, to any location inside the human body through blood microvessels to make clinical treatments [33].

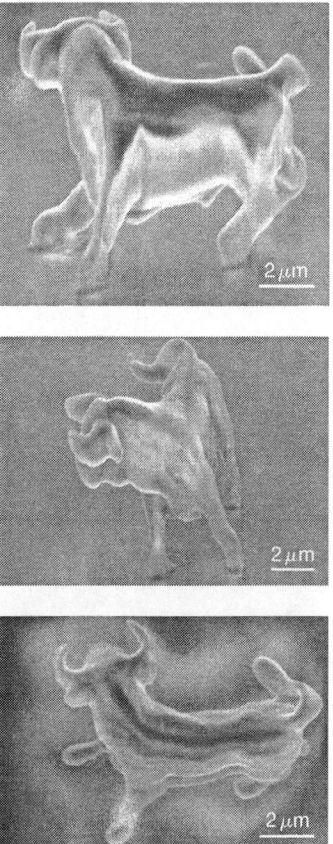

Figure 12   SEM image of the microbull sculpture

The optical setup used for the fabrication of the microbull is shown in Figure 13. In this setup the laser beam is steered in horizontal directions by a set of Galvono mirror and in vertical direction by a piezo stage. Compared with a 3D piezo-stage system, this setup features fast movement and fast stabilization of the beam. After beam expansion with a beam expander, the laser light is guided to the high-NA (NA ~ 1.4) objective lens. The typical parameters for laser fabrication are 1-msec exposure time and 50-nm step length in all three dimensions.

The micro-bull, as shown in the Figure 12, is just a sculpture. A microgearwheel and an icosahedron, as shown in Figure 14, are potentially more useful structures. They are fabricated using fluorescent dye (LD490, Exciton Inc.)-doped resin, which permits internal micro-diagnosis by means of two-photon confocal imaging. With optical sectioning, the proper shaping, positioning,

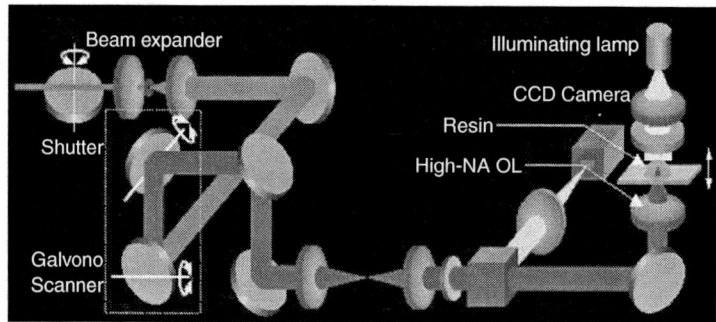

Figure 13   A typical setup for pinpoint two-photon photopolymerization

Figure 14   Reconstructed optical microscopic images of fluorescent micromechanical devices: (a) a microgearwheel; and (b) an icosahedron

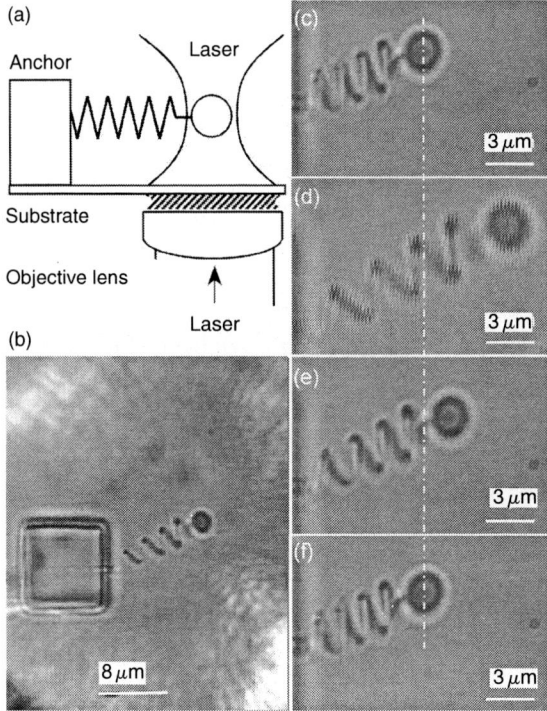

Figure 15 A functional micro-oscillator system. (a) A designed scheme for driving the oscillator using the laser trapping force. (b) A photograph of a fabricated micro-oscillator, where the ended bead is being trapped by the laser. The micro-spring is (c) in its natural state, (d) pulled by a length, (e) released, and (f) recovered to its original states 20 sec after the release

and connecting could be checked [34]. Additionally, the series of 2D optical images could be utilized for reconstructing an entire 3D image of objects. Both Figure 14(a) and 14(b) are the reconstructed optical microscopic images.

Micromechanical devices are expected to work under a certain driving mechanism [35]. We fabricated a 300-nm cord diameter and 2-μm coil diameter spring with its one end attached with a 2-μm diameter bead. When the focus of a laser used to drive the micro-oscillator is carefully adjusted (Figure 15[b]), the bead gets trapped and freely manipulated in three dimensions. The spring is pulled by moving the trapped bead and then released by blocking the laser; an oscillation is thus initiated. The spring is observed to be prolonged (Figure 15[d]) from its original length (Figure 15[c]), and restored (Figure 15[e]) to its original state after the laser is turned off (Figure 15[f]). An elongation up to 7 μm in multi-operation does not cause any elasticity failure, as evidenced by the fact that the spring always restores to its original length. From the oscillation curve, the spring constant is estimated to be 10^{-9} N/m [35]. Such a soft microspring is expected to be useful, for example, for detecting extremely weak interaction between the macromolecules like protein molecules.

PHOTONIC CRYSTALS

The artificial photonic materials called photonic crystals (PhCs) bring novel opportunity for deep understanding on light–matter interactions in a tailored electromagnetic circumstance, and for developing diversified optoelectronic and photonic devices with performance that is otherwise not accessible [36]. PhCs working at the visible and the near-infrared (NIR) wavelengths require a lattice period of several hundreds of nanometers, which is not as small as those well-developed semiconductor expitaxial growth technologies that could be economically utilized, and is not so large as to be accomplishable by mechanical processing [37]. A number of approaches have been employed for PhC fabrication, for example, self-organization of colloidal particles [38,39], layer-by-layer packing of 2D semiconductor meshes [40,41], electrochemical etching [42], and vacuum deposition on structured substrates [43].

Pursuit of more appropriate nanofabrication technologies remains critical to further progress of PhC research and application. First, inclusion of materials of versatile characteristics to PhC structures becomes essential for the purpose of realizing device functions like high-efficiency light emission, large photonic bandgap (PBG) tunability, enhancement of nonlinear phenomena, and so forth. This is sometimes a harsh requirement to the currently existing methods since they are generally applicable to a specific material system. Furthermore, theoretical work has predicted that complex lattice structures may possess stronger PBG effect, like distorted diamond [37], face-centered cubic (FCC) with non-spherical atoms [44], and spiral structures [45]. These theoretically ideal structures are considered impractical due to the difficulty of fabrication by means of conventional nanofabrication tools.

Pinpoint two-photon writing has an intrinsic 3D processing capability, which is particularly suitable for fabricating complicated photonic structures. Figure 16 shows a $\langle 100 \rangle$-orientated diamond lattice PhC structure, where the period is $\Lambda = 2.5$ μm [46]. The distance between the rod-ended balls, or intuitively called photonic atoms, is 1.1 μm. The rods and the photonic atoms have diameters of 500 and 580 nm, respectively. A smooth surface of rods and its entire structure can be clearly seen in Figure 16, showing a high quality fabrication. FTIR measurement shows PhC structures of 2.5- and 3.0-μm lattices giving rise to transmission minima at 3790 cm^{-1} ($\lambda = 2.6$ μm) and 3100 cm^{-1} ($\lambda = 3.2$ μm), respectively. They occur at the identical normalized frequency of 1.05, as defined by λ/Λ, showing the linear scaling performance of the diamond lattices. The single-layer attenuation is around 35%, much better than those from previously reported logpile lattices [47].

The pinpoint writing is characteristic of individual addressing capability, namely any feature inside a structure including defects could be designed and depicted. However, the fabrication speed is slow, generally on a time scale of hours. As an alternative laser fabrication technology, multi-beam interference patterning is liable to produce large-volume perfect 3D

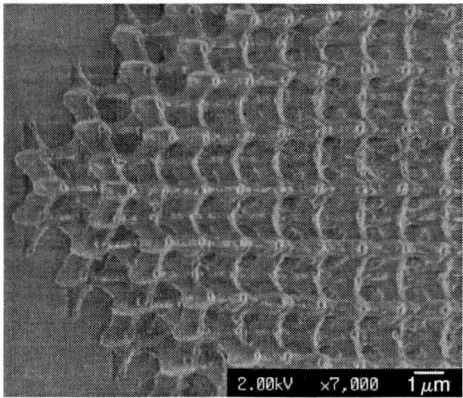 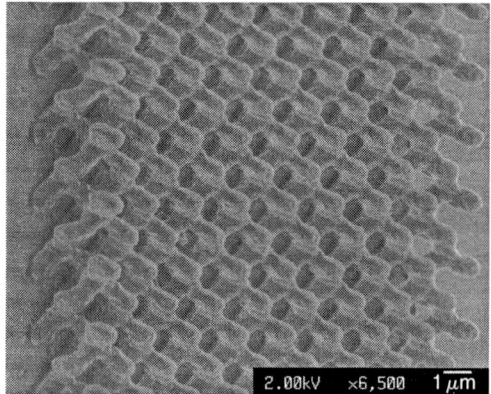

Figure 16   SEM image of a diamond-lattice PhC structure

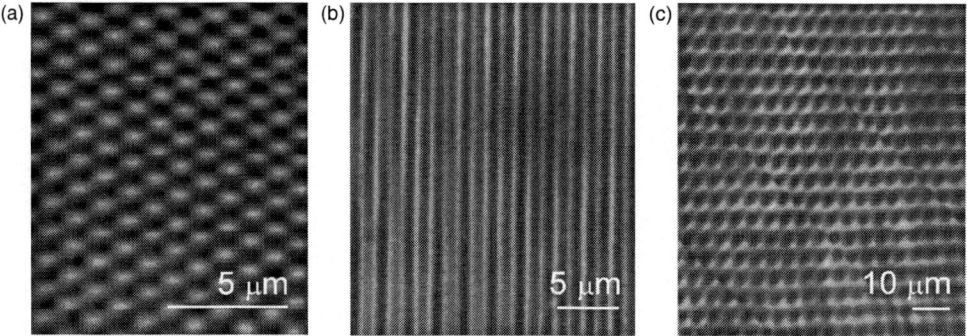

Figure 17   SEM images of 3D PhC structures created by two-step holographic lithography. (a) and (b) are the same structure from the first 3-beam interference, and (c) taken after completion of the additional 2-beam interference

periodic structures within the time range of seconds. The basic idea of the holographic lithography is introducing the interference pattern created by several coherent beams into a photopolymer. The resin at sites of maxima of light intensity is solidified while interval liquid resin is washed out in the postfabrication development. We utilized a two-step exposure scheme [48]. In the first step, three beams (442 nm, He–Cd CW laser), all symmetrical to the sample surface normal, interfere with each other and interact with the photopolymer spin-coated on a substrate, and produce 2D triangular rod array. In the top view SEM image (Figure 17[a]), the bright spots correspond to the tip end of the micro-rods. The planar lattice constant is 1.0 μm. The rods are grown from the bottom to the top of a sample cell, and are 150 μm in length (Figure 17[b]), distributed in an area of 500-μm diameter.

The structure acquired from the first step three-beam interference has the translation symmetry along the rod axis. Hence, there is no periodicity in this direction. For fabricating 3D PhCs, two coherent beams were introduced from the cell top and bottom in the second step of the fabrication, whereby the rods were vertically intersected by 150 cross-sectional layers (Figure 17[c]). Thus by combining the sequential three-beam and two-beam interferences, a 3D periodic structure is produced, where we see diversified possibilities of lattice design by freely changing beam arrangement.

TERABIT OPTICAL MEMORY

Conventional optical recording media like compact disk (CD) and magneto-optical (MO) disk register binary bits, spots with optically modified phase status, or refractive index, in an active layer of the disk, by which information of about 10^8 cm^{-2} was recordable using visible light at the diffraction limit. The information capacity in a disk volume can be expanded by means of a multi-layer recording [49,50]. A simple scheme is of focusing laser at a series of different depths in an optically thick active media, and at each depth, a bit plane is recorded. Estimated from the diffraction-limit-defined volume, a storage density as high as 10^{12} bits cm^{-3} is possible. However, in linear recording-reading (single-photon absorption for both processes), the same amount of photon energy is absorbed in each plane transverse to the optical axis since nearly the same amount of photon flux crosses them. This strongly contaminates the planes above and below a particular focal plane to be addressed, causing the issue of crosstalk. Therefore it is quite difficult to realize the multilayer recording strategy by single photon process. The problem was solved by using multiphoton absorption. First, the intrinsic nature of deep penetration allows a laser to address a certain depth inside materials without power dissipation; and second, the excitation depends roughly on the n-order of intensity so that net excitation per distant plane falls off sharply, where "n" means n-photon absorption. This enables recordings in well-separated layers and minimizes the crosstalk.

Photochromic molecules [51] are characteristic of existence of two chemically stable forms, and the two isomers are inter-switchable by photochemical reactions after absorbing light of different wavelengths. This implies a rewritable optical memory. The two isomers differ in their absorption, refractive index, fluorescence wavelength, and even molecular orientation-induced polarizations, permitting recorded binary bits of one isomer status embedded in a matrix of the other isomer status. As a good 3D storing medium, the material should have high sensitivity and fast response to excitation, stable isomers at both states, and high resistance to fatigue during cyclic writing and erasing. Three classes of molecules, spirobenzopyran, diarylethene, and azobenzene and their derivatives are found promising for this purpose. Particularly, diarylethene derivatives with heterocyclic rings exhibit no thermochromicity till 200°C; their colored close-ring form are stable for more than 3 months at 80°C; and no significant fatigue has been observed even after 10^4 cyclization/ring open reaction cycles [51]. As an example of multilayer writing and reading, Figure 18 shows absorption spectra of two-form 1,2-dicyano-1,2-bis(2,4,5-trimethyl-3-thienyl)ethane (B1536) (Figure 18[a]) and several two-photon recorded bit patterns out of 26 sequential bit planes (Figure 18[b]) [50,52]. Bits were recorded by exciting the 380-nm absorption of open-ring isomers (red color) using 760-nm femtosecond laser. The bits exist mostly in close-ring isomers (yellow). The recording layer and bit intervals are 5 and 2 μm, respectively. The refractive index change around 10^{-4} is distinguished and readout by a reflection confocal microscope.

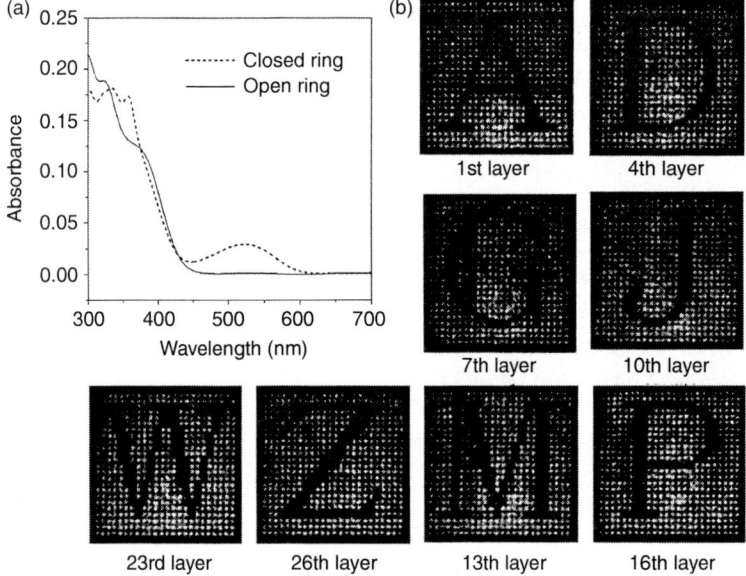

Figure 18 Photochromic materials for 3D optical memory. (a) Absorption spectra of open-ring and closed-ring diarylethene derivative B1536. (b) Bit patterns written by femtosecond two-photon absorption and readout using a reflection confocal microscope

MOLECULAR IMAGING BY NONLINEAR VIBRATIONAL SPECTROSCOPY

The nonlinear process in two- or multi-photon absorption requires that the sum of the energy of several photons is equal to or larger than the energy level for electron transition. Here we show the excitation of molecular vibration by matching the energy difference of photons from different incident beams with the molecular vibrational energy levels. This is another type of multi-photon process. Assuming that a molecule has a vibrational level of $\omega_1 - \omega_2$, where ω_1 and ω_2 are the frequencies of the two pumping laser beams, it may emit the light of frequency $\Omega = 2\omega_1 - \omega_2$ through Raman scattering. This process is called coherent anti-Stokes Raman scattering (CARS) as illustrated in Figure 19. CARS spectrum of Ω obtained by scanning ω_1 could be utilized to analyze molecular vibration and determination of chemical species. As a third order nonlinear process, the CARS signal is proportional to ω_1^2. It is not difficult to realize sub-diffraction-limited resolution when the near-infrared ultrashort laser pulse is tightly focused. In addition, the signal frequency has a blue shift relative to the excitation laser wavelength; therefore, it is easy to separate it from the fluorescence signal.

Figure 20 shows the CARS microscope that we have built. Two mode-locked Ti : Sapphire picosecond lasers are utilized as the light source. The lasers are directed to an optical microscope and focused into a sample. Very recently, a rotating micro-lens array has been installed to the above system, by which a real-time CARS imaging has been enabled. Shown in Figure 21 are two CARS images of C–H vibrations in amide group of yeast (Figure 21[a]) and polystyrene ball (Figure 21[b]).

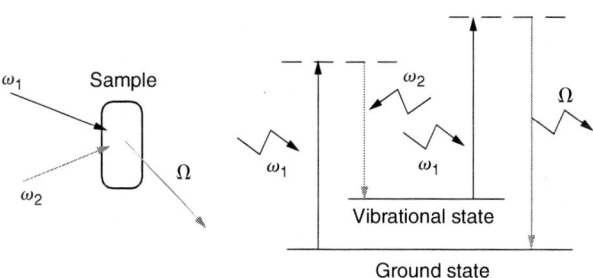

Figure 19   Nonlinear spectroscopy: CARS

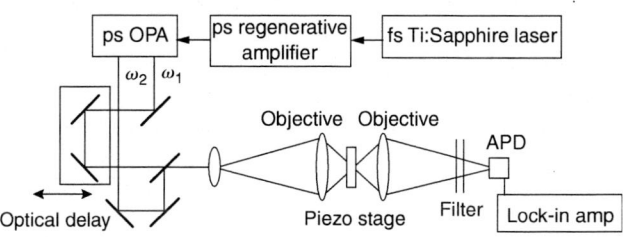

Figure 20   Schematic representation of a CARS microscope

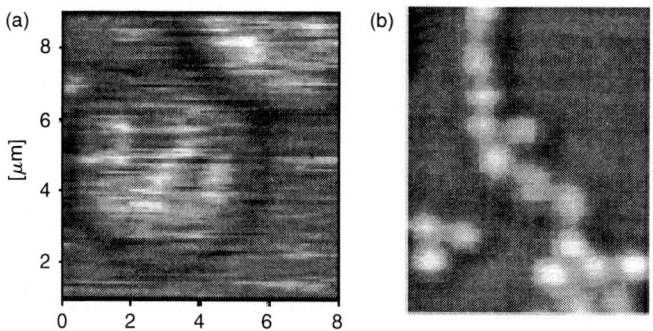

Figure 21   CARS imaging. (a) C–H bonds in an amide group of yeast ($\omega_1 - \omega_2 = 1215$ cm^{-1} and (b) a polystyrene ball ($\omega_1 - \omega_2 = 980$ cm^{-1}).

NEAR FIELD NONLINEAR SPECTROSCOPY AND MOLECULAR IMAGING

The CARS, a third-order nonlinear optical phenomenon, can be induced by the localized surface plasmon polaritons under the metallic tip. Since probability of two- or three-photon process is in proportion to square or cubic photon density, the spatial response of the nonlinear phenomenon is smaller than that of linear phenomenon. Accordingly, the combination of near field optics with nonlinear optics provides us further confinement of photons and higher spatial resolution for molecular or chemical imaging.

Figure 22(a) shows the tip-enhanced near field CARS image of DNA network of poly(dA-dT)-poly(dA-dT). Observed is the DNA network consisting of bundles of DNA double-helix aligned parallel to the glass substrate. The frequency difference of the two excitation lasers for tip-enhanced CARS is set to 1337 cm^{-1} corresponding to a Raman mode of adenine (ring breathing mode of diazole) by tuning the excitation frequencies ω_1 to 12,710 cm^{-1} (λ_1: 786.77 nm) and ω_2 to 11,373 cm^{-1} (λ_2: 879.25 nm). Tip-enhanced CARS image obtained at off-resonant frequencies shows no contrast of DNA bundles as shown in Figure 22(b). The CARS signals are not detected after removing the tip from the sample at the on- and off-resonant frequencies, which confirms that the CARS is effectively induced in the presence of the tip. Figure 22(c) shows one-dimensional line profiles at $y = 270$ nm, which were acquired with ~5 nm step. The cross-section of far-field CARS acquired without the silver tip is also added for comparison. Only the TE-CARS in the on-resonant condition has peaks at the positions of $x \sim 370$ nm and $x \sim 700$ nm where adenine molecules exist in the DNA double helix, while the other line profiles do not sense the existence of the molecules. The FWHM of the narrower peak at $x \sim 700$ nm is ~15 nm. This spatial resolution can be attributed to the spatial confinement effect of the CARS generation due to the third-order non-linearity. The intensity enhancement factor for each electric field is estimated to be ~100.[1,2] This value is quite realistic and reasonable, as compared to the previous numerical results [6,53], although this estimation is very subject to the changes in each parameter with high-order

---

[1] In order to estimate the enhancement factor we first measured far-field CARS intensity of a DNA cluster with different excitation power and repetition rate, then calculate the efficiency of the CARS emission per unit volume, peak power, and repetition rate.

[2] In order to estimate the excited volume we assume the tip-enhanced local excitation spot to be a sphere with 20-nm diameter centered at the tip end in which the excitation efficiency is uniform. The locally excited volume is defined as the volume overlapped by the excitation spot and the sample. In these assumptions, the excited volume is approximately a column with 2.5-nm height and 20-nm diameter.

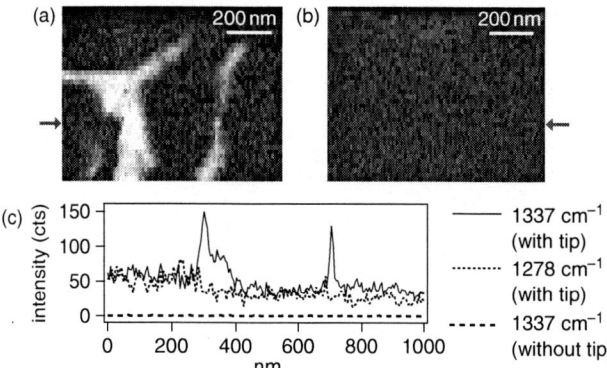

Figure 22 Tip-enhanced CARS images of the DNA network (a) at on-resonant frequency (1337 cm^{-1}) and (b) at off-resonant frequency (1278 cm^{-1}). (c) Line profile of the row indicated by the solid arrows. The scanned area is 1000 nm by 800 nm. The number of photons counted in 100 msec was recorded for one pixel. The acquisition time was ~12 min for the image. The average powers of the ω_1 and ω_2 beams were 45 μW and 23 μW at the 80-MHz repetition rate.

dependency. We also estimated the size of the locally excited volume of the DNA structure to be ~1 zeptolitter.[2] The smallest detectable volume under the current experimental condition is estimated to be ~$\lambda/4$, which is derived from the signal-to-noise ratio of ~15 : 1 in Figure 22(c) and the quadratic dependence of the CARS intensity on interaction volume. This indicates that our TE-CARS microscope is capable of sensing a vibrational-spectroscopic signal from an enormously small volume of subzeptolitter.

CONCLUSION

Intuitively optical nanotechnology implies use of light of nanometer wavelength, which covers from extreme UV to x-ray region. Light of this wave range is actually quite difficult to generate and difficult to use. In this report, we review how photons can bypass the restriction of diffraction limit so that the visible and near-infrared laser could be utilized for nanoprocessing. Two technologies are summarized. In near-field optics, the nanometric resolution relies on the photon tunneling of an evanescent wave field. The evanescent field could be constructed either on the tip end (aperture-type probe) or on the sample surface (apertureless probe). In the latter case, a sharp metallic tip has been utilized for tunneling the evanescent wavefield, which provides not only high resolution and high throughput of signal, but also enhances signals through several mechanisms. An alternative approach to break the diffraction limit is nonlinear optics. The multi-photon process narrows the response curve to light excitation. In addition, thresholding effects of materials provide further possibility to reduce excitation volume to sub-diffraction limit range.

   With the further development of nanophotonics, we wish to observe, analyze, control, modify nano-objects like atoms, molecules, protein molecules, cells, and quantum structures and devices. It is expected that nanophotonics, as an interdiscipline research field, would be utilized for nanofabrication, nano-recording, nano-analyzer, nano-operation, nano-light source, and nanomachines. Nanophotonics will not only be a subject of intense scientific research, but also a foundation for the near-future industrial application.

REFERENCES

1.   S. Kawata, M. Ohtsu, and I. Masahiro, Eds., *Near-field Optics and Surface Plasmon Polaritons* (Springer-Verlag, Heidelberg, 2001).

2. S. Kawata, M. Ohtsu, and M. Irie, Eds., *Nano-Optics* (Springer-Verlag, Heidelberg, 2002).

3. I. Newton, *Opticks: or a Treatise of the Reflections, Inflections and Colours of Light*, 2nd ed. (Dover, Mineola, NY, 1952).

4. S. Zhu, A. W. Yu, D. Hawley, and R. Roy, "Frustrated total internal reflection — a demonstration and review," *Am. J. Phys.*, 54, 601–606 (1986).

5. M. Born and E. Wolf, *Principles of Optics* (Pergamon Press, Frankfurt, 1975).

6. H. Furukawa and S. Kawata, "Local field enhancement with an apertureless near-field-microscope probe," *Opt. Commun.*, 148, 221–224 (1998).

7. H. Hatano and S. Kawata, "Applicability of deconvolution and nonlinear optimization for reconstructing optical images from near-field optical microscope images," *J. Microsc.*, 194, 230–234 (1999).

8. H. Furukawa and S. Kawata, "Analysis of image formation in a near-field scanning optical microscope: effects of multiple scattering," *Opt. Commun.*, 132, 170–178 (1996).

9. H. Furukawa and S. Kawata, "Near-field optical microscope images of a dielectric flat substrate with sub-wavelength strips," *Opt. Commun.*, 196, 93–102 (2001).

10. E. H. Synge, "A suggested method for extending the microscopic resolution in the ultramicroscopic region," *Philos. Mag.*, 6, 356–362 (1928).

11. E. A. Ash and G. Nicholls, "Super-resolution aperture scanning microscope," *Nature*, 237, 510 (1992).

12. D. W. Pohl, W. Denk, and M. Lanz, "Optical stethoscopy — image recording with resolution lambda/20," *Appl. Phys. Lett.*, 44, 651–653 (1984).

13. E. Betzig, J. K. Trautman, T. D. Harris, J. S. Weiner and R. L. Kostelak, "Breaking the diffraction barrier — optical microscopy on a nanometric scale," *Science*, 251, 1468–1470 (1991).

14. S. Kawata, Y. Inouye, and T. Sugiura, "Near-field scanning optical microscope with a laser trapped probe," *Jpn. J. Appl. Phys.*, 33, L1725–L1727 (1994).

15. T. Sugiura, T. Okada, Y. Inouye, O. Nakamura, and S. Kawata, "Gold-bead scanning near-field optical microscope with laser-force position control," *Opt. Lett.*, 22, 1663–1665 (1997).

16. R. E. Collin, *Foundations for Microwave Engineering* (McGraw Hill, London, 1966).

17. T. Nakano and S. Kawata, "Numerical aperture analysis of the near-field diffraction pattern of a small aperture," *J. Mod. Opt.*, 39, 645–661 (1992).

18. A. Ashkin, J. M. Dziedic, J. Bjorkhlm, and S. Chu, "Observation of a single-beam gradient force optical trap for dielectric particles," *Opt. Lett.*, 11, 288–290 (1986).

19. T. Sugiura, S. Kawata, and T. Okada, "Fluorescence imaging with a laser trapping scanning near-field optical microscope," *J. Microsc.*, 194, 291–294 (1999).

20. Y. Inouye and S. Kawata, "Near-field scanning optical microscope with a metallic probe tip," *Opt. Lett.*, 19, 159–161 (1994).

21. F. Zenhausern, M. P. O'Boyle, and H. K. Wickramasinghe, "Apertureless near-field optical microscope," *Appl. Phys. Lett.*, 65, 1623–1625 (1994).

22. R. Bachelot, P. Gleyzes, and A. C. Boccara, "Near-field optical microscope based on local perturbation of a diffraction spot," *Opt. Lett.*, 20, 1924–1926 (1995).

23. Y. Inouye and S. Kawata, "Reflection-mode near-field optical microscope with a metallic probe tip for observing fine structures in semiconductor materials," *Opt. Commun.*, 134, 31–35 (1997).

24. E. Betzig and R. J. Chichester, "Single molecules observed by near-field scanning optical microscope," *Science*, 262, 1422–1425 (1993).

25. N. Hayazawa, Y. Inouye, and S. Kawata, "Evanescent field excitation and measurement of dye fluorescence in a metallic probe near-field scanning optical microscope," *J. Microsc.*, 194, 472–476 (1999).

26. R. Menzel, *Photonics: Linear and Nonlinear Interactions of Laser Light and Matter* (Springer-Verlag, Berlin, 2001).

27. Y. H. Pao and P. M. Rentzepis, "Laser-induced production of free radicals in organic compounds," *Appl. Phys. Lett.*, 6, 93–94 (1965).

28. S. Maruo, O. Nakamura, and S. Kawata, "Three-dimensional microfabrication with two-photon-absorbed photopolymerization," *Opt. Lett.*, 22, 132–134 (1997).

29. G. Odian, *Principles of Polymerization*, 3rd ed. (John Wiley & Sons, New York, 1991).

30. P. J. Flory, *Principles of Polymer Chemistry* (Cornell University Press, New York, 1952).

31. T. Tanaka, H.-B. Sun, and S. Kawata, "Rapid sub-diffraction-limit laser micro/nanoprocessing in a threshold material system," *Appl. Phys. Lett.*, 80, 312–314 (2002).

32. H.-B. Sun and S. Kawata, "Two-photon laser precision microfabrication and its applications to micro-nano devices and systems," *J. Lightwave Technol.*, 21, 624–633 (2003).

33. S. Kawata, H.-B. Sun, T. Tanaka, and K. Takada, "Finer features for functional microdevices," *Nature*, 412, 697–698 (2001).

34. H.-B. Sun, T. Tanaka, K. Takada, and S. Kawata, "Two-photon photopolymerization and diagnosis of three-dimensional microstructures containing fluorescent dyes," *Appl. Phys. Lett.*, 79, 1411–1413 (2001).

35. H.-B. Sun, K. Takada, and S. Kawata, "Elastic force analysis of functional polymer submicron oscillators," *Appl. Phys. Lett.*, 79, 3173–3175 (2001).

36. C. M. Soukoulis, Ed., *Photonic Crystals and Light Localization in the 21st Century*, NATO Science Series, Series 3, Vol. 563 (Kluwer Academic Press, Dordrecht, 2000).

37. C. M. Soukoulis, Ed., *Photonic Band Gap Materials*, NATO Asi Series. Series E, No. 315, (Kluwer Academic Press, Dordrecht, 1995).

38. Y. N. Xia, B. Gates, and Z. Y. Li, "Self-assembly approaches to three-dimensional photonic crystals," *Adv. Mater.*, 13, 409–413 (2001).

39. H.-B. Sun, J. F. Song, Y. Xu, S. Matsuo, H. Misawa, G. T. Du, and S. Y. Liu, "Growth and property characterizations of photonic crystal structures consisting of colloidal microparticles," *J. Opt. Soc. Am. B*, 17, 476–480 (2000).

40. S. Noda, A. Chutinan, and M. Imada, "Trapping and emission of photons by a single defect in a photonic bandgap structure," *Nature*, 407, 608–610 (2000).

41. S. Y. Lin, E. Chow, V. Hietala, P. R. Villeneuve, and J. D. Joannopoulos, "Experimental demonstration of guiding and bending of electromagnetic waves in a photonic crystal," *Science*, 282, 274–276 (1998).

42. U. Gruning, V. Lehmann, and C. M. Engelhardt, "2-dimensional infrared photonic bandgap structure based on porous silicon," *Appl. Phys. Lett.*, 66, 3254–3256 (1995).

43. S. Kawakami, "Fabrication of submicrometre 3D periodic structures composed of Si/SiO_2," *Electron. Lett.*, 33, 1260–1261 (1997).

44. Z.-Y. Li, J. Wang, and B.Y. Gu, "Full band gap in fcc and bcc photonic band gaps structure: non-spherical atom," *J. Phys. Soc. Jpn.*, 67, 3288–3291 (1998).

45. O. Toader and S. John, "Proposed square spiral microfabrication architecture for large three-dimensional photonic band gap crystals," *Science*, 292, 1133–1135 (2001).

46. K. Kaneko, H.-B. Sun, X. M. Duan, and S. Kawata, "Submicron diamond-lattice photonic crystals produced by two-photon laser nanofabrication," *Appl. Phys. Lett.*, 83, 2091–2093 (2003).

47. H.-B. Sun, S. Matsuo, and H. Misawa, "Three-dimensional photonic crystal structures achieved with two-photon-absorption photopolymerization of resin," *Appl. Phys. Lett.*, 74, 786–788 (1999).

48. S. Shoji and S. Kawata, "Photofabrication of three-dimensional photonic crystals by multibeam laser interference into a photopolymerizable resin," *Appl. Phys. Lett.*, 76, 2668–2670 (2000).

49. J. H. Strickler and W. W. Webb, "3-dimensional optical data storage in refractive media by 2-photon point excitation," *Opt. Lett.*, 16, 1780–1782 (1991).

50. S. Kawata and Y. Kawata, "Three-dimensional optical data storage using photochromic materials," *Chem. Rev.*, 100, 1777–1788 (2000).

51. M. Irie, "Diarylethenes for memories and switches," *Chem. Rev.*, 100, 1685–1716 (2000).

52. A. Toriumi, S. Kawata, and M. Gu, "Reflection confocal microscope readout system for three-dimensional photochromic optical data storage," *Opt. Lett.*, 23, 1924–1926 (1998).

53. J. T. Krug II, E. J. Sánchez, and X. S. Xie, "Design of near-field optical probes with optimal field enhancement by finite difference time domain electromagnetic simulation," *J. Chem. Phys.*, 116, 10895–10901 (2002).

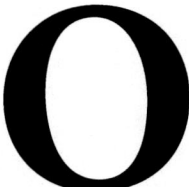

OPTICAL COUPLING IN WAVEGUIDES

Kazuhito Furuya

The power transfer, that is, the optical or mode coupling is caused by any coupling mechanism between two waves. If such couplings, are not there the multimode fibers would be used as single-mode transmission media by using only one mode to transmit a narrow pulse without any broadening. However, in practice, mode couplings are unavoidable. On the other hand, functional optical devices, such as directional-couplers, switches, etc., use the coupling as their operational principle. The basics of the coupling are described here.

Let us consider optical Waves 1 and 2 of angular frequency ω propagating in the +z direction. Waves 1 and 2 may be waves propagating either along two parallel waveguides 1 and 2 or as two Modes 1 and 2 in a single waveguide. Their amplitudes are $a_1(z)$ and $a_2(z)$, while $P_1(z) = |a_1(z)|^2$ and $P_2(z) = |a_2(z)|^2$ are their powers. When there is no coupling, amplitudes are given as

$$a_1(z) = a_{10} \exp(-i\beta_1 z) \qquad a_2(z) = a_{20} \exp(-i\beta_2 z),$$

where β_1 and β_2 are propagation constants of waves. When there is coupling, the equation can be written as

$$\frac{da_1}{dz} = -\beta_1 a_1 + C_{12} a_2 \qquad \frac{da_2}{dz} = -\beta_2 a_2 + C_{21} a_1, \tag{1}$$

where C_{12} is the coupling coefficient representing the amplitude transfer from Wave 2 to Wave 1 while C_{21} is that from Wave 1 to Wave 2. Assuming lossless waveguide, that is, $(d/dz)(P_1 + P_2) = 0$, condition $C_{12} = -C_{21}^*$ is derived from Eq. (1). In the following sections, weak coupling thoery, periodic coupling, random coupling, and microbending loss in fibers are described.

WEAK COUPLING THEORY

Assuming that $|C_{12}| \ll \beta_1, \beta_2$ (weak coupling) and constant C_{12}, substituting $a_1(z) = A\exp(-i\beta z)$ and $a_2(z) = B\exp(-i\beta z)$ into Eq. (1) we obtain solutions for β, $a_1(z)$, and $a_2(z)$ as follows:

$$\beta = \beta_a \pm \beta_b,$$

where $\beta_a = (\beta_1 + \beta_2)/2$ (average propagation constant), $\beta_b = \sqrt{\beta_d^2 + |C_{12}|^2}$ (beat propagation constant) and $\beta_d = (\beta_1 - \beta_2)/2$.

$$a_1(z) = \exp(-i\beta_a z)\left\{\left(\cos\beta_b z - i\frac{\beta_d}{\beta_b}\sin\beta_b z\right)a_1(0) + \frac{C_{12}}{\beta_b}\sin\beta_b z a_2(0)\right\},$$

$$a_2(z) = \exp(-i\beta_a z)\left\{\left(\cos\beta_b z + i\frac{\beta_d}{\beta_b}\sin\beta_b z\right)a_2(0) + \frac{C_{21}}{\beta_b}\sin\beta_b z a_1(0)\right\}.$$

Substituting $a_1(0) = 1$ and $a_2(0) = 0$ as the initial condition, we obtain

$$P_1(z) = 1 - F\sin^2\beta_b z \quad P_2(z) = F\sin^2\beta_b z, \quad \text{where } F = \frac{1}{1 + \left\{(\beta_1 - \beta_2)/2|C_{12}|\right\}^2}.$$

Factor F is the maximum ratio of the power transfered from one wave to the other, to the total power carried by the waves. The maximum transfer takes place at the distance of the coupling length $\pi/2\beta_b$ multiplied by an integer. Between the two waves, fractional power is transferred back and forth as shown in Figure 1. When two propagation constants coincide, $\beta_1 = \beta_2$, F is unity (100% power transfer). A relation between F and the mismatch in the propagation constant is shown in Figure 2. This is useful in designing directional coupler type devices.

PERIODIC COUPLING

When $\beta_1 \neq \beta_2$ and the coupling coefficients are periodic functions with the period $2\pi/|\beta_1 - \beta_2|$ as

$$C_{12} = iC\exp\left(-i(\beta_1 - \beta_2)z\right) \qquad C_{21} = iC\exp\left(i(\beta_1 - \beta_2)z\right),$$

100% power transfer takes place. These satisfy the lossless condition $C_{12} = -C_{21}^*$. This is shown by substituting the previous and the following equation

$$a_1(z) = A_1(z)\exp(-i\beta_2 z) \qquad a_2(z) = A_2(z)\exp(-i\beta_2 z) \quad \text{into Eq. (1) to obtain}$$

$$a_1(z) = a_1(0)\cos Cz\exp(-i\beta_1 z) \qquad a_2(z) = ia_1(0)\sin Cz\exp(-i\beta_2 z).$$

RANDOM COUPLING

In multimode waveguides, for example, multimode fibers, even if single mode is excited, after the propagation, the power distributes over multiple modes by the mode coupling. The cause of the coupling is imperfect homogeneity of the waveguide structure along the axial direction [1,2], for example, as a result of bending of the fiber axis. Such a coupling mechanism varies at random along the axial direction.

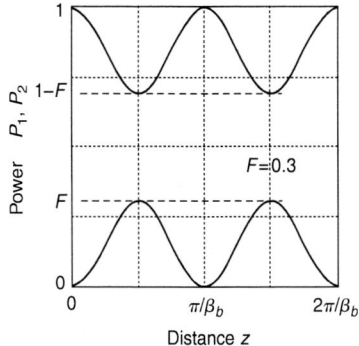

Figure 1   Powers in Waves 1 and 2 versus propagation distance

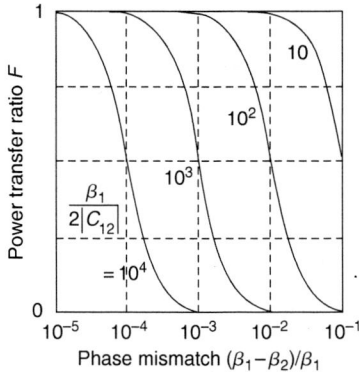

Figure 2   Power transfer ratio versus mismatch in propagation constant

Let us consider that only Wave 1 carries the power at $z = 0$, $\beta_1 \neq \beta_2$ and, therefore, the power of Wave 2 is always much smaller than that of Wave 1. Neglecting the second term in the right-hand side of the first equation in Eq. (1) we get

$$a_1(z) = a_1(0)\exp(-i\beta_1 z).$$

Substituting this into the second equation in Eq. (1)

$$a_2(z) = \exp(-i\beta_2 z)\int_0^z C_{21}(z')a_1(0)\exp\left(i(\beta_2 - \beta_1)z'\right)\,dz'.$$

Since $C_{21}(z')$ is a statistical function of z', any deterministic analysis is unavailable. However, we can obtain $P_2(z)$ as follows:

$$P_2(z) = |a_2(z)|^2$$

$$= |a_1(0)|^2 \int_0^z\int_0^z C_{21}(z')C_{21}^*(z'')\exp\left(i(\beta_2 - \beta_1)z'\right)\exp\left(-i(\beta_2 - \beta_1)z''\right)\,dz'\,dz''$$

$$= z|a_1(0)|^2 \int_{-\infty}^{\infty} R(u)\exp\left(i(\beta_1 - \beta_2)u\right)\,du$$

$$= z|a_1(0)|^2 S(\beta_1 - \beta_2),$$

where $R(u)$ is the autocorrelation function of the function $C_{21}(z')$, that is,

$$R(u) = \lim_{z \to \infty} \frac{1}{z} \int_0^z C_{21}(z') C_{21}^*(z' + u) \, dz'.$$

$R(u)$ is symmetric with respect to the origin and decreasing function with $|u|$. A characteristic length, such as l in $R(u) = \exp(-(u/l^2)$ for example, is called as the correlation length. Here we assume $z \gg l$ in the derivation of the earlier formula. $S(\beta)$ is the Fourier transform of $R(u)$ and is equal to the power spectrum (Wiener–Khinchin's theorem). Finally, the power of Wave 2 is proportional to the distance, and the power spectral density at $\beta_1 - \beta_2$ is the statistic function of the coupling coefficient $C_{21}(z)$. Function $S(\beta)$ is determined by the statistical property of the coupling mechanism. In many cases, it is a decreasing function with the argument. That is, when $\beta_1 - \beta_2$ is larger, the power transfer is smaller. Therefore, to suppress the optical coupling between waves, $\beta_1 - \beta_2$ should be large so that $S(\beta_1 - \beta_2)$ is small. To suppress the coupling between two orthogonal polarization in a single-mode fiber, anisotropy in the refractive index is formed by inroducing the anisotropic stress.

MICROBENDING LOSS IN FIBER—AN EXAMPLE OF RANDOM COUPLING

The optical axis of the fiber is not straight but has minute bends. Such microbendings cause a transmission loss. Microbendings are unavoidably introduced during manufacturing, that is, jacketing, stranding, and cabling processes. Therefore, structural parameter of optical fibers should be designed taking the microbending loss into account. Microbendings couple the guided mode with radiation modes to cause a radiation loss. The coupling coefficient varies statistically along the distance as discussed in Random coupling section. The identification of the power spectrum function and parameter dependencies of the microbending loss [3] are discussed here. Figure 3 shows increases in loss caused by the jacketing and cabling process.

From this process, Gaussian power spectral function well explains measured spectrum of the loss increase caused by the jacketing and cabling processes. Using Gaussian function, theoretical microbending loss is given as

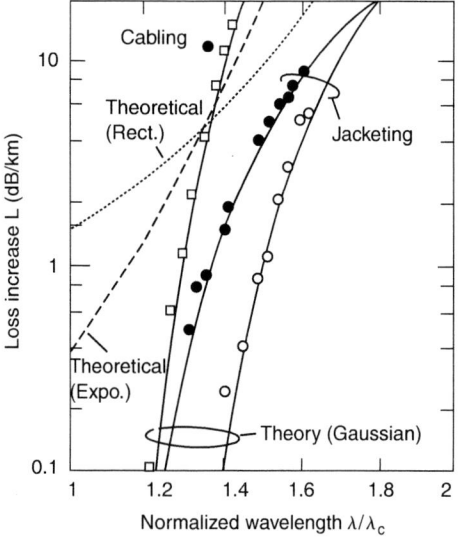

Figure 3   Theoretical and measured spectral loss increase caused by jacketing and cabling

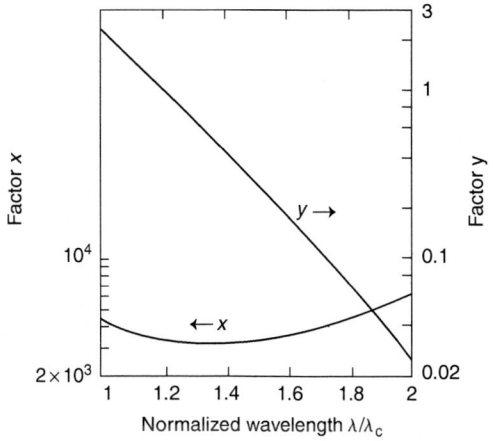

Figure 4 Factors x and y

$$L = xN\left(\frac{1}{R_1}\right)^2 \overline{W}^2 \frac{1}{\Delta}\exp\left(-y\left(\frac{\overline{W}n_{\text{core}}}{\lambda_c}\right)^2\Delta^2\right)(\text{dB/km})$$

where x and y are constants as given in Figure 4, N is the average number of bends per meter, $(1/R_1)^2$ is the mean square of the curvature of the fiber axis, $\overline{W}$ is the correlation length of the curvature of the fiber axis, $\Delta (= (n_{\text{core}} - n_{\text{clad}})/n_{\text{core}})$ is the relative refractive index difference between the core and the cladding of the fiber, and λ_c is the cutoff wavelength of the higher-order mode.

REFERENCES

1. D. Marcuse, "Mode conversion caused by surface imperfactions of a dielectric slab waveguide," *Bell Syst. Tech. J.*, 48, 3187–3215, 1969 and D. Marcuse, "Radiation losses of dielectric waveguides in terms of the power spectrum of the wall distortion function," *Bell Syst. Technol. J.*, 48, 3233–3242, 1969.
2. D. Marcuse, *Light Transmission Optics*, Van Nostrand-Reinhold, New York, 1972, and D. Marcuse, *Theory of Dielectric Optical Waveguides*, Academic Press, New York, 1974.
3. K. Furuya and Y. Suematsu, "Random-bend loss in single-mode and parabolic-index multimode optical fiber cables," *Appl. Opt.*, 19, 1493–1500, 1980.

Optical Coupling of Laser and Fiber

Kenichi Iga and Yasuo Kokubun

The optical coupling efficiency of laser-to-fiber is defined by the overlap integral of the optical field at the facet of optical fiber and the focused light from laser. In general, the coupling efficiency of input optical field $E_1(x, y)$ to guided mode $E_2(x, y)$ is express by

$$\eta = \frac{\left| \iint E_1(x,y) \cdot E_2^*(x,y) dx\, dy \right|^2}{\left[\iint |E_1(x,y)|^2 dx\, dy \right] \cdot \left[\iint |E_2(x,y)|^2 dx\, dy \right]} \tag{1}$$

where * denotes the complex conjugate. The integral at the position behind the focusing lens, can be easily calculated as shown in Figure 1, since the far field of optical fiber is calculated by

$$\Psi(x',y') = \frac{j}{\lambda l_2} e^{-jkl_2} \cdot \iint \Psi_2(x,y) \cdot \exp\left[-jk \frac{(x-x')^2 + (y-y')^2}{2 l_2} \right] dx\, dy, \tag{2}$$

where the arguments x' and y' are the cordinates in the plane behind the focusing lens and k is the propagation constant in a vacuum. In such a case, the coupling efficiency is written as: [1]

$$\eta = \frac{\left| \iint \Psi_1(x',y') \cdot L \cdot \Psi_2^*(x',y') dx'\, dy' \right|^2}{\left[\iint |\Psi_1(x',y')|^2 dx'\, dy' \right] \cdot \left[\iint |\Psi_2(x',y')|^2 dx'\, dy' \right]} \tag{3}$$

where L is the transfer function of the focusing lens expressed in terms of the focal length f given by

$$L = \exp\left[jk \frac{x'^2 + y'^2}{2f} \right]. \tag{4}$$

A put-in microconnector, a method of self-aligned optical coupling scheme, is proposed for efficient coupling of laser and fiber (Figure 2).

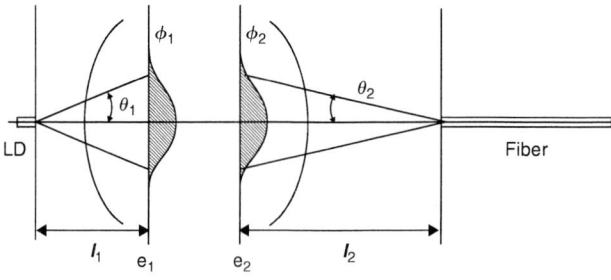

Figure 1   Analytical mode of light coupling from laser-to-fiber

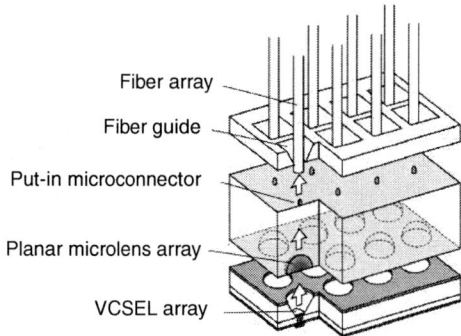

Figure 2 A light coupling scheme using a put-in microconnector array and vertical cavity surface emitting laser array

REFERENCES

1. R. E. Wagner and W. J. Tomlinson, "Coupling efficiency of optics in single-mode fiber component," *Appl. Opt.*, 21, 2671–2688 (1982).
2. K. Iga, M. Oikawa, S. Misawa, J. Banno, and Y. Kokubun, *Appl. Opt.*, 21, 3456 (1982).
3. A. Sasaki, T. Baba, and K. Iga, "Put-in microconnectors for alignment-free coupling of optical fiber arrays," *IEEE Photonics Technol. Lett.*, 4, 908–910 (1992).

OPTICAL DISK PICKUP

Kenya Goto

WHY NEEDED?

Since CD players were in the market by October 15, 1982, various kinds of optical pickups have been produced in the world. An optical pickup is the key device in a CD player, which consists of a high NA (numerical aperture) objective lens, a laser diode chip, photo diode chips, a two-axis mechanical actuator for focusing and tracking of the objective lens, optical systems for focusing error signal detection, and tracking error signal detection. As the target of the optical disk system is to focus the laser beam as small as possible for larger memory capacity, keeping the wave front aberration within the Maréchel criterion from the laser facet to the objective lens is a very important technology. The grade of this aberration technology is the same as in the high NA optical microscope. In order to make the pickup production cost lower, many kinds of integrated optics including waveguide optics have been studied; however, problems of the wave front aberration of the laser diode through the objective lens to the disk substrate and larger dynamic movement of both focusing and tracking prevent the integrated optical pickup from adoption in real players. The conventional optical pickup however adopts diffractive optical elements in only signal detection devices for both the focusing and the tracking error detection optical systems because there is no aberration problem, where several optical components combined into a DOE (diffractive optical element) or a HOE (holographic optical element).

Recently a high-density optical disk system of super-parallel optical heads [1,2], using a two-dimensional VCSEL (vertical cavity surface emitting laser [3]) array was studied for the higher data transfer rate and technological capability. Optical heads of the VCSEL array and microlens array play a key role to get higher evanescent light from a small aperture for the optical disk system, of which the disk surface is coated with lubricant and a protective film on the flat recording medium surface in order to keep the gap between the super-parallel optical head and the disk surface within 20 nm [1]. Higher throughput efficiency has been obtained in the near-field semiconductor optical probe array head [4]. However, the obtained evanescent light power is about 10 μW from the 100 nm probe aperture using a VCSEL 1 mW laser output power, which is still not enough to write bits on the phase change optical disk medium. One solution for improvement of the writing power to get 20 μW evanescent light for 30 nm bit size is to develop a special nano-fabricated corrugated thin metal film for higher throughput efficiency by surface plasmon polariton (SPP) enhancement [5,6]. A metal fine grating fabrication method to get evanescent light wave resonantly enhanced has been studied with the finite-difference time-domain (FDTD) simulation results.

PRINCIPLE

Diffractive Optical Element

Figure 1 shows an integrated optical unit for the optical signal detection (digital bit signal, focusing and tracking error signal for both CD and DVD) in recent CD/DVD optical pickup heads. Conventional pickup heads have each CD and DVD unit in one pickup. Using a DOE, that is a grating, a laser beam splits into three beams by diffraction effect. The three beams reflected back from the optical disk bit-pattern split again into nine beams. Combined with multi-divided photo diodes these nine beams can make focusing and tracking error signals. A DOE consists of a glass plate and a thin plastic film engraved by the mold, which is fabricated with the electron beam lithography. A wave plate operates as one of the tracking signal dividers from the focusing signal. To get these error signals there should be signal subtraction or signal adder using multi-divided photodiode chips and electronic circuits.

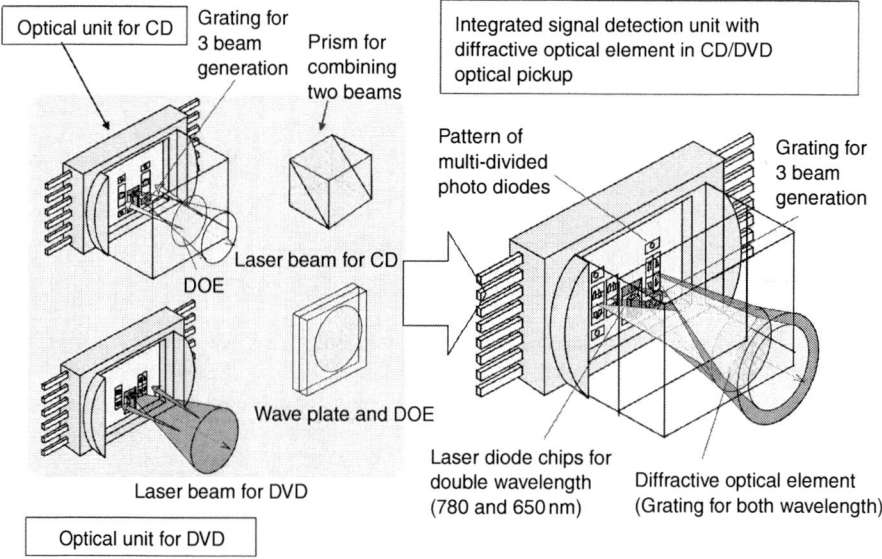

Figure 1   Integrated optical unit in a pickup for error signal detection of both focussing and tracking

Principle of Three-Dimensional Integrated Optical Head

The largest optical-memory capacity will be about 100 GB per surface in a 120 mm diameter optical disk if the objective lenses such as more than 0.85 NA are used. Much larger capacity cannot be realized as far as the lens is adopted. In recent years, it is required to realize an ultra-high density optical memory for the new trends of digital applications for moving pictures and the rapid increase of data capacity in personal computers. Among many approaches to increase the memory capacity in the optical data storage, a near-field optical system has been considered as one of candidates for next generation data storage [7]. However, optical power of an evanescent wave at the nano-aperture is very weak due to the cutoff wavelength and low optical throughput of the near-field probe, resulting in many difficulties to apply it to the real optical data storage. To improve the optical efficiency of nano-aperture probe and realize fast data transfer rate, the parallel optical array head has been proposed and prepared using a VCSEL [3] array and a flat-tip microprobe array [1,2,4–6,8–11]. Since the two-dimensional array system is based on a multi-beam recording with a small spot size with an aid of high efficiency microprobe design, it has advantages for both approaches of fast data-transfer rate and high-memory capacity [1,2]. Figures 2 to 4 show the schematic diagram of two-dimentional array recording system and the integrated head structure. As shown in Figure 4, the VCSEL array is inclined at a specific angle (e.g., 0.57° for 100 × 100 array head) to the track direction to align all light sources on separate data tracks [1,2]. In this array head, all lasers are manipulated simultaneously to record the data on and read them from multi-tracks. The realization of small laser spot with sufficient power is an important requirement for increasing the memory capacity in this parallel optical system. The integrated structure of VCSEL and flat-tip microprobe arrays has also been studied to produce more compact optical array head [4–6,8–11], as shown in Figure 2. The light from VCSEL is focused using microlens and guided to the nano-aperture of microprobe flat-tips. The optical throughput was increased with the use of semiconductor materials of high refractive index and flat-tip structure for higher intensity of laser beam at the aperture, resulting in the improved optical throughput of 1.25% with the 150 nm aperture in the

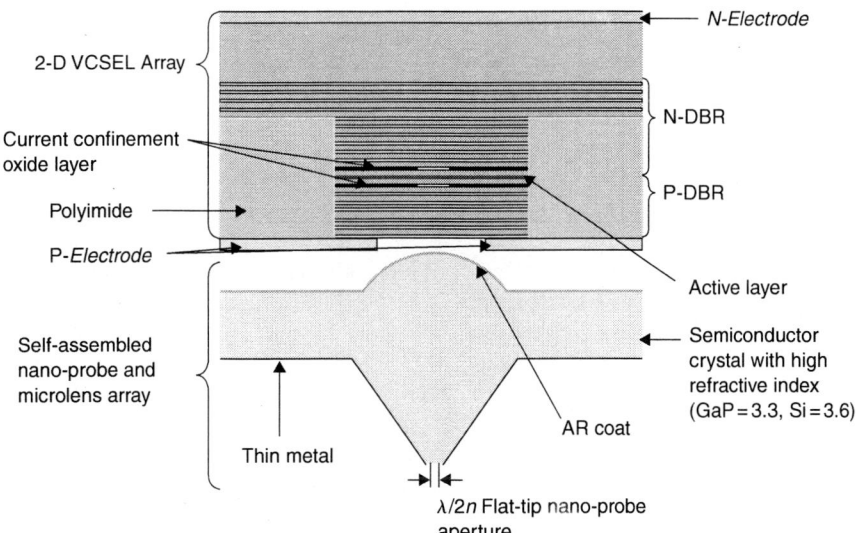

Figure 2   Only one element of the near-field optical head is shown for the schematic near-field semiconductor optical probe with high throughput consisting of VCSEL and lens

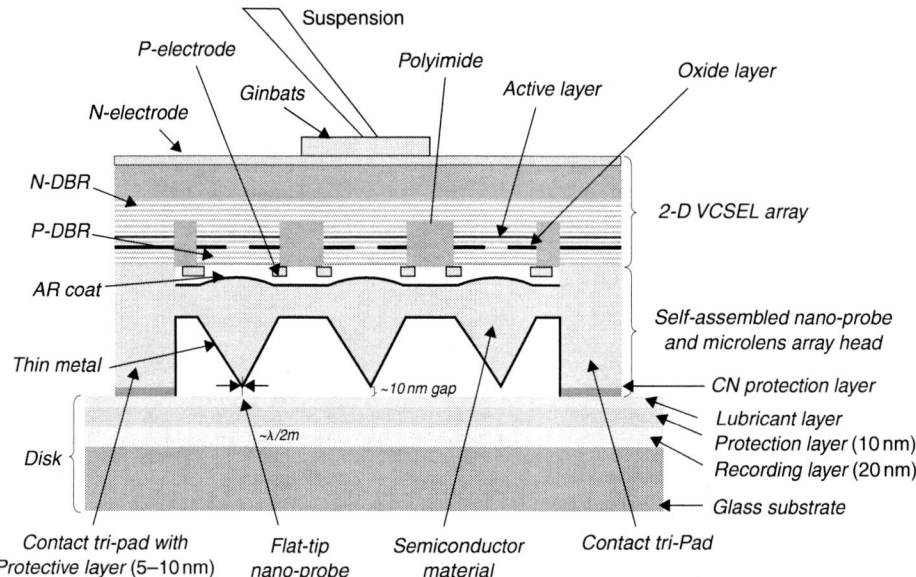

Figure 3   Cross sectional concept of the two-dimensional optical array head. A part of the head is included with VCSEL, microlens, and near-field probe array. The pitch of the element is 25 μm in this case

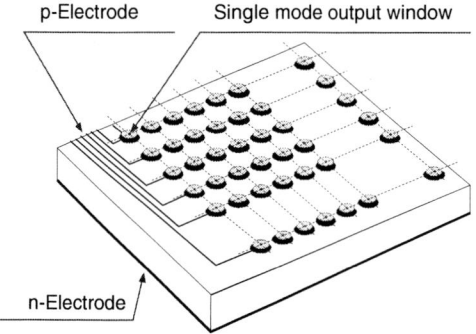

Figure 4   Schematic figure of a 2D array super-parallel head, which is inclined to the disk tangential direction as shown in Reference 1

GaP semiconductor microprobes [4,12]. However, it is not enough to record the 150 nm or less marks on the conventional phase change optical media because the near-field optical power from 1 mW single mode VCSEL is only 12.5 μW for a 150 nm aperture in the experiment [4,12], and finally 100 μW evanescent power is needed to write on the disk. Thus, the enhancement design of optical throughput in the nano-aperture head is still required. There are some efforts to enhance the near-field optical throughput for the nano-apertures, including metal structure modifications and propagation mode control methods [13–16]. From the FDTD simulation, the optical throughput was improved with the buried type microprobe with asymmetric metal-coated structure since it is a better structure for the coupling efficiency between propagating wave and surface plasmon polarition [10]. The "Scoop" type of metal coating on the microprobe (metal coating on just three side planes) also improved the light field intensity at the metal aperture and decreased the beam spot size. Theoretical calculation of high optical power throughput using the "C"-aperture design was reported with the optimization of the aperture structure design and metal thickness control to increase the resonant transmission up to ~1000 times power throughput [13].

In this study, we have developed new head structure using the microlens and metallic film grating for the surface plasmon enhancement [17,18]. The integrated optical head of VCSEL and microprobe array has been modified to increase the optical throughput with the patterned two-dimensional metal grating by the surface plasmon resonance effect. The FDTD numerical simulation was conducted to find the best structural design of metal grating, including the grating pitch, grating width, and metal thickness. Finally, basic experimental approach will be discussed to realize the integrated array head and apply it to the real optical data storage.

DEVICE CONCEPT, DESIGN, AND FABRICATION OF THE INTEGRATED OPTICAL NEAR-FIELD PROBE

Figure 2 shows the integrated optical near-field probe consisting of the VCSEL, the high NA micro-lens, and the self-aligned NF-Probe [16] (Pyramidal Prism Probe: PPP). This figure is only one part of the integrated array head. The VCSEL consists of a multi-quantum active layer and a current confinement oxide layer sandwiched between two distributed Bragg reflector (DBR) multi-layers. An Ohmic contact layer and electrode layers are formed on both n-DBR layer and p-side substrate.

Figure 2 is one part of the fabricated conventional near-field optical disk head, in which the laser light comes from VCSEL and is focused through the microlens on the aperture of the GaP probe. With an aid of the focused light and high refractive index of GaP, the throughput of this probe is about 1% for the 100 nm aperture according to the experiment [4,12]. The output window size of a VCSEL is about 8 μm for a 10 μm column size VCSEL. The pitch between columns is 25 μm as shown in Figure 3, which also shows only a part of the cross-sectional view of the arrayed head. The two-dimensional array is schematically shown in Figure 4. The surface profile of the microlens in Figure 2 is the semi-sphere, which is fabricated with ion beam shower made from continuous etching mixed gas (Cl_2 + Ar) concentration during either the ion milling process or the reactive ion etching process. The beam entrance surface of the semiconductor lens is anti-reflection coated with ECR sputtering using mixed gas with oxygen and nitrogen. Most important technology of this probe array fabrication is the development of the self-alignment method [16] between the focal point of the lens and the aperture central position of the semiconductor near-field optical probe. First fabrication of the 100×100 VCSEL array is carried out using an epi-wafer processed in a VCSEL wafer manufacturer and the microlens array is formed using a thin semiconductor wafer (Si or GaP) for the combined devices of lens and probe array with use of the same photolithographic mask. After completing the VCSEL array and the lens array (lens substrate will still remain plane), a special infrared sensitive photo-film material layer and a photo-lithographic resist layer are coated on the focal plane of the lens substrate (as shown in Figure 5). The visible VCSEL or the infrared VCSEL must each emit laser light and the light beam focuses on the focal plane of the semiconductor surface. With the proper laser light exposure the infrared sensitive photo-film material turns black. After developing and fixing the IR sensitive material using the hypo liquid the UV-light for the photo-resist layer is exposed from the bottom of the plate. As the resolution of the infrared sensitive photo material is not small to fit the NF tip, the black size is bigger than the estimated tip aperture size of the PPP; so it is good for the self-aligned fabrication mask like SiO_2, which can be formed using the conventional photo-lithography for the PPP. With the mask made of such SiO_2 combined with subtle etching technology, the PPP fabrication and the thin metal

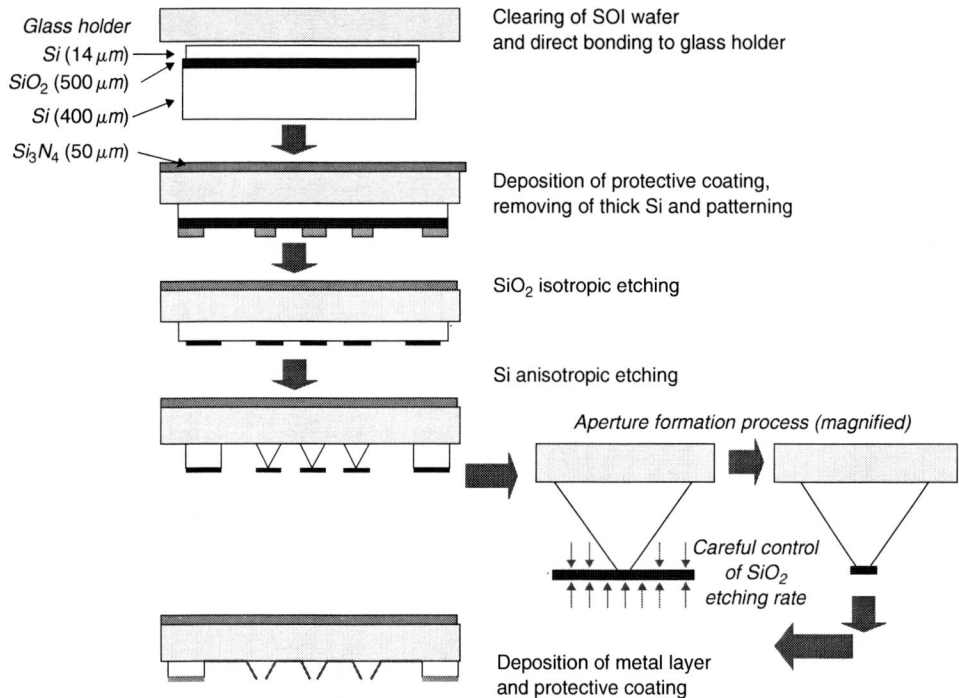

Figure 5 Fabrication method of 2D semiconductor near-field probe array made of Ge or GaP material and each has the same probe height

coating were both down before getting rid of the SiO$_2$ mask completely. This process is described in Figure 5 as a conventional PPP fabrication method. This is the reason why the height of all PPPs is exactly the same as that of the plate thickness of the lens and the PPP material. This same-height feature is very important for the optical-contact head where the only contact of the tri-pads to the disk is through the lubricant over the cover layer of the recording medium [1,2,8], and all the PPP tips are apart from the lubricant layer with about 10 nm gap. The disk rotation speed is rather slow because the contact head consists of 2500 near-field probes (when the array is of 50×50 elements). The roughness of the microlens surface is 17.65 nm [12] (rms), which is quite smooth after getting rid of the contamination deposited on its surface by chemicals resulting out of the dust from the photo-resist etching and the chlorine-active gas in the RIE chamber. By the ion beam irradiation for removing the contamination, using oxygen shower after forming the microlenses array on the surface of GaP or Si as shown later in Figure 9 which was taken using AFM.

The self-alignment process is shown in Figure 6. and the microlens fabrication process in Figure 7. For the first step the conventional photo-resist layer is formed by means of the spray coating method on the surface of the thin semiconductor substrate. Column patterning is formed by the photo-lithographical method. When the substrate is heated up to about 200°C, the photo-resist columns become semi-spherical in shape due to surface tension. After heat treatment and being exposed in the ion beam chamber the Si or GaP plate is etched and the shape of the semi-sphere is transferred to the semiconductor material. Thus the spherical curvature can be fabricated as shown in the SEM picture of Figure 8. The curvature of the lens array has been measured by the AFM as shown in Figure 9. This curvature can be calculated by the ray tracing method using the

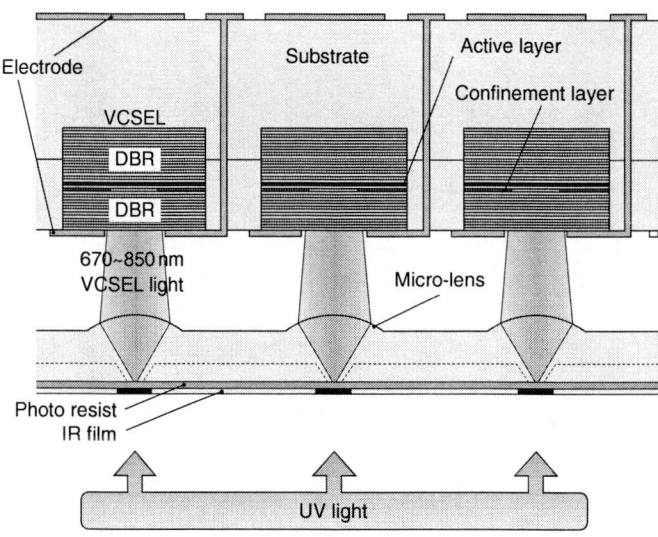

Figure 6 Fabrication method to obtain the nanometer controlled adjustment between focal points of each VCSEL light and each aperture of the probe array, where photo-resist and IR film resin are specially used

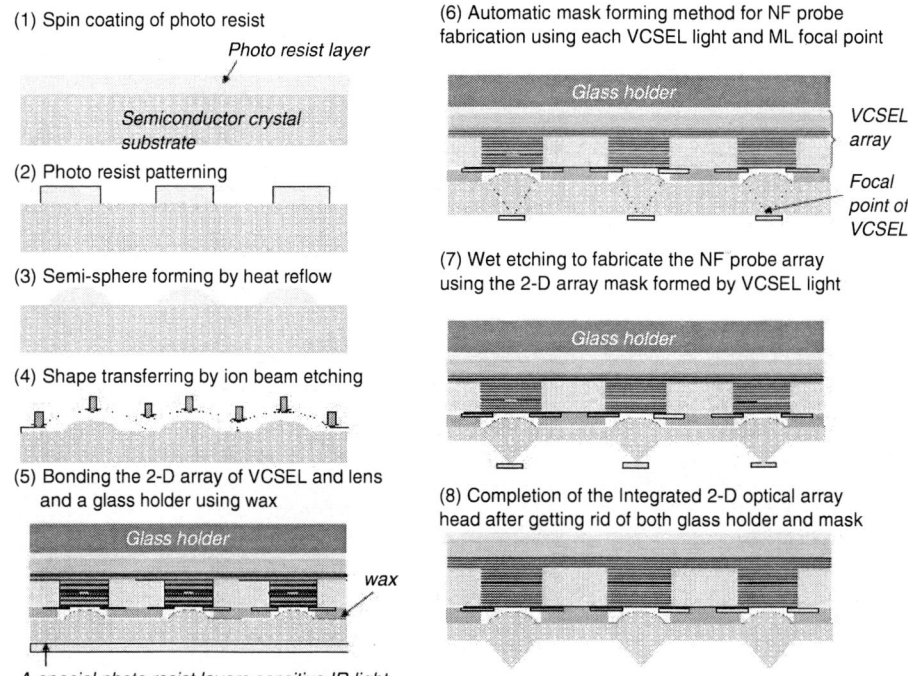

Figure 7 Nanometer controlled 2D microlens array and N–F probe array fabrication method using a self-alignment technology with VCSEL

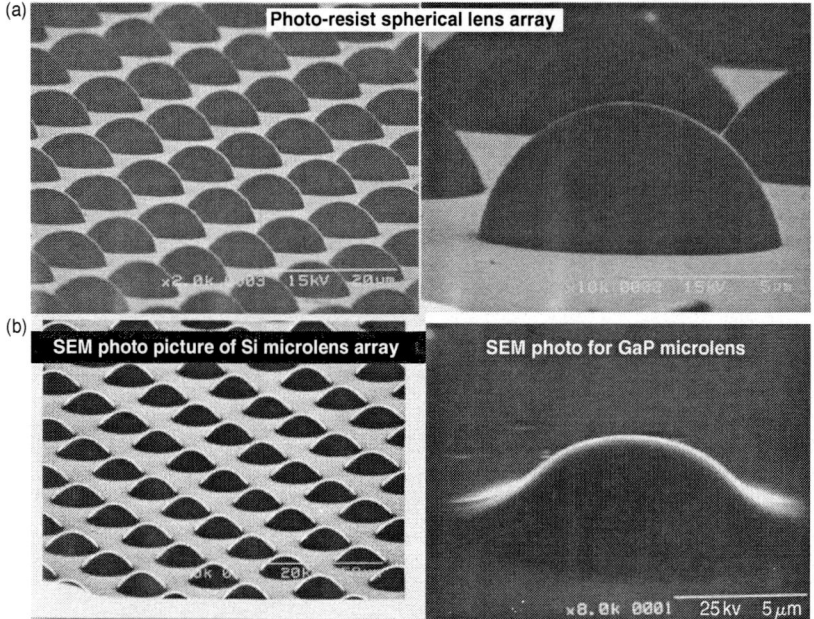

Figure 8 (a) Shows the semi-sphere resists lens array and (b) shows the transferred semiconductor microlens array

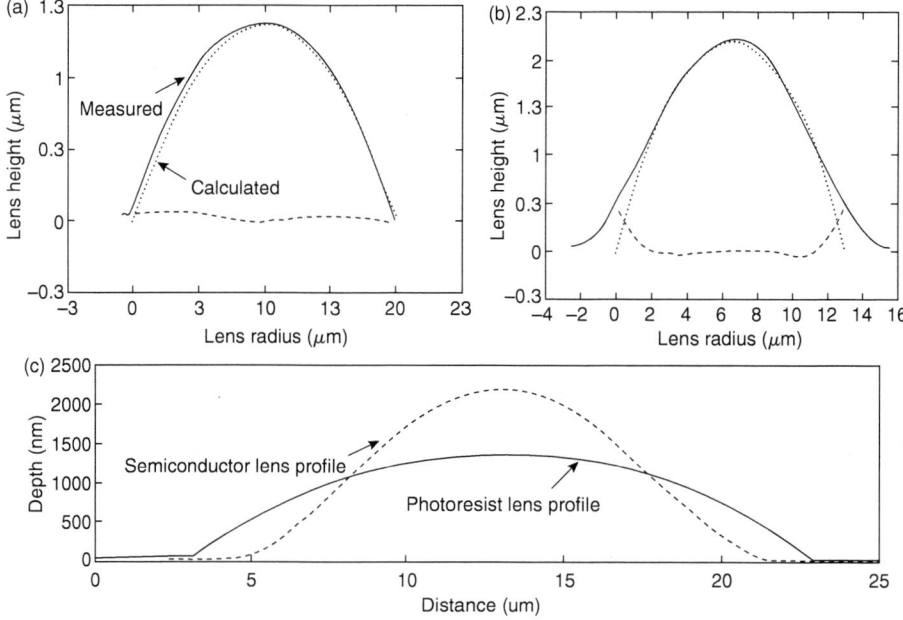

Figure 9 Semiconductor microlens profile measured with AFM (a) spherical photo-resist lens, (b) GaP lens, and (c) comparison between photo-resist lens and semiconductor GaP lens. They are all spherical curves with about $10\,\mu$m diameter

Snell's equation with both conditions of the exact focal point on the $x - y$ co-ordinates and the diffraction angle of the VCSEL output with its wavelength. The calculated throughput efficiency of this probe is about 3.5% when the aperture of the PPP is 100 nm and the refractive index of the GaP probe is 3.3 for 670 nm wavelength and the NA of the spherical lens is 0.5. This figure of 3.5% is the case of the whole Gaussian beam of the VCSEL light, however the peak power of the evanescent light just beneath the aperture of the PPP will be extracted more than that. Perhaps it will be 10% of the input power. The required light power from the top aperture of the PPP is about 100 μW when the aperture size is 100 nm. This means that each VCSEL output power should be about 1.5 mW for the case of 3.5% and can be estimated to about 0.6 mW for the case of 10% throughput. If the PPP aperture size is 30 nm, then the needed light power from the PPP for the phase change optical disk writing will be about 10 μW. In the latter case the VCSEL output power will be only 100 μW when the throughput efficiency is 0.1% in this narrow aperture case, then the total VCSEL power for 50×50 array will be 250 mW, which will be reasonable value for the VCSEL array head. Figure 9 shows maximum deviation of 40 nm between calculated and measured curvatures of the microlens. The focal length of microlens prepared in this research was measured. It was found that the self-alignment between the focal position of the microlens and the exact aperture position of the near-field optical probes can be adjusted using the self-alignment technology (Figure 6). The lens diameter of about 10 μm shows any difference in beam waist size between spherical and aspherical microlens.

NEAR-FIELD PROBE THROUGHPUT IMPROVEMENT BY CORRUGATED THIN METAL WITH SURFACE PLASMON

There will be more improvement for the probe throughput efficiency by applying the SPP effect to the probe outer surfaces with thin Au metal-film coating. The surface plasmon polariton resonance can make the power enhancement even if the aperture is narrower than the cutoff wavelength. If there is a phase matching condition between the surface plasmon polariton wave of transversal and the light wave of longitudinal, then the power of the not decreasing surface plasmon wave will be converted into useful light wave for writing the bits on the optical disk surface. Figure 10 shows FDTD calculation results of PPP partly covered with gold metal film. Figure 11 shows the electric field intensity of the light out of the PPP apertures for the various type of the thin metal cover. Light power density profiles out of the probe aperture are changed for various sides covered with metal film [10].

High throughput efficiency as shown in Figure 12 had been obtained using microlens between VCSEL array and near-field semiconductor optical probe array. Figure 13 is one of the PPP array fabricated in this experiment. Before assembling those two types of optical arrays into one integrated unit for the writing and reading experiments through-hole technology for electrodes to each VCSEL is needed to use the limited space between the VCSEL array and the lens /PPP array which is apart from each other with only 25 μm pitch. After finishing the assembling of the two optical units, that is the Microlens array and the NF probe array, writing and reading experiment will be performed at higher efficiency in the near future.

SOLUTION BY NEW OPTICAL HEAD DESIGN COVERED WITH CORRUGATED GOLD THIN FILM

From the results of Figure 13, one of the solution using PPP is proposed [17,18]. The head surface consists of gold grating thin-film inside the GaP base materials, where a patterned width and depth of gold film are 10 and 30 nm, respectively. The other side of a GaP material also works as a microlens array to focus the VCSEL light beams on both the aperture and the fine

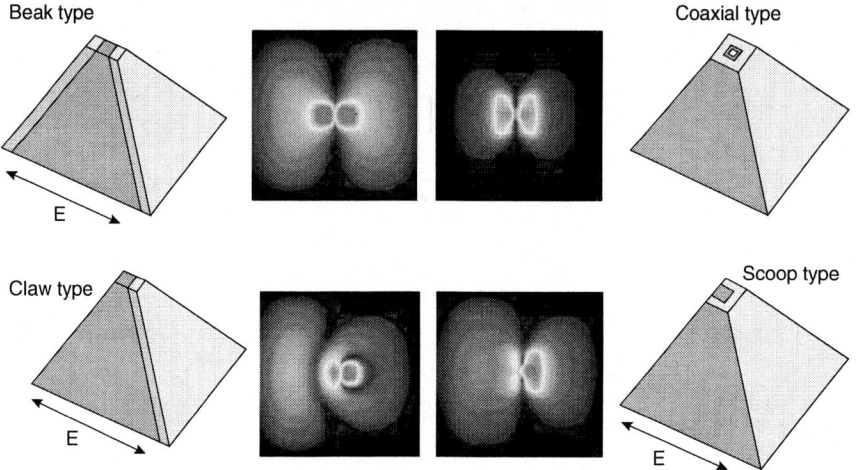

Figure 10   Various types of PPP covered with thin metal film

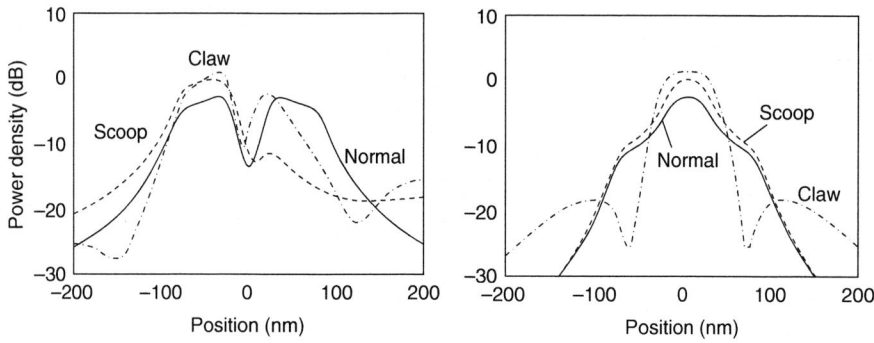

Figure 11   Simulation result by FDTD method [8]

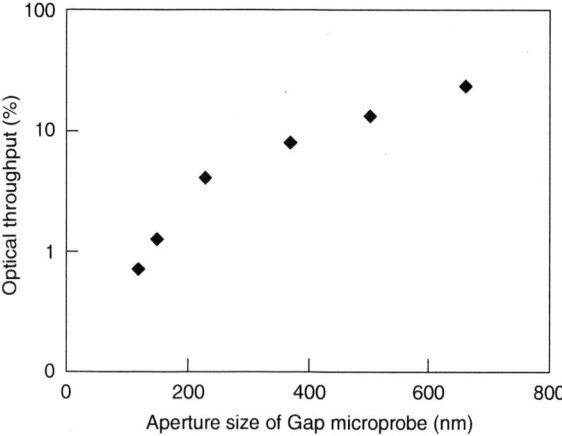

Figure 12   Experimental result of higher throughput efficiency in the near-field semiconductor optical probe array which is shown

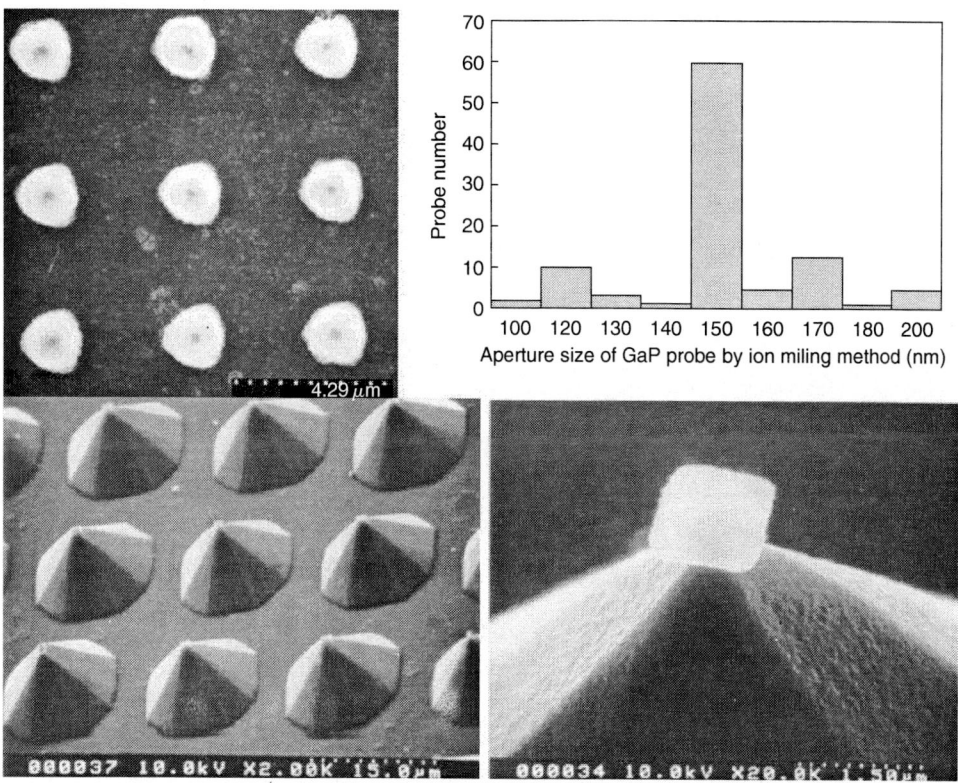

Figure 13 One of the examples for the near-field optical probe array fabricated in this experiment, where the aperture size distribution is also shown

grating tooth around it. This type of head design requires the micro- and nano-fabrication using a non-doped semiconductor and a gold metallic thin-film.

The schematic diagram of optical devices with periodically corrugated gold thin film having tiny nano apertures is shown as in Figure 14. It is required to improve the evanescent light throughput efficiency from a small aperture of 30 nm diameter up to 1% to realize the evanescent light of 10 μW using 1 mW VCSEL output power.

Figure 15 is a schematic figure of a part of the head shown in Figure 14. In this figure only one element of the integrated optical head is three-dimensionally shown, but the VCSEL element is not shown. We can see the nano-aperture (30 nm diameter) surrounded by fine gold grating structure (corrugated gold thin film). The formation of a fine grooving pattern on the GaP substrate and the filling method of a nano metal particles into the groove have been tried using an ultrahigh resolution electron beam lithography and a selective chemical bonding method between the groove inside and the metal particles with a colloid liquid of gold particles for filling gold metals inside the nano groove of 10 μm width and 30 nm depth. Figure 16(a) and 16(b) show two SEM photos of the patterned grooves on the GaP substrate. The EB pattern is 10 nm lines and 235 nm spaces in this fabrication (it is not a match to the exact design mentioned above). Etching for the fine nano grooves will be performed with CF4-RIE. The microlens fabrication method is reported earlier [16].

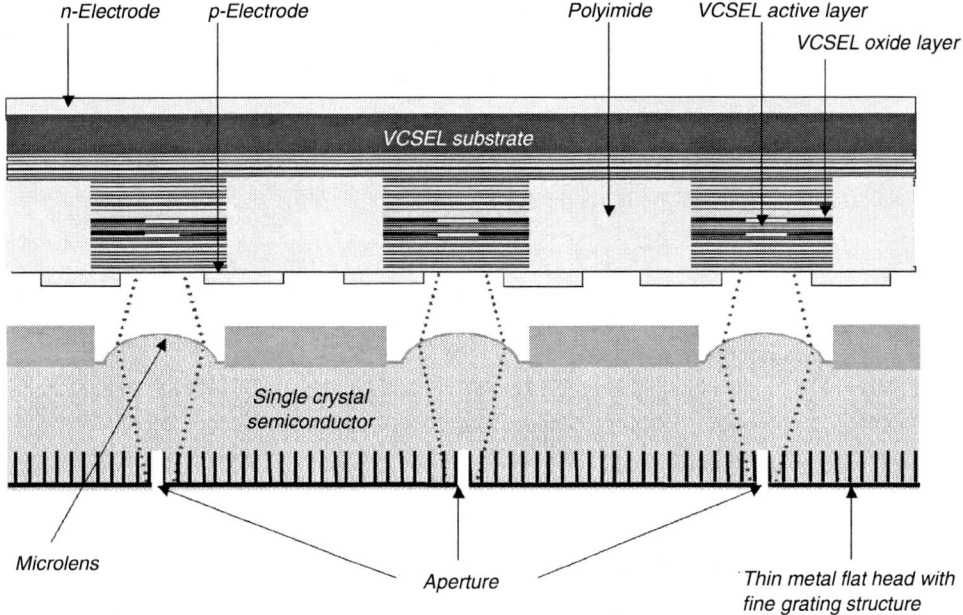

Figure 14 A flat top near-field optical head with an aperture array and periodically corrugated Au metal thin film covered a semiconductor GaP microlens base, in which each laser beam from a VCSEL array irradiated by focusing with the each microlens

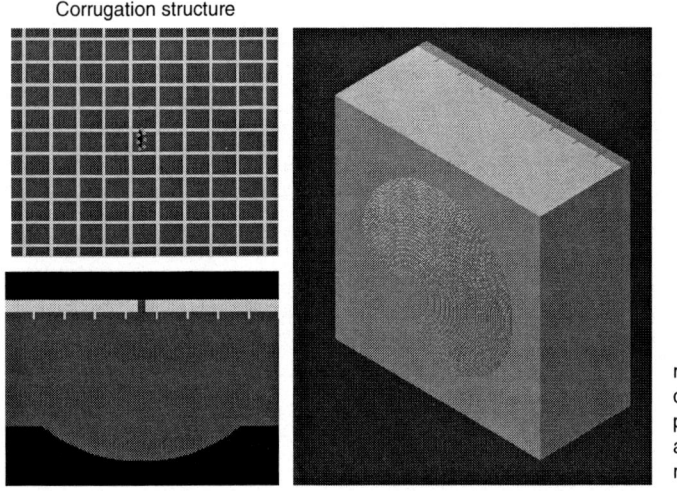

metal width: 10 nm,
depth: 30 nm,
pitch: 120 nm,
aperture size: 20 nm,
metal thin film thickness: 50 nm

Figure 15 Schematic figure of the corrugated metallic thin film. In this case only one element of the integrated optical head is three-dimensionally shown without VCSEL. We can see the nano-aperture surrounded by fine gold grating structure

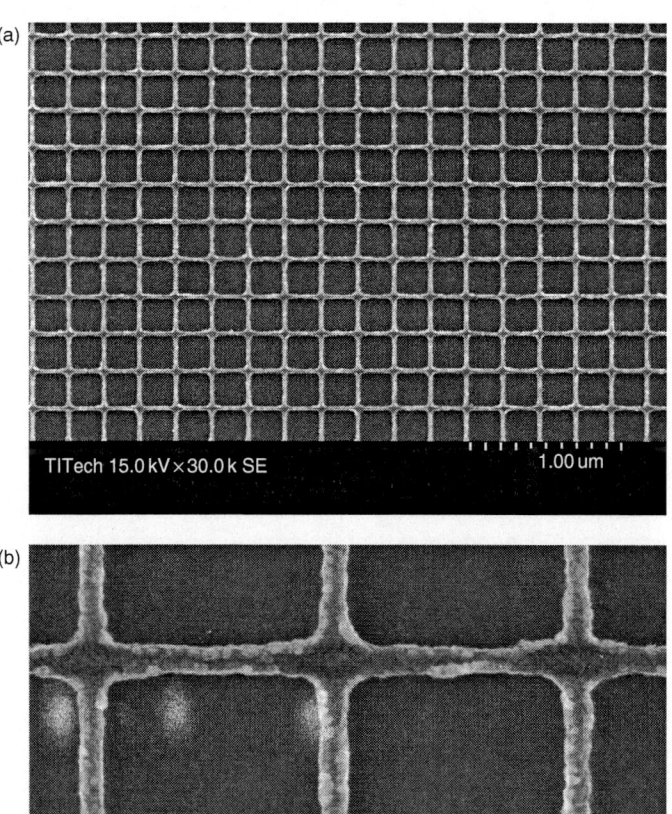

Figure 16 Observation for the lift-off patterned groves (a) SEM and (b) enlarged SEM. Chromium of 30 nm thickness had been evaporated, of which line width and spacing are 10 and 235 nm, respectively. However the obtained width was about 25 nm

RESULT OF FDTD SIMULATION FOR NEW INTEGRATED OPTICAL HEAD

Ebbesen et al. [15,19] reported the highly unusual transmission properties of metal film perforated with a periodic array of sub-wavelength holes. Though light in the visible to infrared range can not be related to the surface plasmons' enhancement on the metal–air interface, it is possible to couple the SPP to metal surface if a periodic structure of sub-wavelength holes is prepared on the metal film. Thus, we tried to add the periodic structural concept on the original integrated VCSEL microprobe array-head to enhance the optical throughput, as shown in Figure 14. A periodic metal grating of sub-wavelength pitch is introduced on the bottom of semiconductor materials with the nanometer size apertures. The revised optical array-head also includes VCSEL and microlens arrays so that the light from VCSEL is focused on the nano-apertures, which are aligned to the center of microlens. In order to increase the optical power of the evanescent wave from the super-parallel two-dimensional VCSELs, the head structure consisting of the microlens and a 30 nm diameter aperture array on a flat face has been prepared with a thin and fine corrugated gold metal as mentioned earlier. Since the weak evanescent waves resonantly excite SPP waves with the fine metal corrugation, this kind of head structure is designed where the resonant pitch of the light waves propagate inside the non-doped GaP material. All parameters are calculated using a FDTD method. In this study, the near-field apertures are prepared on the flat surface and thus, the focal points of the microlenses are located at the same distance. The irradiated spots in the grating plane are larger than expected focal spot size.

By calculations, it is observed that very high evanescent light power were enhanced resonantly when the grating pitch is coincident with a half wavelength and the focused laser light is located inside the semiconductor head, as shown in Figures 17 to 20. In order to obtain this resonantly enhancement result, we varied the width and depth of the fine gold metal corrugation. While calculating we found that the maximum enhancement occurs when the gold metal width is about 10 nm and the height is 30 nm. With these parameters of width and height constant it is calculated that the evanescent wave electric field enhancement phenomena with the variation of aperture diameter. Figures 17 and Figure 18 show some of these results. From the view point of evanescent wave enhancement in Figure 18, it is understood that the smaller aperture size makes the larger evanescent light power. As for the power increment with an aid of SPP, it is realized that the aperture size of the evanescent light for writing on the surface of the optical disk should be <30 nm diameter. Figure 19 is the calculation result of the grating pitch variation for the aperture size of 30 nm.

The wavelength is 780 nm and the refractive index of GaP is 3.3. The enhanced peak evanescent light occurs when the light inside GaP is in resonance with the gold grating pitches. The resonance can be seen in every half wavelength in GaP crystal. The irradiated laser light inside GaP crystal is focused around fine gold grating. Wave vector of the irradiated light and grating vector are resonant with each other. From Figure 20 it is decided that the aperture surface metal thickness should be thicker than 30 nm in order to prevent the leaky far-field light as shown in the figure. However the inside of the aperture holes should be filled with the same material used as the substrate.

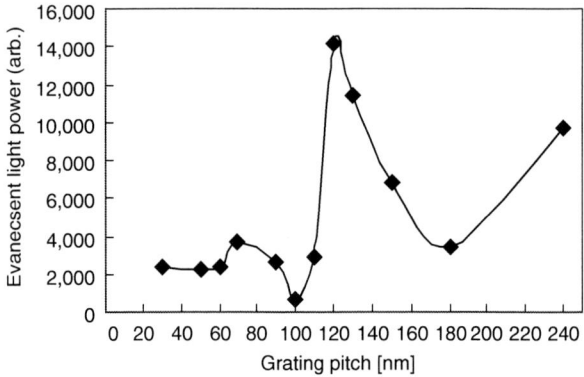

Figure 17   Resonant evanescent light power enhancement with the grating pitch of a half wavelength of the light inside the gold corrugated surface inside the GaP substrate had been observed through FDTD calculation. The enhanced evanescent light power ratio to the power without a resonance shows an around 500 times difference

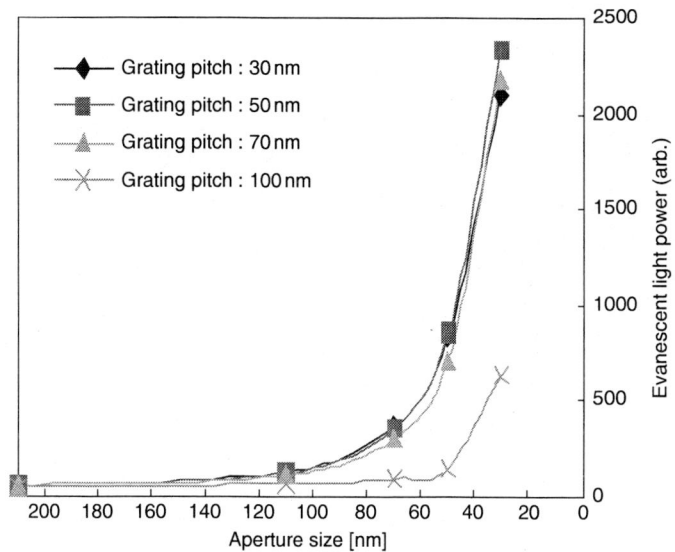

Figure 18 Calculation results of the evanescent light power versus grating pitch variation. The power in this figure means square of the electric field of the evanescent wave

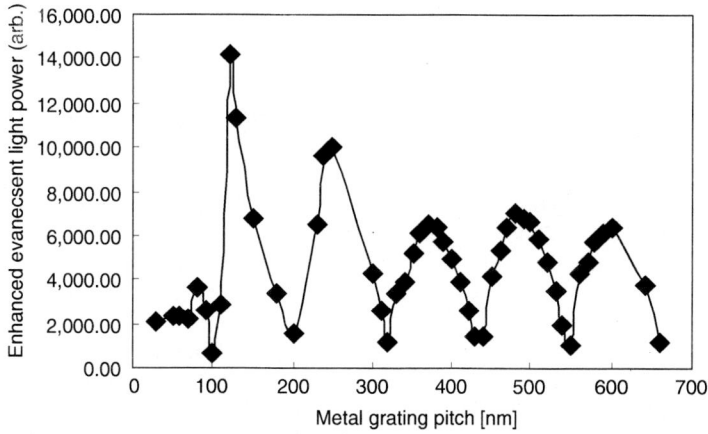

Figure 19 Calculation results of the evanescent light power versus grating pitch variation. The power in this figure means square of the electric field of the evanescent wave

PERFORMANCES

The interface between the gold film and the non-doped GaP crystal may support the charge density oscillations with proportion to the discontinuity of the electric field component of the evanescent wave normal to the interface. An incident electromagnetic wave excites an SPP, if the wave vector component parallel to the non-doped semiconductor and metal interface matches with the propagation vector of the SPP. Since the SPP wave vector is usually larger than the wave vector incident in a non-doped semiconductor or dielectric media adjacent to the metal film, the

Figure 20 Even in non-resonant periodicity there is a large difference between periodic grating gold surface and no periodic corrugation. In both cases the gold film thickness is only 30 nm, then there we can observe leaky light field over the surface. Electric field E^2 distribution (the evanescent light wave power) near the aperture when the pitch is 118 nm is observed

exciting field is usually an evanescent field produced by the grating. In the experimental conventional array head [4–6,8–11] the refractive index of the GaP is as high as 3.3 of light (780 nm wavelength) beam waist size inside the semiconductor probe is almost 500 nm in diameter. Evanescent light from the 100 nm aperture on the top tip of the GaP probe is reduced to 10 μW, even though the throughput efficiency is as high as 1% in those experimental cases. The required light power for writing bits on the surface of the optical disk is 100 μW for a 100 nm aperture, and 12 μW for a 30 nm aperture. This means that at least ten times evanescent light power increment for a 30 nm aperture with the aid of the SPP is required.

The most important technology in this new head structure is that the head has the grating shape on the surface structure where the arrayed apertures are prepared by engraving the gold film surface to excite the SPP wave inside the non-doped GaP substrate. The non-doped semiconductor crystal surface covered with a thin gold film, which exhibits a negative ε at the frequency of the incident 780 nm light shows effective SPP excitation by an incident light wave if the wave vector component parallel to the non-doped semiconductor and metal interface matches with the propagation vector of the SPP. Since the weak evanescent wave can excite the SPP waves resonantly with fine metal grating, new optical array head was designed to confirm the resonant behavior between the input light and metal grating. The interface between the gold film and GaP substrate is related to the charge density oscillation with the electric field component of evanescent wave normal to it. This occurs at a different frequency from the bulk plasmon oscillation and is confined on the interface of metal and dielectric layer. Then only TM-polarized light can excite the SPP and the metal layer must exhibit a negative value of real part of ε at the frequency of incident light. Since the negative dielectric permittivity is required to support the SPP resonance, the best metal at the optical frequency range is gold or silver. An incident electromagnetic wave excites the SPP wave if the wave vector component parallel to the interface of semiconductor and metal is coincident with the propagation vector of the SPP. Since the wave vector of SPP is usually larger than that of the incident light on the semiconductor or dielectric medium adjacent to the metal layer, the excited field is an evanescent field enhanced by the metal grating. Since the surface plasmon enhancement is dependent on the polarization direction of laser beam, the evanescent field enhancement will be limited to one direction. When the polarized direction of input light is parallel to the grating direction in the x-direction, the enhancement occurs in the grating pattern through the x-direction. However, there is little enhancement in the perpendicular grating pattern through the y-direction. The simulation results show that the surface plasmon enhancement strongly depends on the polarization direction and incident angle of focused input laser. In this research, the VCSEL arrays [3] are used for light source since it has many advantages over edge emitting lasers, including the two-dimensional array structure on the wafer, circular beam shape, and single longitudinal mode. However, it shows some limitations due to the multi-transverse mode behavior and polarization instabilities [20]. Although our system requires the polarization control of input light and the current VCSEL does not show the exact polarization control, it is believed that the polarization control in VCSEL will be realized in the near future. A lot of approaches are now studied to control the polarization behavior in VCSEL and some good results are reported by many researchers [20] including the growth on non-(1 0 0) substrate, design of non-cylindrical reso-nators, the use of polarization selective mirrors, and the method of asymmetric current injection.

IMPACT TO SYSTEM

Since the VCSEL microprobe array head reported in the previous paper [4,12] does not satisfy the required recording power for conventional phase change optical media, a new structure of parallel optical array head had been studied. The new structure was designed to enhance the optical throughput using the surface plasmon resonance between the incident light and the thin metal grating. The theoretical analysis and fabrication process for the new integrated array head are discussed with the emphasis on the FDTD simulation about metal grating structure. About 500 times resonant power enhancement of the evanescent light is observed when the grating periodicity (minimum pitch is 118 nm) is equal to half the wavelength of incident wavelength (780 nm) inside the GaP crystal. The calculated data of the evanescent light wave increment have been shown. In the resonant case with the corrugated thin gold metal surface, there is larger evanescent wave enhancement compared to the case of no corrugation metal surface.

It is interesting that enormous power enhancement can be observed with every half wavelength in the non-doped GaP crystal. After establishment of fine engraved metal periodical corrugation with nano-fabrication and assembling technology the new flat type head will be developed in the near future, using the corrugated gold metal film on the top of the super-parallel near-field inte-grated optical head as shown in Figure 14. The parallel optical array head has advantages for real-izing both fast data transfer rate (>10 Gbpsec) and high data capacity (several Tera byte in 120 mm disk size) since it is based on a multi-beam recording and a small spot size using the VCSEL and microlens arrays. Currently it is attempting to develop new integrated VCSEL array head and eval-uate the optical properties for realizing the parallel near-field optical data storage. After the estab-lishment of fine engraved metal periodic grating with nano-fabrication process, the new head will be completed with the attachment process to VCSEL array. The evanescent light power enhance-ment between the incident light and metal grating will be a guideline for the future near-field optical data storage application.

REFERENCES

1. K. Goto, "Proposal of ultrahigh density optical disk system using a vertical cavity surface emitting laser array," *Jpn. J. Appl. Phys.*, **37**, 2274–2278 (1998).
2. Kenya Goto, "Two-dimensional near-field optical memory head," U.S. Patent No. 6,084,848 (Date of Patent July 4, 2000).
3. Kenichi Iga, "Surface emitting laser-Its birth and generation of new optoelectronics field," *IEEE JSTQE*, **6**, 1201–1215 (2000).

4. K. Kurihara, Y.-J. Kim, and K.Goto, "High-throughput GaP microprobe array having uniform aperture size distribution for near-field optical memory," *Jpn. J. Appl. Phys.*, **41**, 2034–2039 (2002).

5. Ivan D. Nikolov, K. Kurihara, and Kenya Goto, "Nano Focusing Probe for a Near-Field Optical Head," paper presented in TNT2003.

6. Kenya Goto, Takayuki Kirigaya, and Yoshiki Masuda, "Design and experiments of a near-field optical disk head with very high efficiency," in *Proceedings of the APNFO-4*, Taroko, Hualien, Taiwan (October 13–16, 2003), p. 17.

7. E. Betzig, J. K. Trautman, R. Wolfe, E. M. Gyorgy, P. L. Finn, M. H. Kryder, and C. H. Chang, "Near-field magneto-optics and high density data storage," *Appl. Phys. Lett.*, **61**, 142–194 (1992).

8. K. Goto, Y. Masuda, Y.-J. Kim, T. Ono, "3D-FDTD simulation and related experiment for evanescent wave enhancement by nanometer sized corrugated metal," in *Proceedings of 10th Microoptics Conference*, M-5, Friedrich-Schiller University, Jena, Germany (September 1–3, 2004).

9. Y.-J. Kim, K. Suzuki, and K. Goto, "Parallel recording array head of nano-aperture flat-tip probes for high density near-field optical data storage," *Jpn. J. Appl. Phys.*, **40**, 1783–1789 (2001).

10. Satoshi Mitsugi, Young-Joo Kim, and Kenya Goto, "Finite difference time domain analysis for electro magnetic field distribution on near-field optical recording probe head," *Opt. Rev.*, **8**, 120–125 (2001).

11. Shu-YingYe, Satoshi Mitsugi,Young-Joo Kim, and Kenya Goto, "Numerical simulation of readout using optical feedback in the integrated vertical cavity surface emitting laser micro-probe head," *Jpn. J. Appl. Phys.*, **41**, 1636–1637 (2002).

12. K. Goto, Y.-J. Kim, S. Mitsugi, K. Suzuki, K. Kurihara, and T. Horibe, "Microoptical 2D devices for the optical memory head of an ultrahigh data transfer rate and density system using VCSEL array," *Jpn. J. Appl. Phys.*, **41**, 4835–4840 (2002).

13. X. Shi and L. Hesselink, "Mechanisms for enhancing power throughput from planar nano-apertures for near-field optical data storage," *Jpn. J. Appl. Phys.*, **41**, 1632–1635 (2002).

14. R. Grober, R. Schoelkopf, and D. Prober, "Optical antenna: Towards a unity efficiency near-field optical probe," *Appl. Phys. Lett.*, **70**, 1354 (1997).

15. T. W. Ebbesen, H. J. Lezec, H. F. Ghaemi, T. Thio, and P. A. Wolff, "Extraordinary optical transmission through sub-wavelength hole arrays," *Nature*, **391**, 667–669 (1998).

16. K. Goto, H. Maruyama, K.Suzuki, and Y.-J. Kim, "Self alignment method of near-field optical probe and micro-lens with VCSEL array," in *Proceedings of the ISOM/ODS 2002*, WC-5, Hawaii (July 7–11, 2002), pp. 293–295.

17. K. Goto, T. Kirigaya, and Y. Masuda, "New optical disk head structure of near-field evanescent wave with writable high power density to a terabyte disk surface using VCSEL," in Technical Digest, Optical Data Storage 2003 (ODS'03), TuD3, Vancouver, Canada (May 11–14, 2003), pp. 129–131.

18. K. Goto, Y.-J. Kim, T. Kirigaya, and Y. Masuda, "Near-field evanescent wave enhancement with nano-meter-sized metal grating and microlens array in parallel optical recording head," *Jpn. J. Appl. Phys.*, **43**, 5814–5818 (2004).

19. H. Ghaemi, T. Thio, D. Grupp, T. Ebbesen, and H. Lezec, "Surface plasmons enhance optical transmission through subwavelength holes," *Phys. Rev. B*, **58**, 6779–6982 (1998).

20. M. Camarena, G. Verschaffelt, M. Moreno, L. Desmet, H. Unold, R. Michalzik, H. Thienpont, J. Danckaert, I. Veretennicoff, and K. Panajotov, in *Proceedings of the Symposium IEEE/LEOS Benelux Chap.*, Amsterdam (2002), p. 103.

OPTICAL FIBER

Yasuo Kokubun

INTRODUCTION

Optical fiber are the most important invention of the signal transmission line during the 20th century. As shown in Figure 1, the transmission loss of silica fiber is extremely low, typically 0.2 dB/km at the wavelength region of 1.55 μm. In addition, a huge transmission capacity can be achieved by the dispersion compensation and dispersion management technologies of single-mode silica fibers. Therefore, in the large capacity (high-bit-rate) and long distance transmission system, the most common optical fiber is the single-mode fiber made of silica glass. Thus, this section is concentrated on the transmission characteristics of single-mode optical fiber.

The structure of optical fiber for telecommunication use is standardized by ITU (formerly CCITT) as shown in Figure 2. The diameter of the glass part (cladding) is 125 μm. The core

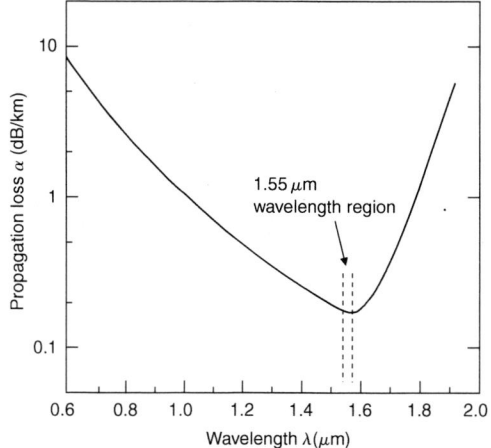

Figure 1   Propagation loss of silica optical fiber

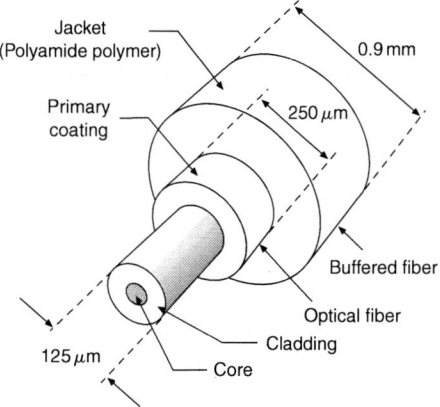

Figure 2   Structure of silica optical fiber for telecommunication use

diameter is determined so that the mode field diameter (see below) is equal to the standardized number, typically 8 to 10 μm.

DISPERSION CHARACTERISTICS AND MODE LABELS

Let us define the co-ordinate system and the parameters as shown in Figure 3. The propagation constant β of round optical fiber is expressed by the relation between the normalized parameters V and b, and is illustrated in Figure 4. The normalized parameters are defined by the following equations.

$$V = k_0 n_1 \sqrt{2\Delta}, \tag{1}$$

$$b = \frac{(\beta / k_0)^2 - n_2^2}{n_1^2 - n_2^2}, \tag{2}$$

$$\Delta = \frac{n_1^2 - n_2^2}{2n_1^2}, \tag{3}$$

where k_0 is the propagation constant in free space.

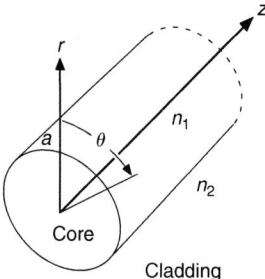

Figure 3   Definition of co-ordinate system and structure of step-index round optical fiber

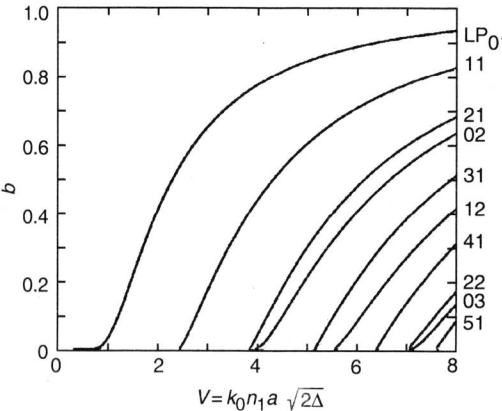

Figure 4   Dispersion curve of step-index round optical fiber

Table 1
Correspondence between LP mode and HE, EH, TE, and TM modes

LP Mode	Hybrid Modes
$LP_{0,1}$	$HE_{1,1}$
$LP_{1,1}$	$HE_{2,1}$, $TE_{0,1}$, $TM_{0,1}$
$LP_{2,1}$	$HE_{3,1}$, $EH_{1,1}$
$LP_{0,2}$	$HE_{1,2}$
$LP_{3,1}$	$HE_{4,1}$, $EH_{2,1}$
$LP_{1,2}$	$HE_{2,2}$, $TE_{0,2}$, $TM_{0,2}$

The solution of exact eigenvalue equations derived from Maxwell's equations are classified into EH_{lm} and HE_{lm} modes [1], where the first mode label l corresponds to the number of nodes of electric z-component in the azimuthal direction and m corresponds to the mode label in the radial direction. It should be noted that l starts from "0" while m starts from "1" in the case of optical fiber.

When the refractive index difference between the core and the cladding is very small compared with the core index itself, the propagation constants of EH and HE modes with the azimuthal mode label satisfying the following relation are almost degenerate [2]. These quasi-degenerate modes are called LP (linear polarized) modes. The correspondence between exact (EH, HE) modes and quasi-degenerate LP modes are shown in Table 1 for several small mode numbers.

$$v = \begin{cases} l-1: & HE_{l,m} \text{ mode,} \\ l+1: & TE_{0,m}, TM_{0,m}, EH_{l,m} \text{ mode.} \end{cases} \tag{4}$$

In the step-index round optical fiber, the field profile is analytically expressed in terms of Bessel functions. On the other hand, for generalized distributed index profile in the core, some numerical calculation methods are needed. For this purpose, the matrix method [3] , finite element analysis method [4] , variational method [5] , and direct integration method [6] have been proposed.

To obtain the single mode condition of distributed-index fibers, simple formula [7,8] can be used instead of numerical modal analysis method.

MODE FIELD DIAMETER

In most optical fiber including distributed-index profile, the lateral field profile $E_x(r)$ of fundamental mode (LT_{01} mode) in the core can be approximated by the Gaussian function as follows:

$$E_x(r) = A \exp\left[-\left(\frac{r}{w_0}\right)^2\right]. \tag{5}$$

Here the intensity profile is proportional to $|E_x(r)|^2$. $2w_0$ is called the "mode field diameter," and w_0 is called spot size. Some formulae for calculating the spot size have been proposed [9,10], and the following Petermann's formula [9] is standardized by ITU:

$$w_0 = \left[\frac{2\int_0^\infty [E_x(r)]^2 \, r \, dr}{\int_0^\infty \left([d \, E_x(r)]/dr\right)^2 \, r \, dr} \right]^{1/2}. \qquad (6)$$

PULSE BROADENING AND DISPERSION

The pulse width transmitted in a single mode fiber is broadened due to the wavelength dependence of group velocity as shown in Figure 5. This wavelength dependence (or frequency dependence) of group velocity is called dispersion. The dispersion is classified into the material dispersion σ_m resulting from the wavelength dependence of refractive index of material and the waveguide dispersion σ_w, which is proportional to the normalized second derivative [11]:

$$V \frac{d^2(Vb)}{dV^2}.$$

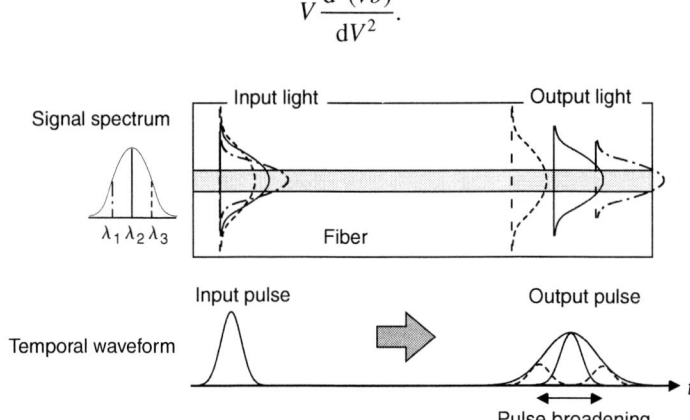

Figure 5   Intuitive consideration on pulse broadening in dispersive fiber

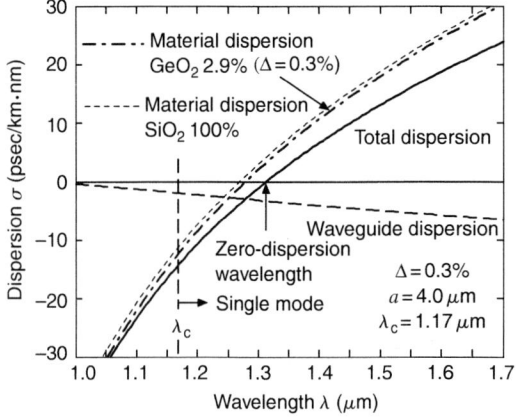

Figure 6   Material, waveguide, and total dispersions of ordinary single-mode fiber

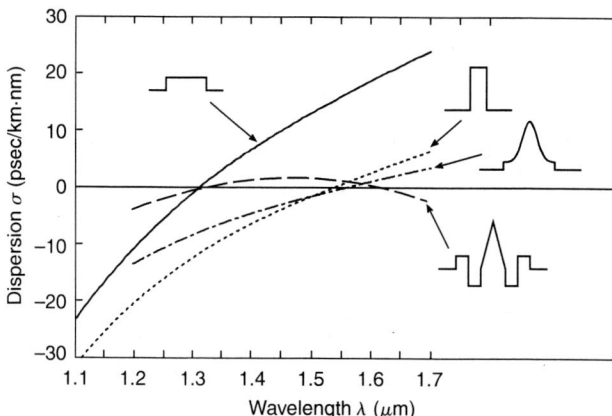

Figure 7   Some examples of dispersion shifted fibers

The total dispersion is expressed by the sum of these two dipsersions, that is, $\sigma_{total} = \sigma_m + \sigma_w$. The wavelength dependences of material dispersion, waveguide dispersion, and the total dispersion are shown in Figure 6.

Since the material dispersion can be compensated for by the appropriate control of waveguide dispersion, some dispersion shifted fibers have been developed as shown in Figure 7. Using these dispersion shifted fibers, a long distance large capacity transmission becomes possible.

REFERENCES

1. E. Snitzer, "Cylindrical dielectric waveguide modes," *J. Opt. Soc. Am.*, 51, 491–498, 1961.
2. D. Gloge, "Weakly guiding fibers," *Appl. Opt.*, 10, 2252–2258, 1971.
3. T. Tanaka and Y. Suematsu, "An exact analysis of cylindrical fiber with index distribution by matrix method and its application to focusing fiber," *Trans. IECEJ*, E59, 1–8, 1976.
4. K. Okamoto and T. Okoshi, "Vectorial wave analysis of inhomogeneous optical fibers using finite element method," *IEEE Trans. Microwave Theory Technol.*, MTT-26, 109–114, 1978.
5. K. Okamoto and T. Okoshi, "Computer-aided synthesis of the optimum refractive index profile for a multimode fiber," *IEEE Trans. Microwave Theory Technol.*, MTT-25, 213–221, 1977.
6. Y. Kokubun and K. Iga, "Mode analysis of graded-index optical fibers using a scalar wave equation including gradient-index terms and direct numerical integration," *J. Opt. Soc. Am.*, 70, 388–394, 1980.
7. Y. Kokubun and K. Iga, "Formulas for TE_{01} cutoff in optial fibers with arbitrary index profile," *J. Opt. Soc. Am.*, 70, 36–40, 1980.
8. Y. Kokubun and K. Iga, "Single-mode condition of optical fibers with axially symmetric refractive index distribution," *Radio Sci.*, 17, 43–49, 1982.
9. K. Petermann, "Constraints for fundamental-mode spot size for broadband dispersion-compensated single-mode fibres," *Electron. Lett.*, 19, 712–714, 1983.
10. K. Hayata, M. Koshiba, and M. Suzuki, "Modal spot size of axially nonsymmetrical fibers," *Electron. Lett.*, 22, 127–129, 1986.
11. D. Gloge, "Dispersion in weakly guiding fibers," *Appl. Opt.*, 10, 2442–2445, 1971.

OPTICAL FILTER SYNTHESIS

Christi K. Madsen

Optical filters perform important signal processing functions in optical communication systems, providing bandwidth management for wavelength division multiplexed (WDM) channels and signal conditioning, such as gain equalization and dispersion compensation for high-bitrate, long-distance transmission. The focus of this chapter is on linear, interference filters characterized in the frequency domain by their magnitude and phase response. The inverse Fourier transform of the frequency response yields the time domain behavior, which is called the impulse response. There are many types of optical interference filters; however, they are broadly distinguished based on the length of their impulse response. The two most general filter types are finite impulse response (FIR) and infinite impulse response (IIR). The basic optical filter discussed in this chapter is an interferometer that splits an incoming signal into multiple paths, each path has an associated delay and an amplitude weight, and the paths are recombined coherently to yield a desired frequency response. FIR filters are characterized by feed-forward interference while IIR filters incorporate feedback interference by either standing wave etalon-based cavities or traveling wave ring resonators. A lumped element approach for describing the splitters and delays is combined with digital filter and signal processing [1–3] concepts to simplify the design and analysis of optical filters.

Although they share a common mathematical description, there are many physical implementations for optical filters including planar waveguide filters, thin-film filters composed of alternating low and high index layers, Bragg or long-period gratings and diffraction gratings. While the first application of digital filter techniques to optical filters was for optical fiber implementations [4–6], the focus of this chapter is on planar waveguide implementations since they offer an interferometrically stable platform with lithographically defined dimensions and wafer-level batch processing for large-scale integration. We begin with a discussion of the simplest implementation of FIR and IIR optical filters, so-called single-stage filters. Their magnitude and phase response, including dispersion, are examined using lumped elements and Z transforms. Then, filter architectures and synthesis techniques for higher-order filter responses are investigated separately for the two general filter types along with practical issues relevant to planar waveguide implementations and applications in WDM systems.

BASIC TYPES OF INTERFEROMETRIC FILTERS

Optical filters based on coherent interference consist of splitters, delay lines, and combiners. Since most splitters are reciprocal, in practice, they may also be used as combiners. Interference occurs when light having the same frequency and polarization propagates along two or more distinct paths and recombines. The maximum difference in path length propagation taken must be shorter than the inverse bandwidth of the incoming signal for coherent interference.

Mathematically, a phasor representation of the dominant transverse electric (TE) (or transverse magnetic, TM) field component is employed. For waveguides, it is convenient to describe the fundamental modes of TE or TM field by an amplitude coefficient and a propagation constant, where the underlying spatial distribution of the mode and the time dependence ωt are suppressed in the notation. Propagation of the electric field for a given polarization and frequency along a lossless delay line of length L and propagation constant β is represented by the input–output relationship $E_{out} = E_{in} e^{-j\beta L}$.

A splitter divides the incoming power among multiple output paths. A simple example of a splitter is a directional coupler. Two waveguides are brought in close proximity so that the evanescent field of one waveguide overlaps with the core of the other waveguide as indicated in Figure 1(a). Coupled mode analysis shows that energy is transferred between the waveguides, and that the coupling depends on several factors including the waveguide confinement, separation, and length of the coupling region. These details may be lumped together to define an overall coupling strength, θ, for our purposes. Since the device has two inputs and outputs, it is convenient to describe it with a 2×2 matrix. For identical waveguides, the matrix is given as follows [7]:

$$\begin{bmatrix} Y_1 \\ Y_2 \end{bmatrix} = \begin{bmatrix} \cos\theta & -j\sin\theta \\ -j\sin\theta & \cos\theta \end{bmatrix} \begin{bmatrix} X_1 \\ X_2 \end{bmatrix} = \mathbf{C} \begin{bmatrix} X_1 \\ X_2 \end{bmatrix}, \tag{1}$$

where the input and output fields for each port are represented by X and Y, respectively. If the coupler is lossless, then the coupling matrix $\mathbf{C}$ is unitary. For later use with multistage filters, it is convenient to define a short-hand notation as follows: $c \equiv \cos\theta$ and $s \equiv \sin\theta$. In practice, the coupling coefficient may be wavelength and polarization dependent. However, for introducing filter design with the least notation, we shall treat θ as a constant. This is a good approximation for intra-channel filters; however, for larger wavelength ranges a more detailed analysis must be considered. Besides directional couplers, other important integrated splitters include multimode interference (MMI) couplers [8] and star couplers [9].

A diagonal matrix is convenient for describing the propagation along multiple, parallel delay lines as depicted in Figure 1(b). Let the relative delay length between parallel paths be an integer multiple of a so-called unit delay ΔL, such that the delay relative to the common path length is $L_n = n\Delta L$ for the nth path. The propagation matrix for two paths is described by a 2×2 diagonal matrix as follows:

$$e^{-j\beta L_C} \begin{bmatrix} 1 & 0 \\ 0 & e^{-j\beta n\Delta L} \end{bmatrix} \rightarrow \mathbf{D} \equiv \begin{bmatrix} 1 & 0 \\ 0 & z^{-n} \end{bmatrix} \tag{2}$$

where L_C is the common path and the substitution $\exp(j\beta\Delta L) \rightarrow z$ is made. The unit delay is given by $T = n_g \Delta L /c$, where n_g is the waveguide's group index. Deviations in the optical path length that are on the order of a wavelength, $\delta(n_e L)$, where n_e is the waveguide's effective index, are treated separately as complex, quasi-frequency independent phase shifts $\phi = 2\pi\delta(n_e L) f/c$ where f is the optical frequency and c is the speed of light. Note that a common delay may be taken outside the matrix to simplify the expression and to render the matrices unitary. Unitary

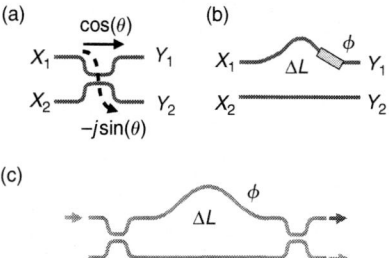

Figure 1   (a) Directional coupler, (b) parallel delay lines with phase shifter, and (c) Mach–Zehnder interferometer

matrices may be described in terms of Caley-Klein parameters α and β, as shown in Eq. (3), where $|\alpha|^2 + |\beta|^2 = 1$. For a directional coupler, $\alpha = \cos\theta$ and $\beta = -j\sin\theta$.

$$\begin{bmatrix} \alpha & \beta \\ -\beta^* & \alpha^* \end{bmatrix}. \tag{3}$$

The overall description of the first-order feed-forward filter shown in Figure 1(c) follows by concatenating these coupling and delay building blocks as described mathematically in Eq. (4). The resulting filter is the well-known Mach–Zehnder interferometer (MZI) [10]. In a folded configuration with a mirror at the midpoint, it is a Michelson interferometer [11].

$$\begin{bmatrix} Y_1 \\ Y_2 \end{bmatrix} = \mathbf{CDC} \begin{bmatrix} X_1 \\ X_2 \end{bmatrix} = \begin{bmatrix} c_2 & -js_2 \\ -js_2 & c_2 \end{bmatrix} \begin{bmatrix} z^{-1} & 0 \\ 0 & 1 \end{bmatrix} \begin{bmatrix} c_1 & -js_1 \\ -js_1 & c_1 \end{bmatrix} \begin{bmatrix} X_1 \\ X_2 \end{bmatrix}. \tag{4}$$

The transfer matrix is defined as follows:

$$\begin{bmatrix} H_{11}(z) & H_{12}(z) \\ H_{21}(z) & H_{22}(z) \end{bmatrix} = \begin{bmatrix} c_1c_2z^{-1} - s_1s_2 & -j\left(s_1c_2z^{-1} + c_1s_2\right) \\ -j\left(s_1c_2 + c_1s_2z^{-1}\right) & c_1c_2 - s_1s_2z^{-1} \end{bmatrix} = z^{-N/2}\begin{bmatrix} A(z) & -B^R(z) \\ B(z) & A^R(z) \end{bmatrix}. \tag{5}$$

So, the transfer function from the X_1 input to the Y_1 output is $H_{11}(z) = c_1c_2z^{-1} - s_1s_2$. Each input/output combination has a transfer function that is a first-order polynomial in z. Just as with the Caley-Klein form of a unitary matrix, the coefficients of the polynomials on the diagonal (and off-diagonal) are related. By reversing the order of $H_{11}(z)$ coefficients, we find an expression for $H_{22}(z)$, and similarly for $H_{12}(z)$ and $H_{21}(z)$. So, the transfer matrix is composed of two unique polynomials $A(z)$ and $B(z)$ and their reverse polynomials designated by the superscript R. For complex coefficients, the reverse polynomial is given by the para-Hermitian conjugate of the forward polynomial, $A^R(z) \equiv A^*(1/z^*)$ [7]. For a single-stage FIR filter, Eq. (5) shows that the reverse polynomial has its root reflected about the unit circle from the original, or so-called forward polynomial. This symmetry in root locations has important implications for the magnitude and the phase response of the various ports, as discussed later. The frequency response is obtained by evaluating z on the unit circle, that is, $z = e^{j\omega}$ for $0 \le \omega < 2\pi$. The output ports are power complementary for a lossless filter, so $|H_{11}(e^{j\omega})|^2 + |H_{21}(e^{j\omega})|^2 = |H_{12}(e^{j\omega})|^2 + |H_{22}(e^{j\omega})|^2 = 1$. The radian frequency ω is proportional to the optical frequency f normalized with respect to the filter's periodicity in frequency, known as its Free Spectral Range (FSR). Typically, a normalized frequency is defined relative to an optical center frequency f_0 such that $\omega/2\pi = (f - f_0)/$ FSR. The FSR is inversely related to the unit delay and effective group index n_g, FSR $= c/n_g\Delta L$, where c is the speed of light. The root of the polynomial is called a "zero" of the transfer function. When the root's magnitude is unity, for example, when $c^2 = s^2 = 0.5$ for $H_{11}(z)$, there is zero transmission for that port at the frequency where $e^{-j\omega} = 1$, or $\omega = 0$. The root is complex if the path length varies slightly from the unit length, indicating that the zero occurs at a frequency $\omega \ne 0$.

Physically realizable filters are stable and causal. A causal filter produces no output before an input occurs. A stable filter means that for any finite energy input, the output will also be finite energy. For FIR filters, causality means that the impulse response has only nonzero terms in z^{-n}, where $n \ge 0$ are physically realizable and stability is guaranteed by the finite sum of square magnitudes of the impulse response coefficients.

The Z transform and frequency response may also be determined by summing the contributions from each path traveled, and is known as the "sum of paths" approach [12]. Consider the ring resonator in Figure 2(a), the "sum of paths" for port X_1 to Y_1 yields an infinite series, which can then be simplified to produce a first-order polynomial in the denominator as follows:

$$\frac{Y_1(z)}{X_1(z)} = -s^2\sqrt{z^{-1}}\left[1 + c^2 z^{-1} + c^4 z^{-2} + \quad\right] = \frac{-s^2\sqrt{z^{-1}}}{1 - c^2 z^{-1}}. \tag{6}$$

The unit delay for IIR filters is the roundtrip delay for the feedback path in contrast to the differential path length for FIR filters. The root of the denominator polynomial is called a "pole". For IIR filters, a stable and causal filter are obtained if the pole magnitude is less than unity. For filters without gain, the pole has a magnitude less than unity. Filter responses with only poles (i.e., no zeros) are minimum-phase filters, a terminology that is explained in more detail in the discussion on phase responses. The amplitude and the phase response of a minimum-phase filter are related by a Hilbert transform, similar to the Kramers-Kronig relations (see reference 7 for a more detailed discussion).

Optical loss due to propagation along the unit delay path is easily incorporated in the Z transform expressions by replacing z^{-1} with γz^{-1}. Loss changes the location of the poles and zeros, and thus the frequency response. The impact of roundtrip loss in Eq. (6) is to both broaden the passband response and decrease the peak transmission. For the port combination X_1 to Y_2, the sum of optical paths yields a transfer function with both a numerator and a denominator polynomial as follows:

$$\frac{Y_2(z)}{X_1(z)} = c - cs^2 z^{-1}\left[1 + c^2 z^{-1} + c^4 z^{-2} + \quad\right] = \frac{c\left(1 - z^{-1}\right)}{1 - c^2 z^{-1}}. \tag{7}$$

This response has a pole and a zero; however, we do not have independent control over their locations. The transfer functions for ring resonators with unequal coupling coefficients are given in reference 7.

Allpass filters are a subclass of IIR filters that have a unity magnitude response when the feedback path is lossless. A ring resonator with a single coupler and single input and output port, as shown in Figure 2(b), is an allpass filter. The etalon realization of an allpass filter is the well-known Gires–Tournois interferometer [13]. The transfer function is given as follows:

$$\frac{Y(z)}{X(z)} = c - s^2 z^{-1}\left[1 + cz^{-1} + c^2 z^{-2} + \quad\right] = \frac{c - z^{-1}}{1 - cz^{-1}} = \frac{D^R(z)}{D(z)}. \tag{8}$$

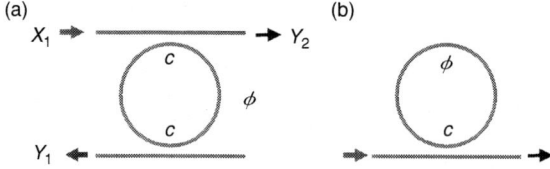

Figure 2 (a) A ring resonator with two identical couplers and (b) a ring resonator-based allpass filter with a single coupler

The numerator is the reverse polynomial of the denominator. Once the poles are determined in the design process, the zeros are automatically defined. A finite roundtrip loss causes the amplitude to vary across the FSR instead of being unity. The peak-to-peak loss depends on the ratio of the roundtrip loss to the coupling coefficient. Critical coupling occurs when $\gamma = c$ and the zero lies on the unit circle causing a perfect null in the transmission spectrum.

The amplitude response is given by the square magnitude of the transfer function evaluated on the unit circle, $|H(e^{j\omega})|^2$, while the phase response is defined as follows:

$$\Theta(\omega) = \tan^{-1}\left\{\frac{\mathrm{Im}\left[H\left(e^{j\omega}\right)\right]}{\mathrm{Re}\left[H\left(e^{j\omega}\right)\right]}\right\},\tag{9}$$

where $\mathrm{Re}[H(e^{j\omega})]$ and $\mathrm{Im}[H(e^{j\omega})]$ denote the real and imaginary parts of the transfer function evaluated on the unit circle. The group delay, normalized to the unit delay, is given by

$$\frac{\tau(\omega)}{T} = -\frac{d\Theta}{d\omega}.\tag{10}$$

The utility of the Z-transform is that simple expressions are available for the magnitude, phase response, and group delay in terms of poles and zeros [14]. Let the pole and the zero locations be denoted by $p_i = r_{pi}\exp(j\phi_{pi})$ and $z_i = r_{zi}\exp(j\phi_{zi})$, respectively. Then, the magnitude response for a filter with M zeros and N poles is given by

$$|H(\omega)|^2 = \frac{|\Gamma|^2 \prod_{i=1}^{M}\left\{1 - 2r_{zi}\cos(\omega - \phi_{zi}) + r_{zi}^2\right\}}{\prod_{i=1}^{N}\left\{1 - 2r_{pi}\cos(\omega - \phi_{pi}) + r_{pi}^2\right\}}\tag{11}$$

where $|\Gamma|^2$ is a constant gain. For an optical filter without gain, the maximum transmission is unity, so $|H(\omega)|^2 \le 1$. The group delay is given by the sum of contributions from all of the poles and zeros as follows:

$$\frac{\tau(\omega)}{T} = \sum_{i=1}^{N}\frac{r_{pi}\left\{\cos(\omega - \phi_{pi}) - r_{pi}\right\}}{\left\{1 - 2r_{pi}\cos(\omega - \phi_{pi}) + r_{pi}^2\right\}} + \sum_{i=1}^{M}\frac{r_{zi}\left\{r_{zi} - \cos(\omega - \phi_{zi})\right\}}{\left\{1 - 2r_{zi}\cos(\omega - \phi_{zi}) + r_{zi}^2\right\}}\tag{12}$$

Zeros at r_{zi} and $1/r_{zi}$ have the same magnitude response since they are the same distance from the unit circle ($|z| = 1$); however, they have different group delay responses. A zero with $r_{zi} > 1$ is called maximum phase while $r_{zi} < 1$ is called minimum phase. This terminology describes the energy distribution in the impulse response. The impulse response is obtained by taking the inverse Z transform of the transfer function. The inverse can typically be found by inspection, given that the inverse Z transform of z^{-1} is the Kronecker delta function $\delta(n)$. The discrete time response is related to the continuous impulse response using the substitution $t = nT$. For a single-stage FIR filter with a transfer function given by $H(z) = 1 - r_z z^{-1}$, the impulse response has two terms, $h(n) = \delta(n) - r_z\delta(n-1)$. The maximum-delay term has the largest amplitude for $r_z > 1$, while the minimum-delay term is larger for $r_z < 1$. The amplitude and group delay responses for a single-stage FIR, minimum-phase IIR, and allpass filter are compared in Figure 3. Note that the sum of the maximum- and minimum-phase group delay responses for

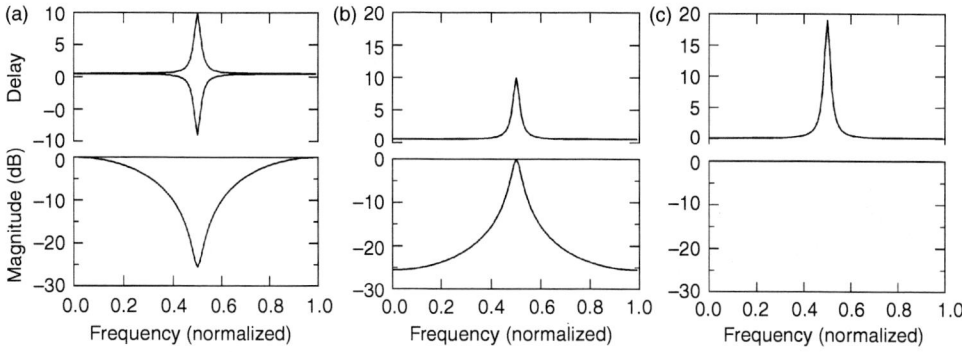

Figure 3 Magnitude and group delay for (a) maximum- and minimum-phase FIR, (b) minimum-phase IIR, and (c) allpass IIR filters

the FIR filter produce a constant group delay as evidenced in Figure 3(a). The impulse response for an FIR and an IIR filter are shown in Figure 4. For an FIR filter with a symmetric or antisymmetric impulse response, that is, $h(n) = \pm h^*(N - n)$, the roots are either located on the unit circle or occur in pairs about the unit circle. With this symmetry (see Figure 4[a] and [b]), the group delay is constant and the filter is dispersion-less. The exponentially decreasing terms of the IIR filter illustrate its minimum-phase nature in the time domain. Note that the impulse response of an IIR filter with a root far away from the unit circle decreases in amplitude very quickly and becomes similar to an FIR filter response in which only a finite number of terms are of interest.

Dispersion is defined as the second derivative of the phase response with respect to frequency, so a dispersion-less filter has a constant group delay. For optical fiber, dispersion varies gradually with wavelength and is quoted in units of ps/nm-km so that the cumulative path dispersion is proportional to the path length. By contrast, a filter's dispersion is given in ps/nm (or ps^2) and may be very frequency dependent as seen by differentiating the group delay curves in Figure 3.

$$D = \frac{d\tau_g}{d\lambda} = -\frac{2\pi c}{\lambda^2} \frac{d^2\Theta}{d\omega^2}. \tag{13}$$

In normalized delay (τ_n) and frequency (ν) units, the normalized dispersion is $D_n \equiv (d\tau_n/d\nu)$. Filter design may be performed using normalized units and converted to physical units as follows:

$$D = -c\left(\frac{T}{\lambda}\right)^2 D_n \tag{14}$$

Note that the filter dispersion is proportional to the square of the unit delay. For very small unit delays, the dispersion will be small. For example, an FSR = 50 GHz ($T = 20$ ps) has $D = 50$ ps/nm for $D_n = 1$ and $\lambda = 1550$ nm. For an FSR = 500 GHz, the dispersion decreases by a factor of 100 for the same normalized dispersion value.

Filter specifications are provided as a desired frequency response such as passband width and flatness and stopband rejection for a bandpass filter. Since single-stage filters do not typically meet all the specifications, multistage filters are required. The filter FSR, number of stages, and nominal values for coupling coefficients and path lengths are determined in the

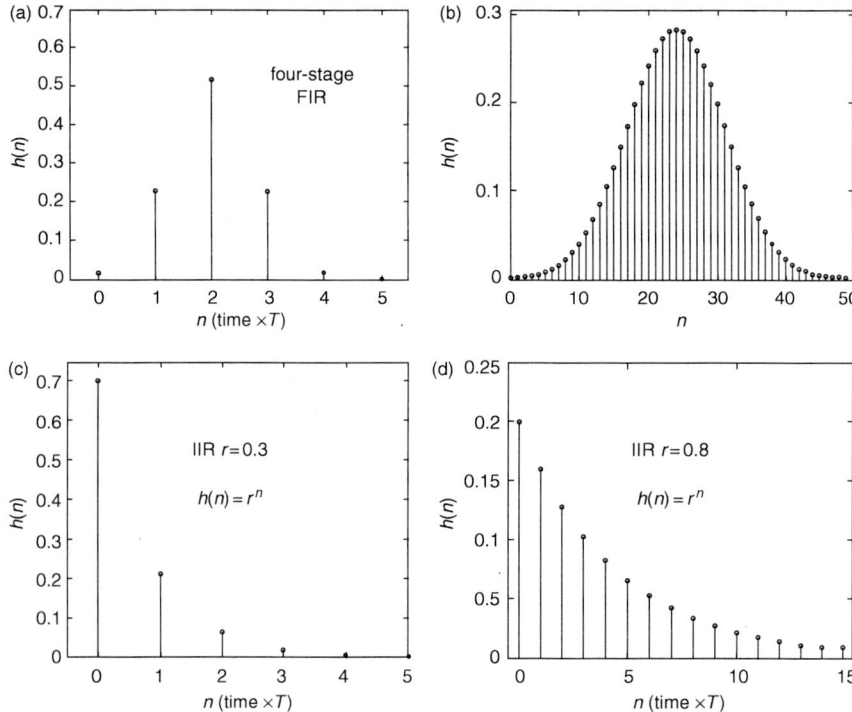

Figure 4   Impulse response for (a) 4-stage and (b) 50-stage FIR filters, and IIR filters with pole magnitudes of (c) 0.3 and (d) 0.8

design process. Least squares or minimum/maximum criterion may be applied in an iterative algorithm to determine an optimal design; then, the sensitivity to variations in the optical parameters is determined to assess the filter's tolerance to fabrication variations in cumulative path lengths and coupling ratios. The maximum deviation of the response from the desired behavior is typically specified.

Tunable filters are often obtained using thermo-optic phase shifters in dielectric and polymer-based planar waveguides [15]. In these devices, a thin-film chromium, resistive heater is deposited on the upper cladding along a section of a waveguide. An optical phase change occurs over the heater length due to the refractive index dependence on temperature change at the waveguide core. For silica waveguides, the change in refractive index with respect to temperature (the so-called dn/dT) is 1e-5/°K. Other tuning mechanisms are available in semiconductors such as the free carrier effect for silicon waveguides and the electro-optic effect in lithium-niobate and indium-phosphide waveguides. Tunable couplers are realized by either varying the coupling ratio directly, for example, by changing the propagation constant of one waveguide relative to the other within the coupling region, or using a phase shifter within a symmetric MZI to vary the effective coupling [16]. A symmetric MZI means that $\Delta L = 0$ in Figure 1(c).

FIR FILTERS

A multistage filter may be realized by cascading single stages so that the output frequency response is a product of the individual responses. Other architectures for multistage FIR filters offer advantages over the cascade architecture including lower loss and power complementary responses. We discuss implementations and applications for transversal, lattice, and phased array multistage FIR architectures in this section.

The transfer function of a multistage FIR filter is described by a finite-order polynomial. An Nth order FIR filter has N zeros, or roots, of the polynomial. Instead of describing the polynomial by its roots and a gain term, it may be described by $N + 1$ coefficients as follows: $H(z) = a_0 + a_1 z^{-1} + \ + a_N z^{-1}$. The relationship between the roots of the polynomial or its coefficients and the optical parameters, such as the coupling coefficients and optical phases, is different for each of the architectures.

An optical transversal filter is illustrated in Figure 5 [17]. The transversal filter architecture has the advantage of either independent control of each filter coefficient by the optical parameters, or a simplified relationship between them, in contrast to the lattice filters that are discussed shortly. Applications include bandpass filters [18], adaptive filters for matched filtering of optical codes [19], and header recognition for packet processing [20,21]. The disadvantage of the transversal filter is that there is typically only a single input and output port as opposed to multiple power complementary outputs of other architectures.

Multistage FIR lattice filters are formed by interconnecting MZIs as shown in Figure 6. Various differential delay lengths may be chosen between the arms for each stage; however, they are typically chosen as a multiple of the unit delay for simplicity. The transfer matrix has the form of Eq. (5) where each polynomial is Nth order. For a lossless filter, the determinant of the transfer matrix is unitary since it is a product of unitary matrices. The determinant yields the following equation that relates the roots of the forward and reverse polynomials.

$$A(z)A^R(z) + B(z)B^R(z) = 1. \tag{15}$$

The consequence of Eq. (15) is that, given one of the polynomials, say $A(z)$, the second polynomial is not uniquely determined. There are N realizations of $B(z)$, or spectral factorizations, that depend on choosing either the minimum- or maximum-phase root from each root pair. When one or more of the relative delays is an integer multiple of the unit delay, the filter order (in z)

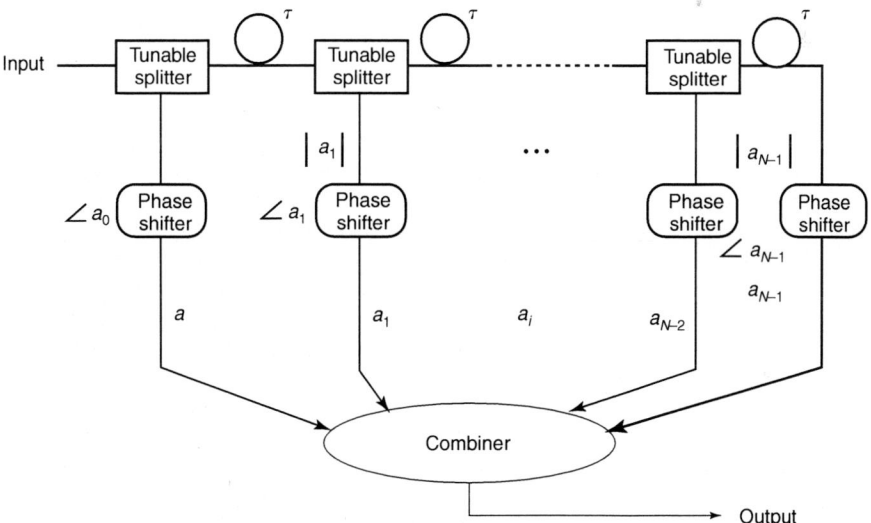

Figure 5 Schematic of an optical transversal filter (From K. Sasayama, M. Okuno, and K. Habara, *J. Lightwave Technol.*, 9, 1225–1230, 1991. With permission. Copyright 1991 IEEE.)

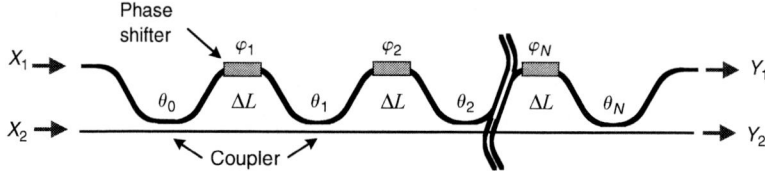

Figure 6 An optical FIR lattice filter

will be larger than the number of stages. The relationship between the optical parameters, that is, the coupling coefficients, and polynomial coefficients is nonlinear in general; however, a layer-peeling algorithm described by Jinguji and Kawachi [22] may be used to directly translate between the optical parameters and filter coefficients. This translation algorithm allows the filter designer to optimize the polynomial roots (or polynomial coefficients) and then translate to the optical parameters or to optimize on the optical parameters directly. Lattice filters have been used in numerous applications including gain equalizers [23,24], delay line interferometers for differential phase-shift-keyed (DPSK) systems [25], dispersion compensators [26–28], intersymbol interference equalizers [29,30], and interleavers [31]. Long-period fiber gratings [32] and codirectional grating-assisted couplers such as acousto-optic filters [33] are FIR lattice filters where propagation in different waveguide modes with different propagation constants determines the differential path lengths instead of dividing the signal among multiple paths. Z transforms have been applied to their analysis as well [34].

Interleavers are important in dense WDM systems where the channel spacing is sufficiently small to make it advantageous to deinterleave neighboring channels to increase the effective channel spacing for further processing. For example, channels on a 50 GHz grid may be deinterleaved to provide two outputs each having 100 GHz channel spacing. Then, reconfiguration such as adding or dropping of particular channels may be accomplished on each output with filters having a 100 GHz channel spacing, which are typically easier to build and less expensive. A particular implementation where the filter dispersion is minimized by cascading stages in a compensating manner is shown in Figure 7. The ports are interconnected so that the transfer function for each path is a product of a forward and reverse polynomial as described in Eq. (5). Since the forward and reverse polynomial have mirror image roots, their dispersion cancels. Interleavers are from a class of filters with special symmetry properties called half-band filters [35]. A dispersion-less interleave filter has been demonstrated at the expense of a small excess loss with a combination of lattice and transversal filters [36].

An architecture that uses 2×2 filters for polarization-dependent filtering is shown schematically in Figure 8(a). A polarization beam splitter (PBS) separates the incoming TE and TM polarized light into two paths. A 90° polarization rotator is inserted in one arm so that the two polarizations are subsequently copolarized and can interfere coherently. A 2×2 filter, such as an FIR lattice filter, is inserted between pairs of PBSs and rotators. Instead of minimizing the filter's polarization dependence, this architecture applies different filter functions depending on the incoming signal's polarization. Any 2×2 filter, including IIR filters, can be inserted in this architecture. Applications include polarization mode dispersion (PMD) compensation [37,38] and polarization demultiplexing [39]. An example of a PMD compensating FIR filter with a single, fixed differential delay line is shown in Figure 8(b) [40]. When polarization-independent filtering is needed, identical filters may be inserted in each arm between PBSs and rotators.

While lattice filters are serial devices that become more complex as the number of stages increase, phased-array filters have the advantage of implementing a large number of feedforward stages in parallel. Diffraction gratings are a classic example of phased arrays where an incoming wavefront is sampled by each grating line and delayed, in this case by the path length

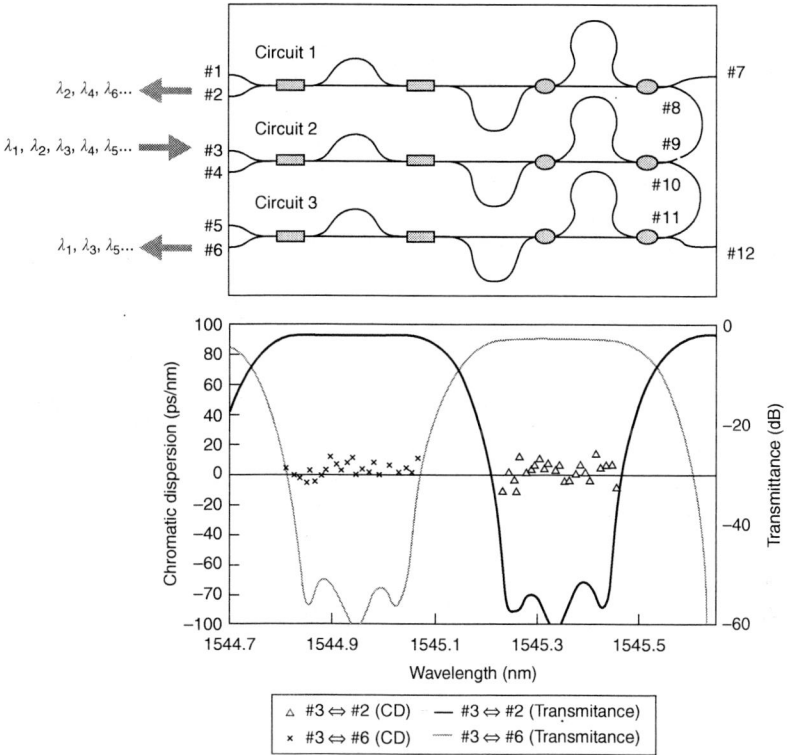

Figure 7 An FIR interleaver filter and spectral response (From T. Chiba, H. Arai, K. Ohira, H. Nonen, H. Okano, and H. Uetsuka, in *Proceedings of the Optical Fiber Communication Conference*, Anaheim, CA, 2001, p. WB5. With permission. Copyright 2001 IEEE.)

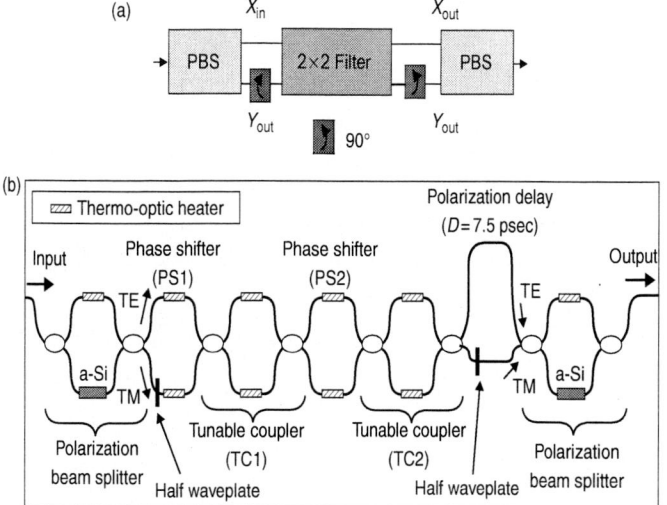

Figure 8 (a) Basic architecture for a polarization filter and (b) a single-stage PMD compensating filter (From T. Saida, K. Takiguchi, S. Kuwahara, Y. Kisaka, Y. Miyamoto, Y. Hashizume, T. Shibata, and K. Okamoto, *IEEE Photonics Technol. Lett.*, 14, 507–509, 2002. With permission. Copyright 2002 IEEE.)

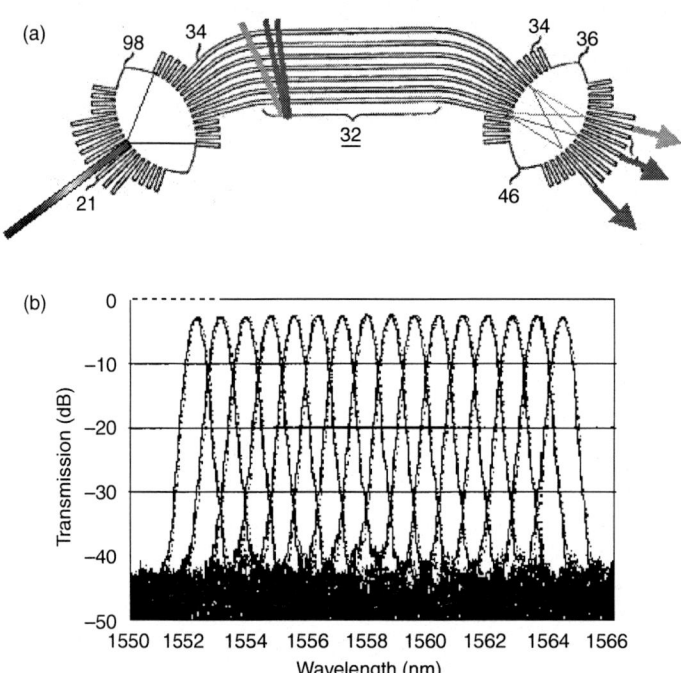

Figure 9 An AWG router (From Y. Hibino, *IEEE J. Select. Top. Quantum Electron.*, 8, 1090–1101, 2002. With permission. Copyright 2002 IEEE.)

differences arising from reflection off the grating surface. For implementation in planar waveguides, echelle gratings are an integrated diffraction grating where deep etching and reflective coatings are required in the fabrication process [41,42]. A device performing an equivalent operation that uses more standard fabrication processes for planar waveguides is the arrayed waveguide grating (AWG) router. An integrated optical phased array was first proposed by Smit [43], combined with a slab coupler [44] and extended to $N \times N$ operation [45] to yield the device in Figure 9(a) known as an AWG, or waveguide grating router (WGR) or PHASAR for phased-array device. The first demonstrations were reported within a year by three different labs [46–48]. An incoming array of waveguides transitions into a slab diffraction region. When a single input waveguide is illuminated, its far-field diffraction pattern excites the grating array of waveguides, which have path length differences of ΔL between adjacent waveguides and a pitch of a at the slab interfaces. The linear, frequency-dependent phase front is focused by the second slab coupler to a wavelength-dependent position along the output array. For the mth-diffraction order, the condition for constructive interference is given by the grating equation [43]:

$$m\lambda = n_e \Delta L + n_s a \left(\theta_i + \theta_j \right), \tag{16}$$

where the grating waveguide and slab effective indices are n_e and n_s, and the far-field plane-wave angle of incidence for the input and output waveguides on the grating array are denoted by θ_i and θ_j. Using the paraxial approximation for the excitation of N grating waveguides, the frequency response is given by the discrete Fourier transform as follows [12]:

$$H_{i,j}(f) = \sum_{n=1}^{N} \left| g_{i,n} h_{n,j} \right| e^{-jn\varphi(f)}, \tag{17}$$

where the amplitude coupling coefficients between the nth-grating waveguide and the ith-input or jth-output waveguide are given by $g_{i,n}$ and $h_{n,j}$, respectively. The phase associated with the unit optical path length through the device is $\varphi(f) = 2\pi f[n_e \Delta L + n_s a(\theta_i + \theta_j)]/c$. The FSR is inversely proportional to the unit optical path length difference, which is dominated by ΔL but also depends on the input and output waveguides.

The AWG demonstrates apodization of the filter coefficients via Gaussian excitation of the array grating as illustrated by the impulse response in Figure 4(b). This tapering of the filter coefficients reduces the sidelobes and thereby improves the stopband rejection so that crosstalk from neighboring channels is minimized in a WDM system. The symmetry in the impulse response implies a linear-phase response; so, AWGs are ideally dispersion-free. By properly designing the amplitude and the phase of the coefficients, a flat passband can be achieved [49]. Other techniques, using spatial filtering at the input/output waveguides are, also used for pass-band flattening including parabolic waveguide horns [50] and MMI couplers [51]. As in RF and microwave phased-array antenna applications, apodization is widely used to suppress out-of-band sidelobes in optical filter applications including fiber long-period and Bragg gratings and thin-film filter design.

The output spectrum for a 16 channel 100 GHz-spacing AWG made with 1.5% index-contrast silica waveguides is shown in Figure 9(b) [52]. Using dopant-rich cladding, the polarization-dependent wavelength shift is < 10 pm. Two-stage tandem AWG multiplexer and demultiplexers with 1000 channels have been demonstrated. Because of their ability to process a large number of channels in parallel, AWGs form the building block for many applications including channelized gain equalizers [53], reconfigurable add/drop multiplexers [54], wave-length selective switches [55], and periodic dispersion compensators [56]. Banded filters have also been realized with AWGs [57]. For a more detailed description of AWG design and applications, see References 58 and 59.

IIR FILTERS

Ring resonators and Fabry–Perot [60] interferometers are well-known single-stage optical IIR filters that contain an all-pole response and two distinct responses containing a coupled pole-zero pair. For approximating a desired frequency response, multistage IIR filters are needed. Because of the narrow band nature of the all-pole response, lattice architectures as shown in Figure 10(a) are favored over cascading single stages to maximize the passband transmission. Bragg gratings and contra-directional grating-assisted couplers are also IIR lattice filters and may be analyzed in a similar fashion to the ring resonators with Z transforms [61]. For arbitrary design control over the poles and zeros, other IIR architectures must be explored. A powerful technique referred to as allpass filter decomposition is presented for the synthesis of higher-order IIR filters.

Lattice ring architectures, as shown in Figure 10(a), have the advantage of power comple-mentary outputs and low theoretical loss over a cascade of single-stage rings. Marcatili [62] first proposed the use of a ring resonator as a bandpass filter; however, for high-bitrate and spec-trally-efficient systems, a filter response with a flat passband and sharper transitions is needed. Similar to the single-stage transfer functions in Eqs. (6) and (7), multistage coupled ring filters have an all-pole response $H_{11}(z) = H_{22}(z)$ and two distinct responses with both poles and zeros, $H_{12}(z)$ and $H_{21}(z)$.

$$H_{11}(z) = H_{22}(z) = \frac{\Gamma\sqrt{z^{-N}}}{D(z)},$$

$$H_{21}(z) = \frac{Y_2(z)}{X_1(z)} = \frac{N(z)}{D(z)} \quad \text{and} \quad H_{21}(z) = \frac{Y_1(z)}{X_2(z)} = \frac{N^R(z)}{D(z)},$$

(18)

where $N(z)$ and $D(z)$ are Nth-order numerator and denominator polynomials. The numerator polynomials are reverse polynomials of each other; so, their dispersion is opposite in sign. Analysis of multistage IIR lattice filters using Z transforms and cascading 2×2 transfer matrices has been reported for both etalon and ring implementations [63–65]. A layer-peeling algorithm may be used to determine the optical parameters from the polynomial coefficients [65]. Multistage filter design using coupled mode theory in time [66], which turns out to be more similar to a Laplace transform than Z transform approach, has also been reported [67,68].

Regarding dispersion, the IIR filter coefficients may be optimized to reduce dispersion across the passband; however, dispersion contributed by the poles cannot be completely cancelled. The all-pole response is minimum phase, so its phase response may be directly calculated from knowledge of the amplitude response [7]. The filter analysis problem is to determine the optical parameters from a measurement of the frequency response. The filter coefficients cannot be uniquely determined from the all-pole response since no information is obtained about $N(z)$ in this case. For a lossless filter, power conservation allows the possible zero

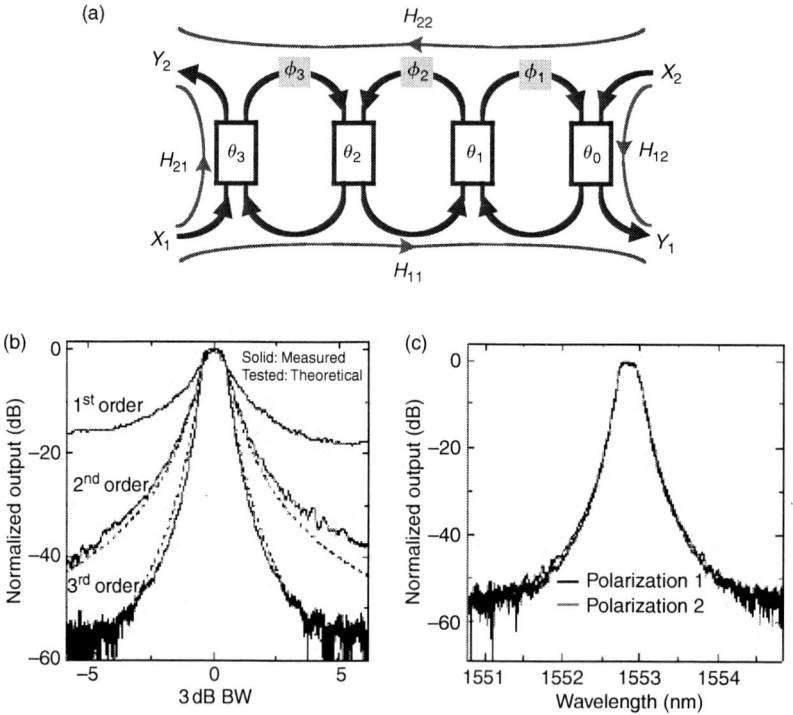

Figure 10 (a) Schematic of a three-ring IIR lattice filter. Experimental results (b) for filters up to third-order, and (c) polarization dependence for a third-order filter (From B. Little, in *Proceedings of the Optical Fiber Communications Conference*, Atlanta, GA, 2003, paper ThD1. With permission. Copyright 2003 IEEE.)

locations to be calculated from knowledge of the pole locations. The particular group delay response of the pole-zero responses can then be used to distinguish the particular solution [69].

The first experimental demonstration of a multistage ring filter employed silica-on-silicon waveguides with millimeter bend radii [65]. For interchannel filtering, larger FSRs are needed. The feedback path length must be reduced, which requires smaller bend radii, and consequently, a larger refractive index contrast between the core and the cladding. Single-stage micro-ring filters with bend radii of a few microns and FSRs over 20 nm were first demonstrated using silicon [70] and indium-phosphide [71] waveguides. The first multistage lattice micro-ring filters were demonstrated in InGaAs [72]. The high mode confinement requires the fabrication of smaller gaps for the directional couplers as illustrated in Figure 11(a) and mode transformers [73] for reducing the coupling loss to standard single-mode fiber. To alleviate the critical fabrication of the gap, vertical coupling [74,75] has been employed as shown in Figure 11(b). Fabrication results for multistage lattice filters in excellent agreement with theoretical predictions have been achieved with very low polarization dependence as shown in Figure 10(b) to 10(c) [76]. Vernier operation may be used to increase the effective FSR without increasing the index contrast. In the simplest form, rings with different FSR are cascaded so that their passbands overlap only once in several periods. Several coupled ring architectures utilizing Vernier operation have been proposed and demonstrated [77–79].

For realizing add and drop functionality where more input and output ports are needed than afforded by the 2×2 lattice filter discussed earlier, the cross-bar architecture shown in Figure 12(a) [80] has been proposed and demonstrated. Other architectures including ring resonators and $N \times M$ couplers have also been investigated theoretically [81]. Active rings employing gain in the feedback path have been demonstrated using semiconductor waveguides [82,83]. High-speed phase shifters using nonlinear optical polymer waveguides [84] are also being pursued for wavelength-dependent resonant-enhanced modulators as indicated schematically in Figure 12(b). Enhanced nonlinear switching has also been investigated using rings [85,86].

Allpass filters are well suited for phase response design since their magnitude response is ideally unity. One application for high-bitrate communication systems is for chromatic dispersion compensators. The first ring resonator used for dispersion compensation was single stage and fiber based [87]. The delay for a single-stage allpass filter with respect to normalized radian frequency is given in Eq. (19).

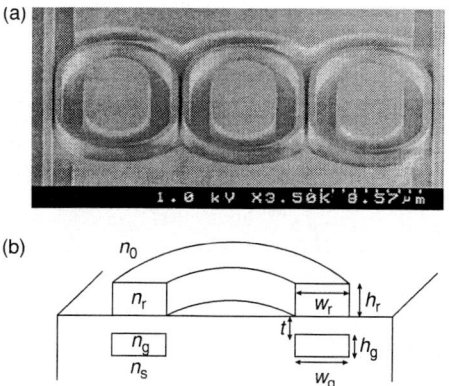

Figure 11 Ring fabrication with (a) horizontal coupling (From J. Hryniewicz, P. Absil, B. Little, R. Wilson, and P.-T. Ho, *IEEE Photonics Technol. Lett.*, 12, 320–322, 2000. With permission. Copyright 2000 IEEE.) and (b) a schematic of vertical coupling (From B. Little, S. Chu, W. Pan, D. Ripin, T. Kaneko, Y. Kokubun, and E. Ippen, *IEEE Photonics Technol. Lett.*, 11, 215–217, 1999. With permission. Copyright 1999 IEEE.)

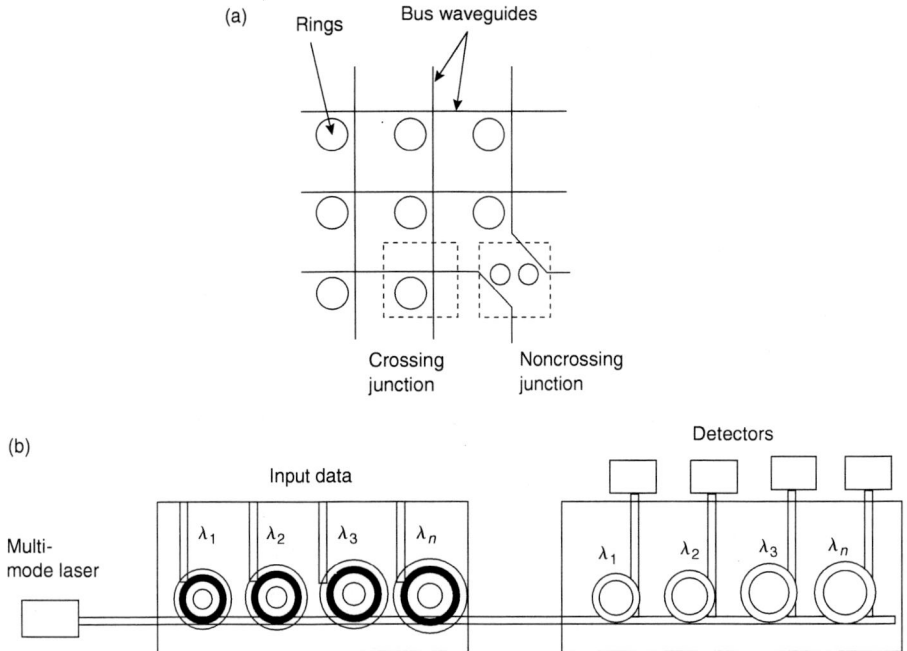

Figure 12 (a) A cross-bar ring add/drop configuration (From B. Little, S. Chu, and Y. Kokubun, *IEEE Photonics Technol. Lett.*, 12, 323–325, 2000. With permission. Copyright 2000 IEEE.) and (b) polymer waveguide modulator application (From P. Rabiei, W. Steier, C. Zhang, and L. Dalton, *J. Lightwave Technol.*, 20, 1968–1975, 2002. With permission. Copyright 2002 IEEE.)

$$\frac{\tau(\omega)}{T} = \frac{\left(1 - \rho^2\right)}{1 + \rho^2 - 2\rho \, \cos(\omega - \phi)}. \tag{19}$$

Two parameters must be specified per stage, the pole magnitude ρ and its phase ϕ. To improve the bandwidth and achieve maximum dispersion, multistage allpass filters are required [88]. Cascades of single-stages shown in Figure 13(a) are advantageous since each stage is independent and the overall delay is the sum of the individual delays. A multistage allpass filter can approximate any arbitrary phase response. The synthesis of a cascaded four-stage response for constant delay and dispersion is shown in Figure 13(d) and 13(e), respectively. For an allpass response, the passband is defined as the frequency range over which the desired phase response is approximated. The bandwidth utilization is defined as the ratio of the passband to the FSR. The peak dispersion, bandwidth utilization and approximation error are optimized for a given number of stages in the design process. The approximation error is defined as the deviation of the filter response from the desired response across the passband. An equiripple design is shown in Figure 13(d) and 13(e).

In practice, both the ring's resonant frequency and the coupling to the feedback path must be precisely controlled for each stage. The coupler lengths contribute to the overall delay length, therefore, long couplers will reduce the FSR. While tuning the coupler directly by changing the propagation constant of one arm relative to another (e.g., by heating) is possible, another approach uses symmetric and asymmetric Mach–Zehnder's within the ring as shown in

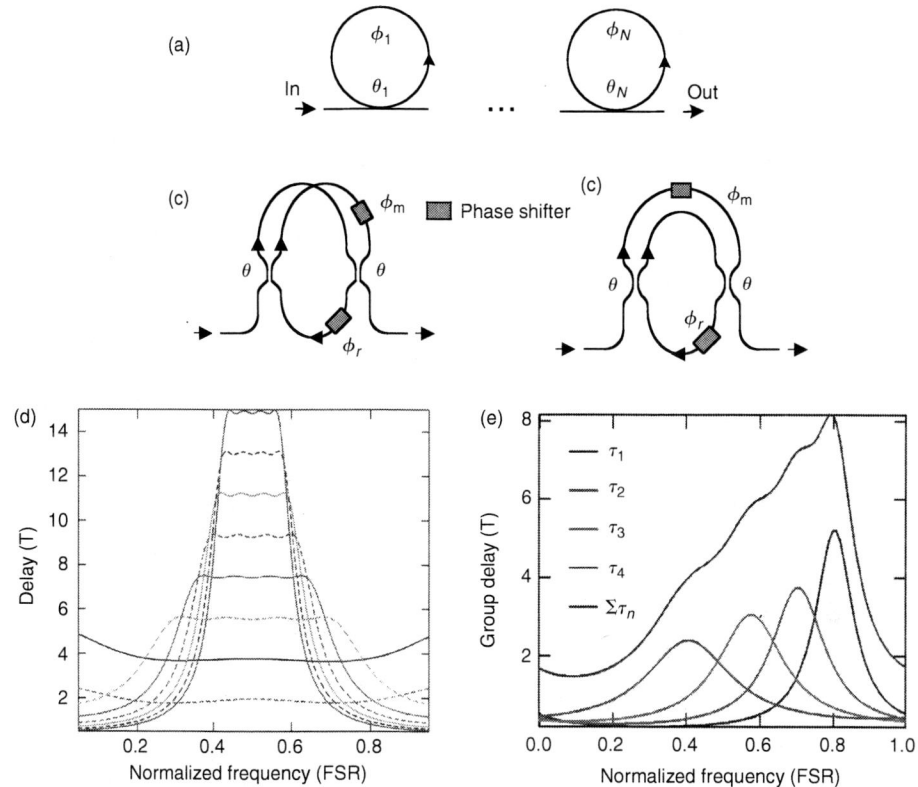

Figure 13   (a) Cascaded multistage allpass filter. Rings incorporating (b) symmetric, and (c) antisymmetric MZIs. Group delay responses for (d) variable delay and (e) tunable dispersion compensating applications

Figure 13(b) and 13(c) to achieve tunable couplers using phase shifters. Since the filtering is intrachannel for dispersion compensators, FSRs equal to the channel spacing are employed that require smaller core-to-cladding index contrasts than interchannel filters requiring FSRs that cover many channels, discussed previously for bandpass filter applications. Tunable dispersion compensators using ring resonators-based allpass filters have been reported in both Ge-doped silica [89] and SiON [90] waveguides. System tests at 10 Gb/sec with a tuning range of 4000 psec/nm [91] and 40 Gb/sec with a tuning range over 200 psec/nm [92] have been performed. More general allpass filter architectures with lattice structures have been proposed [88,93] and demonstrated with etalon implementations [94].

For bandpass filter applications, a steeper transition-band rolloff is achieved if control over both the pole and zero locations is available. A multistage IIR filter architecture with arbitrary pole and zero locations is shown in Figure 14(a) [95]. The stages are coupled making it difficult to precisely set the optical parameters. The simplified architecture in Figure 14(b) places multistage allpass filters within the interferometer arms [96,97]. The resulting transfer functions are sums and differences of allpass filter responses as indicated in Eqs. (20) and (21). This architecture also has many applications in digital filters [98,99].

$$G(z) = \frac{1}{2}\left[A_1(z) + A_2(z)\right] = \frac{P(z)}{D(z)}, \tag{20}$$

$$H(z) = \frac{1}{2}\left[A_1(z) - A_2(z)\right] = \frac{Q(z)}{D(z)}. \tag{21}$$

The two distinct output responses share a common denominator but have different numerator polynomials. The polynomial roots are coupled as a consequence of power conservation, resulting in an equation similar to Eq. (15) for FIR lattice filters [7]. Filter design is accomplished by decomposing the desired response into appropriate allpass filter responses, $A_1(z)$ and $A_2(z)$. Bandpass filters with optimal passband and stopband amplitude characteristics can be directly implemented, such as Butterworth, Chebyshev, and elliptic filter responses [96]. A maximally-flat response was first demonstrated using a ring resonator [100]. A fifth-order elliptic response is shown in Figure 14(c). The allpass decomposition architecture also lends itself to etalon-based implementations [96,101]. In general, IIR bandpass filters require substantially fewer stages to realize a given passband flatness, transition width, and stopband rejection compared with FIR filters as shown in the comparison of Figure 15. The drawback for IIR filters is the dispersion introduced by the poles. An allpass filter may be cascaded with the IIR bandpass

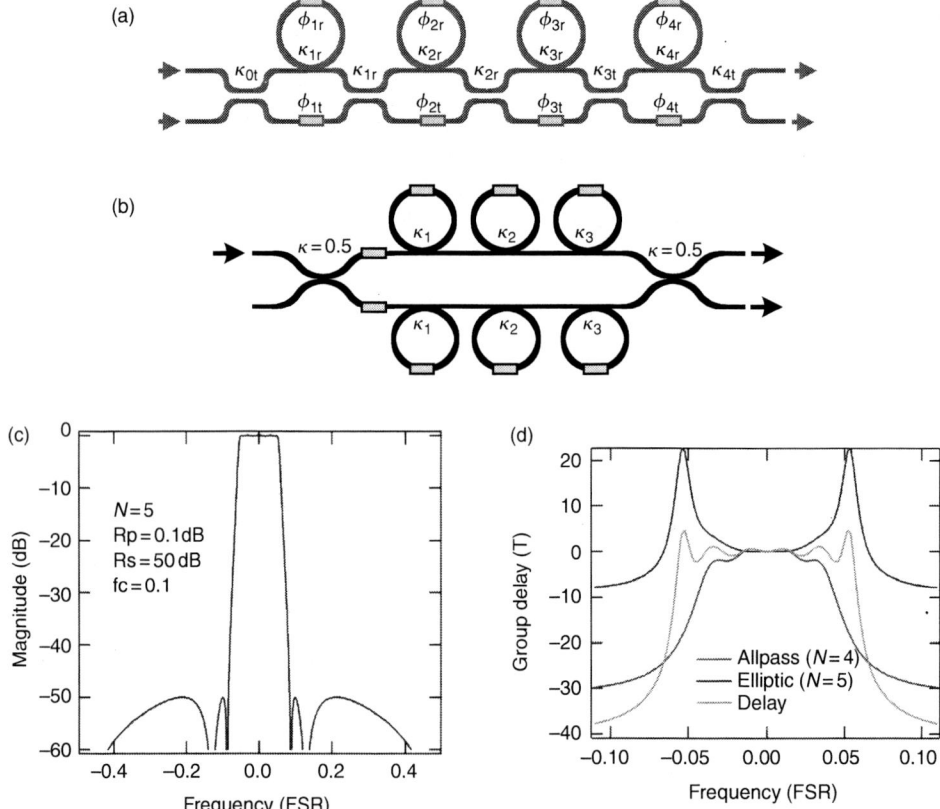

Figure 14 (a) General IIR filter (From K. Jinguji, *J. Lightwave Technol.,* 14, 1882–1898, 1996. With permission. Copyright 1996 IEEE.) and (b) architecture using allpass filter decomposition (From C. Madsen, *IEEE Photonics Technol. Lett.,* 10, 1136–1138, 1998. Copyright 1998 IEEE.) A fifth-order elliptic filter bandpass (c) magnitude and (d) delay response with compensating allpass response (From C. Madesen and G. Lenz, *IEEE Photonics Technol. Lett.,* 10, 994–996, 1998. Copyright 1998 IEEE.)

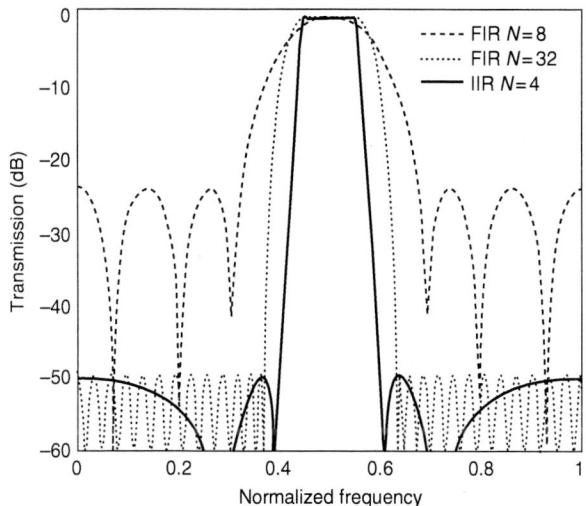

Figure 15 Comparison of bandpass FIR ($N = 8$ and $N = 32$) and IIR ($N = 4$) filter designs

filter to mitigate its dispersion as shown in Figure 14(d); however, this increases the total number of stages and complexity of the implementation. For very large FSRs, the dispersion may become negligible given the physical scaling with the unit delay as indicated in Eq. (14). An arbitrary 2×2 unitary filter, that is, two coupled magnitude and phase responses, can also be realized with a cascade of multistage allpass filters interconnected with couplers and has been applied to PMD emulation/compensation [39]. Notch filters may also be realized using allpass filter decomposition [102,103].

SUMMARY

Optical filters are, clearly, important building blocks for signal processing in optical systems. A great deal of progress has been made in the theory, design, and fabrication of adaptive optical filters within the last few years. A broad range of applications has been demonstrated from reconfigurable optical add/drops multiplexers for the bandwidth management of a large number of optical channels to tunable dispersion compensators for high bitrate optical networks. As per channel bitrates and spectral efficiency increase and systems drive toward optical mesh networks instead of point-to-point links, more ideal filters with agile responses and adaptive filters for dispersion and PMD compensation will be needed.

REFERENCES

1.   A. Oppenheim and R. Schafer, *Digital Signal Processing*. Englewood, NJ: Prentice-Hall, Inc., 1975.
2.   L. Jackson, *Digital Filters and Signal Processing*. Boston, MA: Kluwer Academic, 1986, pp. 145–150.
3.   J. Proakis and D. Manolakis, *Digital Signal Processing: Principles, Algorithms, and Applications*, 3rd ed. Upper Saddle River, NJ: Prentice Hall, 1996.
4.   M. Tur, J. Goodman, B. Moslehi, J. Bowers, and H. Shaw, "Fiber-optic signal processor with applications to matrix-vector multiplication and lattice filtering," *Opt. Lett.*, 7, 463–465, 1982.
5.   B. Moslehi, J. Goodman, M. Tur, and H. Shaw, "Fiber-Optic Lattice Signal Processing," *Proc. IEEE*, 72, 909–930, 1984.

6. K. Jackson, S. Newton, B. Moslehi, M. Tur, C. Cutler, J. Goodman, and H. Shaw, "Optical fiber delay-line signal processing," *IEEE Trans. Microwave Theory Tech.*, 33, 193–209, 1985.

7. C. Madsen and J. Zhao, *Optical Filter Design and Analysis: A Signal Processing Approach.* New York, NY: John Wiley & Sons, 1999.

8. L. Soldano and E. Pennings, "Optical Multi-Mode Interference Devices Based on Self-Imaging: Principles and Applications," *J. Lightwave Technol.*, 13, 615–627, 1995.

9. C. Dragone, "Efficient NXN star coupler based on fourier optics," *Electron. Lett.*, 24, 942–944, 1988.

10. L. Mach, *Zeitschr.f. Instrkde*, 12, 89, 1892; L. Zehnder, *Zeitschr.f. Instrkde*, 11, 275, 1891.

11. J. Michelson, *Am. J. Sci.*, 3, 120, 1881.

12. Y. Li and C. Henry, "Silicon optical bench waveguide technology," in *Optical Fiber Telecommunications IIIB*, I. Kaminow and T. Koch, Eds. San Diego, CA: Academic Press, 1997, pp. 319–376.

13. F. Gires and P. Tournois, "Interferometer useful for pulse compression of a frequency-modulated light pulse," *C. R. Acad. Sci*, 258, 6112–6115, 1964.

14. A. Deczky, "Synthesis of recursive digital filters using the minimum p-error criterion," *IEEE Trans. Audio Electroacoust.*, 20, 257–263, 1972.

15. M. Okuno, N. Takato, T. Kitoh, and A. Sugita, "Silica-based thermo-optic switches," *NTT Rev.*, 7, 57–63, 1995.

16. K. Jinguji, N. Takato, A. Sugita, and M. Kawachi, "Mach–Zehnder interferometer type optical waveguide coupler with wavelength-flattened coupling ratio," *Electron. Lett.*, 16, 1326–1327, 1990.

17. K. Sasayama, M. Okuno, and K. Habara, "Coherent Optical Transversal Filter Using Silica-Based Waveguides for High-Speed Signal Processing," *J. Lightwave Technol.*, 9, 1225–1230, 1991.

18. K. Sasayama, M. Okuno, and K. Habara, "Photonic FDM Multichannel Selector Using Coherent Optical Transversal Filter," *J. Lightwave Technol.*, 12, 664–669, 1994.

19. P. Prucnal, M. Santoro, and T. Fan, "Spread spectrum fiber-optic local area network using optical processing," *J. Lightwave Technol.*, 4, 547–554, 1986.

20. K. Kitayama and N. Wada, "Photonic IP Routing," *IEEE Photonics Technol. Lett.*, 11, 1689–1691, 1999.

21. T. Saida, K. Okamoto, K. Uchiyama, T. Takiguchi, T. Shibata, and A. Sugita, "Integrated optical digital-to-analogue converter and its application to pulse pattern recognition," *Electron. Lett.*, 37, 1237–1238, 2001.

22. K. Jinguji and M. Kawachi, "Synthesis of coherent two-port lattice-form optical delay-line circuit," *J. Lightwave Technol.*, 13, 72–82, 1995.

23. Y. Li, C. Henry, E. Laskowski, H. Yaffe, and R. Sweatt, "Monolithic optical waveguide 1.31/1.55 Mm WDM with -50 DB crosstalk over 100 Nm bandwidth," *Electron. Lett.*, 31, 2100–2101, 1995.

24. B. Offrein, F. Horst, G. Bona, R. Germann, H. Salemink, and R. Beyeler, "Adaptive gain equalizer in high-index-contrast SiON technology," *IEEE Photonics Technol. Lett.*, 12, 504–506, 2000.

25. A. Gnauck, G. Raybon, S. Chandrasekhar, J. Leuthold, C. Doerr, L. Stulz, A. Agarwal, S. Banerjee, D. Grosz, S. Hunsche, A. Kung, A. Marhelyuk, D. Maywar, M. Movassaghi, X. Liu, C. Xu, X. Wei, and D. Gill, "2.5 Tb/s (64 × 42.7 Gb/s) transmission over 40 × 100 Km NZDSF using RZ-DPSK format and all-Raman-amplified spans," in *Proceedings of the Optical Fiber Communications Conference*, 2002, postdeadline paper FC2.

26. K. Takiguchi, K. Okamoto, and K. Moriwaki, "Dispersion compensation using a planar lightwave circuit optical equalizer," *IEEE Photonics Technol. Lett.*, 6, 561–564, 1994.

27. K. Takiguchi, K. Jinguji, K. Okamoto, and Y. Ohmori, "Dispersion compensation using a variable group-delay dispersion equalizer," *Electron. Lett.*, 31, 2192–2194, 1995.

28. K. Takiguchi, K. Okamoto, and K. Moriwaki, "Planar lightwave circuit dispersion equalizer," *J. Lightwave Technol.*, 14, 2003–2011, 1996.

29. M. Bohn, G. Mohs, C. Scheerer, C. Glingener, C. Wree, and W. Rosenkranz, "An adaptive optical equalizer concept for single channel distortion compensation," in *Proceedings of the European Conference on Optical Communications*, 2001, paper Mo.F.2.3.

30. C. Doerr, S. Chandrasekhar, P. Winzer, L. Stulz, and A. Chraplyvy, "Simple multi-channel optical equalizer for mitigating intersymbol interference," in *Proceedings of the Optical Fiber Communication Conference*. Atlanta, GA, March 23–28, 2003, paper no. PD11.

31. T. Chiba, H. Arai, K. Ohira, H. Nonen, H. Okano, and H. Uetsuka, "Novel Architecture of Wavelength Interleaving Filter with Fourier Transform-Based MZIs." in *Proceedings of the Optical Fiber Communication Conference*, Anaheim, CA, 2001, p. WB5.

32. A. Vengsarkar, P. Lemaire, J. Judkins, V. Bhatia, T. Erdogan, and J. Sipe, "Long period fiber gratings as band rejection filters," *J. Lightwave Technol.*, 14, 58–65, 1996.

33. D. Smith, R. Chakravarthy, Z. Bao, J.J. Baran, J.L., A. d'Alessandro, D. Fritz, S. Huang, X. Zou, S. Hwang, A. Willner, and K. Li, "Evolution of the acousto-optic wavelength routing switch," *J. Lightwave Technol.*, 14, 1005–1019, 1996.

34. R. Feced and M. Zervas, "Efficient inverse scattering algorithm for the design of grating-assisted codirectional mode couplers," *J. Opt. Soc. Am. A*, 17, 1573–1582, 2000.

35. K. Jinguji and M. Oguma, "Optical Half-Band Filters," *J. Lightwave Technol.*, 18, 252–259, 2000.

36. T. Mizuno, T. Kitoh, T. Saida, M. Oguma, T. Shibata, and Y. Hibino, "Dispersionless interleave filter vased on transversal form optical filter," *Electron. Lett.*, 38, 1121–1122, 2002.

37. T. Ozeki and T. Kudo, "Adaptive equalization of polarization mode dispersion," in *Proceedings of the Optical Fiber Communication Conference*, 1993, pp. 143–144.

38. C. Madsen, "Optical allpass filters for polarization mode dispersion compensation," *Opt. Lett.*, 25, 878–880, 2000.

39. C. Madsen and P. Oswald, "Optical filter architecture for approximating any 2×2 unitary matrix," *Opt. Lett.*, 28, 534–536, 2003.

40. T. Saida, K. Takiguchi, S. Kuwahara, Y. Kisaka, Y. Miyamoto, Y. Hashizume, T. Shibata, and K. Okamoto, "Planar lightwave circuit polarization-mode dispersion compensator," *IEEE Photonics Technol. Lett.*, 14, 507–509, 2002.

41. J. Soole, A. Scherer, H. Leblanc, N. Andreadakis, R. Bhat, and M. Koza, "Monolithic InP-based grating spectrometer for wavelength-division multiplexed systems at 1.5 microns," *Electron. Lett.*, 27, 132–134, 1991.

42. S. Janz, M. Pearson, B. Lamontagne, L. Erickson, A. Delage, P. Cheben, D.-X. Xu, Gao, A. Balakrishnan, J. Miller, and S. Charbonneau, "Planar waveguide echelle gratings: an embeddable diffractive element for photonic integrated circuits," in *Proceedings of the Optical Fiber Communications Conference*, Anaheim, CA, March 19–22, 2002.

43. M. Smit, "New focusing and dispersive planar component based on an optical phased array," *Electron. Lett.*, 24, 385–386, 1988.

44. C. Dragone, "Optimum design of a planar array of tapered waveguides," *J. Opt. Soc. Am. A*, 7, 2081–2093, 1990.

45. C. Dragone, "An N × N optical multiplexer using a planar arrangement of two star couplers," *IEEE Photonics Technol. Lett.*, 3, 812–815, 1991.

46. H. Takahashi, S. Suzuki, K. Kato, and I. Nishi, "Arrayed-waveguide grating for wavelength division multi/demultiplexer with nanometre resolution," *Electron. Lett.*, 26, 87–88, 1990.

47. C. Dragone, C. Edwards, and R. Kistler, "Integrated optics NxN multiplexer on silicon," *IEEE Photon. Technol. Lett.*, 3, 896–899, 1991.

48. A. Vellekoop and M. Smit, "Four-channel integrated-optic wavelength demultiplexer with weak polarization dependence," *J. Lightwave Technol.*, 9, 310–314, 1991.

49. K. Okamoto and H. Yamada, "Arrayed-waveguide grating multiplexer with flat spectral response," *Opt. Lett.*, 20, 43–45, 1995.

50. K. Okamoto and A. Sugita, "Flat spectral response arrayed-waveguide grating multiplexer with parabolic waveguide horns," *Electron. Lett.*, 32, 1661–1663, 1996.

51. M. Amersfoot, C. de Boer, F. van Ham, M. Smit, P. Demeester, J. van der Tol, and A. Kuntze, "Phased-array wavelength demultiplexer with flattened wavelength response," *Electron. Lett.*, 30, 300–302, 1994.

52. Y. Hibino, "Recent advances in high-density and large-scale AWG multi/demultiplexers with higher index-contrast silica-based PLCs," *IEEE J. Select. Top. Quantum Electron.*, 8, 1090–1101, 2002.

53. C. Doerr, K. Chang, L. Stulz, R. Pafchek, Q. Guo, L. Buhl, L. Gomez, M. Cappuzzo, and G. Bogert, "Arrayed waveguide dynamic gain equalization filter with reduced insertion loss and increased dynamic range," *IEEE Photonics Technol. Lett.*, 13, 329–331, 2001.

54. C. Doerr, L. Stulz, J. Gates, M. Cappuzzo, E. Laskowski, L. Gomez, A. Paunescu, A. White, and C. Narayanan, "Arrayed waveguide lens wavelength add-drop in silica," *IEEE Photonics Technol. Lett.*, 11, 557–559, 1999.

55. C. Doerr, L. Stulz, D. Levy, M. Cappuzzo, E. Chen, L. Gomez, E. Laskowski, A. Wong-Foy, and T. Murphy, "Silica-waveguide 1×9 wavelength-selective cross connect," in *Proceedings of the Optical Fiber Communication Conference*, Anaheim, CA, 2002, paper FA3.

56. C. Doerr, L. Stulz, S. Chandrasekhar, L. Buhl, and R. Pafchek, "Multichannel integrated tunable dispersion compensator employing a thermooptic lens," in FA6. *Proceedings of the Optical Fiber Communications Conference*, Anaheim, CA, March 19–22, 2002.

57. C. R. Doerr, R. Pafchek, and L. W. Stulz, "Integrated band demultiplexer using waveguide grating routers," *IEEE Photonics Technol. Lett.*, 2003.

58. K. Okamoto, *Fundamentals of Optical Waveguides*. New York: Academic Press, 2000.

59. C. Doerr, "Planar lightwave devices for WDM," in *Optical Fiber Telecommunications IVA*, I. Kaminow and T. Li, Eds. New York: Academic Press, 2002, pp. 405–476.

60. C. Fabry and A. Perot, *Ann. Chim. Phys.*, 7, 115, 1899.

61. R. Feced, M. Zervas, and M. Muriel, "An efficient inverse scattering algorithm for the design of nonuniform fiber bragg gratings," *IEEE J. Quantum Electron.*, 35, 1105–1115, 1999.

62. E. Marcatili, "Bends in optical dielectric guides," *Bell Syst. Technol. J.*, 48, 2103–2132, 1969.

63. E. Dowling and D. MacFarlane, "Lightwave lattice filters for optically multiplexed communication systems," *J. Lightwave Technol.*, 12, 471–486, 1994.

64. R. Orta, P. Savi, R. Tascone, and D. Trinchero, "Synthesis of multiple-ring-resonator waveguides," *IEEE Photonics Technol. Lett.*, 7, 1447–1449, 1995.

65. C. Madsen and J. Zhao, "A general planar waveguide autoregressive optical filter," *J. Lightwave Technol.*, 14, 437–447, 1996.

66. H. Haus and W. Huang, "Coupled-mode theory," *Proc. IEEE*, 79, 1505–1518, 1991.

67. B. Little, S. Chu, H. Haus, J. Foresi, and J.-P. Laine, "Microring resonator channel dropping filters," *J. Lightwave Technol.*, 15, 998–1005, 1997.

68. B. Little, S. Chu, J. Hryniewicz, P. Absil, and P. Ho, "Filter synthesis for periodically coupled microring resonators," *Opt. Lett.*, 25, 344–346, 2000.

69. C. Madsen and J. Zhao, "Post-fabrication optimization of an autoregressive planar waveguide lattice filter," *J. Appl. Opt.*, 36, 642–647, 1997.

70. B. Little, J. Foresi, G. Steinmeyer, E. Thoen, S. Chu, H. Haus, E. Ippen, L. Kimerling, and W. Greene, "Ultra-compact Si-SiO2 microring resonator optical channel dropping filters," *IEEE Photonics Technol. Lett.*, 10, 549–551, 1998.

71. D. Rafizadeh, J. Zhang, S. Hagness, A. Taflove, K. Stair, and S. Ho, "Nanofabricated waveguide-coupled 1.5-Um microcavity ring and disk resonators with high Q and 21.6-Nm free spectral range." in *Proceedings of the CLEO Conference*, Baltimore, MD, May 18–23, 1997, pp. CPD23-2.

72. J. Hryniewicz, P. Absil, B. Little, R. Wilson, and P.-T. Ho, "Higher order filter response in coupled microring resonators," *IEEE Photonics Technol. Lett.*, 12, 320–322, 2000.

73. T. Shoji, T. Tsuchizawa, T. Watanabe, K. Yamada, and H. Morita, "Spot-size converter for low-loss coupling between 0.3-Um-square Si wire waveguides and single-mode fibers," in *Proceedings of the LEOS Annual Meeting*. Glasgow, Scotland, November 10–14, 2002, paper TuU3.

74. S. Suzuki, K. Shuto, and Y. Hibino, "Integrated-optic ring resonators with two stacked layers of silica waveguide on Si," *IEEE Photonics Technol. Lett.*, 4, 1256–1258, 1992.

75. B. Little, S. Chu, W. Pan, D. Ripin, T. Kaneko, Y. Kokubun, and E. Ippen, "Vertically coupled glass microring resonator channel dropping filters," *IEEE Photonics Technol. Lett.*, 11, 215–217, 1999.

76. B. Little, "A VLSI photonics platform," in *Proceedings of the Optical Fiber Communications Conference*, Atlanta, GA, 2003, paper ThD1.

77. K. Oda, N. Takato, and H. Toba, "A wide-FSR waveguide double-ring resonator for optical FDM transmission systems," *J. Lightwave Technol.*, 9, 728–736, 1991.

78. G. Barbarossa, M. Armenise, and A. Matteo, "Triple-coupler ring-based optical guided-wave resonator," *Electron. Lett.*, 30, 131–133, 1994.

79. Y. Ja, "Vernier operation of fiber ring and loop resonators," *Fiber Integrated Opt.*, 14, 225–244, 1995.

80. B. Little, S. Chu, and Y. Kokubun, "Microring resonator arrays for VLSI photonics," *IEEE Photonics Technol. Lett.*, 12, 323–325, 2000.

81. D. MacFarlane, E. Dowling, and V. Narayan, "Ring resonators with N × M couplers," *Fiber Integrated Opt.*, 14, 195–210, 1995.

82. K. Djordjev, S.-J. Choi, S.-J. Choi, and P. Dapkus, "Active semiconductor microdisk devices," *J. Lightwave Technol.*, 20, 105–113, 2002.

83. D. Rabus, M. Harmacher, U. Troppenz, and H. Heidrich, "Optical filters based on ring resonators with integrated semiconductor optical amplifiers in GaInAsP-InP," *IEEE J. Select. Top. Quantum Electron.*, 8, 1405–1411, 2002.

84. P. Rabiei, W. Steier, C. Zhang, and L. Dalton, "Polymer micro-ring filters and modulators," *J. Lightwave Technol.*, 20, 1968–1975, 2002.

85. J. Heebner and R. Boyd, "Enhanced all-optical switching by use of a nonlinear fiber ring resonator," *Opt. Lett.*, 24, 847–849, 1999.

86. V. Van, T. Ibrahim, P. Absil, F. Johnson, R. Grover, and P.-T. Ho, "Optical signal processing using nonlinear semiconductor microring resonators," *IEEE J. Select. Top. Quantum Electron.*, 8, 705–713, 2002.

87. S. Dilwali and G. Pandian, "Pulse response of a fiber dispersion equalizing scheme based on an optical resonator," *IEEE Photonics Technol. Lett.*, 4, 942–944, 1992.

88. C. Madsen and G. Lenz, "Optical All-Pass filters for phase response design with applications for dispersion compensation," *IEEE Photonics Technol. Lett.*, 10, 994–996, 1998.

89. C. Madsen, G. Lenz, A. Bruce, M. Cappuzzo, L. Gomez, and R. Scotti, "Integrated tunable allpass filters for adaptive dispersion and dispersion slope compensation," *IEEE Photonics Technol. Lett.*, 11, 1623–1625, 1999.

90. F. Horst, C. Berendsen, R. Beyeler, G.-L. Bona, R. Germann, H. Salemink, and D. Wiesmann, "Tunable ring resonator dispersion compensators realized in high-refractive-index contrast SiON technology," in *Proceedings of the European Conference on Optical Communications*, 2000, paper PD2.2.

91. C. Madsen, S. Chandrasekhar, E. Laskowski, K. Bogart, M. Cappuzzo, A. Paunescu, L. Stulz, and L. Gomez, "Compact integrated tunable chromatic dispersion compensator with a 4000 Ps/Nm tuning range." in *Proceedings of the Optical Fiber Communication Conference*, Anaheim, CA, 2001, p. PD9.

92. C. Madsen, S. Chandrasekhar, E. Laskowski, M. Cappuzzo, J. Bailey, E. Chen, L. Gomez, A. Griffin, R. Long, M. Rasras, A. Wong-Foy, L. Stulz, J. Weld, and Y. Low, "An integrated tunable chromatic dispersion compensator for 40 Gb/s NRZ and CSRZ," in *Proceedings of the Optical Fiber Communications Conference*, Anaheim, CA, March 19–22, 2002, PD-FD9.

93. G. Lenz and C. Madsen, "General optical all-pass filter structures for dispersion control in WDM systems," *J. Lightwave Technol.*, 17, 1248–1254, 1999.

94. D. Moss, S. McLaughlin, G. Randall, M. Lamont, M. Ardekani, P. Colbourne, S. Kiran, and C. Hulse, "Multichannel tunable dispersion compensation using all-pass multicavity etalons," Optical Fiber Communications Conference. Anaheim, CA, March 19–22, 2002, paper TuT2.

95. K. Jinguji, "Synthesis of coherent two-port optical delay-line circuit with ring waveguides," *J. Lightwave Technol.*, 14, 1882–1898, 1996.

96. C. Madsen, "Efficient architectures for exactly realizing optical filters with optimum bandpass designs," *IEEE Photonics Technol. Lett.*, 10, 1136–1138, 1998.

97. C. Madsen, "A multiport band selector with inherently low loss, flat passbands and low crosstalk," *IEEE Photonics Technol. Lett.*, 10, 1766–1768, 1998.

98. P. Vaidyanathan, P. Regalia, and S. Mitra, "Design of doubly-complementary IIR digital filters using a single complex allpass filter, with multirate applications," *IEEE Trans. Circuits Syst.*, 34, 378–389, 1987.

99. P. Regalia, S. Mitra, and P. Vaidyanathan, "The digital all-pass filter: A versatile signal processing building block," *Proc. IEEE*, 76, 19–37, 1988.

100. K. Oda, N. Takato, H. Toba, and K. Nosu, "A wide-band guided-wave periodic multi/demultiplexer with a ring resonator for optical FDM transmission systems," *J. Lightwave Technol.*, 6, 1016–1022, 1988.

101. B. Dingel and M. Izutsu, "Multifunction optical filter with a michelson-Gires-Tournois interferometer for wavelength-division-multiplexed network system applications," *Opt. Lett.*, 23, 1099–1101, 1998.

102. C. Madsen, "General IIR optical filter design for WDM applications using allpass filters," *J. Lightwave Technol.*, 18, 860–868, 2000.

103. P. Absil, J. Hryniewicz, B. Little, R. Wilson, L. Joneckis, and P.-T. Ho, "Compact microring notch filters," *IEEE Photonics Technol. Lett.*, 12, 398–400, 2000.

OPTICAL INTERCONNECT

Kenichi Iga

The idea of optical interconnect is to optical technology to transmit or connect devices, components, and subsystems instead of metal wiring. The advantage of optical interconnect includes its high speed capability basically with no limit, light weight, and low power consumption. Another important issue is parallel lightwave systems including numerous optical fibers. By taking this advantage, the optical interconnect is considered to be inevitable also in the computer technology. Some parallel interconnect schemes and new concepts are being researched. Vertical optical interconnect of LSI (Large Scale Integration) chips and circuit boards may be another interesting issue. The two-dimensional arrayed configuration of surface emitting lasers and planar optics will give way to a new era of opto-electronics.

Massively integrated parallel optical devices are becoming important for use in future parallel electronics such as optical routing systems, optical data transfer, optical image processing, parallel optical recording [1].

REFERENCE

1.   Kenichi Iga, "Surface emitting laser—its birth and generation of new optoelectronics field," *IEEE J. Select. Top. Quantum Electron.*, 6, 1201–1215, 2000.

OPTICAL PARALLEL PROCESSORS

Kenichi Iga

Optical parallel processing is a way of information processing by using essential parallelism and high speed of light, especially for image information processing. An optical parallel processor composed of two-dimensional (2D) microlens array is considered [1]. Computer simulation and experiments showed a possibility of ultra-parallel optical image processors.

Several other schemes for optical computing have been considered; however, one of the bottle necks may be a lack of suitable optical devices, in particular, 2D vertical cavity surface emitting lasers (VCSEL) and surface operating switches. Very low threshold surface emitting lasers have been developed, and stack integration together with 2D photonic devices is actually considered [2].

REFERENCES

1.   Takeo Katayama, Takanori Takahashi, and Kenichi Iga, "Optical pattern recognition experiments of Walsh spatial frequency domain filtering method," *Jpn. J. Appl. Phys.*, 39, 1576–1581, 2000.
2.   Kenichi Iga, "Surface emitting laser—its birth and generation of new optoelectronics field," *IEEE J. Select. Top. Quantum Electron.*, 6, 1201–1215, 2000.

OPTICAL PARAMETRIC AMPLIFIER

Kyo Inoue

BACKGROUND

Optical parametric amplification (OPA) is a phenomenon induced by nonlinear interaction among two or three lightwaves, in which signal light is amplified and light having a new frequency, called the "idler," is generated from the pump light (Figure 1). In general, optical nonlinearity is so weak that extremely high pump power is required to observe this phenomenon. In optical fiber, a lightwave is confined to a small area and thus its intensity can be high because of the waveguide structure, and the interaction length can be long owing to the fiber low-loss property. These factors of high intensity and long interaction length, are preferable for nonlinear interaction. Pioneering work on fiber parametric amplifiers was done in the early 1980s [1], but a high-power laser source was necessary even with such advantages. The situation changed drastically in the early 1990s with the advent of erbium-doped fiber amplifiers (EDFAs), which make it possible to obtain high optical power from laser diode sources. Fiber nonlinearity has been extensively studied since then; for parametric amplifiers in particular, the demonstration of signal gain as high as 49 dB in 2000 has stimulated research activity [2].

OPTICAL PARAMETRIC INTERACTION IN FIBER

Optical parametric amplification in fiber originates from third-order nonlinearity in glass materials. When lightwave is incident on a material, polarization P is induced by electrical field E. The induced polarization is basically proportional to the electrical field but deviates slightly from the linear relationship. This deviation is expressed by a Taylor expansion with respect to the electrical field as $P = \chi_1{:}E + \chi_2{:}EE + \chi_3{:}EEE + \cdots$, where χ denotes susceptibility. In glass materials, which are isotropic media, the second term vanishes because of symmetry, and the third term is the lowest-order nonlinearity. Let us suppose that two lightwaves of different frequencies f_p and f_s are incident on the fiber and they are expressed as $E = A_p e^{i2\pi f_p t} + A_s e^{i2\pi f_s t} + (c.c.)$. Several frequency components are induced via the third term, such as $f_i \pm f_j \pm f_k$ ($i, j, k = p$ or s). A component with a frequency of $2f_p - f_s$ is one of them, from which a new light field is generated.

The newly generated light and the originally incident ones interact with each other while propagating along the fiber length. The behavior of the interaction is described by the following coupled wave equations, which are obtained from Maxwell's equations with the nonlinear polarization term:

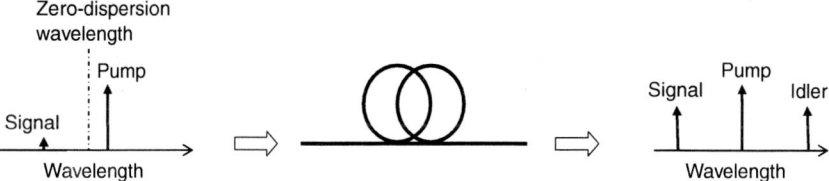

Figure 1   Parametric amplification in fiber

$$\frac{dE_p}{dz} = i\gamma\left[\left(|E_p|^2 + 2|E_s|^2 + 2|E_i|^2\right)E_p + 2E_p^*E_sE_i\exp(i\Delta k_0 z)\right], \tag{1a}$$

$$\frac{dE_s}{dz} = i\gamma\left[\left(2|E_p|^2 + |E_s|^2 + 2|E_i|^2\right)E_s + E_p^2E_i^*\exp(-i\Delta k_0 z)\right], \tag{1b}$$

$$\frac{dE_i}{dz} = i\gamma\left[\left(2|E_p|^2 + 2|E_s|^2 + |E_i|^2\right)E_i + E_p^2E_s^*\exp(-i\Delta k_0 z)\right], \tag{1c}$$

where E is the light amplitude (complex), z the propagation direction, γ the nonlinear coefficient, and Δk_0 linear phase mismatch given by $\Delta k_0 = k_s + k_i - 2k_p$ (where k is the propagation constant). Subscripts p, s, i denote components of frequencies f_p, f_s, f_i, respectively. By expressing the amplitude as $E = \sqrt{P}e^{i\phi}$ (where P is the light power, ϕ is the phase) and substituting it into Eq. (1), we obtain

$$\frac{dP_p}{dz} = -4\gamma\sqrt{P_p^2 P_s P_i}\,\sin\theta, \tag{2a}$$

$$\frac{dP_{s,i}}{dz} = 2\gamma\sqrt{P_p^2 P_s P_i}\,\sin\theta, \tag{2b}$$

$$\frac{d\theta}{dz} = \Delta k_0 + \gamma\left\{2P_p - P_s - P_i + \left(\sqrt{P_p^2 P_i/P_s} + \sqrt{P_p^2 P_s/P_i} - 4\sqrt{P_s P_i}\right)\cos\theta\right\}, \tag{2c}$$

where $\theta = \Delta k_0 z + \phi_s + \phi_i - 2\phi_p$. These equations show the power exchange among the three lightwaves, which depends on θ. For $\theta > 0$, P_s and P_i increases and P_p decreases. Note that $dP_s = dP_i$ and $dP_s + dP_i = -dP_p$, which means the power is transferred equally from P_p to P_s and P_i. Thus, under this condition, f_s and f_i lights are amplified while consuming f_p light. Conventionally, f_s light (originally incident) is called the "signal," f_i light (newly generated) the "idler," and f_p light (source of amplification) the "pump."

The power is most efficiently transferred when $\theta = \pi/2$. In the fiber input region, θ is automatically $\pi/2$. This is because the idler generated in the input region is $dE_i = i\gamma E_p^2 E_s^* \exp(-i\Delta k_0 z)dz$ (from Eq. [1c]), whose phase is $\phi_i = 2\phi_p - \phi_s - \Delta k_0 z + \pi/2$; thus, $\theta = \pi/2$ at the input port. As the lights propagate, θ changes according to Eq. (2c). In the particular case of $\Delta k_0 + \gamma(2P_p - P_s - P_i) = 0$, $d\theta/dz = 0$ and then θ holds $\pi/2$. In this case, the signal and the idler continuously grow along the fiber length. The condition $\Delta k_0 + \gamma(2P_p - P_s - P_i) = 0$ is called the "phase-matching condition," and $\Delta k = \Delta k_0 + \gamma(2P_p - P_s - P_i)$ the "phase mismatch," which represents the deviation from the phase-matching condition. Signal amplification is obtained when the phase-matching condition is satisfied.

To describe amplification behavior exactly, one should carry out numerical calculations. An analytical solution is obtained under the condition that $|E_p| \gg |E_s|, |E_i|$ and pump power is so large that its decrease through parametric interaction (which is called "pump depletion") is negligible. In this condition, the pump field is written from Eq. (1a) as $E_p(z) = E_p(0)\exp(i\gamma|E_p(0)|^2 z)$. Then, Eqs. (1b) and (1c) can be analytically solved with this pump and the boundary condition of $E_i(0) = 0$ as

$$E_s(z) = \{\cosh(gz) + i(\Delta k/2g)\sinh(gz)\}E_s(0)e^{-iqz}, \tag{3a}$$

$$E_i^*(z) = -i\sqrt{1-(\Delta k/2g)^2}\sinh(gz)\}E_s(0)e^{-iqz}, \qquad (3b)$$

where $g^2 = (\gamma P_0)^2 - (\Delta k/2)^2$, $q = (\Delta k_0 - 2\gamma P_0)/2$, P_0 is the pump power, and $\Delta k = \Delta k_0 + 2\gamma P_0$ (phase mismatch). These equations indicate that the signal and the idler power increase with $\exp(2gz)$ for $\exp(2gz) \gg 1$, meaning that $2g$ is the effective gain coefficient in OPAs. The gain is maximum for $\Delta k = 0$, that is, the phase-matching condition is satisfied. In the phase-matched condition, the signal gain is $\sim \exp(2\gamma P_0 L)$ (where L is the fiber length). The nonlinear coefficient is ~ 2 $W^{-1}km^{-1}$ in dispersion-shifted fiber (DSF) and thus, for example, signal gain of ~ 40 dB can be obtained using 2.5 km DSF with 1 W pump power.

ISSUES FOR IMPLEMENTATION: STIMULATED BRILLIOUN SCATTERING

To obtain parametric gain, high pump power should be incident on the fiber. However, the incident power is limited by stimulated Brillioun scattering that scatters light in the backward direction. Suppressing such scattering is crucial for implementing fiber OPAs. A typical way to suppress it is to spread the frequency spectrum of the incident light, because the Brillioun band-width is as narrow as 20 to 30 MHz. External phase modulation or direct frequency modulation to a pump laser diode is usually employed in fiber OPAs.

Spreading the pump spectrum necessarily results in the idler spectrum broadening because $f_i = 2f_p - f_s$, which is not good for some applications, such as wavelength conversion. There are several counter measures to avoid idler spectrum spread, such as the use of two pump lights, for which phase modulation or frequency modulation is imposed in an opposite way (described later), and binary phase modulation of $\{0, \pi\}$ onto pump light.

Usually, EDFAs are used to boost pump power in OPAs. Eliminating amplified spontaneous emission (ASE) from EDFAs is important, in practice, because it can consume pump power via parametric amplification and as a result sufficient signal gain is not obtained.

GAIN SPECTRUM

The gain spectrum or bandwidth is an important characteristic for amplifiers. In OPAs, the gain spectrum is determined by phase mismatch $\Delta k = k_s + k_i - 2k_p + \gamma(2P_p - P_s - P_i)$, which depends on the pump and the signal wavelengths because of fiber chromatic dispersion. For incident lights around the fiber zero-dispersion wavelength λ_0, Δk can be expressed by expanding k with respect to light frequency around the zero-dispersion frequency [3], such that

$$\Delta k = -\frac{2c\pi}{\lambda^2}\frac{dD_c}{d\lambda}(\lambda_p - \lambda_s)^2(\lambda_p - \lambda_0) + \gamma(2P_p - P_s - P_i), \qquad (4)$$

where c is the light velocity, λ_i the light wavelength ($i = $ pump, signal), and D_c the fiber chro-matic dispersion. This expression indicates that the first term, that is, the linear phase mismatch Δk_0, is 0 for $\lambda_p = \lambda_0$ and negative for $\lambda_p > \lambda_0$. Since $\gamma(2P_p - P_s - P_i)$ is a positive value, the phase-matching condition $\Delta k = 0$ is satisfied when the pump light is positioned at a wavelength somewhat longer than the zero-dispersion wavelength (Figure 1). The amount of the wavelength

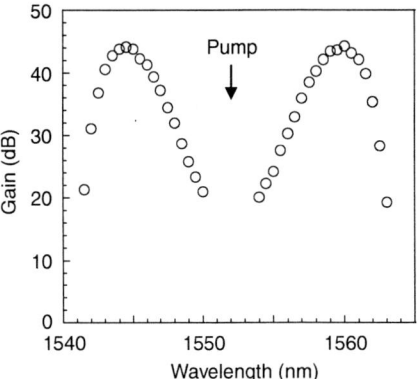

Figure 2   Signal gain spectrum

shift is dependent of the light power such that the pump wavelength should be longer for a higher power. The gain peak wavelength is derived from $\Delta k = 0$ with $P_p \gg P_s$, P_i:

$$\lambda_s = \lambda_p \pm \sqrt{\frac{\gamma P_p}{(2c\pi / \lambda^2)(\mathrm{d}D_c / \mathrm{d}\lambda)(\lambda_p - \lambda_0)}} \tag{5}$$

This expression indicates that there are two gain peaks at symmetrical wavelengths with respect to the pump. Figure 2 shows an example of the gain spectrum, which was measured in an OPA using 2.5 km dispersion-shifted fiber and 1.3 W pump power.

The gain bandwidth is dependent of the dispersion slope $\mathrm{d}D_c/\mathrm{d}\lambda$, as indicated in Eq. (4). For a small dispersion slope, Δk does not deviate much from 0 as the signal wavelength shifts from the gain peak, that is the phase matched wavelength, and thus the gain does not decrease much. This consideration suggests that the gain bandwidth is large for a small dispersion slope. The fiber length also affects the bandwidth. As discussed in the previous section, when $\Delta k \neq 0$, the phase relation among the propagating lights θ deviates from the optimum value ($\pi / 2$) along the fiber length. In short fiber, the signal reaches the fiber end before the deviation becomes serious, and thus the signal gain is not largely decreased from the phase matched condition. As a result, wide bandwidth is obtained in short fiber, while the gain itself is small as a penalty. Highly non-linear fiber (described later) is useful for compensating for the penalty of small gain. Short fiber is also preferable in terms of uniformity of chromatic dispersion. Generally, the fiber dispersion is assumed to be uniform along the fiber length. However, it is not uniform in the actual fiber due to fluctuations in the fabrication process. In fiber with nonuniform chromatic dispersion, the phase-matching condition cannot be perfectly satisfied along the whole length and the bandwidth is narrower than that in uniform fiber. Short fiber has small nonuniformity and thus provides wide bandwidth.

GAIN SATURATION

As the signal power increases, the pump power is depleted and the phase-matching condition shifts from the optimum via parametric interaction. As a result, the signal gain is reduced, that is, gain saturation occurs. Since the response time of the fiber nonlinearity is quite fast (~fsec), the signal gain instantly changes in accordance to the signal input power when an OPA is gain saturated. Channel crosstalk in multichannel amplification and extinction degradation occur in that condition.

A unique property of gain saturation in OPAs is that, as the gain saturation proceeds, the direction of power transfer reverses, that is, from pump → {signal and idler} to {signal and idler}→ pump. As the signal and the idler powers increases and the pump power decreases, the power balance of $2P_p - P_s + P_i$ changes. As a result, the phase relation θ in Eq. (2) shifts from the amplification condition of $\pi / 2$ to a negative value and the power turns to be transferred from the signal and idler to the pump.

NOISE FIGURE

As in other optical amplifiers, the quantum limited noise figure is 3 dB in OPAs, which originates from spontaneous parametric fluorescence. In addition to this inherent noise, several other factors also cause signal fluctuation, such as pump fluctuation and ASE from EDFAs used for boosting the pump power. Recent reports show, a noise figure of 4.2 dB [4].

POLARIZATION DEPENDENCE

A serious problem in OPAs is their inherent polarization dependency. In the previous section, we treated the light field as scalar. In fact, however, light field is a two-dimensional vector representing the state of polarization and the nonlinear susceptibility χ_3 is a tensor. The induced polarization P depends on the state of the polarization of incident lights, and so does the parametric process. For relatively long fiber, the polarization dependence can be expressed as $d\hat{E}_s \propto [\hat{E}_p \cdot \hat{E}_i^*]\hat{E}_p$ and $d\hat{E}_i \propto [\hat{E}_p \cdot \hat{E}_s^*]\hat{E}_p$ where $\hat{E}$ denotes a two-dimensional vector and $[\cdot]$ denotes the inner product of two vectors [5]. This expression indicates that the parametric gain is maximum when pump and signal have an identical state of polarization, and minimum when they have orthogonal states.

For practical applications, polarization insensitive operation is desired, since signal light transmitted through fiber transmission lines has various states of polarization. A fiber loop configuration with a polarization beam splitter (PBS) can compensate for the polarization dependency. Signal light is incident to a PBS, which divides the signal into vertical and horizontal components. The divided components propagate through a fiber loop in clockwise and counter-clockwise directions, respectively, and then return to the PBS. Pump light is also incident to the loop such that equal powers propagate in the two directions. Parametric amplification occurs in both directions with an equal gain, and, at the PBS output, the amplified vertical and horizontal components are summed up, which is independent of the incident signal polarization.

TWO-PUMP CONFIGURATION

The previous sections described one-pumped OPAs. It is also possible to obtain parametric amplification in fiber into which two pump lights are incident. When the phase-matching condition $\Delta k = k_s + k_i - k_{p1} - k_{p2} + \gamma(P_{p1} + P_{p2} - P_s - P_i) = 0$ is satisfied, power is transferred from pump 1 and pump 2 to the signal and the idler $(f_i = f_{p1} + f_{p2} - f_s)$ and the signal is amplified and the idler is generated (Figure 3). Though the configuration is complicated, two-pump schemes offer some advantages.

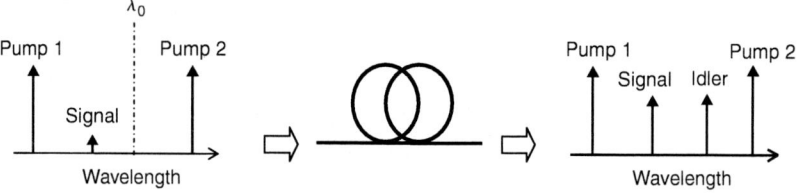

Figure 3 Two-pumped parametric amplification in fiber

First, the gain bandwidth can be wide. The phase mismatch for a two-pump scheme is written as

$$\Delta k = -\frac{c\pi}{\lambda^2}\frac{dD_c}{d\lambda}(\lambda_{p1} + \lambda_{p2} - 2\lambda_0)(\lambda_{p1} - \lambda_s)(\lambda_{p2} - \lambda_s) + \gamma(P_{p1} + P_{p2} - P_s - P_i). \qquad (6)$$

From $\Delta k = 0$ with $P_{p1}, P_{p2} \gg P_s, P_i$,

$$(\lambda_s - \lambda_{p1})(\lambda_s - \lambda_{p2}) = \frac{2\gamma P_p}{(c\pi/\lambda^2)(dD_c/d\lambda)(\lambda_{p1} + \lambda_{p2} - 2\lambda_0)}. \qquad (7)$$

The left-hand side is a parabolic function with respect to signal wavelength λ_s and the right-hand side is a constant with respect to λ_s. The solutions for λ_s are obtained by graphically looking at cross-points of the parabolic curve of the left-hand side, which crosses the zero line at $\lambda_s = \lambda_{p1}$ and λ_{p2}, and the horizontal linear line of the right-hand side. For $\lambda_{p1} + \lambda_{p2} - 2\lambda_0 > 0$, the right-hand side is positive and the solutions are $\lambda_s = \lambda_{p1} - \Delta$ and $\lambda_{p2} + \Delta$ (where Δ is a positive constant and $\lambda_{p1} < \lambda_{p2}$ is assumed). For $\lambda_{p1} + \lambda_{p2} - 2\lambda_0 < 0$, the right-hand side is negative and the solutions are $\lambda_s = \lambda_{p1} + \Delta$ and $\lambda_{p2} - \Delta$. The phase-matching condition is satisfied and the signal gain is maximum at these wavelengths. In the latter case ($\lambda_{p1} + \lambda_{p2} - 2\lambda_0 < 0$), the two gain peaks can be closely located. By appropriately overlapping the two gain peaks, one can have a wide gain spectrum in a two-pump configuration.

Another advantage of two-pump schemes is that polarization-insensitive operation is possible. For a two-pump scheme, polarization dependence is expressed as $d\hat{E}_s \propto [\hat{E}_{p2} \cdot \hat{E}_i^*]\hat{E}_{p1} + [\hat{E}_{p2} \cdot \hat{E}_i^*]\hat{E}_{p2}$ and $d\hat{E}_i \propto [\hat{E}_{p2} \cdot \hat{E}_s^*]\hat{E}_{p1} + [\hat{E}_{p1} \cdot \hat{E}_s^*]\hat{E}_{p2}$. When the two pumps have orthogonal states of polarization as $\hat{E}_{p1} = (E_{p1}, 0)$, and $\hat{E}_{p2} = (0, E_{p2})$, those expressions are decomposed to two sets of expressions:

$$\left\{ dE_{s(y)} \propto (E_{p1}E_{i(x)}^*)E_{p2}, \, dE_{i(x)}^* \propto (E_{p2}^*E_{s(y)})E_{p1}^* \right\}$$

and

$$\left\{ dE_{s(x)} \propto (E_{p2}E_{i(y)}^*)E_{p1}, \, dE_{i(y)}^* \propto (E_{p1}^*E_{s(x)})E_{p2}^* \right\},$$

where subscripts (x) and (y) denote the x- and y-components, respectively. These sets indicate that $\left\{ E_{s(y)}, E_{i(x)}^* \right\}$ and $\left\{ E_{s(x)}, E_{i(y)}^* \right\}$ are respectively amplified. The gains for each set is equal, provided that the two pump powers are the same. Thus, x- and y-components obtain an identical gain, and polarization insensitive operation is achieved as a whole.

Two-pump schemes are also useful to suppress idler spectrum spread. As described in the previous section, a spectrum-broadened pump is usually used in fiber OPAs in order to suppress the stimulated Brillioun scattering, which broadens the idler spectrum. In two-pumped parametric amplifiers, the idler phase (including the frequency term) is $2\pi(f_{p1} + f_{p2} - f_s)t + \phi_{p1} + \phi_{p2} - \phi_s + \pi/2$ in the phase-matched condition. When the frequencies or phases of the two pump lights are oppositely dithered, the deviations are compensated and no spectrum spread is induced in the idler light.

APPLICATIONS TO FUNCTIONAL DEVICES

While OPAs are available for simple signal amplification, they can also be applied to all-optical functional devices. A straightforward application is wavelength conversion. Idler light is newly generated from signal and pump lights, which can be regarded as wavelength conversion from the signal to the idler wavelength. A feature of this wavelength conversion scheme is that it is a coherent process. Thus, signal is converted independent of the modulation format. Multichannel conversion is also possible owing to the property of the coherent process.

The idler is generated when the signal and the pump are simultaneously incident into a fiber. This can be utilized as an all-optical gate or time demultiplexer controlled by pump pulses. The fast response time of fiber nonlinearity is favorable for this application.

PHASE-SENSITIVE AMPLIFIER (NOISELESS AMPLIFIER)

A unique feature of OPAs is that noiseless amplification is possible when the pump and the signal frequencies are identical. In this case, the signal and the idler are degenerate and Eq. (2b) in the phase-matched condition is rewritten as $dP_s/dz = 2\gamma P_p P_s \sin\theta$ with $\theta = 2(\phi_s - \phi_p)$. The signal is amplified for $\theta = \pi/2$, which is satisfied when $\phi_p = \phi_s - \pi/4$. This means that the signal whose phase is synchronized to the pump phase is amplified. The term "phase-sensitive amplifier" comes from this property.

The signal amplification is accompanied by spontaneous emission generated due to the quantum mechanical uncertainty. This spontaneous emission is amplified in the same way for the signal, that is, spontaneous light whose phase is synchronized to the pump phase is amplified. As a result, the signal and the ASE have the same phase. The classical picture of amplifier noise is that interference between signal and ASE causes level fluctuations (signal-spontaneous beat noise). Here, the phases of signal and ASE are the same, thus, the interference causes no fluctuation and noise-free amplification is achieved.

Though noise-free amplification is attractive, implementation of a phase-sensitive amplifier is challenging mainly due to the difficulty of phase synchronization.

HIGHLY NONLINEAR FIBER

To lower the pump power or shorten the fiber length, high nonlinearity is desired for fiber used in OPAs. The fiber nonlinearity is represented by $\gamma = (2\pi/\lambda)(n_2/A_{eff})$, where n_2 is the nonlinear refractive index and A_{eff} is the mode field area. Although dispersion-shifted fiber, which has $\gamma \sim 2\ W^{-1}\ km^{-1}$, is often used in fiber OPAs, fiber with a larger γ has been developed for non-linear applications. A way to achieve large γ is to fabricate fiber with a small mode field area by properly designing the fiber waveguide structure. The small mode field area increases light intensity, which results in large nonlinearity. Increasing the GeO_2 doped into the core is also effective, which increases n_2 and, thus, enhances γ. With these techniques, γ of ~20 W^{-1} km^{-1} has been achieved.

REFERENCES

1.   R. H. Stolen and J. E. Bjorkholm, "Parametric amplification and frequency conversion in optical fibers," *IEEE J. Quantum Electron.*, QE-18, 1062–1072 (1982).
2.   J. Hansryd and P. A. Andrekson, "Broadband CW fiber optical parametric amplifier with 49 dB gain and wavelength conversion efficiency," in *Proceedings of the Optical Fiber Conference*, Baltimore, MD, PD3 (2000).
3.   K. Inoue, "Four-wave mixing in an optical fiber in the zero-dispersion wavelength region," *J. Lightwave Technol.*, 10, 1553–1561 (1992).
4.   J. Blows and S. French, "Low-noise-figure parametric amplifier with a continuous-wave frequency-modulated pump," *Opt. Lett.*, 27, 491–193 (2002).
5.   K. Inoue, "Polarization effect on four-wave mixing efficiency in a single-mode fiber," *IEEE J. Quantum Electron.*, 28, 883–894 (1992).

OPTICAL RESONATOR

Kenichi Iga

The optical resonator is a medium or space that can confine optical waves in three-dimensional space to achieve laser oscillation or amplification including the active medium. In vertical cavity surface emitting laser (VCSEL), for example, a couple of multilayer Bragg reflectors are used in its Fabry-Perot cavity formation [1].

REFERENCE

1.   K. Iga and F. Koyama, *Surface Emitting Laser*, Kyoritsu Pub. Co., Tokyo, 1999.

OPTICAL SWITCH

Renshi Sawada

If a high switching rate is required, switches based on EO (Electro-Optical) effect such as Pockels and Kerr effect are frequently used. LN (Lithium Niobate) is particularly used. A high cost, however, is the weakest point. SOA (Semiconductor Optical Amplifier) switches, in which nonreflective coating is applied to both end faces of normal LD (Laser Diode), have a high threshold current if they are used independently and practically do not cause oscillation. The amount of light that passes through can be controlled by putting external light in and varying the current in this LD.

With regard to mechanical types, prismatic drive and fiber direct drive with solenoid are principal systems. Currently, a method using the micromachining technology is reviewed vigorously instead of the existing method by assembling discrete components and adjusting them for configuration.

Expectations for the micro electro-mechanical systems (MEMS) switches increase as WDM (Wavelength Division Multiplexing) spreads because the MEMS switches, particularly optical switches with a micromirror array have small wavelength dependence [1]. Since former switching performs EO conversion, an increased number of channels causes nonnegligible power capacity supplied to the EO elements. Also, if the loss of $N_{ch} \times N_{ch}$ switches is estimated temporarily for a system where normal 1×2 switches with relatively low normal loss are connected serially, olive switches for moving bubbles are expanded or AWG (Arrayed Waveguide Grating) is used, the number of channels N_{ch} that can reduce the loss to 7 dB is at most 30. With regard to the device size, if TO (Thermo Optic) switches with quartz type waveguide are used, the size reaches 4 in. for 8×8 channels. With the consideration of wiring, only the MEMS switches can be actually used as switches for more than 100 channels and these MEMS switches attract attention for solving these problems. Table 1 shows principal optical switches that have been reported.

Although we think mechanical switches perform slow switching, microswitches are not so. On–off switches, particularly, perform rapid switching. In the case of MARS (Mechanical Anti-Reflection Switch) [2,3] (Figure 1) that does not require any stopper, the switching rate is

Table 1
Principal optical switches

Mechanical type/ micromachine type	Fiber		
	Optical fiber drive		Drive of fiber with solenoid (Lorentz force)
	Loading/unloading of micromirror/shutter		
	Rotation of three-dimensional mirror (around orthogonal two axes)		
	Movement of linear stage with fiber guide		Use of linear stage
	Movement of microlens		Many types of mirror loading/unloading drive systems such as rotation of torsion spring, rotation with bending, and linear movement can be used.
	Planar waveguide		
	Waveguide drive		Use of electrostatic force other than Lorentz force is also available.
	Bubble/matching oil drive		Use of thermal capillarity, ink jet type
	Micromirror loading/unloading		Many methods of mirror loading/ unloading are available like fibers.
	Coupling with evanescent light		
	Optical film		
	Variable gap of opposed mirrors		MRS
	Movement of diffraction grating film		GLV
	Variable diffraction grating pitch		
Nonmicromachine type	Prismatic drive (mechanical type)		Optical path switching by moving mini-prism
	EO type		Optical path switching by varying refraction index with electric field
	Use of thermo optic effect and interference		TO switch: Use of MZI
	Plasma switch		Use of plasma effect
	Use of magneto-optic effect (MO effect)		Use of variation of polarized light direction with magnetic field

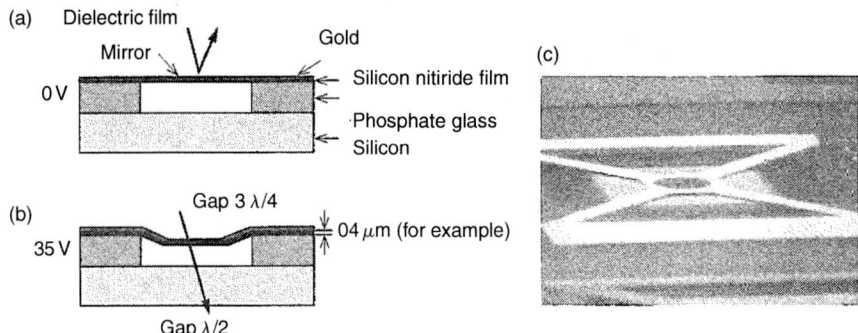

Figure 1   High-speed MARS switch (From K. W. Goosen et al., *IEEE Photonics Technol. Lett.*, 6, 1119–1121 (1994). With permission.)

high. In the case of a rotary mirror type such as DMD (Digital Micromirror Device) described later, the switching time is a little longer and is the order of μ. Since the switching time of most MEMS switches is the order of 1/10 to 1 msec, the rate is not so high but is sufficiently practical because the switching time of former TO switches is the order of millisecond. Since the mass and second moment of inertia are small and the spring and the cantilever can be shortened relatively, the resonance frequency can be very high. For application of mirrors, development of attenuators that block a part of optical transfer path with mirrors, as well as scanners is performed actively.

The difficulty of switching, particularly, switching that uses three-dimensional mirrors, consists in the necessity of instant fixation of a preset mirror rotation angle with high accuracy unlike scanners or oscillator sensors.

ON/OFF MEMS SWITCH SUCH AS GLV, MARs SWITCH, AND DMD APPLICATION SWITCH

Grating Light Valve (GLV) that uses a diffraction grating composed of six silicon nitride movable ribbons is used for switches as well as projectors [4]. As shown in Figure 2(a), six ribbons of 3 μm width, 100 μm long, and 150 nm thick correspond to one DMD micromirror described later. The ribbons are arranged at 4 μm intervals and displaced up and down alternately to change the light direction (same Figure 2[b]). When all ribbons are placed on the same plane, the reflected light returns along the incident light path. When the movable ribbons in every other position are pulled down with electrostatic force, the reflected light intensity decreases gradually and the diffracted light intensity increases. When the movable ribbons lower by $\lambda/4$, the diffracted light is the most intense and the reflected light is the least intense. In addition, the MARS switch [3,5] moves the silicon nitride membrane up and down to vary the gap in the Fabry–Perot etalon and to vary the intensity of reflected and transmitted light accordingly (Figure 1)[6]. This switch is suitable for use as an on/off switch, and the switching time is approximately 40 nsec. Also, this switch is used as a variable attenuator by converting the displacement to an analog value. An example where a DMD mirror [7] is inserted between two input/output fibers for application as a switch is shown in Figure 3.

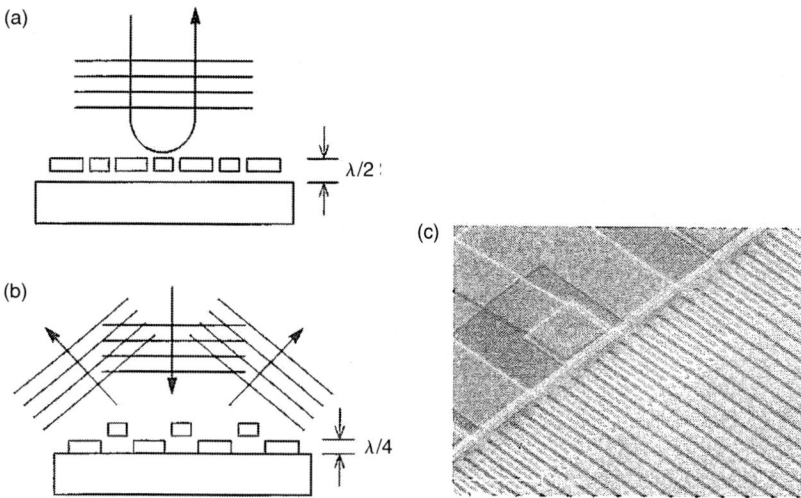

Figure 2 GLV switch

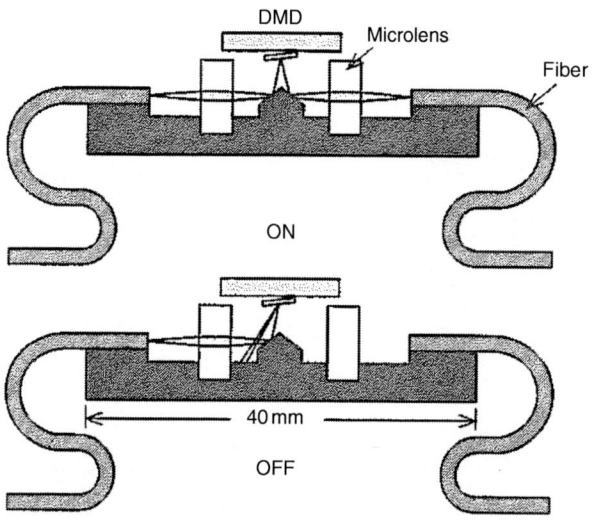

Figure 3 Optical switch with DMD mirror (From R. M. Boysel, T. G. McDonald, G. A. Magel, G. C. Smith, and J. L. Leonard, *Proc. SPIE*, 1793, 34–39 (1992)

SWITCH MADE ONLY BY REACTIVE ION ETCHING (RIE)

Switching is performed by loading and unloading a mirror that is formed with comb electrode actuators that are the most reliable among MEMS actuators (Figure 4) [8–10]. This switch is made of an Si active layer on SOI (Silicon On Insulator) substrate only by RIE (Reactive Ion Etching). Use of a side face to which dry etching is applied as a mirror deserves attention [11]. Also the comb electrode actuators have no self-holding force; however, a ratchet mechanism that uses buckling is formed to hold the loading and unloading status of the mirror. The roughness of the side face is reduced to 40 nm or less. Since thick thermal oxidation is performed selectively at the projecting portions of surface roughness, thermal oxidation and elimination of oxide film with hydrofluoric acid are repeated to reduce the roughness of the side face. This reduction of side face roughness with thermal oxidation is based on the Gibbs energy and related closely to the thermal oxidation temperature. The performance obtained is the insertion loss of 0.5 to 1 dB, crosstalk of 50 dB or more, switching time of 1 msec or less, and power supply voltage of 5 V (70 mW).

FIBER AND WAVEGUIDE DRIVE SWITCH

There are many switches that move an optical fiber and an optical waveguide to transfer light to other fibers and waveguides (Figure 5) [12–15]. Figure 6 shows a switch with the self-holding

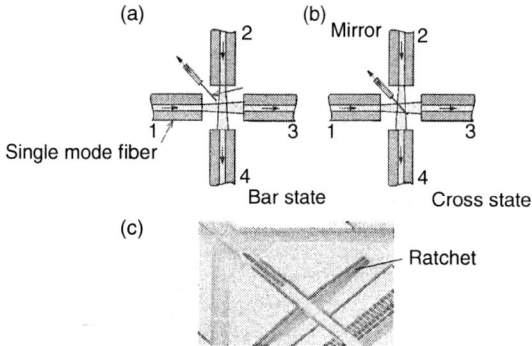

Figure 4 2 × 2 optical switch made only with RIE

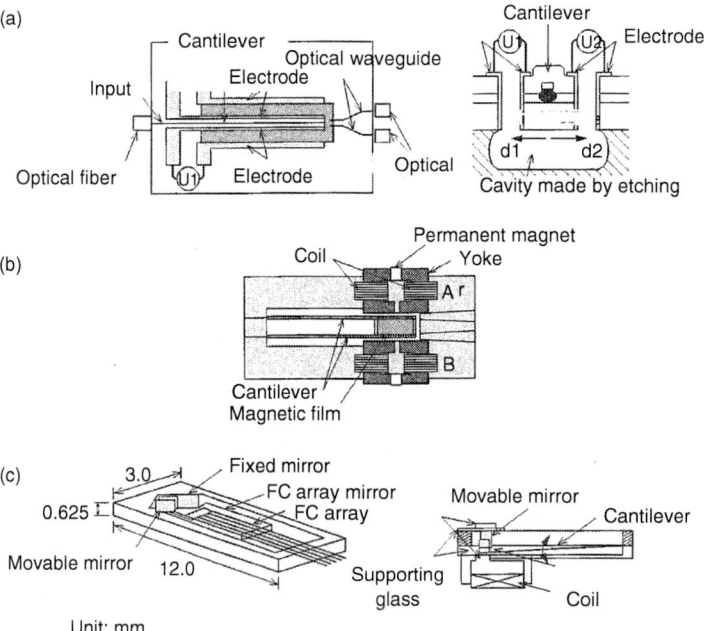

Figure 5 Waveguide movable switch (a) electrostatic force drive, (b) and (c) electromagnetic force (Lorentz force) drive

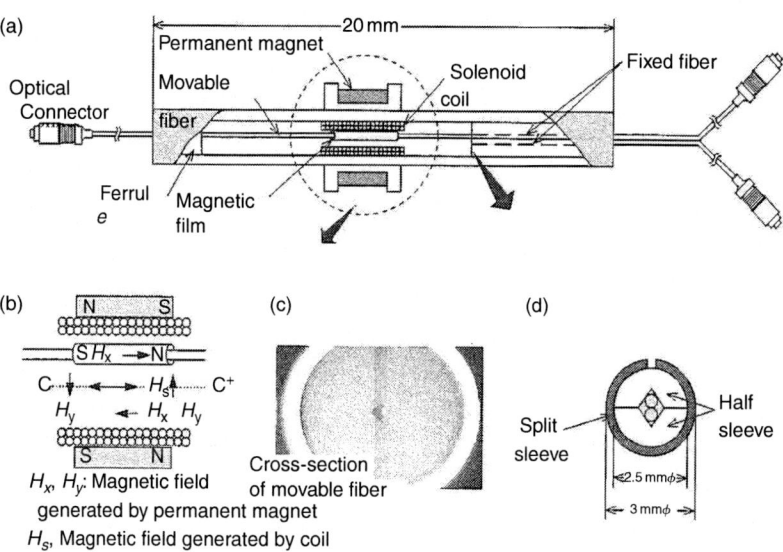

Figure 6   Movable fiber switch (From S. Nagaoka, *IEICE Trans. Electron.*, E80-C, 149–153 (1997). With permission.)

function that attracts a movable fiber on which magnetic film is applied with a permanent magnet and emits current to the external coil at the time of switching to move the fiber toward the permanent magnet at the opposite side with stronger force than the attraction force of the permanent magnet [16,17]. A switch of this type has insertion loss of 1 dB and switching time of approximately 1 msec. In the magnet type, a high integration degree is impossible because problems such as interference with adjacent switches in the magnetic field occur.

THREE-DIMENSIONAL AND MATRIX MIRROR ARRAY SWITCH

For configuration of a large-scale switch that uses micromirror arrays, two types are available: matrix switch and three-dimensional mirror switch (Figure 7) [18–27]. A switch in which an input fiber and an output fiber are arranged so that each optical axis crosses each other and a mirror is placed on that crosspoint is called matrix switch [22]. With regard to a three-dimensional mirror switch, rotation around two axes can be performed independently and light can be reflected in any direction. The role of mirror control is, in both switches, to put the most intense light into the output fiber. In the case of the matrix switch or the N_{ch} channels, only on/off switching is needed for the mirrors and therefore control is easy only if calibration of manufacturing error is performed at the beginning. The number of mirrors is, however, N_{ch}^2. In the case of the three-dimensional mirror array switch, although the number of required mirrors is $2N_{ch}$, high-precision angle setting for N_{ch} types including no switching is required.

   A mirror is raised by moving the lower part with a scratch drive actuator (SDA) driven by the electrostatic force. In the mirror, holes for engaging to prevent return of the mirror from the mounted position are formed [21–25]. Another feature of the matrix mirror switch is high-precision manufacturing of relative positions of the guide groove for alignment of the fiber and the mirror with the photolithography technology. Consequently, optical axis alignment that is normally difficult can be performed easily. The mirror in Figure 8 is a good example of this. A vertical mirror and a V-groove guide for fixing a fiber are made together by wet anisotropic

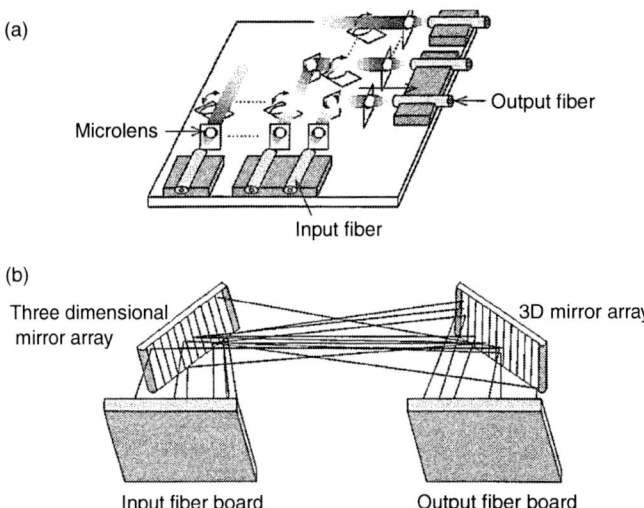

Figure 7 Large-scale switch with micromirrors (a) matrix switch and (b) three-dimensional mirror switch

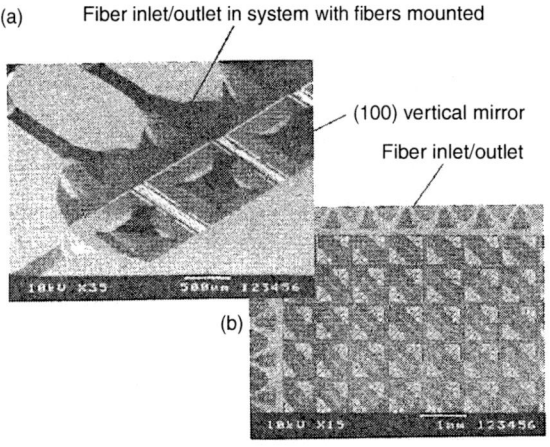

Figure 8 16 × 16 switch in which mirrors and V-grooves for fiber guide are made together

etching of (100) Si substrate with KOH solution [26,28,29]. A permalloy magnet piece of 100 μm thick is mounted to the back side of the vertical mirror, and attraction drive is performed with an external coil. A permanent magnet is mounted together to keep holding power even after the coil is turned off. Good alignment precision and reflecting surface precision give a good characteristic of insertion loss of 0.5 dB in both the reflection and the transmission modes. Figure 9 shows a pencil type switch. A number of switches equivalent to the number of channels are arranged for use. It is difficult to say that this system is a micromachine. Switching between any input and output fibers is performed by adjusting the angle of the movable mirror mounted on the front end.

In a three-dimensional mirror array switch, a part of the light (e.g., 10%) that passes through the output fiber after switching must be monitored (branched) and feedback must be given so that the most intense light enters the output fiber. Although monitoring of the output fiber is required only at the first calibration if mirror operation is stable and has good reproducibility, it must be performed at each switching. Also, monitoring the variation of electrostatic capacity is tried for easy and low-cost control of mirror rotation angle. Figure 10 shows a typical three-dimensional mirror made with the surface micromachining technology in which multiple layers of polycrystalline silicon, insulating film, and metal film are accumulated, patterning is performed, and a part of it is eliminated chemically to form multiple units of movable sections and actuators simultaneously [18,30]. The polycrystalline silicon film of 1.5 to 3.5 μm thick in the mirror section includes a grain boundary and therefore causes bad smoothness of the surface (although the roughness depends on the accumulation conditions and the thickness, the maximum roughness R_{max} is approximately 0.1 μm). Variation of warpage and large intrinsic strain occur due to the heat treatment process. Warpage affects aberration, and surface roughness affects scattering, namely reflectivity. Use of this film in a section that is exposed to direct light such as a mirror and etalon, therefore, cause problems. Figure 11 shows a mirror that is made with high aspect ratio machining (machining for making narrow and high structure) of substrate and bulk micromachining technology [27]. For this a movable mirror is formed in the section of active layer of SOI substrate for bonding with another substrate of step electrode [31]. Features of this mirror are small warpage of the order of 10 nm and smooth surface roughness of a few tens of nanometers that can be obtained by using monocrystalline silicon substrate and

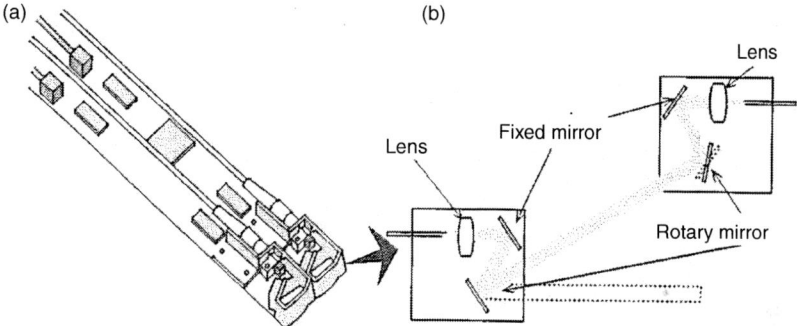

Figure 9   Pencil type switch

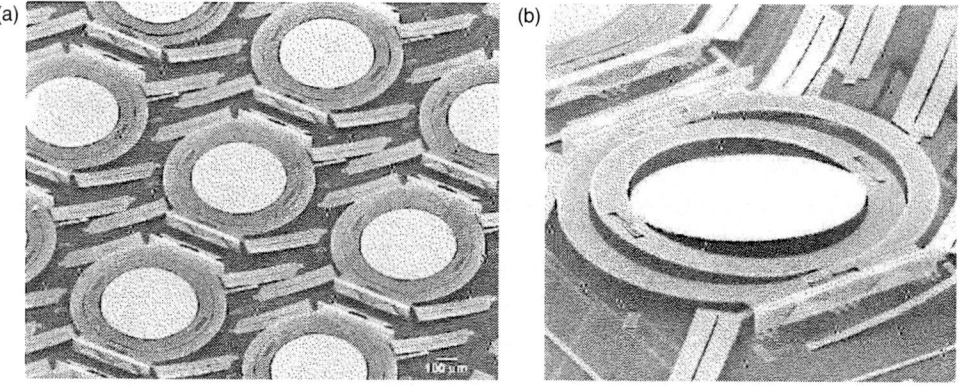

Figure 10   Typical three-dimensional mirror (rotation around two axes) made with surface micromachining (19 × 32)

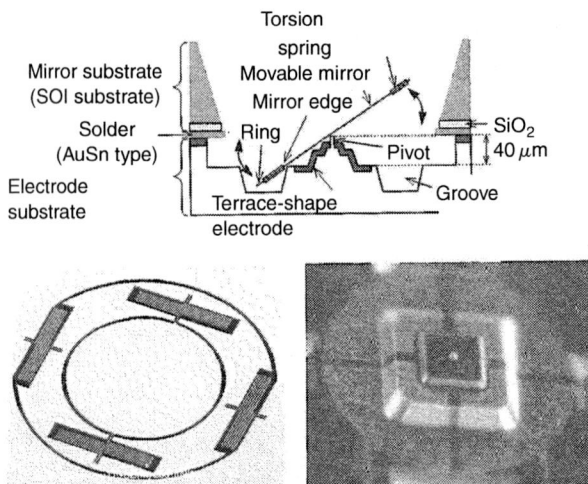

Figure 11   Three-dimensional made with bulk micromachining

forming reflection layer symmetrically on the upper and the lower surfaces of the movable mirror (diameter of 500 μm or more) in the same conditions.

For driving rotation of a micromirror array, electrostatic force is used, and for its restoring force, a torsion spring similar to DMD is used. The mirror rotates in any direction with nonuniform electric field between the mirror section that rotates around two axes and the lower electrode. The mirror size is normally 400 to 600 μm. Difficulty in controlling the three-dimensional mirrors is caused by nonlinear relation between drive voltage V and mirror rotation angle θ as well as variation of the electrode shape in each mirror. Since a large angle rotation can be obtained and linear approximation can be used in the area with a high drive voltage in the $V - \theta$ curve, control with this linear area by adding bias voltage to the mirror voltage is performed. If doughnut-shaped electrodes are formed without any electrode at the center, the drive voltage increases largely but pull-in is not easily generated, resulting in a large rotation angle. For control of the mirror in the linear area, addition of bias V_{bias} to the mirror voltage or the like is tried. Practically, in addition to the bias voltage, calibration is required separately. If variation of torsion spring shape, gap, etc., due to manufacturing error occurs, additional calibration must be performed for each mirror.

LENS DRIVE SWITCH

The direction of collimated light can also be changed by displacing the lens instead of rotating the micromirror [32]. If the focal length is supposed to be f and the collimated light angle is [rad], lens displacement d can be expressed.

The light angle can be changed freely by moving the lens in the two-dimensional plane. For example, rotation angle θ of 6° (approximately 0.1 rad) or more can be realized with a movable microstage with the maximum displacement of 60 μm and a microlens with focal length of 600 μm. Displacement of its stage can be reduced by mounting a lens for enlarging the angle of deflection of light beam. Although this stage gives resolution of 7 nm and allows movement of a few hundred, micrometer, the response time is the order of 100 msec, longer than the three-dimensional mirror switch by one digit.

MACH–ZEHNDER INTERFERENCE (MZI) APPLICATION SWITCH (TO SWITCH)

There is a TO switch in which a MZI is configured with two quartz type waveguide paths (Figure 12) [33]. A heater formed near a part of the optical path of either waveguide is used to heat to vary the phase of light with photothermal refraction index effect and to generate interference with a directional coupler for use as a switch. The loss including fiber connection loss is approximately 1 dB, and the switching time is approximately 1 msec.

BUBBLE MOVEMENT SWITCH

A groove is formed obliquely at some midpoint in the waveguide, and if the groove is filled with air, light is reflected completely to change the direction. If the groove has the same refraction index as the waveguide, although light is diffused and spread a little, it goes straight and enters the waveguide. Liquid is moved by thermal capillarity (Figure 13) or ink-jet system [34]. The basic configuration is a kind of matrix switch. The ink-jet system can be regarded as explosion phenomenon with boiling by heating. The switching time is approximately 10 msec in both types.

Heating one side of the tube causes the liquid to move from the high-temperature to the low-temperature because of unbalance of surface tension. If bubbles exist at the crosspoint of the waveguide, light is reflected completely to change the direction. This state is called cross state and the reflection loss in this state is 1.5 dB. Also, if the liquid exists at the crosspoint, light goes straight and enters the next waveguide. This state is called bar state and the loss in this state is 0.1 dB. In the case of 1×8 switch, for example, the cross loss is 7×0.1 dB (transmission loss) + 1.5 dB (reflection loss) = 2.2 dB. For connection with optical fibers on both ends, connection loss of 0.2 dB on both ends is also added. At the crosspoint, if light is reflected from high reflection index material to low index material like a total reflection side mirror on the side face of

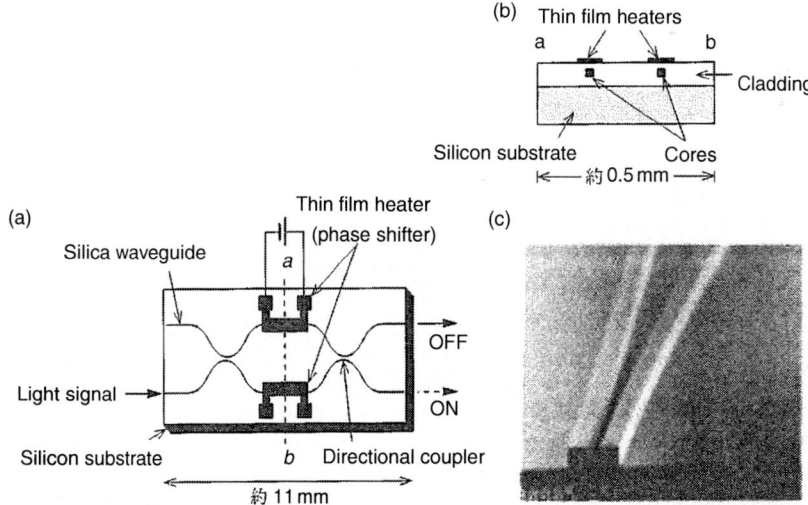

Figure 12 Quartz waveguide switch made by applying MZI

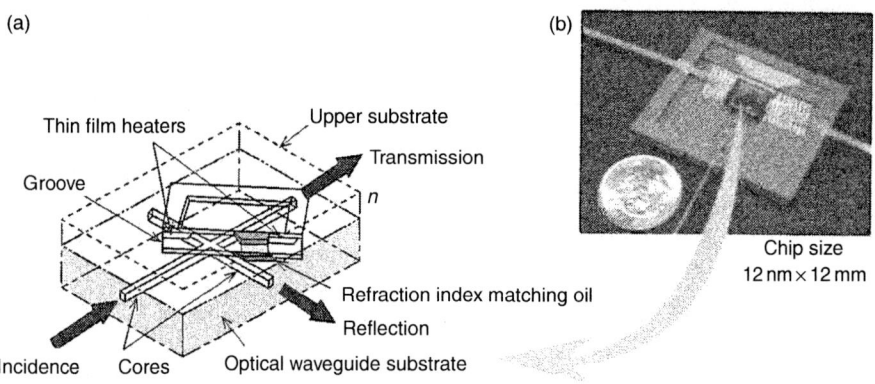

Figure 13   Bubble movement switch (Olive switch)

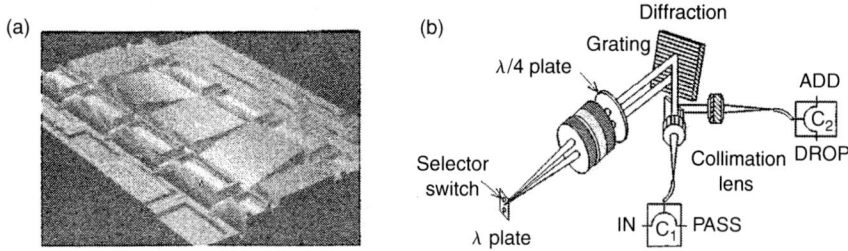

Figure 14   Example of application of single-axis rotation mirror to add-drop multiplexer

glass waveguide, the wavelength of light becomes relatively short, λ/n, and therefore the influence of the roughness of the side face is large. Although the processing technology for glass
etching is not established like Si etching that can produce small side face roughness, almost
vertical etching of 89.5° with groove width of 5 μm and depth of 45 μm can be performed.

Quartz glass is made by deposition with flame hydrolysis reaction, sintering and fusion or
EB evaporation. Basically, similar to the manufacturing method of optical fibers, it is not deposited around an axis but deposited on a plane. Although an organic polyimide can easily give
verticality and side face smoothness, it is not currently applied to communication because light
intensity ratio of TE and TM (polarization characteristic) changes during transmission, and
reliability is checked less sufficiently than quartz glass.

SINGLE-AXIS ROTATING MIRROR

Single-axis rotation micromirror arrays are applicable to WDM routers (branching), wavelength selective crossconnect (WSXC), and so on. Large rotation from low voltage is expected.
For example, in an add-drop multiplexer that has functions of adding and dropping multiplexing
information on the way, a switch for changing the functions is required (Figure 14), along with
high speed. Mirrors of 137×120 μm arranged at 150 μm intervals are formed as an array.
Moving the mirrors with vertical comb electrode actuators allows a large rotation angle of 6° to
be obtained even at a low voltage of 9 V. The linear fill-factor that indicates the ratio of the area
occupied by the mirrors is 91%, and the switching time is relatively short among the MEMS
switches, a few tens of nanosecond.

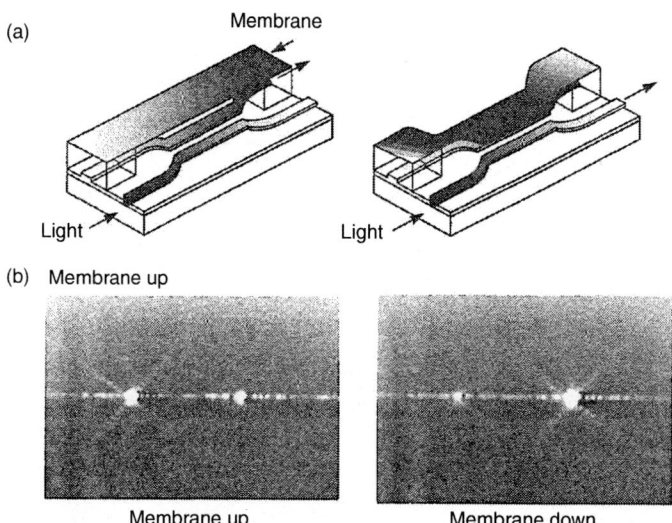

Figure 15   Evanescent coupling switch (From G. A. Magel, *Proc.* SPIE, 2686, 54–63 (1999). With permission.)

EVANESCENT COUPLING SWITCH

Inside an optical waveguide, total reflection is repeated for propagation. Evanescent waves are generated on the total reflection surface, and like directional couplers used in TO switches, if the core of another waveguide is brought close, a part of light transfers to that waveguide. The light intensity of transfer depends largely on the gap and the length of the waveguide that is brought close, as characteristics of evanescent wave indicated in the directional couplers. This system is used as a switch by moving the waveguide to vary the total coupling length and the gap [35–37].

Also, there are switches that turn on or off transmission of light at the extinction ratio of 65 dB/cm by bringing material with a higher refraction index than the core section, for example, bringing movable Si cladding close to the waveguide section of Si_3N_4 core and moving it away (Figure 15) [35].

REFERENCES

1.   J. A. Walker, Topical Reviews. "The future of MEMS in telecommunications network," *J. Micromech. Microeng.* 10, R1–R7 (2000).
2.   J. Ford et al., "Micromechanical fiber-optic attenuator with 3 ms response," *J. Lightwave. Technol.*, 16, 1663–1670 (1998).
3.   K. W. Goossen et al., "Silicon modulator based on mechanically active anti-reflection layer with 1 Mbits/sec capability for fiber-in the loop applications," *IEEE Photonics Technol. Lett.*, 6, 1119–1121 (1994).
4.   O. Solgaard et al., "Deformable grating optical modulator," *Opt. Lett.*, 17, 688–690 (1992).
5.   C. Marxer "Megahertz opto-mechanical modulator," *Sensors Actuators*, A52, 46–50 (1996).
6.   J. Ford et al., "Wavelength add/drop switching using tilting micromirrors," *IEEE J. Lightwave Technol.*, 17, 904–911 (1999).
7.   R. M. Boysel et al., "Integration of deformable mirror devices with optical fibers and waveguides," *Proc. SPIE*, 1793, 34–39 (1992).

8. C. Marxer et al., "Vertical mirrors fabricated by reactive ion etching for fiber optical switching applications," *Proceedings of the Tenth Annual International Workshop on MEMS'97* (Nagoya, Japan), pp. 49–54 (1997).

9. C. Marxer et al., "Micro-opto mechanical 2×2 switch for single-mode fibers based on plasma-etched silicon mirror and electrostatic actuation," *J. Lightwave Technol.*, 17, 2–6 (1999).

10. C. Marxer et al., "Vertical mirrors fabricated by deep reactive ion etching for fiber optic switching applications," *J. Microelectromech. Syst.*, 6, 277–285 (1997).

11. L. Dellmann et al., "4×4 Matrix switch based on MEMS switches and integrated waveguides," paper presented at Transducers'01 (Munich, Germany, June, 10–14), pp. 1332–1335 (2001).

12. M. Horino, "Development of prototype micromechanical optical switch," JSME Int. J., series C, 41, 978–982 (1998).

13. E. Ollier, et al., *Electron. Lett.*, 32, 2007 (1996).

14. P. Kopka et al., "Coupled U-shaped cantilever actuators for 1×4 and 2×2 optical fiber switches," *J. Micromech. Microeng.*, 10, 260–264 (2000).

15. K.E. Burcham et al., "Freestanding, micromachined, multimode silicon optical waveguides at $1 = 1.3$ m for micromechanical systems technology," *Appl. Opt.*, 37, 8397–8399 (1998).

16. S. Nagaoka, "Compact latching type single-mode fiber switches and their applications in subscriber loop network," *IEICE Trans. Electron.*, E80-C, 149–153 (1997).

17. S. Nagaoka, "Compact latching type single-mode fiber switched by a fiber-micromachining technique and their practical applications," *IEEE J. Select. Top. Quantum Electron.*, 5, 36–45 (1999).

18. D. J. Bishop et al., "The lucent lambda router: MEMS technology of the future here today," *IEEE Commun. Mag.*, March, 75–79 (2002).

19. P. De Dobbelaere et al., Digital MEMS for optical switching, *IEEE Commun. Mag.*, March, 88–95 (2002).

20. P. B. Chu et al., "MEMS: The path to large optical crossconnects," *IEEE Commun. Mag.*, March, 80–87 (2002).

21. L. Y. Lin et al., "Free-space micromachined optical switches with sub-millisecond switching time for large-scale optical crossconnect," *IEEE Photonics Technol. Lett.*, 10, 525–527 (1998).

22. L. Y. Lin et al., "Free-space micromachined optical switches for optical networking," *IEEE J. Select. Top. Quantum Electron.*, 5, 4–9 (1999).

23. L. Y. Lin et al., "On the expandability of free-space micromachined optical cross connects," *J. Lightwave Technol.*, 18, 482–489 (2000).

24. L. Y. Lin, "Integrated signal monitoring and connection verification in MEMS optical crossconnects," *IEEE Photonics Technol. Lett.*, 1, 885–887 (2000).

25. L. Y. Lin et al., "Angular-precision enhancement in free-space micromachined optical switches," *IEEE Photonics Technol. Lett.*, 11, 1253–1255 (1999).

26. H. Maekoba et al., "Self-aligned vertical mirror and V-grooves applied to an optical-switch: Modeling and optimization of bi-stable operation by electromagnetic actuation," *Sensors Actuators*, A87, 172–178 (2001).

27. R. Sawada et al., "Single crystalline mirror actuated electrostatically by terraced electrodes with high-aspect ratio torsion spring," in *IEEE/LEOS Proceedings of the International Conference on Optical MEMS* (Okinawa, Japan), pp. 23–24 (2001).

28. P. Helin et al., "Self-aligned micromachining process for large-scale, free-space optical cross-connect," *J. Lightwave Technol.*, 18, 16–22 (2000).

29. P. Helin et al., "Single crystal silicon, vertical mirrors arrays with improved integration density for optical crossconnects," in *IEEE/LEOS Proceedings of the International Conference on Optical MEMS* (Okinawa, Japan), pp. 83–84 (2001).

30. D. T. Neison, et al., "Fully provisioned 112×112 micro-mechanical optical crossconnect with 35.8 Tb/s demonstrated capacity," in *Proceedings of the OFC2000*, FD12-1 (2000).

31. R. Sawada, et al., "Single Si crystal 1024ch MEMS mirror based on terraced electrodes and a high-aspect ratio torsion spring for 3D cross-connect switch," in *IEEE/LEOS Proceedings of the Optical MEMS* (Lugano, Switzerland), pp. 11–12 (2002).

32. H. Toshiyoshi et al., "Surface micromachined 2D lens scanner array," in *IEEE/LEOS Proceedings of the Optical MEMS* (Kauwai, Hawaii), pp. 11–12 (2000).

33. N. Takato, "Silica-based single-mode waveguides on silicon and their application to guide-wave optical interferometers," *J. Lightwave Technol.*, 6, 1003–1010 (1988).
34. M. Makihara et al., "Micromechanical optical switches based on thermo-capilarity integrated in waveguide substrate," *J. Lightwave Technol.*, 17, 14–18 (1999).
35. G. A. Magel, "Integrated optic devices using micromachined metal membranes," *Proc. SPIE*, 2686, 54–63 (1999).

OPTICAL TAP

Yasuo Kokubun

Optical tap is used for monitoring the signal transmitted in the busline to control certain characteristics of the transmission line such as the polarization. Some amount of light power transmitted in the transmission line is dropped to the drop port and the rest of the input power is transmitted to the through port. Therefore, this device must have at least one input port and two output ports, that is, the through port and the drop port. In the *in situ* measurement, an optical tap is needed to detect the signal transmitted from the transmission line or from an optical device. To measure the reflected signal from the transmission line or a device, an optical circulator is used.

For stable and accurate monitoring and measurement, the tap coefficient, which is defined by the power ratio of input power to the drop power, must be polarization independent and wavelength independent. In addition, the tap coefficient must be stable against time and ambient temperature.

A half mirror, directional coupler, and fused fiber coupler are used as the optical tap device.

OPTOELECTRONIC INTEGRATED CIRCUIT

Osamu Wada

INTRODUCTION

Discrete optoelectronic devices based on III–V semiconductor materials have been a key to optical communication, data processing and memory, and sensoring systems. Their functions, however, cannot be utilized without being linked with other electronic and optical elements such as driving and data-processing integrated circuits (ICs), waveguide devices, and optical fiber circuits. Integration of these elements on a common substrate is a very important technique in order to realize compact, highly reliable devices exhibiting high performance, many different functions, and high manufacturability at low cost. Optoelectronic integrated circuit (OEIC) is defined as, in a broad sense, an integrated circuit that consists of different types of optoelectronic and electronic devices on a common substrate, and, in a narrower sense, a monolithic chip integrating optoelectronic and electronic devices on a common semiconductor substrate. Integrating optoelectronic devices with other different devices on a common substrate was proposed by S. E. Miller [1] in 1969 and discussed later by P. K. Tien [2] with a focus on its capability of sophisticated signal processing functions. Experimental research of optoelectronic integration was begun only in the late 1970s by A. Yariv and his group under his research project named "integrated optoelectronics circuits [3]." Figure 1 illustrates the structure of their early optical repeater chip composed of an AlGaAs/GaAs laser and GaAs metal–semicondutor field-effect transistors (MESFETs) monolithically integrated on a semi-insulating GaAs substrate [4]. Further extensive developments on integration, particularly for the application to optical communications, have been performed by many other research groups including Japanese MITI project teams for OEICs [5,6] and DARPA project teams in the United States [7]. The term of OEICs has been commonly used worldwide since then. OEIC for long-wavelength optical communications was first demonstrated by an InGaAs/InP PIN-photodiode/FET optical receiver chip as reported by R. Leheny et al. [8]. Figure 2 shows the circuit diagram and structure of InP-based PIN/FET OEIC receiver.

The advantages of OEICs include improvement of various properties of optoelectronic devices such as performance, functionality, compactness, reliability, manufacturability, and cost-effectiveness [9,10]. Reduction of parasitic reactance by integration leads to high-speed and low-noise performance. Complicated signal processing functions can be added by integration to the simple optoelectronic conversion function in discrete optoelectronic devices. Functions to be added are not limited to signal processing functions by electronic circuits, but a variety of photonic functions such as optical switching and wavelength filtering and conversion can be merged by using waveguide-based photonic circuits. The later category of integration is

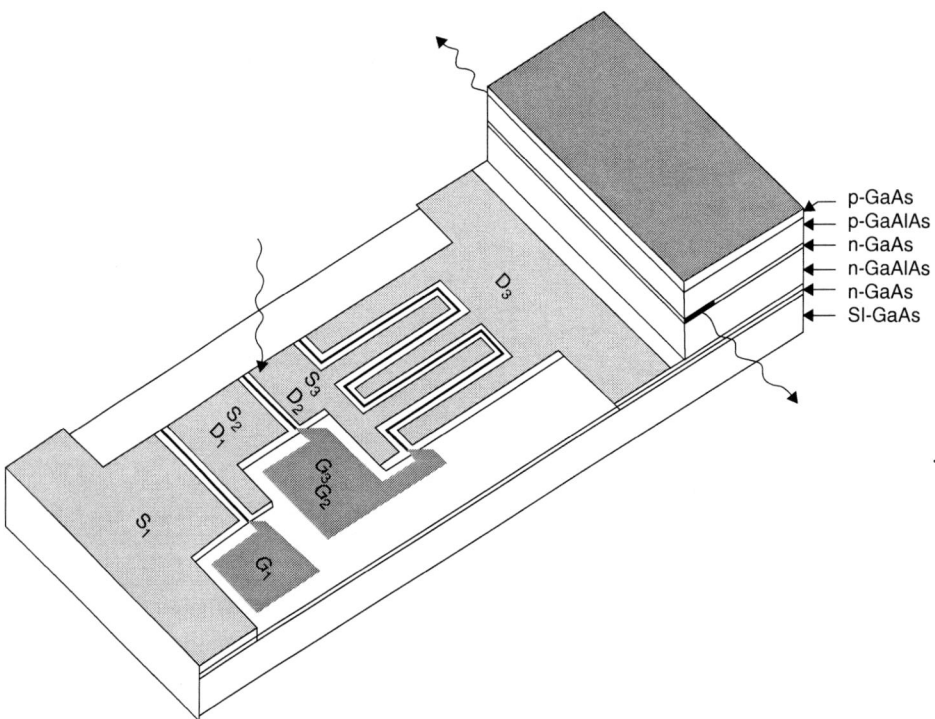

Figure 1 Structure of GaAs-based laser/MESFET OEIC repeater (After J. Shimada, *Optoelectronics* [Maruzen, Tokyo, 1989]. With permission.)

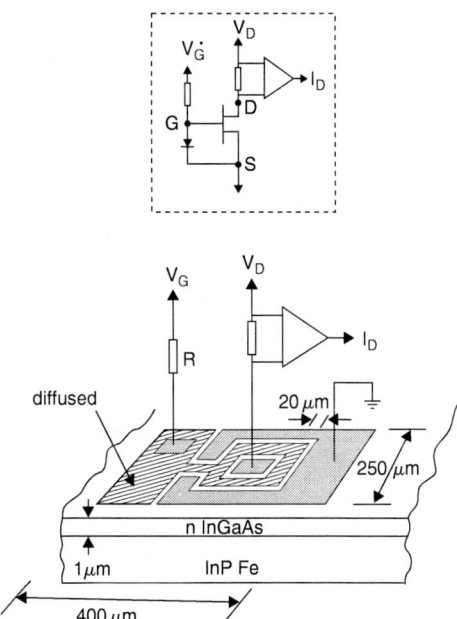

Figure 2 Structure of InP-based PIN/FET OEIC receiver (After R. F. Leheny, R. E. Nahory, M. A. Pollack, A. A. Ballman, E. D. Beebe, J. C. Dewinter, and R. J. Martin, *Electron. Lett.*, 16, 353–355 [1980]. With permission.)

often called photonic integrated circuits (PICs) [11]. The integration provides compact components, reduces system component count, simplifies assembly, stabilizes optical coupling, enhances component reliability and manufacturability, and thus can eventually lower component cost. Such properties are important in a variety of application fields such as optical communications based on both the high-speed time-division and wavelength-division multiplexing schemes, optical interconnections, and optical data storage, processing and computing.

Materials used for fabricating OEICs include III–V compound semiconductors such as GaAs- and InP-based systems. Although Si itself is not an efficient light emitter, Si and SiGe materials grown on Si substrates can be used in optical receivers at short wavelengths and optical modulators even at long-wavelengths [12]. Heterogeneous materials such as InP-based material on Si substrate can be prepared by heteroepitaxial growth [13], direct wafer bonding [14], and epitaxial lift-off techniques [15]. This mixed material system enables us to optimize independently the operating wavelength and circuit performance, which can be often very difficult to achieve in a homogeneous material system. Adding to such monolithic scheme, hybrid integration scheme using, for example, flip-chip bonding technique can be practically more useful for urgent system applications [16].

In the following sections, some representative examples of OEICs are introduced with a focus on the application to optical communication and interconnection systems, in order to illustrate the capability already demonstrated and technical issues to be challenged in the future.

SHORT-WAVELENGTH OEICs

AlGaAs/GaAs material system has been used widely for the application to optical communications and interconnections at the wavelength near 0.8 μm. The most difficult issue in fabricating

OEICs on single semiconductor substrates has been the fabrication process to merge different devices such as lasers, photodiodes, and transistors, which have vastly different layer thicknesses and structures. A variety of process techniques have been developed for overcoming this problem.

Figure 3 shows (a) a structure and circuit diagram and (b) a chip micrograph of 4-channel OEIC transmitter consisting of AlGaAs/GaAs quantum well (QW) lasers with microcleaved facets, monitor photodiodes, and GaAs MESFET laser driver circuits integrated on a semi-insulating (SI)-GaAs substrate [17]. The introduction of low-threshold QW lasers with cavity facets formed without wafer cleavage has enabled monolithic, multi-channel array of transmitters exhibiting uniform characteristics. Figure 4 shows the structure and circuit diagram and a surface micrograph of four-channel OEIC receiver incorporating GaAs metal–semiconductor–metal photodiodes (MSM-PDs) and GaAs MESFET preamplifier circuits, all monolithically integrated on a SI-GaAs substrate. Completely planar structure has been achieved by the introduction of MSM-PDs with structures compatible with MESFETs, giving rise to high uniformity of circuit characteristics and stable fabrication yield [18]. These OEIC transmitter and receiver have been combined with a GaAs-IC cross-point switch to form a compact 4×4 optical matrix switch module as is shown in Figure 5. The whole optical switch module has been demonstrated to operate CW at the bit rate of 400 Mb/sec [19].

On the basis of matured GaAs MESFET-IC technology, fairly large-scale circuits can be integrated with optoelectronic devices. GaAs-based OEICs so far fabricated include a laser transmitter involving a MESFET signal multiplexer as well as a laser driver circuit [7], and four-channel MSM-PD-based optical receiver including amplification, decision, and demultiplexing circuits as well as a clock recovery circuit composed of 8000 devices [20]. The latter has been

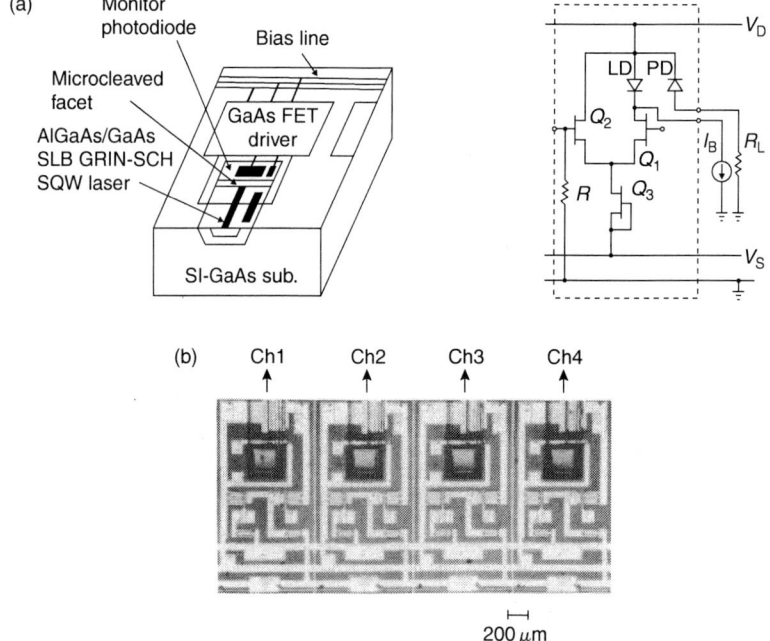

Figure 3   (a) Structure and circuit diagram and (b) chip micrograph of four-channel OEIC transmitter

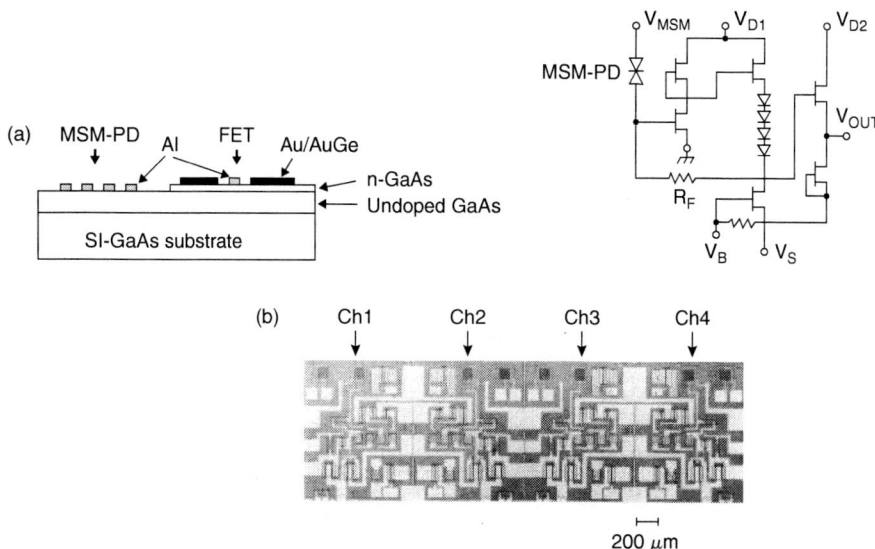

Figure 4 (a) Structure and circuit diagram and (b) chip micrograph of four-channel OEIC receiver

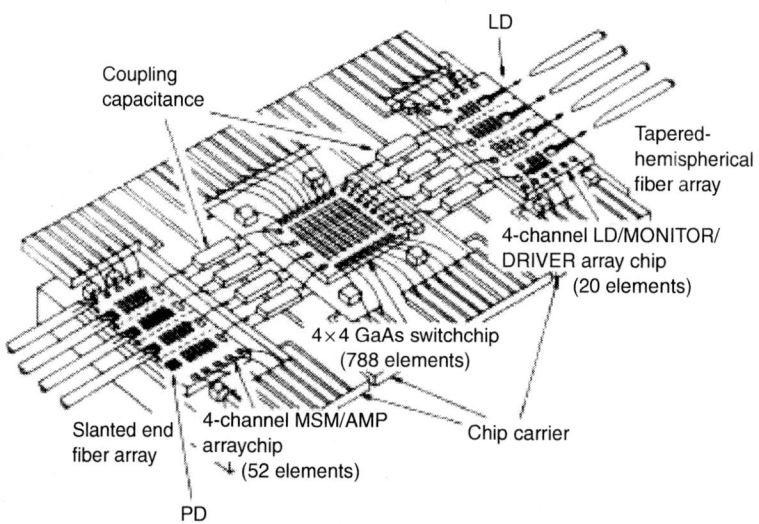

Figure 5 Structure of 4 × 4 OEIC switch module

applied to 32-channel board-to-board optical parallel link modules operating at the signal bit rate of 0.5 Gb/sec/channel [21]. Recent optical link modules often use vertical cavity surface emitting laser (VCSEL) arrays for easy and uniform optical coupling [21].

Si-based receiver OEICs composed of photodiodes and bipolar junction transistor circuit have been developed and commercialized for optical data links [22] and read/write head-modules for optical disk storage systems with the use of GaAs-based light sources [23]. Si-based photodiodes are limited in speed mostly below 100 Mb/sec primarily due to persistent diffusion of photo-generated carriers and also low carrier saturation velocity and low

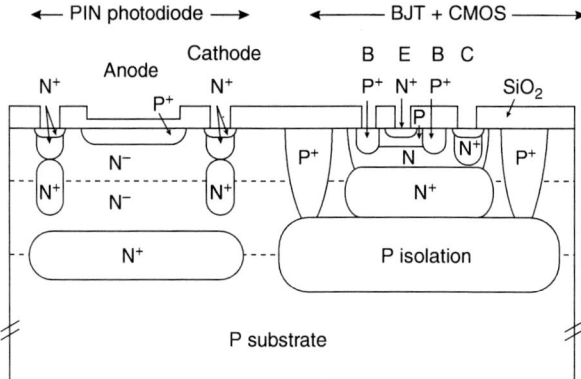

Figure 6 Structure of Si PIN/BiCMOS receiver OEIC (After R. Swoboda and H. Zimmermann, *IEEE J. Select. Top. Quantum Electron.*, 9, 419–424 [2003]. With permission.)

photoabsorbance in Si. Recent effort devoted to device structure design and fabrication techniques has lead to the development of high-speed Si OEIC receivers. Figure 6 illustrates the cross-section of such a Si OEIC receiver consisting of a PIN-PD and 0.6-μm BiCMOS circuit exhibiting excellent low-noise performance at the bit rate of 1.5 Gb/sec [24].

LONG-WAVELENGTH OEICs

InP-based material systems including InGaAsP and InAlGaAs compounds lattice-matched to InP substrate are used for long-wavelength lasers and photodetectors. Since the Schottky barrier height to *n*-type material in this material system is too low to suppress the leakage current, electronic devices required for this material system are heterojunction devices such as InAlAs/ InGaAs high electron mobility transistors (HEMTs) and heterojunction bipolar transistors (HBTs). A transmitter OEIC that integrates a distributed feedback (DFB) single-mode laser [25] and an HEMT laser driver circuit, and also a receiver OEIC that consists of a PIN-PD and a preamplifier circuit [26] have been demonstrated so far. One of the most recent advances in long-wavelength receivers is achievement of very high speed operation in cooperation with monolithic microwave circuit (MMIC) technology. Figure 7 shows (a) a circuit diagram based on coplanar waveguide MMIC design and (b) a chip cross-section of a PIN-PD/HEMT receiver OEIC which has shown operation at the bit rate of 50 Gb/sec [27]. A waveguide photodiode (WGPD) with a fairly thin InGaAs photoabsorption layer has been adopted, so that the carrier transit time is reduced with the quantum efficiency kept high due to long enough waveguide.

Heterogeneous or hybrid integration techniques are particularly important in this wavelength region, because material systems most suitable for optical devices (e.g., InGaAs) and electronic circuits (e.g., Si or GaAs) are independently selected and optimized. Heteroepitaxial growth and epitaxial lift-off techniques will become extremely powerful in OEIC fabrication, once reliable processes are developed [14]. They are under development toward the realization of stable, defect-free heterointerfaces. Figure 8(a) illustrates the cross-sectional structure of receiver OEIC consisting of an InGaAs/InP PIN-PD with monolithic lens and a Si bipolar transistor preamplifier circuit using flip-chip bonding technique [16]. Figure 8(b) shows an SEM image of lensed PIN-PD flip-chip bonded on a GaAs-preamplifier circuit chip. Such a monolithic lens formed on the rear side of PIN-PD by ion-beam etching technique enables the signal light beam to focus on an extremely small junction with the diameter of 5 to 10 μm. Flip-chip

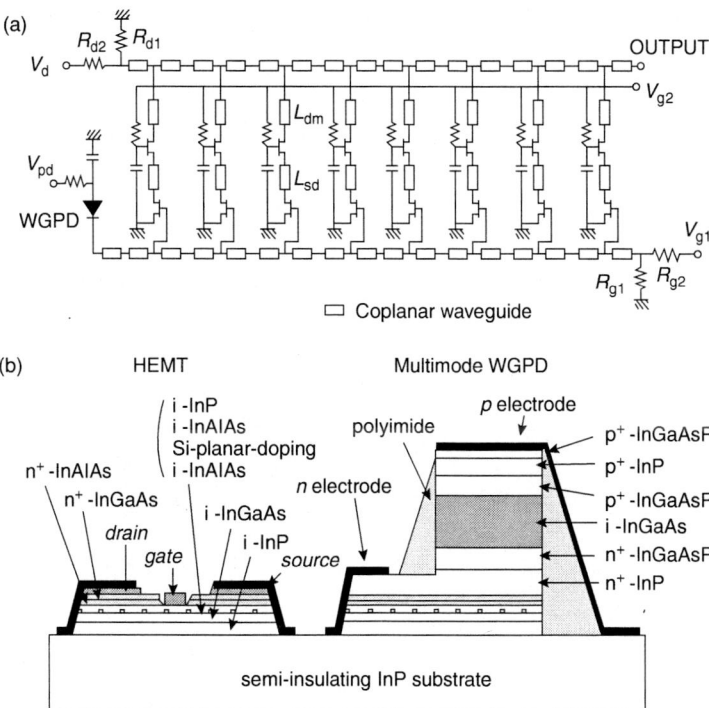

Figure 7 (a) Circuit diagram and (b) cross-sectional structure of very high-speed PIN-PD/HEMT receiver OEIC (After K. Takahata, Y. Muramoto, H. Fukano, K. Kato, A. Kozen, S. Kimura, Y. Imai, Y. Miyamoto, O. Nakajima, and Y. Matsuoka, *IEEE J. Select. Top. Quantum Electron.*, 9, 31–37 [2003]. With permission.)

integration provides shortest interconnection between the PIN-PD and front-end HEMT, so that the input capacitance is minimized for enabling low-noise operation. OEIC receiver using an InGaAs/InP avalanche photodiode (APD) and Si preamplifier circuit has exhibited very large gain-bandwidth product (80 GHz) [28] and excellent sensitivity characteristics at 10 Gb/sec.

CHALLENGES IN OPTOELECTRONIC INTEGRATION

In traditional OEICs as described above, circuit functions are limited in the range of electronic signal processing. Incorporation of photonic functions sensitive to, not just limited to the light intensity, but the wavelength, phase, and polarization can provide a lot of different signal processing capabilities very useful in high-speed time-division multiplexing (TDM), wavelength-division multiplexing (WDM), and also ultrafast all-optical time-division multiplexing (OTDM) techniques. Figure 9 shows structure of high-speed light source in which a DFB laser is integrated with an electro-absorption waveguide modulator on an InP substrate. Such PIC light source can minimize the wavelength chirping under the modulation rate as high as 10 Gb/sec [29] and even 40 Gb/sec in the most recent report [30], which would have not been achieved by direct modulation of lasers.

Figure 10 illustrates the structure of an eight-channel WDM transmitter incorporating eight DFB lasers with incrementally different emission wavelengths, a multimode interferometer (MMI) optical combiner, and an semiconductor optical amplifier (SOA) [31]. Many different

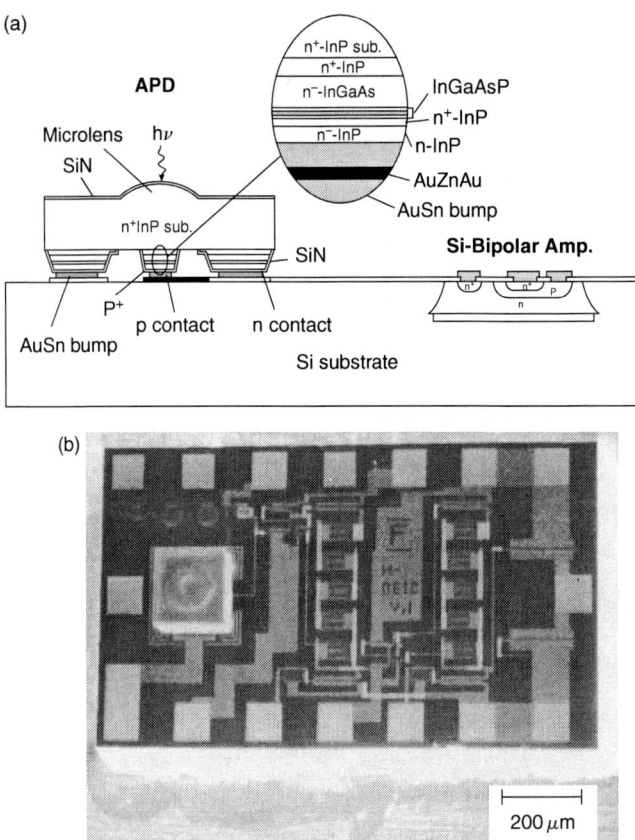

Figure 8 (a) Cross-sectional structure of PIN-PD/Si preamplifier receiver and (b) SEM image of optical receiver involving a PIN-PD flip-chip integrated on a GaAs preamplifier circuit

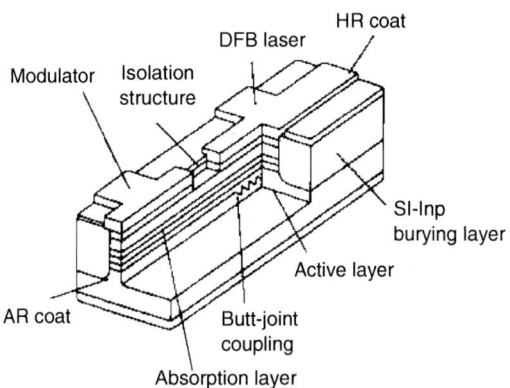

Figure 9 Structure of high-speed light source integrating a DFB laser and EA-modulator on an InP substrate (After H. Soda, M. Furutsa, K. Sato, N. Okazaki, Y. Yamazaki, H. Nishimoto, and H. Ishikawa, *Electron. Lett.*, 26, 9–10 [1990]. With permission.)

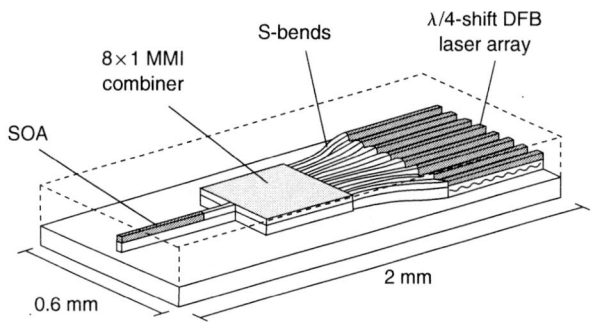

Figure 10 Structure of eight-channel wavelength light source (After M. Bouda, M. Matsuda, K. Morito, S. Hara, T. Watanabe, and Y. Kotaki, Technical Digest Conference on Optical Fiber Communication [OFC'00], [March 7, 2000], pp. 178–183. With permission.)

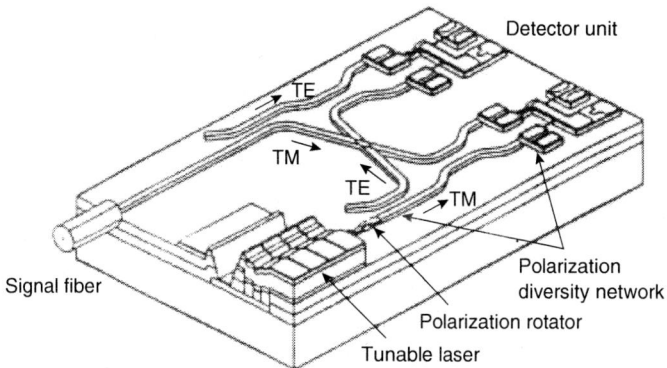

Figure 11 Structure of monolithic optical heterodyne receiver (After P. Kaiser, D. Trommer, H. Heidrich, F. Fidorra, S. Malchow, D. Franke, W. Passenberg, W. Rehbein, H. Schoeter-Janssen, R. Stanzel, and G. Unterboersch, Technical Digest 5th Optoelectronics Conference [OEC'94], Tokyo, PD-II-1 [1994])

PICs have been demonstrated so far, which include monolithic wavelength tunable lasers, multi-electrode monolithic bistable laser switches, monolithic colliding pulse mode-locked lasers for ultrashort pulses in femtoseconds region, Mach–Zehnder interferometer type all-optical switches incorporating SOAs, WDM demultiplexing receivers consisting of an arrayed waveguide grating (AWG) and a photodiode array, and also optical heterodyne receivers [32]. To illustrate technical feasibility of photonic integration, an example of heterodyne receiver OEIC, which integrates a local oscillator DFB laser, polarization sensitive waveguide couplers, and photodiode receiver circuits, all on an InP substrate is illustrated in Figure 11 [33]. OEIC and PIC techniques will become more important in future advanced WDM and OTDM systems, where compact, stable, low-cost components are required.

One of the most promising application of OEICs is believed to be in optical interconnection at various levels in the system, covering the frame, board, module, and chip levels. Figure 12 illustrates a variety of possible optical interconnection component structures at different system levels, where optoelectronic integration techniques can be utilized extensively to achieve most effective interconnection scheme at each system level. Optical interconnection modules for the board level applications have been commercialized and the most important issue there is the

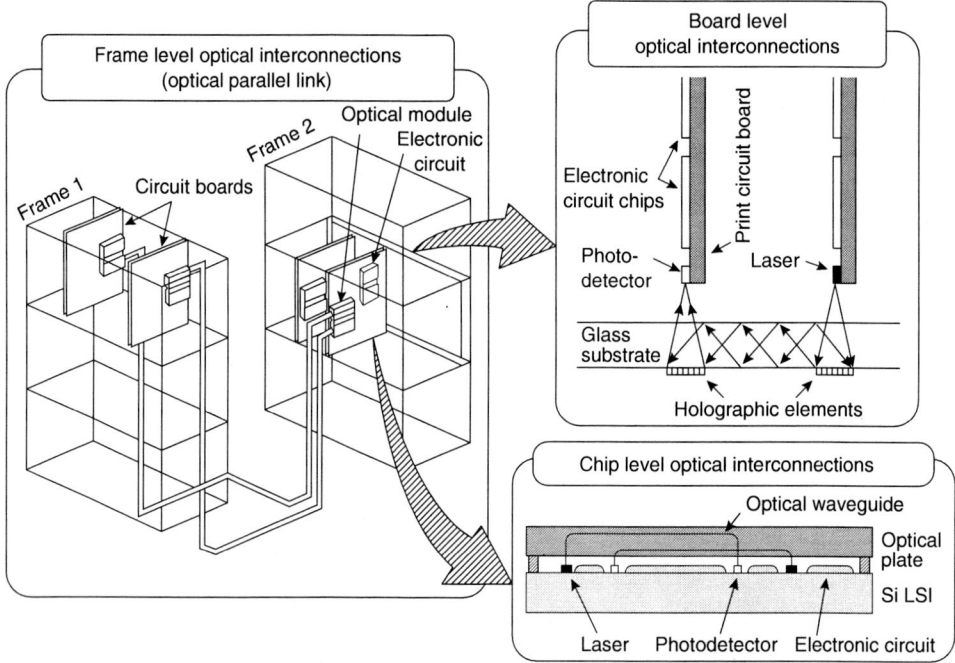

Figure 12   Various structures of optical interconnections at different system levels: frame, board, and chip levels

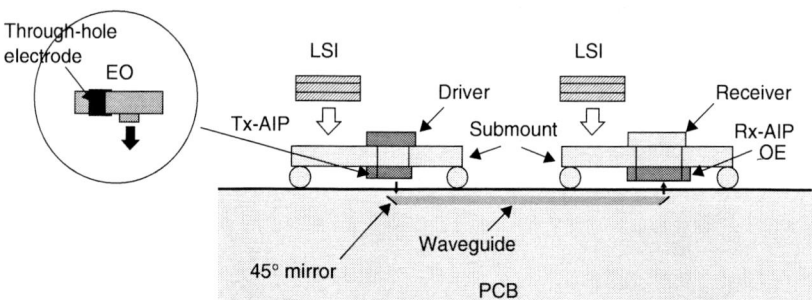

Figure 13   Packaging structure of active interposer incorporating waveguide network board, optical transmitter and receiver circuits, and LSIs (After T. Mikawa, M. Kinoshita, K. Hiruma, T. Ishitsuka, M. Okabe, S. Hiramatsu, H. Furuyama, T. Matsui, K. Kumai, O. Ibaragi, and M. Bonkohara, *IEEE J. Select. Top. Quantum Electron.*, 9, 452–459 [2003]. With permission.)

cost reduction. Current development is focused on the module to chip levels [34]. Multi-chip module, which incorporates an optical waveguide board for providing global interconnecion network and optical transmitters and receivers composed of optoelectronic devices together with integrated LSIs, would be one of the most plausible schemes of optical interconnections in real systems. For example, active interposer (AIP), which enables such an interconnection network system, has been proposed recently. The integration structure of active interposer is shown in Figure 13, and the target overall throughput is 800 Gb/sec for 64 channels [35].

SUMMARY

OEIC technique has been developed and applied in high-speed TDM and WDM optical communication systems as well as in optical storage systems due to its advantages of enhancing performance, functions, reliability, and cost-effectiveness. Application of OEICs to optical interconnections and signal processing systems will become even more important in future optical systems, in which the speed problem of conventional electrical interconnections must be resolved at every system level. In order to develop practical OEICs, the development must target to fulfill high performance and, simultaneously, low cost through the innovation of fabrication process technology as well as breakthrough in device principles and system design.

REFERENCES

1. S. E. Miller, "Integrated optics," *Bell Syst. Tech. J.*, 48, 2059–2069 (1969).
2. P. K. Tien, "Integrated optics and new wave phenomena in optical waveguides," *Rev. Mod. Phys.*, 49, 361–420 (1977).
3. A. Yariv, "The beginning of integrated optoelectronic circuits," *IEEE Trans. Electron. Devices*, ED-31, 1656–1661 (1984).
4. M. Yust, N. Bar-Chaim, S. Izadpanah, S. Margalit, I. Ury, D. Wilt, and A. Yariv, "A monolithically integrated optical repeater," *Appl. Phys. Lett.*, 35, 795–797 (1979).
5. J. Shimada, *Optoelectronics*, Maruzen, Tokyo (1989) [in Japanese].
6. I. Hayashi, "'OEIC': Its concepts and prospects," Technical Digest 4th IOOC (Tokyo, 1983), p. 170.
7. J. K. Carney, M. J. Helix, and R. M. Kolbas, "Gigabit optoelectronic transmitters," Technical Digest GaAs IC Symposium Phoenix, AZ (1983), pp. 48–51.
8. R. F. Leheny, R. E. Nahory, M. A. Pollack, A. A. Ballman, E. D. Beebe, J. C. Dewinter, and R. J. Martin, "Integrated $In_{0.53}Ga_{0.47}As$ p–i–n F. E. T. photoreceiver," *Electron. Lett.*, 16, 353–355 (1980).
9. M. Dagenenais, R. Leheny, and J. Crow, Eds., *Integrated Optoelectronics*, Academic Press, New York (1994).
10. O. Wada, Ed., *Optoelectronic Integration*, Kluwer Academic Publishers, Boston, MA (1994).
11. T. L. Koch and U. Koren, "Semiconductor photonic integrated circuits," *J. Quantum Electron.*, QE-27, 641–653 (1991).
12. A. Irace, G. Coppola, G. Breglio, and A. Cutolo, *IEEE J. Select. Top. Quantum Electron.*, 6, 14–18 (2000).
13. T. H. Windhorn and G. M. Metze, "Room temperature operation of GaAs/AlGaAs diode lasers fabricated on a monolithic GaAs/Si substrate," *Appl. Phys. Lett.*, 47, 1031–1033 (1985).
14. H. Wada and T. Kamijoh, "1.3 mm InP–InGaAsP lasers fabricated on Si substrates by wafer bonding," *IEEE J. Select. Top. Quantum Electron.*, 3, 937–951 (1997).
15. E. Yablonovitch, T. Gumitter, J. P. Harbison, and R. Bhat, "Extreme selectivity in the lift-off of epitaxial GaAs films," *Appl. Phys. Lett.*, 51, 2222–2224 (1987); and also N. M. Jokerst, M. A. Brooke, S.-Y. Cho, S. Wilkinson, M. Vrazel, S. Fike, J. Tabler, Y. J. Joo, S.-W, Seo, D. S. Wills, and A. Brown, "The heterogeneous integration of optical interconnections into integrated microsystems," *IEEE J. Select. Top. Quantum Electron.*, 9, 350–360 (2003).
16. O. Wada, M. Makiuchi, H. Hamaguchi, T. Kumai, and T. Mikawa, "High-performance, high-reliability InP/GaInAs p–i–n photodiodes and flip-chip integrated receivers for lightwave communications," *J. Lightwave Technol.*, 9, 1200–1207 (1991).
17. O. Wada, H. Nobuhara, T. Sanada, M. Kuno, M. Makiuchi, T. Fujii, and T. Sakurai, "Optoelectronic integrated four-channel transmitter array incorporating AlGaAs/GaAs quantum-well lasers," *J. Lightwave Technol.*, 7, 186–197 (1989).
18. O. Wada, H. Hamaguchi, M. Makiuchi, T. Kumai, M. Ito, K. Nakai, T. Horimatsu, and T. Sakurai, "Monolithic four-channel photodiode/amplifier receiver array integrated on a GaAs substrate," *J. Lightwave Technol.*, LT-4, 1694–1703 (1986).
19. T. Iwama, T. Horimatsu, Y. Oikawa, K. Yamaguchi, M. Sasaki, T. Touge, M. Makiuchi, H. Hamaguchi, and O. Wada, "4 × 4 OEIC switch module using GaAs substrate," *J. Lightwave Technol.*, 6, 772–778

(1988); and also O. Wada and J. Crow, "Current status of optoelectronic integrated circuits," in M. Dagenenais, R. Leheny, and J. Crow, Eds., *Integrated Optoelectronics*, Academic Press, New York, Chap. 12, pp. 447–488 (1994).

20. D. J. Crow, "Optoelectronic integrated circuits for high-speed computer networks," Technical Digest Conference Optical Fiber Communication (OFC'89) p. 83; and also D. J. Crow, Technical Digest IOOC (IOOC'89) Kobe, Vol. 4, p. 86.

21. J. Crow, "Packaging integrated optoelectronics," in M. Dagenais, R. Leheny, and J. Crow, Eds., *Integrated Optoelectronics,* Academic Press, New York, Chap. 16, pp. 627–644 (1994).

22. H. Nagao and M. Yamamoto, IEICE Tech rep. 89, 7 (1989) [in Japanese].

23. O. Matsuda, N. Nishi, and T. Mizuno, "Optical pick-up devices for disk storage systems," in H. Cho, Ed., *Opto-Mechatronic Systems Handbook – Techniques and Applications*, CRC Press, New York, pp. 21/1–21/33 (2002).

24. R. Swoboda and H. Zimmermann, "A low-noise monolithically integrated 1.5 Gb/s optical receiver in 0.6 μm BiCMOS technology," *IEEE J. Select. Top. Quantum Electron.*, 9, 419–424 (2003).

25. H. Y. Lo, P. Grabbe, M. Z. Iqbal, R. Bhat, J. L. Gimlett, J. C. Young, P. S. D. Lin, A. S. Gozdz, M. A. Koza, and T. P. Lee, "Multigigabit/s 1.5 μm λ/4-shifted DFB OEIC transmitter and its use in transmission experiments," *IEEE Photon. Technol. Lett.*, PTL-2, 673–674 (1990).

26. O. Wada, H. Nobuhara, M. Makiuchi, H. Hamaguchi, S. Sasa, and T. Fujii, "AlInAs/GaInAs HEMT application for high performance OEIC receivers," *J. Crystal Growth*, 95, 378–381 (1989).

27. K. Takahata, Y. Muramoto, H. Fukano, K. Kato, A. Kozen, S. Kimura, Y. Imai, Y. Miyamoto, O. Nakajima, and Y. Matsuoka, "Ultrafast monolithic receiver OEIC composed of multimode waveguide p–i–n photodiode and HEMT distributed amplifier," *IEEE J. Select. Top. Quantum Electron.*, 9, 31–37 (2003).

28. Y. Kito, H. Kuwatsuka, T. Kumai, M. Makiuchi, T. Uchida, O. Wada, and T. Mikawa, "High-speed flip-chip InP/InGaAs avalanche photodiodes with ultralow capacitance and large gain-bandwidth products," *IEEE Photon. Technol. Lett.*, 3, 1115–1116 (1991).

29. H. Soda, M. Furutsu, K. Sato, N. Okazaki, Y. Yamazaki, H. Nishimoto, and H. Ishikawa, "High-power and high-speed semi-insulating BH structure monolithic electroabsorption modulator/DFB laser light source," *Electron. Lett.*, 26, 9–10 (1990).

30. H. Kawanishi, Y. Yamauchi, N. Mineo, Y. Shibuya, H. Murai, K. Yajima, and H. Wada, "EAM-integrated DFB laser modules with more than 40-GHz bandwidth," *IEEE Photon. Technol. Lett.*, 13, 954–956 (2001).

31. M. Bouda, M. Matsuda, K. Morito, S. Hara, T. Watanabe, and Y. Kotaki, "Compact high-power wavelength selectable lasers for WDM applications," Technical Digest Conference on Optical Fiber Communication (OFC'00), pp. 178–183 (March 7, 2000).

32. P. Kaiser, D. Trommer, H. Heidrich, F. Fidorra, S. Malchow, D. Franke, W. Passenberg, W. Rehbein, H. Schoeter-Janssen, R. Stanzel, and G. Unterboersch, "Polarization diversity heterodyne receiver OEIC on InP:Fe substrate," Technical Digest 5th Optoelectronics Conference (OEC'94), Tokyo, PD-II-1 (1994).

33. P. Kaiser and H. Hiedrich, "Optoelectronic/photonic integrated circuits on InP between technological feasibility and commercial success," *IEICE Trans. Electron.*, 85-C, 970–981 (2002).

34. Special issue on Optical Interconnect: *IEEE J. Select. Top. Quantum Electron.*, 9, 347–676 (2003).

35. T. Mikawa, M. Kinoshita, K. Hiruma, T. Ishitsuka, M. Okabe, S. Hiramatsu, H. Furuyama, T. Matsui, K. Kumai, O. Ibaragi, and M. Bonkohara, "Implementation of active interposer for high-speed and low-cost chip level interconnects," *IEEE J. Select. Top. Quantum Electron.*, 9, 452–459 (2003).

P

PERIODIC STRUCTURES

Toshiaki Suhara

Periodic structures or gratings in a waveguide are one of the most important elements for integrated optics [1–4]. They can perform various passive functions on guided waves, such as deflection and reflection [5], input and output coupling [6], mode conversion [7], wavelength filtering [8], wavelength dispersion, wavefront conversion [2], phase matching in directional coupling between waveguides [9], and phase matching for nonlinear-optic interactions [4]. Periodic modulation in refractive index produced through electro-optic and acousto-optic effects, that is, dynamic gratings, can provide an effective means for guided-wave control. Examples of passive waveguide grating elements are illustrated in Figure 1. The grating structures are fabricated by patterning based on photolithography, electron-beam lithography, or holographic interference recording, associated with dry etching (reactive ion etching and ion-beam etching, etc.) to form the grating grooves. Fabrication techniques include holographic interference recording or holographic contact printing using a phase mask in waveguides of a photopolymer or an ultraviolet (UV)-light-sensitive material such as Ge-doped silica, and direct electron-beam writing, to produce periodic modulation in the refractive index. Periodic structures for integrated optics include fiber gratings, which are fabricated by UV-light-induced refractive index change of an optical fiber.

GRATING DESCRIPTION AND PHASE MATCHING

Consider an optical waveguide in the yz plane. It is described by a cross-sectional distribution of the relative dielectric permittivity $\varepsilon(x, y)$. A mathematical expression of a waveguide grating is given by the modification in the relative permittivity $\Delta\varepsilon(x, y, z)$ superimposed on $\varepsilon(x, y)$. Since $\Delta\varepsilon$ is periodic with respect to y and z, it can be written in a Fourier series as

$$\Delta\varepsilon(x, y, z) = \sum_{q} \Delta\varepsilon_q(x)\exp(-jq\mathbf{Kr}), \qquad \mathbf{r} = y\mathbf{e}_y + z\mathbf{e}_z, \tag{1}$$

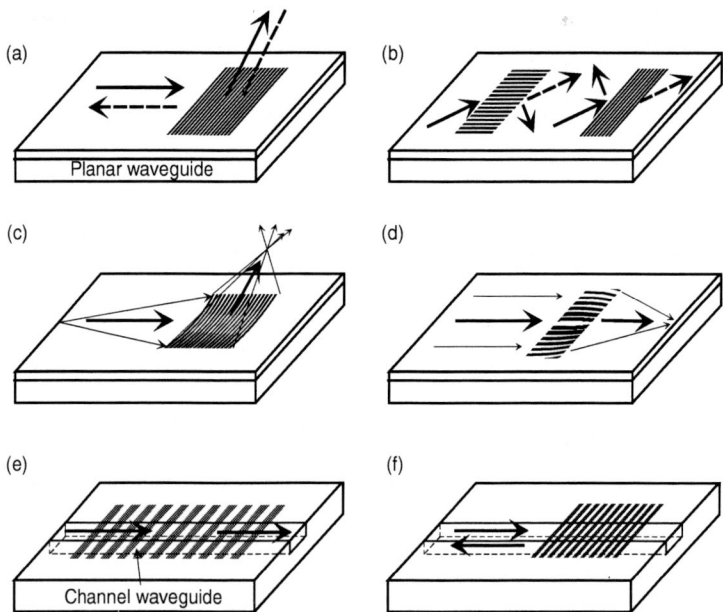

Figure 1   Passive waveguide grating elements for integrated optics (a) input/output coupler, (b) deflector/polarization splitter, (c) focusing grating coupler, (d) grating lens, (e) mode converter, and (f) reflector/wavelength filter

where $K = K_y e_y + K_z e_z$, is the grating vector normal to the grating lines, and $\Delta \varepsilon_q(x)$ is the qth Fourier amplitude $(\Delta \varepsilon_q = \Delta \varepsilon_q^*$ since $\Delta \varepsilon$ is real). The fundamental period Λ is correlated with K by $|K| = K = 2\pi/\Lambda$.

Optical functions of periodic structures can be interpreted in terms of coupling between optical modes. When an optical wave of a mode characterized by a wave vector β_a (a wave of field amplitude having a space dependence in a form of $\exp(-j\beta_a r)$, β_a is assumed to have the z component) is incident in the grating region, wave components (space harmonics) of wave vectors $\beta_a + qK$ are induced. The harmonic can propagate as a mode, provided that the wave vector coincides with a wave vector β_b of a mode in the structure. This means that qth order coupling between modes a and b takes place when

$$\beta_b = \beta_a + qK, \quad q = \pm 1, \pm 2, \dots. \tag{2}$$

Actually, Eq. (2), called Bragg condition, gives the coupling condition for cases where the waves and the grating are extended in infinite space. In integrated optics, however, $\Delta \varepsilon$ is nonzero only near the waveguide plane (yz plane). Therefore, in a planar (channel) waveguide, coupling takes place even if the x component (x and y components) of Eq. (2) is not satisfied. Each component of Eq. (2) is called phase matching condition. The Bragg condition and the phase matching condition can be depicted as a wave vector diagram using β_a, β_b, and K.

COUPLED-MODE EQUATIONS

The characteristics of waveguide gratings can be analyzed based on the coupled-mode theory [1,3,10]. For simplicity, consider two guided modes propagating along the z axis in the waveguide without grating, and let $E_a(x, y)$, $E_b(x, y)$ be the normalized mode profiles, and β_a, β_b be the propagation constants. The grating is assumed to have a grating vector K parallel to the z axis ($K = K e_z$).

The optical field in the waveguide grating structure is expressed approximately by a super-position of these modes, and may be written as:

$$E(x, y, z) = A(z)E_a(x, y)\exp(-j\beta_a z)B(z) + E_b(x, y)\exp(-j\beta_b z). \tag{3}$$

The mode fields $E_a(x, y)\exp(-j\beta_a z)$ and $E_a(x, y)\exp(-j\beta_a z)$ satisfy the Maxwell equations with waveguide permittivity $\varepsilon(x, y)$, and the total field $E(x, y, z)$ satisfies those with $\varepsilon(x, y) + \Delta\varepsilon(x, y, z)$ for the waveguide with grating. Starting from these Maxwell equations, and using the orthonormal relation of the modes, one can deduce the equations that describe the spatial evolution of the mode amplitudes $A(z)$ and $B(z)$:

$$\pm\frac{d}{dz}A(z) = -j\kappa^* B(z)\exp(-j\,2\Delta z), \tag{4a}$$

$$\pm\frac{d}{dz}B(z) = -j\kappa A(z)\exp(+j\,2\Delta z), \tag{4b}$$

$$2\Delta = \beta_b - (\beta_a + qK). \tag{5}$$

As for the $\pm$ in Eq. (4a) (Eq. [4b]), $+$ and $-$ should be taken for $\beta_a > 0$ and $\beta_a < 0$ ($\beta_b > 0$ and $\beta_b < 0$), respectively. Equation (4) are coupled-mode equations, and 2Δ given by Eq. (5) describes the deviation from the phase matching condition.

The parameter κ, called coupling coefficient, is given by

$$\kappa = \frac{\omega\varepsilon_0}{4}\iint E_a^*(x, y)\Delta\varepsilon_q(x)E_b(x, y)\,dx\,dy, \tag{6}$$

where ω is the angular frequency of the optical wave.

COUPLING BETWEEN GUIDED MODES

Collinear Coupling

Coupling between guided modes propagating along the same axis (collinear coupling) is classified into codirectional and contradirectional couplings. They are illustrated in Figure 2(a) and 2(b) with the wave vector diagrams.

Codirectional coupling: Consider coupling between two different modes propagating in the same directions ($\beta_a > 0$, $\beta_b > 0$, $\beta_a \neq \beta_b$). The solution of Eq. (4) for boundary conditions $A(0) = 1$, $B(0) = 0$ indicates periodic transfer of the guided mode power from mode a to b, which implies that the grating functions as a mode converter. The efficiency of the power transfer in a grating of a length L is given by

$$\eta = \left|\frac{B(L)}{A(0)}\right|^2 = \frac{\sin^2\sqrt{|\kappa|^2 + \Delta^2}L}{1 + \Delta^2/|\kappa|^2}. \tag{7}$$

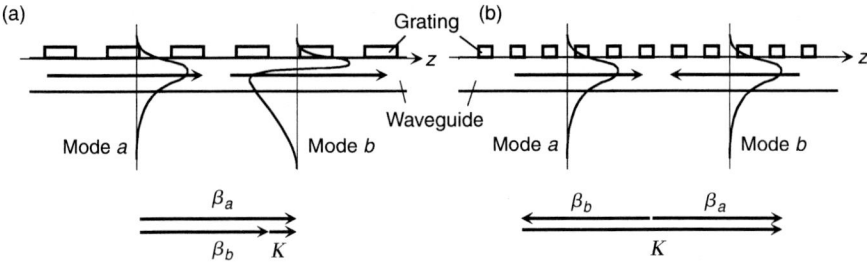

Figure 2   Collinear coupling of guided modes (a) codirectional coupling and (b) contradirectional coupling

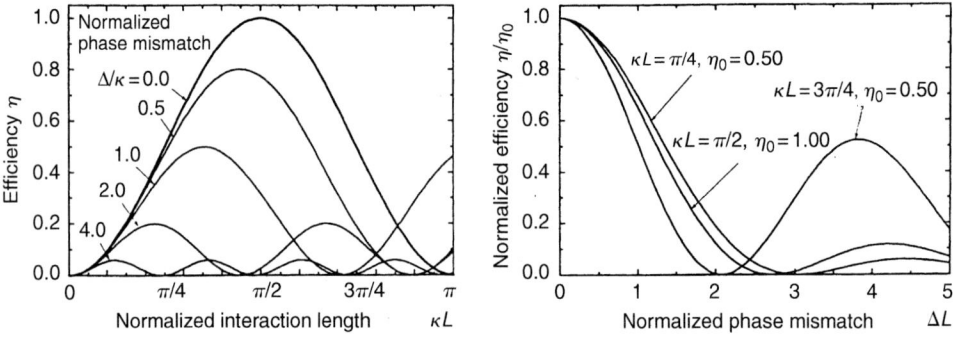

Figure 3   Dependence of the efficiency on the interaction length and the phase mismatch for codirectional coupling

The efficiency under exact phase matching ($\Delta = 0$) is given by $\eta_0 = \sin^2|\kappa|L$. Complete power transfer takes place when L equals to odd integer multiple of the complete coupling length $L_c = \pi/2|\kappa|$. The dependence of η on κL and that of η/η_0 on ΔL are plotted in Figure 3. For $L = \pi/2|\kappa|$, $\eta/\eta_0 = 0.5$ at $\Delta L \approx \pm 1.25$.

Contradirectional coupling: Consider coupling between modes propagating in the opposite directions ($\beta_a > 0$, $\beta_b < 0$) in a waveguide grating of length L. Equation (4) with $A(0) = 1$, $B(L) = 0$ gives a solution that indicates monotonous power transfer. This implies that distributed reflection of forward mode a into backward mode b takes place. The mode b may be the same lateral mode as a ($\beta_b = -\beta_a$, reflection without mode conversion), or may be a different mode ($\beta_b \neq -\beta_a$, reflection associated with mode conversion). The grating is called distributed Bragg reflector (DBR). The efficiency, or the reflectivity, is given by

$$\eta = \left|\frac{B(0)}{A(0)}\right|^2 = \left[1 + \frac{1 - \Delta^2/|\kappa|^2}{\sinh^2\sqrt{|\kappa|^2 - \Delta^2}L}\right]^{-1}. \tag{8}$$

Under exact phase matching ($\Delta = 0$), the efficiency is given by $\eta_0 = \tanh^2|\kappa|L$, and most of the power is transferred ($\eta > 0.84$) when $L > \pi/2|\kappa|$. The dependence of η on κL and that of η/η_0 on ΔL are plotted in Figure 4. For $L = \pi/2\kappa$; $\eta/\eta_0 = 0.5$ at $\Delta L \approx \pm 2.5$.

Equations (7) and (8) indicate that a high efficiency is obtained only when phase matching ($\Delta = 0$) is satisfied, and the mismatch gives rise to a reduction of the efficiency. The propagation constant is given by $\beta = N(\lambda)k = N(\lambda)(2\pi/\lambda)$ with the optical wavelength λ and the mode index

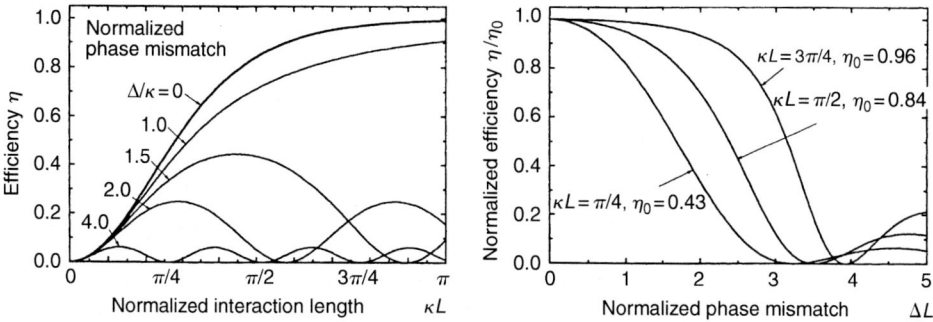

Figure 4 Dependence of the efficiency on the interaction length and the phase mismatch for contradirectional coupling

$N(\lambda)$, and the matching condition is $\Delta = 0$ with Δ given by Eq. (5). Therefore, for a given waveguide grating, a high efficiency is obtained at or near a phase matching wavelength λ_0, and the efficiency is reduced with deviation of the wavelength from λ_0. Thus, the mode conversion and the reflection exhibit wavelength selectivity, and the grating can be used as a wavelength filter. Let $\delta\lambda = \lambda - \lambda_0$ be the wavelength deviation. Assuming $N(\lambda) \approx N(\lambda_0)$, we have $2\Delta \approx (N_a - N_b)(2\pi\delta\lambda/\lambda^2)$ for codirectional coupling, and $2\Delta \approx 2N(2\pi\delta\lambda/\lambda^2)$ for contradirectional coupling of $\beta_b = -\beta_a$. Combining these relations with the bandwidths in terms of ΔL, previously given, the wavelength bandwidths are given by $2\delta\lambda/\lambda \approx 2.5\lambda/\{\pi|N_a - N_b|L\} = 2.5\Lambda/\pi|q|L$ and $2\delta\lambda/\lambda \approx 2.5\lambda/(\pi NL) = 5\Lambda/\pi|q|L$, for co- and contra-directional couplings, respectively. Since the grating period Λ for contradirectional coupling is much shorter, contradirectional coupling (DBR) exhibits wavelength selectivity much sharper than that of codirectional coupling.

Coupling Coefficient

The coupling coefficient κ is an important parameter needed to design the grating structure and predict the performances. κ is calculated by substituting the normalized mode field profiles E_a and E_b and the Fourier amplitude $\Delta\varepsilon_q$ of the grating into Eq. (6). The magnitude of κ depends on the order and polarizations of the coupling modes, and coupling order q, as well as the type and the configuration of the grating. Here, approximate expressions of κ for collinear coupling in a planar waveguide are presented.

Index modulation gratings: Consider a grating consisting of periodic modulation in the refractive index of the guiding layer, as shown in Figure 5(a). The refractive index modulation can be written as:

$$\Delta n(x, z) = \sum_{q \geq 0} \Delta n_q(x)\cos(qKz + \phi_q), \quad (\phi_0 = 0). \tag{9}$$

Using a relation $\varepsilon + \Delta\varepsilon = (n + \Delta n)^2 \approx n^2 + 2n\Delta n$, Eq. (9) can be converted into the permittivity modulation $\Delta\varepsilon$ in the form of Eq. (1). For a grating with uniform modulation within the guiding layer of thickness T, $\Delta\varepsilon$ can be written as:

$$\Delta\varepsilon_q(x) = \Delta\varepsilon_q = n_f\Delta n_{|q|}\exp(-j\phi_{|q|}), \quad -T < x < 0. \tag{10}$$

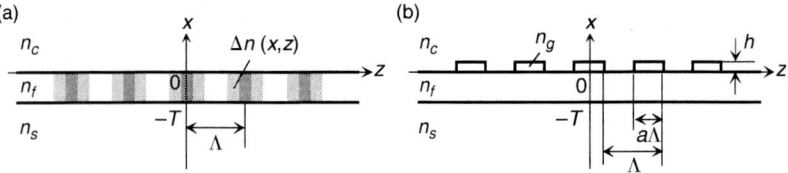

Figure 5   Cross-sectional structures of waveguide gratings (a) index modulation and (b) surface relief gratings

Then the coupling coefficient, calculated from Eq. (6), can be written as

$$|\kappa| = \frac{\pi \Delta n_{|q|}}{\lambda} F, \tag{11}$$

where the first factor $\pi \Delta n_{|q|}/\lambda = \kappa_b$ is the coupling coefficient in bulk, λ the optical wavelength, and the second factor F is an overlap integral between modes and index modulation describing the effect of mode confinement in the waveguide. Since modes are orthogonal, $|\kappa|$ for uniform index modulation and well-guided modes can be written as $|\kappa| \approx \kappa_b \delta_{ab}$. Thus, substantial coupling is limited to contradirectional coupling (reflection) without mode conversion. F for the coupling without mode conversion takes slightly different values for TE–TE (Transverse Electric) and TM–TM (Transverse Magnetic) couplings. F is close to unity if the guided waves are well guided, and is reduced if the guide waves approach to cutoff.

Surface relief gratings: Consider a grating consisting of periodic surface relief on the guiding layer, as shown in Figure 5(b). Let h be the depth of the grating groove and a ($0 < a < 1$) be the duty ratio of the grating tooth. The permittivity modulation $\Delta \varepsilon$ is given by Eq. (1) with

$$\Delta \varepsilon_q(x) = \Delta \varepsilon_q = (n_g^2 - n_c^2) \frac{\sin(qa\pi)}{q\pi}, \quad (q \neq 0, -h/2 < x < h/2). \tag{12}$$

The coupling coefficient κ can be calculated by using Eq. (6). Assuming that the groove depth is much smaller than the thickness of the guiding layer ($h \ll T$), the field amplitudes in the integral can be approximated by the value at the surface ($x = 0$). Then κ for TE–TE coupling can be written as

$$\kappa \approx \frac{\omega \varepsilon_0}{4} E_{y,n}(0) E_{y,m}(0) h \Delta \varepsilon_q$$

$$\approx \frac{2\pi}{\lambda} \frac{\sin(qa\pi)}{q\pi} \frac{n_g^2 - n_c^2}{n_f^2 - n_c^2} \frac{h}{\sqrt{T_{\text{eff},n} T_{\text{eff},m}}} \frac{\sqrt{(n_f^2 - N_n^2)(n_f^2 - N_m^2)}}{\sqrt{N_n N_m}} \tag{13}$$

where m and n denote the mode orders, N the mode index, T_{eff} the effective guide thickness. κ for TM–TM coupling is given in a similar form. Coupling in a grating of $a = 0.5$ is limited to that of odd q. In contrast to an index grating, a relief grating allows coupling associated with mode conversion ($m \neq n$). The coupling coefficient is larger for higher-order modes.

Coplanar Coupling

In a planar waveguide (in yz plane) with a grating (of a length L in z direction) of appropriate orientation, coplanar coupling takes place between guided waves of different propagation directions.

This type of coupling is also called diffraction or deflection, since it involves a change of propagation direction. The propagation direction(s) of the diffracted beam(s) is determined by the y component of Eq. (2). Coplanar coupling is classified into two categories using a parameter defined by $Q = K^2 L/\beta$.

In a thin grating of $Q < 1$, several beams appear corresponding to different diffraction orders q, because coupling may take place even without exact phase matching in the z component. This type of diffraction is called Raman–Nath diffraction. The solution of the coupled-mode equations can be written by using Bessel functions, and the qth order diffraction efficiency is given by $J_q^2(2\kappa L)$. The first-order efficiency $\eta_{\pm 1}$ takes the maximum 0.339 at $2\kappa L = 1.84$. The grating does not exhibit sharp angular and wavelength selectivity.

In a thick grating of $Q \gg 1$, strong coupling takes place only when the Bragg condition (both y and z components of Eq. (2)) is satisfied between the incident and diffracted waves. Therefore, a diffracted wave of a single order appears, as illustrated in Figure 6. This type of diffraction is called Bragg diffraction. The wave vector diagram to determine the diffraction angle is shown in Figure 7, where the phase mismatch in the z direction is denoted as 2Δ. The coupled-mode equations for the amplitudes of the incident and the diffracted waves, $A(z)$ and $B(z)$, are given by

$$\cos \theta_i \frac{d}{dz} A(z) = -j\kappa^* B(z) \exp(-j2\Delta z), \qquad (14a)$$

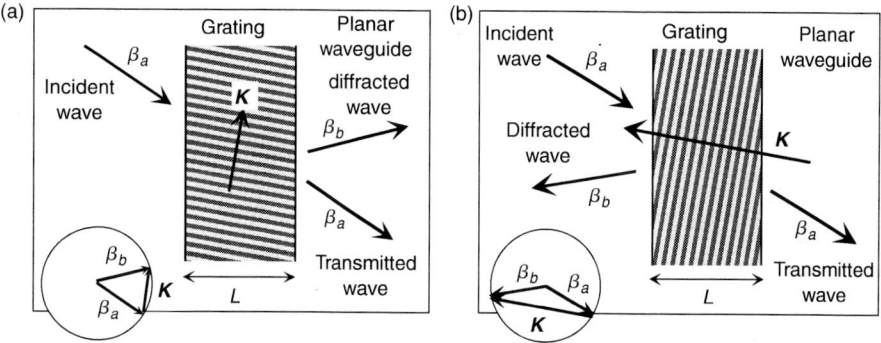

Figure 6   Bragg diffraction of guided wave in a planar waveguide (a) transmission and (b) reflection type

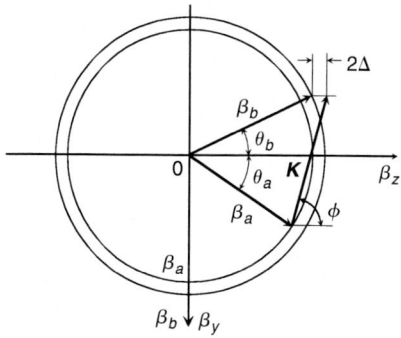

Figure 7   Wave vector diagram for coplanar coupling

$$\cos\theta_d \frac{d}{dz} B(z) = -j\kappa A(z)\exp(+j2\Delta z), \tag{14b}$$

where θ_i and θ_d are incident angle and diffraction angles, respectively. The phase mismatch 2Δ is correlated with the deviation of the incident angle from the Bragg angle. When the incident angle is fixed, a wavelength shift results in a deviation from the Bragg condition.

Transmission grating: Noting that $\cos\theta_i > 0$, $\cos\theta_d > 0$, Eqs. (14) are solved with the boundary conditions $A(0) = 1$, $B(0) = 0$. The diffraction efficiency η can be written as:

$$\eta = \left| \frac{B(L)}{A(0)} \right|^2 = \frac{\sin^2\sqrt{v^2 + \xi^2}}{1 + \xi^2/v^2}, \quad v = \frac{|\kappa|L}{\sqrt{\cos\theta_i \cos\theta_d}}, \quad \xi = \Delta L. \tag{15}$$

The dependence of η on v and ξ are same as the dependence of η on κL and ΔL shown in Figure 3. The efficiency under the Bragg condition ($\xi = 0$) is given by $\eta_0 = \sin^2 v$ and takes a maximum of 100% at $v = \pi/2$. The efficiency decreases with deviation from the Bragg condition. For $v = \pi/2$, $\eta/\eta_0 = 0.5$ at $\xi \approx 1.25$, and the angular and the wavelength selectivity can be evaluated by combining $\Delta L \approx 1.25$ with the relations between 2Δ and the angular or the wavelength deviation.

Reflection grating: Noting that $\cos\theta_i > 0$, $\cos\theta_d < 0$, Eqs. (14) are solved with $A(0) = 1$, $B(L) = 0$. The diffraction efficiency η can be written as:

$$\eta = \left| \frac{B(0)}{A(0)} \right|^2 = \left[1 + \frac{1 - \xi^2/v^2}{\sinh^2\sqrt{v^2 - \xi^2}} \right]^{-1}, \quad v = \frac{|\kappa|L}{\sqrt{\cos\theta_i |\cos\theta_d|}}, \quad \xi = \Delta L. \tag{16}$$

The dependence of η on v and ξ are same as the dependence of η on κL and ΔL shown in Figure 4. The efficiency under the Bragg condition ($\xi = 0$) is given by $\eta_0 = \tanh^2 v$ and increases monotonously with v. The efficiency is 84.1% at $v = \pi/2$ and larger than 99.3% for $v > \pi$. For $v = \pi/2$, $\eta/\eta_0 = 0.5$ at $\xi = \Delta L \approx 2.5$.

The coupling coefficient for coplanar coupling, that is, κ in Eqs. (14) to (16), depends upon polarizations of the incident and the diffracted waves. It is given by $\kappa_{TE-TE}\cos\theta_{di}$ and κ_{TM-TM} for TE–TE and TM–TM couplings, respectively, where κ_{TE-TE} and κ_{TM-TM} are the coupling coefficients for collinear coupling, and $\theta_{di} = \theta_d - \theta_i$ denotes the deflection angle. When $\theta_{di} = \pi/2$, TE–TE coupling does not take place, since the polarizations are perpendicular to each other. A grating of $\theta_{di} = \pi/2$ serves as a TE–TM mode splitter. For TE–TM coupling, the coefficient can be written as $\kappa_{TM-TE}\sin\theta_{di}$. Note that TE–TM mode conversion may occur when $\theta_{di} \neq 0$, although κ_{TE-TM} is considerably small when compared with κ_{TE-TE} and κ_{TM-TM}.

BRILLOUIN DIAGRAM

Optical coupling in a waveguide grating and the wavelength dispersion can be illustrated in a Brillouin diagram, that is, $\omega/c - \beta$ diagram as shown in Figure 8. ($\omega/c = k = 2\pi/\lambda$ is wavenumber in vacuum.) Figure 8(a) shows the dispersion of a waveguide without grating. The bold solid curves indicate guided modes, and radiation modes occupy the shaded region. The curves for the qth order space harmonics produced by grating are obtained by shifting the guided-mode

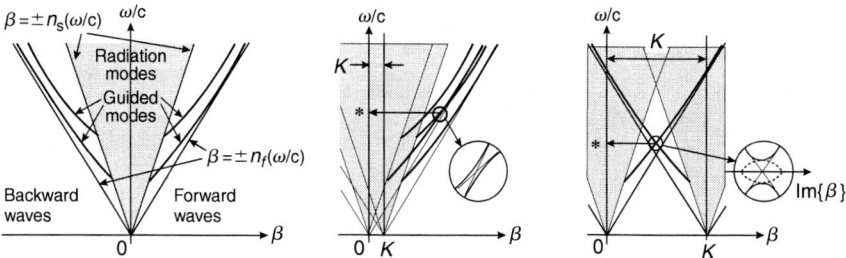

Figure 8 Brillouin diagrams for guided wave coupling by a grating. (a) wave dispersion, (b) codirectional coupling, and (c) contradirectional coupling. The coupling occurs at a wavelength corresponding to ω/c indicated by *

curves by qK along the β axis. Figure 8(b) and 8(c) show the Brillouin diagrams for waveguide grating structures, where the interactions are the first-order codirectional and contradirectional couplings, respectively. Here the interactions of the curves (phase matching points), as a result of mode coupling, the curves for the normal mode are modified as shown in the insets. Codirectional coupling can be interpreted as interference between two normal modes with close β values represented by two close curves. For contradirectional coupling, the curves are separated into upper and lower branches. In the gap between them, β is not real and has an imaginary part as shown by the dotted curve in the inset of Figure 8(c). In this stop band, the wave cannot propagate substantially and is reflected. Since optical coupling occurs only at or near the wavelength corresponding to $\omega/c = k = 2\pi/\lambda$ indicated in the figure by *, a grating in waveguide can be used as wavelength filters.

GUIDED MODE–RADIATION MODE COUPLING

Output Coupling

Figure 9 illustrates coupling between a guided mode and radiation modes in a planar waveguide with a grating. Coupling takes place between waves satisfying phase matching for the z component. When a guided wave of propagation constant $\beta_0 = Nk$ is incident, the qth harmonics are radiated into air and substrate at angles determined by

$$n_c k \, \sin \theta_q^{(c)} = n_s k \, \sin \theta_q^{(s)} = \beta_q = Nk + qK. \tag{17}$$

The number of radiation beams equals the number of real values for $\theta_q^{(c)}$ and $\theta_q^{(s)}$. Figure 9(a) shows multibeam coupling where more than three beams are yielded, and Figure 9(b) shows two-beam coupling where only a single beam for the fundamental order ($q = -1$) is yielded in both air and substrate. Another possibility is one-beam coupling where a beam radiates only into the substrate. The amplitude of the guided and the radiation wave decays as $g(z) = \exp(-\alpha_r z)$ due to the power leakage by radiation. Since the attenuation of the guided power corresponds to the power transferred to the radiation modes, the qth order output coupling efficiency of a grating of length L can be written as

$$\eta_{\text{out}} = P_q^{(i)}\{1 - \exp(-2\alpha_r L)\}, \quad \alpha_r = \sum_{q,i} \alpha_q^{(i)}, \ P_q^{(i)} = \alpha_q^{(i)} / \alpha_r, \tag{18}$$

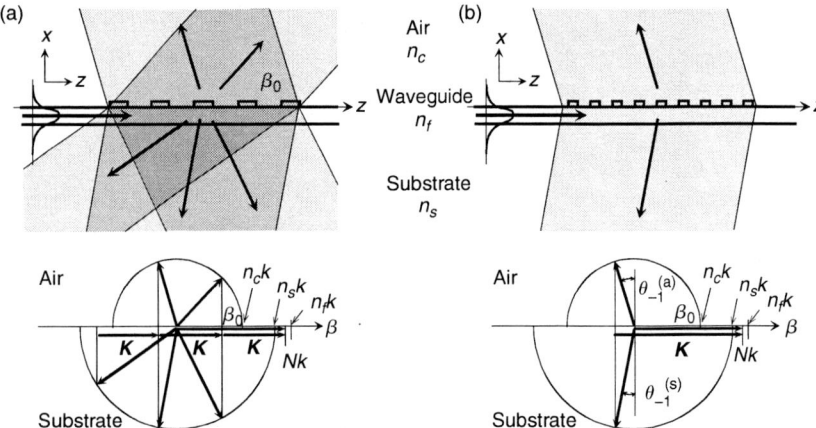

Figure 9 Coupling between guided mode and radiation modes in grating coupler. (a) Multibeam coupling and (b) two-beam coupling

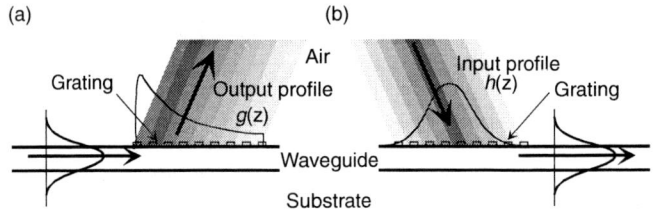

Figure 10 Output and input couplings in a grating coupler. (a) output and (b) input coupling

where $\alpha_q^{(i)}$ and α_r are radiation decay factors, $P_q^{(i)}$ is a ratio for power distribution to q–(i) radiation beam, and $i = a$ or s.

Input Coupling

A guided mode can be excited through input coupling of an external beam incident on a grating at an angle satisfying Eq. (17). Figure 10 compares output and input couplings. A reciprocity theorem analysis shows that the input coupling can be written as [3]

$$\eta_{in} = P_q^{(i)} \cdot I(g,h), \quad I(g,h) = \frac{\left| \int g(z)h(z)dz \right|^2}{\int |g(z)|^2 \, dz \int |h(z)|^2 \, dz} \tag{19}$$

where $h(z)$ is the input beam profile. The overlap integral $I(g,h)$ takes the maximum of unity, when the input profile $h(z)$ is proportional to the output profile $g(z)$. A high efficiency can be accomplished under conditions $a_r L \gg 1$, $P_q^{(i)} \approx 1$ for a single q–(i) beam, and $h(z) \approx g(z)$. For a Gaussian input beam the maximum of $I(g,h)$ is 0.801, and the maximum input efficiency is 80.1%.

Radiation Decay Factor

The radiation decay factors $\alpha_q^{(i)}$ and α_r can be calculated by various methods. In the coupled-mode analysis, $\alpha_q^{(i)}$ is correlated with the coupling coefficient $\kappa_q^{(i)}$ by $\alpha_q^{(i)} = \pi |\kappa_q^{(i)}|^2$, and $\kappa_q^{(i)}$ can be calculated by substituting the guided-mode and radiation-mode profiles into Eq. (6) [3]. Other methods include numerical analysis to calculate the complex propagation constant of the normal modes by space harmonics expansion based on Floquet's theorem [11], analyses based on Green function approach [12], transmission-line approach [13], and Beam Propagation Method (BPM) [14]. For grating couplers of relief type with groove depth h, the decay factor α_r increases in proportion to h^2 in the region where h is small. For larger h, α_r saturates because of the limited penetration of the evanescent tail of the guided mode into the grating layer.

High-Efficiency Grating Couplers

One-beam coupling is desirable for achieving high input and output efficiencies. Such coupling can be realized in a short-period grating, which allows only fundamental backward coupling into substrate $(\theta_{-1}^{(s)} < 0)$. Two-beam couplers as shown in Figure 9(b) are more practical, but they involve a drawback that the output power is divided into air and substrate. A high directionality into air $(P_{-1}^{(a)} \approx 1)$ for eliminating this drawback can be accomplished by inserting a reflection layer on the substrate, by using the Bragg effect in a thick index-modulation grating, or using the blazing effect in a relief grating having an asymmetric triangular cross-section.

WAVEGUIDE GRATING ELEMENTS WITH WAVEFRONT CONVERSION FUNCTION

Various wavefront conversion functions based on the principle of holography can be incorporated in mode coupling in a planar waveguide (either guided-to-guided or guided-to-radiation mode coupling) by spatially modulating the grating pattern (forming a curved and chirped grating) [2]. A useful wavefront conversion is the lens function for focusing, collimating, and imaging. Waveguide grating elements with lens function are illustrated in Figure 1(c) and 1(d). Suppose that $\Phi_i(y, z)$ and $\Phi_o(y, z)$ are the phase distribution functions, on the yz waveguide grating plane, of the incident wave and the desired output wave, respectively. The desired wavefront conversion is accomplished by giving the input wavefront a phase modulation corresponding to the phase difference $\Delta\Phi = \Phi_o - \Phi_i$. The waveguide grating (waveguide hologram) for such phase modulation consists of grating lines described by

$$\Delta\Phi(y,z) = \Phi_o(y,z) - \Phi_i(y,z) = 2m\pi, \quad (m = ..., -2, -1, 0, +1, +2, ...). \tag{20}$$

The grating patterns can be generated and fabricated by computer-controlled electron-beam writing technique.

REFERENCES

1. A. Yariv and M. Nakamura, "Periodic structures for integrated optics," *IEEE J. Quantum Electron.*, **QE-13**, 233–253 (1977).

2. T. Suhara and H. Nishihara, "Integrated optics components and devices using periodic structures," *IEEE J. Quantum Electron.*, **QE-22**, 845–867 (1986).
3. H. Nishihara, M. Haruna, and T. Suhara, *Optical Integrated Circuits*, McGraw-Hill, New York (1989).
4. T. Suhara and M. Fujimura, *Waveguide Nonlinear-Optic Devices*, Springer-Verlag, Berlin (2003).
5. Y. Handa, T. Suhara, H. Nishihara, and J. Koyama, "Microgratings for high-efficiency guided-beam deflection fabricated by electron-beam direct writing techniques," *Appl. Opt.*, **19**, 2842–2847 (1980).
6. M. L. Dakss, L. Kuhn, P. F. Heidrich, and B. A. Scott, "Grating couplers for efficient excitation of optical guided waves in thin films," *Appl. Phys. Lett.*, **16**, 523–525 (1970).
7. K. Ogawa, W. S. C. Chang, B. L. Sopri, and F. J. Rosenbaum, "Grating mode coverter/directional coupler for integrated optics," *J. Opt. Soc. Am.*, **63**, 478–480 (1973).
8. D. C. Flandars, H. Kogelnik, R. V. Schmidt, and C. V. Shank, "Grating filters for thin film optical waveguides," *Appl. Phys. Lett.*, **24**, 194–196 (1974).
9. J. M. Hammer, R. A. Bartolini, A. Miller, and C. C. Nail, "Optical grating coupling between low-index fibers and high-index film waveguides," *Appl. Phys. Lett.*, **28**, 192–194 (1976).
10. A. Yariv, "Coupled-mode theory for guided-wave optics," *IEEE J. Quantum Electron.*, **QE-9**, 919–933 (1973).
11. S. T. Peng, T. Tamir, and H. L. Bertoni, "Theory of periodic dielectric waveguides," *IEEE Trans. Microwave Theory Technol.*, **MTT-23**, 123–133 (1975).
12. K. Ogawa, W. S. C. Chang, B. L. Sopri, and F. J. Rosenbaum, "A theoretical analysis of etched grating couplers for integrated optics," *IEEE J. Quantum Electron.*, **QE-9**, 29–42 (1973).
13. T. Tamir and S. Peng, "Analysis and design of grating couplers," *Appl. Phys.*, **14**, 235–254 (1977).
14. S. F. Helfert and R. Pregla, "Efficient analysis of periodic structures," *J. Lightwave Technol.*, **16**, 1694–1702 (1998).

PHOTONIC CRYSTAL

Toshihiko Baba

INTRODUCTION

Photonic crystals (PCs) are artificial multidimensional periodic structures whose period is of the order of optical wavelength, as shown in Figure 1. Fundamentally, they are based on a concept extended from conventional diffraction gratings and have a unique analogy to solid state crystals. This enables us to use solid state physics theory for the analysis of PCs. For example, one can calculate photonic bands, photonic bandgaps (PBGs), impurity, defect and surface states, etc., for any PCs. This allows the accurate prediction of light propagation in PCs, and expands the possibility of various novel photonic devices based on PCs.

The history of PCs originates from the concept of the photonic band by K. Ohtaka in 1978. This concept has been established as a new theory in physics and many structures exhibiting PBGs have been investigated in the late 1980s to the early 1990s [1,2]. The study on their device applications started since 1994 to 1995, and at present are one of the key technologies for future opto-electronic devices.

This section reviews important theoretical background, fabrication methods, and device applications. Here, focus is on two- and three-dimensional (2D and 3D) periodic structures in order to clarify the difference from conventional one-dimensional (1D) gratings and the uniqueness arising from the multidimensionality. Since many important references are shown in Reference 3, the citation in this section will be limited to those not included in it and some very recent works.

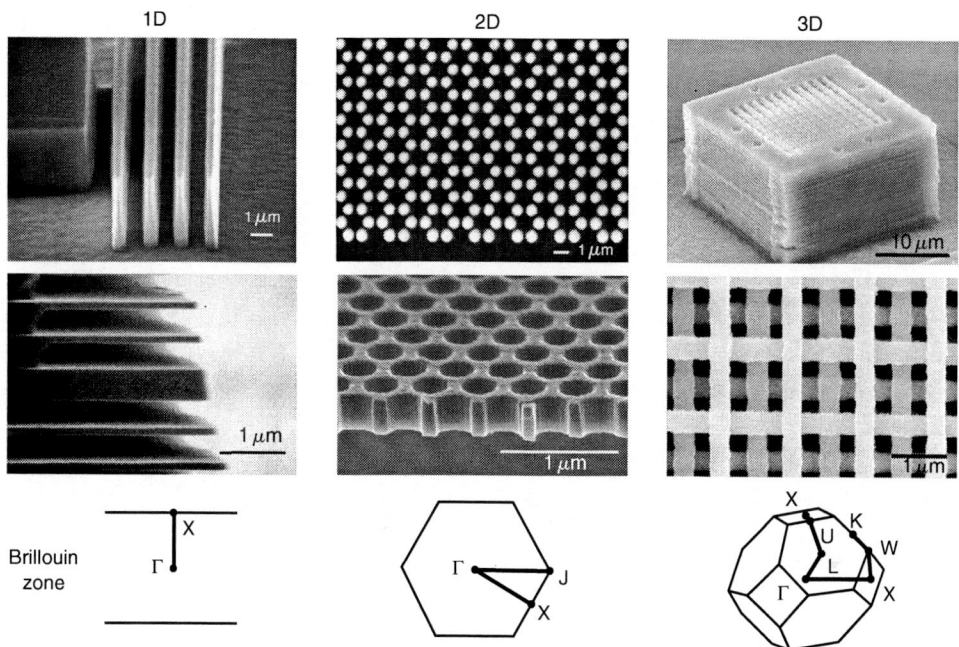

Figure 1   Different dimensional PCs and corresponding Brillouin zones (From Baba et al., *Jpn. J. Appl. Phys.*, 39, 3407 (2003). With permission © 2000 IPAP; from Baba et al., *Nature Materials*, 2, 118 (2003). With permission © 2003 Nature Publishing Group; from Baba, *IEEE Sel. Top. Quantum Electron.*, 3, 816 (1997). With permission © 1997 IEEE; from Baba et al., *Nature Materials*, 2, 119 (2003). With permission © 2003 Nature Publishing Group)

FUNDAMENTAL THEORY

The photonic band is the dispersion relation between the normalized frequency $\omega a/2\pi c$ ($= a/\lambda$) and the wave number k of light, where a is the lattice constant, and ω, c, and λ are the angular frequency, the vacuum velocity, and the wavelength of light respectively. As shown in Figure 2(a), photonic bands are categorized into three frequency ranges, that is, the PBG range, frequencies higher than the PBG (the wavelength is shorter than the lattice constant), and frequencies lower than the PBG (the wavelength is longer than the lattice constant).

PBG Structures

As there are no photonic bands in the PBG, light in this frequency range cannot exist in the PC. Therefore, external light incident on the PC will be completely reflected. Most of the PBGs in 2D and 3D PCs have been calculated by early researchers. As shown in Figure 2(a) and 2(b), circular holes consisting of low refractive index medium, which are arranged in a high index medium in a close-packed triangular lattice and in a honeycomb triangular lattice, exhibit a 2D PBG for arbitrary polarizations. However, the close-packed PC is more widely studied because its PBG is robust against structural imperfections and is easier to apply for devices. Other studies investigated PBGs for one polarization and one direction. The current interest of researchers is on the use of these PBGs. As for 3D PCs, many structures exhibit PBGs because the 3D periodicity achieves an isotropic Brillouin zone more easily than the 2D one. Fundamentally, diamond structures and asymmetric face-centered-cubic structures open a PBG. Recent studies concentrate on some structures that can

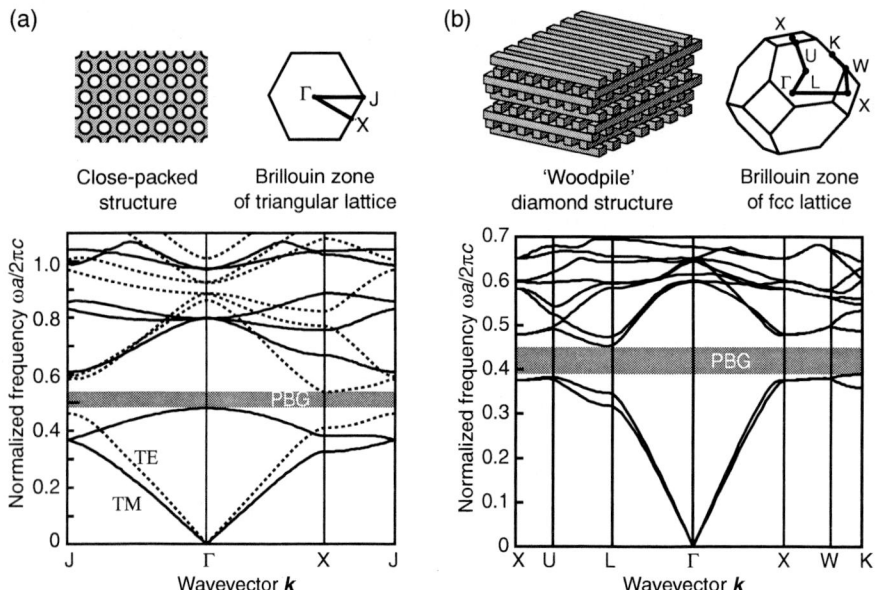

Figure 2 Examples of photonic band diagram. (a) 2D PC consisting of close-packed circular airholes in a triangular lattice. (b) 3D PC called woodpile, which is constructed by stacking square rods

be easily fabricated, for example, the woodpile structure (sometimes called layer-by-layer structure) formed by stacking square rods by shifting their position by half period, as shown in Figure 2(b). For the fabrication and device applications, it is still important to investigate the dependence of optical characteristics on the structural details of these PCs. Before starting the research on device applications, attention should be given to the restriction that a PBG does not appear in 2D and 3D PCs when the index ratio between two media is less than two.

Unique Phenomena in Light-Conductive Frequency Range

In general, complex photonic bands appear in the frequency range higher than the PBG. In this range, light transmits through the PC; however, it shows a peculiar transmission characteristic. The photonic band displayed by contour plots in a Brillouin zone (Figure 3), is called the dispersion surface. It is well known that the derivative of the frequency with respect to the wave vector, which corresponds to the slope of the photonic band, gives the group velocity of light, and light power in the PC propagates toward the gradient of the dispersion surface. This surface is isotropic in free space, while it is complex and anisotropic in PCs. It causes a prism effect and a collimation effect, such that the direction of light propagation is strongly and weakly dependent on the wavelength and the incident angle, respectively. When the dispersion surface is almost flat, a small group velocity enhances the light-matter interaction. The second derivative of the dispersion surface corresponds to the dispersion coefficient. Therefore, one can design a large positive or negative dispersion coefficient and the zero-dispersion condition.

Though unique, it is a composite phenomenon. If any of these are used an effective design should be found that suppresses the other phenomena. One example is the negative refraction phenomenon near the center (Γ point) of the Brillouin zone; light refracts in the negative direction independent of the wavelength and the incident angle. At frequencies much higher than the PBG, the composite phenomenon becomes evident due to many overlapping bands. Therefore, only two or three bands are discussed for device applications.

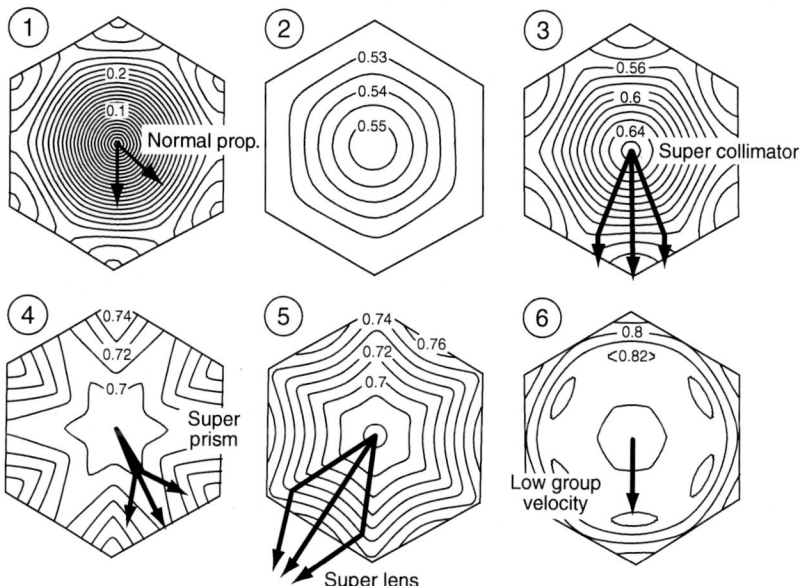

Figure 3   Dispersion surfaces of 2D PC consisting of close-packed circular airholes in a triangular lattice. The number denotes the order of photonic bands

Low Frequency Limit

When the normalized frequency is very low (close to 0), the lattice constant is much smaller than the wavelength. This situation resembles light in a solid state crystal consisting of periodically arranged atoms. It can be easily understood that such a structure exhibits the birefringence, which in an artificial periodic structure is known as the form birefringence. It is expressed by a rigorous formula for 1D PCs; however, is difficult to calculate for 2D and 3D PCs in the classical optic theory. In the photonic band theory, it is precisely calculated from the slope of the lowest order band curve at low frequency limit and enables us to design a large birefringence structure.

Slab Structure and Light Cone

The structure that consists of airholes in a high index slab (Figure 2[b]) is called PC slab. It is widely used for various device applications, because it confines light by the PBG effect in the slab plane, and by the total internal reflection (TIR) in the out-of-plane direction. For this type of structure, the projected photonic band is normally used, in which the k vector is projected to the slab plane. The projected band diagram is separated by a boundary called light line expressed by the relation $\omega a/2\pi c = k/(2\pi n/a)$ for a cladding medium of index n sandwiching the slab. Below the light line, the optical field becomes the evanescent wave in the cladding and is perfectly confined around the slab. Above the light line (sometimes called the light cone), light can leak into the cladding. The amount of the leakage depends on the photonic band and the wave number. The second Γ point is the special case that absolutely suppresses the leakage. However, one should use photonic bands below the light line, when using light confined around the slab.

ANALYSIS AND SIMULATION METHODS

Since early studies, a useful approximate solution for light propagation characteristics in PCs has been desired. However, the structural design still requires numerical calculations such as the photonic band calculation and the simulation of lightwaves, which are discussed in the following sections (at least, plane wave expansion method and FDTD method are indispensable).

Plane Wave Expansion Method

It can be used for the simulation, but is mainly used for the calculation of photonic bands and dispersion surfaces. In this method, the PC structure is expanded by a finite number (ideally infinite number) of plane waves and eigen frequencies for each wave vector is numerically calculated. It rigorously takes the vector nature of light into account, and uses no approximations. When the unit cell of the PC has a simple form, precise bands are obtained within a short time. Compared with the band calculation by the FDTD method shown here, the calculation of the dispersion surface is also easy due to the simple calculation procedure. However, for a complex structure unit cell, a large number of plane waves is necessary for the precise expansion of the structure. The amount of calculation is proportional to the third power of the number of plane waves. Therefore, this method cannot be used for very complex structures.

FDTD Method

It is an abbreviation of the finite difference time domain method. Since this method is based on simple finite difference expressions of Maxwell's equations, it requires a large amount of computer memory and greater calculation time. However, it is widely used in personal computers due to the recent progress in speed and memory capacity. The most common use of the FDTD method is the simulation of lightwaves in a finite size structure. It gives fundamental characteristics such as electromagnetic field distributions, transmission spectrum, time response, etc. It is not only applied for isotropic, nondispersive and linear media, but also for anisotropic, dispersive, and nonlinear media with special algorithms. By using the periodic boundary condition, it can also be used for the photonic band calculation. Here, the production of the band diagram needs complex tasks such as the Fourier transform of the time response and the detection of resonant peaks. However, since the calculation time is proportional to the model size, it can still be used for very complex structures, which cannot be modeled in the plane wave expansion method. This method was originally developed in the field of microwaves, for example, to design the antenna. Therefore, many software tools based on this method are commercially available (recently, those specialized for PCs appeared). They easily serve graphical and reasonable results but sometimes cause serious errors, when users do not understand about the restrictions of this method. There are many special techniques to obtain precise results within short calculation time.

Other Methods

The transfer matrix method and the scattering matrix method are used as simulation methods of light in finite size PCs. The former calculates the transmission spectrum by combining the finite element analysis of the cross-sectional structure and the transfer analysis in the direction of light propagation. The latter uses the expansion of the PC structure and lightwaves by cylindrical waves. In many 2D PCs, the unit cell consists of a circular structure. For such PCs, the scattering matrix method gives precise solutions of electromagnetic fields within one-tenth of the calculation time to that of the FDTD method. In recent years, however, they are rarely used when compared with the FDTD method.

FABRICATION METHODS

These are artificial methods, self-organized methods, and their intermediate ones. The artificial methods can form arbitrary lattices and defects, but require expensive facilities for the process. On the other hand, the flexibility of the self-organized methods is limited and are attractive because a large-scale PC is formed by a simple equipment.

Fabrication of PC Slab by Dry Etchings

Airholes of the PC slab are formed by lithography and dry etching against a high index slab (e.g., semiconductor) sandwiched by low index media. The typical thickness of the slab is <300 nm, and the typical diameter of the airholes is 200 nm. Therefore, the fabrication of the PC slab is not very difficult if the present mature techniques are used. For Si and GaAs slabs, reactive ion etching (RIE) and electron-cyclotron-resonance (ECR) etching are widely used. For the InP system, which is much more difficult to process, a fine structure is formed by using chemically assisted ion beam etching (CAIBE) and inductively coupled plasma (ICP) etching. At present, formed structures have sufficient accuracy so that they exhibit PBGs predicted in the photonic band calculation.

Formation of Deep 2D PC by Chemical Etchings

By using the anodic oxidation of Al and the anodic porous etching of Si, deep airholes with an aspect ratio of 50 to 100 are formed. If an initial pattern is formed on the substrate by another etching and molding, 2D PCs of deep airholes showing good photonic band characteristics are automatically formed. At present, however, there are two problems. One is the nonuniformity in the vertical direction and the other is the self-repair of defects, which disturbs the introduction of intentional defect elements. This method is expected to provide a large-scale thick 2D PC that can be a bulk optic component. For this purpose, a very deep (e.g., over several 100 μm) etching is required.

Production of 2D and 3D PCs by Bias Sputtering

It is a method of depositing multilayer film on a shallow substrate, maintaining the initial pattern. Since large-scale 2D and 3D PCs can be formed by a single process, it is useful for the production of commercial devices. This method restricts the formation of structures, and so it is difficult to form a PBG structure. However, additional formation of asymmetric airhole array on the PCs will open a full PBG. Based on this technique, a polarization filter has been developed as a commercial device.

Formation of 3D PC by Wafer Fusion

The wafer fusion technique has been studied for the purpose of combining different semiconductor materials with a large difference in lattice constant. However, no studies have been successful in forming a stable fused interface due to the mismatching of thermal expansion coefficient. On the other hand, the fusion of the same material is stable and useful for constructing complex 3D structures. This technique has succeeded to construct the woodpile 3D PC and to show the full PBG in the lightwave frequency range. Due to the electrical conductivity at fused interfaces, it is a promising PC for future functional devices.

2D and 3D Polymer PCs formed by Scanning or Interference Exposure

The 3D polymerization by the scanning exposure of ultraviolet light is effective to construct complex PCs. However, due to the spatial resolution of the order of 100 μm, its application is restricted to PCs operating at microwaves. To overcome this limit, the two-photon exposure was proposed and demonstrated, in which strong pulsed laser light at longer wavelength range is focused on a polymer. Since only the focal point is exposed by the two-photon absorption phenomenon, a high resolution of the order of sub-micron is realized, which is suitable for the formation of PCs at lightwave frequencies. The interference exposure forms large-scale 2D and 3D PCs very quickly. By using 4 to 5 light beams, various 3D PCs are formed. A problem with these PCs is about using formed structures, which cannot have a PBG because of the low index contrast between the polymer and the air. Recently, studies are done to transfer the structure into another high index medium.

Self-Organization of Opal 3D PCs

When a large number of uniform nano-spheres are inserted into a beaker filled with liquid, they are precipitated and automatically organized into a 3D periodic structure. After the evaporation of the liquid and the curing process, a large-scale 3D PC called opal is formed. Simple opal structures cannot have a PBG, whereas an inverted structure can. This inverted structure called inverse opal is possible by burying the opal with a high index medium such as TiO_2. Owing to the improvement of growth process of opals, the PBG is well observed in the experiment. The lattice structure can be changed if a template is placed on the bottom of the beaker, which is expected to be a simple production method of large-scale PCs. Recently, their application as a color fixing agent in a pigment was examined.

Construction of 3D PCs by Micro-Manipulation

This is a method of putting, moving, and stacking micro-elements by a nano-probe in a vacuum chamber. The woodpile 3D PC has been experimentally demonstrated by stacking 1D PCs by this technique and is not suitable for mass-production in the present form. However, due to the flexibility of constructing arbitrary structures, it has a potential of forming ideal 3D PCs.

PC Fibers

The fibers are fabricated by drawing the pure silica preform with an airhole array. The preform is formed by bundling silica rods or drilling airholes into a silica plate and by drawing the preform. Earlier studies have investigated the transmission characteristics in the cross-sectional direction. Presently, however, most studies are concerned with PC fibers. The features of PC fibers will be discussed later.

APPLICATIONS

The device applications of PCs are summarized in Figure 4. Most of them studied now utilize 2D PCs, which controls the light propagation inside the 2D plane. Therefore, the means of coupling external light into the 2D plane and confining light inside the 2D plane are important issues for the development of the devices.

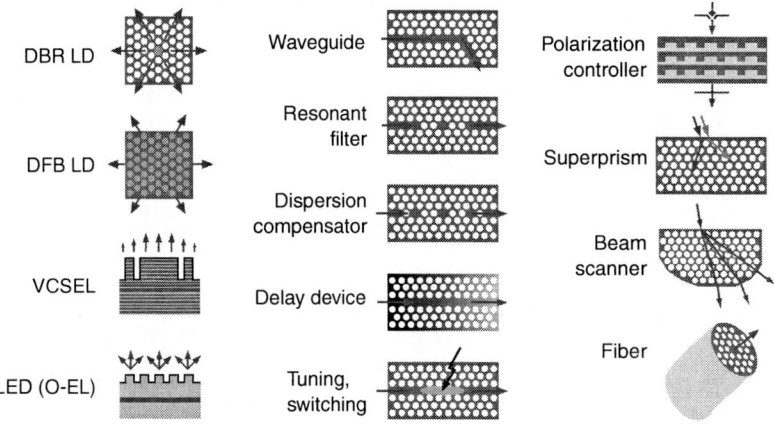

Figure 4 Various applications of PCs

Light Sources

They are categorized into the following four types (1) low threshold microlaser consisting of point defect as a small cavity, (2) high power laser utilizing the whole area of a PC, (3) high power vertical cavity surface emitting laser (VCSEL) with holey structure, and (4) light-emitting diode (LED) with a PC lattice exhibiting high light extraction efficiency.

Low threshold microlaser is the first application discussed for PCs in 1987. The cavity QED effect that enhances the spontaneous emission rate in the PBG microcavity has attracted much attention. The first lasing operation was achieved by a single point defect cavity in a semiconductor PC slab in 1999, and henceforth, various modified structures were also demonstrated (Figure 5). In addition, the large enhancement of the spontaneous emission rate was experimentally observed at room temperature [4]. Due to the small size of the device, the light output is low and the high efficiency current injection is essentially difficult. Thus, it is not expected as a high power light source but as a special device such as a single photon source for the quantum communication. This kind of device is not only realized in a normal PC slab but also in a quasiperiodic PC [5]. Different from normal PCs, this structure has a unique potential to obtain the light localization and the spontaneous emission control in a uniform structure.

High power laser is an extension of conventional distributed feedback (DFB) laser. The modal behavior is not only analyzed by the photonic band, but also well understood by the coupled mode theory. The most unique feature is the coherent lasing over the wide area of the PC. The lasing at various photonic band edges and corresponding far field patterns were observed. Particularly, the second Γ point is expected to achieve a high power vertical emission with very narrow output beam. Currently, two problems faced are the high threshold arising from the wide area pumping and the different directions between the oscillation and the light extraction.

The VCSELs is the most recent topic that are commercially available devices, which are used as light sources in local area networks. However, low single mode output power of <5 mW is a problem. The holey VCSEL was proposed for the purpose of stabilizing the lateral single mode in a large diameter device by the same concept as those in PC fibers. A record high output power of 7 mW was reported not directly based on the PC fiber concept but on the unique control of in-phase oscillating fundamental mode in the periodic structure [6]. If the mode control mechanism and the structural optimization are clarified, this structure will be widely used in various wavelength range VCSELs.

A LED is the device widely studied because of its huge market potential for displays and illuminations in the near future. In LEDs, light extraction efficiency dominates the total

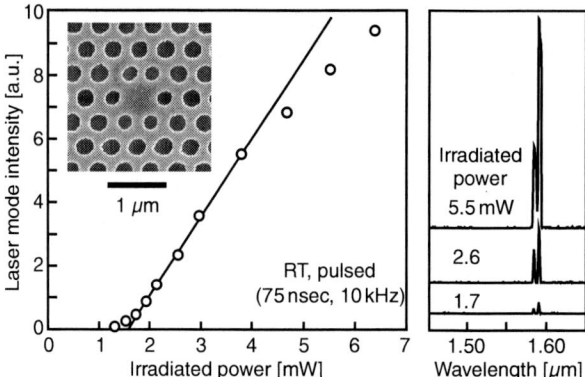

Figure 5   The single point defect cavity formed into a GaInAsP PC slab and its lasing characteristics at room temperature (From Baba et al., *Appl. Phys. Lett.*, 85, 3989 (2004). With permission © 2004 AIP)

efficiency. As discussed earlier, the PC slab has a light cone region, which effectively scatters the internal light to the air. By using this characteristic, a very high efficiency over 30% was experimentally demonstrated. However, there are several problems, which are particularly serious for low cost LEDs, such as the complex fabrication process, the difficulty of forming the electrode, and nonradiative effects induced in the fabrication process. Another simpler structure is the 2D PC formed on top of the emitting material wherein, the efficiency is improved by a factor of 3 to 4 in a semiconductor LED and 1.5 to 2.0 in an organic LED [7]. This device is close to the practical application, because of the shallow PC, large lattice constant (four to five times larger than the wavelength in the material), the robustness, and adaptability to arbitrary materials.

Optical Waveguides

The PC waveguide, in which the line defect put into a uniform PC with the PBG, was first discussed in 1994 and demonstrated by using the PC slab in 1999. In earlier studies, the leakage loss by the light cone was very large. However, once the relation between the waveguide mode and the light cone were clarified by the photonic band theory, the loss-less mode was designed and the experimental loss was rapidly reduced. Figure 6 shows the waveguide formed on a silicon-on-insulator substrate. The first observation of light propagation estimated a large loss of over 100 dB/cm, while recent typical value is <10 dB/cm [8]. In addition, the coupling loss from the single mode silica fiber to the tiny core of $(0.3 \ \mu m)^2$ cross-sectional area has been greatly reduced to <1 dB/port by a spot size converter [8].

The next issue is the realization of sharp bends, branches, and directional couplers that operate in a wide spectral range. However, these waveguide components have already been obtained by a high index contrast waveguides such as the Si photonic wire. Therefore, it will be appropriate to choose the combination of functional devices based on PC waveguides and the optical wiring based on high index contrast waveguides. As mentioned later, the ultralow group velocity near the band edge and the ultrasmall point defect cavity are unique applications of the PC waveguide.

One important issue is the airbridge structure of the PC slab, which is considered necessary for low loss light propagation that causes the fragility of the structure and disturbs the electrical control and the heat sinking. However, such an airbridge (or high index contrast) structure is not necessarily required for some active applications utilizing the low group velocity and, hence, should be designed for each purpose. Another interesting example is the line defect waveguide

Figure 6 Fabricated line defect PC waveguide in an airbridge PC slab, which was formed into a silicon-on-insulator substrate (From Baba et al., *IEEE J. Sel. Top. Quantum Electron.*, 10, 489 (2004). With permission © 2004 IEEE)

formed in a 3D PC without PBG where the light is guided by the effective index confinement arising from the heterostructure. It does not allow a sharp bend, but has a unique dispersion characteristic and the flexibility of various devices can be done by a simple process.

Fibers

As mentioned previously, these fibers have many airholes along the core. This was first proposed and fabricated in 1995. The rapid development of production methods and the discoveries of many unique characteristics, has made it commercially available. There are mainly two types: the dielectric core type and the air core type. The former is based on the index confinement of light and has two unique characteristics. One, is the single mode propagation of light in a wide frequency range and in a wide range of the core size, which arise from the flexible effective index of the PC cladding. Another, is the flexible design of dispersion characteristics; the zero dispersion condition and a very strong positive and negative dispersion conditions can be designed at any wavelength. The propagation loss has been rapidly reduced to 0.2 dB/km, which is almost the same as the lowest limit of the present single mode fiber [8]. It will provide a strong impact to the network infrastructure, if it achieves a loss of 0.1 dB/km. In addition, by the large optical power density in a small core, this type of fiber achieves the strong polarization maintaining propagation by the high index contrast boundaries, and light amplification with the rare earth metal doping, Raman amplification, four wave mixing, etc. One of the most attractive phenomenon is the generation of super continuum light in a wide spectral range almost covering the visible wavelengths. On the other hand, the air core type is expected to achieve a low power density in a large air core, which is effective for the large number of channels in next generation wavelength division multiplexing (WDM) networks. This type uses the Bragg reflection at the PC cladding for the light propagation and the loss has also been reduced rapidly from dB/m order to 2 dB/km recently [9]. There occurs a problem of the narrow transmission window however,

Various lattice structures are investigated to solve this. Another unique feature is the guidance of various media such as nano-particles and liquids.

Filters

The diffraction and the resonant are the two types of a narrow band WDM filter that are studied. The superprism phenomenon at frequencies higher than the PBG is applied for the diffraction type filter. In general, the zone-folding characteristic feature, which is one of dispersion characteristics in diffraction gratings is utilized as a filter. In PCs, dispersion characteristics are modified and deformed by the multidimensionality and therefore, the filtering characteristic is tuned and improved by an optimum design. To obtain a high resolution, the dispersion characteristic slightly apart from the deformed dispersion characteristic is desired. However, the optimum design so far reported still has almost the same performance as that of the present silica-based arrayed waveguide grating (AWG). In simple diffraction gratings and AWGs, a high diffraction order improves the wavelength resolution. On the other hand, the superprism cannot use a high diffraction order due to the composite characteristic, as mentioned earlier. As the filter uses only lower order bands, the performance of the superprism is limited. However, other design principles are being investigated to overcome this limit [10].

As resonant type filters, point defects placed in-between and beside the PC waveguides have been studied. Since the point defect is the ultimately small cavity for lightwaves, it achieves the widest free spectral range (FSR). To achieve the add/drop function, two methods were proposed and demonstrated: one is to use another waveguide, and the other is to use the light radiation and detection in the vertical direction. By controlling the defect size, wide tunability of the resonant wavelength is obtained whereas, it must be controlled very precisely (sub-nanometer order) to obtain a target wavelength. Practically, some postprocess such as the trimming will be indispensable. Recently, the cavity Q and the transmission efficiency have been improved; a remarkable high Q over 10^5 and a high drop efficiency over 80% have been achieved in a cavity volume of the order of λ^3 [11]. In addition, the control of the filtering function to a box-like shape and the integration of multichannel filters are being investigated.

Regarding the polarization filter, a device formed by using the bias-sputtering technique has reached practical performance, and so is partly available commercially. A unique feature of this device is the integration of multiple filters on a substrate. Simplification of the present polarization analyzing system and to provide novel parallel optic systems is required [12].

Dispersion and Group Delay Control Devices

As described earlier, PC fibers allow various unique dispersion characteristic (very large or small dispersion coefficient, flat spectral characteristic, large shift of zero dispersion condition, etc.) and the basic performance of the fibers have been greatly improved recently. Therefore, they have potentials not only as a dispersion compensation device but also as an almighty fiber for various applications. On the other hand, a small dispersion compensation device based on PC line defect waveguides, which has a much larger dispersion coefficient and a tunability, is also considered. For a coupled cavity waveguide, a large dispersion coefficient of the order of ps/nm/mm was theoretically calculated for a wide spectral range and was partly demonstrated in a 1D PC of stacked films. In a normal line defect waveguide, a large dispersion occurs near the band edge. It is not used for the dispersion compensation because of the large wavelength dependence.

Such a band edge characteristic is more useful as a group delay device in which case problems such as strong wavelength dependence and large dispersion that deforms the short optical pulse occur. However, this can be solved by a directional coupler structure of two chirped PC waveguides [13]. In this device, a structural parameter of the waveguide is chirped so that any wavelength is delayed at the corresponding positions by the ultralow group velocity. In addition, two waveguides are designed so that they have upward and downward band profiles and their band edges are always equal. Then, the input optical pulse is expanded into the frequency space by the chirping and equally delayed by the band edge. It is moved to another waveguide due to the same frequency and the same wave number, and finally forwarded in the waveguide. If the two band profiles are symmetric, the input pulse shape is completely reformed. Thus, the group delay device for the short optical pulse is achieved. The delay time is simply proportional to the chirp length; a 1 psec pulse will be delayed by 1 nsec in a 1 mm-long device. If the chirp length is externally controlled, it obtains the tunable group delay device (or in other words, optical buffer memory).

Light Control Device

Among various opto-electronic devices, the two most desired devices are the optical switch and the wavelength tuning device; the optical amplifier is also investigated as a multifunctional device having these functions. However, the discussion on these devices-based PCs arse limited. Fundamentally, the control of light propagation requires the linear or nonlinear change of the index and absorption/amplification coefficient, which are not directly related with PCs. However, the low group velocity and the strong light localization in point defect cavities are effective for the enhancement of the light-matter interaction, as mentioned earlier. Since such an effect is also obtained in 1D PCs such as Fabry–Perot etalons, care should be taken about the real effectiveness of higher dimensional PCs. In such structures, the effect is enhanced by the low group velocity and by the large cavity Q, while the usable wavelength band is limited. This issue may be solved by using the chirped structure, as described in reference 13. In a simple point defect cavity, high transmission is only obtained at the center of the Lorentzian spectral function. Some of the coupled cavity structures are being studied to properly expand the transmission band. Other more practical advantage of 2D PCs are the integration of the light control device and input/output waveguides, high optical power density in the tiny core, and the design flexibility for low transmission loss. The tunability of a defect cavity, which is integrated with input/output waveguides, has been demonstrated by the photopumping [14]. Recently, studies are done on the thermal tuning and switching of the cavity using current injection. Besides, the bistability and the all-optical switching based on the enhancement of the nonlinearity were numerically calculated.

Harmonic Wave Generation, Wavelength Conversion

In PC fibers, the strong optical confinement in a small core strongly enhances the nonlinearity. Particularly, the combination with the wide band single mode propagation generates highly coherent supercontinuum light, in which case high efficiency of four-wave mixing is also required. For other PCs, high efficiency second harmonic generation (SHG) is expected due to the following two reasons: the multi-dimensional phase matching, and the enhancement of internal field by the low group velocity and strong localization of light. The PC for the phase matching has a large lattice constant of several micrometers to 10 μm, and is easily fabricated. The high efficiency SHG in some multiple directions has been observed in a $LiNbO_3$ device. On the other hand, the enhancement of the internal light is a similar effect to that in a cavity etalon. Therefore, the low group velocity band and the point defect cavity are used for this purpose, where the lattice constant is of the same order as the wavelength in the medium. It will be an important device, if the phase matching and field enhancement conditions are satisfied simultaneously with the large tolerance.

FUTURE OUTLOOK

The roadmap on future researches of PCs was published in 2003, (as shown in Reference 3) that has been edited from 2000. In the renewing process, the editors found that the advance of many applications overpaced the prediction. This chapter indicated many current issues of PC devices. However, they are expected to be solved in the near future.

REFERENCES

1. C. M. Bowden and J. P. Dowling, Eds., Development and applications of materials exhibiting photonic band gaps, *J. Opt. Soc. Am. B,* 10-2, 283–413, 1993.
2. J. D. Joannopoulos, R. D. Meade, and J. N. Winn, *Photonic Crystals*, Princeton University Press, Princeton, NJ, 1995.
3. S. Noda and T. Baba, Eds., *Roadmap on Photonic Crystals*, Kluwer Academic, New York, 2003.
4. T. Baba, D. Sano, K. Nozaki, K. Inoshita, Y. Kuroki, and F. Koyama, "Observation of Purcell effect in photonic crystal nanocavity at room temperature," in *Proceedings of the Conference on Laser and Electro-Optics*, no. CPDB3, May 2004.
5. K. Nozaki and T. Baba, Quasiperiodic photonic crystal microcavity lasers, *Appl. Phys. Lett.*, 84, 4875–4877, 2004.
6. A. Furukawa, S. Sasaki, M. Hoshi, A. Matsuzono, K. Moritoh, and T. Baba, High-power single-mode VCSELs with triangular holey structure, in *Proceedings of the Conference on Laser and Electro-Optics*, San Francisco, CA, no. CPDB1, May 2004.
7. H. Ichikawa and T. Baba, Efficiency enhancement in surface grating type two-dimensional photonic crystal light emitting diode, *Appl. Phys. Lett.*, 84, 457–459, 2004.
8. K. Tajima, J. Zhou, K. Kurokawa, and K. Nakajima, Low water peak photonic crystal fibers, Technical Digest European Conference Optical Communications, pp. 42–43, September 2003.
9. B. J. Managan, L. Farr, A. Langford, P. J. Roberts, D. P. Williams, F. Couny, M. Lawman, M. Mason, S. Coupland, R. Flea, H. Sabert, T. A. Birks, J. C. Knight, and P. St. J. Russel, Low loss (1.7 dB/km) hollow core photonic bandgap fiber, Technical Digest Conference Optical Fiber Communication, no. PDP24, February 2004.
10. T. Matsumoto and T. Baba, Photonic crystal *k*-vector superprism, *J. Lightwave Technol.*, 22, 917–922, 2004.
11. Y. Akahane, T. Asano, B.-S. Song, and S. Noda, High-Q photonic nanocavity in a two-dimensional photonic crystal, *Nature*, 425, 944–947, 2003.
12. S. Kawakami, "Optical parallel processing using photonic crystal array: New scheme of ellipsometry," Technical Digest Photonic and Electromagnetic Crystal Structures, no. Mo-C6, March 2004.
13. D. Mori and T. Baba, Dispersion-controlled optical group delay device by chirped photonic crystal waveguides, *Appl. Phys. Lett.*, 85, 2004.
14. T. Baba, M. Shiga, and K. Inoshita, Carrier plasma shift in GaInAsP photonic crystal point defect cavity, *Electron. Lett.*, 39, 1516–1518, 2003.

PLANAR LIGHTWAVE CIRCUIT (PLC)

Tohru Maruno

INTRODUCTION

The rapid and global spread of the Internet is accelerating the growth of optical communications networks. Photonic networks based on wavelength division multiplexing (WDM) systems and fiber-to-the-home (FTTH) systems [1,2] have played a key role in increasing the capacity and the flexibility of these communication networks. IP traffic is growing at a higher rate than expected, making further increases in backbone network capacity a critical issue. In response to this, the WDM channel number has been increased and the bit rate in time division multiplexing (TDM) systems has been raised from 2.5 to 10 Gbit/sec, and a rate of as high as 40 Gbit/sec will soon become practical. The capacity increase by WDM and TDM has made it technically and economically difficult to electrically process network operations. This includes the resumption of operations after failures, which in turn, has made the need to construct photonic networks that are free from electrical processing an urgent development issue. Processing optical signals in photonic networks requires optical add/drop multiplexing (OADM) and optical cross-connect (OXC) systems. Just as optical multi/demultiplexing filters are the key parts of WDM, optical switches are basic elements in OADM and OXC systems. New optical devices such as multi/demultiplexers and optical switches are, therefore, essential if we wish to construct large-capacity photonic networks.

This chapter reports on the recent progress of planar lightwave circuit (PLC) type devices with advanced functions developed for the photonic networks. We begin by describing silica-based PLC fabrication techniques and device characteristics of arrayed waveguide gratings (AWG) and thermo-optic PLC switches (TO PLC-SWs), followed by a detailed description of polymer-based PLCs, such as Y-branch PLC-SWs and TO tunable filters.

SILICA-BASED PLC

Silica-based PLCs have excellent physical and chemical stability and good compatibility with optical fibers. In addition, PLCs are manufactured by a combination of flame hydrolysis deposition (FHD) and reactive ion etching (RIE) [3]. FHD and RIE are used in optical fiber manufacturing and LSI technologies, respectively. This fact allows us to fabricate low-loss single-mode channel waveguides very precisely, and manufacture compact and economical devices with advanced functions.

Fabrication

Silica-based optical waveguides are fabricated on silicon substrates as shown in Figure 1. The first step is to use FHD to deposit two successive glass particle layers that serve as the under cladding and the core. After deposition, the substrate with these layers is heated to about 1300°C for consolidation. The waveguide core ridges are then formed by photolithography and RIE. Finally, FHD is again used to cover the core ridges with an over cladding. The typical waveguide characteristics for PLCs are listed in Table 1. The waveguide core is usually 6×6 μm in size, and the relative refractive index difference D between the core and cladding of the waveguide is around 0.7%. The advantages of this high-D (HD) waveguide are its low-loss and low-coupling loss with optical fiber. Specifically, we have achieved a propagation loss <0.01 dB/cm in a 10-m

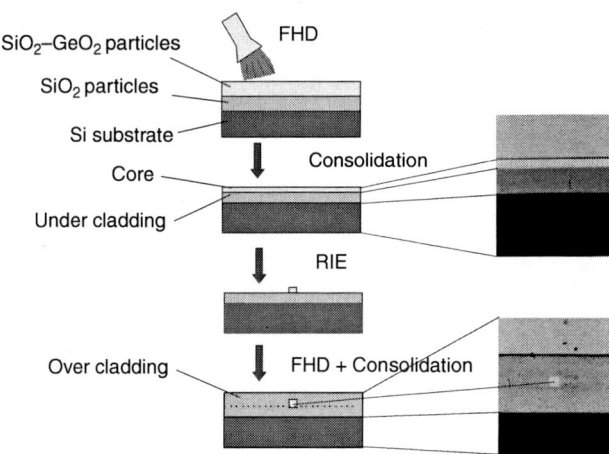

Figure 1   Fabrication process of silica-based PLCs

Table 1
Waveguide parameters using Silica-based PLCs

	HΔ	SHΔ
Relative refractive index difference (%)	0.75	1.5
Core size (μm)	6×6	4×4
Propagation loss (dB/cm)	<0.01	<0.05
Minimum bending radius (mm)	5	2
Optical fiber coupling loss (dB/point)	0.4	>1

long waveguide [4]. To realize higher degree of integration, we have been developing super-HD waveguides (SHD, $D > 1\%$) with a minimum bending radius of 2 mm [5]. These waveguides with a well-defined structure and a low loss enable us to design complicated circuits by using various numerical simulation techniques. Moreover, the reliability of the PLC devices has been confirmed with reference to the Telcordia reliability requirements for passive optical devices [6].

AWG Characteristics

The AWG multi/demultiplexer is one of the most successful optical filters, and is a key component of photonic networks. The configuration of an $N \times N$ AWG multiplexer is shown in Figure 2(a). The multiplexer consists of N input/output waveguides, two focusing slab waveguides, and arrayed waveguides with a constant path length difference DL between neighboring waveguides. The input light is launched into the first slab waveguide and then excites the arrayed waveguides. After traveling through the arrayed waveguides, the light beam interferes constructively at one focal point in the second slab. The location of the focal point depends on the signal wavelength l because the relative phase delay in each arrayed waveguide is given by DL/l. The slab and the array waveguides act as lens and grating, respectively, as shown in Figure 2(b).

We have fabricated various kinds of AWG multiplexers ranging from a 15 nm spacing 8-channel AWG to a 25 GHz spacing 256-channel AWG. The grating parameters and achieved performance of the fabricated AWGs are listed in Table 2. All the AWGs, except the 25 GHz

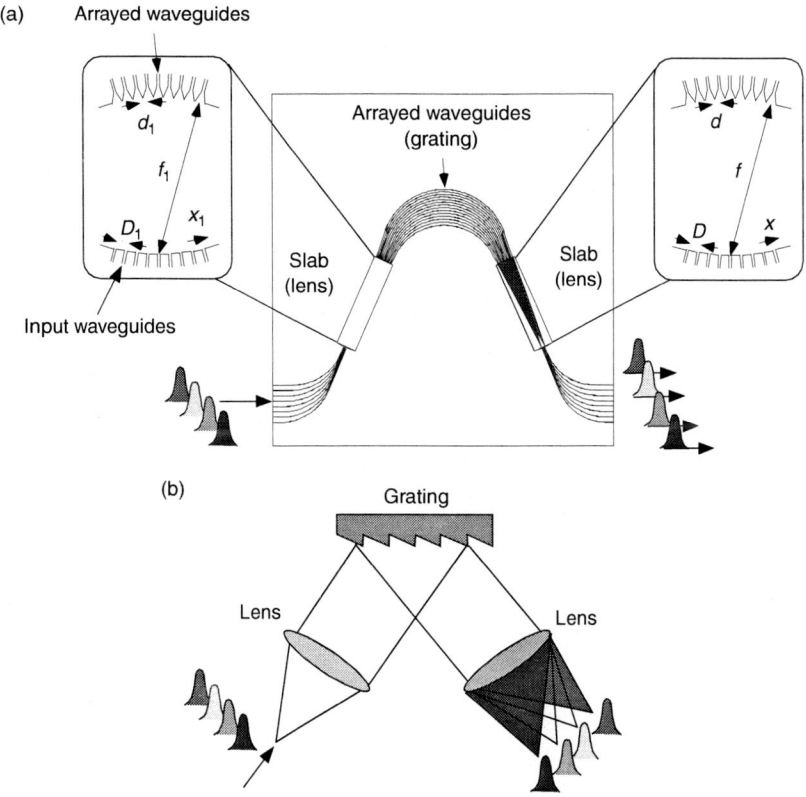

Figure 2 Configuration of $N \times N$ AWG multiplexer. (a) AWG configuration and (b) Micro-optics configuration with the same function as that of the AWG

Table 2
Grating parameters and achieved performance of AWGs

Parameters	Experimental Results					
	15 nm	**2 nm**	**0.8 nm**	**0.4 nm**	**0.2 nm**	**0.2 nm**
Channel spacing			**(100 GHz)**	**(50 GHz)**	**(25 GHz)**	**(25 GHz)**
Number of channels	8	8	32	64	128	256
Center wavelength			1.55 μm			
Path difference (μm)	12.8	50.3	63.0	63.0	63.0	27.7
Focal length (mm)	2.38	5.68	11.35	24.2	36.3	41.1
Diffraction order	12	47	59	59	59	26
On chip loss (dB)	2.4	6.1	2.1	2.8	3.5	2.7
3-dB Bandwidth	6.3 nm	0.74 nm	40 GHz	19 GHz	11 GHz	14.4 GHz
BPM simulation	6.3 nm	0.75 nm	37 GHz	21 GHz	9.5 GHz	12.5 GHz
Channel cross talk (dB)	<−28	<−29	<−28	<−27	<−16	<−33

spacing 256-channel AWG, were fabricated using HD waveguides. The experimentally obtained 3-dB bandwidths agree very well with the theoretical values calculated by using the beam propagation method (BPM). The 25 GHz spacing 256-channel AWG was fabricated using SHD waveguides [7]. Transmission spectra of the 25 GHz spacing 256-channel AWG are shown in Figure 3. The on-chip loss at the center port of the AWGs is as low as 2.7 dB.

The center wavelength of a conventional AWG shifts to a longer wavelength by about 0.1 nm/°C. This is because the refractive index of a silica-based waveguide depends on temperature. The AWG temperature must therefore be controlled with a heater or a Peltier device to stabilize the center wavelength. By contrast, an athermal AWG has also been shown to expand the application area and reduce the cost [8]. Figure 4 shows transmission properties of the athermal AWG. The temperature dependent optical path difference in silica-based waveguides is compensated for by employing a triangular groove filled with silicone adhesive, which has a negative thermal coefficient. Since the negative optical path change in the silicone is several tens of times larger than in silica-based waveguides, the negative optical path change in the silicone can compensate for the positive optical path changes in the silica thereby achieving temperature insensitive operation with small silicone gaps, which cause only a slight loss increment. Although the wavelength shift observed over a 0 to 85°C temperature range was 0.95 nm for a conventional AWG, it is only 0.05 nm for the athermal AWG. This result confirms that the athermal AWG is sufficiently stable for practical use.

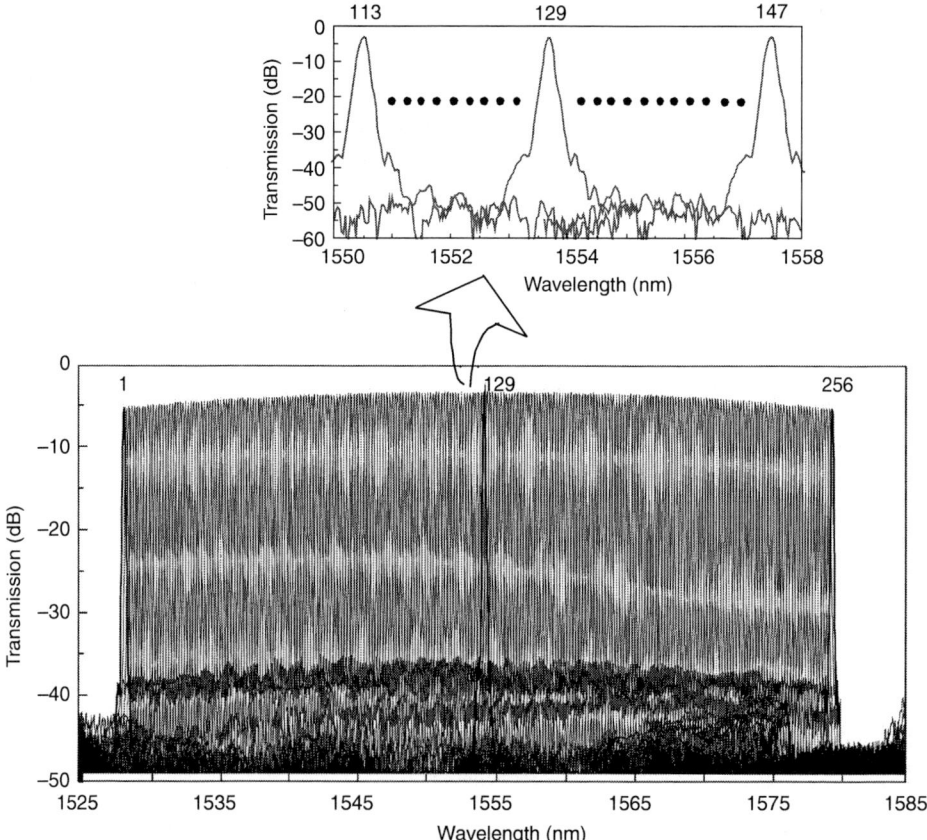

Figure 3   Transmission spectra of the 25 GHz spacing 256-channel AWG

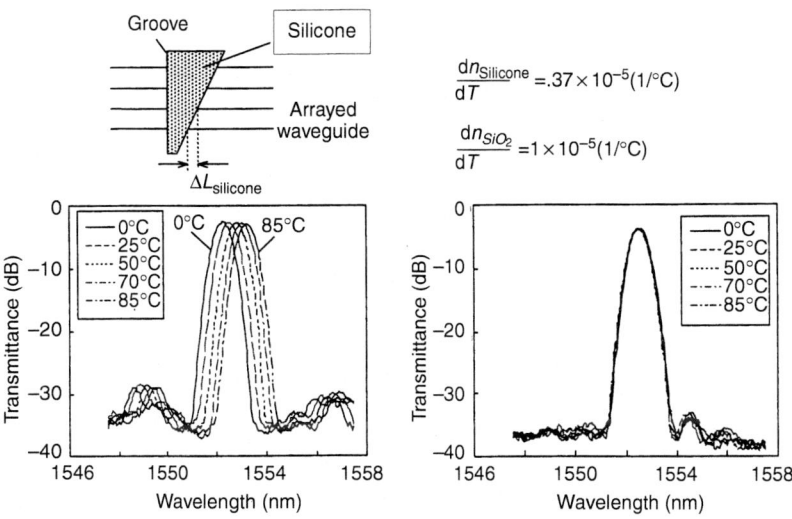

Figure 4   Temperature dependence of transmission spectra for (a) conventional and (b) athermal AWG

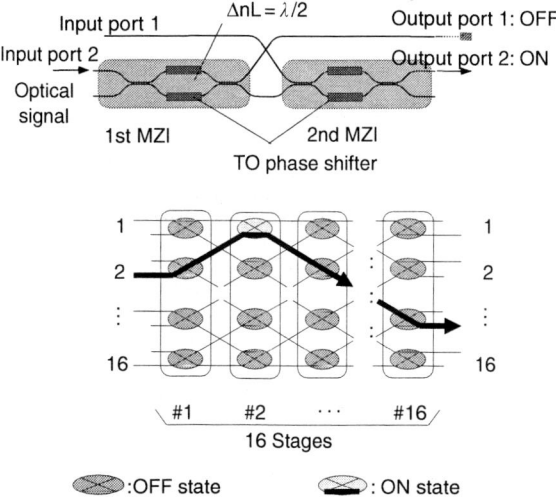

Figure 5   Switching unit with a double-MZI switch configuration and 16 × 16 matrix switch configuration

TO PLC-SW Characteristics

Silica-based TO PLC-SW unit can be made by a combination of a Mach–Zehnder interferometer (MZI) consisting of two 3-dB directional couplers and two waveguide arms equipped with a thin film heater, which operates as a TO phase shifter [9]. The unit is in the cross state when no heating power is supplied. It can be switched to the bar state by adjusting the heating power supplied to the TO phase shifter. The switching time is <5 msec. This performance is acceptable for the majority of WDM applications, such as OADM or cross-connect switches. Strictly nonblocking matrix switches with scales up to 16 × 16 [10] have been fabricated using silica-based PLCs. The configurations of the switching unit and the $N \times N$ matrix switch are shown in Figure 5.

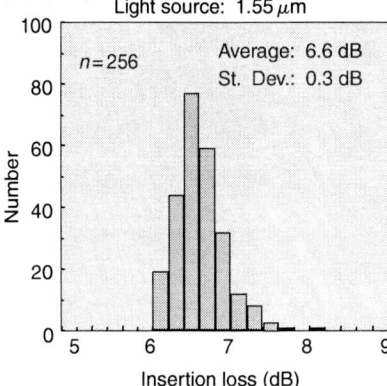

	16×16 TOSW
Insertion loss (dB)	6.6
Extinction ratio(dB)	53
Power (W)	17
Switching time (ms)	<5
Chip size (mm)	100×10^7

Figure 6   Characteristics of 16 × 16 TOSW

A double-gate switching unit, which consists of two asymmetrical MZIs with TO phase shifters and an intersection, is used to improve the extinction ratio. Theoretically, this unit can double the extinction ratio of a single MZI unit. The logical arrangement of an $N \times N$ matrix switch is a path-independent insertion loss arrangement. It is useful for reducing the insertion loss because the total waveguide length is shorter than that in a conventional crossbar matrix switch. The measured losses and extinction ratios of the 16 × 16 matrix switch for all 256 possible input and output connection patterns are shown in Figure 6. The insertion loss ranges from 6.0 to 8.0 dB with an average value of 6.6 dB. The extinction ratio ranges from 40 to 63 dB with an average value of 53 dB. The extinction ratio was considerably improved by adopting a double-gate switching unit. The electric power needed to operate this switching unit is about 1.1 W. The total power for the 16 × 16 matrix switch is 17 W.

Optical Splitter Characteristics

Silica-based optical splitters have been developed for PON systems in FTTH networks. Figure 7(a) shows schematic configuration of PLC-type 1 × 4 and 1 × 8 optical splitters and Figure 7(b) shows the loss spectra for the 1 × 4 splitters. We obtained low insertion losses of <7 dB in the 1.25–1.70 μm range. The insertion loss uniformities were as low as <0.3 dB. A flat wavelength response was realized in the 1.25–1.70 μm wide wavelength range. The reliability has been investigated based on Telcordia GR-1221. Change in the insertion loss was kept sufficiently low during damp heat storage at 75° and 90% RH for 2000 h, which is one of the most severe tests for PLC modules. These low and uniform loss characteristics confirm that the PLC-type splitter is very promising for realizing durable PON systems.

POLYMER-BASED PLC

Polymer waveguides have attracted much attention because they have a variety of optical functions and their fabrication process is simple and cost effective [11]. Moreover, the fact that polymers have large TO coefficient (10 times higher than SiO_2s) enables us to develop TO devices with low power consumption. Device characteristics for two types of polymer-based TO devices have also been described [12,13].

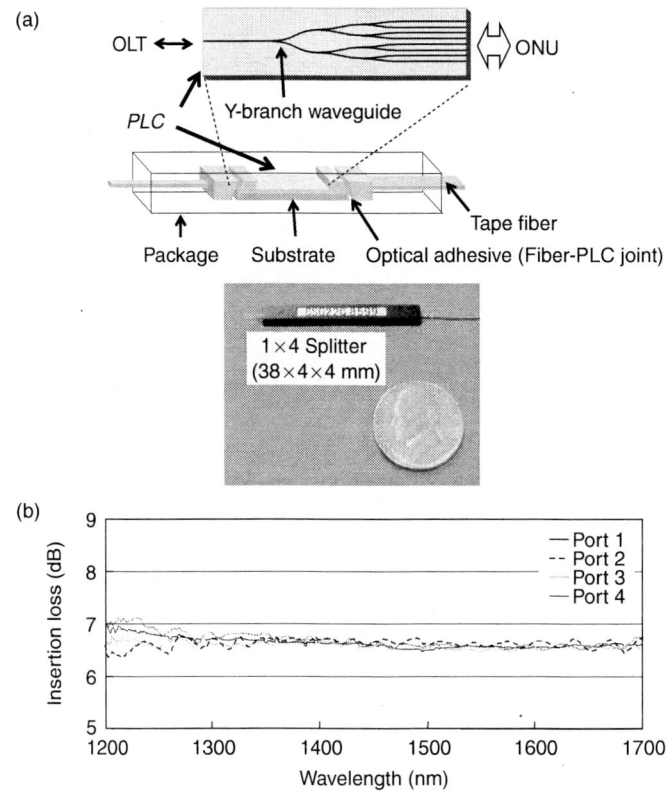

Figure 7   Configuration and transmission properties of (a) $1 \times 4 / 1 \times 8$ PLC optical splitters and (b) 1×4 optical coupler spectral loss

Fabrication

We fabricated the waveguide using a thermosetting silicone resin with low birefringence, a low propagation loss, and good environmental stability [14]. Silicone resin layers for the under-cladding and the core were formed on a Si substrate by spin coating and thermal curing at 250°C. A core ridge was then formed by photolithography and RIE techniques. An over-cladding layer was formed in the same way as the under-cladding layer. The thin film heaters were formed by the sputtering deposition of an Au alloy followed by a dry etching patterning process. Input and output fibers were fixed to the polished waveguide endfaces with a UV adhesive after the fibers had been actively aligned with the waveguide cores.

TOSW Characteristics

The configuration of the 1×8 TO switch chip is shown in Figure 8. We measured the transmission using a 1550 nm laser diode by changing the electric power supplied to three heaters located on a selected path. The three heaters were driven by the same electric power. Figure 9 shows the fiber-to-fiber transmission of the selected (ON) and other (OFF) ports as a function of total supplied power. At 390 to 450 mW, the insertion loss was <2.9 dB and the crosstalk was <41 dB. Since the waveguide propagation loss was 2.0 dB and the fiber coupling loss was about 0.1 dB, the excess loss for each element switch was estimated to be 0.3 dB. The excess loss became large when the power of the heater was >450 mW. This was caused by a leaked lightwave. We tested

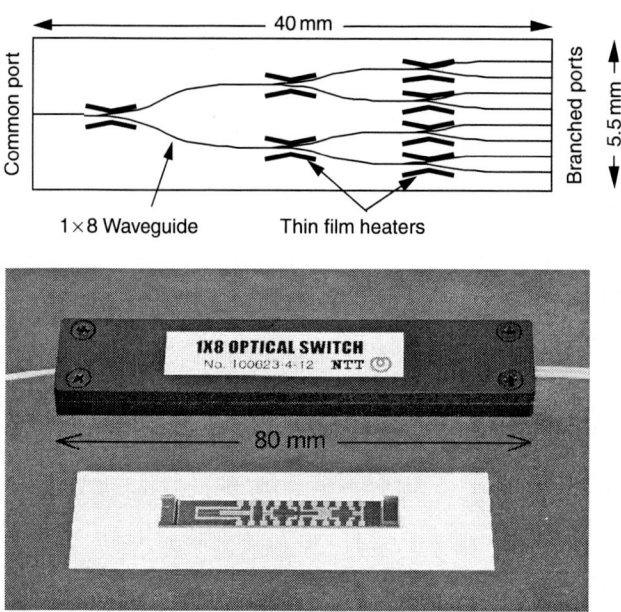

Figure 8   Schematic configuration of the 1 × 8 digital TO switch

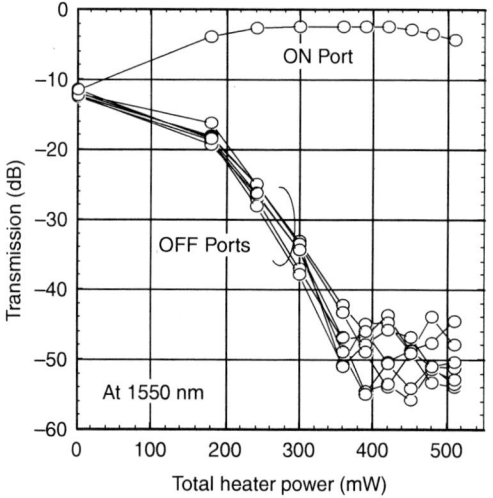

Figure 9   Fiber-to-fiber transmission of the 1 × 8 digital TO switch at 1550 nm as a function of total heater power

all the switched states of the 1 × 8 switch chip at a driving power of 450 mW. The results are summarized in Table 3.

Figure 10 shows the transmission spectra of the ON and OFF port with the worst crosstalk. The ON port transmission corresponds to the absorption spectrum of the silicone resin. In addition to this absorption, the OFF transmission increased at short wavelengths because higher mode propagation in the branching region degraded the extinction. We achieved a low insertion loss

Table 3
Transmission characteristics of the 1 × 8 digital TO switch at a 450 mW driving power

	ON Port (dB)	OFF Ports (dB)
Transmission	>3.0	<43
PDL	<0.1	about 3

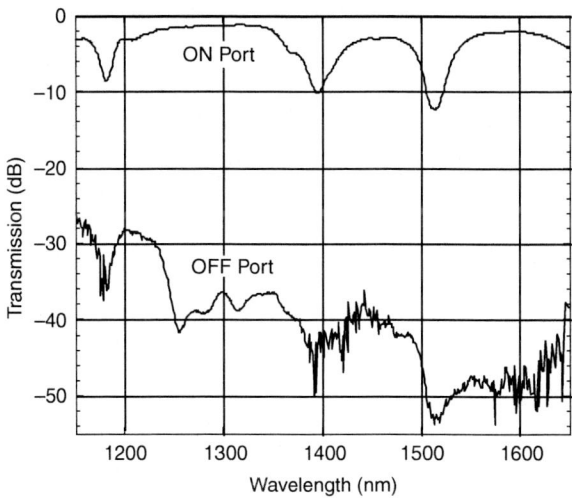

Figure 10   Transmission spectra of the 1 × 8 digital optical switch at a 390 mW driving power

of <4 dB in the 1540 to 1650 nm wavelength band, which includes the bands of both conventional and gain-shifted EDFAs.

We measured the temperature dependence of the switch characteristics in dry conditions and the humidity dependence at 25°C. The OFF port transmission increases at low temperature and the ON port transmission decreases slightly at high temperature. The TO coefficient of the silicone polymer has a positive temperature dependence [12] and so the switch operations at high and low temperature correspond to the operations with large and small heater powers, respectively. The reason for the excess loss at high temperature was the same as that for the excess loss at too large a heater power. The humidity dependence was small.

TO Wavelength Tunable Filter Characteristics

We have developed a polymer TO wavelength tunable filter with a pair of triangular phase shifters on arrayed waveguides (Figure 11), which provide a fast response and wideband tunability. Figure 12 shows the transmission spectra as a function of supplied power to the heater. This filter showed a crosstalk of <–30 dB, a tunability wider than 20 nm, and a response time of 2 to 60 msec. We have already demonstrated its stable operation by testing it in the feedback system of a filter with a 9 nm tuning range and its applicability to LAN systems for selecting any WDM channel from among WDM channels.

CONCLUSION

This chapter has reported the recent progress on PLC devices for WDM-based photonic networks, which are expanding rapidly and globally. Each device introduced in this chapter has its own

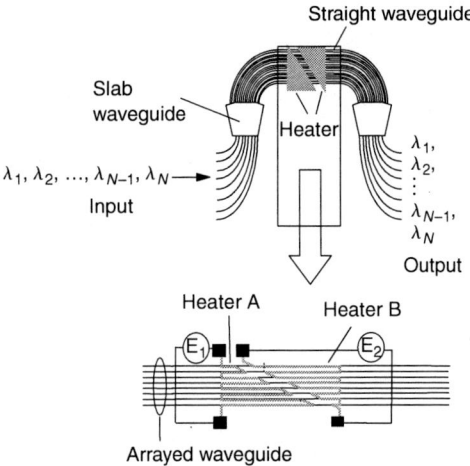

Figure 11   Configuration of TO wavelength tunable filter

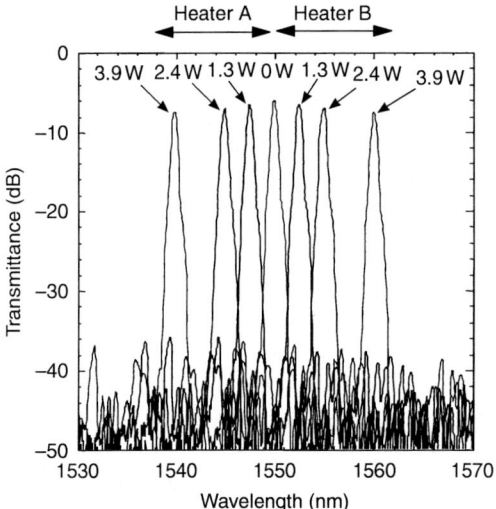

Figure 12   Tunability of TO wavelength tunable filter

advantages to construct flexible and large-capacity networks, and has been or will be installed in a suitable application in WDM systems. The next step is to meet the demands for lower cost, larger scale, and expanded functions. Further advances in optical components and related technologies will greatly contribute to the construction of photonic networks.

REFERENCES

1. H. Toba, K. Oda, K. Nakanishi, N. Shibata, K. Nosu, N. Takato, and M. Fukuda, "A 100-ch optical WDM transmission/distribution at 622 Mbits/s over 50 km," *J. Lightwave Technol.*, 8, 1396–1401 (1990).
2. I.P. Kaminow, "A wideband all-optical WDM network," *IEEE J. Sel. Areas Commn.*, 14, 780–799 (1996).

3. M. Kawachi, "Silica waveguides on silicon and their application to integrated components," *Opt. Quantum Electron.*, 22, 391–416 (1990).

4. Y. Hida, Y. Hibino, H. Okazaki, and Y. Ohmori, "10 m long silica-based waveguide with a loss of 1.7 dB/m," in *Proceedings of the IPR*, Dana Point, paper IthC6 (1995).

5. S. Suzuki, M. Yanagisawa, Y. Hibino, and K. Oda, "High-density integrated planar lightwave circuits using SiO$_2$–GeO$_2$ waveguides with a high refractive index difference," *J. Lightwave Technol.*, 12, 790–796 (1994).

6. Y. Hibino, F. Hanawa, H. Nakagome, M. Ishii, and N. Takato, "High reliability optical splitters composed of silica-based planar lightwave circuits," *J. Lightwave Technol.*, 13, 1728–1735 (1995).

7. Y. Hibino, Y. Hida, A. Kaneko, M. Ishii, M. Itoh, T. Goh, A. Sugita, T. Saida, A. Himeno, and Y. Ohmori, "Fabrication of silica-on-Si waveguide with higher index difference and its application to 256 channel arrayed-waveguide multi/demultiplexer," Technical digest of OFC2000, paper WH2 (2000).

8. Y. Inoue, A. Kaneko, F. Hanawa, H. Takahashi, K. Hattori, and S. Sumida, "Athermal silica-based arrayed-waveguide grating multiplexer," *Electron. Lett.*, 33, 1945–1946 (1997).

9. M. Okuno, N. Takato, T. Kitoh, and A. Sugita, " Silica-based thermo-optic switches," *NTT Rev.*, 7, 57–63 (1995).

10. T. Goh, M. Yasu, K. Hattori, A. Himeno, M. Okuno, and Y. Ohmori, "Low-loss and high-extinction ratio strictly nonblocking 16×16 thermooptic matrix switch on 6-inch wafer using silica-based planar lightwave circuit technology," *J. Lightwave Technol.*, 19, 371–379 (1999).

11. T. Watanabe, N. Ooba, Y. Hida, and M. Hikita, "Influence of humidity on refractive index of polymers for optical waveguide and its temperature dependence," *Appl. Phys. Lett.*, 72, 1533–1535 (1998).

12. N. Ooba, S. Toyoda, and T. Kurihara, "Low crosstalk and low loss polymeric 1×8 digital optical switch," *Jpn. J. Appl. Phys.*, 39, 2369–2371 (2000).

13. S. Toyoda, N. Ooba, A. Kaneko, M. Hikita, T. Kurihara, and T. Maruno, "Wideband polymer thermo-optic wavelength tunable filter with fast response for WDM systems," *IEEE Electron. Lett.*, 36, 658–660 (2000).

14. T. Watanabe, N. Ooba, S. Hayashida, T. Kurihara, and S. Imamura, "Polymeric optical waveguide circuits formed using silicone resin," *J. Lightwave Technol.*, 16, 1049–1055 (1998).

POLARIZATION

Kenichi Iga

It is characterized by the direction of an electric field of light. In a planar waveguide, it is defined as TE (Transverse Electric) and TM (Transverse Magnetic) as shown in Figure 1 [1,2]. Sometimes the polarization is expressed as s-wave and p-wave when it is incident to a plane surface as shown in Figure 2. When electrified, the direction and the amplitude changes from time to time and is called elliptically polarized light having angular momentum.

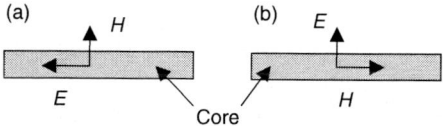

Figure 1   Polarizations of planar waveguide: (a) TE and (b) TM

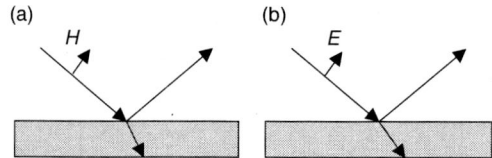

Figure 2   Polarizations of incident plane wave: (a) s-wave and (b) p-wave

REFERENCES

1. Y. Suematsu and K. Iga. *Introduction to Optical Fiber Communication.* Ohm-sha, Tokyo, 2004.
2. Y. Suematsu and and A. R. Adams, *Handbook of Semiconductor Lasers and Photonic Integrated Circuits.* Ohm-sha, Chapman & Hall, New York, 1994, Chap. 3.

POLARIZATION CONTROL

Antao Chen and Fred Heismann

INTRODUCTION

Fixed or adjustable polarization transformers have been known for many decades and are typically used to convert a well-defined input state of polarization (SOP) of a quasi-monochromatic light beam into another well-defined output SOP. Optical polarization controllers, on the other hand, are relatively new and were developed for a substantially more difficult task, namely to transform an *unknown* and even *time-varying* input SOP continuously and automatically into a prescribed output SOP. Such adaptive polarization transformation is often referred to as automatic polarization stabilization or automatic polarization control [1,2]. As we shall see later, automatic polarization controllers are significantly more complex than conventional polarization transformers.

The need for automatic polarization controllers arose with the introduction of single-mode optical fibers in telecommunication systems and in optical sensors. In general, single-mode fibers do not preserve the SOP of the transmitted light, except for some specially designed polarization-maintaining fibers. Hence, the polarization state of the output light of a fiber-optic cable is usually unknown and may even fluctuate with time [3]. These polarization fluctuations do not cause any problems as long as all optical elements connected to the fiber output operate independent of their input SOP. A polarization-dependent optical component, however, would convert these polarization fluctuations into power fluctuations and, hence, would cause temporary fading of the optical signal. Such signal fading can be avoided by placing an adaptive polarization controller in the input of the polarization-dependent component, as shown in Figure 1 [4].

APPLICATIONS AND REQUIREMENTS

It is essential in many applications that the polarization controller generates the desired polarization transformation continuously and without any interruptions or significant degradations of the

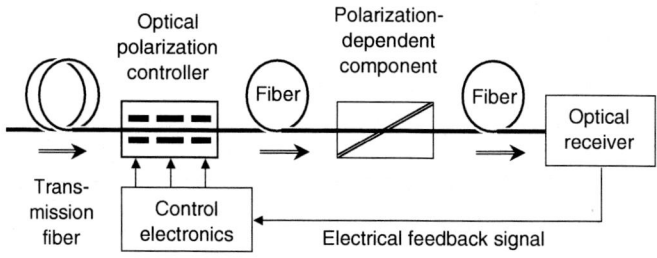

Figure 1   Typical application of a polarization controller with automatic feedback loop

transmitted optical signal. It is often further required that the controller automatically searches for the desired transformation using just a single electrical feedback signal, which typically is proportional to the optical amplitude or power of the output light in the desired SOP. This feedback signal is generated by an optical receiver downstream the optical path, as shown in Figure 1, and the polarization controller is the part of an automatic feedback loop that generates its error signals by dithering the various adjustable parameters of the controller slightly about their current values. Then an electronic circuit monitors the effects of these dither variations on the feedback signal and readjusts the control parameters for maximal signal level.

Automatic polarization controllers must, therefore, be capable of transforming any general input SOP into the prescribed output SOP. However, it is usually advantageous if the controller is able to transform any general input SOP into any general output SOP, in particular, when its output ports are connected to an optical fiber, which may randomly alter the output SOP. More importantly, the controller must have an effectively unlimited transformation range as well as sub-millisecond control speed to allow continuous compensation of the random and potentially large polarization fluctuations in the transmission fiber, which is known as "endless" polarization control [5,6].

It turns out that electro-optic polarization controllers are readily capable of sub-millisecond control speeds, especially when they are implemented with compact single-mode waveguides [7]. Moreover, it has been shown that certain combination cascades of three or more controllers exhibit inherently unlimited transformation ranges [8]. Therefore, integrated-optic polarization controllers in lithium niobate have become the preferred choice for many applications where fast and endless polarization control is required [9]. Thus far, however, electro-optic polarization controllers are not as widely used as conventional amplitude or phase modulators, but they have found a few important applications, especially in high-speed fiber-optic transmission systems, which we shall briefly review in the following paragraphs.

Coherent receivers: The first major application for automatic polarization controllers came about when coherent homodyne or heterodyne detection schemes were considered for fiber-optic communication systems [10]. In these coherent receivers, the optical signal emerging from the fiber link is mixed with the output light of a local reference laser to produce an electrical homo- or heterodyne beat signal in a quadratic photo-detector. However, the amplitude of this beat signal strongly depends on how well the SOP of the transmitted optical signal is matched to that of the reference laser signal when the two beams are mixed. In fact, the beat signal may completely disappear if the transmitted optical signal and the local reference laser are in orthogonal polarization states. Such undesired signal fading can be avoided if an adaptive polarization controller is employed to automatically match the polarization states of the two signals [11]. This controller may be employed either in the transmitted signal beam to compensate for the polarization fluctuations in the transmission fiber or in the local oscillator path to track the polarization changes that occur in the signal beam [12]. In either case, the feedback signal for the polarization controller is usually proportional to the amplitude of the received electrical beat signal [13].

Polarization demultiplexing: Although the interest in coherent optical detection has faded away with the advent of erbium-doped fiber-optic amplifiers (EDFA), polarization controllers have, nevertheless, found other important applications in fiber-optic transmission systems. In polarization-multiplexed transmission systems, for example, adaptive controllers are needed to demultiplex the orthogonally polarized signal components at the receiving end of the transmission link, for example, by transforming the fluctuating polarization states of the two components into the eigenstates of a polarization splitter [14]. Likewise, polarization controllers may be employed in the optical transmitter of these systems to align the two independently modulated signals in orthogonal polarization states.

Compensation of polarization mode dispersion (PMD): Recently, the interest in polarization controllers has increased significantly with the emerging need for mitigating PMD in high-speed fiber-optic transmission links [15]. First-order PMD causes the same effects as a simple birefringent waveguide, that is, it introduces a differential time delay between two orthogonally polarized signal components, which are known as the principal states of polarization (PSP). This PMD-induced time delay may cause substantial broadening of the transmitted optical pulses, especially at data rates above 10 Gb/sec [16]. However, unlike ordinary birefringence, the differential time delay of PMD is not constant and may vary randomly with time and optical frequency. Likewise, the two PSPs, between which the time delay occurs, may also fluctuate with time and optical frequency. In recent years, various optical and electronic schemes have been devised to mitigate or compensate the potentially deleterious effects of first-order PMD [15,17].

Optical compensators for first-order PMD typically employ a birefringent element with fixed or adjustable differential time delay, which is connected to the output of the transmission fiber via an automatic polarization controller. This controller transforms the varying PSPs of the transmission fiber continuously into the fixed polarization eigenstates of the birefringent element, in such a way that the PSP with the shortest time delay in the fiber is coupled into the eigenstate of the birefringent element with the longest time delay. As a result, the PMD-induced differential time delay in the transmission fiber is compensated for by the fixed or adjustable differential time delay in the birefringent element [18].

The effects of higher-order PMD can be mitigated with more complex compensators, which typically employ several polarization controllers, birefringent elements, and sometimes even a polarizer [19]. It is possible to integrate most of these elements in a single integrated-optic device by implementing the polarization controller with a highly birefringent waveguide [20]. However, these controllers operate only over a relatively narrow band of optical frequencies and are, therefore, not well suited for general applications [21].

Polarization scrambling: Another important class of applications for optical polarization controllers involve rapid randomization of a fixed or slowly varying SOP, which is known as polarization scrambling. This random polarization modulation has the effect of "depolarizing" the optical signal on the timescale of the polarization changes, which often helps to mitigate polarization-dependent effects in the fiber-optic transmission line. One example of such an effect is the so-called polarization hole burning in EDFA, which impairs the optical signal-to-noise ratio for polarized optical signals. Although this effect is small in a single EDFA, it can rapidly grow in long strings of concatenated amplifiers and, hence, cause severe impairments in transoceanic fiber-optic cables [22]. It turns out that rapid scrambling of the polarization state of the launched optical signal with just a single polarization controller efficiently suppresses polarization hole burning throughout the entire chain of amplifiers [23,24].

Moreover, polarization controllers are often used in laboratory transmission experiments, where a sequence of optical pulses is repeatedly transmitted through a relatively short loop of fiber-optic cable to investigate the effects of long-haul transmission systems. In these applications, the controller randomly changes the SOP of the optical signal after each roundtrip through the fiber loop to simulate the random changes of the SOP in a real fiber cable [25].

POLARIZATION CONTROL SCHEMES

Optical polarization controllers have been realized in various bulk-optic materials as well as in optical fibers and integrated-optic waveguides. Table 1 summarizes the most popular control schemes and their main features. It is clear that most mechanically adjustable polarization control-lers, such as rotatable half-wave plates (HWP) and quarter-wave plates (QWP) as well as rotat-able fiber-optic coils, are far too slow for applications in deployed fiber-optic systems [11,26]. However, these devices are often very useful in laboratory experiments as well as in certain test equipment, where polarization changes are expected to be slow. Liquid-crystal-based polariza-tion controllers are significantly faster and offer the additional advantage of being relatively inexpensive; however, their response time is inherently limited to above a few milliseconds, which often prohibits applications in long-haul fiber-optic links [27,28]. Fiber-optic polarization controllers based on piezo-electric fiber squeezers or stretchers, on the other hand, exhibit suffi-ciently fast response times and very low insertion losses, but their long-term reliability is still an open question [5,6,29].

Therefore, it is not surprising that electro-optic and magneto-optic polarization controllers have gained the greatest attention, since they offer very fast response times and do not employ any mechanically moving parts [2,30]. However, electro-optic polarization controllers imple-mented with bulk-optic crystals usually suffer from high drive voltages, which severely limit their control speeds [31,32]. Integrated-optic polarization controllers in lithium niobate or III–V semi-conductors, on the other hand, offer significantly lower drive voltages (by about a factor of 10) while also allowing low-loss coupling to optical fibers [4,33]. Lithium niobate ($LiNbO_3$) polar-ization controllers are usually preferred, because they offer lower insertion losses and because they employ the same mature technology as used in conventional phase and amplitude modula-tors, which are widely deployed in fiber-optic telecommunication systems [34]. In fact, the design of $LiNbO_3$ polarization controllers is very similar to that of conventional phase modulators, except that they employ a different crystal orientation and usually do not require high-speed traveling wave electrodes [35]. Polarization controllers, however, comprise several concatenated electrode sections, as shown in Figure 2, and thus require multiple sets of drive voltages.

Table 1
Schemes for optical polarization control and their main attributes

Principle	Implementation	Control Mechanism	Insertion Loss	Speed
Cascaded rotatable wave plates	Bulk-optic crystals	Mechanic	Medium	Very slow
	Fiber-optic loops	Mechanic	Low	Slow
	Bulk-optic crystals	Electro-optic	Medium	Fast
	Liquid crystals	Electro-optic	Medium	Medium
	I/O waveguides	Electro-optic	Medium	Very fast
Cascaded phase retarders	Fiber squeezers	Mechanic	Low	Fast
	Fiber stretchers	Mechanic	Low	Fast
	Liquid crystals	Electro-optic	Medium	Medium
	I/O waveguides	Electro-optic	Medium	Very fast
Faraday rotators	Fiber coils	Magneto-optic	Low	Fast

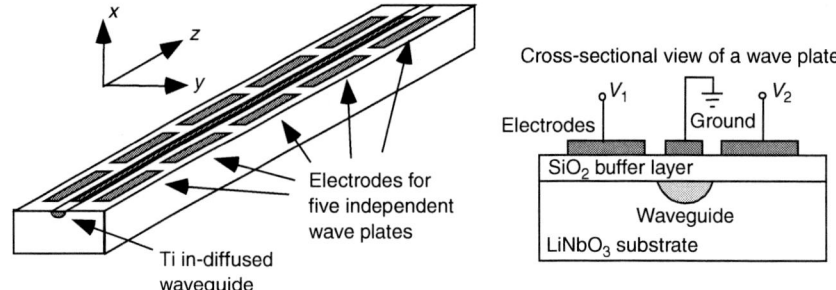

Figure 2 Schematic diagram of a LiNbO$_3$ waveguide polarization controller with five independent electrode sections

The LiNbO$_3$ polarization controller depicted in Figure 2 is designed to emulate the polarization transformations of a combination cascade of five independently rotatable zero-order QWP. It can, therefore, be operated over a wide optical bandwidth [7,8]. Moreover, each electrode section of the controller induces linear phase retardation that rotates endlessly without requiring a reset in the drive voltages. As a result, this controller exhibits an inherently unlimited transformation range [36]. Polarization controllers employing variable optical phase retarders with fixed angular orientations, such as fiber-optic squeezers, are also capable of providing unlimited transformation ranges but usually require occasional resets of the control parameters of the individual elements [5,6]. However, it has been recently shown that with a properly designed control algorithm one can avoid such reset cycles [37].

The amount of differential phase retardation that needs to be generated by the various electrode sections depends on the total number of sections and may be chosen from a relatively wide range [6]. It is well known, however, that a three-section controller emulating an endless rotatable QWP followed by a HWP and another QWP allows reset-free polarization transformations from an arbitrarily varying input SOP into an arbitrarily varying output SOP [6]. The overall polarization transformation of this controller is described by the Jones matrix,

$$
\mathbf{T} = \begin{bmatrix} A - jB & C - jD \\ -C - jD & A + jB \end{bmatrix},
\tag{1}
$$

where

$$
\begin{aligned}
A &= \cos[2\tilde{\gamma}(t)] \cdot \cos[\alpha(t) - \beta(t)], \\
B &= -\sin[2\tilde{\gamma}(t)] \cdot \sin[\alpha(t) - \beta(t)], \\
C &= \cos[2\tilde{\gamma}(t)] \cdot \sin[\alpha(t) - \beta(t)], \\
D &= \sin[2\tilde{\gamma}(t)] \cdot \cos[\alpha(t) - \beta(t)],
\end{aligned}
\tag{2}
$$

and $\tilde{\gamma}(t) = \gamma(t) + [\alpha(t) - \beta(t)]/2$, with t denoting time, $\alpha(t)$ the angular orientation of the first QWP, $\gamma(t)$ that of the HWQ, and $\beta(t)$ the orientation of the second QWP [36]. In theory, only two independent variables are needed to achieve endless polarization control; however, in practice, it is preferred to adjust all three parameters independent of one another, which results in a more robust operation with lower sensitivity to wavelength changes or voltage drifts [34]. In fact, it is even advantageous to add more independent (i.e., redundant) elements to the polarization controller, as shown in the example of Figure 2, which employs a total of five independently driven electrode sections.

IMPLEMENTATION OF LiNbO$_3$ POLARIZATION CONTROLLERS

The LiNbO$_3$ polarization controller in Figure 2 employs a standard titanium in-diffused waveguide that runs nearly parallel to the z-axis of the crystal. This particular orientation has the advantage that the transmitted light does not experience the large material birefringence of LiNbO$_3$, thus, allowing optically broadband polarization transformations via simple uniform electrodes [7]. Each section of the controller employs three separate electrodes that run parallel with the waveguide, one of which is centered on top of the waveguide and the other two are placed symmetrically on both sides of the waveguide. A dielectric buffer layer separates the electrodes from the waveguide to avoid polarization-dependent losses [35].

The two outer electrodes are driven by voltages $V_1(t)$ and $V_2(t)$ relative to the center electrode, such that the difference between $V_1(t)$ and $V_2(t)$ induces a horizontal electrical field E_y in the waveguide, whereas their sum induces a vertical electrical field E_x in the waveguide [35]. Moreover, applying sinusoidal voltages of the form

$$V_1(t) = V_s \cdot \cos[2\theta(t)] + V_c \cdot \sin[2\theta(t)] + V_t + V_0,$$
$$V_2(t) = -V_s \cdot \cos[2\theta(t)] + V_c \cdot \sin[2\theta(t)] - V_t + V_0, \tag{3}$$

with constant amplitudes, V_s and V_c, and fixed offsets, V_t and V_0, induces an endless rotatable electric field in the waveguide, whose angular orientation is determined by the variable electrical phase $2\theta(t)$. The y-component of this field then generates a differential phase shift between the TE- and TM-polarized optical modes of the waveguide via the r_{12} and r_{22} electro-optic coefficients, whereas its x-component generates coupling between these two modes via the r_{61} electro-optic coefficient, where $r_{12} = r_{61} = -r_{22} \approx 3.4 \times 10^{-12}$ m/V [7]. The total differential phase retardation, $\psi(t)$, generated in the waveguide is given by

$$\psi(t) = \frac{2\pi}{\lambda_0} L \sqrt{[(n_{TE} - n_{TM}) + C_y r_{22} n_0^3 E_y(t)]^2 + [C_x r_{22} n_0^3 E_x(t)]^2} \tag{4}$$

and its angular orientation by

$$\phi(t) = \frac{1}{2} \tan^{-1} \left[-\frac{C_x r_{22} n_0^3 E_x(t)}{(n_{TE} - n_{TM}) + C_y r_{22} n_0^3 E_y(t)} \right], \tag{5}$$

where λ_0 denotes the wavelength in free space, L the length of the particular electrode section, $n_{TE} - n_{TM}$ the modal birefringence in the waveguide, n_0 the average index of the TE and TM polarized modes, and C_x and C_y are dimensionless constants between 0 and 1 [38].

The differential bias voltage V_t compensates for any residual modal birefringence in the waveguide by generating a constant index change of $C_y r_{22} n_0^3 E_{y,t} = -(n_{TE} - n_{TM})$, whereas the common bias voltage V_0 compensates for the effects of lateral misalignments between the centerline of the electrodes and that of the optical waveguide [35,39]. The voltage amplitudes V_s and V_c can then be adjusted in such a way that $\psi(t)$ is always constant and equal to the desired retardation, for example, $\psi(t) = \pi/2$ for a QWP and $\psi(t) = \pi$ for a HWP [8].

The voltage amplitudes V_s and V_c as well as the bias voltages V_t and V_0 in any given device may differ from section to section but are otherwise constant for any given wavelength and temperature. Thus, the only adjustable parameter in the two control voltages is the electrical phase $2\theta(t)$, which determines the angular orientation of the induced phase retardation, that is, $\phi(t) \equiv \theta(t)$. Since $\theta(t)$ is continuously adjustable over an infinite range without requiring a voltage reset, it follows that the phase retardation ψ can be rotated endlessly, similar to an endless rotatable fractional-wave plate [36]. The drive voltages $V_1(t)$ and $V_2(t)$, however, are always bound to the range $\pm\left[|V_t|+|V_o|+\sqrt{V_c^2+V_s^2}\right]$.

The required voltage amplitudes, V_s and V_c, and the bias voltages, V_t and V_0, are determined experimentally for each device and individual electrode section by means of a complicated procedure, in which the variations in the output SOP of the controller are monitored with a polarization analyzer while the various components of the drive voltages are swept individually for various input SOPs. This procedure was designed for devices that are connected to short lengths of optical fibers, in which case the input and the output SOPs of the controller cannot be directly monitored [38].

If an electrode section is not correctly biased at the prescribed voltage offsets V_t and V_0, then the induced linear phase retardation and its angular orientation will deviate from their desired values and, hence, the polarization controller may fail to operate as desired. Unlike in the case of LiNbO$_3$ amplitude modulators, where the bias point can be monitored and adjusted continuously during normal operation, no practical procedure has been devised thus far to monitor and readjust the bias voltages in the various sections of a polarization controller without taking the device out of service. In practice, V_s, V_c, V_t, and V_0 are therefore permanently set in the driver electronics of the controller, with possible adjustments for known wavelength or temperature changes. The requirements for the long-term stability of the bias voltages in polarization controllers are therefore much more stringent than those for amplitude modulators. Over the entire lifetime of the device, the drift in V_t and V_0 should be less than the voltage required to induce a phase retardation of $\pi/8$, whereas the bias voltage of an amplitude modulator is usually allowed to drift over a range that corresponds to total phase shift of 2π [40].

A major challenge for the design of LiNbO$_3$ polarization controllers, therefore, is the elimination of a significant long-term drift in the bias voltages, V_t and V_0. An efficient way of reducing this drift is to minimize the bias voltages themselves, since the drift is generally proportional to the magnitude of the applied voltage [41]. While V_0 can be kept small through tight control of the fabrication process, V_t depends on the modal birefringence in the waveguide and can be of the order of 50 V or more [35]. It turns out, however, that this modal birefringence can be compensated for by some of the material birefringence of LiNbO$_3$, which can be accomplished simply by introducing a small angle of about 2° between the orientation of the waveguide and the z-axis of the crystal [39]. The resulting reduction in V_t is demonstrated in Figure 3, which displays the bias voltage as a function of the offset angle for four slightly different waveguide designs. With proper design and tight fabrication control it is therefore possible to keep V_t and V_0 consistently below 10 V.

DEVICE PERFORMANCE AND RELIABILITY

The voltage amplitudes V_s and V_c depend on the length of the electrodes and typically are <20 V for a QWP section. As shown in Figure 4, they are not very sensitive to temperature variations and vary slightly with wavelength. The bias voltages V_t and V_0, on the other hand, exhibit somewhat

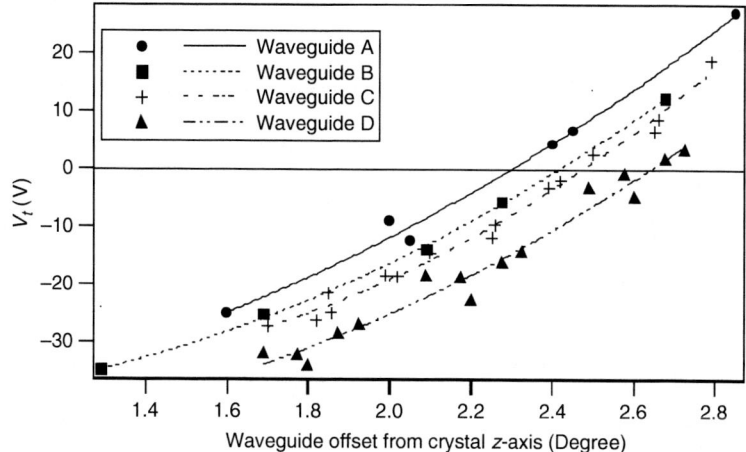

Figure 3 Bias voltage V_t versus waveguide orientation for various waveguide designs

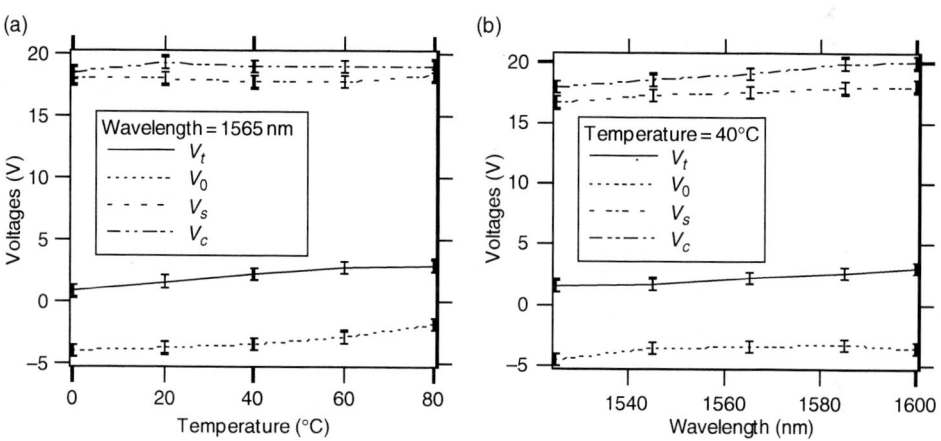

Figure 4 Variations in the required voltage offsets and amplitudes of a LiNbO$_3$ polarization controller with (a) temperature and (b) wavelength

larger variations with temperature, which are attributed to stress resulting from differential thermal expansion of the LiNbO$_3$ substrate and the buffer and the electrode layers. However, these temperature variations are very consistent and may therefore be automatically compensated for in the electronic driver circuit. For this purpose, a temperature sensor may be placed in the device package. The wavelength variations of V_t and V_0 are also small and may be compensated for in the driver circuit.

The maximum speed of the polarization controller is limited by the electrical bandwidth of the individual electrode sections. Figure 5 displays measurements of the small-signal frequency response of a LiNbO$_3$ polarization controller with conventional lumped electrode design. This particular device exhibits a 1 dB bandwidth of more than 100 kHz and a 3 dB bandwidth of about 1 MHz. It has been shown that such a controller is capable of tracking polarization changes at speeds of more than 4000 rad/sec, which is at least ten times faster than natural polarization fluctuations in optical fibers, including rapid manual twisting and handling of the fiber [42].

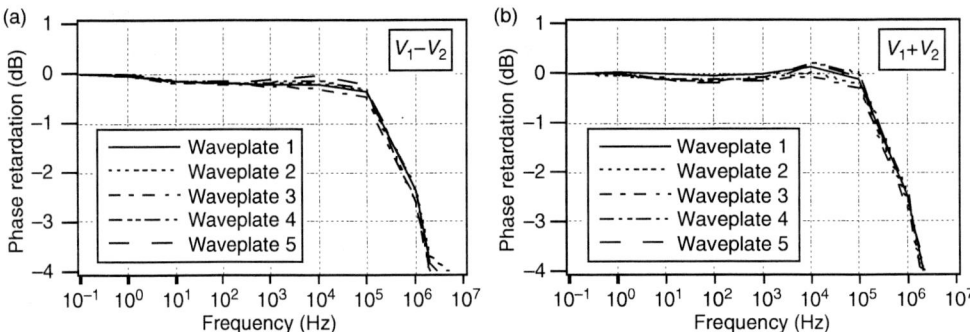

Figure 5 Measured small-signal frequency response of a LiNbO$_3$ polarization controller, showing the phase retardations induced by voltages (a) $V_1(t) - V_2(t)$ and (b) $V_1(t) + V_2(t)$

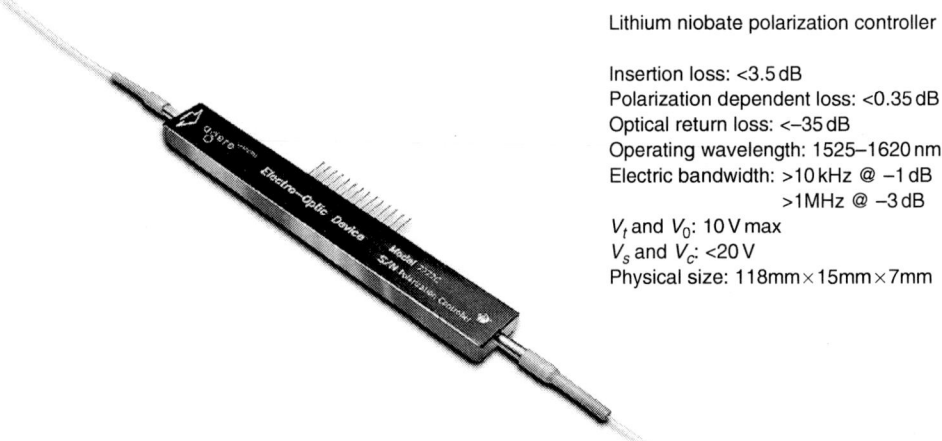

Lithium niobate polarization controller

Insertion loss: <3.5 dB
Polarization dependent loss: <0.35 dB
Optical return loss: <−35 dB
Operating wavelength: 1525–1620 nm
Electric bandwidth: >10 kHz @ −1 dB
                                >1 MHz @ −3 dB
V_t and V_0: 10 V max
V_s and V_c: <20 V
Physical size: 118 mm × 15 mm × 7 mm

Figure 6 Hermetically packaged lithium niobate polarization controller and its performance specifications

Figure 6 shows a photograph of a fully packaged commercial polarization controller along with a list of its main performance specifications. The input and the output ports of this device are permanently connected to standard single-mode fibers and the entire package is hermetically sealed for environmental stability. The controller meets the Telcordia reliability standard GR-486-CORE and exhibits remarkable stability in accelerated aging tests. As shown in Figure 7, the drift in V_t and V_0 is <5 V even when the device is operated for >1000 h at a temperature of 125°C, corresponding to a change in induced phase retardation of <$\pi/16$. The lifetime of the controller is therefore estimated to be at least 33 years when operated at a temperature of 60°C, or 99 years when operated at 50°C, thus exceeding the requirements for commercial telecommunication equipment.

CONTROL ALGORITHM

The drive algorithm for the polarization controller is an important element in automatic control systems and shall be briefly discussed here. In most applications, a feedback loop generates the

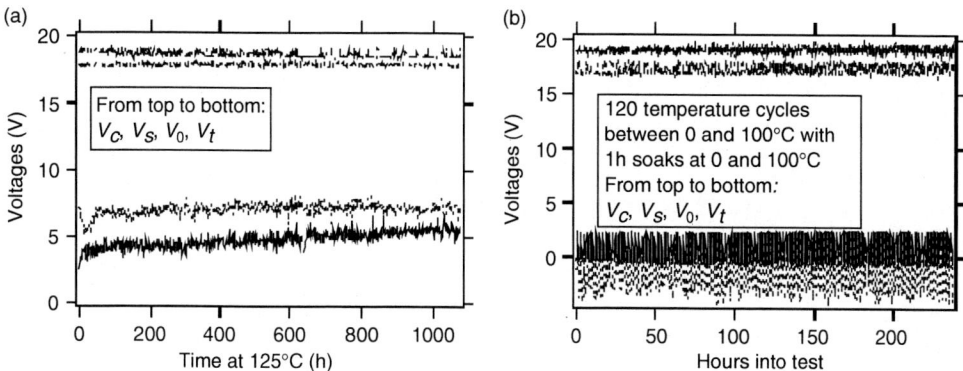

Figure 7   Measured bias voltage drifts in a LiNbO$_3$ polarization controller during (a) accelerated aging and (b) temperature cycling

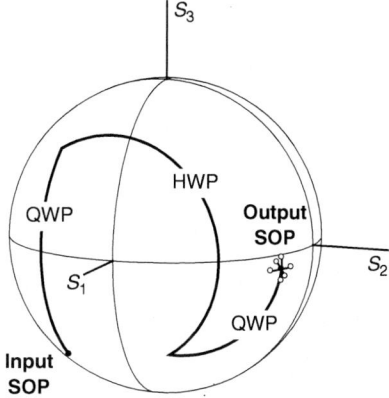

Figure 8   Poincaré sphere representation of a general polarization transformation generated by a QWP–HWP–QWP polarization controller. The bold line shows the evolution of the SOP in the three controller sections, whereas the open circles near the output SOP display the dither excursions caused by small variations in the three control parameters

error signals for the polarization controller by dithering the various adjustable phase angles $2\theta_i(t)$ slightly about their current values, either independent of one another or in certain combinations [36]. An electronic circuit then monitors the effects of these phase variations in the feedback signal and readjusts the phase angles in the various sections of the polarization controller for maximal signal level. The magnitude of the required phase excursions, $2\Delta\theta_i$, depends on the sensitivity of the feedback signal and is typically of the order of 0.1 rad [42].

The operation of this control loop can be readily visualized on the Poincaré sphere, as shown in Figure 8, which displays a general polarization transformation generated by a three-section QWP–HWP–QWP controller. This diagram depicts the evolution of the SOP in the three sections as well as the dither excursions in the output SOP that result from small variations in the parameters $\alpha(t)$, $\gamma(t)$, and $\beta(t)$. In the example of Figure 8, the output SOP is dithered symmetrically along three different directions on the Poincaré sphere. Obviously, the output SOP needs to be probed symmetrically along at least two orthogonal directions to allow continuous tracking of a randomly varying input or output SOP. For this purpose, it is advantageous to keep the parameter $\tilde{\gamma}(t) = \gamma(t) + [\alpha(t) - \beta(t)]/2$, constant while dithering $\alpha(t)$ or $\beta(t)$ [36].

However, there exist certain transformations, in which all three phases dither the output SOP essentially along the same direction. In these cases, large variations in the drive voltages are needed to move the output SOP in a direction orthogonal to these dither variations [36]. Thus, changing the output SOP into such a direction requires many more adjustment cycles for the controller than usual. The particular conditions causing such problems can be identified by expressing the changes in the output SOP, $\hat{e}_{out}$, as a Taylor expansion in the dither excursions $2\Delta\theta_i$, with $\theta_i = \alpha, \tilde{\gamma}, \beta$

$$\hat{e}_{out}\left(\theta_i + \Delta\theta_i\right) = \hat{e}_{out}\left(\theta_i\right) + \sum_{n=1}^{\infty} \frac{\Delta\theta_i^n}{n!}\left(\frac{\partial^n \mathbf{T}}{\partial\theta_i^n}\right)\mathbf{T}^{-1}\hat{e}_{out}\left(\theta_i\right), \tag{6}$$

where $\mathbf{T}$ is the Jones matrix of Eq. (1) and $\hat{e}_{in} = \mathbf{T}^{-1}\hat{e}_{out}$ is the input SOP. The output SOPs that exhibit the smallest dither response to the phase variations $2\Delta\theta_i$ are the eigenstates of the matrix $(\partial\mathbf{T}/\partial\theta_i)\mathbf{T}^{-1}$[43], which are easily found to correspond to the six Stokes vectors

$$\vec{S}_\alpha = \pm\begin{vmatrix} \sin(4\tilde{\gamma})\cdot\cos(2\beta) \\ \sin(4\tilde{\gamma})\cdot\sin(2\beta) \\ -\cos(4\tilde{\gamma}) \end{vmatrix}, \quad \vec{S}_{\tilde{\gamma}} = \pm\begin{vmatrix} -\sin(2\beta) \\ \cos(2\beta) \\ 0 \end{vmatrix}, \quad \vec{S}_\beta = \pm\begin{bmatrix} 0 \\ 0 \\ 1 \end{bmatrix}, \tag{7}$$

where the subscripts denote the corresponding phase dither. These vectors define three rotation axes on the Poincaré sphere, about which the SOP can be dithered linearly with $\Delta\theta_i$. In general, these three axes do not share a common plane, so that even when the output SOP is in one of the above states, it still can be dithered symmetrically along two significantly different (and sometimes orthogonal) directions. However, if $\tilde{\gamma} = m\pi/4$, with $m = 0, \pm1, \pm2, ...$, then the output SOP may be dithered only about $\vec{S}_{\tilde{\gamma}}$ and $\vec{S}_\beta$. Thus, when the output SOP is now parallel to either $\vec{S}_{\tilde{\gamma}}$ or $\vec{S}_\beta$, it can move only along one direction. Moreover, even when the output SOP merely has the same longitude as $\vec{S}_{\tilde{\gamma}}$, it cannot be dithered symmetrically along two orthogonal directions, since in this case small rotations about $\vec{S}_{\tilde{\gamma}}$ cause nearly the same changes in the output SOP as small rotations about $\vec{S}_\beta$.

The likelihood of these undesired situations to occur can be reduced by dithering the three parameters alternately in different combinations or, alternatively, by adding redundant transformer sections to the polarization controller, as shown in the example of Figure 2. However, none of these measures completely solves the problem. A more general solution is to deliberately force a gradual change in at least one of the various control parameters while maintaining the same overall polarization transformation, so that the output SOP is no longer in or near one of the eigenstates of $(\partial\mathbf{T}/\partial\theta_i)\mathbf{T}^{-1}$. It should be noted that the dither problems described here are not specific to fractional-wave-plate transformers but affect all types of endless polarization controllers, although in slightly different ways [44].

SUMMARY

Integrated-optic polarization controllers in $LiNbO_3$ have demonstrated control speeds of >4000 rad/sec and are, therefore, attractive devices for automatic polarization stabilization in fiber-optic telecommunication systems and sensors. Their inherently unlimited transformation range readily allows continuous automatic polarization control with a relatively simple drive algorithm. Moreover, $LiNbO_3$ polarization controllers are based on the same mature fabrication and packaging technologies as used for conventional amplitude and phase modulators and, as a result, are able to meet the stringent reliability requirements of commercial telecommunication systems. It is therefore expected that these electro-optic polarization controllers will play an important role in future fiber-optic communication systems, especially in the areas of PMD compensation and polarization demultiplexing.

REFERENCES

1. R. Ulrich. "Polarization stabilization on single-mode fiber." *Appl. Phys. Lett.* 35, pp. 840–842 (1979).
2. Y. Kidoh, Y. Suematsu, and K. Furuya. "Polarization control on output of single-mode optical fibers." *J. Quantum Electron.* QE-17, pp. 991–994 (1981).
3. T. Imai and T. Matsumoto. "Polarization fluctuations in a single-mode optical fiber." *J. Lightwave Technol.* 6, pp. 1366–1375 (1988).
4. F. Heismann, A. F. Ambrose, T. O. Murphy, and M. S. Whalen. "Polarization-independent photonic switching system using fast automatic polarization controllers." *IEEE Photon. Technol. Lett.* 5, pp. 1341–1343, (1993).
5. R. Noé, H. Heidrich, and D. Hoffmann. "Endless polarization control systems for coherent optics." *J. Lightwave Technol.* 6, pp. 1199–1208 (1988).
6. N. G. Walker and G. R. Walker. "Polarization control for coherent communications." *J. Lightwave Technol.* 8, pp. 438–458 (1990).
7. S. Thaniyavarn. "Wavelength independent, optical damage immune Z-propagation $LiNbO_3$ waveguide polarization converter." *Appl. Phys. Lett.* 47, pp. 674–677 (1985).
8. F. Heismann and M. S. Whalen. "Broadband reset-free automatic polarisation controller." *Electron. Lett.* 27, pp. 377–379 (1991).
9. A. Chen, R. W. Smith, and W. E. Derbyshire. "High performance lithium niobate polarization controller for PMD compensators." In *Proceedings of 15th Annual Meeting IEEE Lasers and Electro-Optics Society.* Glasgow, Scotland, 2002, pp. 885–886.
10. D. W. Smith. "Techniques for multigigabit coherent optical transmission." *J. Lightwave Technol.* LT-5, pp. 1466–1478 (1987).
11. T. Okoshi. "Polarization-state control schemes for heterodyne and homodyne optical fiber communications." *J. Lightwave Technol.* LT-3, pp. 1232–1237 (1985).
12. R. Noé, H. J. Rodler, A. Ebberg, G. Gaukel, B. Noll, J. Wittmann, and F. Auracher. "Comparison of polarization handling methods in coherent optical systems." *J. Lightwave Technol.* 9, pp. 1353–1366 (1991).
13. M. C. Brain, M. J. Creaner, R. C. Steele, N. G. Walker, G. R. Walker, J. Mellis, S. Al-Chalabi, J. Davidson, M. Rutherford, and I. C. Sturgess. "Progress towards the field deployment of coherent optical fiber systems." *J. Lightwave Technol.* 8, pp. 423–437 (1990).
14. F. Heismann, P. B. Hansen, S. K. Korotky, G. Raybon, J. J. Veselka, and M. S. Whalen. "Automatic polarization demultiplexer for polarization multiplexed transmission systems." *Electron. Lett.* 29, pp. 1965–1966 (1993).
15. H. Kogelnik, R. M. Jopson, and L. E. Nelson. "Polarization-mode dispersion." In I. P. Kaminow T. Li, Eds., *Optical Fiber Telecommunications IV B.* San Diego, CA: Academic Press, 2002, pp. 725–861.
16. H. Sunnerud, M. Karlsson, C. Xie, and P. A. Andrekson. "Polarization-mode dispersion in high-speed fiber-optic transmission systems." *J. Lightwave Technol.* 20, pp. 2204–2219 (2002).
17. H. Bülow and S. Lanne. "Optical and electronic PMD compensation." In *Technical Digest Optical Fiber Communication Conference.* Atlanta, 2003, p. 541, Tutorial Notes ThP1-1–ThP1-24.
18. F. Heismann, D. A. Fishman, and D. L. Wilson. "Automatic compensation of first-order polarization mode dispersion in a 10-Gb/s transmission system." In *Proceedings of 24th European Conference on Optical Communication Conference.* Madrid, 1998, pp. 529–530.
19. H. Sunnerud, C. Xie, M. Karlsson, R. Samuelsson, and P. A. Andrekson. "A comparison between different PMD compensation techniques." *J. Lightwave Technol.* 20, pp. 368–378 (2002).
20. R. Noé, D. Sandel, S. Hinz, M. Yoshida-Dierolf, V. Mirvoda, and G. Feise, H. Herrmann, R. Ricken, W. Sohler, F. Wehrmann, C. Glingener, A. Schöpflin, A. Färbert, and G. Fischer. "Integrated optical

LiNbO$_3$ distributed polarisation mode dispersion compensator in 20 Gbit/s transmission system." *Electron. Lett.* 35, pp. 652–654 (1999).

21. F. Heismann. "Integrated-optic polarization transformer for reset-free endless polarization control." *J. Quantum Electron.* 25, pp. 1898–1906 (1989).

22. M. G. Taylor. "Observation of new polarization dependence effect in long haul optically amplified system." *IEEE Photon. Technol. Lett.* 5, pp. 1244–1246 (1993).

23. F. Heismann, D. A. Gray, B. H. Lee, and R. W. Smith. "Electrooptic polarization scramblers for optically-amplified long-haul transmission systems." *IEEE Photon. Technol. Lett.* 6, pp. 1156–1158 (1994).

24. F. Heismann. "Compact electro-optic polarization scramblers for optically amplified lightwave systems." *J. Lightwave Technol.* 14, pp. 1801–1814 (1996).

25. Y. Sun, A. O. Lima, I. T. Lima, J. Zweck, L. Yan, C. R. Menyuk, and G. M. Carter. "Statistics of the system performance in a scrambled recirculating loop with PDL and PDG." *IEEE Photon. Technol. Lett.* 15, pp. 1067–1069 (2003).

26. T. Matsumoto and H. Kano. "Endlessly rotatable fractional-wave devices for single-mode fibre optics." *Electron. Lett.* 22, pp. 78–79 (1986).

27. S. H. Rumbaugh, M. D. Jones, and L. W. Casperson. "Polarization control for coherent fiber-optic systems using nematic liquid crystals." *J. Lightwave Technol.* 8, pp. 459–464 (1990).

28. T. Chiba, Y. Ohtera, and S. Kawakami. "Polarization stabilizer using liquid crystal rotatable waveplates." *J. Lightwave Technol.* 17, pp. 885–890 (1999).

29. H. Shimizu, S. Yamazaki, T. Ono, and K. Emura. "Highly practical fiber squeezer polarization controller." *J. Lightwave Technol.* 9, pp. 1217–1224 (1991).

30. J. Prat, J. Comellas, and G. Junyent. "Experimental demonstration of an all-fiber endless polarization controller based on Faraday rotation." *IEEE Photon. Technol. Lett.* 7, pp. 1430–1432 (1995).

31. H. Shimazu and K. Kaede. "Endless polarization controller using electro-optic waveplates." *Electron. Lett.* 24, pp. 412–413 (1988).

32. K. Hirabayashi and C. Amano. "Variable and rotatable waveplates of PLZT electrooptic ceramic material on planar waveguide circuits." *IEEE Photon. Technol. Lett.* 14, pp. 956–958 (2002).

33. F. Rahmatian, N. A. F. Jaeger, R. James, and E. Berolo. "An ultrahigh-speed AlGaAs–GaAs polarization converter using slow-wave coplanar electrodes." *IEEE Photon. Technol. Lett.* 10, pp. 675–677 (1995).

34. F. Heismann, S. K. Korotky, and J. J. Veselka. "Lithium niobate integrated optics: selected contemporary devices and system applications." In I. P. Kaminow and T. L. Koch, Eds. *Optical Fiber Telecommunication III B*. San Diego, CA: Academic Press, 1997, pp. 377–462.

35. T. Kawazoe, K. Satoh, I. Hayashi, and H. Mori. "Fabrication of integrated-optic polarization controller using z-propagating Ti-LiNbO$_3$ waveguides." *J. Lightwave Technol.* 10, pp. 51–56 (1992).

36. F. Heismann. "Analysis of a reset-free polarization controller for fast automatic polarization stabilization in fiber-optic transmission systems." *J. Lightwave Technol.* 12, pp. 690–699 (1994).

37. M. Martinelli and R. A. Chipman. "Endless polarization control algorithm using adjustable linear retarders with fixed axes." *J. Lightwave Technol.* 21, pp. 2089–2096 (2003).

38. A. J. P. van Haasteren, J. J. G. M van der Tol, M. O. van Deventer, and H. J. Frankena. "Modeling and characterization of an electrooptic polarization controller on LiNbO$_3$." *J. Lightwave Technol.* 11, pp. 1151–1157 (1993).

39. A. Donaldson and K. K. Wong. "Phase-matched mode converter in LiNbO$_3$ using near-Z-axis propagation." *Electron. Lett.* 23, pp. 1378–1379 (1987).

40. A. Mahapatra and E. J. Murphy. "Electrooptic modulators." In I. P. Kaminow and T. Li, Eds. *Optical Fiber Telecommunications IV A*. San Diego, CA: Academic Press, 2002, pp. 258–294.

41. S. K. Korotky and J. J. Veselka. "An RC analysis of long term Ti:LiNbO3 bias stability." *J. Lightwave Technol.* 14, pp. 2687–2697 (1996).

42. F. Heismann and M. S. Whalen. "Fast automatic polarization control system." *IEEE Photon. Technol. Lett.* 4, pp. 503–505 (1992).

43. W. Shieh and H. Kogelnik. "Dynamic eigenstates of polarization." *IEEE Photon. Technol. Lett.* 13, pp. 40–42 (2001).

44. W. H. J. Aarts and G.-D. Khoe. "New endless polarization control method using three fiber squeezers." *J. Lightwave Technol.* 7, pp. 1033–1043 (1989).

Q

QUANTUM WELL

Masahiro Asada

An ultra-thin semiconductor layer with the thickness upto a maximum of about 10 nm is sandwiched between layers with higher band gap energies. The lineup of the conduction and valence band edges forms well potentials, and electrons and holes are confined in the wells. Since the well thickness is as thin as the de Broglie wavelength of electrons in a semiconductor, the energy of electrons in the wells is discrete according to the quantum mechanics. Semiconductor lasers utilizing quantum wells as the active layers are called quantum well lasers. These lasers have superior characteristics, such as ultra-low threshold and less sensitivity to temperature. Today, quantum well lasers are used so widely that semiconductor lasers has become synonymous with quantum well lasers.

Figure 1 shows quantized energy levels and the density of states in a quantum well. The density of states has a step-like feature, and thus, the electron distribution which is the product between the density of states and the Fermi function has sharp peaks at the quantized energy levels. This situation is quite different compared with a conventional bulk semiconductor. Since the stimulated emission and gain are large at these peaks with the same carrier density, the lasing threshold is low compared with a conventional double heterostructure laser, and also less sensitive to temperature. The frequency limit of the direct modulation is also high because of the large differential gain due to the same reason. Since the optical confinement is not effective in a single quantum well, the actual laser uses a separate confinement heterostructure as shown in Figure 2.

As the active layers are very thin in quantum well lasers, a slight lattice mismatch up to around 1% between the quantum well layers and the surrounding layers is allowed. These quantum well lasers are called the strained layer quantum well lasers. Compressive and tensile strains exist according to the direction of mismatching, relative to the surrounding layers. One of the main advantage of the strained quantum well lasers is that the limit of the choice of heterostructure combination is relaxed, by which the lasing wavelength which is impossible in the lattice matched case becomes possible. Another advantage is that the valence band structure is modified due to the strain, resulting in a low threshold (100 A/cm^2 or less in GaInAsP/InP

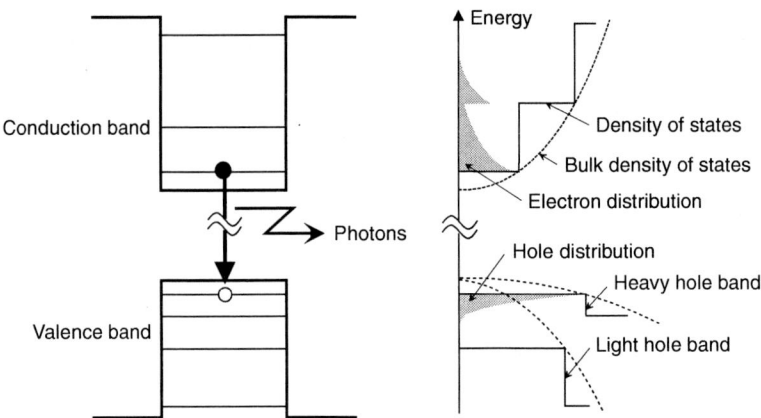

Figure 1 Quantum well structures. (a) Quantized energy levels with electron transitions and (b) density of states and electron distribution

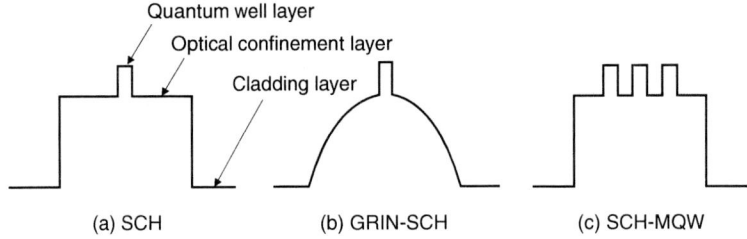

Figure 2 Refractive index profile of quantum well lasers including quantum well layers and optical confinement layers

lasers for optical fiber communication). The effective mass of holes and carrier density that is necessary for the population inversion become small by the valence band mixing due to the strain. The polarization of the output light is the transverse electric (TE) mode for lattice-matched and compressive-strained quantum well lasers, and transverse magnetic (TM) mode for tensile-strained quantum well lasers.

Lasers utilizing transitions only within the levels in the conduction band are also possible in quantum wells. These lasers are called the quantum cascade lasers or intersubband lasers. In these lasers, unipolar carriers are used. The lasing wavelength is determined by the quantum well layer structure, independent of the inherent band gap of used semiconductor materials. Very long wavelength oscillation, such as mid- or far-infrared, have been achieved and short wavelengths are also possible by using deep quantum wells.

In a conventional quantum well laser, electron confinement is made one-dimensional along the direction normal to the heterointerface. The structures with confinement in two and three directions are called the quantum wire and quantum dot structures, respectively. Such a conventional one-dimensional quantum well is called the quantum film. By increasing the degree of electron confinement, further improvement of lasing characteristics are expected because the density of states becomes more sharper. Nowadays fabrication of quantum wire and dot lasers is intensively studied for which, the main techniques used are self-organization of quantum dots during the epitaxial growth and the nanofabrication with electron-beam lithography. Problems such as size fluctuation in array and interface damage causing nonradiative recombinations must be suppressed. The threshold as low as one tenth of conventional quantum film lasers has been reported in self-organized quantum dot array lasers. Quantum wires and dots with low interface damage have also been obtained in electron-beam lithography and etching techniques.

R

RAMAN AMPLIFIER

Shu Namiki

PRINCIPLE

The Raman effect is an inelastic scattering of light with optical phonons of matter, through which light emerging from a matter downshifts in the frequency from the incident light. This phenomenon was first discovered by C. V. Raman in 1928 [1]. Woodbury and Ng proved the possibility of stimulated Raman scattering [2], which can be exploited for amplification of optical signal in optical domain. Through stimulated Raman scattering, a light beam having a frequency coincident with a Raman downshifted frequency of another light beam can receive gain. Figure 1 depicts a quantum-mechanical process of stimulated Raman scattering. During the scattering, an electron of the medium is excited to a virtual state, the lifetime of which is usually very short, and deexcites to a vibrational state of the medium with one optical excited phonon. When the electron deexcites from the upper virtual state, both stimulated and spontaneous emissions are equally possible. The downshift in frequency is called "Stokes shift." If the initial state is the vibrational state, a reverse process of the Stokes shift is possible, which is called "antiStokes shift," in which the frequency upshifts by the same vibrational energy. Usually, antiStokes shift is negligible in room temperature.

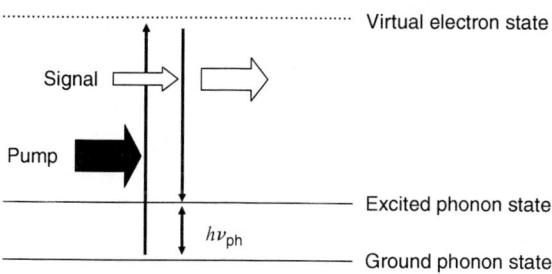

Figure 1 Quantum-mechanical representation of Raman scattering

The total transition probability per unit time, W_R, from the ground to the vibrational state through Raman scattering is expressed as the difference between the emission and the absorption rate, and is proportional to an absolute square of the scattering factors:

$$W_R \propto \left| \left\langle n_m + 1, n_p - 1, n_s + 1 \left| a_s^+ a_p b^+ \right| n_m, n_p, n_s \right\rangle \right|^2 - \left| \left\langle n_m - 1, n_p + 1, n_s - 1 \left| a_s a_p^+ b \right| n_m, n_p, n_s \right\rangle \right|^2$$

$$= n_s (n_p - n_m) + n_p (n_m + 1)$$

$$\approx n_s n_p + n_p (n_m + 1), \tag{1}$$

where the initial number state of molecule optical phonons, pump photons and signal (Stokes) photons was denoted as $\left| n_m, n_p, n_s \right\rangle$, and, a_s^+, a_p^+, b^+, and their Hermite conjugate are the creation and the annihilation operators of signal and pump photons, and molecule optical phonon, respectively. It was assumed that $n_p \gg n_m, n_s$ and found that the first term in the third line of Eq. (1) is related to stimulated Raman scattering that is proportional to the number of pump and signal photons while independent of thermal phonons, and that there are two sources of noise in Raman amplification process corresponding to the second term; one is spontaneous emission and the other is spontaneous emission enhanced by the initial number of phonons. In thermal equilibrium, the mean number of phonons is given by the Bose–Einstein factor, that is, $n_m = (\exp[h v_m / kT] - 1)^{-1}$, where v_m is the frequency of optical phonon, h is the Planck's constant, k is the Boltzmann's constant, and T is the temperature. The proportionality constant is inherently determined by the detailed structures of medium.

FIBER RAMAN AMPLIFIERS

Fiber Raman amplifiers are a kind of optical amplifier that exploits stimulated Raman scattering in fiber for amplifying optical signals propagating through optical fiber. The Raman gain in optical fiber was first measured by Stolen and Ippen [3]. The gain coefficients of silica-based fibers are plotted in Figure 2. Unlike other molecules and crystals and due to amorphous nature of silica, the gain spectra are broad and continuous. This characteristic is suitable for amplifying broadband and wavelength division multiplexed (WDM) signals.

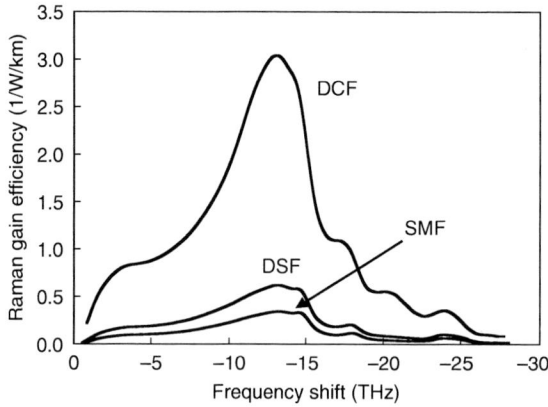

Figure 2 Raman gain efficiency spectra for 1511 nm pump. Raman gain efficiency is defined along with Eqs. (2) and (3). SMF: single mode fiber, DSF: dispersion shifted fiber, DCF: dispersion compensating fiber

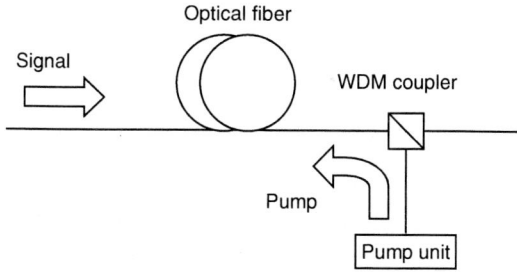

Figure 3   Typical configuration of fiber Raman amplifiers. The figure shows a case of counter pumping

A chief virtue of fiber Raman amplifiers is that gain is supplied in transmission fiber itself. Therefore, not many extra components are necessary to obtain a fiber Raman amplifier. A typical configuration of a fiber Raman amplifier is illustrated in Figure 3 and can vary by changing the pumping orientations, such as forward-, backward-, and bidirectional propagating pump, and the number of stages, and so on. The configuration can also be either distributed or discrete amplifier. A distributed amplifier provides gain distributed along a long transmission fiber line, while a discrete amplifier is placed within a discrete component located in a system rack together with a gain fiber spool.

Because the lifetime of optical phonons in silica is very short due to its amorphous nature, the Raman amplification process can be regarded as nearly a fully inverted process. This also implies that even when the pump rate is high, n_{m} is almost kept in thermal equilibrium. In fact, the approximation from the second to the third line in Eq. (1) is consistent with this observation. As a result, a spontaneous emission factor n_{sp} of Raman amplification process, defined as $n_{\mathrm{sp}} = n_{\mathrm{m}} + 1$, is approximated as 1.1 to 1.2, which is slightly higher than the fully inverted process. Because of this fact, Raman amplifiers can perform extremely well particularly as a distributed amplifier [4].

The equations of motion of the optical powers of pump and signal are derived using Eq. (1) and taking into account the fiber loss is expressed as

$$\frac{dP_s}{dz} = -\alpha_s P_s + \frac{g_R}{\lambda_s A_{\mathrm{eff}}} P_p (P_s + h v_s B[n_{\mathrm{m}} + 1]), \tag{2}$$

$$\frac{dP_p}{dz} = \mp \alpha_p P_p \mp \frac{g_R}{\lambda_p A_{\mathrm{eff}}} P_p (P_s + h v_s B[n_{\mathrm{m}} + 1]), \tag{3}$$

where P_s and P_p are the signal and pump powers, respectively, g_R is the Raman gain coefficient for the wavelengths of signal and pump, λ_s and λ_p are the wavelengths of signal and pump waves, respectively, A_{eff} is the effective area of the optical fiber, α_s and α_p are the attenuation coefficient of the optical fiber at signal and pump wavelengths, respectively. B is the pertinent bandwidth, and v_s is the frequency of the signal. The minus signs on the right-hand side of Eq. (3) correspond to the copumping, and the plus signs correspond to the counter pumping. $g_R / \lambda_s A_{\mathrm{eff}}$ is often called Raman gain efficiency, which scales inversely with the wavelength and effective area. The second term in the parentheses of Eqs. (2) and (3) represent additive white noise due to spontaneous emission.

The main design issues of fiber Raman amplifiers are gain, gain ripple, optical signal-to-noise ratio (SNR), noise figure, the amount of double Rayleigh backscattering, nonlinear phase shift, and pump mediated noise. In order to analyze these issues, the following models are often used, in which the numbers of pump and signal frequencies are more than one [5].

$$\pm\frac{dP_v^{\pm}}{dz} = -\alpha_v P_v^{\pm} + \varepsilon_v P_v^{\mp} + \sum_{\mu>v}\frac{g_{\mu v}}{A_{\mu v}}(P_\mu^{+}+P_\mu^{-})P_v^{\pm}$$

$$+\sum_{\mu>v}\frac{g_{\mu v}}{A_{\mu v}}(P_\mu^{+}+P_\mu^{-})2hv\Delta v(1+n_m) - \sum_{\mu<v}\left(\frac{v}{\mu}\frac{g_{v\mu}}{A_{v\mu}}\right)P_v^{\pm}(P_\mu^{+}+P_\mu^{-}) \tag{4}$$

$$-\sum_{\mu<v}\left(\frac{v}{\mu}\frac{g_{v\mu}}{A_{v\mu}}\right)P_v^{\pm}4h\mu\,\Delta\mu(1+n_m),$$

where subscripts μ and v denote optical frequencies, superscripts + and − denote forward- and backward-propagating waves, respectively, P_v is the optical power within infinitesimal bandwidth Δv around frequency v, α_v is the attenuation coefficient, ε_v is the Rayleigh backscattering coefficient, $A_{\mu v}$ is the effective area of optical fiber for a signal frequency v and pump frequency μ, $g_{\mu v}$ is the Raman gain parameter at frequency v due to pump at frequency μ, and $\Delta\mu$ is the bandwidth of each frequency component around frequency μ. The plus and the minus sign on the left-hand side corresponds to the forward- and backward-propagation, respectively.

The first term of Eq. (4) is fiber loss and second represents Rayleigh backscattering from the light propagating in the opposite direction. The third and the fourth are the gain and the spontaneous scattering caused by the light at frequency μ, respectively. The fifth and the sixth correspond to the depletion of the light at a frequency v due to the amplification and the spontaneous scattering to the light at a frequency μ, respectively. The factor of two in the fourth term is because the spontaneous scattering is independent of the polarization of the signal light and has twice efficiency as much as that of the stimulated emission under the assumption that the state of polarization (SOP) of all lights is uncorrelated and randomly changing with respect to each other, while the degree of polarization of each signal is well preserved. In addition to this, another factor of two enters in the sixth term because the spontaneous emission is in both directions.

CHARACTERISTICS OF FIBER RAMAN AMPLIFIER

Small Signal Gain

Because Raman gain efficiency of silica fiber is small, in many Raman amplifiers, the pump power is much larger than the output signal power. In such cases, the second term of Eq. (3) can be neglected. The amplified spontaneous emission (ASE) noise terms in Eqs. (2) and (3) can be negligible compared with the output signal power in most cases. Then, Eqs. (2) and (3) can be solved analytically, and the output signal power at distance z is expressed as

$$P_s(z) = P_s(0)\exp\left(\frac{g_R}{\lambda_s A_{eff}}P_p(0)L_{eff}(z) - \alpha_s z\right), \tag{5}$$

where L_{eff} is the effective interaction length and defined as

$$L_{eff} = \int_0^z \frac{P_p(z')}{P_p(0)}dz' = \frac{1-\exp(-\alpha_p z)}{\alpha_p}. \tag{6}$$

Gain of Raman amplifiers is often expressed in terms of the so-called "on–off gain," which is related to the signal output powers when the pump is on and off. It is defined as

$$G_R \equiv \frac{P_s(L; P_p = \text{on}) - P_N(L; P_p = \text{on})}{P_s(L; P_p = \text{off})} = \exp\left(\frac{g_R}{\lambda_s A_{\text{eff}}} P_p(0) L_{\text{eff}}\right), \tag{7}$$

where L is the physical length of the optical fiber and P_N is the optical power of the ASE noise measured in the signal band by turning off the signal input.

From Figure 2, a typical Raman gain efficiency of a single mode fiber (SMF) is 0.3 1/W/km. A typical high pump power available in market is 400 mW per laser module. If one launches 400 mW pump power into 10 km long SMF, whose L_{eff} is approximately 7.6 km assuming the attenuation coefficient is 0.25 dB/km at 1450 nm. The on–off gain G_R in this case then becomes 2.5 (4 dB). In case of using dispersion compensating fiber (DCF), the Raman gain efficiency is approximately 2.7-1/W/km and the attenuation coefficient at 1450 nm is typically 0.6 dB/km. Then, G_R will be calculated as 349 (25.4 dB).

Effective Noise Figure

It is known that population inversion of fiber Raman amplifiers is effectively close to 100% because of very short lifetime of the optical phonons. In addition, the dominant noise source of a properly designed Raman amplifier can only be ASE [6]. Therefore, the noise figure F of a Raman amplifier may be represented by

$$F = \frac{P_{\text{ASE}}}{G h v_s B}, \tag{8}$$

where P_{ASE} is the ASE power within the bandwidth B, G is net gain, and v_s is a signal frequency. Because a Raman amplifier tends to require relatively long fiber as the gain medium for which the attenuation is nonnegligible, the noise figure of the Raman amplifier is considerably affected by the attenuation. Particularly, in a distributed amplification scheme, the on–off Raman gain is often smaller than the total attenuation of a fiber span. In such a case, the noise figure is dictated by the attenuation as well as Raman gain itself. Because amount of attenuation varies from a span to another, a property called "effective noise figure" is defined in order to separate the effect of attenuation from the noise figure intrinsic in Raman amplification. The Effective noise figure is expressed as:

$$F_{\text{eff}} = \frac{P_{\text{ASE}}}{G \exp(\alpha L) h v B} = \frac{P_{\text{ASE}}}{G_R h v B}. \tag{9}$$

Likewise, the effective noise figure is defined in terms of on–off gain rather than net gain, and can take values <2 (or 3 dB). This value may represent the amount of improvement in SNR due to introducing Raman pump. In other words, it compares SNR between the cases when the pump is on and off, and not the SNR of input and output signal. By definition, this property is only applied to distributed amplifiers.

Effect of Double Rayleigh Backscattering

In optical fiber, there is always some portion of propagating signal reflected back due to the Rayleigh scattering. Because of the relatively long length of gain fiber, Raman amplifiers are likely to suffer from double Rayleigh backscattering, in which the signal undergoes twice the Rayleigh backscattering. This portion of the double Rayleigh backscattered signal acts as noise onto the original signal, because they copropagate with an indefinite time delay between each other. The noise due to DRBS is also called "multiple path interference (MPI)" noise. It is known that this type of noise sets the limit in improvement of SNR in Raman amplifiers [7]. The effect of DRBS generally depends on configurations of Raman amplifiers such as pumping direction and number of amplifier stages. It was shown that bidirectional pumping better suppressed DRBS than either co- or counter pumping scheme [8,9].

Nonlinear Effects

Nonlinear phase shift is a useful measure of nonlinear effects. The nonlinear phase shift Φ_{NL} of the signal is calculated as the path-integrated power times the nonlinear coefficient:

$$\Phi_{NL} = \gamma \int_0^L P_s(z) \, dz, \tag{10}$$

where γ is the nonlinear coefficient describing self-phase modulation term in nonlinear Schrödinger equation [10], and is defined as

$$\gamma \equiv \frac{2\pi n_2}{\lambda_s A_{eff}}, \tag{11}$$

where n_2 is the nonlinear refractive index. The nonlinear coefficient scales inversely with the wavelength and the effective area of the signal. This scaling is the same as the Raman gain efficiency because the Raman process is closely related to the self-phase modulation (a real part of the third-order nonlinear susceptibility) [11].

From Eq. (10), it is obvious that Raman amplification increases the path average power, and hence the nonlinear phase shift. Unlike EDFAs, a fiber length of Raman amplifiers is usually kilometers long. Therefore, the impact on Φ_{NL} in Raman amplifiers usually appear more significant than in EDFAs.

Pump Mediated Noise and Pumping Configurations

Because the Raman interaction is instantaneous, the temporal pattern of the pump power defines the temporal pattern of the gain. One of the typical high power pump lasers, or a fiber Bragg grating (FBG) stabilized multimode laser diode, usually has a larger relative intensity noise (RIN) than a single frequency laser such as DFB laser [12]. In counter-pumping configuration, signal travels over a rapidly varying gain, while in copumping where the relative velocity between signal and pump is small as determined only by the group velocity dispersion; it becomes more likely that the pattern of the temporally varying gain is imprinted as the temporal pattern of the signal. In general, a transfer function can be defined through which the pump RIN is transferred to noise of the signal [13]. There is an effort to develop a high power pump laser with low RIN while the wavelength is stabilized by internal integrated grating [14,15].

Transient Effects

Even though power conversion efficiency of a Raman amplifier tends to be small and it operates as a linear amplifier, the sudden large change in the input signal power can result in a significant

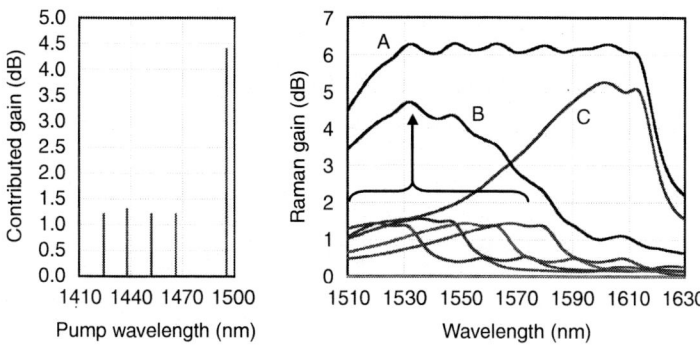

Figure 4   Design scheme of multiple wavelength pumping

transient response of the amplifier and it is often required that the amplifier carries fast electronic control circuits to suppress the transient response [16].

Multiple Wavelength Pumping for WDM Applications

The wavelength of gain in Raman amplifiers is dictated by pump wavelength. In other words, by properly choosing a pump wavelength, one can design gain at any wavelength. If one prepares multiple pump wavelengths, then the gain appears correspondingly at multiple wavelengths. For example, one can design a multiple-wavelength-pumped Raman amplifier to provide a broad and a flat spectral profile of gain, which is suitable for WDM transmissions [5]. Figure 4 illustrates a design scheme of the multiple wavelength pumping. In the figure, five pump wavelengths are used to realize a flat gain over C and L bands. The contributed gain in Figure 4 is the path averaged pump power, and does not have a simple correspondence to the launched pump power because of pump-to-pump interactions [5]. The gray low profiled curves and curve C in Figure 4 are the gain spectra provided by different pump wavelengths, respectively. Curve A is decomposed into curves B and C having opposite slopes.

The multiple wavelength pumping is also attractive in the sense that each pump power at different wavelengths may not be too large; however when combined they provide a sufficient amount of gain. In the configuration of Figure 4, the launched power from each pump laser is no larger than 180 mW in order to achieve a flat 10 dB on–off Raman gain in 80 km SMF over the wavelength range between 1530 and 1610 nm.

PUMP LASER TECHNOLOGIES

There are primarily two kinds of pump laser technologies: one is based on compound semiconductor material and the other is based on fiber laser technologies.

Pump laser diodes are often used in configuration of multiple wavelength pumping. For this purpose, the wavelength of each pump laser has to be well stabilized and narrowed in order to guarantee an efficient and a stable WDM of the pump lasers. The Raman amplifiers operating over C and L band require pump wavelengths ranging from 1400 to 1500 nm. Therefore, the pump laser diodes for such Raman amplifiers are called "14XX nm" pump lasers.

Most of 14XX nm pump lasers available in market have a structure called GRIN-SCH-SL-MQW-BH LD based on InGaAsP/InP material system [17]. Figure 5 describes a photograph and a schematic diagram. This is a Fabry–Perot laser, and is usually stabilized by a FBG written

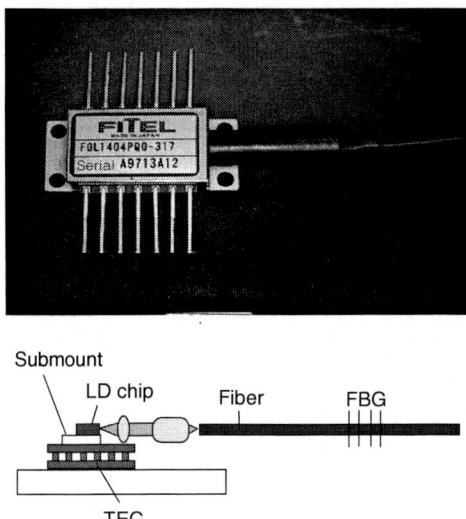

Figure 5   Photograph and schematic diagram of 14XX nm pump laser diode

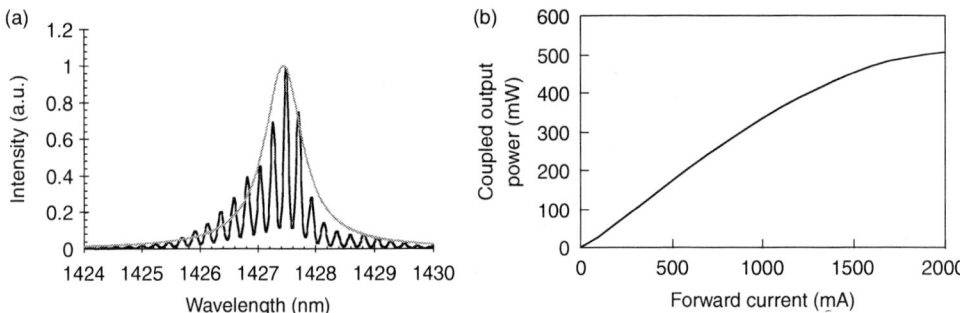

Figure 6   Spectrum and fiber output power versus forward current of 14XX nm pump laser diode

in the output fiber pigtail of the package. The FBG provides a feedback at a selected wavelength to the laser chip to stabilize and narrow the lasing wavelength. Figure 6 shows a spectrum and a fiber output power versus forward current curve of such a laser. A fiber coupled output power more than 400 mW at any wavelength within 14XX nm range is available in market.

Another attractive high power pump laser technology is Raman fiber laser, which is based on the so-called cladding-pumped fiber laser technologies [18]. This technology realizes a wide wavelength tuning as well as multi-watt power output.

APPLICATIONS

Modules

Figure 7 shows a photograph of a multiple wavelength pumping unit for distributed Raman amplifiers. Its schematic diagram is depicted in Figure 8. Figure 9 shows a Raman pumping unit for DCF modules. This pumping unit is embedded in the center of the DCF spool.

Figure 7   Photograph of a multiple wavelength pumping unit for distributed Raman amplifiers (Courtesy of Fitel Products Division, Furukawa Electric Co., Ltd.)

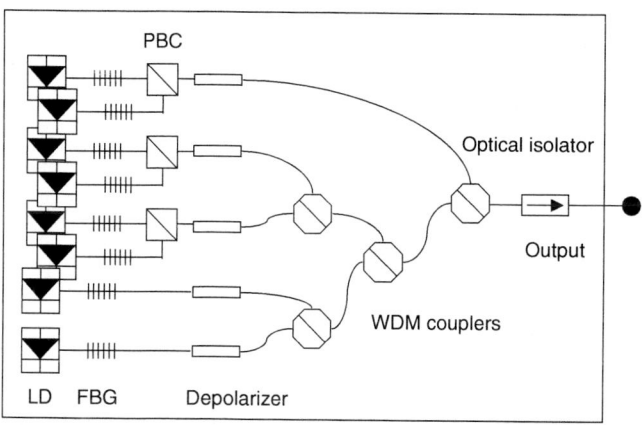

Figure 8   Schematic diagram of a multiple wavelength pumping unit for distributed Raman amplifiers. PBC: Polarization beam combiner

Raman Assisted Transmission Systems

There are a tremendous number of transmission experiments demonstrating high performance of Raman amplifiers. In the early stage when EDFAs had not yet been developed, fiber Raman amplifiers were considered as one of the most promising optical amplifiers. For example, Aoki et al. demonstrated the first signal transmission based on fiber Raman amplifiers in 1985 [19], and almost at the same time, the first experimental demonstration of optical soliton propagation amplified with a fiber Raman amplifier was reported by Mollenauer et al. [20]. It is also noteworthy that the first diode-pumped EDFA was pumped by a 1480 nm laser diode that was originally developed for a light source of optical time domain reflectometer (OTDR) and pumping Raman amplifiers [21,22].

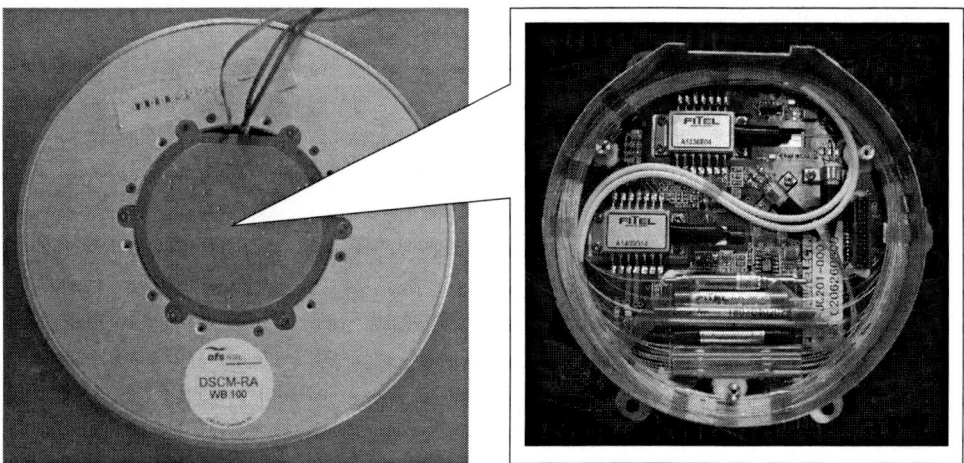

Figure 9   Photograph of a Raman pumping unit for DCF modules (Courtesy of Fitel Products Division, Furukawa Electric Co., Ltd.)

After the pump lasers were developed to achieve a high power that was enough for various Raman amplifiers [17,18], multi-Tera bit/sec long haul dense WDM transmission experiments have been practically successful in employing Raman amplifiers, examples of which are reported in References 23 to 25. It has been widely agreed that such high capacity long haul transmissions would have never been successful without the use of Raman amplifiers.

REFERENCES

1. C. V. Raman, "A new radiation," *Indian J. Phys.*, 2, 387 (1928).
2. E. J. Woodbury and W. K. Ng, "Ruby laser operation in the near IR," *Proc. IRE*, 50, 2367 (1962).
3. R. H. Stolen and E. P. Ippen, "Raman gain in glass optical waveguides," *Appl. Phys. Lett.*, 22, 276–278 (1973).
4. M. Vasilyev, "Raman-assisted transmission: toward ideal distributed amplification," in Technical Digest, *Optical Fiber Communication Conference and Exposition*, Atlanta, Georgia USA, paper WB1 (March 26, 2003).
5. S. Namiki and Y. Emori, "Ultrabroad-band Raman amplifiers pumped and gain-equalized by wavelength-division-multiplexed high-power laser diode," *J. Select. Top. Quantum Electron.*, 7, 3–16 (2001).
6. N. A. Olsson and J. Hegarty, "Noise properties of a Raman amplifier," *J. Lightwave Technol.*, LT-4, 396–399 (1986).
7. P. B. Hansen, L. Eskildsen, A. J. Stentz, T. A. Strasser, J. Judkins, J. J. DeMarco, R. Pedrazzani, and D. J. DiGiovanni, "Rayleigh scattering limitations in distributed Raman pre-amplifiers," *IEEE Photonics Technol. Lett.*, 10, 159–161 (1998).
8. M. Nissov, K. Rottwitt, H. D. Kidorf, and M. X. Ma, "Rayleigh crosstalk in long cascades of distributed unsaturated Raman amplifiers," *Electron. Lett.*, 35, 997–998 (1999).
9. R.-J. Essiambre, P. Winzer, J. Bromage, and C. H. Kim, "Design of bidirectionally pumped fiber amplifiers generating double Rayleigh backscattering," *IEEE Photonics Technol. Lett.*, 14, 914–916 (2002).
10. G. P. Agrawal, *Nonlinear Fiber Optics*, 3rd ed., Academic Press, San Diego, CA, 2001.
11. R. H. Stolen, J. P. Gordon, W. J. Tomlinson, and H. A. Haus, "Raman response function of silica-core fibers," *J. Opt. Soc. Am. B*, 6, 1159–1166 (1989).
12. S. Kado, Y. Emori, S. Namiki, N. Tsukiji, J. Yoshida, and T. Kimura, "Broadband flat-noise Raman amplifier using low-noise bi-directionally pumping sources," in *Proceedings of 27th European Conference on Optical Communications*, Amsterdam, The Netherlands, PD.F.1.8 (October 4, 2001).

13. C. R. S. Fludger, V. Handerek, and R. J. Mears, "Pump to signal RIN transfer in Raman fiber amplifiers," *J. Lightwave Technol.*, 19, 1140–1148 (2001).

14. N. Tsukiji, N. Hayamizu, H. Shimizu, Y. Ohki, T. Kimura, S. Irino, J. Yoshida, T. Fukushima, and S. Namiki, "Advantage of inner-grating-multi-mode laser (iGM-laser) for SBS reduction in co-propagating Raman amplifier," in *Proceedings of the OSA Topical Meeting on Optical Amplifiers and their Applications*, Vancouver, Canada, OMB4 (July 15, 2002).

15. L. L. Wang, R. E. Tench, L. M. Yang, and Z. Jiang, "Linewidth limitations of low noise, wavelength stabilized Raman pumps," in *Proceedings of the OSA Topical Meeting on Optical Amplifiers and their Applications*, Vancouver, Canada, OMB5 (July 15, 2002).

16. C.-J. Chen, J. Ye, W. S. Wong, Y.-W. Lu, M.-C. Ho, Y. Cao, M. J. Gassner, J. S. Pease, H.-S. Tsai, H. K. Lee, S. Cabot, and Y. Sun, "Control of transient effects in distributed and lumped Raman amplifiers," *Electron. Lett.*, 37, 1304–1305 (2001).

17. S. Namiki, N. Tsukiji, and Y. Emori, "Pump laser diodes and WDM pumping," in *Raman Amplifiers For Telecommunications 1*, M. Islam, Ed., Springer-Verlag, New York (2004), Chap. 5.

18. C. Headley, M. Mermelstein, and J. C. Bouteiller, "Raman fiber laser," in *Raman Amplifiers For Telecommunications 2*, M. N. Islam Ed., Springer-Verlag, New York (2004), Chap. 11.

19. Y. Aoki, S. Kishida, K. Washio, and K. Minemura, "Bit error rate evaluation of optical signals amplified via stimulated Raman process in an optical fibre," *Electron. Lett.*, 21, 191–193 (1985).

20. L. F. Mollenauer, R. H. Stolen, and M. N. Islam, "Experimental demonstration of soliton propagation in long fibers: Loss compensated by Raman gain," *Opt. Lett.*, 10, 229–231 (1985).

21. M. Nakazawa, Y. Kimura, and K. Suzuki, "Efficient Er^{3+}-doped optical fiber amplifier pumped by a 1.48 μm InGaAsP laser diode," *Appl. Phys. Lett.*, 54, 295–297 (1989).

22. S. Oshiba, A. Matoba, M. Kawahara, and Y. Kawai, High-power output over 200 mW of 1.3 μm GaInAsP VIPS lasers," *IEEE J. Quantum Electron.*, QE-23, 738–743 (1987).

23. G. Charlet, W. Idler, R. Dischler, J.-C. Antona, P. Tran, and S. Bigo, "3.2Tbit/s (80 × 42.7 Gb/s) C-Band transmission over 9 × 100km of TeraLight™ fiber with 50GHz channel spacing," in *Proceedings of the OSA Topical Meeting on Optical Amplifiers and their Applications*, Vancouver, Canada, PD1 (July 17, 2002).

24. B. Zhu, L. E. Nelson, S. Stulz, A. H. Gnauck, C. Doerr, J. Leuthold, L. Grüner-Nielsen, M. O. Pedersen, J. Kim, R. Lingle, Jr., Y. Emori, Y. Ohki, N. Tsukiji, A. Oguri, and S. Namiki, "6.4-Tb/s (160 × 42.7 Gb/s) transmission with 0.8 bit/s/Hz spectral efficiency over 32 × 100 km of fiber using CSRZ-DPSK format," in *Proceedings of the Optical Fiber Communication Conference and Exposition*, Atlanta, Georgia USA, PD-19 (March 27, 2003).

25. C. Rasmussen, T. Fjelde, J. Bennike, F. Liu, S. Dey, B. Mikkelsen, P. Mamyshev, P. Serbe, P. van der Wagt, Y. Akasaka, D. Harris, D. Gapontsev, V. Ivshin, and P. Reeves-Hall, "DWDM 40G transmission over trans-Pacific distance (10,000 km) using CSRZ-DPSK, enhanced FEC and all-Raman amplified 100 km UltraWave™ fiber spans," in *Proceedings of the Optical Fiber Communication Conference and Exposition*, Atlanta, Georgia USA, PD-18 (March 27, 2003).

RF SPECTRUM ANALYZER

Chen S. Tsai

INTRODUCTION

The purpose of this chapter is to provide a review on guided-wave acousto-optic (AO) and magneto-optic (MO) modulators and the resulting integrated optic modules for radio frequency (RF) signal processing with a focus on real-time spectral analysis. Efficient and wideband AO interactions between guided-optical waves (GOW) [1] and surface acoustic waves (SAW) [2] have facilitated a number of unique applications in signal processing and communications [3].

For example, the resulting wideband planar AO Bragg cell modulators were widely used in the development and realization of micro-optical modules for real-time processing of radar signals, especially, integrated optic RF spectrum analyzers [3]. Subsequently, integrated AO tunable filters were also actively explored for applications in wavelength-division-multiplex (WDM) fiber optic communication systems [4]. In the meantime, MO interactions between GOW and magnetostatic waves (MSWs) have also resulted in tunable wideband planar MO Bragg cell modulators [5]. In this chapter, a review on the science and technology relevant to integrated optic RF spectrum analyzers is given. First, the interaction geometry, physical mechanism, analytical treatment, the key device parameters of a basic planar AO Bragg cell modulator, and four potentially viable substrate materials are presented in "Planar Guided-wave Acousto-optic Bragg Cell Modulator" section. The techniques for realization of efficient and wideband AO Bragg cells using multiple-tilted SAWs and phased-array SAWs are also briefly discussed in this section. In "Planar Guided-wave Magneto-optic Bragg Cell Modulator" section the interaction geometry, physical mechanism, and analytical treatment for MO interactions together with the device parameters and characteristics of the resulting MO Bragg cell modulator are presented. A comparison of AO and MO Bragg cell modulators is also provided in this section. Subsequently, the architecture for a monolithic integrated AO RF spectrum analyzer and that for hybrid counterparts in planar- and spherical-waveguide LiNbO₃ substrates, the architecture for a hybrid integrated MO RF spectrum analyzer, and the fabrication technologies involved are presented in "Integrated Optic RF Spectrum Analyzers" section. Prospects for further advances toward monolithic integration for the integrated optic RF spectrum analyzers are also briefly mentioned in this section. Finally, some concluding remarks are provided.

PLANAR GUIDED-WAVE ACOUSTO-OPTIC BRAGG CELL MODULATOR

Interaction Geometry, Physical Mechanism, and Analytical Treatment

Figure 1 shows the basic interaction geometry for the planar guided-wave AO Bragg cell modulator [3,6]. The optical waveguide material should possess desirable acoustic, optical, and AO properties [3]. The SAW is commonly excited by an interdigital electrode transducer (IDT) deposited on the optical waveguide [3]. If the substrate material is sufficiently piezoelectric, such as Y-cut LiNbO₃, the IDT may be deposited directly on it. Otherwise (e.g., GaAs, InP, and Si), a piezoelectric film such as ZnO [3] must be deposited either beneath or above the interdigital electrode array. The optical waveguide can be either a graded index layer created beneath the substrate or a step index layer deposited on top of the substrate. Propagation of the SAW creates a moving optical grating in the optical waveguide caused by elasto-optic effect. The moving grating in turn causes diffraction of an incident guided light wave. Guided-Wave AO diffraction may be either Raman-Nath, Bragg, or collinear, depending upon the angle between the direction of the incident light and that of the SAW wavefront as well as the acoustic aperture. Raman-Nath diffraction consists of a number of side orders when the AO parameter $Q \equiv 2\pi\lambda_0 L/n\Lambda^2 \leq 0.3$. The symbols λ_0 and Λ designate, respectively, the wavelengths of the guided optical wave (in free space) and the SAW; n is the effective refractive index of the guiding medium; and L designates the aperture of the SAW. When the light wave is incident at the Bragg angle θ_B defined by $\lambda_0/n = 2\Lambda \sin \theta_B$ and $Q > 4\pi$, diffraction is of the Bragg type and consists of only one side order. Here, we shall limit our discussion to Bragg-type diffraction because this type of diffraction is capable of higher acoustic center frequency, wider modulation bandwidth, larger dynamic range, and thus greater versatility in application. Note that in the case of a piezoelectric and electro-optical (EO) substrate such as LiNbO₃ and ZnO, the piezoelectric fields accompanying the SAW can be so large that the induced index changes caused by the EO effect becomes very significant [3].

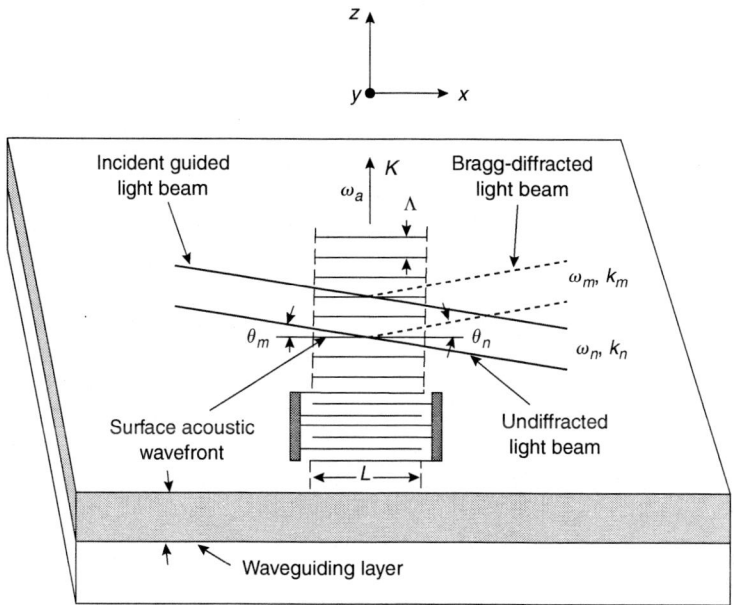

Figure 1   Guided-wave acousto-optical Bragg diffraction from a single-surface acoustic wave

The most common approach for treatment of AO interactions is the so-called coupled-mode technique [7]. This technique has been employed for the analysis of guided-wave AO Bragg diffraction from a single SAW in a LiNbO$_3$ planar waveguide [3]. Such an analysis serves to reveal the physical parameters involved and the key device parameters as well as the performance limitations of the resulting devices. Both the analytical procedures and the methodology for numerical computation developed for this simple case can be conveniently extended to the case involving multiple SAWs [3] as well as other material substrates. The relevant momentum (wave vector) and energy (frequency) conservation relations between the incident guided-light wave, the diffracted guided-light wave, and the SAW are expressed by Eq. (1a,b), where m and n designate the indices of the waveguide modes; k_n, k_m, and K are, respectively, the wave vectors of the diffracted light in the nth mode, the undiffracted light in the mth mode, and the SAW; and ω_n, ω_m, and ω_a are the corresponding radian frequencies:

$$k_n = k_m \pm K \tag{1a}$$

$$\omega_n = \omega_m \pm \omega_a. \tag{1b}$$

The diffracted light may have a polarization parallel or orthogonal to that of the incident light.
    The angles of incidence and diffraction measured with respect to the acoustic wavefront θ_m and θ_n are

$$\sin\theta_m = \frac{\lambda_0}{2n_m\Lambda}\left[1 + \frac{\Lambda^2}{\lambda_0^2}(n_m^2 - n_n^2)\right], \tag{2a}$$

$$\sin\theta_n = \frac{\lambda_0}{2n_n\Lambda}\left[1 - \frac{\Lambda^2}{\lambda_0^2}(n_m^2 - n_n^2)\right], \tag{2b}$$

where n_m and n_n are the effective refractive indices of the undiffracted and the diffracted light waves, and, as defined earlier λ_0 and Λ are, respectively, the optical wavelength in free space and the wavelength of the SAW. Clearly, when the undiffracted and diffracted light propagate in the same waveguide mode ($n_n = n_m$), Eq. (2a,b) both reduce to the well-known Bragg condition in isotropic diffraction, namely, $\sin\theta_n = \sin\theta_m = (\lambda_0/2n_m\Lambda) = (\lambda_0\omega_d/4\pi n_m V_R)$, where $\theta_n = \theta_m$ is the Bragg angle θ_B referred to earlier and V_R designates the propagation velocity of the SAW. Thus, the diffraction angle is identical to the incidence angle, and both angles increase linearly with the acoustic frequency because in practice the latter is much lower than the optical frequency. Note that many guided-wave AO Bragg diffraction experiments using $LiNbO_3$ and GaAs belong to this particular case [3].

Bragg Diffraction Efficiency and Frequency Response

The diffraction efficiency $\zeta(f_a)$, defined as the ratio of Bragg-diffracted light power at the output ($x = L$) and incident light power at the input ($x = 0$) of the interaction region (see Figure 1), as a function of the acoustic frequency f_a, the field distributions of the incident and diffracted light waves and the SAW, and the angular deviation $\Delta\theta$ of the incident light from the Bragg angle θ_B can be calculated [3]. Although the exact expression for $\zeta(f_a)$ is quite complicated, for the case in which the EO contribution to the changes in refractive index is either negligible or proportional to the elasto-optical contribution, $\zeta(f_a)$ can be greatly simplified, however. The results are applicable to nonpiezoelectric materials such as glass, oxidized silicon, As_2S_3, GaAs, and InP. We further assume that the power of the SAW (P_a) is uniformly distributed in a depth of one acoustic wavelength [3] and restrict to the case of low to moderate diffraction efficiency. The resulting expression for the Bragg diffraction efficiency is as follows:

$$\zeta(f_{a0}) \approx \left(\frac{\pi^2}{2\lambda_0^2}\right) M_{2mn}\left(\frac{f_{a0}L}{V_R\cos\theta_m\cos\theta_n}\right)P_a, \tag{3}$$

where f_{a0} is the acoustic center frequency and M_{2mn} is defined as $(n_m^3 n_n^3 P^2/\rho V_R^3)$ in which P and ρ are the relevant photoelastic constants and the mass density, respectively. Equation (3) shows that the diffraction efficiency is proportional to the product of the center frequency and the aperture of the SAW as well as the total acoustic power.

The absolute AO Bragg bandwidth for the case of isotropic interaction involving the optical TE_0 mode is given as follows:

$$\Delta f_{3dB.Bragg} \approx \frac{1.8n_0 V_R^2\cos\theta_{B0}}{\lambda_0 f_{a0}L}, \tag{4}$$

where n_0 and θ_{B0} designate the effective refractive index and the Bragg angle for the TE_0 mode. Equation (4) shows that the absolute AO Bragg bandwidth is inversely proportional to the product of the center frequency and the aperture of the SAW. Thus, the diffraction efficiency and the absolute AO Bragg bandwidth impose conflicting requirements on the acoustic frequency and the acoustic aperture. In fact, the diffraction efficiency-Bragg bandwidth product is a constant that is independent of both the center frequency and the aperture of the SAW.

In summary, because of the complicated spatial distributions of the GOW and the SAW as well as the frequency dependence of the latter, numerical calculations using digital computers are required to determine the diffraction efficiency and the exact frequency response of a guided-wave AO Bragg cell modulator. The procedure is as follows:

1. Obtain appropriate analytical expressions for the field distributions of the GOW and the SAW based on their directions and modes of propagation.
2. Include the frequency dependence of the amplitudes, phases, and penetration depth of the SAW.
3. Identify the relevant photoelastic and EO constants.
4. Calculate the diffraction efficiency versus the acoustic frequency with the acoustic drive power as a parameter.

Note that the frequency response of the SAW transducer can be incorporated in step 2.

Using the above procedure detailed numerical calculations have been carried out for both isotropic and anisotropic Bragg diffraction in Y-cut $LiNbO_3$ waveguides [3]. The results on isotropic Bragg diffraction to be given here involves a He–Ne laser light (0.6328 μm) propagating with the TE_0 mode in the Y-cut $LiNbO_3$ in-diffused optical waveguides and a SAW propagating in the Z(c) direction. Here, the appropriate photoelastic constants are $P_{31} = P_{32}$ and P_{33}. The only relevant EO coefficient is r_{33}. First, the frequency response as determined by the confinements of the optical wave and the SAW is calculated. This is equivalent to calculating the coupling coefficient $C_{mn}^2(f)$ [3] as a function of the acoustic frequency under the assumptions that the transducer bandwidth is so large that it does not introduce any band-limiting effect and that the Bragg condition ($\Delta\theta \equiv 0$) is satisfied at all frequencies of interest. While the first assumption leads to a constant total acoustic power, the second assumption is equivalent to utilization of a sufficiently small acoustic aperture. Subsequently, Bragg diffraction efficiency as a function of the acoustic frequency is calculated with the penetration depth of an optical waveguide mode as a parameter. Figure 2 shows the calculated results for the case involving a TE_0 mode for both the Bragg diffracted and undiffracted light waves. Note that in these calculated plots the total acoustic power is assumed to be a constant, a consequence of the assumption that the transducer bandwidth is sufficiently large so that it does not introduce a band-limiting effect. These plots clearly show that the smaller the optical penetration depth the higher will be the optimum acoustic frequency and the diffraction efficiency. It is seen that at the optical wavelength of 0.628 μm the desirable range of optical penetration depth is from 1.0 to 2.0 μm and that the optimum range of acoustic frequency should be centered around 700 MHz.

Key Device Parameters

The planar AO Bragg cell modulator just presented can be used to scan a guided-light beam in the waveguide plane by changing the frequency of the RF drive. Clearly, it can also be used to analyze the frequency spectrum in the RF drive because each frequency component will result in a diffracted light beam propagating in a corresponding direction. The key device parameters that determine the ultimate performance characteristics of planar AO Bragg cell modulators for RF signal processing and light beam deflection/scanning applications are bandwidth, time-bandwidth product, acoustic and RF drive power, nonlinearity, and dynamic range [3]. A detailed discussion on the first three of these performance parameters is provided here, while some discussion on the last two parameters is provided elsewhere [3],

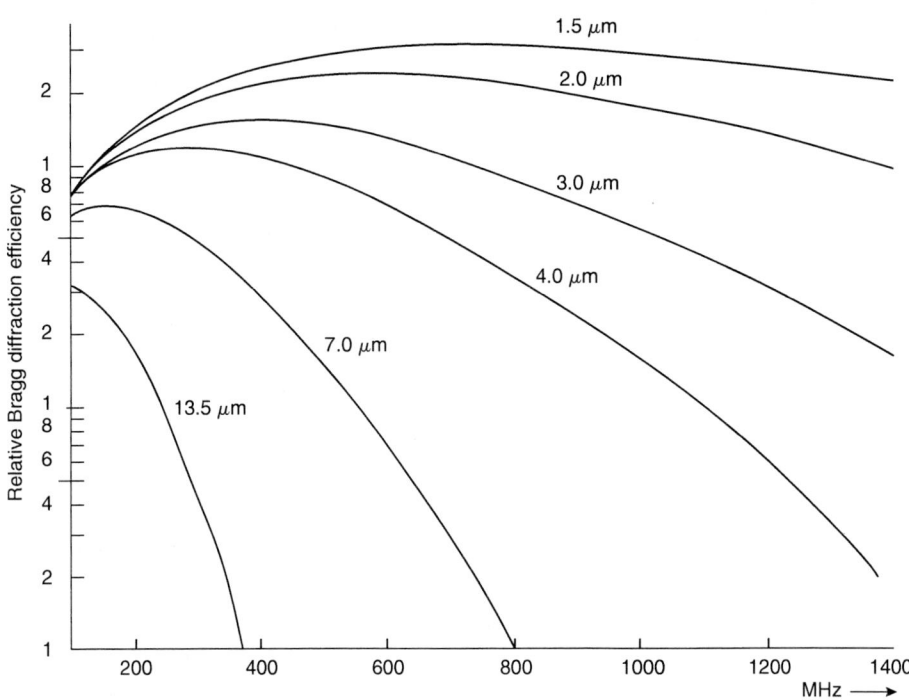

Figure 2 Calculated frequency response based on coupling coefficient alone with penetration depth of TE_0-mode in a Y-cut $LiNbO_3$ waveguide as a parameter (for Z-Propagation SAW). Optimum optical penetration depth $X_t = 1.5\ \mu m$

Bandwidth

As shown in the last subsection the diffraction efficiency-bandwidth product of a planar guided-wave AO modulator that employs isotropic Bragg diffraction and a single periodic ID SAW transducer [3,8] is rather limited. However, if the absolute modulator bandwidth is the sole concern, a large bandwidth can be realized by using either a single *periodic* ID SAW transducer with small acoustic aperture and small number of finger pairs or a single *aperiodic* ID SAW transducer with small acoustic aperture and large number of finger pairs (chirp transducers) [9]. It should be emphasized, however, that in either case the large bandwidth is obtained at a reduced diffraction efficiency because of the very small interaction length. Also, in either case the center frequency of the transducer is set at a relatively high value. Consequently, a higher diffraction efficiency will necessarily require large RF or acoustic drive power. Therefore, it may be concluded that, for applications that require both large bandwidth and high diffraction efficiency, more sophisticated SAW transducer configurations must be employed. It is now possible to realize high-performance planar guided-wave AO Bragg cell modulators with gigahertz center frequency and gigahertz bandwidth using multiple SAW transducer configurations. The design, fabrication, testing, and measured performance figures for a variety of wide-band devices have been reported [3]. A brief discussion on such wide-band transducer configurations is provided in the section "Wideband AO Bragg cell modulator using multiple SAWs."

Time-bandwidth product

The time-bandwidth product of an AO modulator, TB, is defined as the product of the acoustic transit time across the incident light beam aperture and the modulator bandwidth [3]. It is readily shown that this time-bandwidth product is identical to the number of resolvable spot diameters of an AO deflector that utilizes the same modulator, N_R, which is defined as the total angular scan of the diffracted light divided by the angular spread of the incident light. Clearly, it is also identical to the number of resolvable frequency channels when the same modulator is used for RF spectral analysis. Thus the following well-known identities hold:

$$TB = N_R = (D/V_R)\Delta f = \tau \Delta f, \tag{5a}$$

$$\delta f_R = V_R / D, \tag{5b}$$

$$\tau = D/V_R, \tag{5c}$$

where D designates the aperture of the incident light beam, V_R is the velocity of the SAW, Δf is the device bandwidth, τ is the transit time of the SAW across the incident light beam aperture, and δf_R is the incremental frequency change required for deflection of one Rayleigh spot diameter. The acoustic transit time may be considered as the minimum AO switching time if the switching time of the RF driver is sufficiently shorter than the acoustic transit time. The desirable value for N_R depends upon the individual application. For example, in RF signal processing, it is desirable to have this value as large as possible because this value is identical to the processing gain. Thus, for this particular area of application, it is also desirable to have a collimated incident light beam of large aperture. Using a Y-cut LiNbO$_3$ optical wave-guide, a guided-light beam aperture as large as 1.5 cm and good uniformity was demonstrated earlier at the author's laboratory. This light beam aperture resulted in an acoustic transit time of about 4.4 μsec for a Z-propagation SAW ($V_R = 3.488 \times 10^5$ cm/sec) and a corresponding frequency resolution δf_R of 0.232 MHz. Because a bandwidth of up to 1 GHz can be realized using multiple SAW transducers, a time-bandwidth product as high as 4400 is achievable.

Acoustic and RF drive power

The required acoustic drive power at the center frequency f_{a0} and 50% diffraction efficiency is

$$P_a(50\% \text{ diffraction}) = \left(\frac{\lambda_0^2 \cos \theta_m \cos \theta_n}{8} \right) \left(\frac{1}{M_{2mn.\text{eff}}} \right) \left(\frac{1}{L} \right), \tag{6a}$$

where

$$M_{2mn.\text{eff}} \equiv C_{mn}^2(f_{a0})M_{2mn}. \tag{6b}$$

Note that $C_{mn}^2(f_{a0})$ takes a form similar to the overlap integral, with its value depending upon the optical and the acoustic modes of propagation [3], and M_{2mn} has been defined previously. The total RF drive power P_e, is readily found by knowing the electrical-to-acoustic conversion efficiency of the SAW transducer used. Clearly, both the total acoustic and RF drive powers required are inversely proportional to the acoustic aperture.

Finally, it is to be noted that, using some of the wide-band transducer configurations to be discussed in the next section, the AO Bragg cell modulators requiring only milliwatts of electric drive power per megahertz of bandwidth at 50% diffraction efficiency with a bandwidth approaching 1 GHz can be realized [3].

Wideband AO Bragg Cell Modulator Using Multiple SAWs

The results presented in section "Bragg Diffraction Efficiency and Frequency Response" show that the Bragg diffraction efficiency-bandwidth product of a planar AO Bragg cell modulator that utilizes a single SAW is a constant and rather limited. However, a much larger composite bandwidth and thus a much larger diffraction efficiency-bandwidth product can be accomplished by employing a multiple of SAWs that are properly tailored and configured. Multiple SAWs of staggered center frequency and tilted propagation direction (Figure 3) [10] as well as phased-multiple SAWs of identical center frequency and propagation direction (Figure 4) [11] can be used to achieve this objective. A unified treatment has been developed to analyze the AO Bragg diffraction

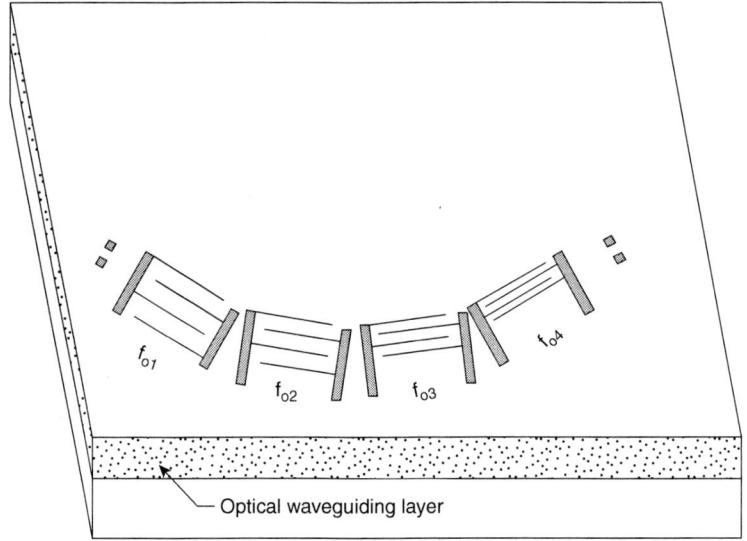

Figure 3   Guided-wave acousto-optic Bragg diffraction from multiple-tilted SAWs

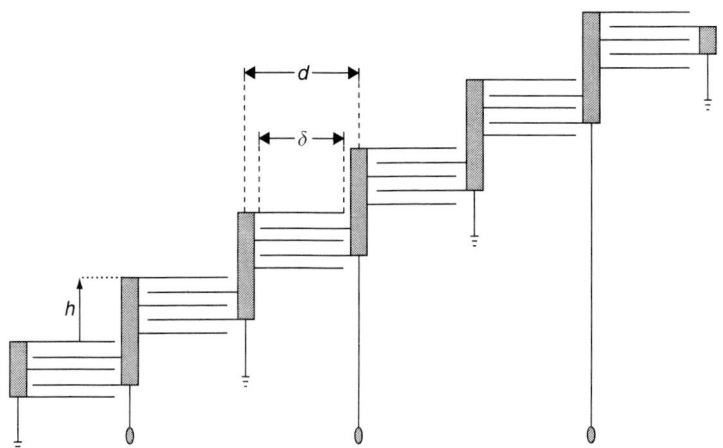

Figure 4   Phased-array multiple SAW transducers

from N number of SAWs generated by N number of transducers [12]. This general approach can be employed to analyze the special cases involving multiple titled and phased SAWs or a combination of both [3].

Using the design procedures and guidelines presented previously, a variety of wideband AO Bragg cell modulators and deflectors have been realized. For example, a modulator (deflector) of 680 MHz composite bandwidth was realized using a Y-cut Ti-diffused $LiNbO_3$ waveguide [13]. The modulator/deflector utilized four tilted transducers with center frequencies of 380, 520, 703, and 950 MHz. The corresponding acoustic apertures were 0.64, 0.47, 0.34, and 0.24 mm, respectively, and the tilt angles between adjacent transducers were 5.9, 8.2, and 11.3 mrad, corresponding to the differences in the Bragg angles at the center frequency of the adjacent transducers. In order to obtain as wide an acoustic bandwidth as possible, the number of finger electrode pairs for each transducer was chosen to be as small as two-and-a-half. The individual transducers were excited in parallel using power dividers. A diffracted light beam of high quality was observed. The measured conversion efficiency of the four element transducers were, respectively, -7.5, -7.0, -10, and -15 dB. The measured diffraction efficiency was 8% at a total RF drive power of 1 W for the entire 680 MHz bandwidth. Since the measured diffraction efficiency of the Bragg cell with only the first three transducers activated was nearly five times higher, a considerably better diffraction efficiency could be expected if the conversion efficiency of the fourth transducer (the one with highest center frequency) were improved to that of the first three.

As an evolvement to Figure 3, we may place all element transducers together on the common substrate, one behind the other in a straight line, and connect them electrically in parallel to result in a "tilted-finger chirp transducer" [13]. Like the conventional chirp transducer the acoustic bandwidth of this composite transducer can be very large. Furthermore, with proper design the wavefront of the SAW generated by this composite transducer can be made to track the Bragg condition for the entire frequency band, and thus result in a large AO Bragg bandwidth. Tilted-finger chirp transducers of both basic design and improved design have been made. For example, the modulator fabricated in a Y-cut $LiNbO_3$ waveguide using a tilted-finger transducer of improved design with a "dog-leg" configuration has provided a bandwidth of 470 MHz at 0.6328 μm light wavelength [14]. The measured diffraction efficiency was 16% at 200 mW RF drive power. This improved modulator was subjected to 1 W of continuous wave (cw) RF drive power without failure. In short, the experimental study has verified that efficient Bragg modulators and deflectors of gigahertz bandwidth can be realized using multiple tilted SAW transducers of staggered center frequency. Specifically, a performance figure of approximately 1 mW electric drive power per megahertz bandwidth with 50% diffraction efficiency and 1 GHz bandwidth should be realizable [3].

Viable Substrate Materials

Based on the AO figure of merit M_{2mn} and the optical, acoustical, and AO properties of the existing materials, the following four viable substrate materials have been identified. A comparison of these substrate types now follows:

LiNbO_3 substrate: Aside from a relatively high AO figure of merit, $LiNbO_3$ also possesses desirable acoustic and optical properties. As a result of large piezoelectricity, a SAW can be generated efficiently by directly depositing the IDT on the substrate. The typical propagation loss of the SAW is 1 dB/cm to 2 dB/cm at 1.0 GHz which is by far the lowest among all AO materials that have been studied. Optical waveguides can be routinely fabricated using the well-established titanium in-diffusion (TI) technique [15]. The measured optical propagation loss is typically 1.0 dB/cm, again the lowest among all existing AO materials. Furthermore, high-quality $LiNbO_3$

crystals of very large size are commercially available. Consequently, $LiNbO_3$ is at present the best substrate material for the realization of wide-band and efficient planar guided-wave AO Bragg cell modulators and related device modules at gigahertz acoustic center frequencies.

SiO_2 As_2, S_3, or SiO_2–Si substrates: The second substrate type is composed of the nonpiezo-electric materials such as fused quartz (SiO_2) [6(a)], arsenic trisulfide (As_2S_3) [16] and oxidized silicon (SiO_2–Si) [17]. Interest in these substrate materials is based on the fact that the first is a common optical waveguide material, the second is an amorphous material with a very large AO figure of merit, and the third may be used to capitalize on the existing silicon technology to further electrical and optical integrations. Even though it is common to deposit a piezoelectric zinc oxide (ZnO) film [3] on such substrate materials for the purpose of SAW generation, ZnO–SiO_2 composite waveguides have also been used to facilitate guided-wave AO interaction because the ZnO film itself also possesses favorable optical waveguiding and AO properties [3]. However, at the present time, measured propagation losses of the SAW in the ZnO–SiO_2–Si composite substrate [18] are much higher than those in the $LiNbO_3$ substrate and also considerably higher than those in the GaAs substrate. This contrast is accentuated as the acoustic frequency goes beyond 200 MHz.

GaAs substrate: GaAs-based substrate can potentially provide the capability for total or mono-lithic integration because both the laser sources [19] and the photodetector arrays as well as the associated electronic devices may be integrated in the same GaAs substrate [20]. The AO Bragg cell modulators that have been realized use a Z-cut [001] GaAs waveguide substrate in which the SAW propagates in $\langle 110 \rangle$ direction [21]. A previous theoretical study [21] has predicted an AO Bragg bandwidth as large as 1.6 and 1.4 GHz for the $\langle 100 \rangle$ and $\langle 110 \rangle$ propagation SAW, respectively, in a Z-cut GaAs–GaAlAs waveguide that supports a TE_0 mode at the optical wave-length of 1.15 μm. For the Bragg cell modulator that used a tilted-finger chirp transducer centered at 485 MHz a –3 dB AO bandwidth of 245 MHz and diffraction efficiency of 5.0% at 1.0 W RF drive power were obtained. The highest acoustic center frequency that has been realized in the GaAs-based Bragg cell modulator was 1.1 GHz. The corresponding diffraction efficiency and AO bandwidth were 19.2% at 1.0 W RF drive power and 78 MHz [21].

InP substrate: As is true of the GaAs substrate, InP substrates can facilitate realization of mono-lithic integrated AO devices and circuits. The prospects for ultimate realization of InP-based inte-grated AO devices has been greatly advanced in light of the continued successes in InP-based photonic integrated circuits (PICs) [22]. Such PICs use the quaternary semiconductor material system of $In_{(1-x)}Ga_xAs_yP_{(1-y)}$ and the computer-controlled growth techniques of metal organic vapor-phase epitaxy (MOVPE) [22], molecular beam epitaxy (MBE) and metalorganic chemical vapor deposition (MOCVD) [23]. By controlling the fractions (x, y) the band gap energy can be tailored in different regions of the InP material substrate, thus enabling monolithic integration of active components such as lasers and photodetectors and passive waveguide regions where AO and EO interactions take place. Furthermore, by varying the fraction (x, y), the refractive index may be tailored to facilitate efficient optical interconnections between the active and passive components.

Figure 5 shows the configuration and geometry for guided-wave AO Bragg diffraction in the Z-cut {001} semi-insulating InP-based composite waveguide substrate that has been studied [24]. The measured photoelastic constants of bulk InP crystals at 1.3 μm wavelength have suggested a large AO figure of merit, which is comparable to that of GaAs [25]. The measured diffraction efficiencies at 1.310 μm wavelength for the TE_0-and TM_0-modes incident light at the acoustic center frequency of 167 MHz and an interaction length of 1.0 mm are 1.41 and 0.95%, respectively, at the acoustic drive power of 0.85 mW. Based on the theoretical predictions and the experimental results reduction in the RF drive power by as much as two orders of magnitude can be achieved by incorporating changes in a new design of the SAW transducer [24].

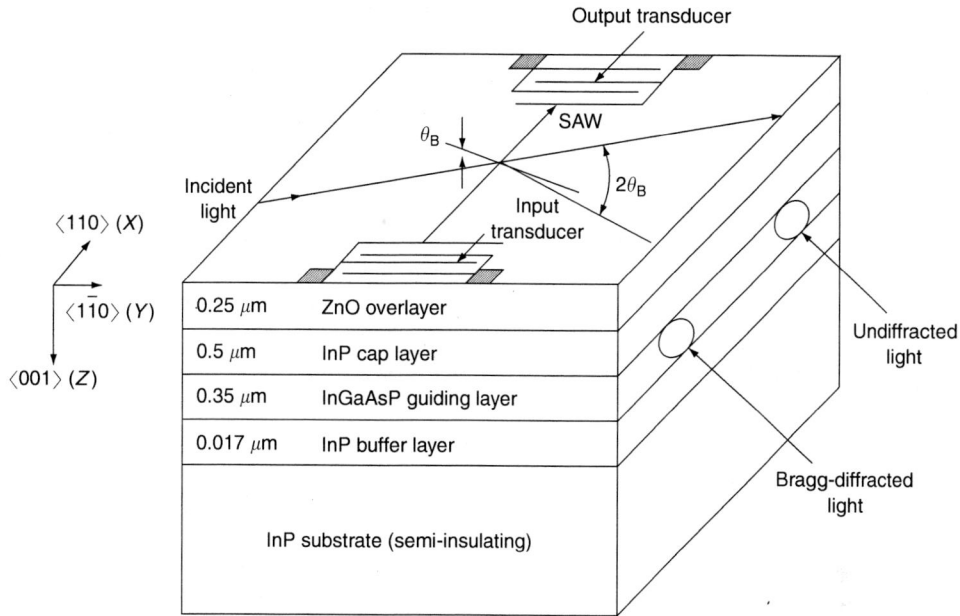

Figure 5   Configuration for guided-wave AO Bragg diffraction in InP/InGaAs/InP substrate

PLANAR GUIDED-WAVE MAGNETO-OPTIC BRAGG CELL MODULATOR

Interaction Geometry, Physical Mechanism, and Analytical Treatment

Significant interactions between the guided-optical waves [1] and the magnetostatic waves (MSWs) [26] can occur through the moving diffraction gratings induced by the latter via Faraday and Cotton–Mouton effects in yttrium iron garnet–gadolinium gallium gannet (YIG–GGG) waveguides [27]. MSWs can be readily and efficiently generated by applying a microwave signal to a microstrip line deposited directly on the YIG–GGG material substrate or brought over it. The physics of guided-wave magneto-optic (MO) Bragg-diffraction in YIG–GGG waveguide has resulted in planar MO Bragg cell modulators [5,28]. The carrier frequency of the MSWs can be tuned, typically from 0.5 to ~40 GHz, by simply varying an external bias magnetic field in synchronism with the carrier frequency of the microwave signal. The corresponding range of wavelength for the MSW is as large as 1000 to 1.0 μm. As in guided-wave AO Bragg interaction the coupled-mode technique [7] is commonly employed for treatment of the MO Bragg diffraction in the interaction geometry of Figure 6 [5]. Similarly, the conservations of carrier frequency and wave vector among the incident and the diffracted light waves, and the MSW must be fulfilled. It should be noted that the wave vector of the MSW not only depends on the carrier frequency, but also the bias magnetic field H_0.

The Bragg diffracted light is scanned in the plane of the waveguide when the carrier frequency for the MSW is varied at a fixed bias magnetic field. The Bragg diffracted light can also be scanned at a fixed carrier frequency of the MSW by varying the bias magnetic field. Note that the latter is not possible with AO Bragg diffraction. The guided-wave MO Bragg diffraction has resulted in Bragg-type planar MO devices such as modulators, scanners, and frequency shifters [29]. Since a very large range of velocity is associated with the MSW, for example, one to three orders of magnitude higher than that of the SAW, the transit time of the MSW across the aperture

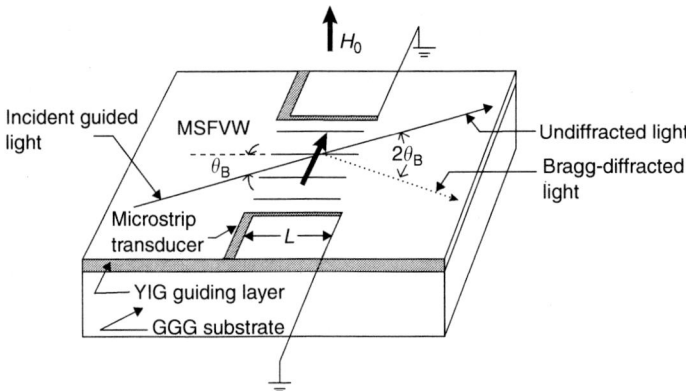

Figure 6   Guided-wave magneto-optic Bragg diffraction from a single magnetostatic wave

of the incident light beam and, thus, the speed of the resulting MO space switches, scanners, and RF spectrum analyzers can be one to three orders of magnitude faster than their AO counterparts.

Bragg Diffraction Efficiency and Frequency Response

Analysis for MO Bragg diffraction has been carried out for the three cases involving magneto-static surface waves (MSSW) [28(a)], magnetostatic backward volume waves (MSBVW) [30], and magnetostatic forward volume waves (MSFVW) [5,28(b), 28(c)] in YIG–GGG waveguides using the coupled-mode technique. The analysis and the results for the third case with the inter-action geometry of Figure 6 are summarized here.

   Similar to AO Bragg diffraction the conservations of frequency and wave vector among the incident and the diffracted light waves, and the MSW must be fulfilled. For the case involving an incident light of the TM_0 mode, the diffracted light of the TE_0 mode, and the MSFVW they are expressed as follows:

$$\vec{\beta}_{TE_0}^{(d)} = \vec{\beta}_{TM_0}^{(u)} \pm \vec{K} \tag{7a}$$

$$\omega_d = \omega_u \pm \Omega, \tag{7b}$$

where $\vec{\beta}_{TE_0}^{(d)}$, $\vec{\beta}_{TM_0}^{(u)}$, and $\vec{K}$ are, respectively, the wave vectors of the Bragg-diffracted light, the undiffracted (incident) light and the MSFVW; and ω_d, ω_u, and Ω are the corresponding radian frequencies. Using the coupled-mode technique the spatial dependence of the diffracted light and the undiffracted light can be determined. The diffraction efficiency η, defined as the ratio of Bragg-diffracted light power at the output ($x = L$) and incident light power at the input ($x = 0$) of the interaction region (see Figure 6) is given as follows:

$$\eta^{\pm}(\Omega) = \sin^2 \left\{ \sqrt{(\kappa^{\pm})^2 + \left(\frac{\Delta}{2}\right)^2} \, L \right\}, \tag{8}$$

$$\kappa^{\pm} \equiv \kappa_1 \mp \kappa_2, \tag{9a}$$

$$\Delta \equiv \left| \vec{\beta}_{TE_0}^{(d)} - \vec{\beta}_{TM_0}^{(u)} \mp \vec{K} \right|, \tag{9b}$$

where the + and − superscript signs designate, respectively, the situations for anti-Stokes (with frequency up-shifted diffracted light) and Stokes (with frequency down-shifted diffracted light) interactions [5], and Δ designates the mismatch in wave vectors.

The coupling coefficients κ_1 and κ_2 are related, respectively, to dynamic Faraday and dynamic Cotton–Mouton coupling coefficients as follows:

$$\kappa_1 = \frac{k_0 |m_x|}{4\sqrt{\varepsilon_r}} f_1, \tag{9c}$$

$$\kappa_2 = \frac{k_0 M_0 |m_y|}{4\sqrt{\varepsilon_r}} \left(2f_{44} + \frac{2}{3} \Delta f \right), \tag{9d}$$

$$\Delta f \equiv f_{11} - f_{12} - 2f_{44}, \tag{9e}$$

$$f_1 \cong \frac{2\sqrt{\varepsilon_r} \phi_r}{k_0 M_0}, \tag{9f}$$

where k_0 is the wave number of the light wave in free space, ε_r is the relative dielectric constant of the YIG film; M_0 is the saturation magnetization; f_{11}, f_{12}, and f_{44} are the three independent components of the linear or second-order magnetic birefringence (Cotton–Mouton effect); f_1 is the circular or first-order magnetic birefringence (Faraday effect); ϕ_r is the Faraday rotation; and m_x and m_y are the RF magnetization components of the MSFVW.

Numerical calculations for the Bragg-diffraction efficiency at the center frequency Ω_0 under perfect phase matching ($\Delta = 0$) and the other frequency Ω ($\Delta \neq 0$) can be carried out using Eq. (8) and all relevant physical parameters and measured MO constants [31]. Furthermore, by inserting the relationship between the RF components, m_x and m_y, of the MSFVW and the corresponding power, and the electrical to MSW conversion efficiency, the RF drive power dependence of the diffraction efficiency can be generated.

In the experimental studies a Bragg diffraction efficiency of 12% was previously measured in a modulator that utilized a 5.0 mm interaction length in a Bi-doped YIG/GGG waveguide and a uniform bias magnetic field [5]. The required RF drive power was 2.0 W. Subsequently, the diffraction efficiency was increased by three- to six-fold using a nonuniform bias magnetic field [32]. Enhancement of Bragg diffraction efficiency by two- to four-fold was accomplished by inserting an electrical feedback loop to the MO Bragg cell modulator [33]. Further increase in Bragg diffraction efficiency using a tilted and nonùniform bias magnetic field is also expected [34].

A Comparison of AO and MO Bragg Cell Modulators

Based on the common characteristics of AO and MO Bragg diffraction gratings, practically all of the RF signal processing functions that can be facilitated by the AO Bragg cell modulator can also be accomplished using the MO Bragg cell modulator. Similar to the AO Bragg cell modulators, the key device parameters of the MO Bragg cell modulators are bandwidth, time–bandwidth

Table 1
Device characteristics—guided-wave AO Bragg cell modulators versus guided-wave MO Bragg cell modulators

	AO	MO
Substrate materials	Ferroelectrics, semiconductors Amorphours	Ferromagnetic Garnets Magnetic semiconductors
Optical wavelength	0.4–12.0 μm	$\geq$1.15 μm
Modulating wave	SAW	MSW
Modulating carrier frequency	Up to around 5 GHz	up to around 40 GHz
Transducer	IDT; rather complex for GHz range	Microstrip line; Rather routine up to upper frequency range
Electronic tunability of carrier frequency	No	Yes, using bias magnetic field
Dispersion	No	Yes
Switching/modulation speed	Microsecond	Nanosecond
Diffraction/modulation efficiency (Per RF driver)	85% per Watt	30% per Watt
Present level of development	High	Moderate
Fabrication technology complexity	Moderate to complex	Moderate
Potential for monolithic integration	High	Moderate

product, magnetostatic and RF drive power, nonlinearity, and dynamic range. A comparison of these two classes of modulators in terms of key device parameters and performance characteristics has shown a number of significant differences (see Table 1). In particular, the MO modulators possess the following unique advantages:

1. A much larger range of tunable carrier frequencies (0.5 to ~40 GHz) may be accomplished by varying the bias magnetic field. Such high and tunable carrier frequencies with MO devices are highly desirable because they allow direct processing at carrier frequency of wideband RF signals and, thus eliminate the need for indirect processing via frequency down-conversion as is required with the AO devices [3]. Indirect processing will require more hardware and incur a higher degree of system complexity, and thus result in higher cost.
2. A large MO bandwidth may be realized by means of a simpler microstrip transducer.
3. Much higher and electronically tunable modulation/switching and scanning speeds are achievable because the velocity of propagation for the MSWs can be higher than that of the SAWs by one- to three-orders of magnitude.
4. A bias magnetic field can be used as a control input. For example, the output angle of MO Bragg diffracted light can be made fixed by synchronous tuning of the carrier frequency of the MSW and the bias magnetic field.
5. The dispersive nature of the MSWs provides potential for implementation of unique signal processing functions.

It is well recognized that the SAW devices in the VHF to UHF region (100 to 2000 MHz) have little competition as broadband RF signal processors. Meanwhile, the MSW devices, when fully developed, will have no competition in the frequency region higher than 2000 MHz. Thus, the related guided-wave AO and MO Bragg cell modulators should compliment each other in the field of real-time wideband RF signal processing at carrier frequencies far beyond 2000 MHz.

INTEGRATED OPTIC RF SPECTRUM ANALYZERS

Acousto-optic RF Spectrum Analyzers

Planar waveguide-based architectures

LiNbO₃-based devices

LiNbO$_3$-based devices
A variety of integrated optic device modules with potential applications to information processing and communications [3,35] have been devised using the basic AO Bragg cell modulator of
Figure 1 and constructed using the substrate materials described in section "Viable substrate
materials" and the fabrication technologies to be presented in this section. The hybrid integrated
AO device modules that have been realized include RF spectrum analyzers, light beam deflectors and scanners, optical frequency shifters, optical space switch arrays, optical correlators,
and matrix multipliers [35]. One of the most active R and D efforts in the 1980s was focused on
the realization of the so-called integrated optic RF spectrum analyzers [3,36]. The basic architecture for a monolithic integrated optic RF spectrum analyzers consists of a solid-state laser
source, a collimation-focusing waveguide lens pair, a wideband AO Bragg cell modulator, a
waveguide photodetector array, and post processing electronic circuits (see Figure 7). For the
LiNbO$_3$ substrate, hybrid modules such as the one shown in Figure 8 in which the diode laser
source and the photodetector array are edge-coupled to the input and output end faces of the
LiNbO$_3$ substrate, respectively, have been realized [3,36]. A review on the performance figures
of all related passive and active components has been reported [3].

It is to be noted that the conventional AO Bragg cell-based RF spectrum analyzers just
described detect only the power and the frequency of the RF signals. Thus, not only the phase
information of the RF signals is lost, but the dynamic range of detection is also severely limited.
In order to alleviate all these disadvantages, an optical heterodyning detection scheme that incorporates a coherent optical reference beam must be used. Accordingly, it is desirable to implement
the interferometric RF receivers, such as the spectrum analyzers, in planar integrated optic architecture format.

Figure 9(a) shows the architecture of a noncollinear interferometric RF spectrum analyzer
module that was constructed in a Y-cut LiNbO$_3$ substrate, $1 \times 8 \times 16$ mm^3 in size [37]. A tilted-
finger chirp SAW transducer having a bandwidth of 205 MHz centered at 350 MHz was used to
facilitate wideband AO Bragg diffraction, and thus produce the optical signal beam. An ion-milled
grating concave lens was added to provide a divergent optical beam via passive Bragg diffraction of the undiffracted light, and thus the optical reference beam for heterodyning detection.

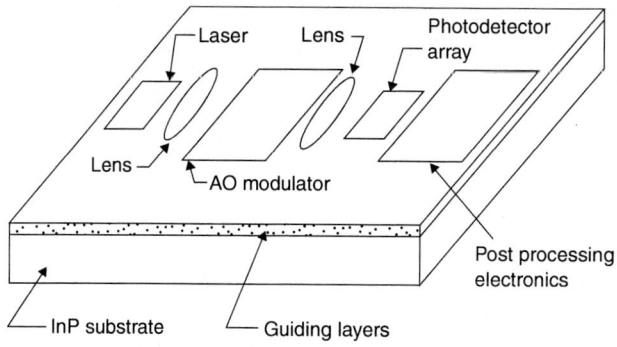

Figure 7   Monolithic integrated acousto-optic RF spectrum analyzer in In$_{1-x}$Ga$_x$As$_y$P$_{1-y}$-based waveguide
substrate

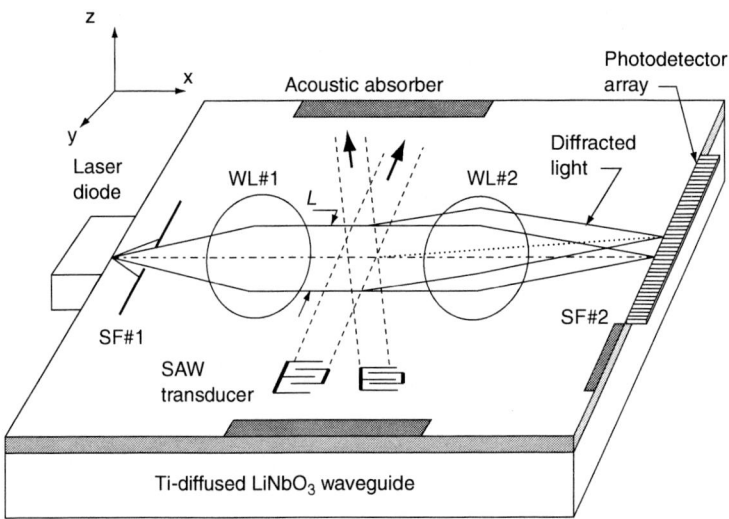

Figure 8   A hybrid integrated acousto-optic RF power spectrum analyzer module using LiNbO$_3$ substrate

Finally, a TIPE waveguide lens was placed at the output region of the substrate so that both focusing and Fourier transform functions were accomplished in the same waveguide substrate. Clearly, for an RF signal applied to the SAW transducer and thus a resulting optical signal beam, a corresponding optical reference beam that is coherent with and propagating in the same direction as the optical signal beam would be provided by the passive ion-milled grating. By design, these two Bragg-diffracted optical beams were made to match with each other and spread across the entire photodetector array, and then efficiently combined by the TIPE lens [38,39]. Therefore, the optical alignment for the entire system is robust. The single-unit (basic) interferometeric spectrum analyzer module just described has demonstrated the capability for simultaneously measuring the amplitude, frequency, and phase of RF signals using a 1 mW single-mode He–Ne laser at the wavelength of 0.6328 μm [37]. Encouraging performance figures including a bandwidth of 205 MHz centered at 350 MHz, a frequency resolution of 3.5 MHz, and single-tone simultaneous and two-tone third-order spurious-free dynamic ranges of 51 and 40 dB, respectively, were measured at the drive power of 50 mW per RF signal input. It should be noted that the measured 51 dB single-tone dynamic range represents a 21 dB enhancement over that measured when the device module was operated as a conventional RF power spectrum analyzer.

Furthermore, a pair of the basic interferometric devices just described were fabricated symmetrically in a Y-cut LiNbO$_3$ planar waveguide, also $1 \times 8 \times 16$ mm^3 in size, to form a dual-unit interferometric RF spectrum analyzer module, as shown in Figure 9(b) [37]. This dual-unit module was successfully used to determine the angle of arrival of RF signals in addition to their frequency, amplitude, and phase, by measuring the phase differences between the heterodyned signals from the two identical basic units of the pair.

GaAs- and InP-based devices
As mentioned previously, GaAs and InP substrates provide the potential capability for total or monolithic integration because both the laser sources and the photodetectors as well as some associated electronic devices can be fabricated in the same substrate of such compound semiconductor materials. For the GaAs-based integrated optic RF power spectrum analyzers, a wideband AO Bragg cell modulator at gigahertz center frequency and a curved-contour hybrid waveguide lens pair in which the gratings lie in a parabolic contour, which is virtually coma-free up to

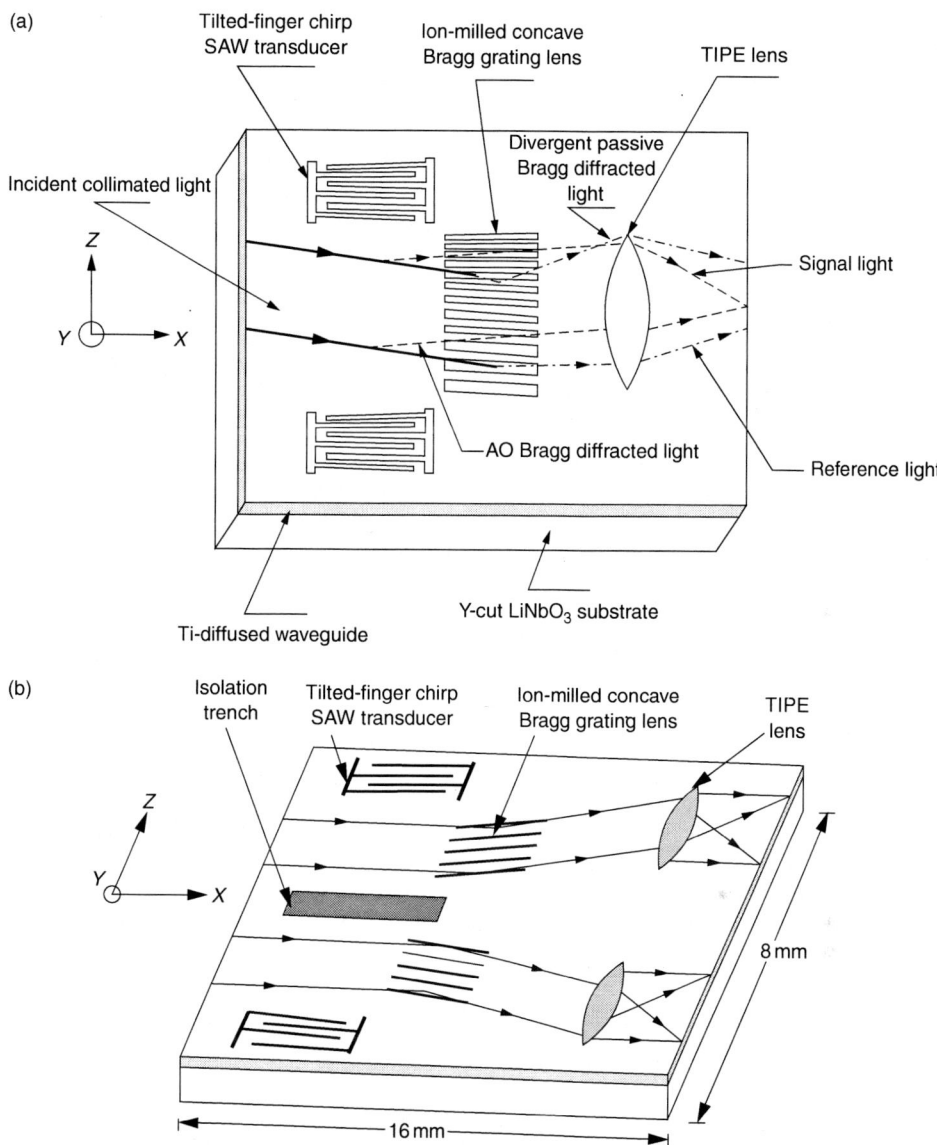

Figure 9 (a) Hybrid integrated AO heterodyne device module using active and passive Bragg diffraction in cascade; (b) Dual-unit hybrid integrated AO heterodyne device module in a Y-cut LiNbO$_3$ planar waveguide

$\pm 4°$ off-axis in a GaAs waveguide, have been realized [40]. This curved-contour waveguide lens is capable of processing RF signals of GHz bandwidth. Such hybrid waveguide lenses were further integrated with a 50-element photodetector array of the InGaAs photoconducting type in the same GaAs waveguide 5×13 mm^2 in size as shown in Figure 10 [41]. The measured cross-talk between adjacent photodetector elements was lower than -14 dB as limited by the side-lobe level of the lens. The InGaAs photoconductive detector array has shown high gain (100 to 1000) at low frequency and a dynamic range of 35 dB, which suggests some potential for use in AO-based devices such as lightbeam switches/scanners in additional to the RF spectrum analyzers.

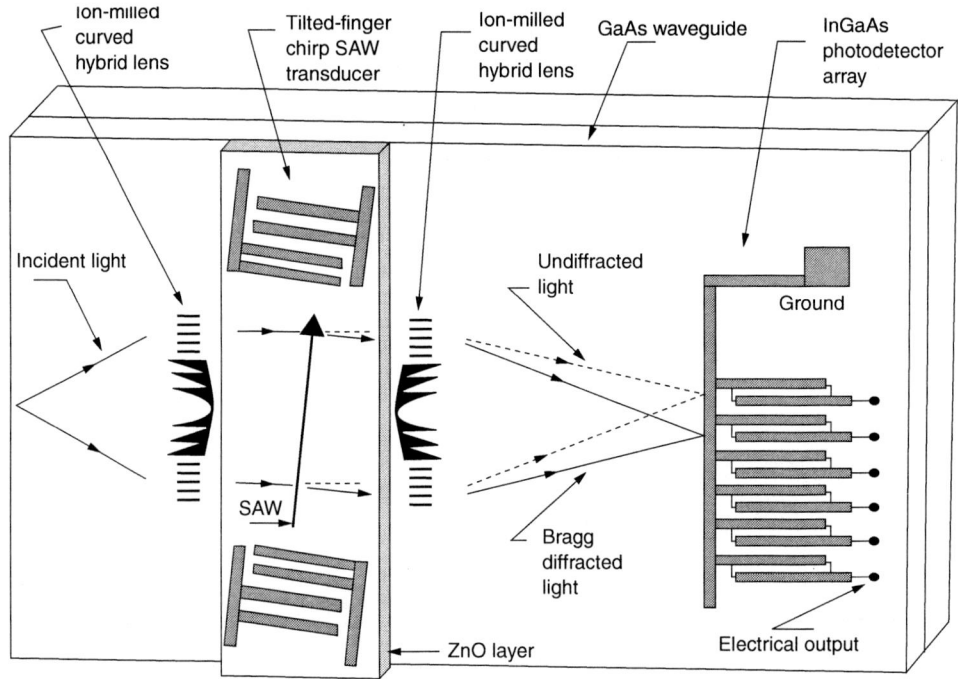

Figure 10 Hybrid integrated acousto-optic RF spectrum analyzer module in a GaAs waveguide

Finally, as mentioned earlier the prospects for realization of monolithic integrated RF spectrum analyzers has been greatly advanced through the continued successes in realization of InP-based PICs in the quaternary semiconductor material system of $In_{(1-x)}Ga_xAs_yP_{(1-y)}$ [22]. For this purpose an AO Bragg cell modulator in a composite InP/InGaAsP/InP planar waveguide has been realized [24].

Spherical waveguide-based architecture

A spherical substrate such as the surface of a $LiNbO_3$ hemispherical block is capable of simultaneously guiding, collimating, and focusing the light beams, and thus serves as an alternate substrate for realization of AO Bragg cell modulator and related device modules such as the integrated optic RF spectrum analyzers [42]. The $LiNbO_3$ spherical waveguide and the AO interaction configuration involved are depicted in Figures 11 and 12, respectively. A focused light beam entering at point P on the rim of the base plane of a hemisphere block is guided, collimated at the top of the hemisphere and refocused at point P' on the opposite side of the rim (see Figure 12). It is clear that the spherical waveguide just described functions like a double-diffraction system in bulk optics. The aberration of this spherical waveguide structure is potentially much smaller than that of a planar waveguide of identical aperture and field angles. Consequently, the diffraction-limited aperture of the light beam can be made considerably larger than that provided by a planar waveguide lens. Accordingly, a larger processing gain or time-bandwidth product can be realized with spherical waveguide AO signal processors. Now if an interdigital transducer of proper orientation and centre frequency is deposited on the top of the hemisphere (see Figure 12), the SAW generated will interact with the incident guided-light wave in a manner similar to the AO interaction in the planar waveguides. In the case of Bragg diffraction, the diffracted light

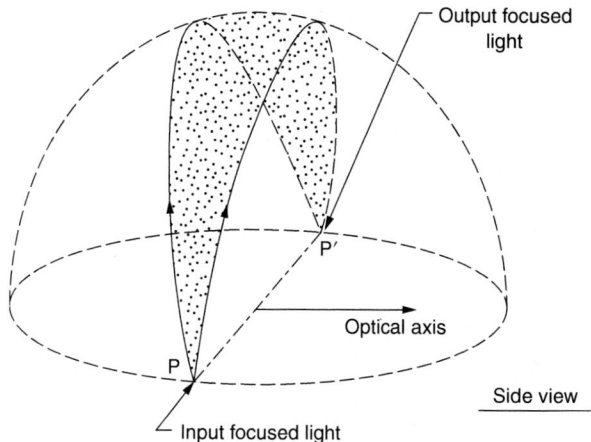

Figure 11   A hemispherical wave guiding surface

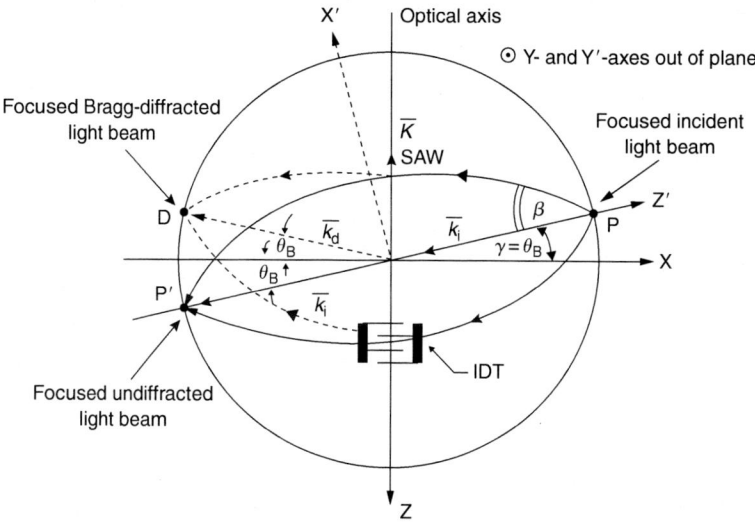

Figure 12   Top view of a hemispherical block for acousto-optic Bragg diffraction

propagates at twice the Bragg angle from the incident light, and focuses at a point D on the rim of the base plane. As the frequency of the SAW is varied, the focal spot of the Bragg diffracted light is scanned along the rim. Thus, the spherical waveguide substrate serves simultaneously as the waveguide and the collimating Fourier transform lens pair, and the resulting spherical Bragg cell modulators can be readily used to implement compact integrated optic device modules for signal processing such as RF spectrum analyzers.

A theoretical analysis for the optical and the acoustic waves using asymptotic expansions and the resulting AO Bragg diffraction using the Green function technique has been carried out [43]. In the experiment the guiding layer on the spherical surface of $LiNbO_3$ hemispherical blocks was formed using the TI method [15], and a pair of identical SAW transducers of 500 MHz centre frequency were fabricated using the lift-off technique [44]. The design parameters of the hemispherical block and the SAW transducer are as follows:

- Spherical waveguide

  - Radius of LiNbO$_3$ hemispherical: 15.5 mm
  - Optical axis of LiNbO$_3$ along the Z-axis on the base plane
  - Incident light wave propagating at Bragg angle off the X-axis

- SAW transducer

  - IDT type: tilted-finger chirp transducer [3] with SAW propagation along the Z- or the optical-axis
  - Acoustic center frequency: 500 MHz
  - Acoustic bandwidth: 250 MHz
  - Finger aperture: 0.5 mm
  - Number of finger pairs: 51

Note that since the dimensions of the IDT SAW transducers are small in comparison to the radius of the LiNbO$_3$ hemispherical block, it was conveniently fabricated on the top of surface of the block. The Bragg diffraction measurements were performed at 0.6328 μm optical wavelength. A laser beam was collimated and focused at the rim of the base plane of the LiNbO$_3$ hemisphere block. The angular spread of the incident light beam was varied by using lenses of different focal length. The incidence angle of the light beam was adjusted to satisfy the Bragg condition. Both the undiffracted light beam and the Bragg-diffracted light beam emerging at the opposite locations along the rim were refocused upon a CCD photo-detector array. A bandwidth of 245 MHz, namely, from 345 to 590 MHz, and a diffraction efficiency of 28% at an RF drive power of 4.4 W were measured. The profile of the undiffracted light beam was a well focused spot with low sidelobe level at the rim and that of the Bragg-diffracted light beam was preserved as well. The Bragg-diffracted light beam was scanned along the rim of the base plane by varying the frequency of the RF drive signal. The diffracted light beams were fully resolved spatially (based on the Rayleigh criterion) at a frequency differential of 4 MHz. In other words, the measured frequency resolution (δf) relevant to applications such as RF spectrum analyzers and light beam scanners was 4 MHz. This measured frequency resolution is to be compared with the calculated value of 3.75 MHz. Finally, the throughput efficiency measured between the input port and output port varied from -9 to -11 dB.

Magneto-optic RF Spectrum Analyzers

A permanent magnet-based MO Bragg cell modulator module was constructed initially [45(a)]. Subsequently, a MO Bragg cell modulator was integrated with a collimation-focusing lens pair in a bismuth-doped YIG–GGG taper waveguide structure, with dimensions of 6.0×16.0 mm^2 (see Figure 13) [45(b)]. As shown in the central region of the figure, the initial thickness of the YIG waveguide layer was 3.25 μm. The two end regions of the taper waveguide, each of 5.0 mm in length, were milled down to 2.68 μm in three steps in order to produce a gradual transition, and thus ensure a high transmission for the light beam. Two identical curved hybrid lenses with 4.0 mm focal length, 0.8 mm aperture, and 300 μm lens length were then fabricated in the two end regions using the ion-milling technique [45(b)].

The MO Bragg cell modulator was constructed by incorporating a pair of identical microstrip line transducers in the central section of the taper waveguide in order to take advantage of the earlier finding that the inherent bandwidth associated with magnetostatic forward volume waves (MSFVW) increases approximately with the thickness of the YIG layer [46]. One transducer was used to convert the microwave drive signal into MSFVW, while the other at a separation of 4.0 mm was used to

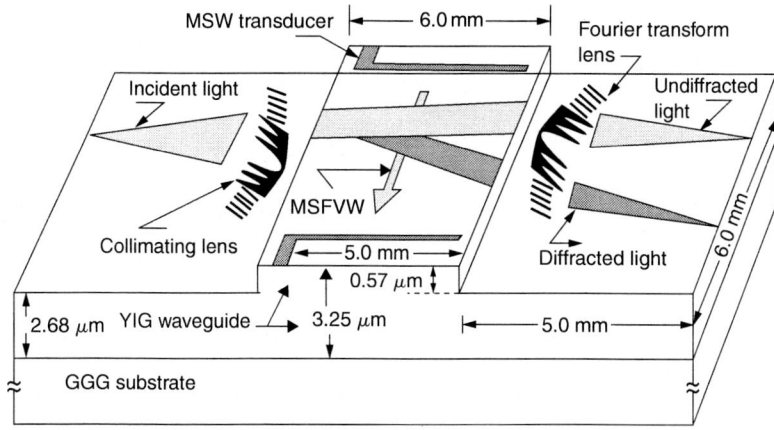

Figure 13 Hybrid integrated magneto-optic RF spectrum analyzer in YIG–GGG taper waveguide structure

convert the MSFVW back to RF signal for measurement of its conversion efficiency, and other propagation characteristics. The length of the waveguide central section of the waveguide was set at 6.0 mm in order to enable usage of maximum MO interaction length, where the microstrip line transducers were configured. In this particular design, the transducers had a strip width of 60 μm and an aperture of 5.0 mm, which was also the MO interaction length. As mentioned previously, the two 5.0 mm end sections were used to accommodate the focal length of the lens pair.

The finished MO Bragg cell modulator sample just described was then inserted into a compact magnetic housing [45] used to provide the required tunable bias magnetic field for saturation of the YIG layer and excitation of wideband MSFVWs. A 1.303 μm laser diode was butt-coupled into the taper waveguide in the TE- or TM-mode for measurement of the performance of the resulting hybrid integrated MO device module. The measured bandwidth was 290 MHz at the center carrier frequency of 9970 MHz and a fixed bias magnetic field of 3600 Oe. The calculated bandwidths for the YIG layer thickness of 3.25 and 2.68 μm are 300 and 250 MHz, respectively. The good agreement between the calculated and the measured bandwidths confirms the capability of the taper waveguide structure to preserve the inherent MO interaction bandwidths. The diffraction efficiency of the modulator was measured to be 5.0% at the carrier frequency of 9990 MHz and the RF drive power of 30.7 dBm. A dynamic range (defined as the range of linear dependence of the measured diffraction efficiency versus the RF drive power) of 25 dB was also measured. A larger dynamic range should be achievable by using a waveguidelens combination with a lower degree of scattering. Scanning of the focused diffracted light spot of orthogonal polarization upon the output edge of the waveguide was accomplished as the carrier frequency of the microwave drive signal was varied from 10.25 to 10.75 GHz at a fixed bias magnetic field of 3700 Oe. The corresponding scanning of the diffracted light spot for the carrier frequency centered at 7.50 GHz and at 2820 Oe bias magnetic field was also accomplished.

Application of the integrated MO Bragg cell modulator to RF spectral analysis was also demonstrated by applying simultaneously two carrier frequencies of 10.315 and 10.350 GHz. A frequency resolution of about 35 MHz was demonstrated. The frequency resolution δf_R, namely, the frequency separation between adjacent resolvable light beam spots or frequency channels is determined by following expression [45]:

$$\delta f_R \approx \left(\frac{V_{msw}}{D} \right) \times \frac{1}{\left\{ 1 - \left(f / V_{msw} \right) \times \left(dV_{msw} / df \right) \right\}}, \qquad (10)$$

where V_{msw} and f designate, respectively, the velocity and the carrier frequency of the MSFVW, and D the light beam width. It is seen that due to the dispersive nature of the MSW the frequency resolution depends upon the carrier frequency, and that the expression for frequency resolution reduces to that for AO Bragg diffraction [3] as the factor corresponding to dV_{msw}/df due to dispersion for the SAW vanishes. The measured frequency resolution was found to be in good agreement with the calculated frequency resolution of 36 MHz at the center carrier frequency of 10.30 GHz. Thus, approximately seven to nine resolvable frequency channels were demonstrated with this preliminary hybrid integrated MO RF spectrum analyzer at the center carrier frequency of 10.30 GHz [45].

Fabrication Technologies and Prospects for Further Integration

Construction of hybrid integrated AO devices involves mainly fabrication of optical waveguide, SAW transducers, and waveguide lenses. For the LiNbO$_3$ substrate, both the titanium-indiffusion (TI) [15] and the proton-exchange (PE) [47] techniques, and their combination (TIPE) [38] have been refined to produce high-quality planar and channel waveguides. These techniques are readily employed to fabricate the planar and the channel waveguide arrays as well as their combinations in a common substrate required for construction of integrated AO device modules and circuits [35]. A variety of waveguide lenses have been explored [48]. As to formation of refractive waveguide lenses in the forms of collimation-focusing lens pairs and microlens arrays, the TIPE technique can also be conveniently employed [39]. Recently, diffractive or grating waveguide lenses were formed in the LiNbO$_3$ waveguides using the ion-milling technique [49]. Finally, formation of interdigital finger electrode transducers on the top of the LiNbO$_3$ waveguide for excitation of the SAW is routinely accomplished through the processing steps including preparation of photomask having the transducer IDT electrode pattern, deposition of metallic thin film, and replication of the IDT electrode pattern by using the liftoff technique [44].

For the GaAs and InAsP materials systems, both the molecular beam epitaxy (MBE) [23(a)], the metal organic chemical vapor deposition (MOCVD) [23(b)], and the MOVPE [22] have become the standard techniques for construction of high-quality planar and channel waveguides. Ion-milling using argon ions has also been developed to fabricate diffractive or grating waveguide lenses in GaAs material substrates [40]. Aspheric waveguide lenses have also been fabricated using selective chemical etching on the InP/InGaAsP material system [50]. With respect to efficient excitation of the SAW, it is necessary to first deposit a piezoelectric thin-film such as zinc oxide (ZnO) on the top of the waveguide and then fabricate the IDT electrode pattern upon it [3]. Finally, the ZnO films are often grown using the RF sputtering technique [3].

As mentioned in section "Viable Substrate Materials," among the variety of potential substrate materials for the integrated AO signal processors such as RF spectrum analyzers ferroelectric LiNbO$_3$ and compound semiconductor GaInAsP material systems continue to be the two most viable ones. Recent progress on integrated lasers and amplifiers in Er-doped LiNbO$_3$ substrates has also significantly enhanced the potential of this substrate material. As to the InGaAsP material system, technologies for construction of high-quality waveguides, lasers, photodetectors, and other active and passive devices in the same substrate have been established through the on-going concerted efforts toward the realization of PIC [22]. Clearly, these same technologies can be employed to implement monolithic integrated AO devices and modules.

As to the substrate materials for planar MO devices, YIG–GGG remains the most viable. A number of significant advances have been made recently in MSW-based guided-wave magneto-optics. These advances include LPE growth of high-quality pure YIG–GGG waveguides, quality improvement in Bi-doped and Ce-doped YIG–GGG wave-guides, design and fabrication of efficient and wideband transducers for MSWs, theoretical analysis on noncollinear coplanar guided-wave MO Bragg diffraction [5], realization of compact MO Bragg cell modulators [5], realization of ion-milled waveguide lenses and their integration with the MO Bragg cell modulator [45], realization of high efficiency modulators using nonuniform bias magnetic field [31] and electronic feedback [32], and demonstration of their applications in RF spectral analysis at X-band and light beam modulation/scanning/switching [5], and wideband optical frequency shifting at the X-band [29]. Thus, planar Bragg-type MO devices constitute a new class of integrated optic devices which are potentially capable of providing desirable features similar to that of the now prevalent AO devices, but at much higher and electronically tunable carrier frequencies.

As to the integrated MO signal processors, prospects for incorporating lasers in the YIG–GGG substrate with doping such as Er ions have started to be recognized. However, it is clear that the technology toward ultimate realization of monolithic integrated AO device modules is considerably ahead of that for its MO counterpart.

CONCLUSION

Advances in the sciences and the technologies of guided-wave AO and MO Bragg diffractions have resulted in planar AO and MO Bragg cell modulator, respectively. The two types of modulators are shown to compliment each other with the second type capable of operating at much higher carrier frequencies. Both modulators possess desirable device characteristics and have demonstrated a variety of applications in wideband real-time RF signal processings and optical communications. Thus, the integrated AO and MO Bragg modulator modules should compliment each other in the field of wideband real-time RF spectral analysis for the range of carrier frequencies as large as subgigahertz to 40 GHz. A number of hybrid integrated AO and MO RF spectrum analyzers have been constructed and tested with encouraging performance figures. At present, $LiNbO_3$ and YIG are, respectively, the most viable material substrates for realization of hybrid AO and MO analyzer modules, while $In_{(1-x)}Ga_xAs_yP_{(1-y)}$ appears to be the most viable material substrate for monolithic AO analyzer modules. The prospects for advances toward further integration are bright for both types of integrated optic RF spectrum analyzers.

REFERENCES

1. See, for example, (a) P. K. Tien, "Light waves in thin films and integrated optics," *Appl. Opt.*, 10, 2395–2413, 1971; (b) H. F. Taylor and A. Yariv, "Guided-wave optics," *Proc. IEEE*, 62, 1044–1060, 1974; (c) H. Kogelnik, "An introduction to integrat356ed optics," *IEEE Trans. Microwave Theory Technol.*, 23, 2–16, 1975; (d) T. Tamir, Ed., *Integrated Optics*, Berlin: Springer-Verlag, 1979; (e) H. P. Nolting and R. Ulrich, Eds., *Integrated Optics*, Berlin: Springer-Verlag, 1985; (f) K. Iga, Y. Kokubun, and M. Oikawa, *Fundamentals of Microoptics*, New York: Academic Press, 1984.
2. See, for example, (a) R. M. White, "Surface elastic waves," *Proc. IEEE*, 58, 1238–1276, 1970; (b) A. A. Oliner, Ed., *Surface Acoustic Waves*, Berlin: Springer-Verlag, 1979; (c) G. W. Farnell and E. L. Adler, "Elastic wave propagation in thin layers," in W. P. Mason and R. N. Thurston, Eds., *Physical Acoustics*, Vol. 9, New York: Academic Press, 1972.
3. C. S. Tsai, Ed., *Guided-Wave Acoustooptic Bragg Diffraction, Devices, and Applications*, Berlin: Springer-Verlag, 1990.
4. See, for example, (a) D. A. Smith et al., "Integrated optic acoustically tunable filters for WDM networks," *IEEE J. Select. Areas Commun.*, 8, 1151–1159, 1990; (b) Y. Yamamoto, C. S. Tsai, K. Esteghamat, and H. Nishimoto, "Suppression of sidelobe levels for guided-wave acoustooptic tunable filters using weighted coupling," *IEEE Trans. Ultrason. Ferroelect. Freq. Control*, 40, 813–818, 1993; (c) A. Kar-Roy and C. S. Tsai, "Integrated acoustooptical tunable filters using weighted coupling," *IEEE J. Quantum Electron.*, 30, 1574–1586, 1994.

5. C. S. Tsai and D. Young, "Magnetostatic forward volume wave-based guided-wave magneto-optic Bragg cells and applications to communications and signal processing," *IEEE Trans. Microwave Theory Technol.*, **MTT-38**, 560–570, 1990.

6. See, for example, (a) L. Kuhn et al., "Deflection of an optical guided-wave by surface acoustic wave," *Appl. Phys. Lett.*, 17, 265–268, 1970; (b) Y. Ohmachi, "Acousto-optical light diffraction in thin films," *J. Appl. Phys.*, 44, 3928–3922, 1973; (c) R. V. Schmidt and I. P. Kaminow, "Acoustooptic Bragg deflection in LiNbO$_3$ Ti-diffused waveguides," *IEEE J. Quantum Electron.*, **QE-11**, 57–59, 1975; (d) C. S. Tsai, Le T. Nguyen, S. K. Yao, and M. A. Alhaider, "High-performance guided-light-beam device using two tilted surface acoustic wave," *Appl. Phys. Lett.*, 26, 140–142, 1975; (e) E. G. Lean, J. M. White, and C. D. W. Wilkinson, "Thin-film acousto-optic devices," *Proc. IEEE*, 64, 779–788, 1976; (f) T. G. Giallorenzi, "Acousto-optic deflection in thin-film waveguides," *J. Appl. Phys.*, 44, 242–253, 1973; (g) V. V. Proklov, "Acousto-optic interactions in planar waveguide," in *Pt.I. Proceedings of the International Symposium on Surface Waves in Solids and Layered Structures*, Novosibirsk, 1986, Vol. 1, pp. I48–I63; D.V. Petrov, "Acousto-optic interactions in a planar optical waveguide," *Pt.II Proceedings of the International Symposium on Surface Waves in Solids and Layered Structures*, pp. I64–I75.

7. A. Yariv, "Coupled-mode theory for guide-wave optics," *IEEE J. Quantum Electron.*, **QE-9**, 919, 1973.

8. R. W. Smith, H. M. Gerard, J. H. Collins, T. M. Reeder, and H. J. Shaw, "Design of surface wave delay lines with integrated transducers," *IEEE Trans. Microwave Theory Technol.*, **MTT-17**, 856–873, 1969.

9. W. R. Smith, H. M. Gerard, and W. R. Jones, "Analysis and design of dispersive interdigital surface-wave transducers," *IEEE Trans. Microwave Theory Technol.*, **MTT-20**, 458–471, 1972.

10. C. S. Tsai, M. A. Alhaider, Le T. Nguyen, and B. Kim, "Wideband guided-wave acoustooptic Bragg diffraction and devices using multiple tilted surface acoustic waves," *Proc. IEEE*, 64, 318–328, 1976.

11. Le T. Nguyen and C. S. Tsai, "Efficient wideband guided-wave acoustooptic Bragg diffraction using phased-surface acoustic wave array in LiNbO$_3$ waveguide," *Appl. Opt.*, 16, 1297–1304, 1977.

12. B. Kim and C. S. Tsai, "High-performance guided-wave acoustooptic scanning devices using multiple surface acoustic wave," *Proc. IEEE (Special Issue on Surface Acoustic Waves)*, 64, 788–793, 1976.

13. C. C. Lee, K. Y. Liao, C. L. Chang, and C. S. Tsai, "Wideband guided wave acoustooptic Bragg deflector using a tilted finger-chirp transducer," *IEEE J. Quantum Electron.*, 15, 1166–1170, 1979.

14. K. Y. Liao, C. L. Chang, C. C. Lee, and C. S. Tsai, "Progress on wideband guided-wave acousto-optic Bragg deflector using a tilted-finger chirp transducer," in *Proceedings of the 1979 Ultrasonics Symposium IEEE Cat. No. 79CH1482-9SU*, 1979, pp. 24–27.

15. R. V. Schmidt and I. P. Kaminow, "Metal-diffused optical wave guides in LiNbO$_3$," *Appl. Phys. Lett.*, 25, 459–460, 1974.

16. T. Suhara, T. Shiono, H. Nishihara, and J. Koyama, "An integrated-optic Fourier processor using an acoustooptic deflector and Fresnel lenses in As$_2$S$_3$ waveguide," *IEEE J.*, **LT-1**, 624, 1983.

17. See, for example, H. Schmidt, M. Weihnacht, and R. Wobst, "A thin-film on-Silicon acoustooptical modulator with multimode behavior," *Proc. IEEE Ultrasonics Symp.*, 847–850, 1995.

18. N. Chubachi, J. Kushibiki, and Y. Kikuchi, "Monolithically integrated Bragg deflector for an optical guided wave made of zincoxide film," *Electron. Lett.*, 9, 193–194, 1973.

19. See, for example, the many references cited in (a) A. Yariv, *Introduction to Optical Electronics*, 2nd ed., New York: Holt, Rinehart and Winston, 1976; (b) H. Kressel, Ed., *Semiconductor Devices for Optical Communication*, 2nd ed., *Top. Appl. Phys.*, Vol. 39 (Springer-Verlag, Berlin, 1982); (c) H. C. Casey Jr. and M. B. Panish, *Heterostructure Lasers*, New York: Academic, 1978; (d) M. Nakamura, *IEEE Trans.*, **CAS-26**, 1055, 1979; (e) Y. Suematsu, "Advances in semiconductor lasers," *Phys. Today.*, 32, 32–39, 1985; (f) L. A. Coldren and S. W. Corzine, *Diode Lasers and Photonic Integrated Circuit*, New York: John Wiley & Sons, 1995.

20. See, for example, (a) J. L. Merz, R. A. Logan, and A. M. Sergent, "GaAs integrated optical circuits by wet chemical etching," *IEEE J. Quantum Electron.*, **QE-5**, 72–82, 1979; (b) A. Yariv, "The beginning of integrated optoelectronic circuits," *IEEE J. Quantum Electron. Devices*, **ED-3**, 1956, 1984; (c) O. Wada, T. Sakurai, and T. Nakagami, "Recent progress in optoelectronic integrated circuits," *IEEE J. Quantum Electron.*, **QE-22**, 805, 1986.

21. See, for example, (a) O. Yamazaki, C. S. Tsai, M. Umeda, L. S. Yap, C. J. Lii, K. Wasa, and J. Merz, "Guided-wave acoustooptic interactions in GaAs-ZnO composite structure," in *Proceedings of the 1982 Ultrasonic. Symposium*, IEEE Cat. No. 82Ch1823-4, pp. 418–421; (b) C. J. Lii, C. S. Tsai, and C. C. Lee, "Wideband guided-wave acoustooptic Bragg cells in GaAs-GaAlAs waveguide," *IEEE J. Quantum Electron.*, **QE-22**,

868–872, 1986; (c) Y. Abdelrazak, C. S. Tsai, and T. Q. Vu, "An integrated optic RF spectrum analyzers in ZnO-GaAs-GaAlAs waveguide," *IEEE Lightwave Technol.*, **8**, 1833–1837, 1990; (d) A. M. Matteo, V. M. N. Passaro, and M. N. Armenise, "High-performance guided-wave acoustooptic Bragg cells in LiNbO$_3$ and GaAs-based structures," *IEEE Trans. Ultrason. Ferroelectr. Freq. Control*, 43, 270–279, 1996.

22. See, for example, T. L. Koch and U. Koren, "Semiconductor photonic integrated circuits," *IEEE J. Quantum Electron.*, **QE-27**, 641–653, 1991.

23. See, for example, (a) W. T. Tsang and A. Y. Cho, "Molecular beam epitaxial writing of patterned GaAs epilayer structures," *Appl. Phys. Lett.*, 32, 491–493, 1978; (b) R. D. Dupus and P. D. Dapkus, "Preparation and properties of Ga$_{1-x}$Al$_x$As–GaAs heterostructures lasers grown by metalorganic chemical vapor deposition," *IEEE J. Quantum Electron.*, **QE-15**, 128, 1979.

24. C. S. Tsai, B. Sun, and A. K. Roy, "Guided-wave acoustooptic Bragg diffraction in indium gallium arsenide phosphide waveguides," *Appl. Phys. Lett.*, 70, 3185–3187, 1997.

25. N. Suzuki and K. Tada, "Elastooptic properties of InP," *Jpn. J. Appl. Phys.*, 22, 441–445, 1983.

26. See, for example, (a) J. D. Adam, "Analog signal processing with microwave magnetics," *Proc. IEEE*, 76, 159–170, 1988; (b) W. S. Ishak, "Magnetostatic wave technology: A review," *Proc. IEEE.*, 76, 171–187, 1988.

27. See, for example, (a) P. K. Tien, R. J. Martin, R. Wolfe, R. C. LeCraw, and S. L. Blank, "Switching and modulation of light in magneto-optic waveguides of garnet films," *Appl. Phys. Lett.*, 21, 394, 1972; (b) K. Ando, N. Takeda, T. Okuda, and N. Koshizuka, "Waveguide mode conversion by magnetic linear birefringence of Bi-substituted iron garnet films tilted from (111)," *J. Appl. Phys.*, 57, 718, 1985; (c) V. J. Fratello and R. Wolfe, "Epitaxial garnet films for nonreciprocal magneto-optical devices," in *Magnetic Thin Film Devices, Handbook of Thin Film Devices*, Vol. 4, J. D. Adam and M. H. Francombe, Eds., Academic Press, 2000, pp. 93–141.

28. (a) C. S. Tsai, D. Young, W. Chen, L. Adkins, C. C. Lee, and H. Glass, "Noncollinear coplanar magneto-optic interaction of guided optical wave and magnetostatic surface waves in yttrium iron garnet-gadolinium gallium garnet waveguides," *Appl. Phys. Lett.*, 47, 651–654, 1985; .(b) C. S. Tsai and D. Young, "Wide-band scanning of guided-light beam and RF spectral analysis using magnetostatic forward volume waves in YIG-GGG waveguide," *Appl. Phys. Lett.*, 54, 196–198, 1989.

29. Y. Pu and C. S. Tsai, "Wideband integrated magnetooptic frequency shifter at X-band," *Appl. Phys. Lett.*, 62, 3420–3422, 1993.

30. Y. Pu, C. L. Wang, and C. S. Tsai, "Magnetostatic backward volume wave-based guided-wave magneto-optic Bragg cells and application to wide-band lightbeam scanning," *IEEE Photon. Technol. Lett.*, 5, 462–465, 1991.

31. R. V. Pisarev, I. G. Sinii, N. N. Kolpakova, and Yu. M. Yakovlev, "Magnetic Birefringence of light in iron garnets," *Sov. Phys. JETP*, 33, 1175, 1971.

32. C. S. Tsai, Y. S. Lin, J. Su, and S. R. Calciu, "High efficiency guided-wave magnetooptic Bragg cell modulator using non-uniform bias magnetic field," *Appl. Phys. Lett.*, 71, 3715–3717, (1997).

33. J. Su and C. S. Tsai, "A magnetostatic forward volume wave oscillator-based magnetooptic Bragg cell modulator," *Appl. Phys. Lett.*, 74, 2878–2880, 1999.

34. W. Zuo, J. Su, G. Q. Liu, and C. S. Tsai, "Guided-wave magnetooptic Bragg diffraction in YIG-GGG waveguide under tilted and non-uniform bias magnetic field," in *Proceedings of the 1997 IEEE Ultrasonics Symposium*, 1997, 757–760, IEEE Cats #97CH36118.

35. (a) C. S. Tsai, "Guided-wave acoustooptic Bragg modulators for wideband integrated optic communications and signal processing," *IEEE Trans. Circuits Syst.*, **CAS-26**, 1072–1089, 1979; (b) C. S. Tsai, "Integrated acoustooptic circuits and applications," *IEEE Trans. Ultrasonics Ferroelectr. Freq. Control*, 39, 529–554, 1992; (c) C. S. Tsai, "Integrated acoustooptic and magnetooptic devices for optical information processing," *Proc. IEEE*, 84, 853–859, 1996.

36. (a) D. B. Anderson, J. T. Boyd, M. C. Hamilton, and R. R. August, "An integrated-optical approach to the Fourier transform," *IEEE J.*, **QE-13**, 268, 1977; (b) C. S. Tsai, "Guided-wave acoustooptic Bragg modulators for wideband integrated optic communications and signal processing," *IEEE Trans. Circuits Syst.*, **CAS-26**, 1072–1089, 1979; (c) M. K. Barnoski, B. Chen, T. R. Joseph, J. Y. M. Lee, and O. G. Ramer, "Integrated-optic spectrum analyzer," *IEEE Trans.*, CAS-26, 1113–1124, 1979; (d) E. Marx, L. D. Hutcheson, and A. L. Keller, "Operational integrated-optic RF spectrum analyzer," *Appl. Opt.*, 19, 3033–3034, 1980; (e) R. L. Davis and F. S. Hickernell, "An integrated optic spectrum analyzer with tin film lenses," in *Proceedings of the International Conference on Integrated Optics and Optical Fiber*

Communication, San Francisco, CA, Paper No. WE-6, 1981; (f) V. Neuman, C. W. Pitt, and L. M. Walpita, "An integrated acousto-optic spectrum analyzer using grating components," in *Proceedings of the First European Conference Integrated Optics*, London, England, 1981, pp. 89–92; (g) T. Suhara, H. Nishihara, and J. Koyam, "A folded-type integrated optic spectrum analyzer using butt-coupled chirped grating lenses," *IEEE J.* **QE-18**, 1057–1059, 1982; (h) R. L. Davis and F. S. Hickernell, "Application of wideband Bragg cells for integrated optic spectrum analyzer," *Proc. SPIE*, **321**, 141–148, 1982; (i) V. M. Ristic, S. A. Jones, and G. R. Dubois, "Evaluation of an integrated acousto-optic receiver," *Can. Elect. Eng. J.*, 8, 59–64, 1983; (j) M. Kanazawa, T. Atsumi, M. Takami, and T. Ito, "High resolution integrated optic spectrum analyzer," in *Proceedings of the International Conference on Integrated Optics and Optical Fiber Communication*, Tokyo, Japan, 1983, Paper No. 30B-3; (k) S. Valette, J. Lizet, P. Mottier, J. P. Jadot, S. Renard, A. Fournier, A. M. Grouillet, P. Gidons, and H. Denis, "Integrated optical spectrum analyzer using planar technology on oxidized silicon substrate," *Electron. Lett.*, **19**, 883–885, 1983 and *IEEE Proc.* H, **131**, 325–331, 1984; (l) T. Suhara, T. Shiono, H. Nishihara, and J. Koyama, "An integrated-optic Fourier processor using an acousto-optic deflector and Fresnel lenses in As_2S_3 wave-guide," *J. Lightwave Technol.*, **LT-1**, 624–630, 1983; (m) C. Stewart, G. Serivener, and W. J. Stewart, "Guided-wave acousto-optic spectrum analysis at frequencies above 1 GHz," in *Proceedings of the International Conference on Integrated Optics and Optical Fiber Communication*, in Technical Digest, 1981, p. 122; (n) E. T. Aksenov, N. A. Esepkina, A. A. Lipovskii, and A. V. Pavenko, "Prototype integrated acousto-optic spectrum analyzer," *Sov Tech. Lett.*, **6**, 519–520, 1980; (o) T. R. Ranaganath, T. R. Joseph, and J. Y. Lee, *Proceedings of the Third International Conference on Integrated Optics and Optical Fiber Communications*, San Francisco, CA, *Techn. Digest*, Paper WH3, 1981.

37. G. D. Xu and C. S. Tsai, "Integrated acousto-optic heterodyning device modules in $LiNbO_3$ substrate," *Appl. Opt.*, **31**, 5259–5268, 1992.

38. M. DeMicheli et al., "Fabrication and characterization of titanium-indiffused proton exchanged (TIPE) waveguide in lithium niobate," *Opt. Commun.*, 42, 101, 1982.

39. D. Y. Zhang and C. S. Tsai, "Titanium-indiffused proton-exchanged waveguide lenses in $LiNbO_3$ for optical information processing," *Appl. Opt.*, 25, 2264–2271, 1986.

40. T. Q. Vu, J. A. Norris, and C. S. Tsai, "Planer waveguide lenses in GaAs using ion milling," *Appl. Phys. Lett.*, 54, 1098–1100, 1989.

41. T. Q. Vu, C. S. Tsai, and Y. C. Kao, "Integration of curved hybrid waveguide lens and photodetector array in a GaAs waveguide," *Appl. Opt.*, 31, 5246–5254, 1992.

42. Q. Li, C. S. Tsai, S. Sottini, and C. C. Lee, "Light propagation and acousto-optic interaction in A $LiNbO_3$ spherical waveguide," *Appl. Phys. Lett.*, 46, 707–709, 1985.

43. C. S. Tsai, W. Chen, P. Le, and S. C. Tsai, "Acousto-optic interactions and devices in a spherical waveguide," *J. Optics A: Pure Appl. Opt.*, 3, S46–S53, 2001.

44. H. I. Smith, F. J. Bachner, and N. Efremow, "A high-yield photolithographic technique for surface wave devices," *J. Electrochem. Soc.*, 118, 822–825, 1971.

45. (a) C. L. Wang, Y. Pu, and C. S. Tsai, "Permanent magnet-based guided-wave magnetooptic Bragg cell modules," *J. Lightwave Technol.*, 10, 624–648, 1992; (b) C. L. Wang and C. S. Tsai, "Integrated magneto-optic Bragg cell modulator in yttrium iron garnet-gadolinium gallium garnet taper waveguide and applications," *J. Lightwave Technol.*, 15, 1708–1715, 1997.

46. Y. Pu and C. S. Tsai, "RF magnetization of magnetosatic forward volume waves in a YIG–GGG layered structure with application to design of high-performance guided-wave magnetooptic Bragg cells," *Int. J. High-Speed Electron.*, 2, 185–208, 1991.

47. J. L. Jackel, C. E. Rice, and J. J. Veslka, "Proton-exchange for high index waveguide in $LiNbO_3$," *Appl. Phys. Lett.*, 47, 607, 1982.

48. See, for example, the variety of other waveguide lenses referred to in Chap. 7 of Ref. #3. C. S. Tsai, Ed., *Guided-Wave Acoustooptic Bragg Diffraction, Devices, and Application*, Berlin: Springer-Verlag, 1990.

49. T. Q. Vu, J. A. Norris, and C. S. Tsai, "Formation of negative index-changes waveguide lenses in $LiNbO_3$ using ion milling," *Opt. Lett.*, 13, 1141–1143, 1988.

50. (a) J. M. Verdiell et al., "Aspheric waveguide lenses for photonic integrated circuits," *Appl. Phys. Lett.*, 62, 808–810, 1993; (b) T. J. Su and C. C. Lee, "Planar fabrication process of a high coupling efficiency interface between optical-waveguides of large index difference," *Appl. Opt.*, 34, 5366–5374, 1995.

S

SECOND HARMONIC GENERATION (WAVEGUIDE)

Kazuhisa Yamamoto

Second Harmonic Generation (SHG) devices based on waveguides achieve high conversion efficiencies because a large nonlinear coefficient is available and the waveguide can attain strong confinement with long interaction length. Utilizing a waveguide SHG, short wave light generation has been realized in $LiNbO_3$, $LiTaO_3$, $KTP(KTiOPO_4)$, and $KNbO_3$ waveguides.

In this chapter, SHG in waveguides and frequency doubling of laser diodes by using SHG devices are described. Descriptions of SHG devices are limited to inorganic ferroelectric crystals, because a major part of waveguide SHG device research has used these crystals.

WAVEGUIDE SHG

Earlier experimental work on waveguide SHG used thin-film planar waveguides [1–3]. In this case, the conversion efficiency is very small. It became evident in the mid-1970s that channel $LiNbO_3$ waveguides were excellent candidates for efficient SHG [4]. Blue Cherenkov-type SHG by frequency doubling of a laser diode was realized in a proton-exchanged waveguide [5]. Waveguide QPM (Quasi-phase-matched) SHG in an $LiNbO_3$ was first demonstrated in 1986 [6]. Today, the QPM technique is widely used because of its possibility of wide conversion wavelength range and high efficiency. A comprehensive review of waveguide SHG is provided in References 3 and 7.

SHG devices in waveguides are attractive, especially for low power fundamental sources as, for example, laser diodes. Waveguide SHG is able to achieve high conversion efficiencies since the waveguide can attain strong confinement with a long interaction length.

When the conversion efficiency is so low that pump depletion is small, the SHG conversion efficiency η is given by

$$\eta = Cd^2L^2(P_\omega / S) \ [\sin(\Delta\beta L / 2) / (\Delta\beta L / 2)]^2,$$

Table 1
Properties of nonlinear materials

Nonlinear material	KTP (KTiOPO$_4$)	KNbO$_3$	LiTaO$_3$	LiNbO$_3$	MgO : LiNbO$_3$
Wavelength (nm)	350~	350~	280~	330~	320~
Nonlinear coefficient (at 852 nm)	d_{33}: 16.6 pm/V	d_{33}: 22.3 pm/V d_{32}: 11 pm/V	d_{33}: 15.1 m/V	d_{33}: 25.7 pm/V	d_{33}: 28.4 pm/V
Optical damage threshold (at 440 nm)	—	—	100 kW/cm^2	20 W/cm^2	>1 MW/cm^2
Crystal size (wafer) (in.)	1~2	1~2	2~3	3~4	3

where d is a nonlinear coefficient, P_ω is the pumping power, $P_{2\omega}$ is the second harmonic power, L is the interaction length, S is the effective cross-sectional area, $\Delta\beta$ is the difference of the propagation phase constant and C is the constant that depends on the refractive index and wavelength. When the phase matching condition is satisfied ($\Delta\beta = 0$), the conversion efficiency η is proportional to the square of the interaction length L in the nondepleted pump approximation.

MATERIALS FOR SHG

As nonlinear materials, in which waveguides have been successfully formed, LiNbO$_3$, LiTaO$_3$, KTP, and KNbO$_3$ are well known. The crystal properties are summarized in Table 1 [8,9].

Among them, large optical crystals, as can be provided in a 3–4 in. size wafer, are grown for LiNbO$_3$ and LiTaO$_3$. LiTaO$_3$ is less susceptible to optical damage due to the photorefractive effect and has a shorter wavelength absorption edge. For LiNbO$_3$, notable optical damage has been known to occur in this material, especially in the visible region. However, it has been reported that Mg doping can greatly reduce the optical damage in LiNbO$_3$ crystals [10].

The KTP(KTiOPO$_4$) and KNbO$_3$ are known to be good materials due to their high optical damage threshold and relatively high nonlinearity. KTP is widely used for frequency doubling of a Nd:YAG laser because of its wide temperature bandwidth and large nonlinearity.

The KNbO$_3$ is most convenient for frequency doubling of a laser diode. This crystal has 90-degree phase matching at a wavelength of 860 nm laser diodes.

WAVEGUIDE STRUCTURE AND FABRICATION PROCESS

For efficient frequency conversion, a low-loss and uniform waveguide of strong modal confinement is necessary. High resistance to optical damage and less reduction of nonlinearity are also essential factors.

Waveguide structures are classified into planar and channel (three-dimensional), as shown in Figure 1. Since channel waveguides have good optical confinement, high efficiency is acquired. The ridge structure is especially effective because of its large refractive index change, but it is subject to making the propagation loss small.

As fabrication techniques of waveguides in a nonlinear material, diffusion, ion-exchange, ion implantation, etc. have been reported. Some typical fabrication methods are introduced below.

Efficient Waveguide SHG in LiNbO$_3$ was first demonstrated by a Ti diffusion waveguide [4]. A Ti-diffused LiNbO$_3$ waveguide [11] has very low loss and can offer both TM and TE modes

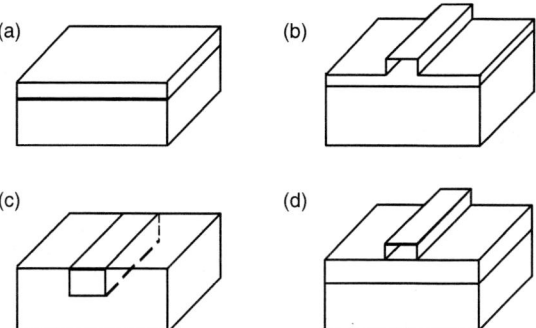

Figure 1 Waveguide structures: (a) planar waveguide; (b) ridge waveguide; (c) buried waveguide; and (d) loaded waveguide

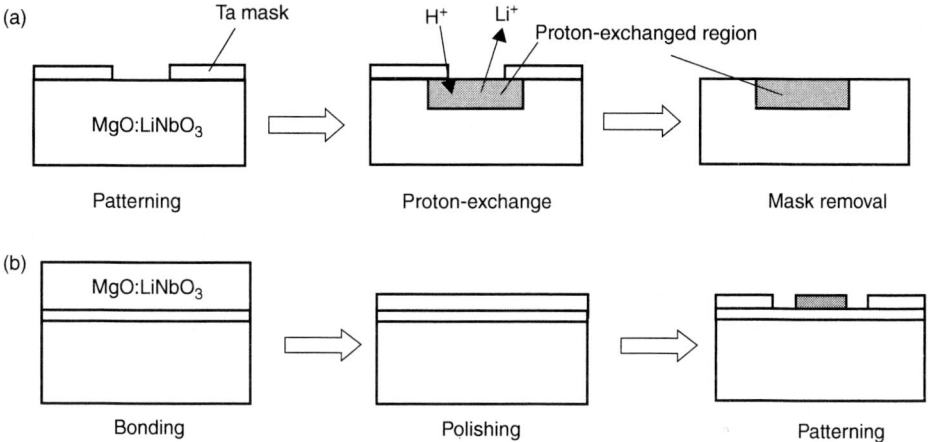

Figure 2 Fabrication procedure of (a) proton-exchange and (b) grinding method

of propagation. However it is prone to photorefractive damage, so it is suitable for the infrared region. For short wavelength conversion, a Zn diffused waveguide is attractive because of its high resistance to optical damage [12].

Proton-exchanged waveguides [13] in $LiNbO_3$ and $LiTaO_3$ are advantageous for use in SHG devices for the visible spectral range because of their high resistance to optical damage. They also make it possible to form a large refractive index change with a step index profile. The fabrication process is shown in Figure 2(a). Li^+ in $LiNbO_3$ is exchanged for H^+ in acid. In the case of a KTP waveguide fabrication, K^+ in KTP is exchanged for Rb^+ in $RbNO_3$ [14].

Using conventional benzoic acid proton-exchange can lead to high propagation loss (2 to 4 dB/cm) in channel waveguides. The propagation loss can be decreased by a thermal annealing technique or pyrophosphoric acid proton-exchange. Some decrease in the nonlinear optical coefficient has been observed. By thermal annealing, some recovery of optical nonlinearity has been reported.

The waveguide fabrication technique by ion implantation is based on physical effects. Ion implantation is applicable to a variety of materials and can be performed at room temperature [15]. Since fabricated waveguides have high propagation loss, an annealing process is necessary. In $KNbO_3$, $LiNbO_3$, $LiTaO_3$, and KTP crystals, ion-implanted waveguides are fabricated.

Waveguides fabricated by liquid phase epitaxy (LPE) or by a grinding technique have step profiles. LPE waveguides with propagation losses <1 dB/cm have been achieved [16]. Waveguides formed by a grinding technique as shown in Figure 2(b) are very attractive [17] because this waveguide is free from impurities and, therefore, a high optical damage threshold and an original nonlinearity of crystal is promised. Moreover, these waveguides have several advantages such as large confinement and large coupling efficiency to the laser diode due to the symmetric mode, compared with annealed proton-exchanged waveguides.

PHASE MATCHING IN WAVEGUIDES

Waveguide devices offer more flexibility in phase-matching, as they not only have larger allowances than bulk resonators, but also allow the generation of wavelengths not possible by the bulk approach. Figure 3 illustrates a representative phase-matching configuration.

Birefringence Phase Matching

Birefringence phase matching as shown in Figure 3(a) is most commonly utilized, combined with temperature adjustments. By using modes of different polarizations at the two wavelengths, the same effective indices in the waveguide at the fundamental and second harmonic wavelengths can be obtained. The phase matching is accomplished with an appropriate crystal cut and angle tuning. For $LiNbO_3$ and KTP waveguides, birefringence phase matching was reported in the green region [4,18]. Green (532 nm) light generation with a normalized conversion efficiency of 15%/W·cm^2 was realized in the KTP ion-exchanged waveguide [18]. Generation of blue light is not possible because the birefringence in KTP, $LiNbO_3$, and $LiTaO_3$ waveguides is too small. For the $KNbO_3$ waveguide, the birefringence phase matching by d_{32} is possible in the blue region [15].

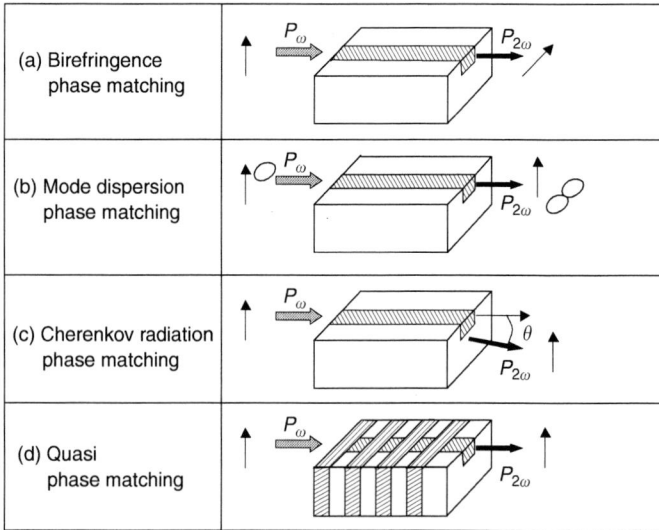

Figure 3 Representative phase-matching configuration: (a) birefringence phase matching; (b) mode dispersion phase matching; (c) Cherenkov radiation phase matching; and (d) quasi phase matching

Mode Dispersion Phase Matching

Waveguide SHG can use mode dispersion phase matching [2]. The mode index depends on the waveguide dimension and structure. Since the overlap of fundamental and harmonic waves is small, the conversion efficiency becomes low.

In the earliest work, amorphous waveguides formed on nonlinear crystals were used for mode dispersion phase matching. After that, some experiments using nonlinear thin films were reported, for example, ZnO or ZnS. Mode dispersion phase matching using d_{33} in $LiNbO_3$ is demonstrated in a Ti diffused waveguide [19]. An average-power of 22 mW at the wavelength of 532 nm was demonstrated.

Cherenkov Radiation Phase Matching

Phase matching can be achieved between the guided mode of the fundamental wave and the radiation mode of the harmonic wave, and the second harmonic generation is observed in the form of Cherenkov radiation [1]. This Cherenkov radiation scheme allows for SHG in a very wide wavelength range, as the phase-matching condition can be satisfied automatically. However, the output second harmonic radiation is difficult to collimate.

The Cherenkov radiation angle θ depends on the refractive index of the fundamental guided mode n_ω and the refractive index of the substrate for SHG $N_{2\omega}$.

$$\cos \theta = n_\omega / N_{2\omega}.$$

The SHG by Cherenkov radiation was first reported in a ZnS planar waveguide on a ZnO substrate with a YAG laser [1]. The proton-exchanged waveguide in $LiNbO_3$ is attractive for the blue SHG device because it provides large optical confinement and low-loss propagation [5].

Using the wide phase-matching wavelength range of the Cherenkov radiation scheme, simultaneous generation of blue, green, and red coherent radiation by sum-frequency and second harmonic generation of laser diodes at wavelengths of 860 and 1300 nm was reported [20].

Quasi-Phase Matching

Quasi-phase matching is an attractive technique for obtaining short-wavelength light, as phase matching of an arbitrary wavelength may be achieved by appropriate periodic modulation of the nonlinear polarization [21,22]. Moreover, the generated light is of diffraction-limited beam quality as a guided mode second harmonic output. The structure is shown in Figure 3(d). Periodically domain-inverted regions are arranged so that the waveguide traverses the regions. The period Λ of the domain-inverted regions are described as

$$\Lambda = 2mLc,$$

where m is the order of QPM ($m = 1, 3, 5, \ldots$), and Lc is the phase-matching length.

$$Lc = \lambda / [4(n_{2\omega} - n_\omega)],$$

where λ is the wavelength of the fundamental wave, and n_ω and $n_{2\omega}$ are the effective indices of the waveguide for the fundamental and second harmonic waves, respectively.

Figure 4 shows the calculated SHG wavelength as a function of the period of first order QPM ($m = 1$) for $LiTaO_3$ and $LiNbO_3$. For 420 nm SHG, the period of QPM is 3.0 μm in $LiNbO_3$ and 3.6 μm in $LiTaO_3$.

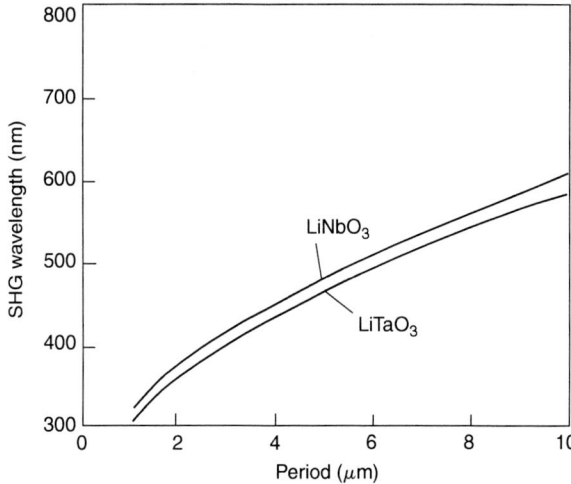

Figure 4 Calculated SHG wavelength as a function of the period of first-order QPM ($m = 1$) for LiTaO$_3$ and LiNbO$_3$

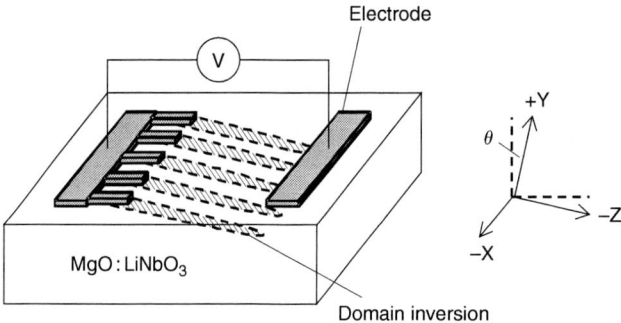

Figure 5 Schematic illustrations of representative electric poling technique for a x-cut waveguide

In the early stage of research, the fabrication of periodically domain-inverted waveguides was demonstrated by using a diffusion domain-inversion technique, such as Ti in-diffusion or Li out-diffusion [23,24]. In the diffusion technique, however, the shape of the domain-inversion was limited to a shallow triangular shape, and so it was difficult to gain effective overlap with the optical waveguides [25]. Since then, many techniques to form deep domain-inversion have been studied. Ion exchange methods have shown excellent results in waveguide devices in KTP [26] and LiTaO$_3$ [27], because of the uniform periodicity and sufficient overlap of the domain-inverted regions with the waveguides. Deep domain-inverted structures have been obtained by electric field poling [28] and also by electron beam writing [29–31] techniques. Recently, the electric field poling technique is mainly used not only for bulk SHG devices, but also for producing waveguide SHG devices because of its superiority in forming uniform and deep domain-inverted regions over a large area. Schematic illustrations of recent representative electric poling techniques for X-cut waveguides are shown in Figure 5. Figure 6 shows a cross-sectional view of the periodically domain-inverted MgO:LiNbO$_3$ with a 1.4 μm period [32].

Following the fabrication procedure of the domain-inversion, channel waveguides are fabricated in the crystals. A low-temperature process is desirable to prevent erasure of the

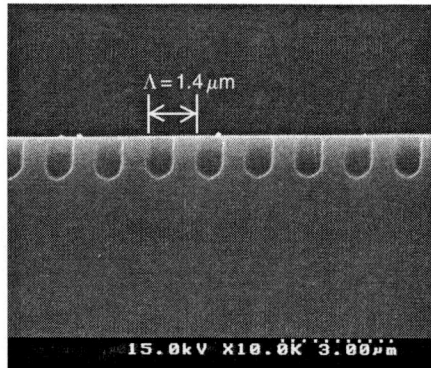

Figure 6 Cross-sectional view of a periodically domain-inverted MgO:LiNbO$_3$ with 1.4 μm period

domain-inverted regions. So, ion exchange or a grinding process is the best choice for waveguide fabrication. Utilizing this method, blue light generation has been achieved in the waveguides of LiNbO$_3$, LiTaO$_3$, and KTP(KTiOPO$_4$). A comprehensive review of QPM–SHG waveguides is given in Reference 7.

IMPROVEMENT OF DEVICE PERFORMANCE

Higher efficiency, shorter wavelength, and wide bandwidth conversion is preferred, as will be discussed below.

High Efficiency QPM–SHG Devices

In LiTaO$_3$, a maximum SHG power of 40 mW, and normalized conversion efficiency of 280%/W was obtained at the wavelength of 435 nm [27]. In KTP, 800%/W·cm^2 (116%/W) normalized conversion efficiency was demonstrated with a periodically segmented waveguide [26]. In LiNbO$_3$ [28], it has long been difficult to form a device with a long interaction length, which may be due to the occurrence of optical damage. After that, periodically domain-inverted regions were successfully formed in MgO:LiNbO$_3$, and highly efficient QPM–SHG devices with high optical damage thresholds have been reported [33–35].

From the viewpoint of achieving a compact SHG blue laser, an X-cut substrate is more attractive than that of a Z-cut substrate, because it couples directly with a conventional TE-mode laser diode [34,35]. Utilizing two-dimensional high-voltage application, deep domains with uniform periodicity can be formed. This waveguide was fabricated by a proton-exchange method in pyrophosphoric acid [36], and clad with a high-index layer to enhance the confinement of the fundamental wave [37]. The SHG device had a 3.2 μm periodic domain and a 2 μm depth waveguide. By a cw AlGaAs laser diode coupled into the waveguide, a peak in harmonic intensity was obtained for a fundamental wavelength of 852 nm to yield a maximum blue light power of 17.3 mW for 55 mW of fundamental power, with a conversion efficiency of 31% [38].

A ridge-type QPM–SHG waveguide device formed by a grinding process [17] has several advantages, such as large confinement, large nonlinearity, and large coupling efficiency to the laser diode due to its symmetric mode, compared with a proton-exchanged waveguide. In this device, periodically domain-inverted regions were fabricated using a multiple short pulse for electric poling to an off-cut MgO : LiNbO$_3$. The MgO : LiNbO$_3$ was bonded on the substrate and processed to a thickness of 3 μm by grinding and polishing. A Channel waveguide was fabricated by grinding, using a diamond blade. The propagation loss was as low as 1 dB/cm and the coupling efficiency to the DBR–LD was over 80%. SHG blue light (411 nm) of 31 mW was gained for a fundamental input power of 90 mW [39]. The ridge-type QPM–SHG waveguide device formed by grinding indicates high conversion efficiency and high resistance for optical damage to give stable high power blue light generation.

Resonant Waveguide SHG Devices

Resonant waveguide SHG devices are constructed by depositing dielectric multilayer films to form high reflection mirrors for fundamental waves on both facets of the waveguides. Precious temperature control is necessary for phase matching because the resonant conditions are critical.

For $Ti:LiNbO_3$ waveguides, 1000%/W normalized efficiency was demonstrated by birefringence phase matching [40]. A resonant QPM device with a proton-exchanged waveguide was accomplished by electro-optic tuning, and a factor of eight was obtained [41].

Improvement of Efficiency with RF Superimposed Laser Diode

In order to gain higher efficiency than those obtained by frequency doubling of continuous wave (cw), laser diodes of high-power pulsed-wave are indispensable in view of the fact that the frequency doubled power is proportional to the square of the fundamental power. The gain-switching method can produce high-power pulse trains of around 20 psec of full width at half maximum (FWHM) at arbitrary repetition rates. Therefore, it is expected to attain increased average conversion efficiency of the SHG device by high-frequency superimposing to the driving current of the laser diode, at the same average light power for cw operation. Figure 7 shows an experimental setup for frequency doubling of a quasi-cw operated laser diode with a grating feedback optical system [42]. The laser was gain-switched by an electrical sine wave of radio frequency (800 MHz). In this experiment, a $LiTaO_3$ QPM–SHG waveguide device was used. While in optical feedback operation, the laser diode oscillated in a single-like longitudinal mode, and the longitudinal mode was stabilized within the acceptable bandwidth of the QPM–SHG device. The average blue output power was four times larger than in the cw operation. Blue light of a 200-mW peak power with 25 psec pulse width was generated.

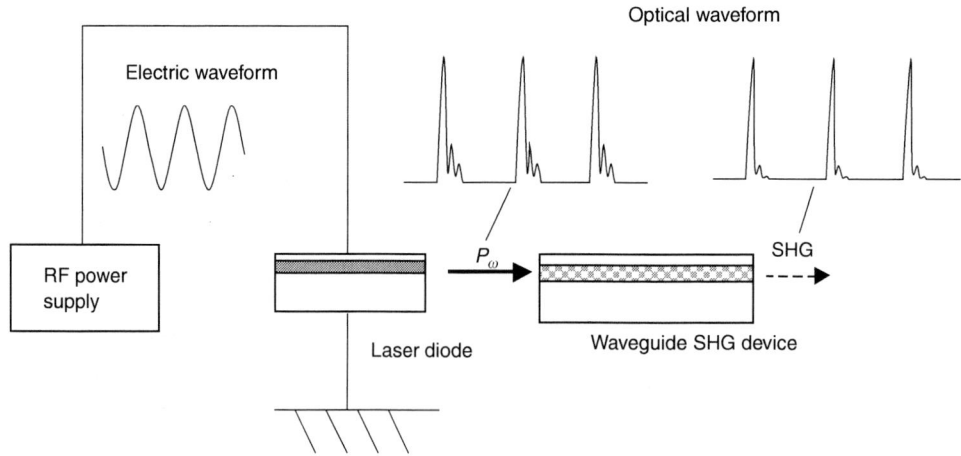

Figure 7 Frequency doubling of a quasi-cw operated laser diode

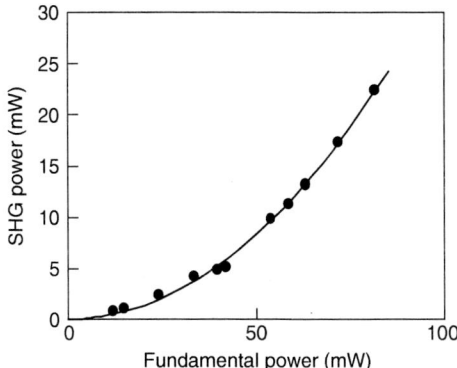

Figure 8 Generated UV-light power as a function of fundamental power in a MgO:LiNbO$_3$ waveguide

Ultraviolet Generation

LiTaO$_3$ has the shortest absorption edge in QPM materials of relatively large nonlinearlity. However, in order to phase match the wavelength region of visible-red lasers (640 to 680 nm), domain-inversion has to be for a short period of <1.7 μm. A first-order periodically poled LiTaO$_3$ by using electric field poling with a precisely controlled technique of domain-inversion was first demonstrated [43]. A UV (ultraviolet) generation by cascaded QPM–SHG–SFG (Sum Frequency Generation) in LiNbO$_3$ [44], first order QPM–SHG in MgO:LiNbO$_3$ [32], and a segmented waveguide in KTP were also reported [45]. As shown in Figure 8, UV-light power of 22 mW at a wavelength of 340 nm was obtained for 81 mW of fundamental power in the MgO : LiNbO$_3$ waveguide [32].

QPM with Wide Bandwidth

As wavelength acceptance of the QPM–SHG device is very narrow, it is very attractive for broadening the QPM bandwidth. One of the methods to broaden the bandwidth is to use a chirped domain-inversion period [46]. It is shown that an extreme improvement of the band-width can be attained.

Segmented QPM structure, pseudo-random polarity reversal, and other methods have also broadened the bandwidth [27,47,48]. Conversion efficiency of an SHG device falls in exchange for broadening at the same waveguide length.

APPLICATIONS OF WAVEGUIDE SHG

Compact laser diode-based second harmonic generation devices hold much potential for use in information fields such as high-density optical disks, laser display and laser-beam printer applications. In this section, SHG lasers using waveguide devices are introduced.

Frequency Doubling of a Wavelength-Locked Laser Diode

To achieve stable and efficient frequency doubling, locking the oscillation wavelength of a laser diode is essential, because wavelength acceptance of the QPM–SHG device is narrow. Various lock-ing methods by use of optical feedback systems have been investigated, such as by using an external grating [42], an optical band-pass filter [49], and distributed-Bragg-reflector (DBR) on an SHG waveguide [50]. Figure 9 shows various kinds of wavelength locking methods for laser diodes.

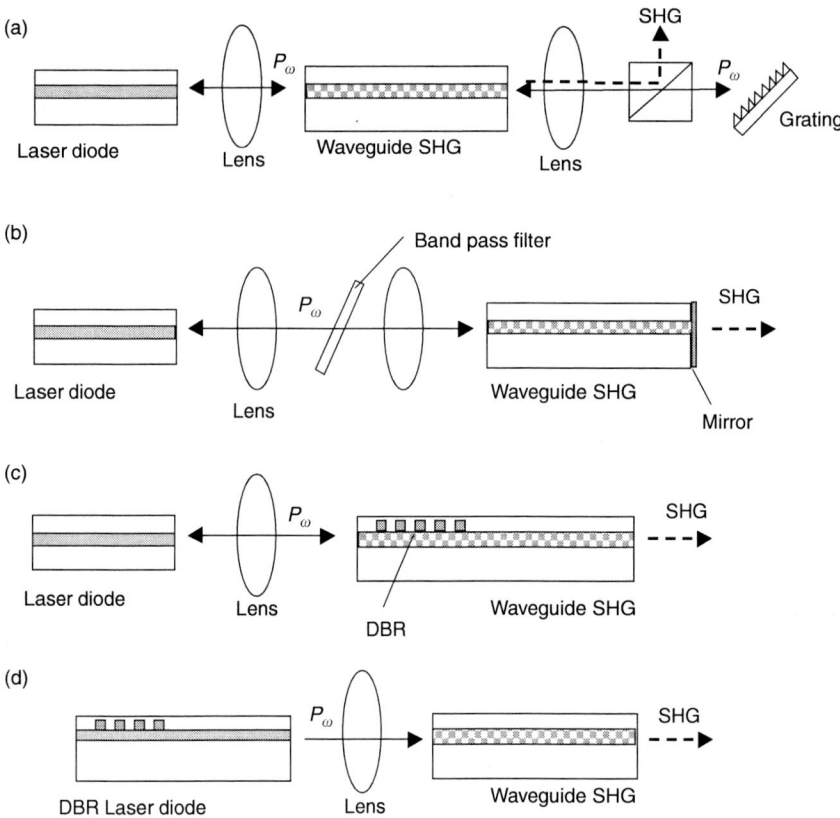

Figure 9 Various kinds of wavelength-locking methods for laser diodes: (a) grating; (b) band-pass filter; (c) DBR on a waveguide; and (d) DBR laser diode

Although a wide tuning range is possible by using the external grating or the confocal optics with an optical band-pass filter, on the other hand, optical systems become large.

A self QPM–DBR technique [50], where the fundamental wave is provided by oscillation at the QPM wavelength by optical feedback from the waveguide DBR on SHG devices, was proposed. In this case temperature control is necessary to obtain stable SHG.

A tunable DBR laser diode is effective as an alternative approach to locking and tuning the wavelength because of its compactness and wide tuning range [51]. The laser has three sections of active, phase-control, and DBR regions. By controlling the current to each section, stable wavelength control and high-speed modulation are attained. The lasing wavelength was 822 nm and the tuning range was 2 nm.

SHG Laser Module

As compact, solid-state sources for blue coherent light, laser diode-based second harmonic generation devices hold great potential for use in high-density optical disk and laser-beam printer applications. The SHG devices have taken advantage of QPM principles as an attractive method for obtaining blue light, because phase-matching of an arbitrary wavelength can be achieved by appropriate periodic modulation of the nonlinear polarization.

Figure 10 shows SHG blue laser modules (a) using a lens and a band-pass filter [49] and (b) using a direct coupling technique and a tunable DBR laser diode [51]. The direct coupling technique without the coupling lens optics is preferable to develop a compact SHG blue laser. When the laser diode was coupled close to the waveguide, a maximum coupling efficiency of 80% was obtained for a waveguide on an X-cut substrate.

Using this coupling technique, the compact SHG blue laser module with a volume of <0.3 cc can be realized. The dimensions of the module as shown in Figure 11 are 5 mm width, 18 mm length, and 3 mm thickness. The obtained continuous output power of the SHG blue light (411 nm) was 31 mW. The blue light is generated in a single longitudinal mode, where the side-mode was negligible because the side-mode suppression ratio for the DBR–LD was larger than 20 dB. Therefore, the relative intensity noise of the blue light was as small as −145 dB/Hz. The parallel and perpendicular far-field pattern indicated the symmetric TE_{00} mode (the Gauss-like mode), and the beam full width at half maximum (FWHM) was 5° and 9°, respectively. When the drive current of the laser diode was modulated, the longitudinal-mode spectrum of the laser diode remained locked, and a peak blue light power of 62 mW was obtained at a modulation frequency of 60 MHz [39]. The blue SHG laser permits compact packaging with high power and direct modulation capability.

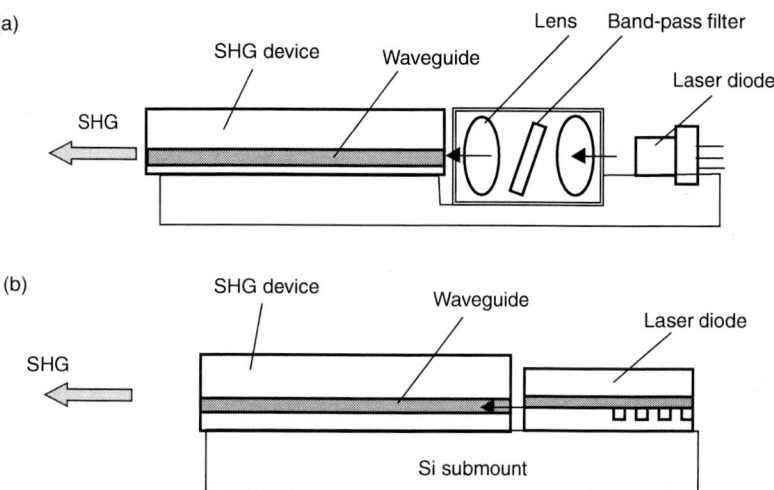

Figure 10   SHG laser modules: (a) using a lens and a band-pass filter and (b) using a direct coupling technique and a tunable DBR laser diode

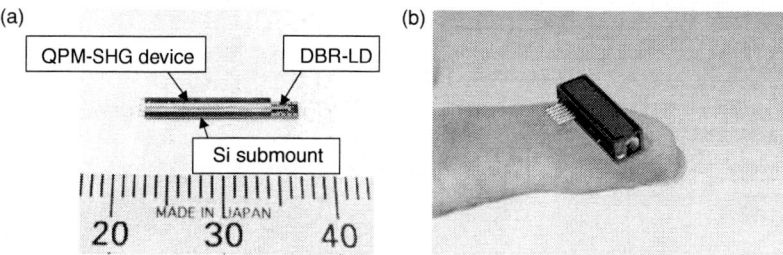

Figure 11   Compact SHG blue laser module consisting of a waveguide QPM–SHG device in MgO:LiNbO₃ and a tunable DBR laser: (a) SHG laser chip and (b) packaged SHG laser (0.3 cc)

REFERENCES

1. P. K. Tien, R. Ulrich, and R. J. Martin, "Optical second harmonic generation in form of coherent Cherenkov radiation from a thin-film waveguide," *Appl. Phys. Lett.*, 17, 447–450 (1970).
2. I. Ito and H. Inaba, "Phase-matched SHG in thin-film nonlinear optical waveguide incorporated with substrate nonlinearity," *IEEE J. Quantum Electron.*, 11, 862 (1975).
3. G. I. Stegeman and C. T. Seaton, "Nonlinear integrated optics," *J. Appl. Phys.*, 58, R57–R78 (1985).
4. N. Uesugi and T. Kimura, "Efficient second-harmonic generation in three-dimensional LiNbO$_3$ optical waveguide," *Appl. Phys. Lett.*, 29, 572–574 (1976).
5. T. Taniuchi, and K. Yamamoto, "Second harmonic generation with GaAs laser diode in proton-exchanged LiNbO$_3$ waveguides," in *Proceedings of the 12th ECOC*, Barcelona , pp. 171–174 (September, 1986).
6. B. Jaskorzynska, G. Arvidsson, and F. Laurell, "Periodic structures for phase matching second haramonic generation in titanium lithium niobate waveguides," *Proc. SPIE*, 651, 221–228 (1986).
7. T. Suhara and M. Fujimura, *Waveguide Nonlinear-Optical Devices*, Springer-Verlag, Berlin (2003).
8. V. G. Dmitriev, G. G. Gurzadyan, and D. N. Nikogosyan, *Handbook of Nonlinear Optical Crystals*, Springer-Verlag, New York (1999).
9. I. Shoji, T. Kondo, A. Kitamoto, M. Shirane, and R. Ito, "Absolute scale of second-order nonlinear-optical coefficients," *J. Opt. Soc. Am. B*, 14, 2268–2294 (1997).
10. D. A. Bryan, R. Gerson, and H. E. Tomaschke, "Increased optical damage resistance in lithium niobate," *Appl. Phys. Lett.*, 44, 847–849 (1984).
11. R. V. Schmidt and I. P. Kaminow, "Metal diffused optical waveguides in LiNbO$_3$," *Appl. Phys. Lett.*, 25, 458–464 (1974).
12. W. M. Young, M. M. Fejer, M. J. F. Digonnet, A. F. Marsgall, and R. S. Feigelson, "Fabrication, characterization and index profile modeling of high-damage resistance Zn-diffused waveguides in congruent and MgO:LiNbO$_3$," *J. Lightwave Technol.*, 10, 1238–1246 (1992).
13. J. L. Jackel, C. E. Rice, and J. J. Veselka, "Proton exchange for high index waveguides in LiNbO$_3$," *Appl. Phys. Lett.*, 41, 607–608 (1982).
14. J. D. Bierlein and H. Vanherzeele, "Potassium titanyl phosphate: properties and new applications," *J. Opt. Soc. Am. B*, 6, 622–633 (1989).
15. D. Fluck and P. Gunter, "Second-harmonic generation in potassium niobate waveguides," *IEEE J. Select. Top. Quantum Electron.*, 6, 122–131 (2000).
16. T. Kawaguchi, K. Mizuuchi, T. Yoshino, M. Imaeda, K. Yamamoto, and T. Fukuda, "Liquid-phase epitaxial growth of Zn-doped LiNbO$_3$ thin films and optical damage resistance for second-harmonic generation," *J. Crystal Growth*, 203, 173–178 (1999).
17. T. Kawaguchi, K. Mizuuchi, T. Yoshino, M. Imaeda, and K. Yamamoto, "New ridge-type waveguide QPM-SHG device fabricated by ultraprecision machining," in *Proceedings of the ISOM 2000*, Th-F-03, (September 2000), pp. 66–67.
18. J. D. Bierlein, D. B. Laubacher, and J. B. Brown, "Balanced phase matching in segmented KTiOPO$_4$ waveguides," *Appl. Phys. Lett.*, 56, 1725–1727 (1990).
19. M. M. Fejer, M. J. F. Digonnet, and R. L. Byer, "Generation of 22 mW of 532 nm radiation by frequency doubling in Ti:MgO:LiNbO$_3$ waveguides," *Opt. Lett.*, 11, 230–232 (1986).
20. K. Yamamoto, H. Yamamoto, and T. Taniuchi, "Simultaneous sum-frequency and second-harmonic generation from a proton-exchanged MgO-doped LiNbO$_3$ waveguide," *Appl. Phys. Lett.*, 58, 1227–1229 (1991).
21. J. A. Armstrong, N. Bloembergen, J. Ducuing, and P. S. Pershan, "Interactions between light waves in a nonlinear dielectric," *Phys. Rev.*, 127, 1918–1939 (1962).
22. S. Somekh and A. Yariv, "Phase matching by periodic modulation of the nonlinear optical properties," *Opt. Commun.*, 6, 301–304 (1972).
23. E. J. Lim, M. M. Fejer, R. L. Byer, and W. J. Kozlovsky, "Blue light generation by frequency doubling in periodically poled lithiumniobete channel waveguide," *Electron. Lett.*, 25, 731–732 (1989).
24. J. Webjörn, F. Laurell, and G. Arvidsson, "Blue light generation by frequency doubling a laser diode light in a lithium niobate channel waveguide," *IEEE Photonics Technol. Lett.*, 1, 316–318 (1989).
25. K. Yamamoto, K. Mizuuchi, K. Takeshige, Y. Sasai, and T. Taniuchi, "Characteristics of periodically domain-inverted LiNbO$_3$ and LiTaO$_3$ waveguides for second harmonic generation," *J. Appl. Phys.*, 70, 1947 (1991).

26. D. Eger, M. Oron, M. Katz, and A. Zussman, "Highly efficient blue light generation in KTiOPO4 waveguides," *Appl. Phys. Lett.*, 64, 3208–3209 (1994).

27. K. Mizuuchi, K. Yamamoto, M. Kato, and H. Sato, "Broadening of the phase-matching bandwidth in quasi-phase matched second harmonic generation," *IEEE J. Quantum Electron.*, 30, 1596–1604 (1994).

28. M. Yamada, N. Nada, M. Saitoh, and K. Watanabe, "First order quasi-phase matched LiNbO3 waveguide periodically poled by applying an external field for efficient blue second-harmonic," *Appl. Phys. Lett.*, 62, 435–436 (1993).

29. H. Ito, C. Takyu, and H. Inaba, "Fabrication of periodic domain grating in LiNbO3 by electron beam writing for application of nonlinear optical process," *Electron. Lett.*, 27, 1221–1222 (1991).

30. M. Yamada and K. Kishima, "Fabrication of periodically domain structure for SHG in LiNbO3 by direct electron beam lithography at room temperature," *Electron. Lett.*, 27, 828–829 (1991).

31. S. Kurimura, I. Shimoya, and Y. Uesu, "Domain-inversion by an electron-beam induced electric field in MgO:LiNbO3, LiNbO3 and LiTaO3," *Jpn. J. Appl. Phys.*, 35, L31–L33 (1996).

32. K. Mizuuchi, T. Sugita, K. Yamamoto, T. Kawaguchi, T. Yoshino, and M. Imaeda, "Efficient 340 nm light generation by a ridge waveguide in a first order periodically poled MgO:LiNbO3," *Opt. Lett.*, 28, 1344–1346 (2003).

33. K. Mizuuchi, K. Yamamoto, and M. Kato, "Harmonic blue light generation in bulk periodically poled MgO:LiNbO3," *Electron. Lett.*, 32, 2091–2092 (1996).

34. K. Mizuuchi, K. Yamamoto, and M. Kato, "Harmonic blue light generation in x-cut MgO:LiNbO3 waveguide," *Electron. Lett.*, 33, 806–807 (1997).

35. S. Sonoda, I. Tsuruma, and M. Hatori, "Second harmonic generation in electric poled x-cut MgO-doped LiNbO3 waveguides," *Appl. Phys. Lett.*, 70, 3078–3080 (1997).

36. K. Yamamoto, K. Mizuuchi, and T. Taniuchi, "Low-loss channel waveguides in MgO:LiNbO3 and LiTaO3 by pyrophosphoric acid proton exchange," *Jpn. J. Appl. Phys.*, 31, 1059–1064 (1992).

37. K. Mizuuchi, H. Ohta, K. Yamamoto, and M. Kato, "Second harmonic generation with a high-index-clad waveguide," *Opt. Lett.*, 22, 1217–1219 (1997).

38. T. Sugita, K. Mizuuchi, Y. Kitaoka, and K. Yamamoto, "31%-efficient blue second-harmonic generation in a periodically poled MgO:LiNbO3 waveguide by frequency doubling of an AlGaAs laser diode," *Opt. Lett.*, 24, 1590–1592 (1999).

39. Y. Kitaoka, K. Kasazumi, A. Morikawa, T. Yokoyama, T. Sugita, K. Mizuuchi, K. Yamamoto, T. Takayama, S. Takigawa, and M. Yuri, "Wavelength stabilization of a distributed Bragg reflector laser diode by use of complementary current injection," *Opt. Lett.*, 28, 914–916 (2003).

40. R. Regener and W. Sohler, "Efficient second-harmonic generation in Ti:LiNbO3 channel waveguide resonators," *J. Opt. Soc. Am. B*, 5, 267–277 (1988).

41. M. Fujimura, M. Sudoh, K. Kintaka, T. Suhara, and H. Nishihara, "Enhancement of SHG efficiency in periodically poled LiNbO3 waveguide utilizing a resonance effect," *Electron. Lett.* 32, 1283–1284 (1996).

42. K. Yamamoto, K. Mizuuchi, Y. Kitaoka, and M. Kato, "Highly efficient quasi-phase-matched second-harmonic-generation by frequency doubling of a high-frequency superimposed laser diode," *Opt. Lett.*, 20, 273–275 (1995).

43. K. Mizuuchi, K. Yamamoto, and M. Kato, "Generation of ultraviolet light by frequency doubling of a red laser diode in a first-order periodically poled bulk LiTaO3," *Appl. Phys. Lett.*, 70, 1201–1203 (1997).

44. K. Kintaka, M. Fujimura, T. Suhara, and H. Nishihara, "Third harmonic generation of Nd:YAG laser light in periodically poled LiNbO3 waveguide," *Electron. Lett.*, 33, 1459–1460 (1997).

45. F. Laurell, J. B. Brown, and J. D. Bierlein, "Simultaneous generation of UV and visible light in segmented KTP," *Appl. Phys. Lett.*, 62, 1872–1874 (1993).

46. T. Suhara and H. Nishihara, "Theoretical analysis of waveguide second-harmonic generation phase matched with uniform and chirped," *IEEE J. Quantum Electron.*, 26, 1265–1276 (1990).

47. M. M. Fejer, G. A. Magel, Dieter H. Jundt, and R. L. Byer, "Quasi-phase-matched second harmonic generation: Tuning and tolerances," *IEEE J. Quantum Electron.*, 28, 2631–2654 (1992).

48. M. L. Bortz, M. Fujimura, and M. M. Fejer, "Increased acceptance bandwidth for quasi-phasemathed second harmonic generation in LiNbO3 waveguides," *Electron. Lett.*, 30, 34–35 (1994).

49. Y. Kitaoka, K. Mizuuchi, K. Yamamoto, and M. Kato, "An SHG blue light source using domain-inverted waveguide device in LiTaO3," *Rev. Laser Eng.*, 23, 788–794 (1995).

50. K. Shinozaki, T. Fukunaga, K. Watanabe, and T. Kamijoh, "Self-quasi-phase matched second harmonic generation in the proton-exchanged LiNbO$_3$ optical waveguide," *Appl. Phys. Lett.,* 59, 510–512 (1991).
51. Y. Kitaoka, T. Yokoyama, K. Mizuuchi, and K. Yamamoto, "Miniaturized blue laser using second harmonic generation," *Jpn. J. Appl. Phys.,* 39, 3416–3418 (2000).

SEMICONDUCTOR OPTICAL AMPLIFIER

Ken Morito

SEMICONDUCTOR OPTICAL AMPLIFIER

Semiconductor Optical Amplifier (SOA) is a kind of optical amplifier of which optical gain is provided by the stimulated emission in semiconductor where the population of free carriers is inverted.

Figure 1 shows the schematic structure of an SOA. The optical signal is coupled into the active waveguide from the input facet of the SOA. The population of free carriers in the active waveguide is inverted by electrical pumping through electrodes. During the propagation of the optical signal in the active waveguide, the optical signal is amplified by the stimulated emission. The amplified optical signal comes out from the output facet of the SOA.

The development of SOA was closely related to that of the semiconductor laser. In 1963, Coupland et al. [1] measured optical amplification in a GaAs laser. In the early stage, the SOAs were laser diodes biased below the threshold, and were called Fabry–Perot (FP) SOAs. The FP SOAs have a deep ripple in the gain spectral due to the optical feedback from facets as shown in Figure 2. The period of the ripple corresponds to the longitudinal mode spacing of the cavity. Since the gain bandwidth of each ripple is very narrow, the resonant peak wavelength must be precisely adjusted to the incoming signal wavelength. This requires the temperature control of the FP-SOA in the order of 0.01 K, which is not suitable for practical use.

To overcome this problem, another type of SOAs called traveling wave (TW) SOA had been developed. In the TW SOAs, the cavity resonance is suppressed by reducing the facet

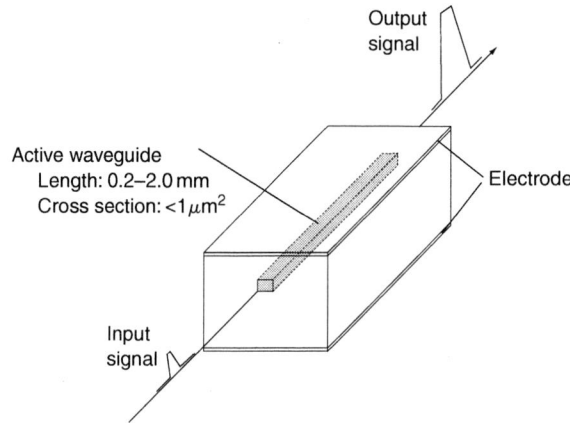

Figure 1 Schematic view of SOA

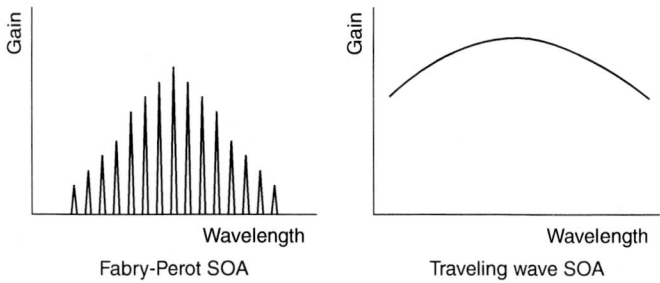

Figure 2 Gain spectrum for FP SOA and traveling wave SOA

reflectivity and the gain ripple can be decreased to a level of <0.5 dB. The TW SOA can have a wide gain bandwidth of several tenth nm to over a hundred nm as shown in Figure 2. Additionally, TW SOAs can have higher saturation output power and lower noise figure as compared to FP SOAs. But we had to wait for TW SOAs until 1980s. In the following sections, the word "SOA" means the TW SOA.

FEATURES OF SOAs

SOAs are compact (typical length ranging from several hundreds of microns to several millimeters), electrically pumped devices. Additionally SOAs can be integrated monolithically with laser diodes, optical modulators, photodiodes, and passive waveguides including multi-mode interference (MMI) couplers and arrayed waveguide (AWG) devices. The integrated device can provide more complex functionalities, for example, all optical wavelength converters, all optical switches, wavelength selectors, and tunable lasers. The features of optical gain characteristics in SOAs are a wide gain bandwidth (several tenth nanometers to over a hundred nanometers), a wide flexibility in the choice of the gain peak wavelength (e.g., 1200 to 1650 nm in InP-based devices), and a fast gain response (several hundreds of picoseconds). Both the wide gain bandwidth and the flexible choice of operation wavelength are attractive features for wavelength division multiplexing (WDM) systems. A fast gain response can provide fast switching devices. However, a fast gain response causes waveform distortion due to the pattern effect in the saturated regime when amplifying signals are modulated in the speed of giga-bit per second (Gbpsec) range. Basically SOAs must be operated in the linear or unsaturated regime when they are used as optical amplifiers for signals modulated in Gbpsec range.

PERFORMANCE CRITERIA OF SOAs

- Chip gain/internal gain: Optical gain obtained in a single propagation through SOA (TW-SOA).
- Fiber-to-fiber gain/module gain: Optical gain including the chip-fiber coupling loss at both input and output sides.
- Gain peak wavelength.
- Gain bandwidth/spectral bandwidth: Typically 3 dB down bandwidth in gain spectrum.
- Gain ripple/spectral ripple: Undulation in gain spectrum caused by cavity resonance due to the residual facet reflectivity.
- Polarization sensitivity in gain/polarization dependent gain: difference in gain between TE polarization and TM polarization.

- Chip saturation output power: Optical output power at which gain decreases 3 dB below the unsaturated gain.
- Fiber coupled saturation output power/module saturation output power: Saturation output power including the chip-fiber coupling loss at output side.
- Chip noise figure: Decrease of the signal-to-noise ratio (SNR) due to the propagation of a quantum noise limited optical signal through SOA.
- Fiber coupled noise figure/module noise figure: Noise figure including the chip-fiber coupling loss at input side.
- Transient time of gain response.
- Power consumption.

TECHNOLOGIES FOR REALIZING HIGH PERFORMANCE SOAs

There existed three technological issues that had to be addressed before SOAs could have acceptable performances for practical applications.

- Reduction of optical feedback
- Reduction of polarization sensitivity in gain
- Reduction of coupling loss between SOA chip and fiber

In this section, main technologies are summarized for each issue.

Technologies for Reducing Optical Feedback

Different approaches have been taken to reduce the facet reflectivity and realize TW SOAs.

Antireflection (AR) coating: The facet reflectivity can be reduced by applying dielectric coating on the facet. Although the facet reflectivity as low as 10^{-3} can be obtained with an optimized AR coating, tight control of the coating layer thickness is required and a low reflectivity can be maintained for a limited wavelength range. Then AR coating is used together with one or both of the following methodologies.

Tilted waveguide: Another way to reduce the facet reflectivity is to tilt the active waveguide from the facet so that the light is reflected away from the active waveguide. Typical angle ranges from 5° to 10°.

Buried facet (Window structure): Another method to reduce the facet reflectivity is to end the active waveguide tips about 20 to 30 μm inside the facets so that less feedback can be coupled back into the waveguide due to diffraction.

Technologies for Reducing Polarization Sensitivity in Gain

The early SOAs had polarization dependent gain. Since the state of polarization of an optical signal randomly changes, however, the SOAs must have polarization insensitive gain for most applications. Different methods have been developed to realize polarization insensitive SOAs. In the section "Development of Polarization Insensitive SOA", the history of the development of polarization insensitive TW SOAs is detailed.

Symmetrical bulk active layer: A square bulk active layer with buried heterostructure can provide both the same optical confinement factor for TE and TM polarizations and the same material gain for TE and TM polarizations. If the square can be realized, perfect polarization insensitivity will be achieved for a wide range of signal wavelength, injection current, and operating temperature. To achieve the square bulk active layer in which only the fundamental transverse mode can exist, however, the active layer width must be less than about 0.5 μm. Then ultra-fine and ultra-accurate fabrication process is required to fabricate the sub-micron square bulk active layer.

Strained MQW active layer with flat waveguide: By introducing the tensile strain into the MQW active layer, the material gain for the TM polarization is enhanced. The polarization insensitivity can be achieved even in the active layer with the flat cross-section like normal laser diodes if a proper amount of tensile strain is chosen. However, it is very difficult to achieve the polarization insensitivity over a wide range of wavelength with the strained MQW SOAs because relatively large strain makes a difference in the shape of the gain spectra between TE and TM mode.

Strained bulk active layer with flat waveguide: It is also possible to achieve the polarization insensitivity in a flat active layer by introducing tensile strain into the bulk active layer. The amount of the tensile strain required is much smaller in the bulk structure than in the MQW structure. This enables to achieve the polarization insensitivity for a relatively wide range of signal wavelength, injection current, and operating temperature.

Technologies for Reducing Coupling Loss between SOA Chip and Fiber

High chip–fiber coupling losses make SOA performances poor because the fiber-to-fiber gain decreases due to the chip–fiber coupling losses at both the input and the output sides, the fiber coupled saturation output decreases due to the chip–fiber coupling loss at the output side, and the fiber coupled noise increases due to the chip–fiber coupling loss at the input side.

Good optical coupling between the SOA and the fiber can be obtained if the mode profiles are made to be equal. Both inserting lens between the SOA and the fiber and integrating spot size converters (SSCs) with SOAs have been used to achieve the best coupling. Additionally, tapering waveguide also decreases the facet reflectivity.

DEVELOPMENT OF POLARIZATION INSENSITIVE SOA

In 1982, Simon [2] demonstrated an AlGaAs SOA with a low polarization sensitivity of 2 dB and proposed that optical confinement factor between two polarizations could be made equal by using a thick bulk active layer or a large optical cavity (LOC) structure together with buried heterostructure. In 1989 first polarization insensitive SOAs, of which polarization sensitivities were within 1 dB for the chip gain larger than 20 dB, were reported [3–6]. However, these SOAs had a limited fiber-to-fiber gain of 8.9 dB [4], 16 dB [5], or 13 dB [6] and a limited fiber-coupled saturation output power of +1.5 dBm [5] due to the high chip–fiber optical coupling losses. Additionally, they still had relatively large gain ripple due to the residual facet reflectivity.

As the technologies matured concerning low facet reflectivity, polarization independent gain, and low chip–fiber coupling loss, the SOA performances have improved. In 1994 and 1995, Doussiere et al. [7,8] reported a polarization insensitive SOA module with a fiber-to-fiber gain of 29 dB and a fiber-coupled saturation output power of +9 dBm at the signal wavelength of 1550 nm. The polarization sensitivity of 0.5 dB was realized by the sub-micron square bulk SOA. The facet reflectivity of the order of 10^{-5} was achieved by 7° tilted waveguide and AR coated facets together with window structures. The coupling loss between the SOA chip and the lensed fiber was reduced to 2 dB per facet by integrating width-tapered SSCs.

On the other hand, other structures, a ridge waveguide structure and a strained MQW active layer structure, were studied to overcome the difficulty in fabricating the sub-micron square bulk SOA.

In 1993, Holtmann et al. realized the first polarization insensitive bulk SOA using a ridge waveguide structure [9] and demonstrated a fiber-to-fiber gain of 27 dB and a fiber-coupled saturation output power of +10 dBm at the signal wavelength of 1310 nm [10]. By controlling the active layer thickness, polarization insensitivity could be achieved with a wide rib of 3 μm.

In the beginning of 1990s, first polarization insensitive SOAs using a strained MQW active layer were reported [11–16]. In 1990, Magari et al. [11,12] proposed the use of lattice-matched InGaAs wells and tensile strained InGaAs barriers and demonstrated the polarization insensitive MQW SOA. In 1991, Joma et al. [13,14] proposed the use of tensile strained InGaAsP wells and demonstrated the polarization insensitive MQW SOA. In 1992, Tiemeijer et al. [15] proposed the use of both tensile strained well and compressively strained well. The SOA exhibited a fiber-to-fiber gain of 16 dB together with a high fiber coupled saturation output power of +13 dBm and a low fiber coupled noise figure of 6.5 dB at the signal wavelength of 1310 nm. In 1996 they reported improved performances of a fiber-to-fiber gain of 36 dB, a high fiber coupled saturation output power of +15 dBm and a low fiber coupled noise figure of 6.5 dB at the signal wavelength of 1310 nm [16].

However, it is difficult to obtain enough gain for a long wavelength of around 1550 nm with the tensile strained well structure. The reason is that the introduction of all the three, the tensile strain, the quantum effect, and the band filling effect, makes the gain peak wavelength much shorter. This puts restrictions on the design of MQW structure for the 1550 nm SOAs, for example, the use of thick tensile strained well [17]. Then the performance of 1550 nm polarization insensitive strained MQW SOAs has been behind that for the 1310 nm polarization insensitive strained MQW SOAs.

In 1996, Emery et al. [18] proposed to use a bulk active layer with a low tensile strain for a 1550 nm polarization insensitive SOA to overcome the problem of both the gain peak shift in the tensile strained MQW SOAs and the stringent fabrication tolerance required for the sub-micron square bulk SOAs. They demonstrated a polarization sensitivity of 0.2 dB for a fiber-to-fiber gain of 29 dB [19]. However, the fiber coupled saturation output power of the SOA still remained at +9 dBm, which was the same with that for the 1550 nm sub-micron square bulk SOAs [8] and 6 dB lower than that for the 1310 nm strained MQW SOA [16]. In 2000, Morito et al. developed a 1550 nm polarization insensitive SOA with a thin tensile strained bulk active layer of 50 nm to increase the saturation output power and achieved a record high fiber-coupled saturation output power of +17 dBm at the signal wavelength of 1550 nm and demonstrated pattern effect free amplification for signals modulated at 10 Gbpsec up to the fiber coupled power of +12 dBm [20]. They also demonstrated that the polarization sensitivity in gain remained <0.3 dB over the wavelength range of 1530 to 1560 nm due to the small strain value of –0.25% [21].

TYPICAL CHARACTERISTICS OF POLARIZATION INSENSITIVE SOA

In this section, characteristics of a 1550 nm high power polarization insensitive SOA [20] are shown as an example. Figure 3 shows the schematic view of the SOA. A 50 nm thick, –0.25% tensile-strained InGaAs bulk active layer is used to achieve the polarization insensitivity. The total SOA length is 1200 μm. The central region of the active waveguide with the constant width of 1.4 μm is 900 μm long. At both input and output sides of the SOA, 150 μm-long polarization-insensitive active-width-tapered SSCs are integrated to improve the chip–fiber coupling. In the SSC region, the active waveguide width is linearly changed from1.4 to 0.4 μm at the tip. The active waveguide mesa is tilted 7° and both facets are AR-coated to reduce a reflectivity.

Figure 4 shows the dependence of the unsaturated module gain on the signal wavelength for both TE and TM polarizations at an SOA injection current of 500 mA. In this wavelength range, the polarization sensitivity of the unsaturated module gain remains <0.3 dB. This small dependence of the polarization sensitivity on the signal wavelength is due to the small tensile strain value required for TE–TM gain matching in the tensile strained bulk active layer.

Figure 5 shows the gain saturation characteristics for both TE and TM polarizations at a current of 500 mA. The module saturation output power is +17 dBm and an unsaturated module

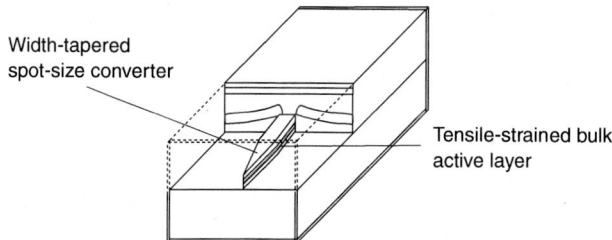

Figure 3 Device structure of polarization insensitive SOA

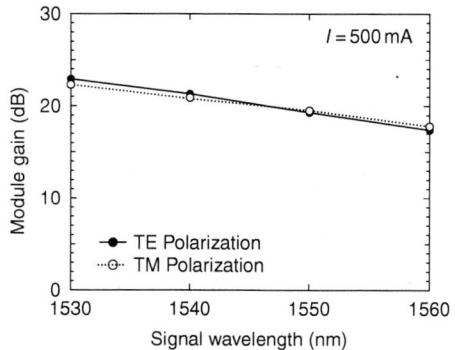

Figure 4 Dependence of module gain on signal wavelength for both polarizations

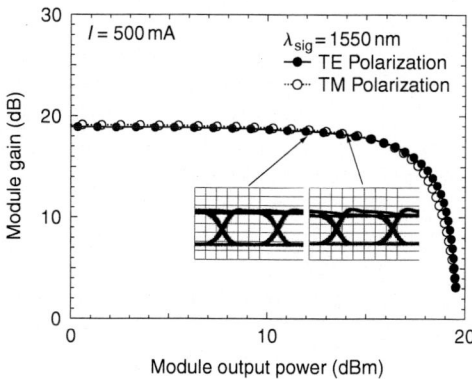

Figure 5 Gain saturation characteristics for both polarizations and eye patterns of amplified outputs modulated at 10 Gbpsec

gain of 19 dB and a polarization sensitivity in gain of 0.2 dB. Since the optical coupling loss between the SOA chip and the fiber is about 2 dB per facet, the chip saturation output power and the unsaturated chip gain are estimated to be +19 dBm and 23 dB. The eye patterns of the amplified signal modulated in 10 Gbpsec NRZ modulated signals with a PRBS of $2^{31}-1$ are also shown. A distortion-free eye pattern is obtained at an average module output power of +12 dBm, that is, 5 dB below the module saturation output power. For an average optical output power of +14 dBm, however, slightly distorted eye pattern due to the pattern effect is observed because the mark level of about +17 dBm reaches the CW saturation output power.

APPLICATION OF POLARIZATION INSENSITIVE SOAs

Polarization insensitive SOAs can be used as linear optical amplifiers, optical gate switches, and nonlinear elements for all optical processing. In this section, the former two applications are described. The examples of the last one are all optical wavelength converters and optical switches. These are discussed in detail in the chapter on "Wavelength Converter" in this book.

SOAs are promising as compact, low-cost optical amplifiers in metropolitan transmission systems. High power 1550 nm polarization insensitive SOAs were used as in-line optical amplifiers in WDM metropolitan transmission experiences [22–27]. Several remediation techniques have recently been proposed to avoid the pattern effect in the saturated regime, such as a large number of WDM signal inputs that statistically decrease the total signal power fluctuation [22], a polarization multiplexing technique [23], or a wavelength multiplexing technique [24]. In 1999, Sun et al. demonstrated the error-free transmission of 32 WDM channels at 2.5 Gbpsec per channel over 125 km using cascaded in-line SOAs [22]. On the other hand, an operating SOA in the (nearly) linear regime [25–27] is attractive because it is applicable for both WDM and single channel operation and no additional or special components are necessary to be used for transmitters and receivers. In 2000, Spiekmann et al. [25] demonstrated the transmission of 8 WDM channels at 20 Gbpsec/ch through 160 km and Wiesenfeld et al. [26] demonstrated the transmission of 32 WDM channels at 10 Gbpsec/ch through 160 km. Additionally, since the linearity can cope with the burst data, this feature will be very attractive in the WDM metropolitan transmission systems where the total signal power varies due to optical add and drop multiplexing (OADM) [27].

The polarization insensitive SOAs are promising to be used as fast optical gate switches with high extinction ratios for WDM cross-connect systems. Loss-less operation can be possible due to the gain. In 1981, Ikeda [28] proposed the use of SOAs as optical switches and the possibilities of integration of SOA array and waveguide. He demonstrated the high on–off ratio with the AlGaAs laser biased below the threshold (FP-SOA). In 1987, Ikeda et al. [29] demonstrated the monolithic integrated 2×2 SOA matrix switches with the crossbar configuration. In 1991, Janson et al. [30] demonstrated the monolithic integrated 2×2 SOA matrix switches with the tree configuration.

Low power consumption is inevitable for large-scale switch matrices. In 1997, Kitamura et al. [30] achieved loss-less operation with the operating current of 10 mA and Ito et al. [32] also demonstrated loss-less operation with the operating current of 7 mA. Since the module saturation output power was −2 dBm, however, the penalty-free-performance for signals modulated at 2.5 Gbpsec NRZ was achieved in the input (= output) range from −20 to −7 dBm [31].

ADVANCED STRUCTURE AND NOVEL MATERIAL OF SOA

Advanced Structure of SOA: Gain Clamped SOAs

Multi-channel amplification is required for inline amplifiers in WDM systems. When SOAs are used as the multi-channel amplifiers, system penalties can be caused due to inter-channel crosstalk and nonlinear distortions in the saturated regime. When SOAs are used for multi-channel signals, the total output must be limited several dB below the saturation output power so that the linear amplification can be guaranteed [25–27]. Then increasing saturation output power of SOAs is very essential. Additionally, optical linearization techniques of SOA gain that utilize the gain clamping mechanism of laser operation have been studied. In 1994 Simon et al. [33] realized the gain clamped (GC) SOA using the TW SOA that was optically coupled to an external diffraction grating through a lens or to a fiber Bragg grating (FBG) [34]. In 1994, Bauer et al. [35] developed the GC SOA using distributed feedback (DFB) laser. The GC SOAs could reduce nonlinear signal distortions and inter-channel crosstalk and increase the dynamic range of the linear regime.

Since the laser light was also coupled to the fiber, however, an optical band pass filter must be placed to prevent the laser light from mixing with the signal. In 1996, Holtmann et al. [36] developed a GC SOA in Mach–Zehnder Interferometer (MZI) configuration and demonstrated that the new GC SOA could spatially separate the laser light and the signal light. In 2001, Francis et al. [37] developed a GC SOA in arrayed vertical cavity surface emitting laser (VCSEL) configuration. This GC SOA, which is called LOA, can also spatially separate the laser light and the signal because the laser light emits in the orthogonal direction.

Novel Gain Material: Quantum Dot

Recently quantum dot (QD) SOAs have received attention due to its unique properties. In 1991, Komori et al. [38] predicted that QD SOAs might have a low noise figure due to almost complete population inversion. In 1990s, significant progress in the fabrication of QDs was achieved by self-assembled technique. In 1999, Borri et al. [39] and Akiyama et al. [40] independently reported the first experimental demonstration of ultra-fast gain response in InAs–GaAs QD SOAs. In 2004, Akiyama et al. [41] demonstrated the pattern effect free amplification up to the record high chip output power of +23 dBm for the 1550 nm signals modulated at 10 Gbpsec in InAs–InP QD SOA with the buried hetero-structure. The QD SOAs are not yet polarization independent and have long chip size (>6 mm) and high injection current. If the effective way to realize the polarization insensitivity can be found and the improvement of the material quality can reduce the chip length and the injection current, however, the QD SOAs will become promising as both linear amplifiers and nonlinear elements.

REFERENCES

1.   M. Coupland, K. H. Hambleton, and C. Hilsum, "Measurement of amplification in GaAs injection laser," *Phys. Lett.*, 7, 231–232, 1963.
2.   J. C. Simon, "Polarization characteristics of a travelling-wave-type semiconductor laser amplifier," *Electron. Lett.*, 18, 438–439, 1982.
3.   S. Cole, D. M. Cooper, W. J. Devlin, A. D. Ellis, D. J. Elton, J. J. Isaac, G. Sherlock, P. C. Spurdens, and W. A. Stallard, "Polarization-insensitive, near-travelling-wave semiconductor laser amplifier at 1.5 μm," *Electron. Lett.*, 25, 314–315, 1989.

4. N. A. Olsson, R. F. Kazarinov, W. A. Nordland, C. H. Henry, M. G. Oberg, H. G. White, P. A. Garbinski, and A. Savage, "polarization-independent optical amplifier with buried facet," *Electron. Lett.*, 25, 1048–1049, 1989.

5. I. Cha, M. Kitamra, H. Honmou, and I. Mito, "1.5 μm band travelling-wave semiconductor optical amplifier with window facet structure," *Electron. Lett.*, 25, 1241–1242, 1989.

6. M. S. Lin, A. B. Piccirilly, Y. Twu, and N. K. Dutta, "Fabrication and gain measurement for buried facet optical amplifier," *Electron. Lett.*, 25, 1378–1380, 1989.

7. P. Doussiere, P. Garabedian, C. Graver, D. Bonnevie, T. Fillon, E. Derouin, M. Monnot, J. G. Provost, D. Leclerc, and M. Klenk, "1.55 μm polarization insensitive semiconductor optical amplifier with 25 dB fiber to fiber gain," *IEEE Photonics Technol. Lett.*, 6, 170–172, 1994.

8. P. Doussiere, F. Pommerau, D. Leclerc, R. Ngo, M. Goix, T. Fillon, P. Bousselet, and G. Laube, "Polarization independent 1550 nm Semiconductor Optical Amplifier Packaged Module with 29 dB Fiber to Fiber Gain," paper presented at OAA1995, paper FA4, 1995.

9. C. Holtmann, P.-A. Besse, T. Brenner, R. Dall'Ara, and H. Melchior, "Polarization Insensitive Bulk Ridge-Type Semiconductor Optical Amplifiers at 1.3 μm Wavelength," paper presented at OAA1993, paper SuB2, 1993.

10. C. Holtmann, P.-A. Besse, T. Brenner, and H. Melchior, "Polarization Independent Bulk Active Region Semiconductor Optical Amplifiers for 1.3 μm Wavelength," *IEEE Photonics Technol. Lett.*, 8, 343–345, 1996.

11. K. Magari, M. Okamoto, H. Yasaka, K. Sato, Y. Noguchi, and O. Mikami, "Polarization insensitive traveling wave type amplifier using tensile-strained multiple quantum well structure," *IEEE Photonics Technol. Lett.*, 2, 556–558, 1990.

12. K. Magari, M. Okamoto, Y. Suzuki, K. Sato, Y. Noguchi, and O. Mikami, "Polarization-insensitive optical amplifier with tensile-strained-barrier MQW structure," *J. Quantum Electron.*, 30, 695–702, 1994.

13. M. Joma, H. Horikawa, C. Q. Xu, K. Yamada, Y. Katoh, and T. Kanijoh, "Polarization Insensitive Amplification at 1.5 μm Wavelength Using Tensile Strained Multiple Quantum Well Structure," paper presented at ISLC1992, Paper E-2, 1992.

14. M. Joma, H. Horikawa, C. Q. Xu, K. Yamada, Y. Katoh, and T. Kanijoh, "Polarization insensitive semiconductor laser amplifiers with tensile strained multiple quantum well structures," *Appl. Phys. Lett.*, 62, 121–122, 1993.

15. L. F. Tiemeijer, P. J. A. Thijs, T. van Dongen, R. W. M. Slootweg, J. M. M. van der Heijden, J. J. M. Binsma, and M. P. C. M. Krijn, "High Performance 1300 nm Polarization Insensitive Laser Amplifiers Employing both Tensile and Compressively Strained Quantum Wells in a Single Active Layer," paper presented at ECOC1992, paper ThPD 2.6, 1992.

16. L. F. Tiemeijer, P. J. A. Thijs, T. van Dongen, J. J. M. Binsma, and E. J. Jansen, "Polarization resolved, complete characterization of 1310 nm fiber pigtailed multiple-quantum-well optical amplifiers," *J. Lightwave Technol.*, 14, 1524–1533, 1996.

17. M. A. Newkirk, B. I. Miller, U. Koren, M. G. Young, M. Chien, R. M. Jopson, and C. A. Burrus, "1.5 μm multiquantum-well semiconductor optical amplifier with tensile and compressively strained wells for polarization-independent gain," *IEEE Photonics Technol. Lett.*, 4, 406–408, 1993.

18. J. Y. Emery, P. Doussiere, L. Goldstein, F. Pommerau, C. Fortin, R. Ngo, N. Tscherptner, J.-L. Lafragette, P. Aubert, and F. Brillouet, "New, Process Tolerant, High Performance 1.55 μm Polarization Insensitive Semiconductor Amplifier Based on Low Tensile Bulk GaInAsP," paper presented at ECOC1996, paper WeD.2.3, 1996.

19. J. Y. Emery, T. Ducellier, M. Bachmann, P. Doussiere, F. Pommerau, R. Ngo, F. Gaborit, L. Goldstein, G. Laube, and J. Barrau, "High performance 1.55 μm polarization insensitive semiconductor optical amplifier based on low-tensile-strained bulk GaInAsP," *Electron. Lett.*, 33, 1083–1084, 1997.

20. K. Morito, M. Ekawa, T. Watanabe, T. Fujii, and Y. Kotaki, "High Saturation Output Power (+17 dBm) 1550 nm Polarization Insensitive Semiconductor Optical Amplifier," paper presented ECOC2000, paper 1.3.2, 2000.

21. K. Morito, M. Ekawa, T. Watanabe, and Y. Kotaki, "High-output-power polarization-insensitive semiconductor optical amplifier," *J. Lightwave Technol.*, 21, 176–181, 2003.

22. Y. Sun, A. K. Srivastava, S. Banerjee, J. W. Sulhoff, R. Pan, K. Kantor, R. M. Jopson, and A. R. Chraplyvy, "Error-free transmission of 32 × 2.5 Gb/s DWDM channels over 125 km using cascaded in-line semiconductor optical amplifier," *Electron. Lett.*, 35, 1863–1865, 1999.

23. B. Banerjee, A. K. Srivastava, B. R. Eichenbaum, C. Wolf, Y. Sun, J. W. Sulhoff, and A. R. Chraplyvy, "Polarization Multiplexing Technique to Mitigate WDM Crosstalk in SOAs," paper presented at ECOC1999, paper PD3-9, 1999.

24. H. K. Kim and S. Chandrasekhar, "Reducing Cross Gain Modulation in the Semiconductor Optical Amplifier by using Wavelength Modulated Signal," paper presented at OAA'2000, paper OTuB3, 2000.

25. L. H. Spiekman, J. M. Wiesenfeld, A. H. Gnauck, L. D. Garrett, G. N. van den Hoven, T. van Dongen, M. J. H. Sander-Jochem, and J. J. M. Binsma, "160 Gbit/s (8 × 20 Gbit/s) DWDM Transmission over 160 km of Standard Fiber using a Cascade of Semiconductor Optical Amplifier," paper presented at ECOC1999, paper PD2-7, 1999.

26. J. M. Wiesenfeld, L. H. Spiekman, A. H. Gnauck, and L. D. Garrett, "32 × 10 Gb/s DWDM Transmission through a 160 km Cascade of Semiconductor Optical Amplifier in Light and Moderate Saturation Regime," paper presented at OAA'2000, paper OtuB4, 2000.

27. A. H. Gnauck, L. H. Spiekmann, J. M. Wiesenfeld, and L. D. Garrett, "Dynamic Add/drop of 8 of −16 10 Gb/s Channels in 4 × 40 km Semiconductor Optical Amplifier based WDM Systems," paper presented at OFC 2000, PD39, 2000.

28. M. Ikeda, "Laser diode switch," *Electron. Lett.*, 17, 899–900, 1981.

29. M. Ikeda, O. Ohguchi, and K. Yoshino, "Monolithic LD Optical Matrix Switches," paper presented at ECOC1987, pp. 227–230, 1987.

30. M. Janson, L. Lundgren, A.-C. Moerner, M. Rask, B. Stoltz, M. Gustavsson, and L. Thylen, "Monolithically Integrated 2 × 2 InGaAsP/InP Laser Amplifier Gate Switch Arrays," paper presented at ECOC/IOOC1991, paper PDP, pp. 28–31, 1991.

31. S. Kitamura, H. Hatakeyama, T. Kato, N. Kimura, M. Yamaguchi, and K. Komatsu, "Very-Low-Operating-Current SOA-Gate Modules for Optical Matrix Switches," paper presented at OAA1997, paper TuC3-1, 1997.

32. T. Ito, N. Yoshimoto, K. Magari, K. Nishi, Y. Kondo, and M. Naganuma, "Extremely low operating current SOA gate for WDM applications," paper presented at OECC1997, paper PDP1-1, 1997.

33. J. C. Simon, P. Doussiere, P. Lamouler, I. Valiente, and F. Riou, "Traveling wave semiconductor optical amplifier with reduced nonlinear distortions," *Electron. Lett.*, 30, 49–50, 1994.

34. L. Lablonde, I. Valente, P. Lamouler, E. Delevaque, S. Boj, and J. C. Simon, "Experimental and Theoretical Investigation of a Gain Clamped Semiconductor Optical Amplifier," paper presented at ECOC1994, paper WeC3.4, 1994.

35. B. Bauer, F. Henry, and R. Schimpe, "Gain stabilization of a semiconductor optical amplifier by distributed feedback," *IEEE Photonics Technol. Lett.*, 6, 182–185, 1994.

36. C. Holtmann, P.-A. Besse, J. Eckner, and H. Melchior, "Monolithically Integrated Gain-Clamped Semiconductor Optical Amplifier Exploiting Mach–Zehnder Interferometer Configuration," paper presented at ECOC1996, paper WeD.3.4, 1996.

37. D. A. Francis, S. P. Dijaili, and J. D. Walker, "A Single-Chip Linear Optical Amplifier," paper presented at OFC2001, paper PD13-1-3, 2001.

38. K. Komori, S. Arai, and Y. Suematsu, "Noise in semiconductor laser amplifiers with quantum box structure," *IEEE Photonics Technol. Lett.*, 3, 39–41, 1991.

39. P. Borri, W. Langbein, J. Mork, J. M. Hvam, F. Heinrichsdorff, M. H. Mao, and D. Bimberg, "Ultrafast Gain and Index Dynamics in Quantum Dot Amplifiers," paper presented at ECOC1999, paper 2, p. 74, 1999.

40. T. Akiyama, T. Shimoyama, H. Kuwatsuka, Y. Nakata, K. Mukai, M. Sugawara, O. Wada, and H. Ishikawa, "Gain Nonlinearity and Ultrafast Carrier Dynamics in Quantum Dot Optical Amplifiers," paper presented at ECOC1999, paper 2, p. 76, 1999.

41. T. Akiyama, M. Ekawa, M. Sugawara, H. Sudo, K. Kawaguchi, A. Kuramata, H. Ebe, K. Morito, H. Imai, and Y. Arakawa, "An Ultrawide-Band (120 nm) Semiconductor Optical Amplifiers Having an Extremely-High Penalty-Free Output Power of 23 dBm Realized with Quantum-Dot Active Layers," paper presented at OFC2004, paper PDP12, 2004.

SINGLE PHOTON SOURCE

Yoshihisa Yamamoto

A standard laser produces a coherent state of light with a Poissonian photon number distribution. Generation of single photons on demand is a key requirement for quantum key distribution and quantum information processing systems based on photonic qubits.

One of the generation schemes for single photons is to use a single semiconductor quantum dot embedded in a monolithic microcavity (see Figure 1). An optical pump pulse or electrical pump pulse generates more than one electron–hole pairs in a quantum dot. Those electron–hole pairs recombine to generate photons one by one via spontaneous emission. The last photon emitted from a quantum dot has a unique wavelength, differing from the other photons, because the multiparticle interaction inside a quantum dot modulates an emission wavelength. We can selectively extract this last photon by a narrow band optical filter, so that single photons are extracted from the system no matter how many electron–hole pairs are injected into the quantum dot initially.

This simple scheme works very well for an InAs self-assembled quantum dot in a GaAs/AlGaAs distributed Bragg reflection (DBR) postmicrocavity. The microcavity enhances a spontaneous emission into a cavity mode, so that both spontaneous emission rate and coupling efficiency are increased. It was demonstrated using a device such that the probability of finding two photons per pulse is <1%, the pulse duration is <100 psec and the coupling efficiency into a single cavity mode is 80% [1].

Moreover, these single photons are quantum, mechanically indistinguishable, so they feature two particle interference effect [2]. Using these properties of single photons, various photonic quantum information systems, such as quantum key distribution [3], postselected entangled photon-pair generation [4], and quantum teleportation gate [5], are demonstrated.

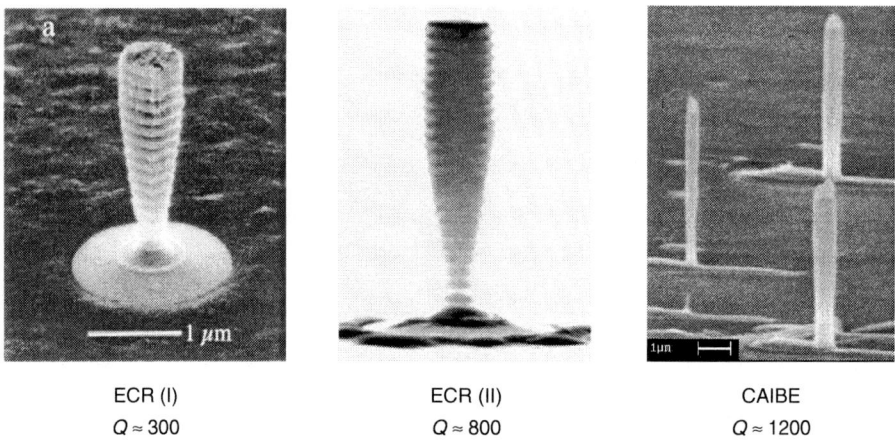

ECR (I)	ECR (II)	CAIBE
$Q \approx 300$	$Q \approx 800$	$Q \approx 1200$

Figure 1   DBR postmicrocavities including a single quantum dot in an optical cavity layer. The first two cavities are fabricated by electron cyclotron resonance (ECR) process and the third one by chemically assisted ion beam etching process. (From G. S. Salomon, M. Pelton, and Y. Yamamoto, *Phys. Rev. Lett.*, 86, 3903–3906 (2001); M. Pelton, C. Santori, J. Vuckovic, B. Y. Zhang, G. S. Solomon, J. Plant, and Y. Yamamoto, *Phys. Rev. Lett.*, 89, 233602-1–233602-4 (2002); J. Vuckovic, D. Fattal, C. Santori, G. S. Solomon, and Y. Yamamoto, *Appl. Phys. Lett.*, 82, 3596–3598 (2003). With permission.)

REFERENCES

1. J. Vuckovic, D. Fattal, C. Santori, G. S. Solomon, and Y. Yamamoto, "Enhanced single-photon emission from a quantum dot in a micropost microcavity," *Appl. Phys. Lett.*, 82, 3596–3598 (2003).
2. C. Santori, D. Fattal, J. Vuckovic, G. S. Solomon, and Y. Yamamoto, "Indistinguishable photons from a single-photon device," *Nature*, 419, 594–597 (2002).
3. E. Waks, K. Inoue, C. Santori, D. Fattal, J. Vuckovic, G. S. Solomon, and Y. Yamamoto, "Quantum cryptography with a photon turnstile," *Nature*, 420, 762 (2002).
4. G. S. Solomon, M. Pelton, and Y. Yamamoto, "Single-mode spontaneous emission from a single quantum dot in a three-dimensional microcavity," *Phys. Rev. Lett.*, 86, 3903–3906 (2001).
5. M. Pelton, C. Santori, J. Vuckovic, B. Y. Zhang, G. S. Solomon, J. Plant, and Y. Yamamoto, "Efficient source of single photons: A single quantum dot in a micropost microcavity," *Phys. Rev. Lett.*, 89, 233602-1–233602-4 (2002).
6. D. Fattal, K. Inoue, J. Vuckovic, C. Santori, G. S. Solomon, and Y. Yamamoto, "Entanglement formation and violation of Bell's inequality with a semiconductor single photon source," *Phys. Rev. Lett.*, 92, 037903-1–037903-4 (2004).
7. D. Fattal, E. Diamanti, K. Inoue, and Y. Yamamoto, "Quantum teleportation with a quantum dot single photon source," *Phys. Rev. Lett.*, 92, 037904-1–037904-4 (2004).

STACKED PLANAR OPTICS

Kenichi Iga

Stacked planar optics, a novel self-aligning scheme that uses a microlens array, has been developed as shown in Figure 1. If we can realize these kinds of two-dimensional arrayed devices available for actual uses, we can expect to align a tremendous number of optical components simultaneously, as in parallel multiplexing lightwave systems. The proposed concept will be a key technology for low cost optical couplers for advanced optical fiber communication systems handling a huge number of fiber and laser arrays. (See Microlens).

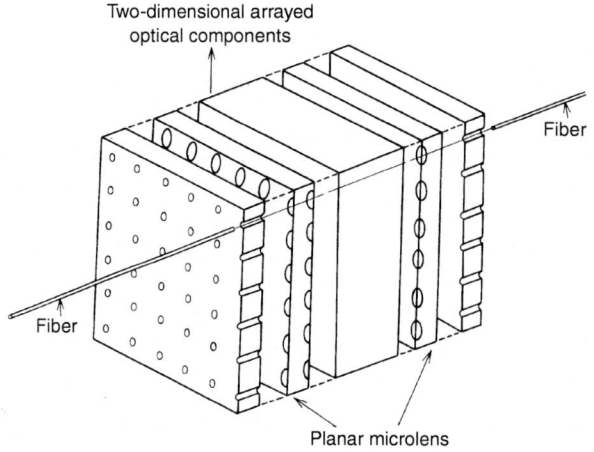

Figure 1 A concept of stacked planar optics

T

THERMO-OPTIC DEVICES

Yoshinori Hibino

INTRODUCTION

The rapid and global spread of the Internet is accelerating the growth of optical communication networks. Optical networks based on wavelength division multiplexing (WDM) systems [1] have played a key role in increasing the capacity and flexibility of these networks. They began as point-to-point WDM transmission systems, and are now evolving into ring and mesh networks. Various optical devices have been developed for WDM-based photonic networks, and some have already been installed in commercial communication systems. Of these, filters and optical switches are two of the most important passive devices used in the networks. Filters are needed wherever signals with different wavelengths propagating in a fiber have to be multiplexed or demultiplexed in WDM networks. We also need various kinds of optical switch for optical add/drop multiplexing (OADM) systems and optical cross connect (OXC) systems [2] to make networks more flexible. Moreover, the rapid progress made on networks has led to a demand for more channels at a lower cost.

Thermo-optic (TO) effects, which are refractive index changes caused by temperature variations in a material, have been used when constructing optical switches. TO switches have been developed for a variety of applications in photonic networks including OADM and OXC systems. These switches are mainly used to provide lightpaths and system protection because the switching speed of passive TO switches is usually only of millisecond order [3]. For these applications, the switches are used inside OXCs to reconfigure them to support new lightpaths. The challenge here is to realize large switches. For protection applications, the switches are used to move the traffic stream from a primary fiber to another fiber should the former fail. Small 2×2 switches are usually sufficient for this purpose. In addition to switching time, other important parameters used to characterize the suitability of a switch for optical networking applications are insertion loss, the extinction ratio between on and off states, crosstalk, polarization-dependent loss (PDL), and wavelength dependence, which will be reviewed in some detail. The state of integration of optical switches is considerably less than that of electric switches as illustrated by the fact that a 16×16 optical switch is currently considered a large switch.

There are two main kinds of TO switch; one is composed of silica-based planar lightwave circuits (PLCs), and the other of polymer waveguides. The following section reviews these two kinds of TO device, which were developed for optical communication networks. PLC-type devices have been playing an important role recently because of their suitability for large-scale integration and mass production.

SILICA-BASED PLC-TYPE TO DEVICES

The PLCs, in which fiber-matched silica-based waveguides are integrated, can provide various key devices for optical networks [4]. This is because they are suitable for large-scale integration, offer long-term stability, and can be mass-produced. The PLC device family includes optical couplers, $1/N$ optical power splitters, TO switches, arrayed waveguide grating (AWG)-type multi/ demultiplexers and lattice filters for high-speed transmission systems. AWGs are key components in high-capacity WDM networks, and we have recently fabricated a 400-channel 25-GHz spacing AWG using high contrast waveguides [5]. The PLC-type TO-switch is one of the most important devices for photonic networks, and is described in detail in subsequent sections.

The advantages of the silica-on-Si waveguide are its low loss and low coupling loss with optical fibers. Specifically, we have realized a propagation loss of <0.02 dB/cm in a 10-m long waveguide. These waveguides with a well-defined structure and a low loss enable us to design various optical integrated circuits by using numerical simulation techniques. In terms of PLC device reliability, fiber-pigtailed 1×8 splitter modules have been the subject of a detailed investigation based on the Telcordia reliability requirements for passive optical devices, which include long-term damp heat and mechanical tests [6]. Moreover, the lifetime of the 1×8 splitter modules was estimated from tests in which temperature and humidity were varied. The results yielded a 30-year failure rate at 25°C–90%RH of below 0.6 FIT.

Fundamentals

PLC-type TO switches consist of a Mach–Zehnder interferometer (MZI) with a phase shifter (thin-film heater) as shown in Figure 1. The MZI consists of two couplers and two optical waveguide

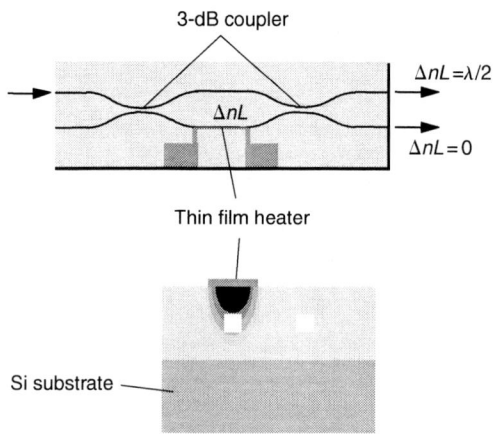

Figure 1   Configuration of Mach–Zehnder interferometer

arms that connect the couplers [7]. These couplers are generally 3-dB (50%) couplers and there are several types including the Y-junction, the directional coupler (DC), and the multi-mode interferometer (MMI) coupler. Various functions can be achieved by controlling the interference conditions using couplers, switches, modulators, wavelength (frequency) filters, and other devices. Of the MZI-type devices fabricated using silica-based PLC technologies, this section focuses on the TO optical switch with a phase shifter whose optical path-length difference is extremely small.

In silica glass, the TO effect is characterized by an increase in the refractive index as the temperature rises. The index change is given by the following equation:

$$n(T) = 0.9 \times 10^{-8} T^2 + 1.02 \times 10^{-5} T + 1.4497. \tag{1}$$

This is an empirically derived equation with T being the temperature of the waveguide in degrees Celsius. This equation tells us that the TO effect in silica waveguides is a nonlinear effect since temperature has a squared term. The coefficient of this term, however, is extremely small, which means that the term can be ignored near room temperature. The TO effect in silica glass therefore, is thought of as an almost linear effect. In silica-based PLC-type optical switches, switching is achieved by driving a thin-film metal heater deposited directly on a waveguide and changing the temperature in the vicinity of the core.

Next, to clarify the way in which an MZI switch operates, we first derive the optical output equations of an MZI. We assume a TO 2×2 switch with a symmetric MZ configuration in which the waveguide arms are equal in length. Also, for the sake of clarity, we limit the input light to port 1 and assume that the two-directional couplers have the same coupling efficiency (k). Here, light incident on the MZI splits at the first-stage DC, and after passing through the waveguide arms, recombines and interferes at the second-stage DC before exiting at output ports. In this case, the output powers are given by the following equations:

$$I_3 / I_1 = (1 - 2k)^2 + 4k(1 - k)\sin^2(\Delta\phi / 2) \tag{2}$$

$$I_4 / I_1 = 4k(1 - k)\cos^2(\Delta\phi / 2). \tag{3}$$

The phase difference $\Delta\phi$ in the above equations is expressed as follows:

$$\Delta\phi = 2\pi\Delta nl / \lambda. \tag{4}$$

Here, Δnl is the difference between the optical path lengths generated between the waveguide arms, Δn is the change in the refractive index due to the TO effect, and l is the length of the thin-film heater.

Equation (2) shows that some light leaks to the through-port (port 1 to 3 or port 2 to 4 in Figure 2) as the coupling efficiency shifts from 0.5 at a phase difference of 0. For example, the DC coupling efficiency must be kept at $50 \pm 5\%$ to hold the leaked light at below 20 dB. By contrast, for a phase difference of π, all light outputs pass to the through-port regardless of the coupling efficiency (from Eq. (3)). This relationship between phase difference and leaked light is a characteristic that must be considered when constructing a low crosstalk switch, and is particularly important when discussing and constructing a low-crosstalk matrix switch.

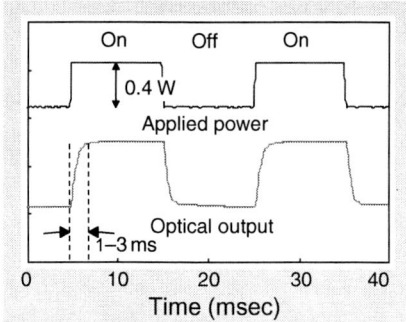

Figure 2   Typical switching power and time of PLC-type MZI switch

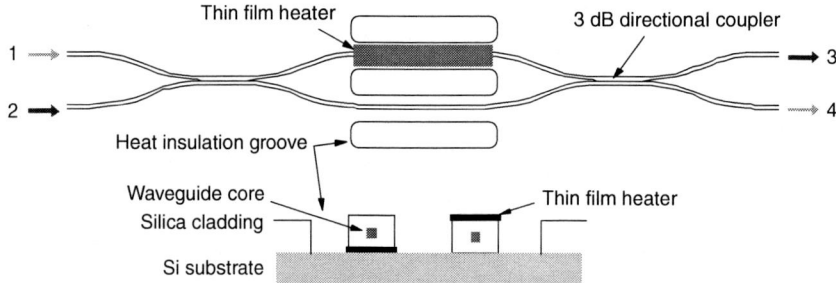

Figure 3   Novel structure of MZI switching unit with heat insulating groove

As for switching power, if the cladding thickness is 50 μm, and the core is located at the center of the cladding, the change in heater temperature and the power consumption at the time of switching would be about 37.5°C and 0.5 W, respectively. The switching time (or speed) is defined as the time required to change the light intensity from 10 to 90%. Figure 2 shows the thermal response characteristics for a 2×2 optical switch. The response characteristics shown at the bottom of Figure 2 depict the voltage applied to the TO phase shifter and those at the top show the corresponding optical output. These are the characteristics for a through-port (port 1 to 3 in Figure 1) in a symmetric MZ switch. The switching time consists of a rise time of 1.1 msec and a fall time of 1 msec. The switching time is mainly determined by cladding thickness and core position. In general, though, the switching time increases with increases in cladding thickness. The switching times for the optical switches are in the 1 to 3 msec range.

Recently, a novel configuration has been demonstrated for reducing the power consumption in a silica-based PL-type TO switch without any insertion loss increase [8]. The schematic structure of the proposed TO switch is shown in Figure 3. This structure consists of a conventional MZI with heat insulating grooves formed by using reactive ion etching (RIE). These grooves are formed symmetrically with respect to the MZI arms to avoid degrading the switching characteristics. The distance between the core center and the heater is set at 15 μm. Heat insulating grooves are also formed to restrict lateral heat diffusion. The fabricated switch needs a switching power of only 90 mW, and this is 80% less than that needed for a conventional TO switch. The insertion losses were as low as 1 dB for both TO switches, and we obtained an extinction ratio of more than 30 dB for the proposed TO switch.

Fabrication of PLC-Type TO Devices

PLC TO switches are fabricated using optical waveguides with a relative refractive-index difference Δ between the core and cladding of either 0.25 or 0.75%. The core size of the 0.25% Δ waveguide is $8 \times 8\ \mu$m, and the core of the 0.75% Δ waveguide is $6 \times 6\ \mu$m. The cladding thickness is about 50 μm and the Si-substrate thickness is 1 mm. Furthermore, a waveguide with a 0.25% Δ has a connection loss to optical fibers not > 0.1 dB/point, thus enabling us to realize a low-loss circuit. However, as Δ is small in this case, the light confinement is weak and the curvature radius of the curved waveguide must be as large as 25 mm or greater. This increases the circuit size, which is not beneficial for a large-scale circuit. By contrast, a waveguide with a 0.75% Δ has strong light confinement allowing us to realize a curvature radius as small as 5 mm. In this case, however, the connection loss with optical fiber is as large as 0.4 dB/point. So, we used a waveguide with a Δ of 0.25% for small-scale circuits such as 2×2 switches and a waveguide with a Δ of 0.75% for large-scale circuits such as $1 \times N$ and $N \times N$ matrix switches.

The process used for fabricating an optical waveguide is shown in Figure 4 [9]. First, glass soot consisting of an under-cladding layer and a core layer is deposited on a silicon (Si) substrate by flame hydrolysis deposition (FHD). The FHD method introduces gas materials such as $SiCl_4$ into an oxy-hydrogen burner thus causing an oxidation reaction within the flame, and then deposits fine particles such as SiO_2 on the substrate. Here, the main component of the under-cladding layer is SiO_2, but traces of B_2O_3 and P_2O_5 are included as dopant materials to lower the softening point temperature below the melting temperature of the Si substrate while keeping the refractive index fixed. GeO_2 is included in the core layer to raise the refractive index slightly. Next, the Si substrate with the deposited soot layers is heated in an electric furnace at temperatures ranging from 1200 to 1300°C to form transparent glass layers. In the next step, the core circuit pattern is formed by using the photolithographic technology employed in the manufacture of large scale integrated (LSI) circuits, and reactive ion etching (RIE). After this, the FHD

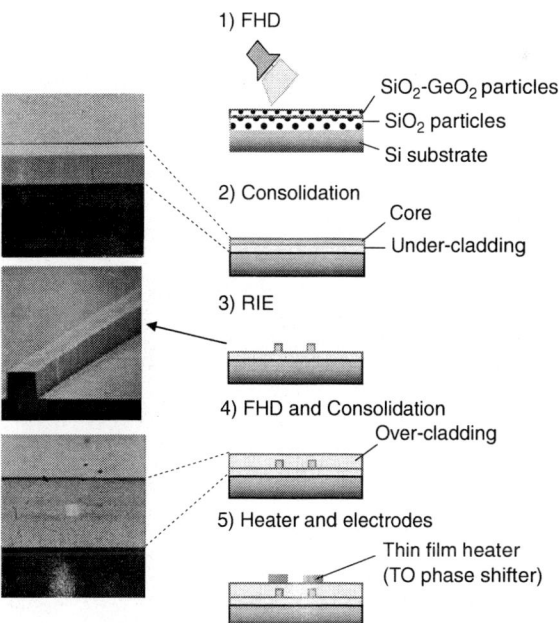

Figure 4 Fabrication process of PLC-type TO switch

method is used again to deposit soot and form an overcladding layer. The components of this layer are the same as those of the under-cladding layer, but the amounts of B_2O_3 and P_2O_5 are increased slightly to reduce the consolidation temperature of the over-cladding to below the temperature used to form the under-cladding layer and prevent core deformation. The over-cladding is also consolidated in an electric furnace to form transparent glass.

To heat the optical waveguides and obtain the TO effects, a thin-film metal heater is deposited on the over-cladding layer. Two types of film are used as thin-film heaters: Ta_2N sputter film and Cr vapor-deposited film. The film thickness is controlled to obtain the resistance desired for the heating function. The heater pattern is transferred to a resist by a photolithographic technique. The Ta_2N sputter heater is superior to the Cr vapor-deposited heater in terms of power durability, and with little fluctuation in resistance caused by oxidation, it is also highly reliable. A standard heater is about 5 mm long and 50 μm wide.

Variable Optical Attenuators

Variable optical attenuators (VOAs) are now playing a more important role in the gain control of linear repeaters for WDM networks and channel power equalization in WDM OADM nodes [10]. Such applications require a relatively fast operating speed of ~1 msec and a compact size. While high-speed nonmechanical VOAs using a Faraday rotator have already been proposed, a silica-based PLC is another candidate for realizing high-speed compact VOAs. Since PLC technologies enable us to integrate arrayed VOAs, a PLC-type VOA is superior in terms of size to a Faraday rotator-type VOA.

A 16-channel VOA array, composed of an asymmetric MZI with stress-releasing grooves, has been fabricated by using silica-based PLC technologies. A cross-path is usually used in the VOA because the attenuation in a through path is strongly affected by the coupling ratio errors that occur in the directional coupler during fabrication. We took care to prevent the fabrication process from affecting the waveguide stress distribution, since the PDL in the silica-based PLC-VOA strongly depends on the waveguide stress. The cladding ridge width was set at 20 μm, and the thin film heater was 20 μm wide and 3 mm long. Δ was 0.75%. The core was 7×7 μm in size, and the under- and over-cladding thicknesses were 15 and 30 μm, respectively. The VOA array chip was 15×30 mm^2.

Figure 5(a) shows insertion loss versus applied electrical power (P_e) superimposed over 16 channels. The insertion losses at $P_e = P\pi$ ranged from 0.97 to 1.08 dB. The average insertion

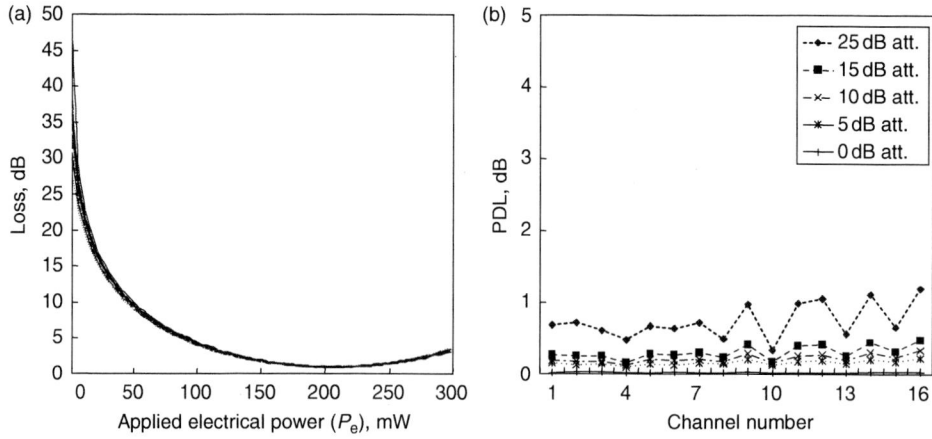

Figure 5   Attenuation and PDL characteristics in PLC-type VOA

loss was 1.03 dB. This insertion loss includes two coupling losses between the single-mode fibers and the input/output waveguides. The maximum attenuation at $P_e = 0$ was larger than 26 dB. The average $P\pi$ value was 203 mW. Figure 5(b) shows the PDL at attenuations of 0, 5, 10, 15, and 25 dB over 16 channels. The PDL values at attenuations of 0, 5, 10, 15, and 25 dB were less than 0.1, 0.2, 0.3, 0.5, and 1.2 dB, respectively, and the average PDL values were 0.0, 0.2, 0.2, 0.3, and 0.7 dB, respectively. Figure 5(b) confirms that the PDL is stably suppressed throughout the 16 channels. This low and uniform PDL results from the precise PLC fabrication technology.

32-Channel 2 × 2 Switch

An optical add/drop multiplexer (OADM), which adds and drops incoming and outgoing channels according to their wavelengths in WDM systems, is important in photonic networks. The OADM was constructed by combining an AWG and PLC-type TO switches. The PLC-type TO switch in the OADM node has two functions. One is high-isolation switching for the add/drop operation. The switch must have a high extinction ratio in order to avoid any crosstalk-induced degradation in the bit error rate. The other requirement is a VOA function for level equalization at each channel. Any power deviation among the WDM channels degrades the signal-to-noise ratio for weaker power channels while high power channels suffer waveform distortion induced by fiber nonlinearity [11].

Figure 6 shows a novel TO switch configuration for an OADM system that was proposed and demonstrated recently. The new configuration consists of four switches in a channel. There are two operation modes, namely the through mode and the add/drop mode, as shown in the figure. In the through mode, the input signal is led to the output port. In the add/drop mode, the input and add signals are led to the drop and output ports, respectively. Figure 6 shows the MZI layout for this function. There are two (#1, #4) and three (#2, #3, #4) MZI switch elements in the add–output and input–output paths, respectively, because a high extinction ratio is required or the signal degradation caused by optical crosstalk would never recover. A high extinction ratio is not required for the input–drop path, and there is only one element (#2) since the electric signal from the photo diode connected to the drop port is disregarded in the through mode. Each switch element is an asymmetric MZI with an optical length difference of a half-wavelength. The shorter arm waveguide is equipped with a thin-film heater, as shown in Figure 3(a). The asymmetric MZI is used to maintain the high extinction ratio of the cross-path (port 1 → 4/port 2 → 3) despite the coupling ratio error of the directional coupler [3].

The switch element is in the "cross" state when the heater is driven, and in the "bar" state when the heater is not driven. When MZI switch elements #2 and #3 are operated, the input signal is led to the output port (through mode). When MZI switch elements #1 and #4 are operated, the input and add signals are led to the drop and output ports, respectively, (add/drop mode). One characteristic of this configuration is that the power consumption is the same for both the through and add/drop modes.

A 32-channel 2 × 2 switch array was fabricated using the PLC technology. The waveguide was 0.75%, and heat-insulating grooves were introduced along the arm waveguides in the MZI

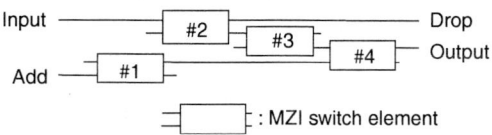

Figure 6 Configuration of PLC-type TOSWs for OADM systems

Figure 7   Photograph of 32 channel OADM switch

to reduce the power consumption needed for integrating 32 units. Figure 7 shows a photograph of the fabricated 32-channel 2×2 switch chip. The chip is 24×56 mm^2 in size and is fitted into a $90 \times 120 \times 16$ mm^3 package that includes cooling fins.

We measured the insertion losses of the input–output path in the through mode, the add–output path in the add/drop mode, and the input–drop path in the add/drop mode. The measured loss includes the coupling loss between the fiber and the waveguide. The average loss was 0.65 dB for all 96 paths (3 paths $\times$ 32 channels). Moreover, we measured the extinction ratio, which is calculated from the insertion losses of the input–output path in the add/drop mode and the add–output path in the through mode. The average crosstalk values for the input–output and add–output paths were 71.4 and 55.1 dB, respectively, which shows that we obtained an extinction ratio as high as 27 dB per MZI element. The average switching power per MZI element was 0.2 W and the total power consumption was 12.8 W (=0.2 W $\times$ 2 $\times$ 32).

The average extinction ratio change per MZI element in a 15 to 80°C temperature range was only 0.6 dB. This is one of the characteristics of the asymmetric MZI switch. We confirmed that these loss fluctuations induced by changes in the chip temperature were very small and that this switch could be deployed in practical systems without chip temperature control.

$1 \times N$ PLC-Type TO Switch

Eight-arrayed 1×8 PLC-type switches have been developed for OXC systems [12]. In the switch, an asymmetric MZI structure with a half-wavelength difference between the two waveguide lengths was chosen to obtain a high extinction ratio. This is because a high extinction ratio is inherently obtained for the cross-port pass in the MZI, and the extinction occurs in the asymmetric MZI at an electrical power of zero. The logical arrangement of the switch is shown in Figure 8. Each switch has bunched MZIs with ports 1 and 4 connected to each other. Thus input light passes though the MZIs when the heater is not activated. The light can be tapped to an arbitrary output port by activating the corresponding tap MZI switch element ("T" in Figure 8). A feature of this structure is that there is very little leakage to the output when the heater is not activated. To obtain a higher extinction ratio, a gate MZI switch element is added at each output port ("G" in Figure 8). There are 128 MZIs but there are only 16 simultaneously activated MZIs for switching.

We used a silica-based waveguide with Δ of 0.75% to form the switch. The core is 7×7 μm^2 The waveguide ridge between the heat insulator grooves is 30 μm wide, and 35 μm high in cross-section. The chip is 110×22 mm^2 as shown in Figure 9. The average switching power was 243 mW per MZI, which is about 60% of that obtained without a heat insulator groove. The total power consumption was about 3.9 (=0.243 $\times$ 16) W.

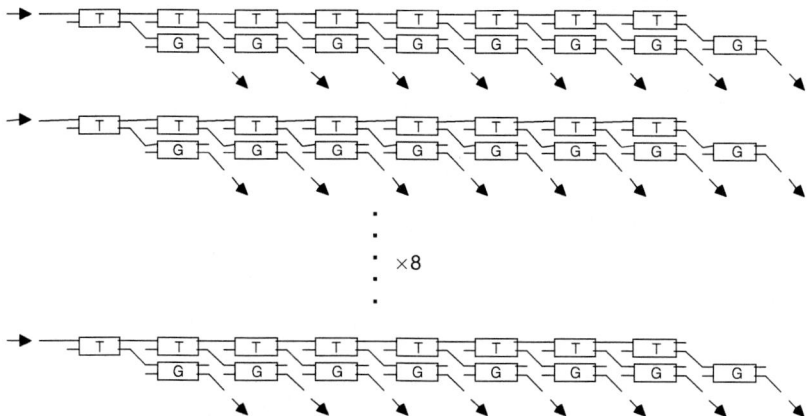

Figure 8 Configuration of eight-arrayed PLC-type 1×8 TO switch

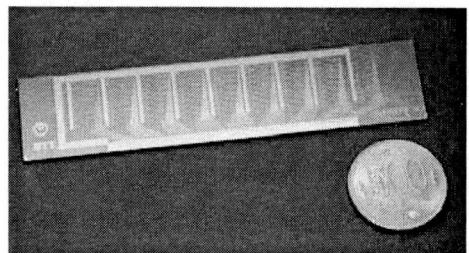

Figure 9 Photograph of eight-arrayed PLC-type 1×8 TO switch

The fiber-pigtailed 8-arrayed 1×8 switch module was equipped with a switch chip that was fixed to fiber ribbons at the input and output waveguides, a ceramic wiring board with a pin-grid array for the power supply, and a cooling fin. We realized very thin modules with a thickness of, for example, 13 mm. We used the switch modules in experimental OXC systems and confirmed that they had suitable characteristics for practical use [13].

We measured the individual and crosstalk suppression for three devices. The crosstalk suppression is defined as the worst suppression at a given output when the switch is set to the remaining seven outputs in turn. The extinction ratio ranged from 21.1 to 36.4 dB and the average value was as high as 28.5 dB because of the small optical length error in the downsized MZI. The crosstalk suppression ranged from 50 to 62 dB, and the average value was 55.7 dB, which is approximately double the MZI extinction ratio. This crosstalk suppression is the highest ever obtained without the use of an optical phase trimming technique. The insertion loss ranged from 0.67 to 2.29 dB with an average value of 1.33 dB. The loss increased from the nearest to the farthest port in steps of about 0.1 dB. That is, the through path excess loss of the MZI is only 0.1 dB.

$N \times N$ Nonblocking Matrix Switch

We fabricated strictly nonblocking matrix switches up to 16×16 using silica-based PLC technologies on a Si substrate [14]. To improve the extinction ratio of the matrix TO switches, we used a double-gate MZI switching unit for the $N \times N$ TO switches. The configuration of the

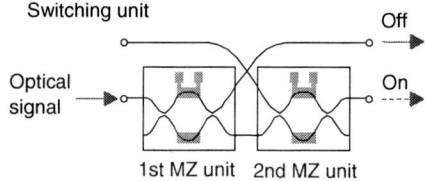

Figure 10   Configuration of double-gate switching unit

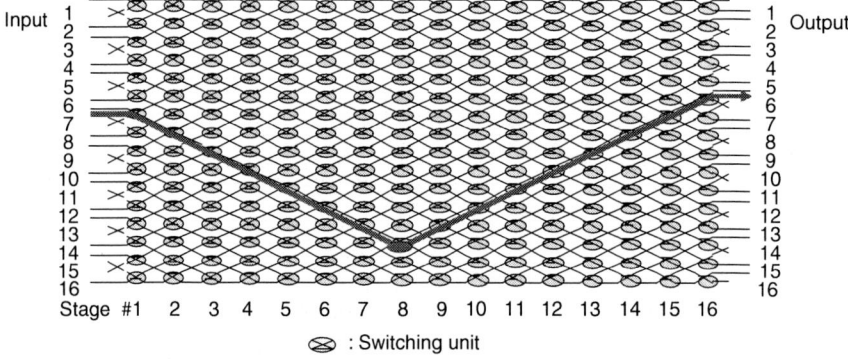

Figure 11   Logical arrangement of 16 × 16 matrix switch with PI-loss structure

double-gate MZI switching unit is shown in Figure 10. The switching unit is composed of two asymmetric MZIs with an optical path length difference of a half-wavelength and TO phase shifters, and an intersection. These asymmetric MZIs are also used to obtain a high extinction ratio. Moreover, the leaked light power to the cross path in the off state is greatly reduced in this proposed configuration because the unwanted light power from the first MZI is blocked by the second MZI. Therefore, the expected extinction ratio of the matrix switch is twice that obtained with a conventional single-MZI switching unit.

Figure 11 shows the logical arrangement adopted for the strictly nonblocking 16 × 16 matrix switch, which has a path-independent insertion loss (PI-loss) characteristic [15]. This is logically equivalent to the conventional crossbar matrix arrangement with a diamond shape. This arrangement requires only N switching unit stages in the $N \times N$ matrix switch, compared with the $2N - 1$ stages required with a conventional crossbar arrangement. This will be useful for reducing waveguide length.

Figure 12 shows the circuit layout of a 16 × 16 matrix switch with the PI-loss configuration on a 6-in. wafer. The 16 × 16 TOSW is four times larger than the 8 × 8 TOSW. The core size was 7×7 μm and the relative refractive index difference between the core and cladding was 0.75%. Sixteen switching unit stages were located along the twisting waveguides to minimize the total waveguide length including the connective waveguides between the stages. One switching unit stage contains sixteen switching units and 256 switching units in total are integrated on the chip. A comparatively short total-circuit length of 66 cm was realized with a bending radius of 5 mm. The chip size was 100×107 mm^2. There are usually fabrication errors [for/as regards?] the optical phase difference between the two MZI arms. Therefore, phase trimming is necessary to obtain high extinction ratios for matrix TO switches. The phase trimming is carried out in the matrix switch using the heaters. When a high power is applied to the heater, a permanent refractive index change can be induced in the core. Phase trimming was performed for all MZIs in the

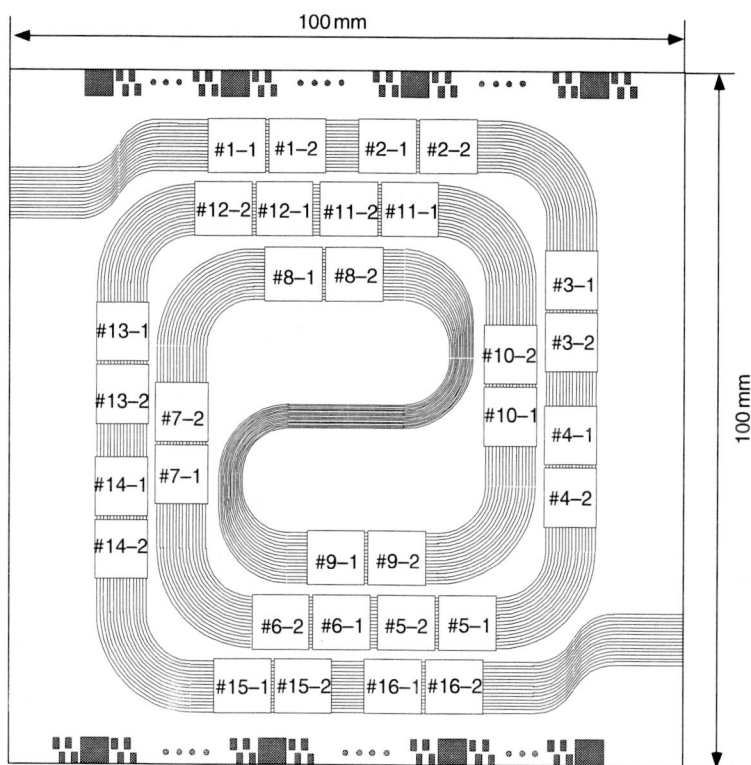

Figure 12   Circuit layout of 16 × 16 matrix switch with PI-loss structure

matrix switch [16]. The phase-error was eliminated and the bias power was adjusted to <3 mW. This accuracy corresponds to 0.6% of the half-wavelength phase change, and is sufficient to obtain a high extinction ratio without bias power. The phase-trimming process caused no loss degradation.

Driving circuits with TTL interfaces were used for the 8 × 8 TOSW module, which require the same number of control terminals as switching units. As for the 16 × 16 matrix switch, a package with newly designed driving circuits has been developed [17]. The new driving circuits are composed of 4 driving ICs with serial interfaces, and they enable us to reduce significantly the required number of control terminals from 256 to only 3. In the module with the new driving circuits composed of four driving ICs, only three control terminals are necessary because of the serial interfaces. The serial control signals are transformed into parallel signals with the driving ICs. In addition, since an arbitrary and independent switching power can be supplied to each switching unit, it is possible to flatten the optical output power level even if the optical input power level fluctuates. This new circuit design can be easily applied to a more densely integrated optical module. Figure 13 shows a photograph of the fabricated module. The TO switch chip was mounted on a multi-layered ceramic substrate on which the driving circuits were integrated. The PLC chip and the driving circuit were connected with gold wires. The terminals of the serial interfaces and those for the switching power supply were all connected to an electrical connector. Two fiber ribbons were butted and fixed to the facets of the input and output ports of the chip. The reverse surface of the ceramic substrate was equipped with a cooling fin. The module was 165 × 160 × 23 mm including the cooling fin.

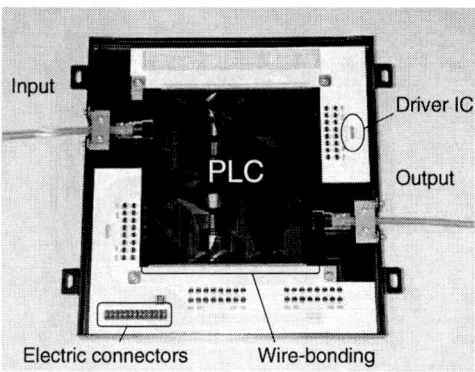

Figure 13   Photograph of 16×16 matrix switch module

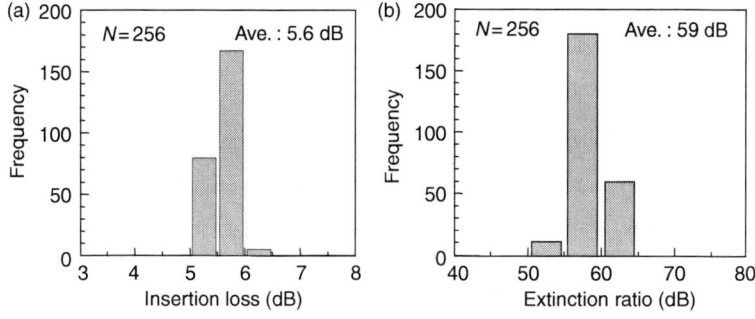

Figure 14   Optical characteristics of 16×16 matrix switch. (a) Insertion loss and (b) extinction ratio

We measured the performance of the fabricated TO switch module at 1.55 μm. Figure 14(a) and (b) show the distributions of the measured insertion loss and extinction ratio, respectively, for all 256 possible optical paths from the input ports to the output ports. We obtained a low insertion loss of 7.3 dB, a high extinction ratio of 60.7 dB, and a low PDL of 0.11 dB. The switching power per switch unit was 0.85 W, corresponding to a total switching power of 13.6 W (= 0.85×16). We realized a switching time of <4.1 msec. This module is a very promising component for realizing large photonic network systems.

We measured the wavelength dependence characteristics by employing 1.55 μm band wavelength light from a tunable laser diode with a high-speed polarization scrambler using a piezo-electric device. Figure 15 shows the wavelength dependence of the extinction ratios and insertion losses for all 64 optical paths. The error bars indicate the minimum and maximum extinction ratios and insertion losses. The worst extinction ratio and insertion loss were 47.6 and 6.4 dB, respectively, in the 1530 to 1560 nm range. This flat response covers the gain band of practical erbium-doped fiber amplifiers. These results show that this device can be practically employed in a wavelength division multiplexing optical communication system.

POLYMER WAVEGUIDE TO DEVICES

Polymeric optical waveguide devices based on the thermo-optic (TO) effect have great potential for use in optical telecommunication systems. This is because polymeric waveguides have a

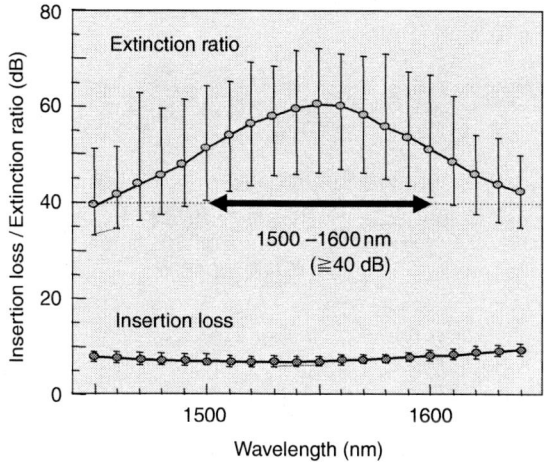

Figure 15   Wavelength dependence of optical characteristics in 8 × 8 matrix switch with double gate switching unit

larger TO coefficient than silica waveguides. Recently, we successfully demonstrated TO polymeric waveguide devices including digital TO switches [18] and a tunable wavelength filter using arrayed-waveguide grating (AWG) multiplexers [19]. As the next step toward practical use these devices must undergo stability and reliability tests. In these studies, however, some of the polymeric waveguide devices have exhibited slightly unstable behavior caused by slight changes in their refractive index. In other words, waveguide devices can function as highly sensitive sensors for detecting refractive index changes. In this paper, we discuss this refractive index change in optical waveguide devices.

Temperature Dependence of Refractive Index

The temperature dependence of the refractive index in polymer is expressed by the Lorentz–Lorentz formula as follows:

$$\frac{\mathrm{d}n}{\mathrm{d}T} = \frac{(n^2 + 2)(n^2 - 1)}{6n}\left(\frac{1}{\alpha}\frac{\mathrm{d}\alpha}{\mathrm{d}T} + \frac{1}{\rho}\frac{\mathrm{d}\rho}{\mathrm{d}T}\right), \tag{5}$$

where, n is the refractive index, T is temperature, α is polarizability, and ρ is density. The $(1/\alpha)(\mathrm{d}\alpha/\mathrm{d}T)$ value is negligible, because the temperature dependence of polymer polarizability is very small. Consequently, the change in density, which is $(1/\rho)(\mathrm{d}\rho/\mathrm{d}T)$, has an influence on the temperature dependent refractive index in polymer.

The temperature dependence of the refractive index of polymethylmethacrylate (PMMA) was measured by using various techniques such as the Fizeau fringe [20] and the deflection of a prism [21]. However, it is difficult to employ these techniques for measurements under humid conditions, because it is impractical to place the apparatus in a humidity chamber.

We measured both the temperature and humidity dependence of the refractive index of deuterated PMMA (d-PMMA), silicone resin and UV-cured epoxy resin by the return loss method [22]. With this approach, a drop of polymer solution is made to adhere to the cleaved end of an optical fiber. Next, this fiber end is placed in a humidity- and temperature-controlled chamber. The return loss L (dB) from the coated end is measured, and the refractive index n of the polymer is estimated from the following relation as

$$10^{-L/10} = \{(n - n_0)/(n + n_0)\}^2,\qquad\qquad(6)$$

where n_0 is the refractive index of the core of the silica optical fiber. Figure 16 shows the temperature and humidity dependence of the refractive index of d-PMMA at a wavelength of 1.3 μm. The refractive index increases with humidity at temperatures below 50°C. However, the humidity dependence of the refractive index changes above 50°C. Figure 17 shows the temperature and humidity dependence of the refractive index of UV-cured epoxy resin at 1.3 μm. Although the humidity dependence of the refractive index of UV-cured epoxy resin is smaller than that of d-PMMA, they have almost the same TO coefficient. Figure 18 compares data for silicone with data for d-PMMA. The effect of humidity on the refractive index is smaller for the epoxy and silicone resins than for d-PMMA.

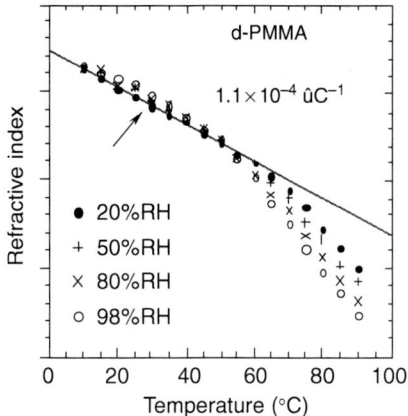

Figure 16   Temperature and humidity dependence of refractive index of d-PMMA at 1.3 μm

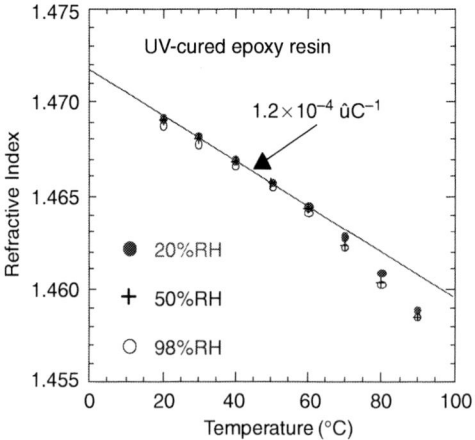

Figure 17   Temperature and humidity dependence of refractive index of UV-cured epoxy resin at 1.3 μm

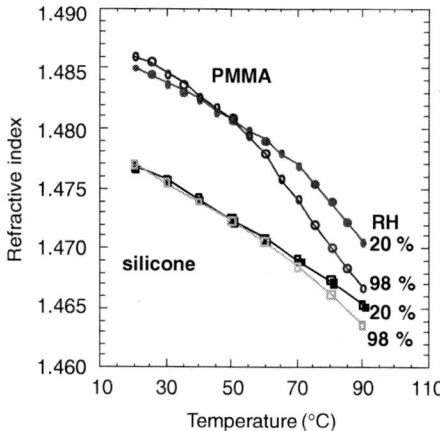

Figure 18 Temperature and humidity dependence of refractive index of silicone resin at 1.3 μm compared with values for d-PMMA

Polymeric Optical Waveguide Devices

The successful fabrication has been reported of highly transparent waveguides with a propagation loss <0.1 dB/cm at 1.3 μm using d-PMMA based polymer, which has a highly controllable refractive index [23]. Subsequently these polymer waveguides were employed in directional couplers, MZIs, ring resonators, MZ interferometer type TO switches [24], and arrayed-waveguide grating multiplexers [4]. Both the refractive indices and waveguide shapes of these lightwave-phase sensitive waveguide devices must be well controlled. These waveguide devices were previously realized in silica glass waveguides.

Polymer TO Switches

As the TO coefficient of d-PMMA is one order of magnitude larger than that of silica glass, it is advantageous in terms of TO switch applications. Hida and Imamura [25] demonstrated low power operation of 4.8 mW with MZ type TO (MZTO) switches, which is two orders of magnitude less than that for silica-based MZTO switches. However, the large humidity dependence of the refractive index sometimes leads to unstable operation [26]. This is because the operation of the MZI is very sensitive to the refractive index difference between its two arms. The refractive index of d-PMMA depends greatly on the circumambient humidity because of its large water absorption. Figure 19 shows the optical loss spectra of d-PMMA under various humidity conditions. A Y-branching type digital TO (YDTO) switch, which is different from the MZ type, has been investigated by Horsthuis et al. [27] and Moosburger et al. [28] with a view to avoiding unstable operation. We described the switching characteristics of a d-PMMA YDTO switch when it was operated under various humidity conditions [29]. Figure 20(a) is a diagram of our YDTO switch and Figure 20(b) shows the humidity dependence of its switching characteristics. The operation was much more stable than with previously fabricated MZTO switches, however, humidity has a slight influence on the extinction ratio of the on–off operation. Recently, we fabricated 1 × 8 YDTO switches using silicone resin, which is much more stable in humid conditions than PMMA based polymers [18].

TO Wavelength-Tunable Filter

As already mentioned, AWG multiplexers are key components for WDM systems. A wavelength-tunable optical filter is also a useful component for these systems because of its multichannnel

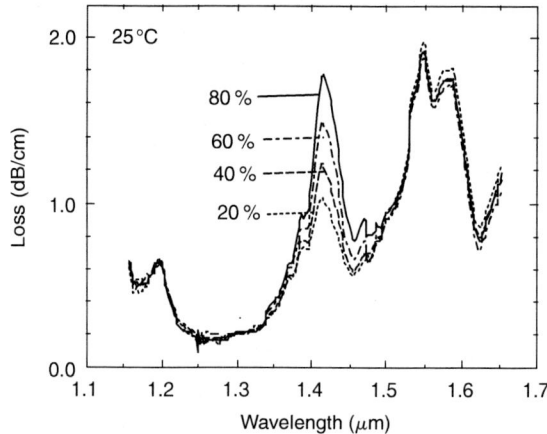

Figure 19   Loss spectra of d-PMMA under different humidity conditions

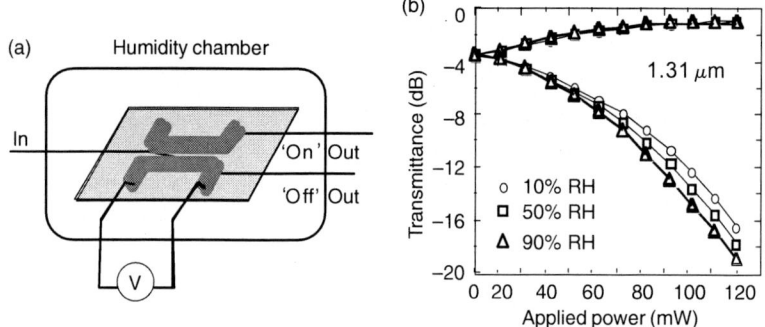

Figure 20   (a) Diagram of digital Y-branching type TO (YDTO) switch. (b) Operation characteristics of YDTO switch at 10, 50, and 90% RH

selectivity. Recently, we fabricated a wavelength-tunable optical filter using a polymeric AWG because AWG multiplexers provide good wavelength resolution [19]. Practical tunable filters require good wavelength resolution, low polarization by controlling the temperature dependence, low levels of crosstalk, and low insertion loss.

AWG multiplexers have already been developed using silica [4], semiconductors [30], and polymers [31] as the waveguide materials. Because of their inherently large TO coefficient, polymers offer excellent potential for the realization of TO tunable wavelength filters. Then, the transmission passband of polymeric AWG multiplexers could be shifted over a wide range by heating. AWG multiplexers have been fabricated using cross-linked silicone. We operated this AWG multiplexer as a thermo-optic tunable filter (TOTF) by controlling the temperature with a Peltier-type heater as shown in Figure 21. The peak wavelength was tuned by controlling the temperature of the AWG chip as shown in Figure 22. However, we found that the device spectrum including the peak wavelength shifts slowly with time over 12 h. This peak shift indicates a refractive index change in the waveguide material. This slow response in refractive index change seems to originate in the volume relaxation of the waveguide polymer. We adopted an optical-output-to-temperature feedback system in order to avoid this unstable behavior [19]. With this feedback system, the device temperature was controlled to stabilize the optical transmission.

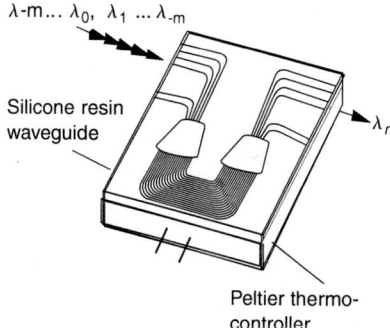

Figure 21 Diagram of thermo-optic tunable filter (TOTF) using AWG multiplexer that operates by controlling the temperature with a Peltier-type heater

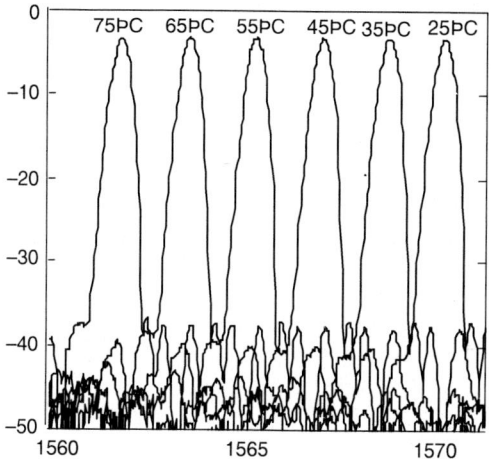

Figure 22 Tunability of polymer TOTF in 20 to 80°C temperature range

CONCLUSION

This chapter has reviewed the development of TO devices for WDM-based photonic networks. Various kinds of optical devices have been developed to make it possible to construct flexible and large-capacity networks, and the optical switch is one of the key components for photonic networks. PLC-type TO switches have been developed to provide lightpaths and protection for such systems as OADM and OXC systems. First we described the basic characteristics and fabrication technologies of PLC-type TO switches. Then, we reported the optical performance and applications of various types of switch including 2×2 switches, a $1 \times N$ switch and $N \times N$ matrix switches. In terms of optical performance, insertion loss, extinction ratio, crosstalk, PDL, and wavelength dependence are important. These characteristics were reviewed here for PLC-type switches. Each device has been or will be installed in a suitable application in WDM systems. The next step is to meet demands that both types of device be made less expensive, on a larger scale and with expanded functions.

In addition, we described the refractive index changes that occur in polymers. We measured both the temperature and humidity dependence of the refractive index for d-PMMA, silicone resin and UV-cured epoxy resin using the newly developed return loss method. We fabricated an MZTO switch and a YDTO switch using photolithographic and dry etching techniques. We observed the refractive index change in these devices that occurred under humid conditions, and found that the YDTO switch operated much more stably than the MZTO switch. We also observed the unstable behavior of the MZTO switch in a fabricated thermo-optic tunable wavelength filter using AWG multiplexers. This operational instability is an indication of the volume relaxation effect generated by a sudden temperature change. The use of a feedback-controlled system overcame this instability. From another point of view, optical waveguide devices are effective as highly sensitive probes for measuring small changes in the refractive index of transparent materials.

REFERENCES

1. I. P. Kaminow, "A wideband all-optical WDM network," *J. Select. Areas Commun.*, 14, 780–799 (1996).
2. M. Koga, Y. Hamazumi, A. Watanabe, S. Okamoto, H. Obara, K. Sato, M. Okuno, and S. Suzuki, "Design and performance of an optical path cross-connect system based on wavelength path connect," *J. Lightwave Technol.*, 14, 1106–1119 (1996).
3. K. Sato, "Photonic transport network OAM technologies," Special issue on "Operation and Management of Broadband Networks," *IEEE Commun. Mag.*, 34, 86–94 (1996).
4. A. Himeno, K. Kato, and T. Miya, "Silica-based planar lightwave circuits," *J. Select. Top. Quantum Electron.*, 4, 913–924 (1998).
5. Y. Hida, Y. Hibino, T. Kitoh, Y. Inoue, M. Itoh, T. Shibata, A. Sugita, and A. Himeno, "400-channel arrayed waveguide grating with 25 GHz spacing using 1.5%-Δ waveguides on 6-inch Si wafer," *Electron. Lett.*, 37, (2001).
6. Y. Hibino, F. Hanawa, H. Nakagome, M. Ishii, and N. Takato, "High reliability optical splitters composed of silica-based planar lightwave circuits," *J. Lightwave Technol.*, 13, 1728–1735 (1995).
7. N. Takato, K. Jinguji, M. Yasu, H. Toba, and M. Kawachi, "Silica-based single-mode waveguides on silicon and their application to guided-wave optical interferometer," *J. Lightwave Technol.*, 6, 1003–1010 (1988).
8. R. Kasahara, M. Yanagisawa, A. Sugita, T. Goh, M. Yasu, A. Himeno, and S. Matsui, "Low-power consumption silica-based 2 × 2 thermooptic switch using trenched silicon substrate," *IEEE Photonics Technol. Lett.*, 11, 1132–1134, 1999.
9. M. Kawachi, "Silica waveguides on silicon and their application to integrated components," *Opt. Quantum Electron.*, 22, 391–416 (1990).
10. Y. Hashizume et al., "Low-PDL 16-channel variable optical attenuator array using silica-based PLC," in *Technical digest OFC2004*, Los Angeles, WC4 (2004).
11. Y. Hashizume, H. Takahashi, T. Watanabe, S. Sohma, T. Shibata, and M. Okuno, "Compact 32-channel 2 × 2 optical switch array based on PLC technology for OADM systems," in *Proceedings of ECOC2003* (2003).
12. H. Takahashi et al., "High performance 8-arrayed 1 × 8 optical switch based on planar lightwave circuits for photonics networks," in *Proceedings of the ECOC2002*, 4.2.6 (2002).
13. K. Koga, A. Watanabe, T. Kawai, K. Sato, and Y. Ohmori, "Large-capacity optical path cross-connect system for WDM photonics transport network," *J. Select. Areas Commun.*, 16, 1260–1269 (1998).
14. T. Goh, M. Yasu, K. Hattori, A. Himeno, and Y. Ohmori, "Low loss and high extinction ratio silica-based strictly nonblocking 16 × 16 thermo-optic matrix switch," *IEEE Photonics Technol. Lett.*, 10, 810–812 (1998).
15. T. Nishi, T. Yamamoto, and S. Kuroyanagi, "A polarization-controlled free-space photonic switch based on a PI-loss switch," *IEEE Photonics Technol. Lett.*, 5, 1104–1106 (1993).
16. K. Moriwaki, M. Abe, Y. Inoue, M. Okuno, and Y. Ohmori, "New silica-based 8 × 8 thermo-optic matrix switch on Si that requires no bias power," in *Tecnical Digest OFC'95*, San Diego, CA, USA, paper WS1, March 1995, pp. 211–212.
17. T. Shibata, M. Okuno, T. Goh, M. Yasu, M. Ishii, Y. Hibino, A. Sugita, and A. Himeno, "Silica-based 16 × 16 optical matrix switch module with integrated driving circuits," in *Technical. Digest OFC'01*, Anaheim, USA, paper WS1, March 2001, pp. 211–212.

18. N. Ooba, S. Toyoda, and T. Kurihara, "Low crosstalk and low loss polymeric 1 × 8 digital optical switch," *Jpn. J. Appl. Phys.*, 2369–2371, 2000.

19. S. Toyoda, N. Ooba, A. Kaneko, M. Hikita, T. Kurihara, and T. Maruno, "Wideband polymer thermo-optic wavelength tunable filter with fast response for WDM systems," *Electron. Lett.*, 36, 658–660 (2000).

20. R. M. Waxler, D. Horowitz, and A. Feldman, *Appl. Opt.*, 18, 101 (1979).

21. J. M. Cariou, J. Dugas, L. Martin, and P. Michel, *Appl. Opt.*, 25, 334 (1986).

22. T. Watanabe, N. Ooba, Y. Hida, and M. Hikita, " Influence of humidity on refractive index of polymers for optical waveguide and its temperature dependence," *Appl. Phys. Lett.*, 72, 1533–1535 (1998).

23. S. Imamura, R. Yoshimura, and T. Izawa, *Electron. Lett.*, 27, 1342 (1991).

24. Y. Hida, S. Imamura, and T. Izawa, *Electron. Lett.*, 28, 1314 (1992).

25. Y. Hida and S. Imamura, *IEEE Photonics Technol. Lett.*, 6, 845 (1994).

26. Y. Hida and S. Imamura, *Jpn. J. Appl. Phys.*, 34, 6416 (1995).

27. W. Horsthuis, B. Hendriksen, M. Diemeer, M. Donckers, T. Hoekstra, M. K. Koerkamp, F. Lipscomb, J. Thackara, T. Ticknor, and R. Lytel, in *Proceedings of the 21st ECOC, Brussels*, Th.L.3.4. (1995), p. 1059.

28. R. Moosburger, G. Fischbeck, C. Kostrzewa, B. Schuppert, and K. Petermann, in *Proceedings of the 21st European Conference on Optical Communications*, Brussels, Th.L.3.5, 1995, p. 1063.

29. Y. Hida, N. Ooba, R. Yoshimura, T. Watanabe, M. Hikita, and T. Kurihara, "Influence of humidity on transmission in a Y-branch thermo-optic switch composed of deuterated fluoromethacrylate polymer waveguide," *Electron. Lett.*, 33, 626–627.

30. H. Tanobe, M. Kohtoku, K. Okamoto, and Y. Yoshikuni, "Semiconductor optical filters for WDM systems," *NTT R&D*, 46, 677–684 (1997).

31. M. B. J. Diemeer, L. H. Spiekman, R. Ramsamoedj, and M. K. Smit, "Polymeric phase array wavelength multiplexer operating around 1550 nm," *Electron. Lett.*, 32, 1132–1133 (1996).

3R (RETIMING, RESHAPING, REGENERATION)

Kazuro Kikuchi

INTRODUCTION

Almost all optical communication systems currently deployed make use of the binary intensity modulation scheme, where the laser light is emitted in bit 1, and not in bit 0. After propagating through a long distance, the signal inevitably suffers from attenuation due to fiber loss and waveform distortion due to fiber dispersion. However, it is possible to make a decision whether the transmitted signal is bit 1 or bit 0 because the signal intensity is digitally modulated.

Through the binary decision process, we can regenerate the signal, removing the waveform distortion and restoring the power level. Such signal regeneration is performed by optical repeaters, which are periodically placed in the fiber link. Since the repeater includes the 3R functions, namely "reshaping," "retiming," and "regeneration," we can maintain the signal quality along the entire length of the link, although certain amounts of decision errors are accompanied with the signal regeneration process.

In this chapter, we deal with the principle of operation of the repeater, and discuss the future direction of research and development.

3R CIRCUIT BASED ON O/E AND E/O CONVERSION

Figure 1 illustrates the construction of the optical repeater based on O/E and E/O conversion. After the transmitted optical signal is received by an O/E converter, an amplifier circuit amplifies the detected signal up to the required voltage level. The signal bandwidth is also limited optimally by a filter so that the signal-to-noise ratio (SNR) is improved, while the intersymbol interference is suppressed. This process is called "reshaping."

The timing of the signal transmitted through an optical fiber fluctuates due to environmental perturbations. In order to perform "retiming" of the signal, we have to extract the clock from the transmitted signal, which determines the optimum instant for binary decision.

A decision circuit that samples the reshaped signal at the optimum instant, compares the sampled value with a threshold, and decides whether it is bit 1 or bit 0. Thus, the binary data is regenerated from the reshaped signal, and this function is called "regeneration."

Owing to the progress in integrated circuit technologies, retiming and regenerating functions have been achieved by a single IC chip called the clock data recovery (CDR) IC. Figure 2 shows the configuration of the CDR IC. The reshaped data are sent to a regenerator consisting of a D-flip flop (DFF) and a phase comparator consisting of an Exclusive OR (XOR) circuit. The phase of the regenerated and the input signals are compared with each other by using the XOR circuit. In order to form the phase-locked loop (PLL), the phase error is led to the voltage-controlled oscillator (VCO), which generates the clock, via a lowpass filter (LPF). The DFF is driven by the clock from the VCO and the input, and the regenerated signal is obtained. The speed of state-of-the-art DFFs using compound semiconductors has reached 50 GHz.

Finally, the intensity of the laser is modulated by the restored binary data through an E/O convertor, and incident on the next span of the fiber link.

BIT-ERROR RATE

The quality of the digital information is maintained over a long distance owing to the 3R functions, because we can eliminate noise and signal distortion at every repeater. However, bit error

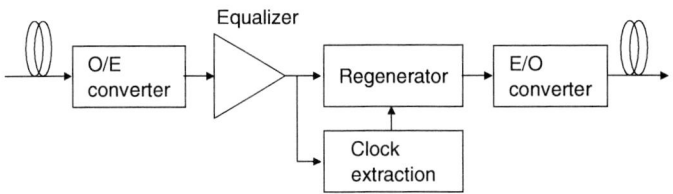

Figure 1 Schematic of O/E/O-based repeaters

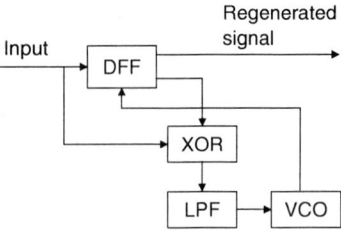

Figure 2 Schematic of CDR ICs

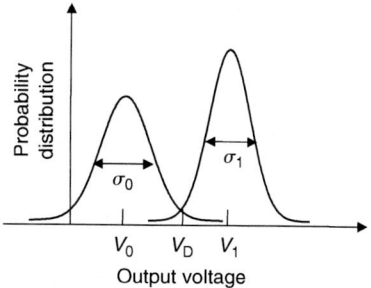

Figure 3   Probability distributions of signals in the mark and the space states

inevitably occurs through the decision process. In this section, we discuss how the bit error occurs through the binary decision process [1].

Let the averages of signal levels incident on the decision circuit be V_1 in the mark state and V_0 in the space, respectively. However, the signal V is fluctuating around these values due to noise and waveform distortion. The distribution function of the level is usually assumed to be Gaussian as shown in Figure 3, where σ_0^2 and σ_1^2 represent variances in the space state and the mark state, respectively. Let the threshold level for binary decision be V_D. In such a case, the decision circuit call the signal bit 1 if $V \geq V_D$ or bit 0 if $V < V_D$. Therefore, bit errors occur in the following cases: $V \geq V_D$ for bit 0 and $V < V_D$ for bit 1.

The bit error rate (BER) is then calculated as

$$P_e = \frac{1}{2}\left(\frac{1}{\sqrt{2\pi}\sigma_1} \int_{-\infty}^{V_D} \exp\frac{-(V-V_1)^2}{2\sigma_1^2} dV + \frac{1}{\sqrt{2\pi}\sigma_0} \int_{V_D}^{\infty} \exp\frac{-(V-V_0)^2}{2\sigma_1^2} dV \right),$$ (1)

where we assume that the mark rate is 1/2. The BER is minimized by optimizing V_D as

$$V_D = \frac{\sigma_0 V_1 + \sigma_1 V_0}{\sigma_0 + \sigma_1}.$$ (2)

Equations (1) and (2) yield

$$P_e = \frac{1}{2}\mathrm{erfc}\left(\frac{Q}{\sqrt{2}} \right).$$ (3)

In this equation, "erfc" means the complementary error function defined as

$$\mathrm{erfc}(x) = \frac{2}{\sqrt{\pi}} \int_x^{\infty} \exp(-y^2)\, dy,$$ (4)

and Q is given by

$$Q = \frac{V_1 - V_0}{\sigma_1 + \sigma_0}.$$ (5)

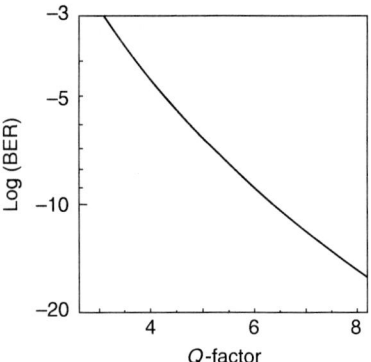

Figure 4   BER as a function of the Q-factor

The Q-factor representing the SNR is widely used for evaluating the quality of the binary signal. Figure 4 shows BER as a function of Q. In order to realize $P_e \simeq 10^{-9}$, for example, we need $Q = 6$.

However, the information theory predicts that by adding redundancy to the signal, we can achieve error correction. Recently, the forward-error correction (FEC) technique has been introduced into practical optical transmission systems in order to improve the BER performance. When we use the Read-Solomon code, for example, the BER of 10^{-6} is reduced to 10^{-16} at the expense of signal-bandwidth increase by 7%.

REPEATER USING ERBIUM-DOPED FIBER AMPLIFIERS

Since erbium-doped fiber amplifiers (EDFAs) have become practical devices, the repeater that employs an EDFA has also been introduced into optical transmission systems. In this type of repeater, the attenuated optical signal is amplified by the EDFA with automatic power control, and the fiber dispersion is usually compensated by a dispersion compensation fiber (DCF); thus, the reshaping function is mainly realized in an all-optical manner.

In present undersea optical transmission systems, two points are connected only by using EDFA-based repeaters placed at ~50 km intervals. Note that the EDFA-based repeater has neither retiming nor regenerating functions; in other words, noise and waveform distortion are accumulated along the entire system length. It is true that the great advantage of the digital communication system has been abandoned in such systems; however, we still have the following merits when using EDFA-based repeaters instead of O/E/O-based repeaters:

1.   The repeater does not require any high-speed electronics, and signal amplification is done in an all-optical manner.
2.   All wavelength-division multiplexed (WDM) channels are amplified simultaneously.
3.   System simplicity reduces the cost.

On the other hand, in terrestrial optical transmission systems, the signal is regenerated by the 3R circuit based on O/E/O conversion after being transmitted through several EDFA-based-repeater spans.

ALL-OPTICAL RETIMING AND REGENERATION

The EDFA-based repeater has realized "1R function (reshaping)." As the next step, all-optical retiming and regeneration technologies have been studied extensively. The all-optical 3R circuit, which includes "reshaping," "retiming," and "regeneration" functions, have the advantage that the speed for signal processing is not limited by electronics.

The all-optical retiming and regeneration can be performed by optical gate switches. Figure 5 depicts the construction of the all-optical 3R circuit. The optical clock pulse train is generated from the clock extraction circuit. The reshaped data signal switches the clock pulse train by using an optical gate switch; and then, the retiming function is realized. When the output power of the gated pulse is a nonlinear function of the power of the input data pulse as shown in Figure 5, the intensity of the gated clock pulse becomes insensitive to the signal intensity fluctuation. In the case of the 2R circuit, where the retiming function is omitted, we only have to launch a CW light on the gate switch instead of the clock pulse.

Note that the optical regeneration technology is still very premature. Most of the optical regenerators presently available are not necessarily the binary decision circuits. Because of this fact, the all-optical regeneration is called "reshaping" in some literatures. Instead, reshaping by EDFAs is referred as "reamplification" in such a case. Therefore, all-optical 3R sometimes mean "reamplification," "retiming," and "reshaping"; however, in this chapter, we do not use these terminologies.

The following explains the retiming and reshaping functions using a nonlinear interferometric switch as a specific example.

All-optical nonlinear interferometric switches (NIS) are very attractive devices to realize the all-optical retiming and regenerating functions. The Mach–Zehnder interferometer (MZI) using semiconductor optical amplifiers (SOAs) [2] and the nonlinear optical loop mirror (NOLM) [3,4] using dispersion-shifted fibers are typical NIS configurations.

Figure 6 illustrates the basic construction of the MZI. The waveguide material has the third-order optical nonlinearity, which means that the refractive index is slightly dependent on the optical power. In the case of SOAs, the mechanism of the third-order optical nonlinearity is as follows: When a light is injected to an SOA, the carrier density reduces because of stimulated emission, which in turn increases the refractive index [5]. On the other hand, the third-order non-linearity of optical fibers is called the Kerr effect, and has a much faster response speed than that of SOAs.

As shown in Figure 6, the input light is divided into two paths by using a 3 dB coupler. After traveling through these paths, two beams interfere with each other by using another 3 dB coupler.

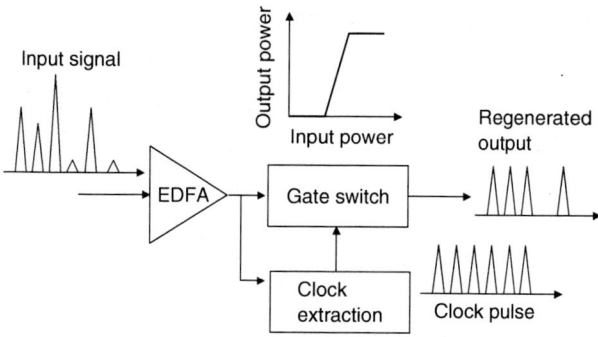

Figure 5   Configuration of all-optical 3R circuits

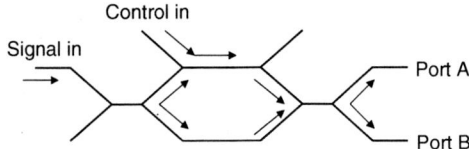

Figure 6   MZI-type optical gate switch

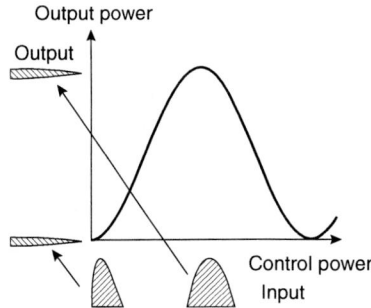

Figure 7   Input/output characteristics of the MZI-type optical gate switch

In such a case, the power P_A of the output light from the port A depends on the phase difference ϕ between the two paths, written as

$$P_A = P_{in}(1 + \cos\phi)/2,\qquad(6)$$

and the power P_B from the output B is expressed as

$$P_B = P_{in}(1 - \cos\phi)/2,\qquad(7)$$

where P_{in} stands for the input power.

When a strong control light is injected into one of the paths, the input light traveling through this path is phase-modulated in proportion to the power P_c of the control light, and ϕ is given as

$$\phi = 2\gamma P_c l,\qquad(8)$$

where γ denotes the nonlinear coefficient of the nonlinear waveguide and l denotes the device length. The output power from the port B is a sinusoidal function of the power P_c of the control light as shown in Figure 7. When $P_c = 0$, the input light is switched off; on the other hand, when $\phi P_c l = \pi/2$, the input light is switched on.

When a noisy data pulse is injected from the control port and a CW light is injected from the input port, the CW light is gated by the data pulse. By adjusting the data power in bit 1 around $\pi/4\gamma l$, the power of the gated signal is stabilized in spite of fluctuations of the data power. Similarly, in bit 0, the gated signal has fluctuations smaller than the data signal. Thus, we find that the nonlinear transmission characteristics of the MZI can improve the SNR of the data pulse, and the regenerating function can be realized.

When we extract a clock pulse train from transmitted data pulses, we can switch the clean clock pulse train by noisy data pulses. In such a case, the retiming function is realized in addition to the above-mentioned regenerating function; thus, we accomplish the all-optical 3R functions.

CONCLUSION

We have described the O/E/O-based repeater, the EDFA-based repeater, and the all-optical 3R circuit.

The EDFA-based repeater has only the function that compensates for signal attenuation; however, because of simplicity and low cost it has been introduced into long-distance transmission systems. The retiming and regeneration of the degraded signal is still performed by O/E/O-based repeaters, whose speed is limited by electronics. The all-optical 3R circuit is investigated to increase the signal processing speed beyond the electronics limit. The all-optical gate switch is the key element to realize such circuit.

REFERENCES

1. G. P. Agrawal, *Fiber-Optic Communication Systems,* John Wiley & Sons, New York, 1997.
2. Y. Ueno, S. Nakamura, and K. Tajima, "Penalty-free error-free all-optical data pulse regeneration at 84 Gb/s by using a symmetric-Mach–Zehnder-type semiconductor regenerator," *IEEE Photonics Technol. Lett.,* 13, 469–471, 2001.
3. T. Sakamoto and K. Kikuchi, "Nonlinear optical loop mirror with an optical bias controller for achieving full-swing operation of gate switching," *IEEE Photonics Technol. Lett.,* 16, 545–547, 2004.
4. S. Watanabe and S. Takeda, "All-optical noise suppression using two-stage highly-nonlinear fibre loop interferometers," *Electron. Lett.,* 36, 52–53, 2000.
5. K. E. Stubkjaer, "Semiconductor optical amplifier-based all-optical gates for high-speed optical processing," *IEEE J. Select. Top. Quantum Electron.,* 6, 1428–1435, 2000.

TRANSMITTER/RECEIVER

Kohroh Kobayashi

A transmitter and a receiver are the most fundamental devices to construct lightwave transmission systems. A transmitter, often called TX, consists of electronic circuits, a driving amplifier, and a light source (Figure 1(a)). The transmitter converts the electrical input signal into the optical output signal. Light sources are semiconductor lasers and light emitting diodes. A combination of a laser and a modulator or a modulator-integrated semiconductor laser is also used as a light source.

A receiver, often called RX, consists of a photo-detector, followed by electrical pre- and post-amplifiers and electronic circuits (Figure1(b)). A semiconductor photo-diode or an avalanche-photo-diode is used as a photo-detector. The receiver converts the input optical signal into the output electrical signal.

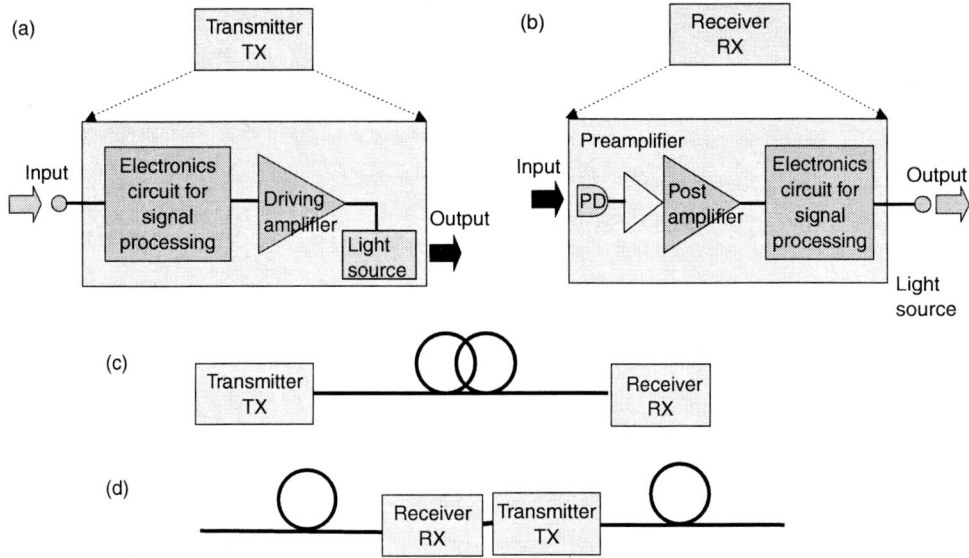

Figure 1 Fundamental configuration of a transmitter, a receiver, and related links (a) an optical transmitter, (b) an optical receiver, (c) a point-to-point optical link, and (d) an optical repeater

A most simple point-to-point optical link can be realized by connecting a transmitter and a receiver through optical fibers (Figure 1(c)). A repeater consists of a receiver followed by a transmitter (Figure 1(d)). The repeater converts the input optical signal to the output optical signal. Inside the repeater, the optical signal is amplified, reshaped, and regenerated.

TRAVELING-WAVE ELECTROABSORPTION

Robert Lewén, Stefan Irmscher, Urban Eriksson,
Urban Westergren, and Lars Thylén

INTRODUCTION

Electroabsorption modulators (EAMs) are suitable for use in compact fiber optical transmitters for high speed and efficiency. Their epitaxial structure is compatible with laser integration and they can be designed to require voltages and currents that can be provided by electronics based on emerging transistor technologies even for speeds above 40 Gb/sec (e.g., References 1 to 19).

An important limitation for high-speed operation of optical modulators, that is, EAMs and electro-optic (EO) modulators is the RC limitation that follows from the device capacitance in combination with the source and the load impedance. A common technique to improve high-speed performance and avoid this RC limitation is to design the modulator to operate in the traveling-wave (TW) regime [11–19]. In this chapter, we will introduce the fundamental operation of TW optical modulators suitable for binary intensity modulation in high-speed fiber optical communications. Microwave properties as well as different implementations of transmission lines (TML) will be discussed. The discussion and presented results on device performance will mainly be focused on EAMs although many aspects of the design on EO modulators are generically similar.

ACTIVE MATERIAL

The optical absorption process in EAMs is either based on the bulk Franz–Keldysh (FK) effect [20–23] or the quantum-confined Stark-effect (QCSE) [22–24]. Recently, there has been an increasing interest in EAMs based on intersub-band transition at the telecommunication wavelength of 1.55 μm [25–28]. The absorption process is discussed in more detail in a different chapter of this book addressing general EAMs while this chapter is devoted to the high-frequency properties of TW-EAMs.

For device optimization, it is useful to approximate the relation between optical absorption and electrical field. The following empirical relation often results in a good model fit compared with various published devices [29]:

$$\Delta\alpha \approx p\left(E - E_0\right)^c = p\left(\frac{V - V_0}{h_i}\right)^c \quad V > V_0,\tag{1}$$

where the proportional parameter (p) and curvature parameter (c) are fitting parameters, and h_i is the intrinsic layer thickness of the p–i–n structure. V and E are the applied voltage and electrical field, respectively. V_0 is the low-field bias that corresponds to a transmitted "1" symbol. The important parameter for device optimization is the curvature parameter (c), which indicates the *nonlinearity* of the absorption curve (Figure 1). From the Stark-shift [24] it is anticipated that the c-parameter value should be close to 2 at a low bias ($E_0 \approx 0$). However, the c-parameter depends on material properties, wavelength detuning, and low-field bias, and usually has a value in the range between 1.5 and 2.5. Equation (1) provides a good model fit for a wide range of voltage, which is independent of the choice of V_0 [29]. However, the c-parameter obtained from the model fit depends on the V_0 we assume as low-field bias. The absorption at this voltage corresponds to the internal insertion loss of the device. For digital communication it is desirable to operate the device with a low bias to minimize the insertion loss but yet at a bias point where

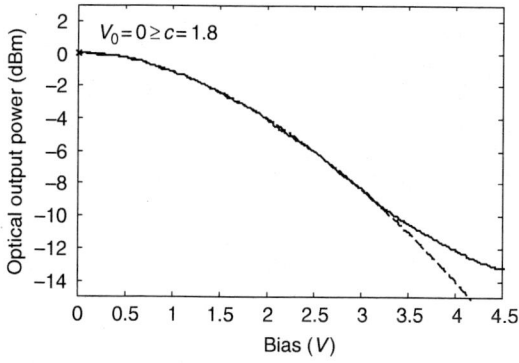

Figure 1 Example of the typical nonlinear absorption curve between applied modulator bias and optical absorption for an InGaAsP/InGaAbP QW device [12] (solid lines) with corresponding model fit (dashed lines) according to Eq. (1) at $V_0 = 0$

an extinction ratio of more than 8.4 dB [30] can be reached with a moderate drive voltage. The value of the curvature parameter is strongly related to a tolerated insertion loss, that is, a high c value is equivalent to the case when the modulator is operating well into the nonlinear region in the transmitting state and results in a lower insertion loss corresponding to a "1" symbol.

The nonlinearity given by the curvature parameter represents a very important difference for device optimization between EO modulators and EAMs. For EO modulators based on the Pockels effect [22] the change in refractive index is proportional to the applied field. This indicates that making the EO modulator twice as long can compensate a scaling of the intrinsic layer thickness by a factor of two. The nonlinear relation between absorption and electrical field for EAMs changes the conditions for this type of scaling. The choice of low-field bias, and the corresponding value of the c-parameter, has a very important influence on the device optimization with respect to geometrical parameters such as length and i-layer thickness. The latter parameter has a pronounced impact on the corresponding microwave parameters such as characteristic impedance and attenuation which are important for the design of TW modulators. This will be further discussed in sections on "Device characteristics" and "High-impedance segmented TW-modulators."

TRAVELING-WAVE MODULATORS

Microwave phase modulation of light operating in a TW regime based on copropagation of the microwave and the light was proposed by Bloembergen in 1960 and some early experimental work on KH_2PO_4 crystals was reported by Kaminow in 1961 [31]. The concept of TW modulators has later been refined, in particular, in $LiNbO_3$-based intensity Mach-Zehnder interferometer (MZI) modulators where (electrical) bandwidths of up to 70 GHz have been demonstrated [32]. The compact size of EAMs has enabled operation in excess of 10 GHz bandwidth using a simple lumped (RC-limited) configuration. Growing interest in optical modulators with 40 GHz bandwidth and above drives the interest in EAMs in TW operation. EAMs with TW electrode configuration were not demonstrated until the mid-1990s [33,34].

Considering the TW modulator illustrated in Figure 2, the amplitude of the optical output is expressed as [35]:

$$A_{out} = A_{in} \exp\left(-\int_{z=0}^{L} \gamma_0(z)\,dz\right), \tag{2}$$

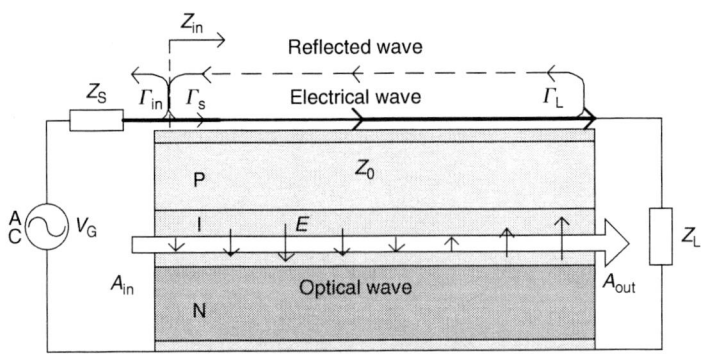

Figure 2   Principle of operation for a TW optical modulator

where A_{in} is the amplitude of the optical input and $\gamma_0(z)$ is the optical complex modal propagation factor, $\gamma_0 = \alpha_0/2 + j\beta_0$. In an optical modulator, the propagation factor is controlled by an external electrical field. For EAMs we are mainly concerned with the change in the intensity absorption, α_0, although the change in refractive index is important for chirp analysis [23,36–37]. The absorption change can be represented by a constant bias term, α_B at the corresponding bias voltage V_B as well as an AC part. For small signal analysis this voltage dependent AC term is approximated as $\alpha_{ac} \approx (d\alpha_B/dV_B)V_{ac}(z)$. In a reference time frame moving with an optical pulse that propagates with the group-velocity n_{go} the AC part of the optical intensity is expressed

$$A_{ac} = A_{in} \exp\left[-\frac{\alpha_B}{2}L - \frac{1}{2}\int_{z=0}^{L}\frac{d\alpha_B}{dV_B}V_{ac}\left(z,t+\frac{n_{go}z}{c_0}\right)dz\right], \tag{3}$$

with sufficiently small modulation amplitude (V_{ac}), A_{ac} is approximated as

$$A_{ac} \approx A_{in}e^{(\alpha_B/2)L}\frac{1}{2}\frac{d\alpha_B}{dV_B}\int_{z=0}^{L}V_{ac}\left(z,t+\frac{n_{go}z}{c_0}\right)dz, \tag{4}$$

In a TW modulator the total voltage is obtained from the sum of the forward and reverse propagating microwave in Figure 2,

$$V_{ac} = (V^+ + V^-)\frac{1}{1+j\omega R_p - C_p}. \tag{5}$$

The last factor in Eq. (5) is a correction that arises as the voltage over the active material which is slightly lower than the voltage on the metal electrode. This effect is due to an internal RC limit formed between the resistance of the cladding material (often dominated by p-type InP) and the active layer capacitance. The origin of this correction will be further discussed in the following section. In the InP material system this effect is mainly of concern for ultra-fast modulator operating at 100 GHz and above but can usually be neglected for devices operating at 40 GHz and below. Using time-harmonic phasor notation [38] the microwave amplitudes on the electrode are obtained from conventional TML theory, for example, Reference 39:

$$V^{\pm}\left(z,t+\frac{n_{go}z}{c_0}\right) = \text{Re}\left\{V_0^{\pm}(\omega)\exp(\mp\gamma_e z)\exp\left[j\omega\left(t+\frac{n_{go}z}{c_0}\right)\right]\right\}, \tag{6}$$

where

$$V_0^+ = \frac{V_0 T}{1-\Gamma_L\Gamma_s e^{-2\gamma_e L}} \quad \text{and} \quad V_0^- = V_0^+\Gamma_L e^{-2\gamma_e L}. \tag{7}$$

$\gamma_e(\omega)$ is the frequency dependent microwave propagation constant for the electrical modulation field, $\Gamma_S = (Z_S - Z_0)/(Z_S + Z_0)$ and $\Gamma_L = (Z_L - Z_0)/(Z_L + Z_0)$ are the internal microwave reflection coefficients at the modulator source and the load ports, respectively. $T = 1 - \Gamma_S$ is the modulator input transmission coefficient.

The set of equations given by Eqs. (4–7) describes the small signal response of a TW-EAM. In the next section the TML properties and their corresponding microwave characteristics, characteristic impedance (Z_0), and propagation constant (γ_e), are discussed. Section on Device characteristics discusses their influence on device performance while in following section we will review the suggested modifications to improve the device performance.

TRANSMISSION-LINE DESIGN AND ANALYSIS

The two typical implementations of an EAM-TML are illustrated in Figure 3 as a buried optical waveguide and a ridge waveguide, respectively [40]. A widely used and attractive solution is the buried optical waveguide with a polymer cladding material (e.g., BCB [41]). The advantage of this configuration is a low waveguide capacitance, which results in a high-characteristic impedance and low electrical attenuation. Due to the high-optical index contrast between the polymer cladding and the InGaAsP core (1.5/3.4) this implementation is usually multimode in the horizontal direction. For single-mode operation one normally uses a semiinsulating (SI) regrown cladding layer or a ridge waveguide (Figure 3, right). Single-mode operation is actually not required for an EAM [22,42]; however, it is feasible for laser integration as the waveguide, as both devices can be etched within the same process step while assuring single-mode operation of the integrated laser.

SI-regrown EAMs with a narrow mesa ($w_a \approx 1\ \mu m$) has proven to be an efficient solution for high-optical power capability and low insertion loss [3,43]. The disadvantages with this configuration are a higher waveguide capacitance because of the higher dielectric constant of InP and potential problems with inter-diffusion between Fe and Zn dopants during the regrowth that may degrade the performance of the device. Recently, Ru-doped SI-InP [19] has been demonstrated as a potential approach of avoiding this problem.

An alternative approach is the ridge-waveguide (Figure 3, right). Besides being optical single mode, this approach is preferred for improved reliability compared with the buried waveguide

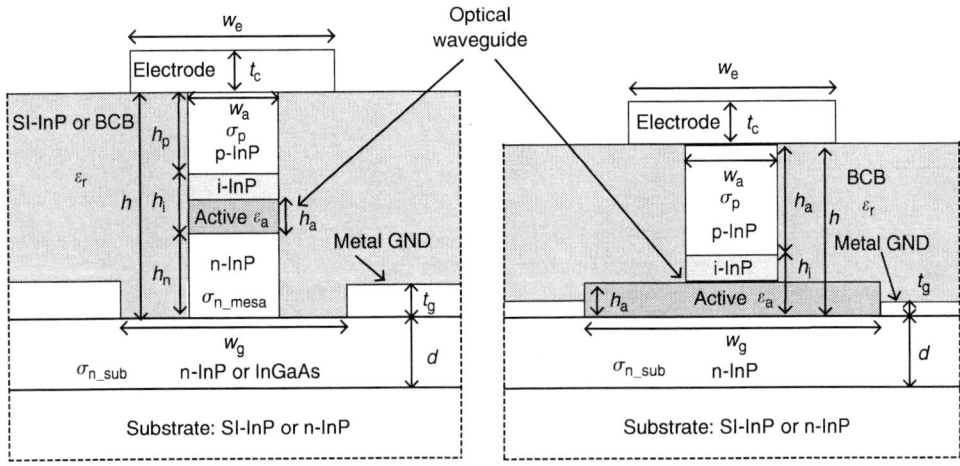

Figure 3 Schematic structures for an EAM–TML. The left figure represents a buried optical waveguide (a.k.a. deeply etched ridge waveguide [22]) with SI-InP or BCB cladding, the right figure represents an optical ridge waveguide [40] (a.k.a. strip loaded waveguide [22])

with polyamide cladding as the potential material damage caused by etching through the active layer is avoided. This can also be an advantage for active layers containing Al alloys as oxidization is avoided. Furthermore, the heat conductivity is improved (compared with a deeply etched waveguide with polymer cladding). The disadvantage is that the ridge waveguide usually shows slightly higher capacitance (10 to 15%) due to a more divergent electrical field (the charge accumulation close to the lower mesa corners is higher), but this capacitance can be reduced by adding a small intrinsic buffer layer on top of the active layer (Figure 3). The influence of this additional intrinsic layer will be further discussed in the next section.

The problem of modeling the microwave properties of a TML with the cross-sectional structure corresponding to Figure 3 has been treated in several publications, for example, References 44 to 47. The low-frequency field solution for the quasi-TEM mode is often referred to as a slow-wave mode [48–50]. This type of mode arises due to the physical nature of the semiconductor material; the skin-depth of the material is typically very large (e.g., $\delta = 8.3\ \mu$m at 100 GHz for n-InP 10^{18}/cm^3) and it can, therefore, be treated as a dielectric material for the magnetic-field, whereas the displacement current in the doped semiconductor is negligible (e.g., $\omega\varepsilon/\sigma = 0.05$ at 100 GHz for p-InP 10^{18}/cm^3) and is therefore treated as a conductor for the electric field. Due to the influence of the semiconductor material, the LC product is much larger than for a conventional TML on an insulating dielectric substrate. The wave propagation is therefore significantly slower than expected from the dielectric constant of the materials used in the structure.

The characteristic microwave properties of the EAM-TML can be estimated from the quasi-static equivalent circuit in Figure 4. Various versions of this circuit implementation have been proposed, for example, References 11, 35, 44, and 46. The illustrated equivalent circuit is motivated in one of our previous papers [44]. Here we also derive an approximate calculation model for estimating the corresponding circuit elements from geometrical and material data. In this circuit implementation, R_C is the conductor resistance, L is the inductance, R_{CG} is the ground metal resistance, R_{SC} represents the induced-current loss in the semiconductor substrate (early currents), C_{int} is the active layer capacitance, while C_{ext} is the fringing field capacitance, and R_p is the cladding layer resistance usually dominated by the p-type semiconductor. I_0 represents the induced photocurrent from the optical absorption. In a small-signal AC approximation this current source can be replaced by a resistor:

$$R_0 = \frac{\hbar\omega}{q}\frac{dV_B}{d\alpha_B}. \tag{8}$$

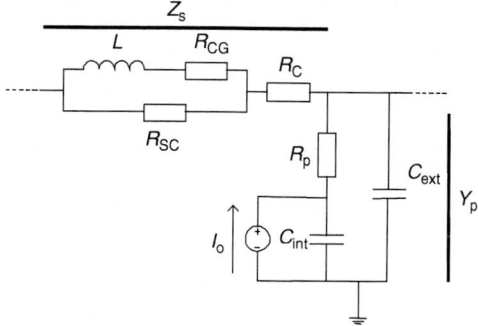

Figure 4   Quasi-static circuit model representing a unit length of the EAM–TML

The resulting equivalent circuit is well suited for circuit level modeling [51,52]. The corresponding TML parameters are expressed as:

$$Z_0(\omega) = \sqrt{\frac{Z_s(\omega)}{Y_p(\omega)}}, \tag{9}$$

$$\gamma_e(\omega) = \alpha_e(\omega) + j\beta_e(\omega) = \sqrt{Z_s(\omega)Y_p(\omega)}, \tag{10}$$

where

$$Z_s(\omega) \approx j\omega L + (R_C + R_{CG}) + \omega^2 \frac{L^2}{R_{SC}}, \tag{11}$$

$$Y_p(\omega) = \frac{j\omega C_{int}}{1 + j\omega C_{int} R_p} + j\omega C_{ext}. \tag{12}$$

To understand how the TML parameters scale with geometry it is useful to approximate Eqs. (11) and (12) with the dominating circuit elements [35]:

$$Z_0 \approx \sqrt{\frac{L}{C_{tot}}}; \qquad \beta \approx \omega\sqrt{LC_{tot}}, \tag{13a}$$

$$\alpha \approx \frac{R_C + R_{CG}}{2Z_0} + \frac{\omega^2}{2}\left(C_{int}^2 R_p Z_0 + \frac{L^2}{R_{SC} Z_0}\right) \approx \frac{R_C + R_{CG}}{2Z_0} + \frac{\omega^2 n_e^2}{2c_0^2}\left(\frac{R_p}{Z_0} + \frac{Z_0}{R_{SC}}\right), \tag{13b}$$

where $C_{tot} = C_{int} + C_{ext}$ is the total TML capacitance.

DEVICE CHARACTERISTICS

Bandwidth Limitations

The small signal response expressed by Eqs. (4–7), derived in section on Traveling-wave modulators, includes the distributed effect from the finite optical propagation velocity (n_0), the electrical TML propagation velocity and attenuation (γ), the electrical reflection at the TML source and termination (Γ_S and Γ_L), and the effect of a voltage drop across the series cladding resistance (R_p). It is essential to take all four effects into consideration when the optical response is determined. In the following sections, the influence of each effect will be discussed.

Velocity mismatch

For optimum performance of a TW device the optical and the electrical waves should propagate at the same velocity. The frequency response in the velocity mismatch limit is given as:

$$A_{ac} \propto e^{j\omega((n_o - n_e(\omega))/2c_0)L} \sin c\left(\omega \frac{(n_o - n_e(\omega))}{2c_0}l\right)e^{j\omega t}, \qquad (14)$$

where l is the length of the modulator, n_e is the phase index for the electrical wave at a single frequency, and n_0 is the group index for the optical wave representing the propagation of a modulated envelope. The attenuation for the p–i–n EAM-TML configuration is high and TW-EAMs are therefore typically short compared with the electrical wavelength ($l < \lambda/2$) within the frequency range of interest. For example, we have previously demonstrated [18] a 450 μm long device with a bandwidth of 43 GHz and an electrical index of $n_e = 5.5$ that results in a ratio of $l - \lambda_{3dB} = 0.35$. This is, to the best of our knowledge, the highest ratio between length and wavelength reported for a TW-EAM based on a p-i-n structure. The name "traveling-wave" can therefore be somewhat misleading as the devices are usually not several wavelengths long. However, it is important to realize that a TW-EAM is generally too long to be modeled as a lumped circuit element and usually show clear distributed effects. The velocity walk-off will degrade the performance slightly, but this is often not the dominating bandwidth limitation.

It is possible to match the propagation velocities of the electrical and the optical wave by designing a lower value of either the capacitance or inductance. The capacitance is, however, normally given by the optical waveguide design, that is, the capacitance is chosen as low as possible without significantly sacrificing the optical modulation efficiency of the device (see next section). This leaves the inductance as the only free parameter to optimize the design with respect to velocity matching. Fundamentally, $n_e = c_0 Z_0 C$ indicates the trade-off between velocity matching and TML impedance. In a previous publication [53] we have experimentally investigated how the microwave properties are changed with the inductance, which is controlled here by the width of the electrode.

TML attenuation

The TML attenuation is usually the dominating bandwidth limitation for TW-EAMs. This is also a limitation of fundamental nature; in contrast to the other limitations discussed in this section, for example, velocity mismatch, the effect of attenuation cannot be completely avoided with a good design. An important difference between EO modulators and EAMs is that optically induced carriers effectively need to be removed from the latter modulators. For devices based on the InP material system the difficulty in making a good p-contact on InP imposes a problem. Usually an InGaAs contact layer is required. Due to the higher refractive index and optical absorption, this contact layer can lead to substantial optical absorption, unless it is sufficiently separated from the active optical waveguide. Usually a cladding layer thicker than 1.5 μm is required. An EAM-TML is often highly attenuating due to this resistive cladding layer, and the effect of the TML attenuation is therefore often dominating over the velocity mismatch for TW-EAMs. With a negligible velocity mismatch and perfect line termination, the f_{3dB} bandwidth is given as:

$$P_{ac}(\omega) = \frac{(1 - e^{-\alpha(\omega)L})}{\alpha(\omega)L} = \frac{1}{\sqrt{2}} \quad \text{or} \quad \alpha(\omega)L = 0.738. \qquad (15)$$

This is equivalent to a TML attenuation of 6.4 dB. A dominating contribution to the attenuation at high frequencies is the second term in Eq. (13), $\alpha \propto \omega^2 C_{int}^2 Z_0$, which indicates that the attenuation is strongly related to the device capacitance. It is easy to get the impression that the TW implementation solves the problem with device capacitance and this is true in the sense that it is possible to significantly exceed the RC-limit. It is, however, important to realize that the capacitance has a major contribution to the TML attenuation and ultimately sets the bandwidth limitation also for the TW configuration. Also note that the attenuation and propagation velocity are related from the cross products between the real and the imaginary parts of Z_s and Y_p (Eqs. (11) and (12)), respectively, as indicated by Eq. (13). Thus, a slow propagation is equivalent to a high attenuation. An intuitive explanation to this phenomenon is that with a shorter wavelength the capacitance is charged more times during the wave propagation through the structure, leading to increased power dissipation in the resistive cladding material. The value of the inductance is an important trade-off between TML attenuation and impedance. This indicates that a high-bandwidth design does inevitably result in low modulator waveguide impedance.

It should also be noted that attenuation limited TW-modulators show a much flatter transfer function than RC-limited modulators at higher frequencies. This means that a peaking of the transfer function caused by a driver design, interconnect, or impedance mismatch (see next section) can have a significant effect on the f_{3dB}-bandwidth. However, this also indicates that we cannot directly compare the performance of an RC-limited modulator to an attenuation limited TW-modulator based only on the value of f_{3dB}.

Impedance mismatch

To truly operate in the TW regime it is important that the electrical wave is properly terminated at the end of the TML, that is, ideally the load should be matched to the characteristic impedance of the TML ($Z_L = Z_0$). The TML termination (Z_L) has a very strong effect on the device bandwidth for a TW-EAM. The reason is that the impedance load is transformed in the TML configuration, and since the modulator length close to f_{3dB} is often $l \approx \lambda/4$, the impedance seen at the input of the modulator is approximately $Z_{in} = Z_0^2/Z_L$ close to that frequency. With $Z_L < Z_0$ the input impedance will increase with frequency, and thereby increase the voltage at the input of the TML at higher frequencies. Intuitively it may be considered that a negative reflection at the TML load would reduce the response, but the voltage peaking at the input from the impedance transformation is often dominating and the net effect is an enhanced high-frequency response. Due to this effect the load impedance has a stronger influence on the bandwidth for the TML-configuration than for a lumped modulator. As a result from the relatively flat frequency response for an attenuation limited modulator, the mismatch effect enables a significant change of f_{3dB} with only a small peaking of the transfer function. (The performance of the modulators in Figure 5 is also affected by the internal RC-penalty discussed in the next section.) A design with low termination impedance flattens the frequency response, and this is a very effective way to increase the bandwidth of the device. However, it is also important to note that with $Z_L > Z_0$ the bandwidth will be significantly reduced due to the same effect. Therefore, it is good to design with slightly lower rather than higher load impedance.

A case study of the effect of the load impedance is illustrated in Figure 5 where the simulations use TML parameters corresponding to a 450 μm long modulator presented in a previously published paper [18]. As illustrated, low termination impedance results in a pronounced peaking of the transfer function. The effect of bandwidth peaking is obviously dependent on the source impedance, for example, if a 50 Ω-source or a dedicated driver design is used. Only if the TML-EAM is matched ($Z_L = Z_0$) the input impedance is almost frequency independent, and the bandwidth is independent of the source impedance. The effect of the bandwidth peaking also depends on the attenuation of the modulator TML. With low TML attenuation, a high mismatch may lead

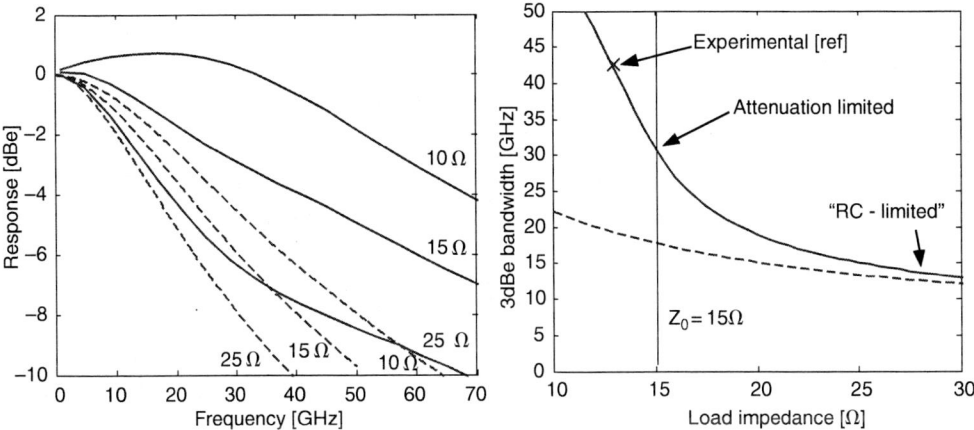

Figure 5   Influence of the load impedance for the TML configuration (solid) compared with the lumped (RC-limited) configuration (dashed). The simulations use TML parameters corresponding to the 450 μm long modulator presented in [18] with a characteristic impedance $Z_0 = 15$ Ω. The left figure shows the optical response as a function of frequency for different values of the load impedance (25, 15, and 10 Ω). The right figure shows the corresponding 3 dB bandwidth as a function of load impedance

to a strong peaking of the transfer function and the reflected wave can result in a nonlinear phase response of the device. This leads to signal distortion if the mismatch is too high.

The important conclusion from this case study is that in order to significantly gain from the TW implementation it is very important that the device load is matched to the characteristic impedance of the EAM–TML. The bandwidth can be further increased by a slight mismatch ($Z_L < Z_0$). However, the bandwidth gain is limited by the tolerated signal distortion.

Internal RC-penalty

Due to the low conductivity of p-InP the voltage over the intrinsic layer is lower than the voltage on the modulator electrode, which indicates an internal RC-constant given by $R_p C_{int}$. This can cause a significant penalty on the transfer function if the intrinsic layer is chosen to be too thin. Ideally, this RC-product usually corresponds to a frequency of 200 GHz or more, although the effect is substantial also at lower frequencies. In a fabrication process where the obtained doping level can be lower than the nominal, or passivation of dopants may occur, this can have pronounced effects at lower frequencies. The frequency response in Figure 5 is notably affected by this internal RC-limit and falls off faster than a TW modulator operating in the attenuation limit. In spite of this effect, the performance is significantly improved compared with the lumped case.

Influence of the Intrinsic Layer Thickness

An important design parameter for high-speed TW-EAM is the thickness of the intrinsic layer. To avoid a pronounced penalty from the internal RC-limit it can be attractive to design this layer slightly thicker than for a conventional lumped EAM. Previous investigations [29] indicate that it is quite difficult to reach bandwidths in the 100 GHz regime with an i-layer much thinner than 350 nm. Note that this layer may deliberately be designed thicker than the active QW package. The thickness of the optical waveguide is limited to maintain single-mode operation, and can be restricted by the number of wells in a QW-design due to built-in strain or with respect to carrier transport. Increasing the thickness of the intrinsic layer (Figure 3) will obviously reduce the electrical field for a fixed applied voltage, and thereby reduce the modulation efficiency.

However, the performance of a TW modulator does not scale with geometry in the same way as a lumped (RC-limited) modulator. This changes the optimum design with respect to the intrinsic layer thickness of the modulator and its corresponding waveguide capacitance (C_{int}). For example: for a lumped modulator the bandwidth is directly proportional to $1/RC_{int}$, and decreasing the value of the capacitance (per unit length) by a factor of two indicates that the device can be made twice as long and still have the same bandwidth. However, as a result from the nonlinear relation between the optical absorption and electrical field discussed in section on Active material we would not gain from using this design. A lumped modulator is therefore often designed with as thin i-layer as possible for a high-optical extinction given a specified drive voltage. For a matched TW-design ($Z_L = Z_0$) the bandwidth is given by the αL product Eq. (15), and with a thin intrinsic layer α is proportional to $C_{int}^{(3/2)}$ at high frequencies Eq. (13). Hence, reducing C_{int} by a factor of two indicates that the device length can be increased by a factor of up to 2.8 without decreasing the device bandwidth. In addition, the penalty for internal RC limitation and velocity mismatch are reduced and the characteristic impedance is increased. This indicates that a TW-design benefits from a design with slightly thicker i-layer.

The thickness of the intrinsic layer has a pronounced effect on the general modulator performance with respect to optical extinction, high-power saturation, and frequency chirp. A detailed optimization is therefore complicated, and it is very difficult to specify an overall optimum design. In general, a thicker intrinsic layer is beneficial for the following reasons:

- Decreased capacitance, which leads to decreased attenuation, increased impedance, and decreased internal RC penalty. The modulator length can be increased without reducing the small signal bandwidth.
- The optical absorption per unit length is reduced; a more distributed optical absorption decreases the effect of high-power saturation related to the carrier escape time from the QW [54], carrier screening [55], and local heating of the active layer [56].
- Decreased free carrier absorption from p-dopants [57].
- Decreased diffusion of p-dopants into the MQW layer in MOVPE growth [11].

The drawback of a long modulator with a thick intrinsic layer is mainly associated with high-insertion loss and frequency chirp as:

- The frequency-chirp parameter ($\alpha_H = \Delta n_0/\Delta k$) [36,37] is usually lower with a high-electrical field so that the chirp is reduced by operating the device with a low curvature value (see section on Active material).
- The insertion loss is increased (depending on the low-field absorption).
- Hole pile-up [58–60] in the bandgap discontinuity may be an issue unless proper bandgap engineering is used.
- The hole drift time is increased due to the thicker i-layer ($v_{sat} \approx 7 \times 10^6$ cm/sec [61]). This may increase the effect of high-power saturation due to carrier screening of the applied field [55].
- Nonintentional residual doping in the intrinsic region may increase the drive voltage and the TML capacitance becomes bias dependent.

HIGH-IMPEDANCE SEGMENTED TW-MODULATORS

A typical EAM–TML structure as illustrated in Figure 3 with an intrinsic layer thickness of 0.35 μm and a mesa width of slightly <2 μm usually results in a characteristic impedance of approximately 25 Ω. Due to the current standard of microwave systems modulator impedances near 50 Ω are preferred. A common practical requirement is that the return loss to a 50 Ω-microwave system should be higher than 10 dB, and some applications require even higher return loss. Higher impedance is also preferable for device packaging as (uncontrolled) parasitic series inductive elements have less effect on the device performance. Therefore, a modulator impedance of 35 Ω or higher is preferred. A widely used technique is to compensate the low TML impedance by adding short TML sections with high impedance at both ends of the modulator [11,12,15,17]. Provided that each TML section is much shorter than the wavelength of the electrical wave the microwave will experience the effective impedance given by the average inductance and capacitance per unit length of the TML. The requirement on a low ratio between TML section length and wavelength indicates that it becomes necessary to divide the modulator into several segments at very high frequencies. The modulator can then be implemented as a segmented TML with active low-impedance modulator sections in between passive TML sections with high impedance [12,62]. Alternatively the modulator is often implemented as a continuous high-impedance passive TML loaded by short capacitive modulator segments, that is, capacitively loaded TML (Figure 7). The idea of dividing the active modulator into several segments was first introduced for electrooptical LiNbO$_3$ modulators by Langmann et al. in 1982 [63]. This concept has later been refined and introduced for electrooptic modulators based on the GaAs [64,65] and InP material systems [66–68] as well as InP TW electroabsorption modulators [12]. A schematic classification of TW modulator implementations based on either segmented or continuous TML is shown in Figure 6.

In the low frequency limit the characteristic impedance, Z_B, and electrical propagation index along the optical waveguide, n_B, are expressed as:

$$Z_B \approx \sqrt{\frac{L_p d_p + L_m d_m}{C_p d_p + C_m d_m}} \tag{16}$$

$$n_B \approx \frac{c_0}{d_{B\text{-optical}}} \sqrt{\left(L_p d_p + L_m d_m\right)\left(C_p d_p + C_m d_m\right)}, \tag{17}$$

where L_p, C_p, L_m, and C_m are the inductance and capacitance per unit length of the passive and active TML, respectively. d_p and d_m are the lengths of each TML segment (Figure 7). $d_{B\text{-optical}}$ is the distance

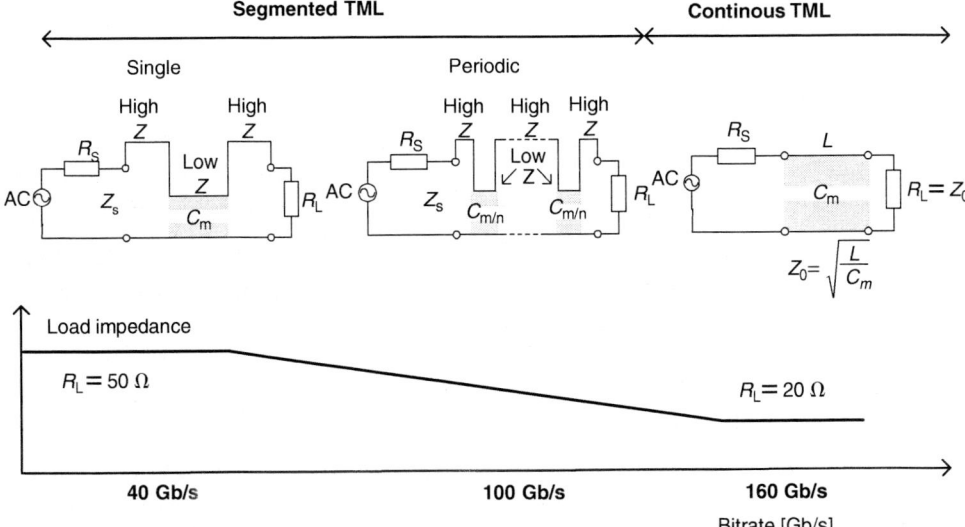

Figure 6 Schematic classification of demonstrated implementations for distributed TML electroabsorption modulators [11–19]. The lower graph schematically reflects our conclusions for the choice of implementation in each bitrate regime

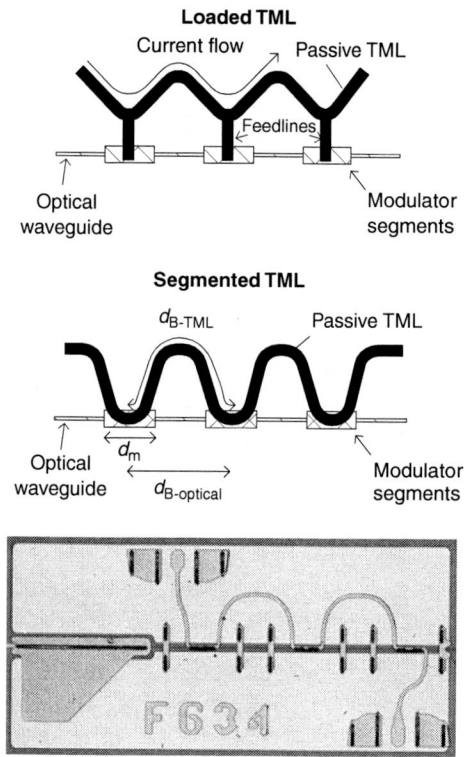

Figure 7 The upper figures schematically show examples of layout-implementations of multisegment modulators as a loaded and segmented TML implementation, respectively. The lower figure shows a microscopic photograph of a segmented TML modulator [12] with GSG microwave pads at the beginning and end of the TML. The large pad on the left is indented to be used as an integrated laser. The total chip length is 700 μm

between two adjacent modulator sections. To meet the design requirement of impedance matching and velocity matching, we need two free design parameters. In a traditional capacitively loaded TML design as presented by for example, Walker [64] these two free parameters are the loading duty-cycle (d_m/d_B in Figure 7) of the TML and the impedance of the passive unloaded TML (Z_p). For some implementations, Z_p may, however, be restricted due to process constraints. In that case a second free parameter can be obtained by letting the electrical TML take a longer path than the optical waveguide (i.e., meandered TML, Figure 7). The design problem is then summarized as follows:

Design goal: Resulting TML impedance for the loaded structure (Z_B), and velocity matching between the optical and the electrical wave, $n_B = n_o$.

Given parameters: Impedance of the active (Z_m) and passive (Z_p) TML sections, and the corresponding propagation constant γ_m, γ_p. Often the propagation constant of the passive waveguide is given by the corresponding effective dielectric constant, $n_e = \sqrt{\varepsilon_{eff}}$.

Free parameters: Loading duty-cycle (d_m/d_B), and the ratio between the electrical TML and optical waveguide lengths ($d_{B\text{-optical}}/d_{B\text{-TML}}$). Z_p is a free parameter for some implementations; however, this value is often limited due to process constraints.

By using the fundamental TML relations, for example, [39], $nZ_0 = c_0L$ and $n = \sqrt{\varepsilon_{eff}} = c_0CZ_0$, and assuming approximately the same inductance per unit length of each TML type, we can rewrite Eqs. (16) and (17) as:

$$\frac{Z_B}{Z_p} \approx \frac{\sqrt{\varepsilon_{eff}}}{n_B} \frac{d_{B\text{-}TML}}{d_{B\text{-}optical}},$$ (18)

$$\left(C_m - C_p\right)\frac{d_m}{d_{B\text{-}optical}} \approx \frac{n_B^2 - \varepsilon_{eff}\left(d_{B\text{-}TML}/d_{B\text{-}optical}\right)^2}{c_0 Z_B n_B}.$$ (19)

By setting $n_B = n_o$ the ratio between TML and optical waveguide length and the value of Z_p is given by Eq. (18) while the loading duty-cycle is given by Eq. (19).

The derived relations given by Eqs. (16–19) are very useful as an approximate design input. Microwave parameters for periodic transmission lines are, however, frequency dependent. It should be emphasized that the relations are only valid at low frequencies and that these equations furthermore do not include the microwave attenuation, which is a very important parameter when estimating the device bandwidth. Once the segment length is comparable to a (quarter of a) wavelength, a stop band will occur for the microwave propagation. Effectively the structure will act as a Bragg mirror which is not a desired property for a TW structure. The frequency dependent characteristic impedance and propagation constant can still be interpreted in terms of Bloch-waves, with the corresponding TML impedance Z_B and propagation constant γ_B. The Bloch-wave formalism is a well established and widely used analysis approach for periodic TW TMLs [39]. With this formalism the periodic TML line can be treated with conventional TML theory, which consequently provides an intuitive understanding of the wave propagation in the structure. Further details on the Bloch-wave analysis implemented on a segmented TW-EAM are found in one of our previous publications [12].

The segmented TML implementation for TW-EAMs is attractive as it enables higher TML impedance without compromising the thickness of the active intrinsic layer. This is important as the relation between applied voltage and optical absorption is nonlinear for EAMs. The periodic TML can be implemented without any pronounced effects of discontinuity parasitics up to at least 110 GHz [69]. The implementation potentially has two drawbacks. First, the required passive optical waveguide between the active modulator segments potentially increases the optical insertion loss. Second, the electrical attenuation normalized to the total active modulator length is increased. This second property is not unique only for the segmented TML implementation, since a similar trade-off needs to be considered also for alternative approaches [70].

The higher impedance that follows from a periodic TML design has obvious advantages due to reduced return loss to a 50 Ω microwave system and reduced influence of interconnect paracitics. However, the trade-off between TML impedance and attenuation indicates that a higher impedance does not necessarily improve the modulation efficiency. Depending on the source impedance, a higher impedance usually leads to a higher voltage applied to the modulator. On the other hand, the higher attenuation leads to a shorter modulator given a specific design bandwidth. This trade-off has to be treated on a case to case basis, given the available drive technology and module interconnect. It is far from obvious that a 50 Ω TML design is the optimum case with respect to modulation efficiency. Assuming the typical nonlinear EAM absorption characteristics with a curvature parameter of $c \approx 2$ in Eq. (1) it is indicated [12] that an impedance close to 35 Ω provides a good trade-off between return-loss and modulation efficiency. It is anticipated that a higher impedance reduces the efficiency. In conclusion, the segmented TML–EAM provides an increased impedance level for TW-EAM structures, resulting in several advantages in designing a high-speed fiber optical transmitter, for example, an excellent return loss figure to a 50 Ω system over a wide frequency range. The segmented TML implementation can improve the modulation efficiency, but there is an optimum with respect to TML impedance. In a 50 Ω system, segmented devices show slightly better high-speed performance when optimized for a device load of 35 Ω.

REFERENCES

1. T. Ido, S. Tanaka, M. Suzuki, M. Koizumi, H. Sano, and H. Inoue, "Ultra-high-speed multiple-quantum-well electro-absorption optical modulators with integrated waveguides," *IEEE J. Lightwave Technol.*, 14, 2026–2034, 1996.

2. Y. Miyazaki, H. Tada, S. Tokizaki et al., "Small chirp 40 Gbps EA modulator with novel tensile-strained asymmetric quantum well absorption layer," in *Proceedings of the European conference on optical communication ECOC 2002*, paper 10.5.6, 2002.

3. D. G. Moodie, A. D. Ellis, P. J. Cannard, C. W. Ford, A. H. Barrell, R. T. Moore, S. D. Perrin, R. I. McLaughin, and F. Garcia, "40 Gbit/s modulator with low drive voltage and high optical output power," in *Proceedings of the European Conference on Optical Communication ECOC 2001*, 2001, pp. 332–333.

4. N. Mineo, K. Nagai, and T. Ushikubo, "Ultra wide-band electroabsorption modulator modules for DC to millimeter-wave band," in *Proceedings Microwave Photonics 2001, MWP '01.* 2002, pp. 9–12.

5. X. Fu, T. Liu, J. Li, and G. Zhang, "Investigation of electroabsorption modulator for 40 Gb/s transmitter application," in *Proceedings of the Optical Fiber Communication (OFC) 2002*, 2002, pp. 718–719.

6. H. Feng, T. Makino, S. Ogita, H. Maruyama, and M. Kondo, "40 Gb/s electro-absorption modulator integrated DFB laser with optimized design," in *Proceedings of the Optical fiber communication (OFC) 2002*, 2002, pp. 340–341.

7. H. Tada, Y. Miyazaki, K. Takagi, T. Aoyagi, T. Nishimura, and E. Omura, "40 GHz modulation bandwidth of electroabsorption modulator with narrow-mesa ridge waveguide," in *Proceedings of the Optical fiber communication (OFC) 2002*, 2002, pp. 722–723.

8. K. Takagi, Y. Miyazaki, H. Tada, E. Ishimura, T. Aoyagi, T. Nishimura, T. Hatta, and E. Omura, "Highly reliable 40 Gb/s electroabsorption modulator grown on InP:Fe substrate," in *Proceedings of the Conference on InP and related materials IPRM 2001*, pp. 432–435.

9. M. Le Pallec, J. Decobert, C. Kazmierski, A. Ramdane, N. El Dahdah, F. Blache, J.-G. Provost, J. Landreau, D. Carpentier, F. Barthe, and N. Lagay, "42 GHz Bandwidth InGaAlAs/InP Electroabsorption Modulator with a Sub-Volt Modulation Drive Capability in a 50 nm Spectral Range," *paper presented at Indium Phosphide and Related Materials (IPRM) 2004.* Kagoshima, Japan, May 2004, pp. 577–580.

10. B. Mason, A. Ougazzaden, C. Lentz, K. Glogovsky et al. (20 authors), "40-Gb/s tandem electroabsorption modulators," *IEEE Photonics Technol. Lett.*, 11, 27–29, 2002.

11. S. Zhang, "Traveling-wave Electroabsorption Modulators," Ph.D. dissertation, Department of Electrical & Computer Engineering, University California at Santa Barbara, CA, 1999.

12. R. Lewén, S. Irmscher, U. Westergren, L. Thylén, and U. Eriksson, "Segmented transmission-line electroabsorption modulators," *J. Lightwave Technol.*, 22, 172–179, 2004.

13. Y.-J. Chiu, H.-F. Chou, V. Kaman, P. Abraham, and J. E. Bowers, "High extinction ratio and saturation power traveling-wave electroabsorption modulator," *IEEE Photonics Technol. Lett.*, 14, 792–794, 2002.

14. Y. Akage, K. Kawano, S. Oku, R. Iga, H. Okamoto, Y. Miyamoto, and H. Takeuchi, "Wide bandwidth of over 50 GHz traveling wave electrode electroabsorption modulator integrated DFB lasers," *Electron. Lett.*, 37, 299–300, 2001.

15. H. Fukano, M. Tamura, T. Yamanaka, H. Nakajima, Y. Akage, Y. Kondo, and T. Saitoh, "Low Driving-Voltage (1.1 Vpp) Electroabsorption Modulators Operating at 40 Gbit/s," *paper presented at Indium Phosphide and Related Materials (IPRM) 2004.* Kagoshima, Japan, May 2004, pp. 577–580.

16. G. L. Li, S. A. Pappert, P. Mages, C. K. Sun, W. S. C. Chang, and P. K. L. Yu, "High-saturation high-speed traveling-wave InGaAsP-InP electroabsorption modulator," *IEEE Photonics Technol. Lett.*, 13, 1076–1078, 2001.

17. M. Shirai, H. Arimoto, K. Watanbe, A. Taike, K. Shinoda, J. Shimizu, H. Sato, T. Ido, T. Tsuchiya, M. Aoki, and T. Tsuji, "Impedance-controlled-electrode (ICE) semiconductor modulators for 1.3-μm 40 Gbit/s transceivers," in *Proceedings of the ECOC 2002*, Vol. 4, paper 9.5.4, 2002.

18. S. Irmscher, R. Lewén, and U. Eriksson, "InP/InGaAsP high-speed traveling-wave electro-absorption modulators with integrated termination resistors," *IEEE Photonics Technol. Lett.*, 14, 923–925, 2002.

19. M. Tamura, T. Yamanaka, H. Fukano, Y. Akage, Y. Kondo, T. Saitoh, "High speed electroabsorption modulators using ruthenium-doped SI-InP: Impact of interdiffusion-free burying technology on E/O modulation characteristics," in *Proceedings of the Indium Phosphide and Related Materials (IPRM'03)*, Santa-Barbara, CA, May 2003, pp. 491–493.

20. W. Franz, "Einfluss eines elektrischen Feldes auf eine Absorptionskante," *Z. Naturforsch.*, 13a, 484–489, 1958.

21. L. V. Keldysh, "The effect of a strong electric field on the optical properties of insulating crystals," *J. Exptl. Theoret. Phys. (U.S.S.R)*, 34, 788–790, 1958.

22. Koichi Wakita, *Semiconductor Optical Modulators*, New York: Kluwer Academic Publishers, 1998.

23. G. L. Li and P. K. L. Yu, "Optical intensity modulators for digital and analog applications," *J. Lightwave Technol.*, 21, 2010–1183, 2003.

24. D. A. B. Miller, D. S. Chemla, T. C. Damen, A. C. Gossard, W. Wiegmann, T. H. Wood, and C. A. Burrus, "Electric field dependence of optical absorption near the bandgap of quantum well structures," *Phys. Rev. B, Condens. Matter*, 32, 1043–1060, 1985.

25. A. Neogi, T. Mozume, H. Yoshida, and O. Wada, "Intersubband transitions at 1.3 and 1.55 μm in a novel coupled InGaAs/AlAsSb double-quantum-well structure," *IEEE Photonics Technol. Lett.*, 11, 632–634, 1999.

26. C. Gmachl, H. M. Ng, S.-N. G. Chu, and A. Y. Cho, "Intersubband absorption at $\lambda = 1.55$ μm in well- and modulation-doped GaN/AlGaN multiple quantum wells with superlattice barriers," *Appl. Phys Lett.*, 77, 3722–3724, 2000.

27. R. Akimoto, Y. Kinpara, K. Akita, F. Sasaki, and S. Kobayashi, "Short-wavelength intersubband transitions down to 1.6 μm in ZnSe/BeTe type-II superlattices," *Appl. Phys Lett.*, 78, 580–582, 2001.

28. P. Janes and P. Holmstrom, "High-speed optical modulator based on intersubband transitions in InGaAs/InAlAs/AlAsSb coupled quantum wells," in *Proceedings of Indium Phosphide and Related Materials (IPRM'03)*, Santa-Barbara, CA, May 2003, pp. 308–311.

29. R. Lewén, "High-Speed Electroabsorption Modulators and p–i–n Photodiodes for Fiber-Optic Communications," Ph.D. dissertation, Department of Microelectronics and Information Technology, Royal Institute of Technology, Sweden, 2003.

30. Telecommunication Standardization Sector of ITU-T G.691–G.693.

31. I. P. Kaminow, "Microwave modulation of the electrooptic effect in KH_2PO_4," *Phys. Rev. Lett.*, 6, 528–529, 1961.

32. K. Noguchi, O. Mitomi, and H. Miyazawa, "Millimeter-wave Ti:LiNbO$_3$ optical modulators," *IEEE J. Lightwave Technol*, 16(4), 615–619, 1998.

33. K. Kawano, M. Kohtoku, M. Ueki, T. Ito, S. Kondoh, Y. Noguchi, and Y. Hasumi, "Polarization-insensitive traveling-wave electroabsorption modulator with over 50 GHz and driving voltage less than 2V," *Electron. Lett.*, 33, 1580–1581, 1997.

34. S. Z. Zhang, Y.-J. Chiu, P. Abraham, and J. E. Bowers, "25 GHz polarization-insensitive electroabsorption modulators with traveling-wave electrodes," *IEEE Photonics Technol. Lett.*, 11, 191–193, 1999.

35. G. L. Li, S. K. Sun, S. A. Pappert, W. X. Chen, and P. K. L. Yu, "Ultrahigh-speed traveling-wave electroabsorption modulator—design and analysis," *IEEE Trans. Microwave Theory Technol.*, MTT-47, 1177–1183, 1999.

36. F. Devaux, Y. Sorel, and J. F. Kerdiles, "Simple measurement of fiber dispersion and of chirp parameter of intensity modulated light emitter," *IEEE J. Lightwave Technol.*, 11(12), 1937–1940, 1993.

37. F. Koyana and K. Iga, "Frequency chirping in external modulators," *IEEE J. Lightwave Technol.*, 6(1), 87–93, 1988.

38. D. K. Cheng, *Field and Wave Electromagnetics*, Reading, MA: Addison-Wesley, 1991, chap. 7.7.

39. R. E. Collin, *Foundation for Microwave Engineering*, 2nd ed., Singapore: McGraw-Hill, 1992.

40. L. A. Coldren and S. W. Corzine, *Diode Lasers and Photonic Integrated Circuits*, New York: John Wiley & Sons, 1995.

41. http://www.dow.com/cyclotene/

42. N. Dagli, "Wide-bandwidth lasers and modulators for RF photonics," *IEEE Trans. Microwave Theory Technol.*, 47, 1151–1171, 1999.

43. K. Wakita, I. Kotaka, S. Matsumoto, R. Iga, S. Kondo, and Y. Noguchi, "Very-high allowability of incidental optical power for polarization-insensitive InGaAs/InAlAs multiple Quantum well modulators buried in semi-insulating InP," *Jpn. J. Appl. Phys.*, 37, 1432–1435, 1998.

44. R. Lewén, S. Irmscher, and U. Eriksson, "Microwave CAD circuit modeling of a traveling-wave electro-absorption modulator," *IEEE Trans. Microwave Theory Technol.*, 51, 1117–1128, 2003.

45. F. Bertazzi, F. Cappelluti, F. Bonani, M. Goano, and G. Ghione, "A novel coupled physics-based electro-magnetic model of semiconductor traveling-wave structures for RF and optoelectronic applications," in *Proceedings of the GaAs 2003*, Munich, Oct. 2003, pp. 239–242.

46. H. H. Liao, K. K. Loi, C. W. Tu, P. M. Asbeck, and W. S. C. Chang, "Microwave structures for traveling-wave MQW electro-absorption modulators for wide band 1.3 μm photonic links," *Proc. SPIE*, 3006, 291–300, 1997.

47. I. Kim, M. R. T. Tan, and S. Y. Wang, "Analysis of a new microwave low-loss and velocity matched III–V transmission line for traveling wave electrooptic modulators," *J. Lightwave Technol.*, 8, 728–738, 1990.

48. Y. R. Kwon, V. M. Hietala, and K. S.Champlin, "Quasi-TEM analysis of 'Slow-Wave' mode propagation on coplanar microstructure MIS transmission lines," *IEEE Trans. Microwave Theory Technol.*, MTT-35, 545–551, 1987.

49. H. Hasegawa, M. Furukawa, and H. Yanai, "Properties of microstrip line on Si-SiO$_2$ system," *IEEE Trans. Microwave Theory Technol.*, MTT-19, 869–881, 1971.

50. Y. Fukuoka, Yi-Chi Shih, and T. Itoh, "Analysis of slow-wave coplanar waveguide for monolithic inte-grated circuits," *IEEE Trans. Microwave Theory Technol.*, MTT-31, 567–573, 1983.

51. F. Cappelluti and G. Ghione, "Self-consistent time-domain large-signal model of high-speed traveling-wave electroabsorption modulators," *IEEE Trans. Microwave Theory Technol.*, MTT-51, 1096–1104, 2003.

52. J. Lim, S. Jeon, J. Kim, and S. Hong, "A circuit model of traveling wave electroabsorption modulators," *Microwave Symp. Dig.*, 3, 1707–1710, 2002.

53. S. Irmscher, R. Lewén, and U. Eriksson, "Influence of electrode width on high-speed performance of traveling-wave electro-absorption modulators," in *13th International Conference on Indium Phosphide and Related Materials (IPRM'01)*, May 2001, Nara, Japan, paper WA3-3, 2001.

54. A. M. Fox, D. A. B. Miller, G. Livescu, J. E. Cunningham, and W. Jan, "Quantum well carrier sweep out: Relation to electroabsorption and exciton saturation," *IEEE J. Quantum Electron.*, 27, 2281–2295, 1991.

55. F. Devaux, S. Chelles, A. Ougazzaden, A. Mircea, and J. C. Harmand, "Electroabsorption modulators for high-bit-rate optical communications: A comparison of strained InGaAs/InAlAs and InGaAsP/InGaAsP MQW," *Semicond. Sci. Technol.*, 10, 887–910, 1995.

56. R. B. Welstand, S. A. Pappert, D. T. Nichols, L. J. Lembo, Y. Z. Liu, and P. K. L. Yu, "Enhancement in electroabsorption waveguide modulators slope efficiency at high optical power," *IEEE Photonics. Technol. Lett.*, 10, 961–963, 1998.

57. C. H. Henry, R. A. Logan, F. R. Merritt, and J.-P Luongo, "The effect of intervalence band absorption on the thermal behavior of InGaAsP laser," *IEEE J. Quantum Electron.*, 19, 947–952, 1983.

58. M. Suzuki, H. Tanaka, and S. Akiba, "High-speed characteristics at high input optical power of GaInAsP electroabsorption modulators," *Electron. Lett.*, 24, 1272–1273, 1988.

59. M. Suzuki, H. Tanaka, and S. Akiba, "Effect of hole pile-up at heterointerface on modulation voltage in GaInAsP electroabsorption modulators," *Electron. Lett.*, 25, 88–89, 1989.

60. D. Meglio, P. Lugli, R. Sabella, and O. Sahlén, "Analysis and optimization of InGaAsP electro-absorption modulators," *IEEE J. Quantum Electron.*, 31, 261–268, 1995.

61. K. Brennan and K. Hess, "Theory of high-field transport of holes in GaAs and InP," *Phys. Rev. B, Condens. Matter*, 29, 5581–5590, 1984.

62. S. Akiyama, S. Hirose, T. Watanabe, M. Ueda, S. Sekiguchi, N. Morii, T. Yamamoto, A. Kuramata, and H. Soda, "Novel InP-based Mach-Zehnder modulator for 40 Gb/s integrated lightwave source," in *Pro-ceedings of the IEEE 18th International Semiconductor Laser Conference 2002*. TuC1, 2002, pp. 57–58.

63. U. Langmann and D. Hoffmann, "Capacitively loaded transmission line for subpicosecond stepped $\Delta\beta$ operation of an integrated optical directional coupler switch," *IEEE MTT-S Dig.*, 82(1), 110–112, 1982.

64. R. G. Walker, "High-speed semiconductor intensity modulators," *IEEE J. Quantum Electron.*, 27, 654–667, 1991.

65. S. R. Sakamoto, A. Jackson, and N. Dagli, "Substrate removed GaAs/AlGaAs Mach-Zehnder electro-optic modulators for ultra wide bandwidth operation," *paper presented* at *International Topical Meeting on Micro-wave Photonics 1999*, November 1999, pp. 13–16.

66. D. Hoffmann, S. Staroske, and K.-O. Velthaus, "45 GHz bandwidth Travelling-Wave Electrode Mach-Zehnder Modulator with Integrated Spot Size Converter," *paper presented at Indium Phosphide and Related Materials (IPRM) 2004.* Kagoshima, Japan, May 2004, pp. 585–588.

67. S. Akiyama, S. Hirose, H. Itoh, T. Takeuchi, T. Watanabe, S. Sekiguchi, A. Kuramata, and T. Yamamoto, "40 Gb/s InP-based Mach-Zehnder Modulator with a driving voltage of 3V$_{pp}$," *paper presented at Indium Phosphide and Related Materials (IPRM) 2004*. Kagoshima, Japan, May 2004, pp. 581–584.
68. B. J. Hawdon, T. Tutken, A. Hangleiter, R. W. Glew, and J. E. A. Whiteaway, "Direct comparison of InGaAs/InGaAlAs and InGaAs/InGaAsP quantum well modulators," *Electron. Lett.*, 29, 705–707, 1993.
69. R. Lewén, S. Irmscher, U. Westergren, L. Thylén, and U. Eriksson, "Traveling-Wave Electrode Electroabsorption Modulators Toward 100 Gb/s," *paper presented at the Conference of Optical Fiber Communication (OFC'03)*, Los Angeles, CA, paper FL1, 2004.
70. Z. Zhuang, Y. Wu, W. S. C. Chang, P. K. L. Yu, S. Mathai, M. Wu, D. Tishini, and K. Y. Liou, "Peripheral coupled waveguide traveling-wave electroabsorption modulators," *Proc. IEEE MTT-S Dig.*, 2, 1367–1370 2003.

TUNABLE SEMICONDUCTOR LASERS

Larry A. Coldren, G. A. Fish, Y. Akulova, J. S. Barton,
L. Johansson, and C. W. Coldren

INTRODUCTION

Tunable lasers have been of interest for some time [1]. Applications range from sources for fiber optic telecommunication systems to broadband sensors. About three or four years ago, the telecom application began to drive significant investments into this field to support the perceived need for dynamic networks and wavelength reconfigurability in wavelength division multiplexing (WDM) systems. Vast reductions in operational costs were predicted for such flexible fiber-optic networks that were thought to be necessary for the rapidly expanding demand for bandwidth. However, as many new companies joined this effort, there was a large overbuild of capacity, and the need for the new networks vanished, or more accurately, was pushed back to at least the present time. The good news for the industry is that the demand for bandwidth continues to nearly double each year.

Although the potential to reduce operational costs with more dynamical networks still exists, the delay in significant network expansion has led to a reappraisal of the value proposition for tunable lasers. Today, the main value for telecom networks appears to be in the areas of inventory reduction, both in the manufacture and operation of WDM systems. With fixed frequency distributed-feedback (DFB) lasers, dozens of different wavelength codes must be manufactured and inventoried, and perhaps more importantly, dozens of different wavelength-specific line cards must be manufactured and inventoried. Since the cost of line cards is measured in multiples of $10k, this can be a significant overhead. Thus, even for this less glamorous application, the savings are finite, but as a result, today's tunable laser solutions are compared to fixed-frequency DFBs for both cost and performance.

Bearing all of this in mind, it is generally agreed that if tunable lasers with the same performance specs as DFBs were available, most systems companies would select them over DFBs for a small price premium. As we will show in this report, some tunable embodiments appear to have reached specification parity with DFBs, so the situation may indeed be favorable for tunables in future WDM networks. By the time one considers the price of a line card, the increased cost of incorporating the tunable laser can be quite small, relatively speaking, and one can gain the functionality of a "universal" line card, which can be programmed to function at any wavelength over the tuning range of the laser [2]. Of course, this is a strong argument for full-band tunability in the laser, because only one part would then be necessary for any slot.

Finally, there is still the compelling argument that the line card can be re-provisioned at some later point in time, should the network architecture evolve to accommodate this, and again, full-band tunability would be desired [3]. The first example application to be mentioned is in reconfigurable optical-add-drop multiplexers (ROADMs). As illustrated in Figure 1 these allow single (or multiple) optical channels to be removed and replaced on a fiber without de-multiplexing, regenerating, and re-multiplexing the entire array of wavelengths contained in the fiber. In applications where this functionality is desired, the ROADM can vastly reduce the cost of dropping and/or adding a relatively small amount of information from or to the fiber.

Tunable lasers are also natural complementary components in optical switches of various kinds. Here they generally are used for the function of wavelength switching or "wavelength conversion," in which an incoming signal on one wavelength is remodulated onto another wavelength on the output [4]. This can be accomplished in numerous ways, the most straightforward of which is to incorporate a tunable laser within a line card or transponder, so that the output can be set to any wavelength value. These "optical–electronic–optical" or OEO components include 3R regeneration to reconstitute the signal to its original form. One can also make "all optical" wavelength converters that use the incoming signal on one wavelength to drive a modulator that applies the signal directly to a second selectable output wavelength generated by a tunable laser. Recently, this function has been demonstrated with a single monolithic chip [5,6]. However, in these "all-optical" approaches 3R or even 2R regeneration of the signal is generally not provided, so that these elements can only work with relatively clean data, and they can only be cascaded a few times before a 3R regeneration is necessary.

Figure 2 shows an all-optical space switch that uses wavelength converters at the input and a passive optical router switching fabric to provide space switching. In this case the input signal is placed on the wavelength that the passive "lambda router" will route to the desired output port. If the signal is to be re-mutiplexed, it would then have to be again converted to the desired wavelength to enter the optical multiplexer. This sort of switching architecture is also currently being investigated by several groups for all-optical packet switching [7,8]. In this case, the tunable lasers in the front-end wavelength converters (shown as line cards with tunable lasers in Figure 2) must switch wavelengths very fast — typically in the nanosecond range. Such a criterion will favor the tunable laser types that are controlled electronically versus the ones that have thermal or mechanical tuning elements.

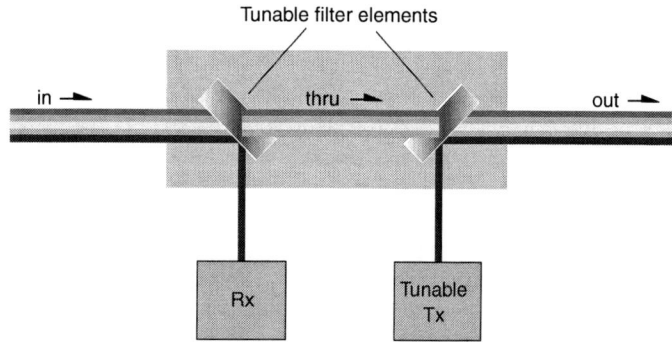

Figure 1   Reconfigurable optical add/drop port. A tunable filter selectively removes (or adds) a single (or several) optical WDM channel from the fiber. A tunable transmitter is needed to insert any desired channel at the add port

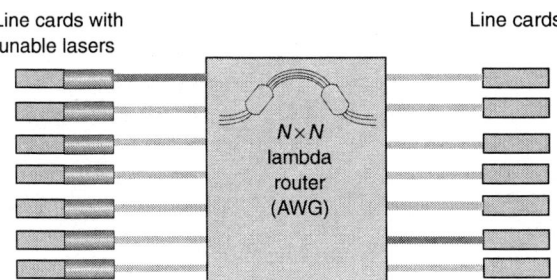

Line cards with
tunable lasers

Line cards

Figure 2 Transparent optical space switch composed of a wavelength converter array and a passive router such as an arrayed-waveguide-router (AWG). Line cards with tunable lasers can more generally be replaced by wavelength converters

The situation in the sensor area is perhaps even more attractive for tunable lasers. Here many sensor types rely upon the ability to sweep the laser frequency over a wide range for their basic functionality, so they are essential. A number of applications are emerging, ranging from the measurement of stress in aircraft to the mapping of pressure in oil wells. A wide tuning range provides increased resolution and accuracy. Rapid tuning allows near real-time processing of the sensor information. Many sensors use some sort of interferometric effect for their measurement. Thus, issues such as noise and linewidth become even more important than in communication systems.

BASIC TUNING MECHANISMS

Figure 3 gives a schematic of a generic tunable laser together with the relative spectra of the necessary filter and gain elements as well as the location of the various cavity modes that all must be properly aligned and translated to create a tunable, single-frequency laser. Of course, in most practical embodiments, the filter, mirror, and phase-shifting elements are combined in some way to create a unique physical structure for the different kinds of tunable lasers. Figure 3 can be used to see how a tunable semiconductor laser evolves from the most basic "Fabry–Perot" laser, which has just the gain and the two simple mirror elements, to a "single-frequency" laser, which adds the mode-selection filter, to a "tunable single-frequency" laser, which adds possible adjustment of the mirror position and the center frequency of the mode-selection filter, as well as adding a new adjustable cavity phase element. For more analytical discussion, the reader is referred to References 9 and 10.

The most common Fabry–Perot laser is composed of a uniform cleaved semiconductor chip that is structured to provide gain for a guided optical mode with the cleaves functioning as the mirrors. The most common single-frequency laser is probably the DFB laser, illustrated in Figure 4(a), in which an index grating is formed near the optical waveguide to provide a continuous reflection that gives both the mirror functionality as well as the mode selection filter. The vertical-cavity surface-emitting laser (VCSEL) as illustrated in Figure 4(b), is also a single-frequency laser, but in this case the cavity is vertical and the grating mirrors sandwich the gain region. Although the distributed-Bragg-reflector (DBR) mirrors are frequency selective, the primary mode selection is done by the finite width of the gain spectrum in this case, because both the mirror spectrum and the mode spacing are made large by the short cavity length — a somewhat different case than that suggested in Figure 3.

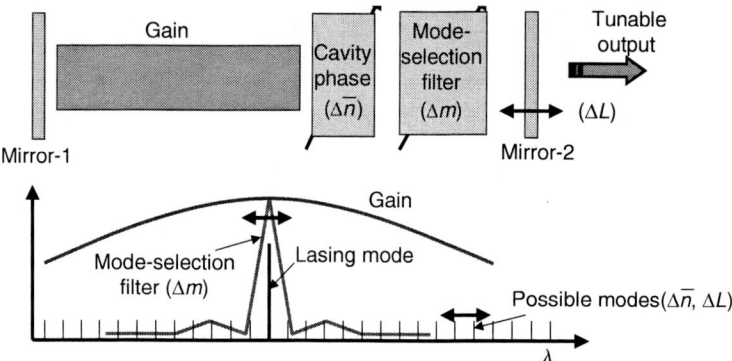

Figure 3   Schematic of generic tunable laser together with relationship of the spectra of each element

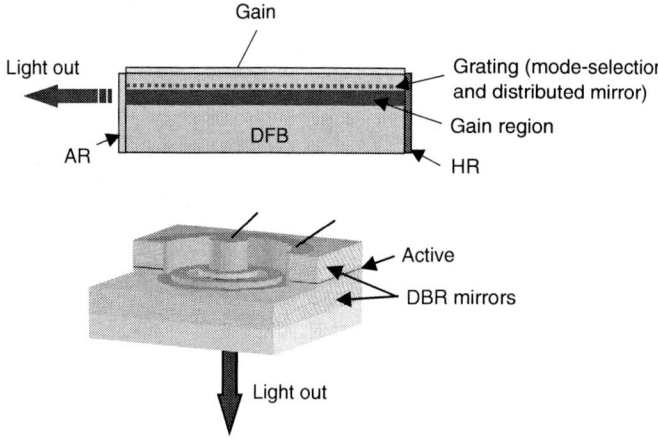

Figure 4   Examples of single-frequency lasers (not tunable): (a) Distributed-feedback (DFB) laser; (b) Vertical-cavity surface-emitting laser (VCSEL)

Equation (1) gives the relationship between the lasing wavelength, λ, and the cavity mode number, m, effective index of refraction seen by the cavity mode, $\bar{n}$, and the effective cavity length, L. Quite obviously, if one changes m, $\bar{n}$, or L, the wavelength must also change. The relative change in wavelength derived from Eq. (1) is given in Eq. (2). As indicated the relative wavelength change is directly proportional to the relative change in either the length, index, or mode number.

$$m\lambda / 2 = \bar{n}L \tag{1}$$

$$\frac{\Delta\lambda}{\lambda} = \frac{\Delta\bar{n}}{\bar{n}} + \frac{\Delta L}{L} - \frac{\Delta m}{m} \tag{2}$$

where $\Delta\bar{n}/\bar{n}$ is tuned by net cavity index change, $\Delta L/L$ is tuned by physical length tuned by mode-selection filter (via index or grating angle).

EXAMPLES OF TUNABLE SEMICONDUCTOR LASERS

Figure 5 shows several different types of tunable single-frequency lasers that have been commercialized. (Since tunable lasers need to be single frequency to be of much use, we will now drop this qualifier.) In the figure we have only included the widely-tunable varieties that are capable of full C- or L-band coverage from a single device.

The first example shows a selectable array of DFB lasers that are combined in a multimode interference coupler. The DFBs are excited one at a time and each is manufactured with a slightly different grating pitch to offset their output wavelengths by about 3 or 4 nm. The chip is then temperature tuned by some 30 to 40°C to access the wavelengths between the discrete values of the array elements. With N-DFB elements, then, a wavelength range of up to about $4N$ nm can

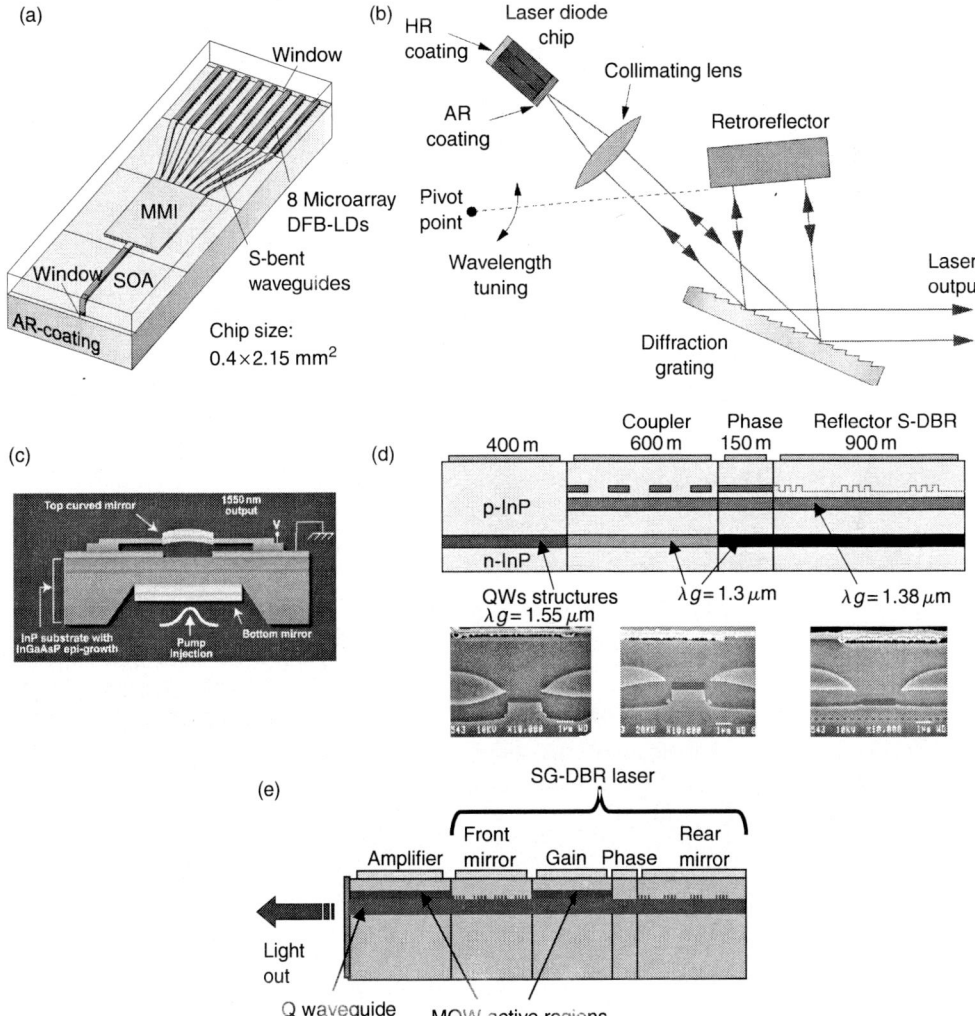

Figure 5 Examples of widely tunable laser types. (a) Selectable DFB array; (b) external-cavity; (c) MEMs/VCSEL; (d) grating-coupled sampled-reflector (GCSR); (e) sampled-grating DBR (SGDBR) with integrated SOA

be accessed, or with 8 to 10 elements the entire C-band can be accessed. The schematic included in Figure 5(a) is from NEC [11]; however, similar work is also being carried out at Fujitsu [12] and other mostly Japanese labs. Santur Corp. uses a similar concept, but with an external MEMs mirror to select which element is coupled to the output fiber [13], thus eliminating the $1/N$ combiner loss, but at the expense of one more element to package and control. In all cases, this approach must deal with the requirement of having a number of closely spaced DFBs all working to tight specifications. The losses in combining, inherent in most varieties, are also significant, and the need to temperature tune over a fairly large tuning range leads to relatively high power dissipation for this approach.

Figure 5(b) is an example of an external-cavity laser. In this case a "gain block" is coupled to external mode-selection filtering and tuning elements via bulk optical elements. The cavity phase adjustment, necessary to properly align the mode with the filter peak and the desired ITU grid wavelength, can be included in one of the several places — for example, on the gain block or by fine tuning the mirror position. In most external-cavity approaches the mode selection filter is a diffraction grating that can also double as a mirror. The so-called Littman–Metcalf cavity arrangement is illustrated. In this case, a retro-reflecting mirror is translated as it is rotated. This combined motion changes the effective cavity length in proportion to the change in center wavelength of the mode-selection filter to track the movement of a single cavity mode. The Littman–Metcalf geometry provides continuous tuning over some range, but due to cavity dispersion, one in general still needs to correct the cavity phase at each ITU channel. This approach has been used by Iolon [14] and New Focus [15] in their products. Other companies tend to just rotate the mirror and let the mode selection filter scan across the modes. This is most common in scientific instruments, where the cavities are quite long and the mode spacing very small. Intel also has reported some research [16] in which the external cavity contains two temperature-tuned etalons with slightly different resonance frequencies, which act in combination to create a widely-tunable filter. A standard external mirror completes the cavity. All of the external cavity approaches appear to provide useable specs for telecommunications, although at this writing we are not aware of any that has completed the full Telcordia qualification exercise. An obvious concern with these structures is their manufacturability and reliability, given the need for assembling numerous micro-optical parts and holding them in precise alignment.

Figure 5(c) shows a tunable VCSEL that is created by mounting one mirror on a flexible arm and using an electrostatic force to translate it up and down. This micro-electromechanical (MEMs) approach has been employed by Coretek [17] — later acquired by Nortel — (as shown in Figure 5[c]) and Bandwidth 9 [18]. In Coretek's case external optical pumping was used, and in Bandwidth 9's case electrical pumping was employed. Both efforts appear to have been discontinued. The Coretek approach used dielectric mirrors for wide reflection bandwidth. Thus, it was able to show full C-band operation; the Bandwidth 9 device had a somewhat smaller tuning range. The use of optical pumping also provides for more power output, although advertised products from Nortel did include an external amplifier to boost the fiber-coupled power to the 20 mW range. A primary appeal for the VCSEL approaches is the wafer-scale manufacturing platform that it appears to provide. The hope here was to make tunable devices for nearly the same cost as the 850 nm VCSELs used in Gigabit Ethernet. However, at 1550 nm VCSEL construction is more difficult, and limited output power together with wide optical linewidth appear to be serious limitations with the VCSEL approaches at 1550 nm.

Figures 5(d) and 5(e) show monolithic widely-tunable semiconductor laser approaches that employ electronic tuning of the index in a single cavity to provide for full C- or L-band wavelength coverage. Both are variations on older DBR laser approaches [19], but both employ concepts to tune the relative wavelength by up to an order of magnitude more than the index of any section can be tuned. In the case of Figure 5(d), the so-called grating-coupled sampled-reflector (GCSR) laser [20], this is accomplished by using a property of a grating-assisted co-directional coupler which has a tuning proportional to the index tuning relative to the *difference* in index between two coupled waveguides, $\Delta n/(n_1 - n_2)$, rather than $\Delta n/n_1$ as in most other filters. However, because the filter is also broad, a back multiple-order sampled-grating reflector is required for good mode selectivity in this case. In the SGDBR of Figure 5(e) [21–22], the wider tuning range filter is provided by the product of the two differently spaced and independently tuned reflection combs of the sampled-grating DBRs at each end of the cavity. This product, R_1R_2, is what appears in the laser cavity loss factors, and the variation in the beating effect between the two different mirror reflection combs is sometimes referred to as the vernier effect. In this case the net mode selection filter wavelength tuning is that of a single grating, $\Delta n/n$, multiplied by $\delta\lambda/\Delta\lambda$, the difference in spacing between the mirror reflection peaks of the two mirrors, $\delta\lambda$, divided by the mean mirror peak spacing, $\Delta\lambda$. Similar physics is involved in the superstructure-grating DBR developed at NTT [23]. In both cases, good side-mode suppression has been demonstrated, and tuning of over 40 nm is easily accomplished, but due to grating losses resulting from current injection for tuning, the differential efficiency and chip output powers can be somewhat limited. In the case of the SGDBR, this is easily addressed by the incorporation of another gain section on the output side of the output mirror, and fiber-coupled powers of up to 40 mW have been reported. In fact, this is the embodiment illustrated in Figure 5(e). Incorporating such a semiconductor-optical-amplifier (SOA) is not as easy for the GCSR, so fiber-coupled powers of typically <10 mW result. The integrated SOA also has other benefits for the SGDBR as will be discussed below.

CHARACTERISTICS OF SGDBR LASERS AND SINGLE-CHIP TRANSMITTERS

Work at UCSB and Agility Communications has aimed to develop widely tunable lasers and transmitters with monolithically integrated modulators. A low-cost "platform technology" that is capable of providing a wide variety of photonic ICs (PICs) without changing the basic manufacturing process has been developed. Figure 6 shows a photograph of a 2 in. InP wafer with arrays of seven-section photonic IC transmitters, each consisting of a full-band-tunable four-section sampled-grating DBR (SGDBR) laser integrated with a monitoring detector, optical amplifier, and modulator. The SEM inset shows one of these mounted on a carrier ready to be inserted into a package. It is important to note that the wafer layer structure and processing procedure used is identical to that developed for the SGDBR laser alone. This same structure and processing procedure is also used in the more complex laser PICs to be discussed below. Note also a key advantage of photonic integration — only one optical coupling to fiber is required, as would be necessary for a simple DFB laser alone.

The basic SGDBR–SOA shown in Figure 5(e) above as well as the integrated SGDBR–SOA–EAM transmitter illustrated in Figure 6 have been productized and Telcordia qualified for

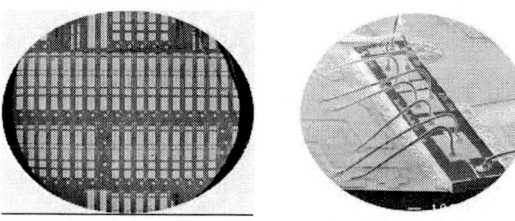

Figure 6   Photo of wafer and SEM of mounted single-chip transmitter

telecom applications [24]. In Figure 7 we give a summary of the characteristics of a 20 mW cw product similar to Figure 5(e) at each of 100 channels spaced by 50 GHz across the C-band. A common quaternary waveguide extends throughout the entire device and offset quantum-well gain layers are included at the laser gain and SOA sections.

Figure 8 shows a schematic cross section of an InP-based transmitter chip [25] as included in the photos of Figure 6. The modulator bias is varied across the 40 nm tuning range to enable efficient modulation and good extinction across this entire range.

Figure 9(a) shows superimposed rf-extinction ratio versus wavelength characteristics for 100 transmitter chips across the C-band, and Figure 9(b) shows the bit-error rate after transmission through 350 km of standard single-mode fiber for two different wavelengths. The data is applied directly to the EAM of the chip. The average modulated output power is about 3 dBm in this case. Error-free operation was observed.

The transmitter illustrated in Figure 6 and Figure 8 and characterized in Figure 9 provides good results at 2.5 Gb/sec for distances up to 350 km. However, for longer distances and/or higher bit rates, some sort of chirp control is necessary. Thus, work at both Agility [6,26] and

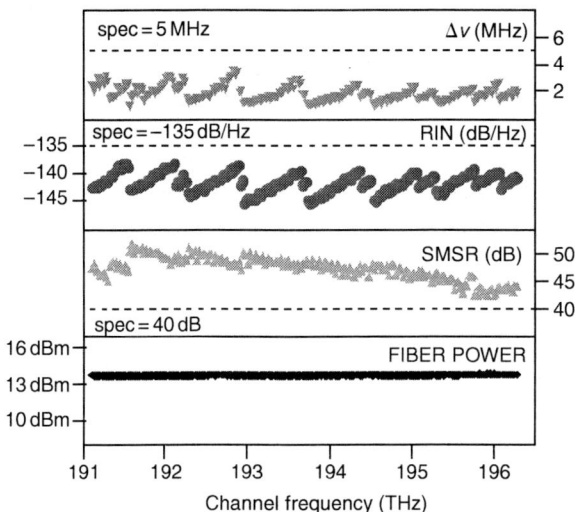

Figure 7 CW characteristics of SGCBR–SOA device for 100 channels — calibrated for 20 mW of fiber power. The linewidth, Δv, relative intensity noise, RIN, and side-mode suppression ratio, SMSR shown for all C-band channels

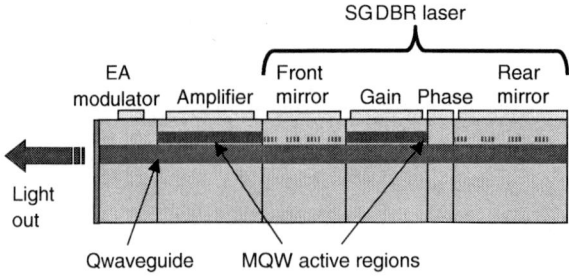

Figure 8 Single-chip widely tunable transmitter schematic showing a SGDBR laser integrated with an SOA and EAM

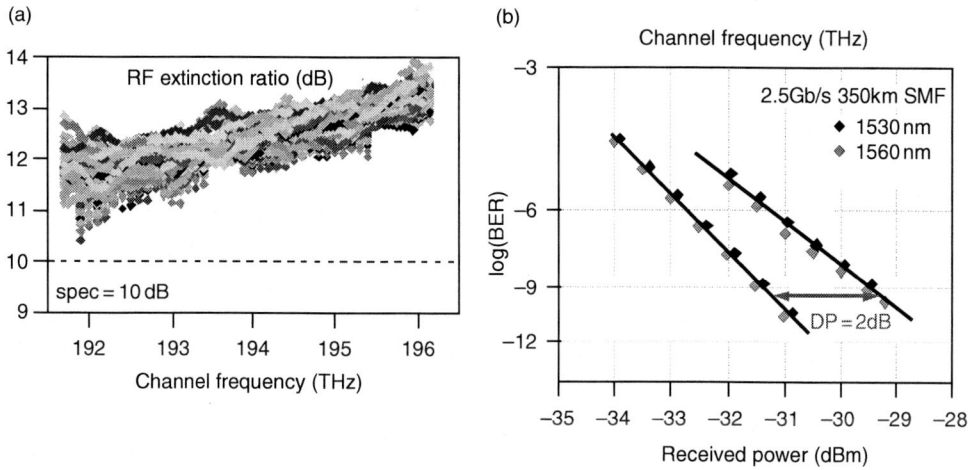

Figure 9 (a) RF extinction ratio for 100 superimposed SGDBR/EAM transmitters across the C-band. (b) Bit-error-rate results after transmission through 350 km of standard fiber at 2.5 Gb/sec

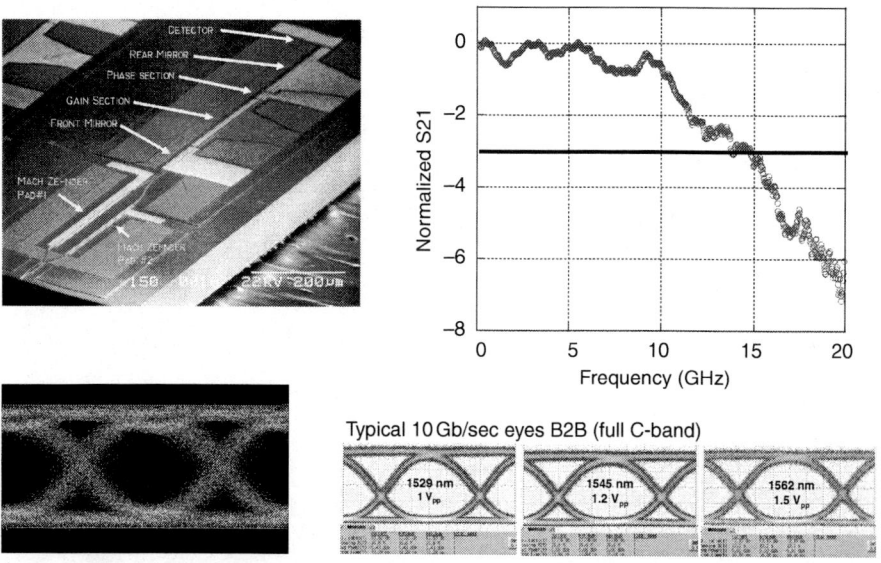

Figure 10 (Top left) SEM photo of UCSB's SGDBR integrated with a Mach–Zehender modulator; (top right) small-signal bandwidth. (bottom left) unfiltered eye; (bottom right) filtered eye-diagrams at 10 Gb/sec for three wavelengths across the band for Agility device

UCSB [27] has explored replacing the EAM with a Mach–Zehnder modulator (MZM) as shown in Figure 10. Such modulators have been used widely for long-haul applications, and they allow negative chirp with only one drive signal, although dual drive of both arms of the MZM are necessary for truly programmable chirp. In the past, researchers have had difficulties in integrating such MZMs directly with lasers because of reflections. However, the UCSB-Agility effort appears to have solved these difficulties. By monolithically integrating the MZM a much smaller footprint and low power dissipation is possible as compared to hybrid packaged or fiber-coupled devices. In addition, the chirp can be tailored for each channel across the wavelength band by adjusting the biases to the two legs of the MZM. Chirp values from +1 to −1 are readily available. Error free transmission over 80 km of standard fiber was demonstrated for all channels at 10 Gb/sec using a negative chirp configuration.

RELIABILITY OF THE SGDBR LASER

Figure 11 summarizes some of the reliability data taken on the 10 mW cw product by Agility [28]. Both the integrated EAM transmitter and the 10 mW cw version have undergone complete Telcordia qualification. Because of the InP single-chip architecture, these PICs can be qualified in much the same way as simple laser chips. Such is not the case with other types of widely-tunable transmitters in which separated optical parts are involved in some sort of hybrid package.

A quantitative model of failure rates was developed for each section of the device by fully characterizing failure modes and determining failure mode accelerants. The activation energy, $E_a = 0.5$ eV was derived assuming an aging rate proportional to $\exp[E_a/kT]$. The current acceleration exponent, $n = 1.5$ was derived assuming the aging rate was also proportional to J^n, where J is the applied current density to the section in question. Mirror drift failure was set to be when the operating point moved half way from the center of a single-mode region toward a mode-hop boundary. For the SGDBRs in question this was equivalent to ±100 pm of allowable open-loop wavelength drift of the mode boundaries. (Of course, with a wavelength locker in operation, the lasing mode wavelength only drifts as much as it does — typically < 1 pm over life.) The aging criteria for the gain and amplifier sections are as for other semiconductor lasers. The same approximate activation energies and current acceleration factors were observed for all sections.

The data indicate that no updating of mirror currents is necessary for a FIT rate of <20 at 15 years. This includes reasonable margins for all device parameters. However, a mirror look-up table updating algorithm has also been developed that both monitors the mirror drift for setting possible alarms as well as updating the table. This mirror-control algorithm improves the FIT rate to <3 at 15 years. The lifetime distribution taken from 200 parts using accelerated aging procedures shows a classical log-normal relationship with a mean lifetime of 186 years for room

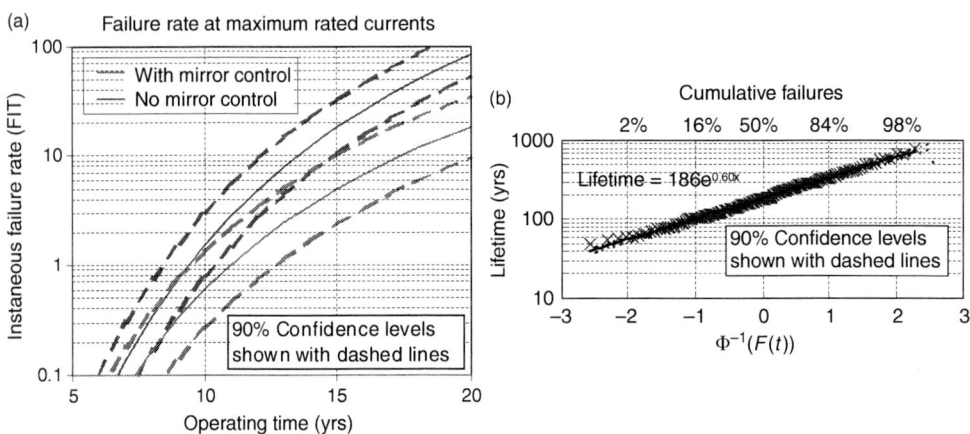

Figure 11   (a) FIT rate versus time, assuming both original mirror biases as well as with bias updating — mirror control. (b) Lifetime distribution of 200 parts tested. Maximum channel currents assumed. Mean lifetime of 186 years shown

temperature, but with maximum channel currents assumed. In a normal WDM system populated with such devices, the channel currents would be distributed over lower values for the various channels, so Figure 11 should be taken as a worst case result that would not occur over any distribution of components in a typical system. Taking a distribution of WDM channels into account, the "no mirror control" FIT rate is estimated to be about 2 at 15 years.

This relatively low wavelength drift for the SGDBR has been ascribed to the relatively small percentage of grating that fills the sampled-grating mirrors. About 90% of the mirror area is free from gratings in a typical design. Studies have shown that this results in much higher material quality within the mirrors [28]. Lack of gratings in most areas permits very high quality regrowth of the InP cladding following grating formation. Not only is the surface more planar and free from defects, it can be composed of InP rather than InGaAsP quaternary waveguide material in the large regions between the grating bursts. Thus, while standard DBR lasers, which contain gratings throughout the mirror tuning sections, continue to have wavelength drift problems, the SGDBR has emerged as being surprisingly stable.

CONTROL OF WIDELY TUNABLE LASERS

The control of multi element tunable lasers, such as those illustrated in Figure 5, has been a roadblock to their general acceptance for some time. Most system engineers are accustomed to incorporating a two-terminal device, such as a DFB laser, in their optical transmitters. Of course, even for the DFB the device temperature is used to fine tune and lock the wavelength in WDM systems. For the widely tunable devices of Figure 5, it seems apparent that we must simultaneously control some additional parameters, although in some cases we may only need to dynamically control the same number as in the DFB to lock the amplitude and wavelength at a given channel. However, there is always a need for a "look-up table" to give the specific set of currents or voltages for each channel to the several sections, and this indeed, does add a complication for the user. To gain more wide-spread acceptance, suppliers of the multiple-section lasers in recent years have provided automatic control systems within the laser module so that the user does not have to deal with the control problem. The wavelength and amplitude are set via a digital command through a common interface. Nevertheless, users are justifiably concerned with the reliability and stability of such systems. So they remain of great interest.

The control system must be capable of two basic functions: (1) staying accurately on the desired wavelength channel and (2) reliably finding a new desired channel when a channel change is requested at some later time, and this time could be near the end of life. To accurately stay on a desired wavelength channel most lasers require a separate wavelength locker. If the wavelength channel plan is relatively coarse, perhaps >100 GHz, this locker may not be necessary. For example, given the low wavelength drift of the SGDBR outlined above, this locker may not be necessary for even 100 GHz channel spacing if a modest FIT rate is tolerable. But, more generally a locker is required. It usually contains an etalon with a free-spectral range roughly equal to the channel spacing, so that it can provide a feedback signal to capture and lock the wavelength within about one-third of a channel spacing on either side of some ITU frequency.

Switching to a new channel after some time is generally a more difficult problem. The immediate question is, will the original look-up table from factory calibration be good enough, or will aging have changed the values? To be able to use the original look-up table, the settings must get us to the correct channel within the capture range of the locker. For embodiments where tuning requires mechanical motion or significant swings in temperature, hysteresis and charging of MEMs elements tend to shift the look-up table. In some DBR structures, changes in carrier lifetime also may result in a shift in wavelength that exceeds the locker capture range. Possible solutions to these problems involve either some sort of global wavelength monitor, a channel

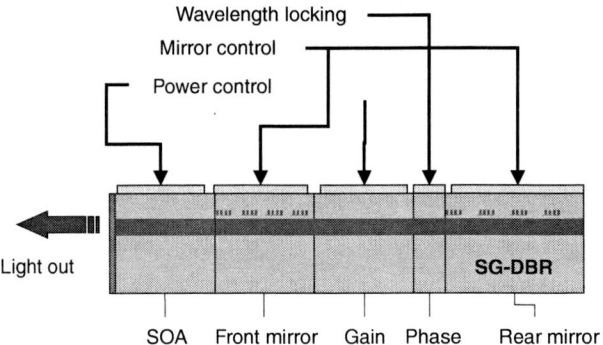

Figure 12   Control signals necessary to operate SGDBR/SOA

counting algorithm, or some means of updating the look-up tables over life. All of these approaches have been demonstrated, but all require a more complex control system.

Figure 12 illustrates the control signals necessary to operate the SGDBR/SOA. An electronic circuit supplies control currents in response to amplitude and wavelength errors derived from the locker. The temperature and the current to the gain section are held constant at factory-set values, so they are not part of the control system. All other currents are contained in a look-up table for each channel. The locker signals are converted to error currents for that are added to the SOA and phase sections. The SOA is used to lock the amplitude and the phase section is used to lock the wavelength. In normal operation no corrections are supplied to the mirrors — this is called the "no mirror control" case referred to in Figure 11. In this mode of operation then, the actual feedback control system is about the same as for the DFB, with the amplitude correction being added to the SOA instead of the DFB gain section and the wavelength correction being added to the phase section instead of the thermoelectric cooler of the DFB. Of course, there is a look-up table to set different initial values for each channel in the SGDBR case, but this involves no dynamic control, just set points.

It may also be seen that the use of an external SOA for amplitude control is desirable in a tunable laser relative to adjusting the gain current in the cavity. This is because the wavelength would also change in response to changing the gain current. In fact, this is one of the primary limitations on wavelength stability in widely-tunable laser embodiments that do not have the external SOA to level the amplitude as the device ages.

For "mirror control" the mirror currents are slowly dithered about their set points and the voltage on the gain section is monitored. Because the wavelength and amplitude locking circuits are operating, there is no change in external optical power or wavelength observed. Second order changes in cavity loss, caused by changing the mirror currents, are also removed in this process. The dithering of the reflectivity peaks of the mirrors cause the gain voltage to change slightly because it monitors the quasi-Fermi level separation in the gain region, and this is proportional to the cavity loss change. Thus, a local minimum in the gain voltage is observed when the mirror peaks are properly aligned with the mode wavelength, where the cavity loss is at a local minimum. The mode, of course, is set by the locker/phase-section feedback circuit to be at the proper ITU grid wavelength. So, it can be seen that this "mirror control" algorithm requires no additional optical elements or electrical connections. Again, this mirror control mode is probably not necessary for reliable device operation according to Figure 11; however, monitoring of the mirror peaks relative to the cavity mode gives one assurance that the device is operating properly.

If the mirror currents must be corrected, then it may be assumed that the currents required to hit other channels must also change. This is the second aspect of control mentioned above — finding

Table 1

Laser	Coarse wavelength	Fine wavelength	Amplitude	VOA
DFB Array/SOA	$V_{\text{array}}(j)$	T	$I_{\text{gain}}(j)$	ΔI_{SOA}
DFB/MEMs	$V_{m1}, V_{m2}(j)$	T	$I_{\text{gain}}(j)$	$V_{m1}, V_{m2}(j)$
SGDBR/SOA	I_{m1}, I_{m2}	I_ϕ	I_{SOA}	ΔI_{SOA}
External cavity/grating	$V_{m\theta}$	V_{mL} or I_ϕ	I_{gain}	V_{mshutter}
External cavity/etalons	$T_{\text{et1}}, T_{\text{et2}}$	V_{mL} or I_ϕ	I_{gain}	—
VCSEL/MEMs	V_{m1}	V_{m1}	I_{gain}	—

a new channel. In the SGDBR case with mirror control the table can be updated dynamically without ever leaving the original channel. This is because the same reduction in carrier lifetime that requires a current increase to maintain a given carrier density and thus index of refraction, is also experienced by all the other channels. Most importantly, it has been verified that this carrier lifetime decrease is due to an increase in nonradiative recombination, and it is well known that this has a linear relationship to carrier density. Since carrier density is predominately determined by the radiative recombination rate, which depends upon the square of the carrier density, we can assume that the shift in the entire look-up table will be linear in the square root of current. Fortunately, extensive measurements have shown that this is indeed the case experimentally [28,29], so updating the table is a valid approach in this case.

Table 1 summarizes the parameters that must be adjusted to enable the amplitude and wavelength of the various types of tunable lasers illustrated in Figure 5 to be set. It also indicates the parameters for variable-optical-attenuator (VOA) operation. This function is desirable both to allow the user to adjust the amplitude as well as to blank the output during tuning between channels. As can be seen most of the widely tunable lasers being considered require several parameters to be set, and in most cases, most of these must be controlled. In the VCSEL/MEMs case there are fewer parameters, but this is an example where changing channels requires some sort of global wavelength monitor or channel counting scheme, because one clearly can not depend upon the look-up table for channel selection, especially after some aging with the MEMs mirror. The case is similar in the other mechanically tuned embodiments.

CONCLUSIONS

Tunable laser issues have been outlined. These include discussions of why tunable lasers might be beneficial, how tunable lasers tune, how they are controlled, and how reliable they can be. Several examples of tunable lasers that have been commercialized were given, and a summary of performance data for the SGDBR type of laser as well as the monolithically integrated SGDBR with both electroabsorption and Mach–Zehnder modulators was given. It was argued that tunable lasers can reduce operational costs, that full-band tunability is desirable for many applications, that rapid tunability is necessary in some applications, that monolithic integration offers the most potential for reducing size, weight, power, and cost, and that sufficient reliability for system insertion has been demonstrated, at least in one case.

REFERENCES

1. L. Coldren, "Monolithic tunable diode lasers," *IEEE J. Select. Top. Quantum Electron.*, 6, 988–, 2000.
2. R.-C. Yu et al., in *Proceedings of the APOC'03*, Wuhan, China, September '03.

3. G. Fish and K. Affolter, "Tunable lasers and their impact on optical Networks," in *Proceedings of the Communications Design Conference*, San Jose, August, 2002.

4. D. J. Blumenthal, B. E. Olsson, G. Rossi, T. Dimmick, L. Rau, M. L. Mašnović, O. A. Lavrova, R. Doshi, O. Jerphagnon, J. E. Bowers, V. Kaman, L. A. Coldren, and J. Barton, "All-optical label swapping networks and technologies," *IEEE J. Lightwave Technol.*, Special Issue on Optical Networks, 18, 2058–2075, 2000.

5. M. L. Mašanović, V. Lal, J. S. Barton, E. J. Skogen, L. A. Coldren, and D. J. Blumenthal, "Monolithically Integrated Mach–Zehnder Interferometer Wavelength Converter and Widely-Tunable Laser in InP," *IEEE Photonics Technol. Lett.*, 2003.

6. L. A. Coldren, "Widely-tunable chip-scale transmitters and wavelength converters," in *Technical Digest Integrated Photonics Research*, paper IMB-1, June 2003.

7. A. Carena, M. D. Vaughn, R. Gaudino, M. Shell, and D. J. Blumenthal, "OPERA: An optical packet experimental routing architecture with label swapping capability," *IEEE J. Lightwave Technol.*, special issue on Photonic Packet Switching Technologies, Techniques, and Systems, 16, 2135–2145, 1998.

8. M. Duelk, J. Gripp, J. Simsarian, A. Bhardwaj, P. Bernasconi, M. Zirngibl, and O. Laznicka, "Fast packet routing in a 2.5 Tb/s optical switch fabric with 40 Gb/d duobinary signal at 0.8 b/s/Hz spectral efficiency," in *Proceedings of the OFC 2003*, Optical Society of Am., postdeadline paper PD8-1, 2003.

9. L. A. Coldren and S. W. Corzine, *Diode Lasers and Photonic Integrated Circuits*, New York: John Wiley & Sons, 1995.

10. M. C. Amann and J. Buus, *Tunable Laser Diodes*, London: Artech, 1998.

11. N. Natakeyama, K. Naniwae, K. Kudo, N. Suzuki, S. Sudo, S. Ae, Y. Muroya, K. Yashiki, S. Satoh, T. Morimoto, K. Mori, and T. Sasaki, "Wavelength — Selectable microarray light sources for S-, C-, and L-band WDM systems," *IEEE Photonics Technol. Lett.*, 15, 903–905, 2003.

12. M. Bouda, M. Matsuda, K. Morito, S. Hara, T. Watanabe, T. Fujii, and Y. Kotaki, "Compact high-power wavelength selectable lasers for WDM applications," in *Proceeding of the OFC 2000*, Vol. 1, March 2000, pp.178–180.

13. J. Heanue, E. Vail, M. Sherback, and B. Pezeshki, "Widely tunable laser module using DFB array and MEMs selection with iternal wavelength locker," in *Proceedings of the OFC 2003*, Vol. 1, 2003, pp. 82–83.

14. D. Anthon, J. D. Brerger, J. Drake, S. Dutta, A. Fennema, J. D. Grade, S. Hrinya, F. Ilkov, H. Jerman, and D. King, "External cavity diode lasers tuned with silicon MEMS," in *Proceedings of the OFC 2003*, 2002, pp. 97–98.

15. Photonics Spectra, January 1999, p. 102. http://www.newfocus.com/Online_Catalog/Literature/Tunable2. pdf

16. http://www.commsdesign.com/story/OEG20030110S0053

17. K. J. Knopp, D. Vakhshoori, P. D. Wang, M. Azimi, M. Jiang, P. Chen, Y. Matsui, K. McCallion, A. Baliga, F. Sakhitab, M. Letsch, B. Johnson, R. Huang, A. Jean, B. DeLargy, C. Pinzone, F. Fan, J. Liu, C. Lu, J. Zhou, H. Zhu, and R. Gurjar, "High power MEMs-tunable vertical-cavity surface-emitting lasers," in *Proceedings of the Advanced Semiconductor Lasers, 2001 Digest of the LEOS Summer Topical Meetings*, 30 July–1 August, 2001.

18. C. J. Chang-Hasnain, "Tunable VCSELs," *IEEE J. Select. Topi. Quantum Electron.*, 6, 978–987, 2000.

19. T. L. Koch, U. Koren, and B. I. Miller, "High-performance tunable 1.55 mm InGaAs/InGaAsP multiple-quantum-well distributed-Bragg-reflector lasers," *Appl. Phys. Letts.*, 53, 1036–1038, 1988.

20. M. Oberg, S. Nilsson, K. Streubel, J. Wallin, L. Backborn, and T. Klinga, "74 nm wavelength tuning range of an InGaAsP/InP vertical grating assisted codirectional coupler laser with rear sampled grating reflector," *IEEE Photonics Technol. Letts.*, 5, 735–738, 1993.

21. V. Jayaraman, A. Mathur, L. A. Coldren, and P. D. Dapkus, "Theory, design, and performance of extended tuning range in sampled grating DBR lasers," *IEEE J. Quantum Electron.*, 29, 1824–1834, 1993.

22. B. Mason, J. Barton, G. A. Fish, L. A. Coldren, and S. P. Denbaars, "Design of sampled grating DBR lasers with integrated semiconductor optical amplifiers," *IEEE Photonics Technol. Lett.*, 12, 762–764, 2000.

23. Y. Tohmori, Y. Yoshikuni, H. Ishii, F. Kano, T. Tamamura, Y. Kondo, and M. Yamamoto, "Broadrange wavelength-tunable superstructure grating 9SSG DBR lasers," *IEEE J. Quantum Electron.*, 29, 1817–1823, 1993.

24. C. Coldren et al., "Workshop on tunable lasers," in *Proceedings of the OFC'03*, Optical Society of America, Atlanta, March 2003.

25. Y. A. Akulova, G. A. Fish, P.-C. Koh, C. L. Schow, P. Kozodoy, A. P. Dahl, S. Nakagawa, M. C. Larson, M. P. Mack, T. A. Strand, C. W. Coldren, E. Hegblom, S. K. Penniman, T. Wipiejewski, and L. A. Coldren,

"Widely tunable electroabsorption-modulated sampled-grating DBR laser transmitter," *IEEE J. Select. Top. Quantum Electron.*, 8, 1349–1357, 2002.

26. T. Wipiejewski, Y. Akulova, G. Fish, P. Koh, C. Show, P. Kozodoy, A. Dahl, M. Larson, M. Mack, T. Strand, C. Coldren, E. Hegblom, S. Penniman, T. Liljeberg, and L. Coldren, in *Proceedings of the 2003 Electronic Components and Technology Conference*, New Orleans, May 2003.

27. J. S. Barton, E. J. Skogen, M. L. Mašanović, S. P. DenBaars, and L. A. Coldren, "Widely-tunable high-speed transmitters using integrated SGDBRs and Mach–Zehnder modulators," *IEEE J. Select. Top. Quantum Electron.* 9, Semi. Lasers issue, 2003.

28. C. Coldren, T. Strand, E. Hegblom, Y. Akulova, G. Fish, M. Larson, and L. Coldren, in *Proceedings of the OFC Technical Digest*, paper MF60, Optical Society of America, Atlanta, March 2003.

29. D. A. Ackerman, J. E. Johnson, George Chu, Sung Nee, Zhang Liming, E. J. Dean, and L. J.-P. Ketelsen, "Assessment and modelling of aging in electro-absorption distributed Bragg reflector lasers," *IEEE J. Quantum Electron.*, 37, 1382–1387, 2001.

V

VERTICAL CAVITY SURFACE EMITTING LASER

Kenichi Iga

INTRODUCTION AND HISTORY

The vertical cavity surface emitting laser (VCSEL) is a semiconductor laser which has a resonant cavity that is vertically formed with the surfaces of the epitaxial layers. The light output is taken from one of the mirror surfaces, as shown in Figure 1. The present author suggested a VCSEL device in 1977 [1]. The first report came out in 1978 [2]. The VCSEL lasing was demonstrated

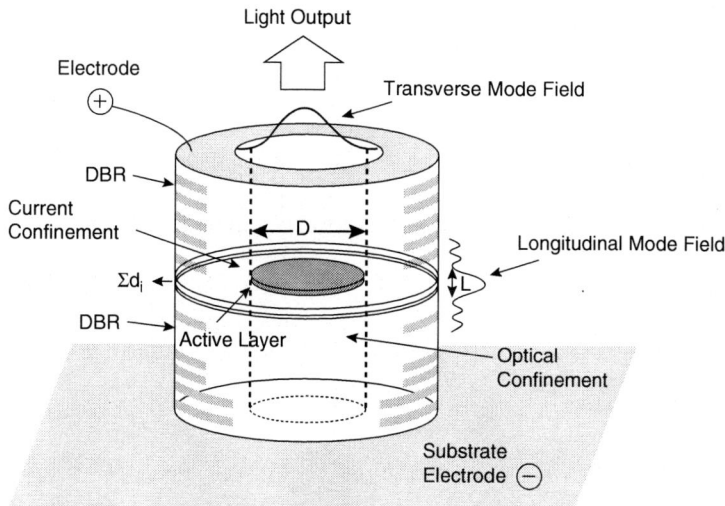

Figure 1 A model of a VCSEL

in 1979, where we used a GaInAsP/InP material for the active region, emitting 1300 nm wavelength light [3]. In 1986, we made a 6mA threshold GaAs device [4]. Then we employed metal organic chemical-vapor deposition (MOCVD) for its crystal growth and the first room temperature continuous wave (CW) laser, using GaAs material was demonstrated in 1988 [5]. Later, in 1989, Jewell and coworkers demonstrated a GaInAs VCSEL exhibiting a 2 mA threshold device [6].

Since 1992, VCSELs based on GaAs have been extensively studied. Some devices exhibiting sub-milliampere thresholds were demonstrated by improving the quality of the active region and the laser cavity [7–9]. Some 980, 850, and 780 nm devices were commercialized and utilized in various optical systems. In 1993, the author's group demonstrated a 1300 nm room temperature CW device [10]. A wafer fusion technique enabled us to operate 1550 nm VCSELs at higher temperatures [11]. In 1993, room temperature high performance CW red color AlGaAs/ GaAs [12] and AlGaInP [13] devices were demonstrated. In 1996, green–blue–ultraviolet device research started [14]. Since 1999, VCSEL-based optical tranceivers have been introduced into Giga-bit/sec Ethernet and high-speed local area networks. The VCSEL is being applied in various optical systems such as optical fiber networks, parallel optical interconnects, laser printers, and high density optical disks [15].

BASIC PRINCIPLE AND SCALING LAWS

Threshold Current

The physical difference between VCSELs and conventional stripe geometry lasers is summarized in Table 1. Of importance here is the cavity length, which for VCSELs is in the order of the wavelength, whereas that of stripe lasers is about 300 μm. The aforementioned structures provide the substantial differences in laser performances [16].

Table 1
Comparison of parameters between stripe lasers and vertical cavity surface emitting lasers

Parameter	Symbol	Stripe Laser	Surface Emitting Laser
Active layer thickness	d	100 Å -0.1 μm	80 Å -0.5 μm
Active layer area	S	3 x 300 μm^2	5 x 5 μm^2
Active volume	V	60 μm^3	0.07 μm^3
Cavity length	L	300 μm	1 μm
Reflectivity	R_m	0.3	0.99–0.999
Optical confinement		3%	4%
Optical confinement (Transverse)	ξ_t	3-5%	50-80%
Optical confinement (Longitudinal)	ξ_l	50%	2x1%x3 (3 QW'S)
Photon lifetime	τ_p	1 psec	1 psec
Relaxation frequency (Low current levels)	f_r	5 GHz	10 GHz

The threshold current I_{th} of surface emitting lasers can be expressed by an equation in the terms of carrier density N_{th} needed for laser oscillation:

$$I_{th} \cong \frac{eB_{eff}}{\eta_i \eta_{spon}} N_{th}^2 V, \tag{1}$$

where e is electron charge, V is the volume of active region, B_{eff} is effective recombination coefficient, η_i is injection efficiency (sometimes referred to as internal efficiency), and η_{spon} is the spontaneous emission efficiency.

As seen from Eq. (1), it is essential to reduce the volume of the active region in order to decrease the threshold current. While assuming that the threshold carrier density does not change significantly, we can decrease the threshold as we can make a small active region. The dimensions of surface emitting and conventional stripe geometry lasers can be compared as shown in Table 1. It is noticeable that the volume of VCSELs can be $V = 0.06 \ \mu m^3$, whereas that for stripe lasers remains $V = 60 \ \mu m^3$. This directly reflects the threshold currents; that is, a typical threshold for stripe lasers is in the milliampere region or higher, but that for VCSELs is far below the milliampere level. It can even be as low as the microampere level by implementing sophisticated carrier and optical confinement structures, as will be discussed later.

The simple estimation discussed earlier has shown that the threshold current can be reduced proportional to the square of the active region diameter. However, there is a minimum value originating from the decrease of the optical confinement factor, which is defined by the overlap of the optical mode field and the gain region when the diameter is reducing. In addition to this, extreme minimization of volume, in particular in the lateral direction, is limited by the optical and carrier losses as a result of optical scattering, lightwave diffraction, nonradiative carrier recombination, and other technical imperfections.

Output Power and Quantum Efficiency

If we use a nonabsorbing mirror for the front reflector, the differential quantum efficiency η_d from the front mirror is expressed as

$$\eta_d = \eta_i \frac{(1/L)\ln(1/R_f)}{\alpha + (1/L)\ln(1/\sqrt{R_f R_r})}. \tag{2}$$

where α is the total internal loss, and R_f and R_r are front and rear mirror reflectivity.

The optical power output is expressed by

$$P_o = \eta_d \eta_{spon} CE_g I \qquad (I \leq I_{th})$$

$$= \eta_d \eta_{spon} CE_g I_{th} + \eta_d E_g (I - I_{th}) \qquad (I \geq I_{th}), \tag{3}$$

where E_g is the bandgap energy, C is the spontaneous emission factor, and I is the driving current. On the other hand, the power conversion efficiency η_P far above the threshold is given by

$$\eta_P = \frac{P_o}{V_b I} = \eta_d \frac{E_g}{V_b} (1 - \frac{I_{th}}{I}) \cong \eta_d \frac{E_g}{V_b} \cong \eta_d, \tag{4}$$

where V_b is a bias voltage and the spontaneous component has been neglected. In the case of a surface emitting laser, the threshold current can be very small, and therefore the power conversion efficiency can be relatively large, that is, higher than 50%. The power conversion efficiency is sometimes called the wall-plug efficiency. However, at the high injection levels of actual devices, the heat generated in or near the active region forces the threshold to increase, and the power output to saturate, showing a so-called roll-off in current-output characteristic.

Modulation Characteristics

The modulation bandwidth is given by

$$f_{3dB} \cong 1.55 f_r,$$
(5)

where f_r denotes the relaxation frequency, which is expressed by the equation

$$f_r = \frac{1}{2\pi\tau_s} \sqrt{\frac{\tau_s}{\tau_p}\left(\frac{I}{I_{th}} - 1\right)}.$$
(6)

The photon lifetime τ_p is given by

$$\tau_p = \frac{n_{eff}/c}{\alpha + \alpha_m}.$$
(7)

When the threshold current I_{th} is negligible compared with the driving current I, f_r can be expressed as

$$f_r \cong \frac{1}{2\pi\tau_s} \sqrt{\frac{\tau_s}{\tau_p} \frac{I}{I_{th}}}$$

$$= \frac{1}{2\pi\tau_s} \sqrt{\frac{\tau_s}{\tau_p} \frac{\eta_i \eta_{spon} I}{eB_{eff} N_{th}^2 V}}.$$
(8)

The relaxation frequency is inversely proportional to the square root of the active volume and can be larger, if the volume is reduced to a minimum.

The photon lifetime is normally in the order of pico-seconds, which can be made slightly smaller than for stripe lasers. Since the threshold current can be very small in VCSELs, the relaxation frequency can be relatively higher than that of stripe lasers even in low driving ranges. The threshold carrier density N_{th} can be expressed in terms of photon lifetime, which represents the cavity loss, and is given using Eqs. (7) and (3) as follows:

$$N_{th} = N_t + \frac{1}{(c/n_{eff})} \frac{1}{(dg/dN)} \frac{1}{\xi} \frac{1}{\tau_p}.$$
(9)

It is noted that the threshold carrier density can be small when we make the differential gain dg/dN, confinement factor ξ, and photon lifetime τ_p large.

The tuning wavelength bandwidth $\Delta\lambda$ of a semiconductor laser is determined by its free spectral range (FSR):

$$\frac{\Delta\lambda}{\lambda} = \frac{(\lambda/2n_{eq})}{L}.$$
(10)

This is inversely proportional to cavity length L, where n_{eq} is the equivalent refractive index of the resonant mode. This means that in a VCSEL it can be as large as almost one wavelength, and that a very wide continuous tuning range is available.

Transverse and Longitudinal Mode

The resonant mode in most surface emitting lasers can be expressed by the well-known Fabry–Perot TEM mode. The near-field pattern (NFP) of the fundamental mode can be given by the Gaussian function

$$E = E_0 \exp\left[-\tfrac{1}{2}(r/s)^2\right], \tag{11}$$

where E is the optical field, r is the lateral distance, and s denotes the spotsize.

The spotsize of normal surface emitting lasers is several microns and is relatively large compared with stripe lasers, for which the spotsize is, say, 2 to 3 μm. In the case of multi-mode operation, the mode behaves like the combination of multiple TEM_{pq}. The associated spectrum is broadened as a result of different resonant wavelengths.

The far-field pattern (FFP) associated with the Gaussian near field can be expressed by a Gaussian function and diffraction angle $\Delta\theta$, and is given by

$$2\Delta\theta = 0.64(\lambda/2s). \tag{12}$$

Here, if $s = 3$ μm and $\lambda = 1$ μm, $\Delta\theta = 0.05$(rad) $\cong 3°$. This kind of angle is narrower than that of conventional stripe lasers.

Polarization Mode

A VCSEL has a linear polarization. This is due to a small amount of asymmetric loss coming from the shape of the device or material. The device grown on a (100)-oriented substrate polarizes in (110) or equivalent orientations. The direction may not be definitely identified and sometimes switches over as a result of spatial hole burning or temperature variation. The switching occurs with relatively high speed (pico-seconds or less). In order to stabilize the polarization mode, some special care should be taken. This will be discussed later.

DEVICE STRUCTURES AND DESIGN

Device Configuration

As shown in Figure 2, the structure common to most of the VCSELs consists of two parallel reflectors, which sandwich a thin active layer. We also show the standing wave of resonant mode. It is clear that the maximum of the standing wave can be allocated exactly to the position of the active layers consisting of quantum wells. The reflectivity necessary to reach the lasing threshold should normally be higher than 99.9%. Together with the optical cavity formation, the scheme for injecting electrons and holes effectively into small volume of active region is necessary for current injection device. The ultimate threshold current depends on how to make the active volume small as introduced in the previous section and how well the optical field can be

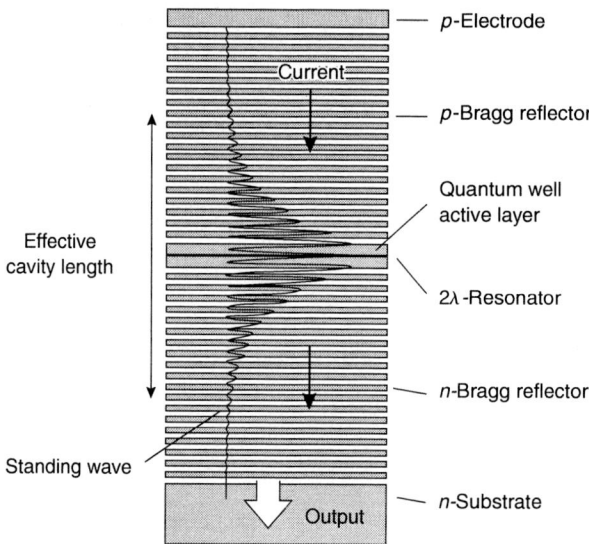

Figure 2   A resonant cavity consisting of multi-layered distributed Bragg reflectors

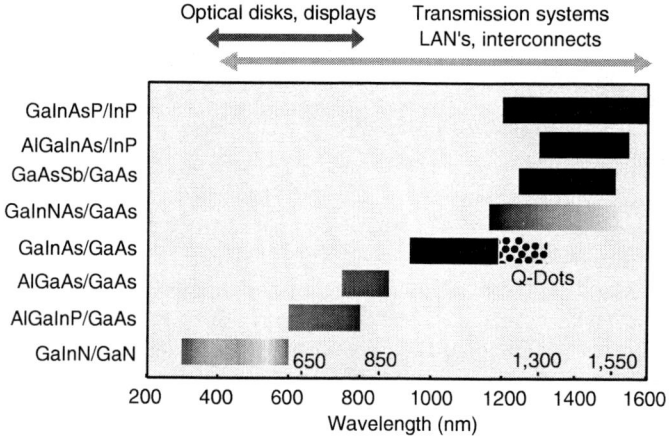

Figure 3   Semiconductor materials for vertical cavity surface emitting lasers in various wavelength bands

confined in the cavity to maximize the overlap with active region. These confinement structures will be presented in latter sections.

Materials

Figure 3 shows the possible materials for VCSELs. Listed below are the problems that should be considered while making vertical cavity VCSELs, discussed in the previous section.

1.   Design of resonant cavity and mode-gain matching.
2.   Multi-layered distributed Bragg reflectors (DBRs) to realize high reflective mirrors.
3.   Optical losses such as free carrier absorption, Auger recombination, intervalence band absorption, scattering loss, and diffraction loss.

4. *p*-type doping to reduce the resistivity in *p*-type materials for CW and high efficiency operation. If we wish to form multi-layer DBRs, then this will become even more severe.
5. Heat sinking for high temperature and high power operation.
6. COD (Catastrophic Optical Damage) level is very important for high power operation.
7. Crystal growth at reasonably high temperatures (e.g., higher than half of melting temperatures).

Current Injection Scheme

Let us consider the current confinement for VCSELs. We show some typical models of current confinement schemes reported so far.

1. *Ring electrode type*: This structure can limit the current flow in the vicinity of ring electrode. The light output can be taken out from the center window. This is easy to fabricate, but the current cannot completely be confined in a small area due to diffusion.
2. *Proton bombardment type*: We make an insulating layer by proton (H+) irradiation to limit the current spreading toward the surrounding area. The process is rather simple and most of the commercialized devices are made by this method.
3. *Buried-heterostructure (BH) type*: We bury the mesa including active region with wide-gap semiconductor to limit the current. The refractive index can be small in the surrounding region, resulting in the formation of an index-guiding structure. This is one of the ideal structure in terms of current and optical confinement. The problem is that the necessary process is rather complicated, in particular, in making a tiny 3D device.
4. *Air-post type*: The circular or rectangular air-post is used to make a current confinement. The simplest way of device fabrication, but nonradiative recombination at the outer wall may deteriorate the performance.
5. *Selective AlAs oxidation type*: We oxidize AlAs layer to make a transparent insulator.
6. *Oxidized DBR type*: The same method is applied to oxidize DBR consisting of AlAs and GaAs. This is one of volume confinement methods and can reduce the nonradiative recombination.

    By developing fine process technology we could reach the laser performance, which is expected from the theoretical limit.

Optical Guiding

Some optical confinement schemes were developed for VCSELs. The fundamental concept is to increase the overlap of optical field with gain region.

1. *Fabry–Perot type*: The optical resonant field is determined by two reflectors that form a plane parallel to the Fabry–Perot cavity. The diffraction loss increases, if the mirror diameter reduces.
2. *Gain-guide type*: We simply limit the field at the region where the gain exists. The mode may be changed at high-injection levels due to spatial hole burning.
3. *BH type*: As introduced in the previous section, an ideal index guiding can be formed.
4. *Selective AlAs oxidation type*: Due to the index difference between AlAs and oxidized region, we can confine the optical field as well by a kind of lens-effect.
5. *Anti-guiding type*: The index is designed to be lower in the surrounding region to make a so-called anti-guiding scheme. The threshold is rather high, but this structure is good for stable mode in high driving levels.

ADVANTAGES OF VCSEL

The initial motivation of surface-emitting laser invention was fully monolithic fabrication of laser cavity. The current issues include, based on this concept, high speed modulation capability at very low power consumption level, reproducible array production, inexpensive moduling, and so on. The VCSEL structure may provide a number of advantages as listed below:

1. Laser devices can be fabricated by a fully monolithic production with a very high yield.
2. Laser cavity can be completed before separation into individual chips enabling wafer level testing.
3. Ultra-low threshold operation is expected from its small cavity volume reaching micro-ampere levels.
4. Dynamic single mode operation is possible at high driving levels.
5. High speed modulation beyond 10 GBits/sec is possible even at low driving ranges.
6. Wide and continuous wavelength tuning is possible.
7. Temperature independent operation is allowable, which yields no power controller operation.
8. Large power-conversion efficiency, that is, >50%.
9. High power and low power devices are subject to design.
10. High device reliability due to completely embedded active region and passivated surfaces.
11. Vertical and circular beam is inherently provided.
12. Easy coupling to optical fibers due to good mode matching from single mode through thick multi-mode fibers.
13. Easy bonding and mounting.
14. Simple modules and package.
15. Densely packed and precisely arranged two-dimensional laser arrays can be formed.
16. Vertical stack integration of multi-thin-film functional optical devices can be made intact to VCSEL resonator, taking the advantage of micro-machining technology (MEMS) providing polarization independent characteristics.
17. Compatible integration together with LSIs.

DEVICE TECHNOLOGY AND PERFORMANCES

In this section, we introduce some representative technologies for VCSEL devices. The original papers may be referred from References 15 and 16. The mostly produced VCSEL devices are from 850 nm wavelength band. In Figure 4, we show a typical device structure of a GaAs/AlGaAs material. The advantage of this system is that we can grow a high-quality quantum-well active layer, and low loss and highly conductive distributed Bragg reflector composed of GaAs/AlGaAs. The selective oxidation of AlAs for current confinement and matured process technology are available here. In Figure 5, we show a typical current–light-output characteristic. Low sub-milliampere thresholds and > 10 mW outputs have been achieved. In extreme cases, the power conversion efficiency of > 50% has been demonstrated. As for the reliability of VCSELs, 10^7 h of room temperature operation is estimated. Life test of oxide-defined devices exhibited higher reliability. Some test results were reported on oxide-defined devices exhibiting no substantial negative failures.

The Gigabit Ethernet has already been in large markets by the use of multimode-fiber-based optical links. This system is being extended to one of 10 Gbits/sec Ethernet standards. The high-speed modulation capability of VCSELs at low driving level is good for low power interconnect applications enabling >10 Gbits/sec transmission or >1 Gbits/sec zero-bias operation. Actually, transmission experiments over 10Gbits/s and zero-bias transmission have been reported.

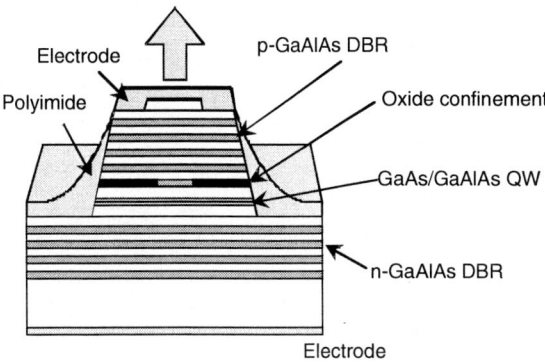

Figure 4 A device structure of a GaAs/GaAlAs VCSEL

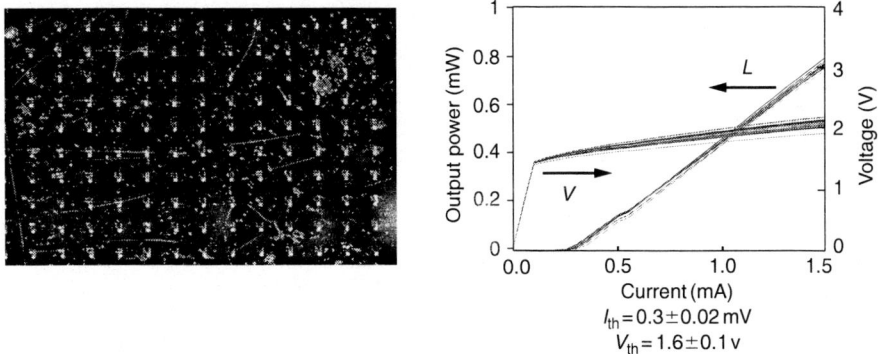

Figure 5 A typical current-output characteristic of a GaAs/GaAlAs VCSEL

The importance of 1300 or 1550 nm devices is recognized for local area networks (LANs) and metropolitan area networks (MAN or METRO). The long wavelength device (1300 nm) is developed for this purpose [17]. The engineered device became available in 2004. A 1550 nm VCSEL with an MEMS tunable mechanism was introduced into a high-end metropolitan area system. Other important technology is to realize long wavelength emitters that can be formed on a GaAs substrate as shown in Figure 6. The highly strained GaInAs/GaAs emitting 980 to 1200 nm spectra is one of the candidates.

Another viable candidate is a GaInNAs system, that also can provide excellent temperature stability and may substantially change the device map in long wavelength region [18,19]. It is found that temperature dependence upon threshold and quantum efficiency could be removed by properly designing the device structure and material. The polarization control technology in VCSELs has been established by using (311)B GaAs substrate as shown in Figure 7. The orthogonal polarization suppression ratio (OPSR) of over 30 dB was obtained even in high-speed modulation condition [20]. The current versus light-output characteristics are shown in Figure 8. A temperature insensitive performance has been achieved.

The VCSEL in 780 nm wavelength can be provided by $Al_xGa_{1-x}As/Al_xGa_{1-x}As$ system with different composition x of Al [12]. We show the design below. If we choose the Al content x to be 0.14 for $Ga_{1-x}Al_xAs$, the wavelength can be as short as 780 nm. The active layer $Ga_{0.86}Al_{0.14}As$ is formed by a super-lattice consisting of GaAs(33.9 Å), and AlAs(5.7 Å), with 14 periods.

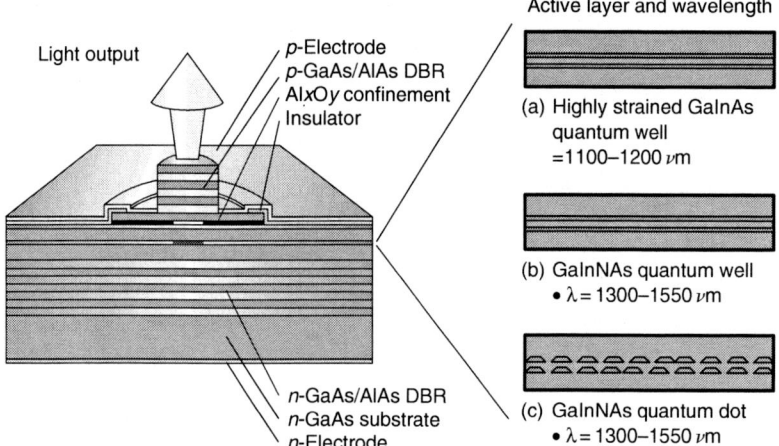

Figure 6 Candidates of long wavelength VCSELs formable on GaAs substrate (After T. Miyamoto, unpublished)

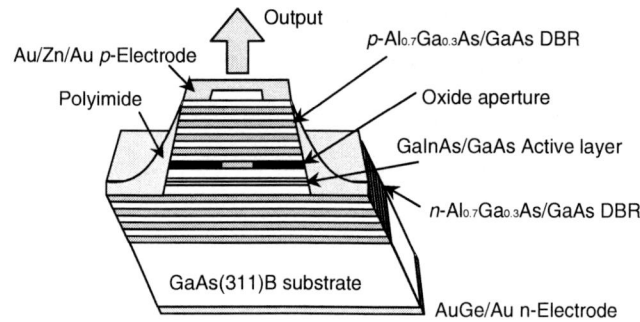

Figure 7 A GaInAs/GaAs VCSEL grown on GaAs (311)B substrate Nobuhiko Nishiyama, Masakaza Arai, Satashi Shinada, koichi Suzuki, Fumio Koyama, and Kenichi Iga, *IEEE Photonics Technical Lett.*, 12, 606–608, 2000. With permission.)

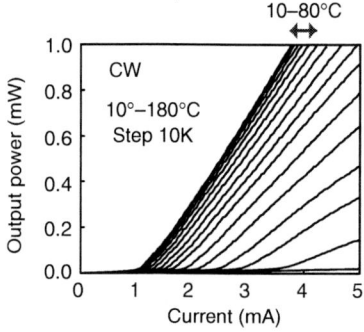

Figure 8 Current versus Light-output characteristics exhibiting single mode and single polarization oscillation Nobuhiko Nishiyama, Masakaza Arai, Satashi Shinada, Koichi Suzuki, Fumio Koyama, and Kenichi Iga, *IEEE Photonics Technical Lett.*, 12, 606–608, 2000. With permission.)

The DBR is made of $AlAs-Al_{0.35}Ga_{0.65}As-Al_{0.3}Ga_{0.7}As-Al_{0.35}Ga_{0.65}As$ as 1 period. The n-DBR has 28.5 pair and p-DBR consists of 22 pairs. The threshold is around low sub-mA and the output was a few milliwatts.

The red color VCSEL emitting 650 nm [13] can match to the low-loss band of plastic fibers. Short-distance data links are considered by using 1 mm diameter plastic fibers having graded-index have been developed. This system provides us with a very easy optical coupling and the VCSEL can very nicely match to this application. Green to UV, VCSELs will be useful in the optoelectronics field as in ultra-high density optical memories [14,21]. The present author proposed a model of optical pickup using VCSEL. Near-field optics scheme is considered to realize high-density optical memories — full color flat displays and large area projectors, illuminations and light-signals, light decorations, UV-lithography, laser processes, medical treatment, and so on.

ARRAYS AND INTEGRATION

The appropriately designed VCSEL for telecom applications can now provide single wavelength, single transverse mode, and stable polarization operation. A wide variety of functions, such as frequency tuning, amplification, and filtering should be integrated.

Moreover, a 2D parallel optical logic system can deal with a large amount of image information with high speed. Micro-machining technology and MEMS will be very helpful. Light beam manipulation technology using MEMS scheme will open up a new field of optical switching and routing in large-scale optical exchanging systems.

A densely packed array has also been demonstrated for the purpose of making high power lasers and coherent arrays. For the purpose of realizing coherent arrays, coherent coupling of these arrayed lasers has been tried by using a Talbot cavity and phase compensation. A wide variety of functions, such as frequency tuning, amplification, and filtering can be integrated along with surface emitting lasers by stacking. Parallel photonic devices and optical subsystems are now being developed and the application areas of these devices are broadened toward realization of high speed LANs, optical interconnects, optical links, optical sensing, and so on. The ultra-parallel and ultra-high speed photonics will also open-up a new era of industry [22]. The possible applications are summarized in Table 2.

Table 2
Applications of vertical cavity surface emitting lasers

Technical fields	Systems
Optical communications	LANs, metropolitan networks, optical links, mobile links, etc.
Computer optics	Computer links, optical interconnects, high speed/parallel data transfer, etc.
Optical memory	CD, DVD, near field, multi-beam, initializer, etc.
Optoelectronic equipments	Printer/copier, laser pointer, mobile tools, home appliances, etc.
Optical information processing	Optical processors, parallel processing, etc.
Optical sensing	Optical fiber sensing, bar code readers, Encoders, etc.
Displays	Array light sources, high efficiency light-sources
Illuminations	Multi-beam search-lights, micro illuminators, Adjustable illuminations, etc.

SUMMARY

The VCSEL is a key laser device in optical high speed networks by taking the advantage of low-power consumption and high-speed modulation capability. This device is also enabling ultra-parallel data transfer in digital photonic equipments and computer systems.

ACKNOWLEDGMENTS

The author would like to thank Prof. Yasuharu Suematsu for his encouragement to research and Prof. Fumio Koyama and laboratory members at Tokyo Institute of Technology for collaborations.

REFERENCES

1. K. Iga, "Surface Emitting Laser," Laboratory Notebook, March 22 (1977).
2. K. Iga, T. Kambayashi, and C. Kitahara, "GaInAsP/InP surface emitting laser (I)," in *Proceedings of the 26th Spring Meeting of Applied Physics Societies*, 27p-C-1 1, 1978.
3. H. Soda, K. Iga, C. Kitahara, and Y. Suematsu, "GaInAsP/InP surface emitting injection lasers," *Jpn. J. Appl. Phys.*, 18, 2329–2330, 1979.
4. Kenichi Iga, Susumu Kinoshita, and Fumio Koyama, "Microcavity GaAlAs/GaAs surface emitting laser with $I_{th} = 6$ mA," in *Proceedings of the 10th International Semiconductor Laser Conference (ISLC'86)*, paper PD-4(post-deadline), Oct. 1986. *Electron. Lett.*, 23, 134–136, 1987.
5. F. Koyama, S. Kinoshita, and K. Iga, "Room temperature cw operation of GaAs vertical cavity surface emitting laser," *Trans. IEICE*, E71, 1089–1090, 1988.
6. J. L. Jewell, S. L. McCall, A. Scherer, H. H. Houh, N. A. Whitaker, A. C. Gossard, and J. H. English, "Transverse modes, waveguide dispersion and 30 ps recovery in submicron GaAs/AlAs microresonators," *Appl. Phys. Lett.*, 55, 22–24, 1989.
7. R. S. Geels and L. A. Coldren, "Sub-milliamp threshold vertical-cavity laser diodes," *Appl. Phys. Lett.*, 57, 1605–1607, 1991.
8. T. Wipiejewski, K. Panzlaf, E. Zeeb, and K. J. Ebeling, "Submilliamp vertical cavity laser diode structure with 2.2 nm continuous tuning," in *Proceedings of the 18th European Conference on Optical Communication* (ECOC'92), no. PDII-4, 1992.
9. D. L. Huffaker, J. Shin, and D.G. Deppe, "Low threshold half-wave vertical-cavity lasers," *Electron. Lett.*, 31, 1946, 1994.
10. T. Baba, Y. Yogo, K. Suzuki, F. Koyama, and K. Iga, "Near room temperature continuous wave lasing characteristics of GaInAsP/InP surface emitting laser," *Electron. Lett.*, 29, 913–914, 1993.
11. D. I. Babic, K. Streubel, R. P. Mirin, J. Pirek, N. M. Margalit, J. E. Bowers, E. L. Hu, D. E. Mars, L. Yang, and K Carey, "Room temperature performance of double-fused 1.54 μm vertical-cavity lasers," IPRM'96, no. ThA1-2 , April 1996; N. M. Margalit, D. I. Babic, K. Streubel, R. P. Mirin, R. L. Naone, J. E. Bowers, and E. L. Hu, "Submilliamp long wavelength vertical cavity lasers," *Electron. Lett.*, 32, 1675, 1996.
12. Y. H. Lee, B. Tell, K. F. Brown-Goebeler, R. E. Leibenguth, and V. D. Mattera, "Deep-red continuous wave top-surface-emitting vertical-cavity AlGaAs superlattice lasers," *IEEE Photonics Technol. Lett.*, 3, 108–109, 1991.
13. K. D. Choquette, R. P. Schneider, M. H. Crawford, K. M. Geib, and J. J. Figiel, " Continuous wave operation of 640–660 nm selectively oxidised AlGaInP vertical cavity lasers," *Electron. Lett.*, 31, 1145–1146, 1995.
14. K. Iga, "Possibility of green/blue/UV surface emitting lasers," in *Proceedings of the International Symposium on Blue Laser and Light Emitting Diodes (ISBLLED'96)*, March 1996.
15. K. Iga, "Surface emitting laser-its birth and generation of new optoelectronics field," *IEEE J. Select. Top. Quantum Electron.*, 6, 1201–1215, 2000.

16. K. Iga, F. Koyama, and S. Kinoshita, "Surface emitting semiconductor laser," *IEEE J. Quantum. Electron.*, QE-24, 1845–1855, 1988.

17. V. Jayaraman, J. C. Geske, M. H. MacDougal, F. H. Peters, T. D. Lowes, and T. T. Char, "Uniform threshold current, continuous-wave, singlemode 1300 nm vertical cavity lasers from 0 to 70°C," *Electron. Lett.*, 34, 1405–1406, 1998.

18. M. C. Larson, M. Kondow, T. Kitatani, K. Nakahara, K. Tamura, H. Inoue, and K. Uomi, "Room temperature pulsed operation of GaInNAs/GaAs long-wavelength vertical cavity lasers," *in Proceedings of the IEEE/LEOS'97*, no. PD1.3, 1997.

19. T. Kageyama, T. Miyamoto, S. Makino, Y. Ikenaga, N. Nishiyama, A. Matsutani, F. Koyama, and K. Iga, "Room temperature continuous-wave operation of GaInNAs/GaAs VCSELs grown by chemical beam epitaxy with output power exceeding 1 mW," *Electron. Lett.*, 37, 225–226, 2001.

20. N. Nishiyama, M. Arai, S. Shinada, K. Suzuki, F. Koyama, and K. Iga, "Multi-oxide layer structure for single-mode operation in vertical-cavity surface-emitting lasers," *IEEE Photonics Technol. Lett.*, 12, 606–608, 2000.

21. T. Someya, K. Tachibana, Y. Arakawa, J. Lee, and T. Kamiya, "Lasing oscillation in InGaN vertical cavity surface emitting lasers," *in Proceedings of the 16th IEEE International Semiconductor Laser Conference*, PD-1, 1998, pp. 1–2. T. Someya, K. Tachibana, J. Lee, T. Kamiya, and Y. Arakawa, "Lasing emission from an In0.1Ga0.9N vertical cavity surface emitting laser," *Jpn. J. Appl. Phys.*, 37, L1424–L1426, 1998.

22. K. Iga and F. Koyama, *Fundamentals and Application of Surface Emitting Laser*, Kyouritsu Pub. Co. Ltd., Tokyo, 1999.

W

WAVEGUIDE BENDS

Kazuhito Furuya

Curved dielectric-waveguides radiate lights to cause losses. The radiation loss of waveguide bends using a slab waveguide model is described. Substituting the electric field in z-direction $E_z = F(r) \exp(-i\nu\theta)$ into Maxwell's equations in cylindrical polar coordinates (Figure 1), the radial function $F(r)$ is determined by

$$\frac{1}{r}\frac{\partial}{\partial r}\left(r\frac{\partial F}{\partial r}\right) + \left(\omega^2\varepsilon_0\mu n^2 - \frac{\nu^2}{r^2}\right)F = 0. \tag{1}$$

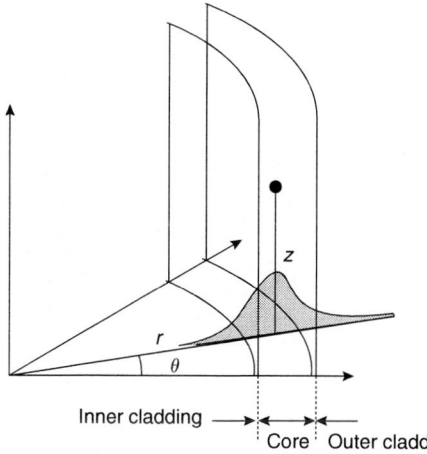

Figure 1 Bending slab dielectric waveguide in (r, θ, z) coordinate

457

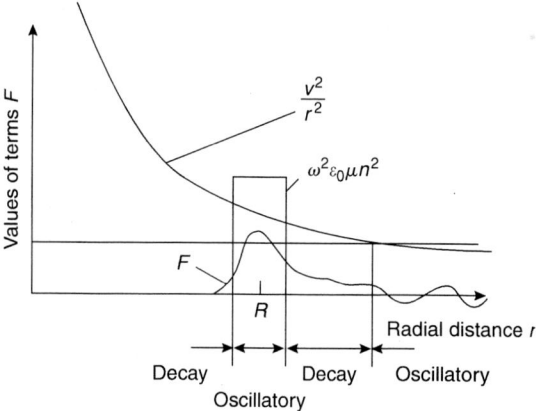

Figure 2   Radial function $F(r)$

The sign of the second term in Eq. (1) depends on the radial distance r as shown in Figure 2. When the sign is positive, the function is oscillatory. It should be noted that $F(r)$ oscillates necessarily in the region where r is large enough and the oscillation outside the core means that the eigenfunction of Eq. (1) has continuous eigenvalue and, therefore, is always in radiation mode. Strictly speaking, no guided mode exists in curved dielectric waveguides. However, in practice, $F(r)$ decays with the increase of r in the region between the outer boundary of the core, and the turning point where the sign changes. Then when the curvature radius R and the length of the region where $F(r)$ decays is large enough, the amplitude of the oscillatory function is very small to be neglected. The radiation mode can be considered substantially as the guided mode and the radiation loss depends on the radius R.

By substituting $u = k_0 n r$, $k_0 = \omega\sqrt{\varepsilon_0 \mu}$, Eq. (1) becomes the Bessel differential equation and its solution is

$$F(r) = \begin{cases} J_\nu(n_{\text{clad}}k_0 r)\exp(-i\nu\theta) & r < R-b, \\ \left\{ A J_\nu(n_{\text{core}}k_0 r) + B N_\nu(n_{\text{core}}k_0 r) \right\}\exp(-i\nu\theta) & R-b < r < R+b, \\ C H_\nu^{(2)}(n_{\text{clad}}k_0 r)\exp(-i\nu\theta) & R+b < r, \end{cases}$$

where n_{core} and n_{clad} are refractive indices of the core and the cladding, respectively, b is half of the core width. $J_\nu(u)$ is the Bessel function, $N_\nu(u)$ is the Neumann function, $H_\nu^{(2)}(u)$ is the Hankel function of the second kind. Boundary conditions determine values A, B, C, and ν Propagation constant ν is a complex number whose imaginary part represents decay constant of the mode. In this manner the loss of the curved waveguide can be calculated. Rather than the numerical calculation of ν, an approximate and analytic expression for the loss is derived as follows (see Reference 1 for the derivation):

$$2\alpha \sim F(T)\frac{1}{b^2 n_{\text{core}}k_0}\exp\left(-\frac{2n_{\text{core}}k_0 R}{3}\left(\frac{\kappa}{n_{\text{core}}k_0} \right)^3 \right) \qquad [\text{Nep./m}],$$

where $F(T)$ is determined solely by $T\left(= k_0 b\sqrt{n_{\text{core}}^2 - n_{\text{clad}}^2}\right)$ and is

$$F(T) = \frac{(\gamma b)^2 \left(T^2 - (\gamma b)^2\right)\exp\left(2\sqrt{T^2 - (\gamma b)^2}\right)}{T^2\left(1 + \sqrt{T^2 - (\gamma b)^2}\right)},$$

where γb is roots of the equation $\left|\cos(\gamma b - \phi)\right| = \gamma b / T$, where ϕ is zero for even modes while $\pi/2$ for odd modes. Numerical radiation losses, 8.6α[dB/m], are shown in Figure 3 and Figure 4.

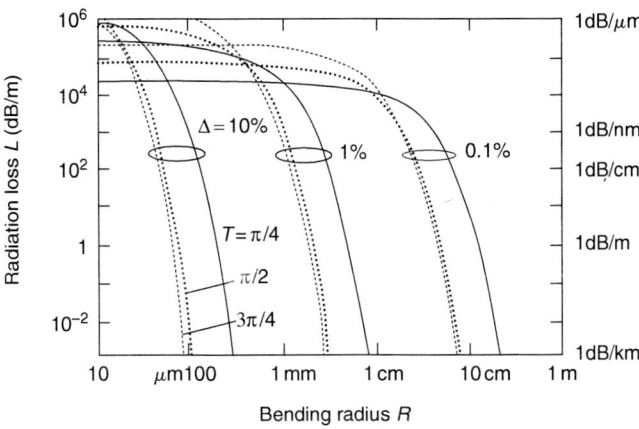

Figure 3 Radiation loss caused by bend. Refractive index of the core is 1.5 corresponding to SiO_2-based waveguides. Wavelength is 1.5 μm

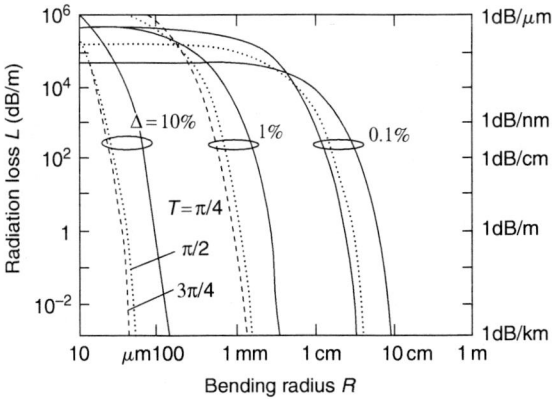

Figure 4 Radiation loss caused by bend. Refractive index of the core is 3.2 corresponding to semiconductor-based waveguides. Wavelength is 1.5 μm

From Figure 3 and Figure 4, the loss abruptly increases when the curvature radius R decrease beyond the critical radius. To suppress the bend loss <0.1 dB/km, for the single-mode fiber of the relative refractive index Δ of 0.2%, the curvature radius should be larger than 1.5 cm.

REFERENCE

1. D. Marcuse, *Light Transmission Optics*, Van Nostrand-Reinhold, New York, 1972.

WAVEGUIDE MODELING

Masanori Koshiba

Recent advances in the field of guided-wave optics, such as fiber optics and integrated optics, have included the introduction of arbitrarily shaped optical waveguides, which, in many cases, also happened to be arbitrarily inhomogeneous, dissipative, anisotropic, and nonlinear. Most of such cases of waveguide arbitrariness do not lend themselves to analytical solutions; hence, computational tools for modeling and simulation are essential for successful design, optimization, and realization of the optical waveguide devices that penetrate into many photonic circuits and systems. For this purpose, various modeling techniques have been developed, and commercialized computer-aided design (CAD) software for photonic component design has become a standard in the industry.

Here, modeling techniques that are in widespread used for the analysis of optical waveguide devices are summarized. Modeling techniques for optical guided-wave propagation could be classified into two groups: time-harmonic (monochromatic CW operation) and time-dependent (pulsed operation) modelings. The most fundamental guiding properties of optical waveguides are expressed in terms of the waveguide modes that can propagate in longitudinally invariant structures as shown in Figure 1(a).

First, the effective index method, the finite difference method, and the finite element method, which are the most important methods employed in the modal analysis, are reviewed. In longitudinally variant structures as shown in Figure 1(b), the guided modes are usually coupled to the

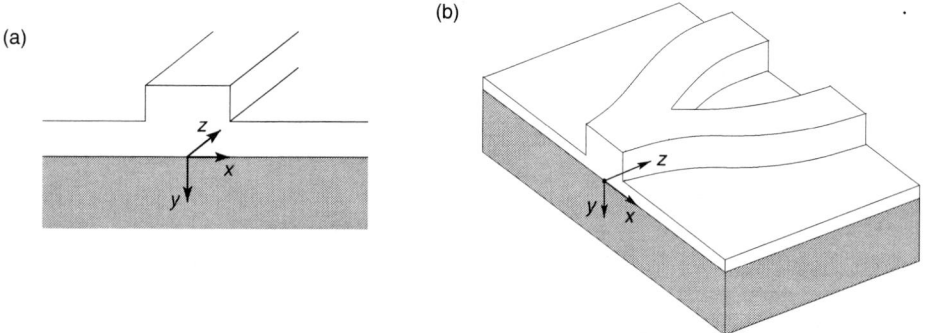

Figure 1 Longitudinally (a) invariant and (b) variant optical waveguides

radiation modes. The beam propagation method, in which the mode coupling and conversion phenomena are automatically included, is then introduced. Lastly, the finite-difference time-domain method and the time-domain beam propagation method, which are applicable to both CW and pulsed operations, are briefly reviewed. Details of numerical implementation can be found in the corresponding references [1–30] and therein.

TIME-HARMONIC MODELING

Basic Equations

We consider a three-dimensional (3D), channel optical waveguide as shown in Figure 1, where x and y are the transverse directions and z is the propagation direction. With a time dependence of the form $\exp(j\omega t)$ being implied, from Maxwell's equations the following vectorial Helmholtz equation is derived:

$$\nabla \times (p\nabla \times \mathbf{\Phi}) - k_0^2 q\mathbf{\Phi} = 0, \tag{1}$$

where ω is the angular frequency, k_0 is the free-space wavenumber, $\mathbf{\Phi}$ denotes either the electric field $\mathbf{E}$ or the magnetic field $\mathbf{H}$, p and q are given by

$$p = 1, \qquad q = n^2 \tag{2}$$

for $\mathbf{\Phi} = \mathbf{E}$, and

$$p = 1/n^2, \qquad q = 1 \tag{3}$$

for $\mathbf{\Phi} = \mathbf{H}$, and n is the refractive index.

An optical waveguide shown in Figure 1 will support the propagation of waves having two possible field configurations, classified as the x-polarized, TE (transverse electric)-like (quasi-TE) and the y-polarized, TM (transverse magnetic)-like (quasi-TM) modes. The main field components of the TE-like modes are E_x and H_y, while those of the TM-like modes are E_y and H_x. Noting that the TE-like and TM-like modes are well approximated by assuming $E_y = 0$ and $H_y = 0$, respectively, the following approximate scalar Helmholtz equation is derived:

$$p\nabla^2\mathbf{\Phi} + k_0^2 q\mathbf{\Phi} = 0, \tag{4}$$

where $\mathbf{\Phi}$ denotes E_x and H_x for the TE-like and TM-like modes, respectively. The remaining field components can be expressed with E_x for the TE-like mode and with H_x for the TM-like mode.

Effective Index Method

The effective index method (EIM) [1] is based on the scalar wave approximation and the field in Eq. (4) is expressed as

$$\Phi(x, y, z) = \phi(x, y)\exp(-j\beta z), \tag{5}$$

where β is the propagation constant of the guided mode. In the EIM, first, a 3D optical waveguide is divided into regions based on structural differences as shown in Figure 2(a). Each region is then considered a two-dimensional (2D), planar waveguide as shown in Figure 2(b). The effective index n_{eff} in each region can be easily obtained by solving the transcendental equation for the 2D waveguide uniform in the x direction. Then, with the effective indexes in the whole regions, the original structure in Figure 2(a) is modeled as a 2D waveguide as shown in Figure 2(c). Finally, the propagation constant β is calculated from the transcendental equation for the 2D waveguide uniform in the y direction.

Although this method is in principle scalar, it enables us to take the polarization effect into account. In the TE-like modes, noting that the main fields are E_x and H_y, first, the effective index n_{eff} is calculated for the TE polarization in Figure 2(b), and then, the propagation constant β is calculated for the TM polarization in Figure 2(c). In the TM-like modes, on the other hand, noting that the main fields are E_y and H_x, first, the effective index n_{eff} is calculated for the TM polarization in Figure 2(b), and then, the propagation constant β is calculated for the TE polarization in Figure 2(c).

The EIM is effective for a waveguiding structure that is slowly varying in the x-direction. The largest errors are for the modes near the cut-off. As the modal confinement increases, the error of gradually reduces to negligibly small level.

Finite Difference Method

In the finite difference method (FDM) [1,2], a waveguide cross-section is divided with a rectangular grid of points which may be of constant or variable spacing, and the Helmholtz equation is discretized by the usual five-point Taylor series formula. Making use of the finite difference representation for the derivatives of $\phi(x,y)$ and of the boundary conditions, we obtain the equation relating the field value in a grid point to the field values in the adjacent points. Assembling these relations for all the grid points, a matrix eigenvalue problem is derived. Sparse matrix routines may be used to obtain the eigenvalues β^2 and the corresponding eigenvectors $\phi(x,y)$.

A waveguide cross-section is usually defined with finite dimensions by enclosing it in a rectangular box and assuming Dirichlet or Neumann conditions on its boundaries, although the absorbing boundary condition (ABC) [3] or perfectly matched layer (PML) boundary condition [4] may also be introduced. The materials in these artificial layers are lossy, and therefore, the matrices become complex, leading to complex eigenvalues, which require more sophisticated numerical techniques.

The errors of FDM are basically implied by two approximations. One is the finite difference approximation for the Helmholtz equation. The other is the finite computational-window size.

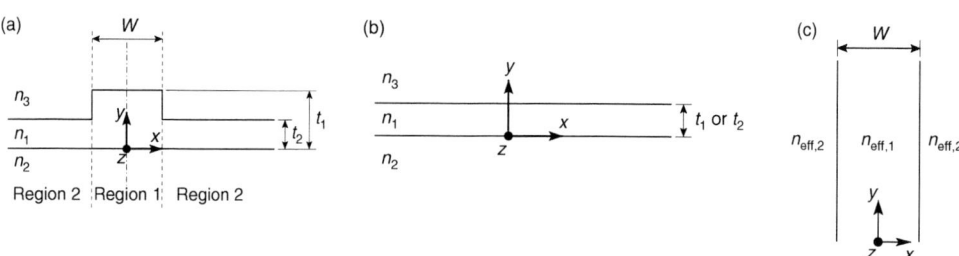

Figure 2 Cross-sections of (a) 3D waveguide, (b) 2D waveguide uniform in the x-direction, and (c) 2D waveguide uniform in the y-direction

To reduce the errors, we need to take enough grid points and to put the computational-window edges far enough from the waveguide core.

For the full-vector analysis, the finite difference approximation is applied to Eq. (1) and the electromagnetic field Φ is expressed as

$$\Phi(x,y,z) = [\phi_x(x,y)i_x + \phi_y(x,y)i_y + \phi_z(x,y)i_z]\exp(-j\beta z), \tag{6}$$

where i_x, i_y and i_z are, respectively, the unit vectors in the x-, y-, and z-directions.

Finite Element Method

In the finite element method (FEM) [2,5], instead of solving the Helmholtz equation (1) or (4), the corresponding functional to which a variational method is applied is set up, where the waveguide cross-section is divided into the so-called elements, an equivalent discretized model for each element is constructed, and then all the element contributions to the whole waveguide cross-section are assembled, resulting in a matrix eigenvalue problem with nodal variables as unknowns. The matrices derived from the FEM are sparse as in the FDM. Elements can have various shapes, such as triangles and rectangles, allowing the modeling of optical waveguides with arbitrarily shaped cross-section.

The most serious difficulty in applying the FEM with conventional nodal elements to the vectorial Helmholtz equation is the appearance of the so-called spurious, nonphysical solutions. The spurious solutions can be eliminated by using the recently developed edge elements impos-ing the continuity of the tangential components of the electric and magnetic fields. Especially, a curvilinear hybrid edge/nodal element [6] is very useful for avoiding the spurious solutions and for modeling curved boundaries with more accuracy and lesser number of degrees of free-dom than rectilinear elements, resulting in low computational cost. For the axial field, ϕ_z, a nodal element with six variables as shown in Figure 3(a), ϕ_{z1} to ϕ_{z6}, is employed, while for the trans-verse fields, ϕ_x and ϕ_y, an edge element with eight variables as shown in Figure 3(b), ϕ_{t1} to ϕ_{t8}, is employed, resulting in significantly fast convergence of solutions. The FEM is useful not only for idealized-model simulations but also for real-model simulations based on actual waveguide structures, and recently, has been applied to modeling fabricated photonic crystal fibers [7].

The accuracy of FEM depends on a proper discretization of the analysis domain. In general, as the mesh is refined, the accuracy of numerical solutions is better but, at the same time, its cost also goes up. Excessive or inefficient refinement of the analysis domain wastes computer resources, and therefore, automatic procedures for adaptive mesh generation are highly desired [8]. In the adaptive mesh generation, individual element weights are, in essence, used to decide which elements should be refined. The procedure continues until a stopping criterion is met and yields a final mesh with local weight (also called local errors) of the same order of magnitude.

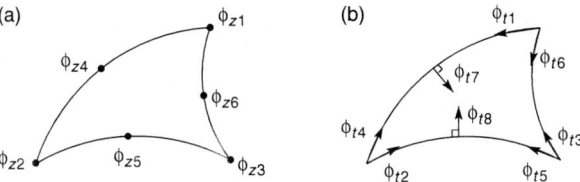

Figure 3   Curvilinear (a) nodal and (b) edge elements

Beam Propagation Method

The beam propagation method (BPM) [1–3,9] is the most widely used propagation technique for modeling integrated and fiber-optic photonic devices, and most commercial software for such modeling is based on it. The BPM automatically includes the effects of both guided and radiation fields as well as mode coupling and conversion.

In the BPM, noting that in typical guided-wave problems, the most rapid variation in the field $\mathbf{\Phi}$ or Φ is the phase variation due to the propagation and assuming that the propagation is predominantly along the z-direction, the so-called slowly varying field $\mathbf{\phi}$ or ϕ is introduced as

$$\mathbf{\Phi}(x, y, z) = \mathbf{\phi}(x, y, z) \exp(-jk_0 n_0 z) \tag{7}$$

$$\Phi(x, y, z) = \phi(x, y, z) \exp(-jk_0 n_0 z), \tag{8}$$

where n_0 is the reference refractive index to be chosen to present the average phase variation of the field. Substituting Eqs. (7) and (8) into Eqs. (1) and (4), respectively, we obtain the equation for the slowly varying field $\mathbf{\phi}$ or ϕ, in which both the first and the second derivative terms in z are included. Now, considering that the variation of $\mathbf{\phi}$ or ϕ with z is sufficiently slow and using the slowly varying envelope approximation (SVEA) that the second derivative terms $\partial^2 \mathbf{\phi} / \partial z^2$ and $\partial^2 \phi / \partial z^2$ can be neglected with respect to the first derivative terms $\partial \mathbf{\phi} / \partial z$ and $\partial \phi / \partial z$ (Fresnel or paraxial approximation), the basic equation for the beam propagation analysis is derived.

Standard version of the BPM based on the fast Fourier transform (FFT–BPM) [10] splits the actual wave propagation into a propagation in a homogeneous medium, followed by a phase correction corresponding to the inhomogeneous index distribution of the waveguide structure (split-step algorithm). The wave propagation in a homogeneous medium is efficiently carried out in the spectrum domain using FFT techniques. However, the FFT-BPM cannot be applied to strongly guiding waveguides and cannot take into account a polarization effect. To overcome these difficulties, BPMs based on FDM (FD-BPM) or FEM (FE-BPM) are also developed, where the waveguide cross-section is discretized with FDM or FEM, and for the propagation direction, as usual, the Crank–Nicholson algorithm is adopted.

By the way, conventional BPMs based on the Fresnel or paraxial approximation limit consideration to fields which propagate primarily along the z-axis. To treat fields propagating at larger angles with respect to the z-axis, the so-called wide-angle BPMs based on the Pade approximation [11] and the two-step finite-difference scheme [12] to solve the second derivative dependence on the propagation direction are also developed.

One of the key issues in implementing BPM to study light propagation in finite spatial domain is the boundary conditions at the computational window edges. For this purpose, various approaches such as the ABC [3], PML [4], and transparent boundary condition (TBC) [13] have been introduced. Especially, PML has been successfully applied to the FD-BPM. Unfortunately, since the PML technique involves a modification of Maxwell's equations based on the splitting of the field components into two subcomponents, these non-Maxwellian equations do not have a desirable form for the FE-BPM. Recently, the PML for anisotropic materials that does not involve the field splitting has been developed [14] and has been incorporated into the full-vector FE-BPM [15].

The BPM is effective also for an eigenmode solver. In the so-called imaginary-distance BPM (ID-BPM), the propagation direction is selected along the imaginary axis and selecting the appropriate propagation step size, we can extract the specific eigenmode from the initial inputted field expressed by arbitrarily superposing the eigenmodes [16]. The main advantages of ID-BPM as an eigenmode solver are as follows: (1) high-efficient calculation algorithms developed for the BPM analysis can be directly utilized; (2) matrices derived from the BPM formulation are essentially complex, so lossy optical waveguides can be easily treated with no additional effort; (3) eigenmodes can be obtained successively from the fundamental to higher-order modes; and (4) employing the appropriate boundary conditions, not only guided modes but leaky modes may be treated because radiation fields are automatically included in the BPM calculation. This BPM-based mode-solving technique has also been incorporated into the full-vector FE-BPM [17].

Since BPMs assume only the forward propagating waves, it is difficult to take backward reflecting waves into account. To handle reflections, various methods, such as the bidirectional BPM with iterative procedure [18], the combined method of BPM and FEM [19], have also been developed.

It is, in general, difficult to check the accuracy of numerical solutions, and so, a comparative exercise between the results from different methods is very useful. Results of benchmark tests for different beam-propagative schemes are reported [20–23].

TIME-DEPENDENT MODELING

Finite-Difference Time-Domain Method

The finite-difference time-domain method (FDTD) [2,24,25] is a widely used propagation solution technique in integrated optics, especially in photonic crystal device simulations. Unlike the BPM, time-domain simulation can model reflections. Since the field variation in time is obtained at an arbitrary position, the frequency response can be readily induced by Fourier transformation of the variation.

The FDTD is based on the difference form of two Maxwell's rotation equations in the time domain given as

$$\nabla \times E = -\frac{\partial B}{\partial t}, \tag{9}$$

$$\nabla \times H = \frac{\partial D}{\partial t}, \tag{10}$$

where D and B are, respectively, the electric flux and magnetic flux densities. These equations are discretized with central differences in time and space. The grid is staggered in time and space that is called Yee mesh or Yee lattice with spatial mesh sizes Δx, Δy, and Δz in the x-, y-, and z-directions, respectively. Each component of the electromagnetic fields of every lattice is calculated at every short time interval Δt. It is well known that the time step size is restricted to

$$\Delta t \leq \frac{1}{c_{max}} \left[\frac{1}{(\Delta x)^2} + \frac{1}{(\Delta y)^2} + \frac{1}{(\Delta z)^2} \right]^{-1/2}, \tag{11}$$

where c_{max} is the maximum light speed in the structure considered. This requirement is called the Courant–Friedrichs–Lewy (CFL) condition. With a given excitation at the input in CW or pulsed form, the excited field may be propagated and finally reaches the computational window edges. To suppress spurious reflections from the computational window edges, Mur's condition [26] and Berenger's PML condition [4] are widely used. Mur's condition requires minimal calculation and memory consumption, while Berenger's PML is superior in the reflection suppression.

The errors of FDTD depend on the mesh sizes. Generally speaking, the FDTD requires less than $\lambda/10$ spatial mesh size with λ being the shortest wavelength in the analysis domain. When the structure includes tilted or curved surfaces, a finer mesh is required because of staircase approximations in FDTD.

Time-Domain Beam Propagation Method

Conventional FDTD simulates the time varying optical fields regardless of their time dependence. As usual, the bandwidth of the optical signal is very narrow relative to the carrier frequency. In this case, much of the intensive computation of FDTD is wasted. Simulation can be performed on the relatively slowly varying envelope of the narrowband signal instead of the raw optical field [27].

Recently, under the condition that the modulation frequency is much lower than the carrier frequency, a simple and efficient propagation algorithm in time domain has been developed and is called the time-domain BPM (TD-BPM) [28–30]. The removal of the fast carrier allows one to track a slowly varying envelope of a pulsed wave directly in time domain and thus, the converged solution could be obtained with moderate time step size.

From Maxwell's equations the following vectorial or approximate scalar wave equation:

$$\nabla \times (p\nabla \times \boldsymbol{\Phi}) + \frac{q}{c^2}\frac{\partial^2 \boldsymbol{\Phi}}{\partial t^2} = 0, \tag{12}$$

$$p\nabla^2 \Phi - \frac{q}{c^2}\frac{\partial^2 \Phi}{\partial t^2} = 0, \tag{13}$$

where c is the light velocity in free space. In the TD-BPM, the so-called slowly varying field $\boldsymbol{\phi}$ or ϕ is introduced as

$$\boldsymbol{\Phi}(x, y, z, t) = \boldsymbol{\phi}(x, y, z)\exp(-j\omega_0 t), \tag{14}$$

$$\Phi(x, y, z, t) = \phi(x, y, z)\exp(-j\omega_0 t), \tag{15}$$

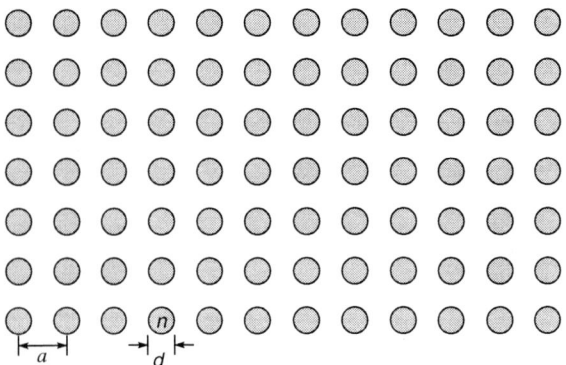

Figure 4 Photonic crystal

where ω_0 is the carrier center angular frequency. Substituting Eqs. (14) and (15) into Eqs. (12) and (13), respectively, we obtain the equation for the slowly varying field ϕ and ϕ, in which the first and the second derivative terms in t are included. Using the SVEA, the Fresnel or paraxial equation (narrowband TD-BPM) is derived. Using the Pade recurrence relation [11], on the other hand, the basic equation for the wideband TD-BPM is derived.

We consider a photonic crystal of dielectric rods in air on a square array with lattice constant a as shown in Figure 4, where the rod diameter and the refractive index are, respectively, $d = 0.36a$ and $n = 3.4$. Assuming the lattice constant to be $a = 0.54$ μm, the crystal has the photonic bandgap for TE modes with the electric fields parallel to the rod axis, which extends over the wavelength range from 1.22 to 1.79 μm, but not for TM modes. Figure 5 and Figure 6 show,

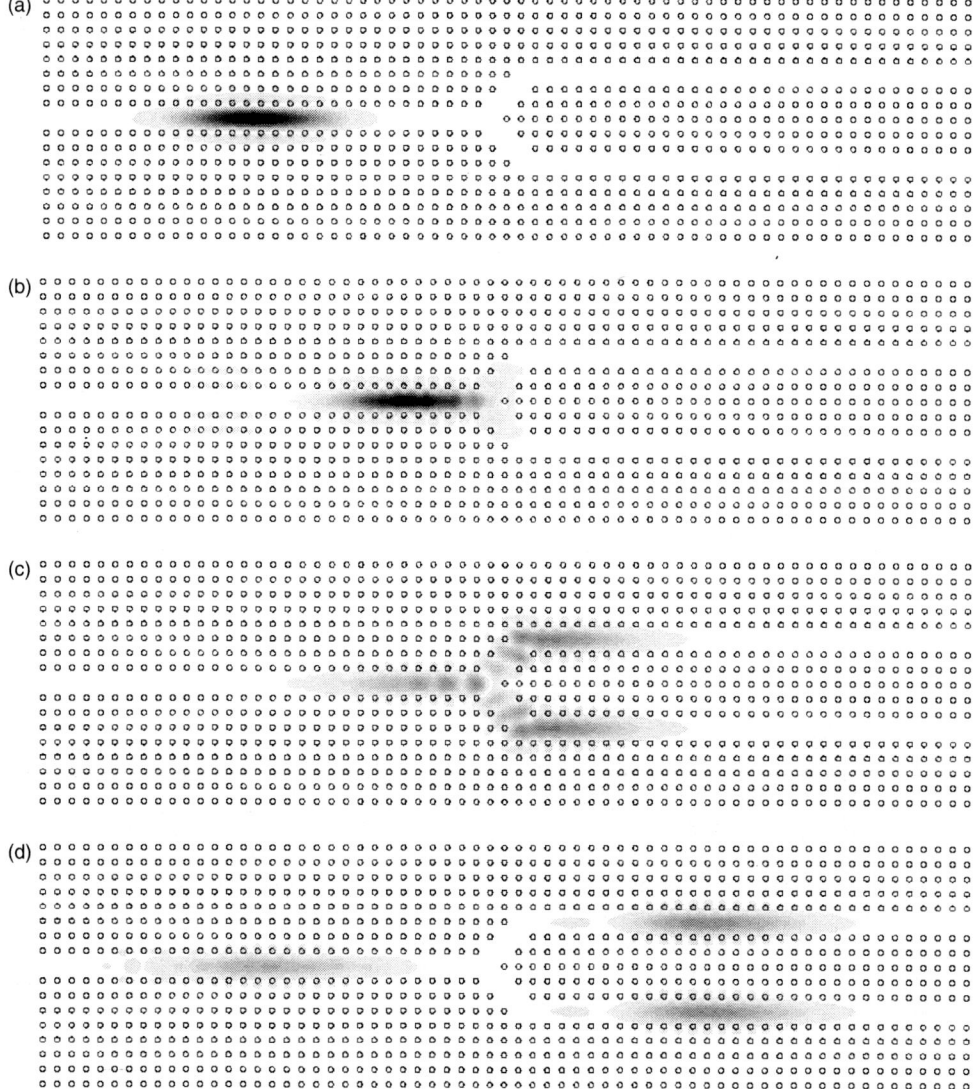

Figure 5   Y-branch with (a) incident Gaussian pulse and electric field patterns at (b) 40 fsec, (c) 80 fsec, and (d) 120 fsec after launching the pulse

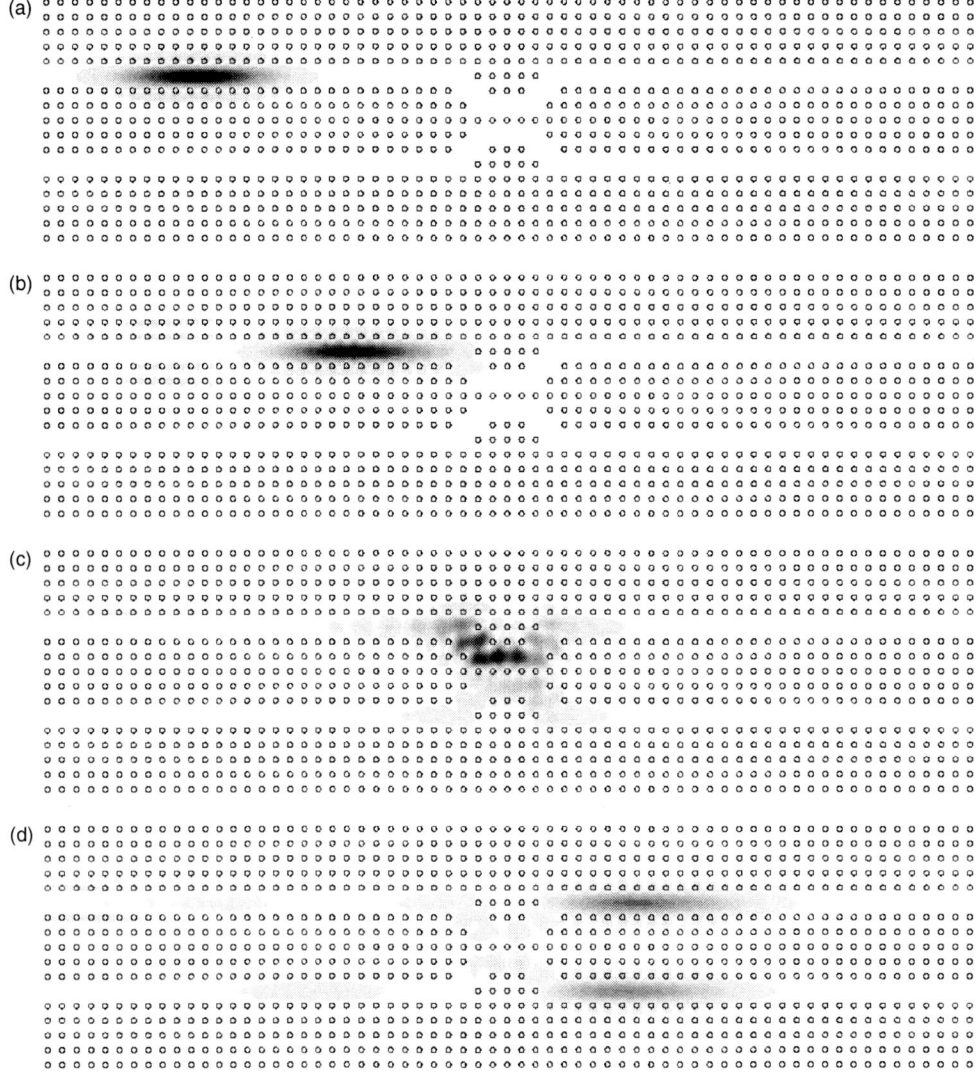

Figure 6 Directional coupler with (a) incident Gaussian pulse and electric field patterns at (b) 40 fsec, (c) 80 fsec, and (d) 120 fsec after launching the pulse

respectively, the electric field patterns in a Y-branch and a directional coupler after launching a Gaussian pulse with the carrier center wavelength of 1.45 μm [30]. The reflected (or transmitted) spectra can be evaluated from the ratio between the Fourier transforms of the reflected (or transmitted) pulse and the incident pulse.

REFERENCES

1. R. Baets, P. Kaczmarski, and P. Vankwikelberge, "Design and modelling of passive and active optical waveguide devices," in *Waveguide Optoelectronics*, J. H. Marsh and R. M. De La Rue, Eds., Dordrechet, The Netherlands: Kluwer, 1992.

2. R. Scarmozzino, A. Gopinath, R. Pregla, and S. Helfert, "Numerical techniques for modeling guided-wave photonic devices," *IEEE J. Select. Top. Quantum Electron.*, 6, 150–162, 2000.

3. J. Sajionmaa and D. Yevik, "Beam-propagation analysis of loss in bent optical waveguides and fibers," *J. Opt. Soc. Am.*, 73, 1785–1791, 1983.

4. J.-P. Berenger, "A perfectly matched layer for the absorption of electromagnetic waves," *J. Comput. Phys.*, 114, 185–200, 1994.

5. M. Koshiba, *Optical Waveguide Theory by the Finite Element Method*, Tokyo, Japan: KTK Scientific/ Kluwer, 1992.

6. M. Koshiba and Y. Tsuji, "Curvilinear hybrid edge/nodal elements with triangular shape for guided-wave problems," *J. Lightwave Technol.*, 18, 737–743, 2000.

7. M. Koshiba and K. Saitoh, "Finite-element analysis of birefringence and dispersion properties in actual and idealized holey-fiber structures," *Appl. Opt.*, 42, 6267–6275, 2003.

8. Y. Tsuji and M. Koshiba, "Adaptive mesh generation for full-vectorial guided-mode and beam-propagation solutions," *IEEE J. Select. Top. Quantum Electron.*, 6, 163–169, 2000.

9. R. März, *Integrated Optics: Design and Modeling*, Norwood, MA: Artech House, 1995.

10. M. D. Feit and J. A. Fleck, Jr., "Light propagation in graded-index optical fibers," *Appl. Opt.*, 17, 3990–3998, 1978.

11. G. R. Hadley, "Wide-angle beam propagation using Pade approximant operators," *Opt. Lett.*, 17, 1426–1428, 1992.

12. H. E. Hernández-Figueroa, "Simple nonparaxial beam-propagation method for integrated optics," *J. Lightwave Technol.*, 12, 644–649, 1994.

13. G. R. Hadley, "Transparent boundary condition for the beam propagation method," *Opt. Lett.*, 16, 624–626, 1991.

14. F. L. Teixeira and W. C. Chew, "General closed-form PML constitutive tensors to match arbitrary bianiso-tropic and dispersive linear media," *IEEE Microwave Guided Wave Lett.*, 8, 223–225, 1998.

15. K. Saitoh and M. Koshiba, "Full-vectorial finite element beam propagation method with perfectly matched layers for anisotropic optical waveguides," *J. Lightwave Technol.*, 19, 405–413, 2001.

16. D. Yevic and B. Hermansson, "New approach to lossy optical waveguide," *Electron. Lett.*, 21, 1029–1030, 1985.

17. K. Saitoh and M. Koshiba, "Full-vectorial imaginary-distance beam propagation method based on a finite element scheme: Application to photonic crystal fibers," *IEEE J. Quantum Electron.*, 38, 927–933, 2002.

18. P. Kaczmarski and P. E. Lagasse, "Bidirectional beam propagation method," *Electron. Lett.*, 24, 675–676, 1988.

19. S. Yoneta, M. Koshiba, and Y. Tsuji, "Combination of beam propagation method and finite element method for optical beam propagation analysis," *J. Lightwave Technol.*, 17, 2398–2404, 1999.

20. H.-P. Nolting and R. März, "Results of benchmark tests for different numerical BPM algorithms," *J. Lightwave Technol.*, 13, 216–224, 1995.

21. J. Haes, R. Baets, C. M. Weinert, M. Gravert, H. P. Nolting, M. A. Andrade, A. Leite, H. K. Bissessur, J. B. Davies, R. D. Ettinger, J. Ctyroky, E. Ducloux, F. Ratovelomanana, N. Vodjdani, S. Helfert, R. Pregla, F. H. G. M. Wijnands, H. J. W. M. Hoekstra, and G. J. M. Krijnen, "A comparison between different prop-agative schemes for the simulation of tapered step index slab waveguides," *J. Lightwave Technol.*, 14, 1557–1569, 1996.

22. M. Koshiba and Y. Tsuji, "A wide-angle finite-element beam propagation method," *IEEE Photon. Technol. Lett.*, 8, 1208–1210, 1996.

23. M. Koshiba, Y. Tsuji, and M. Hikari, "Finite element beam propagation method with perfectly matched layer boundary conditions," *IEEE Trans. Magn.*, 35, 1482–1485, 1999.

24. K. S. Yee, "Numerical solution of initial boundary value problems involving Maxwell's equations in isotropic media," *IEEE Trans. Antennas Propagat.*, AP-14, 302–307, 1966.

25. A. Taflove, *Computational Electrodynamics: The Finite-Difference Time-Domain Method*, Norwood, MA: Artech House, 1995.

26. G. Mur, "Absorbing boundary conditions for the finite-difference approximation of the time-domain electromagnetic-field equations," *IEEE Trans. Electromagn. Compatibility*, EMC-23, 377–382, 1981.

27. F. Ma, "Slowly varying envelope simulation of optical waves in time domain with transparent and absorb-ing boundary conditions," *J. Lightwave Technol.*, 15, 1974–1985, 1997.

28. P.-L. Liu, Q. Zhao, and F.-S. Choa, "Slow-wave finite-difference beam propagation method," *IEEE Photon. Technol. Lett.*, 7, 890–892, 1995.
29. G. H. Jin, J. Harari, J. P. Vicot, and D. Decoster, "An improved time-domain beam propagation method for integrated optics components," *IEEE Photon. Technol. Lett.*, 9, 348–350, 1997.
30. M. Koshiba, Y. Tsuji, and M. Hikari, "Time-domain beam propagation method and its application to photonic crystal circuits," *J. Lightwave Technol.*, 18, 102–110, 2000.

WAVELENGTH CONVERSION

Hiroyuki Uenohara

Wavelength conversion is a technique for changing the wavelength of signal light λ_1 to that of output light λ_2 all optically, that is, with no electrical-to-optical and optical-to-electrical conversion, wherein λ_2 is different from λ_1. Wavelength conversion is important for photonic networks, when the optical signal is transferred from one node to another by assigning optical wavelength to each optical path and changing it in accordance with the destination of the optical signal. In another application, wavelength division multiplexed signals are wavelength converted, where the work path is failed and is dynamically routed to the protection path. Conventionally, wavelength change is performed by an optical-to-electrical converter (a photodetector or an avalanche photodiode), electrical driver, a laser diode, and a modulator ($LiNbO_3$), or an electroabsorption modulator (EAM). With speeding optical data, the operational speed of electric circuit will become the bottleneck for up grading the photonic networks, and electric power consumption will increase drastically with the increasing data rate. The main purpose of the all-optical wavelength conversion is to overcome the difficulties of the operation speed and electric power consumption.

Till date, wavelength conversion has been proposed and verified experimentally by several methods. From the device point of view, it is categorized as follows:

1. Semiconductor optical amplifier (SOA) and lasers (LD) [1–27]
2. Electroabsorption modulator (EAM) [28–31]
3. Electro-optic crystal [32–39]
4. Optical fiber [40–46]

Physically, wavelength conversion can be achieved by the following phenomena:

1. Cross gain modulation (XGM) [1–6]
2. Cross phase modulation (XPM) [7–21,30,31]
3. Four wave mixing (FWM) [22–27,40,41,44–46]
4. Cross absorption modulation (XAM) [28,29]
5. Self-phase modulation (SPM) [42,43]
6. Quasi-phase matching (QPM) [32–39]

In the following sections, principle of operation and experimental approaches are categorized and described from the device point of view.

SOA- AND LD-BASED WAVELENGTH CONVERSION

A SOA is a semiconductor optical device in which optical signal intensity is amplified 10 to 100 times greater than the input signal by stimulated emission. The typical device length is 500 μm to 1.5 mm. For suppressing lasing operation and achieving large optical gain, antireflection coating and window region on both facets are fabricated. Polarization independent amplification can be obtained by introducing a bulk active layer with slightly tensiled strain [47] or tensile-strained multi-quantum well (MQW) structure [48–50]. When signal light with large intensity is injected into a SOA, carrier density in the active region is decreased by stimulated emission. In this situation, optical gain is decreased. This phenomenon is called gain saturation. Under gain saturation, the optical gain and refractive index are modulated simultaneously. In addition, efficient nonlinear effect is expected because third-order nonlinear effect is relatively larger in semiconductors than silica-based optical fibers and electro-optical crystals such as $LiNbO_3$. Thus, wavelength conversion by XGM, XPM, and FWM are achieved by using a SOA.

A LD is also a gain medium with large optical gain and nonlinearity. As a wavelength converter using a LD, FWM technique is mainly used for the operation speed exceeding 40 Gbps.

XGM Wavelength Conversion in a SOA

The XGM is a wavelength conversion technique by modulating optical gain for CW probe light when RF signal light is injected [1–6]. The principle of operation is briefly indicated in Figure 1. CW light is amplified by a SOA. In case the injection current into a SOA is large enough for gain saturation, carrier density is modulated and the optical gain for CW probe light is also modulated. So signal bit stream is wavelength converted to the probe light. The bit sequence is inverted in intensity. That is, the intensity of wavelength conversion signal decreases when signal bit is "1" level (high intensity), and it increases in case of "0" level signal light (low intensity).

There are mainly two factors for operation speed limitation. The fall time for wavelength conversion signal (transition time from high to low intensity) corresponds to the carrier density decrease by stimulated emission by signal light injection. So this is very fast in the order of 10 to 20 psec. On the other hand, the rise time for wavelength conversion signal (transition time

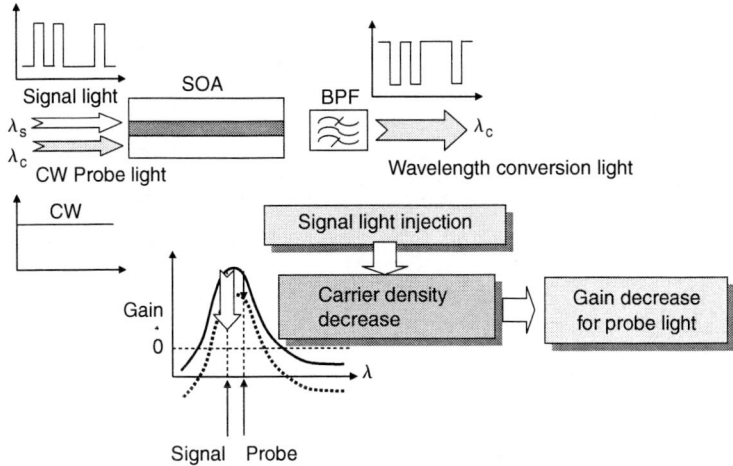

Figure 1 Operation principle of XGM in a SOA

from low to high intensity) corresponds to the carrier recovery. No signal light is injected in the latter case, and, in general, the rise time is larger than the fall time. The typical rise time is about 100 psec. For this reason the rise time is the main limitation in the switching operation. For improving it, several methods have been tried such as (a) high carrier injection, (b) long cavity for large internal optical gain, and (c) third injection CW light with different wavelength from both signal and probe light. In all these methods, the technical point for reducing carrier recovery time is to enlarge stimulated emission effect. By using these techniques, rise time is reduced to <50 psec, and operation speed of >100 Gbps has been achieved [3].

XPM Wavelength Conversion in a SOA

The XPM is a wavelength conversion technique using the optical phase change, or in other words, refractive index change by signal light injection [7–9]. Its basic structure and operation principle is shown in Figure 2. Refractive index change occurs by modulating the carrier density in the active region of a SOA, when CW probe light and RF-modulated signal light are injected at the same time. To convert the refractive index change into light intensity modulation, an interferometer is formed as shown in Figure 2 and Figure 3. Two SOAs are set up in each optical branch in the interferometer. By setting the injection current into the two SOAs at different levels, output signal phase can be cancelled and the normally off operation is possible. When signal light is injected in one of SOA's, optical phase balance is destroyed, and output light intensity increases.

The operation response speed is basically the same as that for XGM. That is, the carrier recovery time is the speed limitation, but operation with up to 40 Gbps can be realized.

One of the applications for XPM wavelength conversion is the all-optical signal regeneration, so-called optical 2R and 3R. 3R is the abbreviation of reshaping, retiming, and regenerating. So far, degraded optical signal due to the dispersion and optical loss in the optical fiber is converted

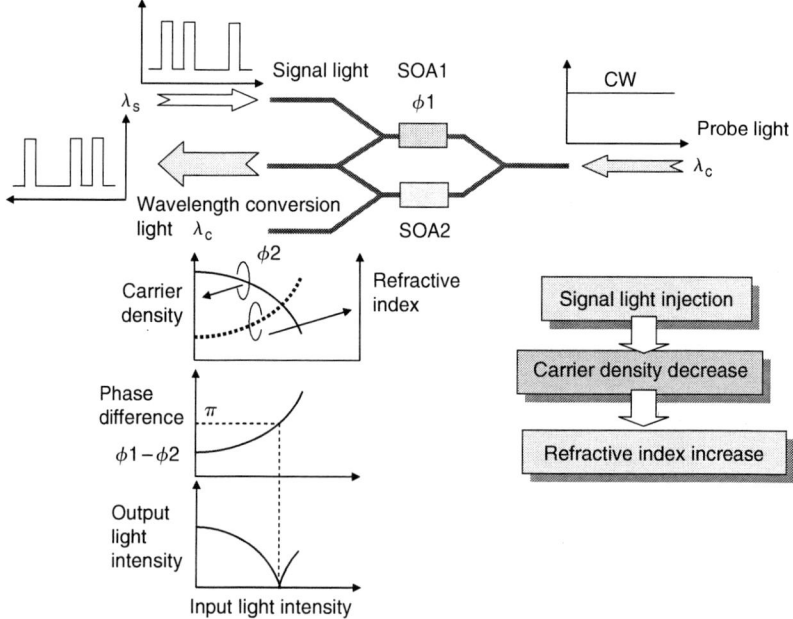

Figure 2 Operation principle of XPM in a SOA

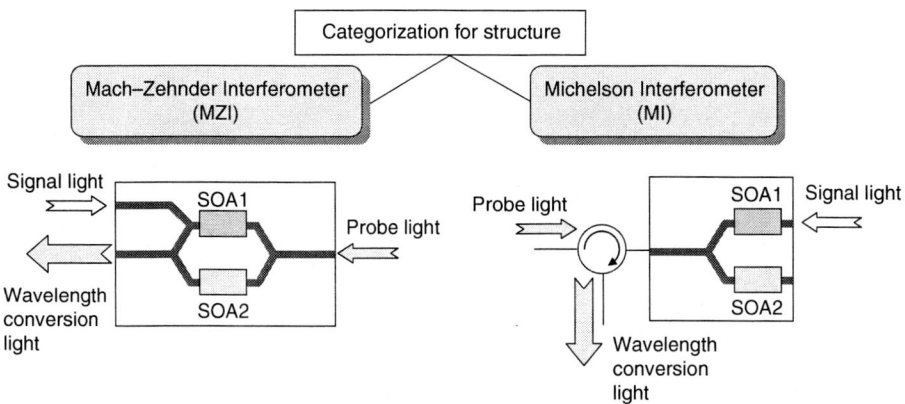

Figure 3 Schematic structure of XPM-based wavelength converters Left: Mach–Zehnder interferometer (MZI). Right: Michelson interferometer (MI)

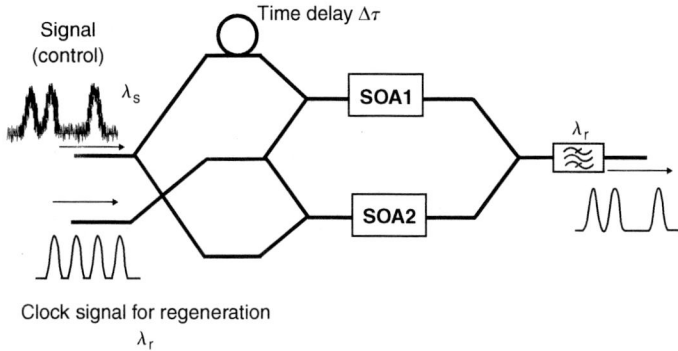

Figure 4 Schematic structure of a SMZ type optical signal regenerator

to the electrical signal. Its logical level ("1" level or "0" level) is discriminated, and is converted to optical signal again. Some issues faced for further speeding up the data rate are the operation response limit of electronic circuits and the increase of electric power consumption. To overcome these problems, all-optical signal regeneration technique has been investigated by many researchers and research institutes.

Wavelength conversion by using a SOA is popular for optical 2R/3R because of the several reasons listed below:

1. High speed capability of >40 Gbps
2. Optical gain performance
3. Polarization independence
4. Large extinction ratio

Figure 4 shows one of the most promising configurations. It is called a symmetric Mach–Zehnder type optical regenerator [10–16]. It is similar to the structure for XPM wavelength converter indicated in Figure 3, however, it has two leading arms connected to both optical branches. One of them has an optical delay with the delay time of optical gating width. The operation

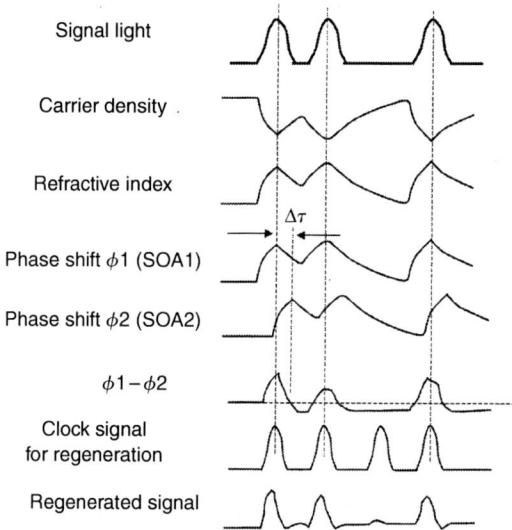

Signal light

Carrier density .

Refractive index

Phase shift $\phi 1$ (SOA1)

Phase shift $\phi 2$ (SOA2)

$\phi 1 - \phi 2$

Clock signal
for regeneration

Regenerated signal

Figure 5 Operation principle of a SMZ type optical signal regenerator

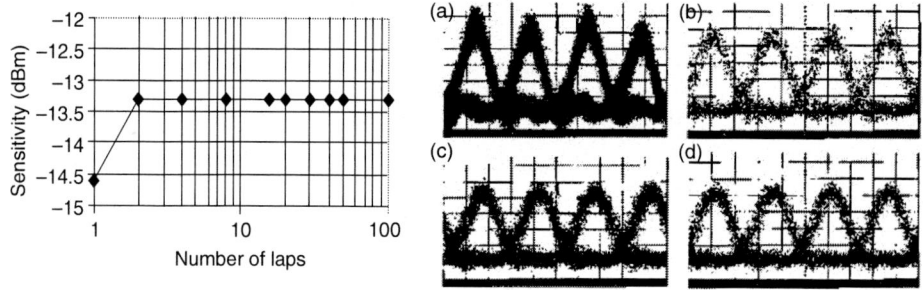

Figure 6 Optical sensitivity as a function of the number of laps for a SMZ type optical signal regenerator (left) and eye diagrams for different laps (right) (after B. Lavigne, P. Guerber, P. Brindel, E. Balmfrezol, and B. Dagens, in *Proceedings of the* 27th *European Conference on Optical Communication* (ECOC2001), Amsterdam, (2001), pp. 290–291. With permission.)

principle is depicted in Figure 5. Optical signal injected into a SOA is amplified due to stimulated emission. This results in refractive index and optical phase change ϕ_1 and ϕ_2 in the two SOAs. The timing of ϕ_1 and ϕ_2 are shifted by the amount of $\Delta\tau$ caused by the optical delay. So the trailing edge of ϕ_1 is cancelled by ϕ_2, and the optical gating window can be obtained. The carrier recovery, the limiting factor for high speed operation, is relatively slow compared with the signal bit rate, but high speed operation exceeding the carrier recovery time is possible because of the operation principle described previously. Optical clock pulse trains are injected into the center port, and the original signal bit streams are regenerated by going through the optical gate. The operation speed of 168 Gbps as a wavelength converter and 336 Gbps as an optical switch for DEMUX have been reported [10,11]. The loop experiments have been tried and no degradation of Q value was reported in spite of the number circulating over 100 as shown in Figure 6 [12].

There is another type of optical 2R/3R. Figure 7 shows an example. It is called a SLALOM (semiconductor laser amplifier in a loop mirror) [17,18]. The operation principle is almost the

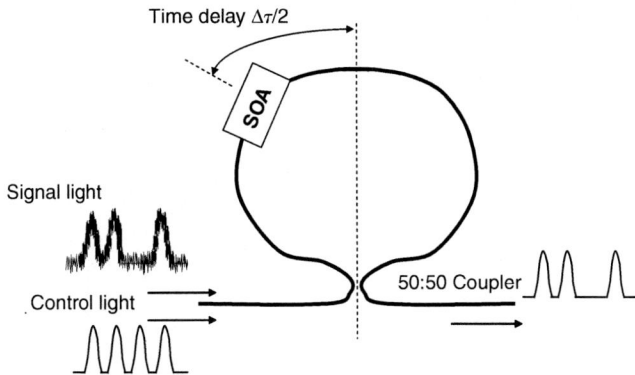

Figure 7 Schematic structure of SLALOM type wavelength converter and optical signal regenerator.

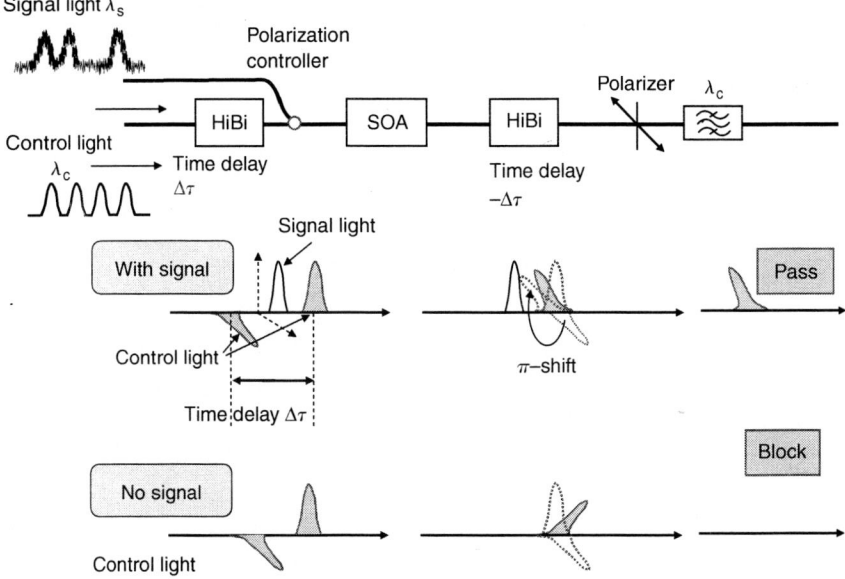

Figure 8 Operation principle of a UNI type optical signal regenerator.

same as SMZ type optical 2R/3R. Fiber loop is used instead of the Mach-Zehnder interferometer (MZI). The SOA is set in the position offset from the center of the fiber loop.

There are different types of optical 2R/3R. The basic principle is similar to SMZ type optical regenerator, but uses the polarization birefringence where architecture is shown in Figure 8. It is called UNI (ultra fast nonlinear interferometer) [19–21]. It has no optical couplers and no interferometer, and uses only one SOA. Control light pulse for regeneration signal is injected into a highly birefrengent medium (HiBi) with polarization angle of 45°. Then the time delay $\Delta\tau$ is generated between two polarization modes because of the polarization mode dispersion. Signal light is injected into a SOA between two polarization modes. The optical phase shift of π for the latter control pulse is caused by XPM. After passing through another HiBi, the split control pulses are recombined into a single pulse with the polarization angle of 90° shift from the original one. So if the polarizer is set at the output port, the gated control signal is extracted as a regenerated signal. When the control light is not injected, no phase shift occurs, and so no optical signal outputs are obtained.

FWM Wavelength Conversion in a SOA and a LD

The FWM is a third-order optical nonlinear effect, and is utilized for wavelength conversion [22–27]. The configuration is very similar to that of XGM as shown in Figure 1, but the converted signal wavelength is different from the XGM wavelength converted signal. Assuming that the frequencies of signal, probe, and wavelength converted light are f_s, f_p, and f_c, respectively, in FWM, relation between each optical frequency is given as:

$$f_c = 2f_p - f_s \tag{1}$$

This means that the wavelength converted signal arises at the frequency at symmetrically opposite side of the signal light against the pump light. Figure 9 indicates the experimental results for FWM wavelength conversion using a LD [24].

The origin of FWM have been reported by several groups of researchers. There are mainly three phenomena; carrier density pulsation, carrier heating effect, and spectral hole burning effect in semiconductors.

The advantages of FWM wavelength conversion are as follows:

1. High speed operation of up to 1 Tbps [26]
2. Several signals with different wavelengths can be wavelength converted simultaneously
3. Large extinction ratio
4. Wavelength converted signal is phase conjugated against original signal light

Because of (2) and (3), FWM wavelength conversion is expected for a dispersion compensator in WDM networks with compact size.

On the other hand, it has a large issue for practical application. The wavelength conversion efficiency is asymmetric for long- and short-wavelength sides of the pump light. It is basically

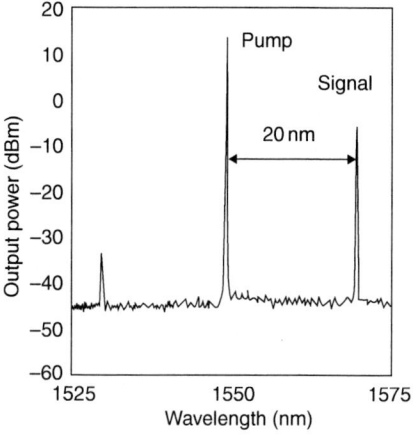

Figure 9 Optical spectrum of FWM wavelength conversion signal (after H. Kuwatsuka, H. Shoji, M. Matsuda, and H. Ishikawa, *J. Quantum Electron.*, 33, 2002–2010 (1997). With permission.)

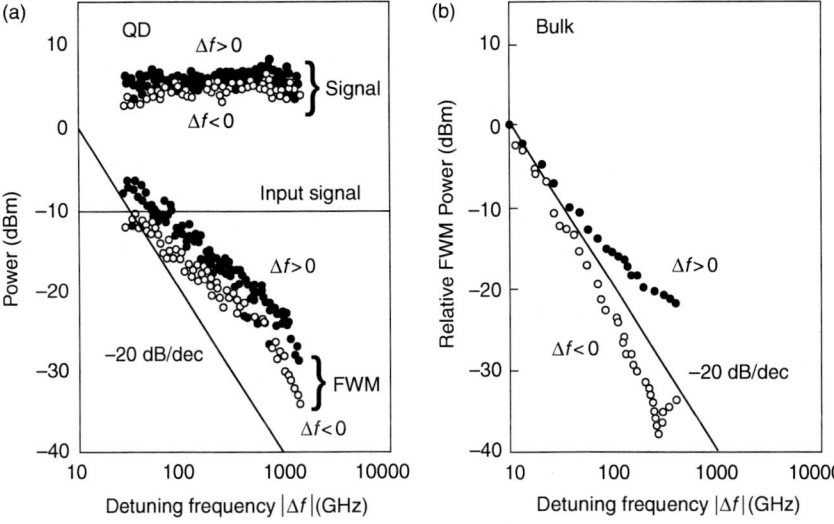

Figure 10 FWM power as a function of frequency detuning for (a) quantum dot and (b) bulk structure (After T. Akiyama, H. Kuwatsuka, N. Haton, Y. Nakata, H. Ebe, and M. Sugawara, *IEEE Photonics Technol. Lett.*, 14, 1139–1141 (2002). With permission.)

caused by the phase difference at the beat frequency between positive and negative detuning between signal and probe light. To improve such characteristic, a quantum-dot type wavelength converter has been proposed [27]. In an ideal quantum-dot structure, gain peak wavelength changes very little in spite of the carrier density change. This means that the refractive index change is very small and the asymmetry in the carrier density pulsation can be suppressed. As shown in Figure 10, symmetric conversion efficiency was achieved [27].

Another issue for FWM is polarization dependent characteristic. In FWM, signal and pump light should interact together inside a SOA or a LD, so polarization for both signal has to be aligned although a polarization independent SOA is used. For realizing polarization independent performance, polarization diversity configuration has been tried.

EAM-BASED WAVELENGTH CONVERSION

An EAM is a semiconductor device which modulates optical signal by absorption. If one wants to use it at wavelength of λ_0, the bandgap energy of EAM should be larger, in other words, the bandgap wavelength λ_1 should be shorter than λ_0. When the reverse bias voltage is applied to the EAM, absorption spectrum shifts toward longer wavelength due to Franz–Keldish effect in case of bulk material, and due to quantum-confined stark effect (QCSE) in case of MQW structure. In this situation, absorption becomes large at around band gap wavelength. So, the input light intensity can be modulated in accordance with applied voltage.

The wavelength conversion in an EAM is achieved by modulating the absorption by the RF optical signal. This phenomenon is called cross absorption modulation (XAM) [28,29]. Operation principle of XAM is shown in Figure 11. CW light with the wavelength λ_p is injected into an EAM. When RF signal light with the wavelength λ_s is injected at the same time, the photo-excited carriers move toward the electrodes. These carriers (electrons and holes) generate internal electric field in the direction opposite to the externally applied reverse bias voltage. In this

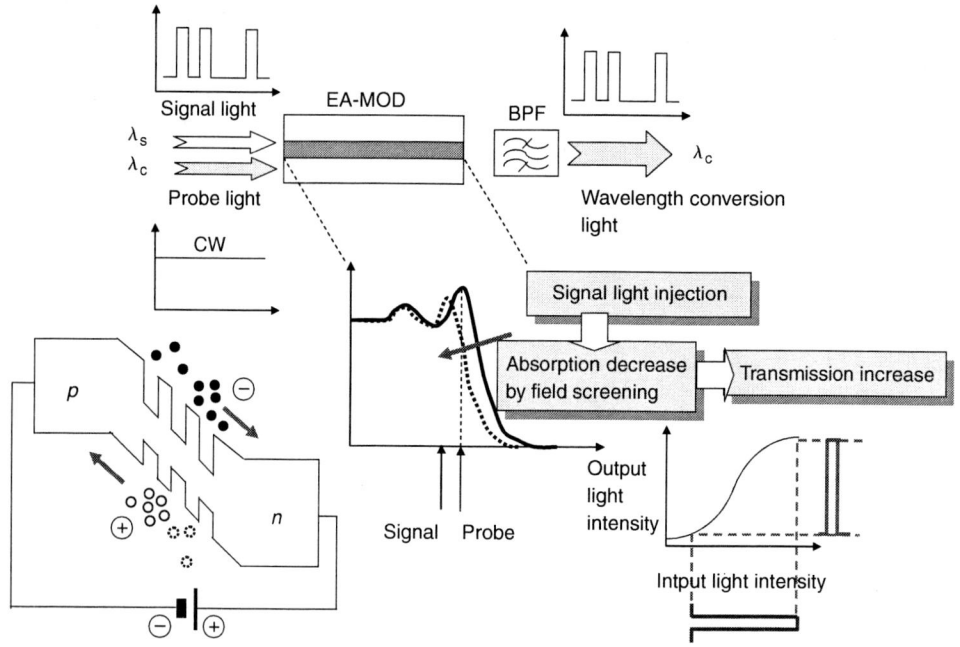

Figure 11 Operation principle of wavelength conversion based on XAM

period, the optical spectrum of the absorption region shifts toward shorter wavelength. So, the absorption close to the bandgap edge decreases. When the photoexcited carriers move toward the cladding layers, the internal electric field diminishes. The response time of the absorption modulation by XAM is dominated by the field screening effect mentioned earlier, and a fast response of >40 Gbps is possible.

XPM as well as XAM occurs in an EAM, therefore, the design of the wavelength converter similar to XPM, SMZ switch, and UNI are available. By using UNI structure with an EAM, 80 Gbps all-optical wavelength conversion and optical signal regeneration have been reported [30,31].

ELECTRO-OPTIC CRYSTAL-BASED WAVELENGTH CONVERTER

Various approaches for wavelength conversion using an electro-optic crystal are being tried of which the most popular crystal is $LiNbO_3$. The mechanism of the wavelength conversion is the second-order nonlinearity. The schematic structure is shown in Figure 12. Signal light with the optical frequency f_s and pump light with the optical frequency f_p are injected into a waveguide formed in the $LiNbO_3$. The second harmonic light is generated in the $LiNbO_3$ waveguide. Then, the differential signal of the second harmonic light and the signal light with the optical frequency $f_c = 2f_p - f_s$ is generated (Figure 13). For achieving this wavelength conversion operation, optical phase matching between the second harmonic light and the signal light are important. Because of this, periodic structure along the propagation direction of the waveguide is formed. For this purpose, polarization inverted domain is fabricated. This structure is called a periodic poled $LiNbO_3$ (PPLN) [32–39].

Recently, the phase and the mode matching technique have been developed, and the wavelength conversion efficiency improved. Simultaneous wavelength conversion of several signals with the operation speed of 40 Gbps have been reported.

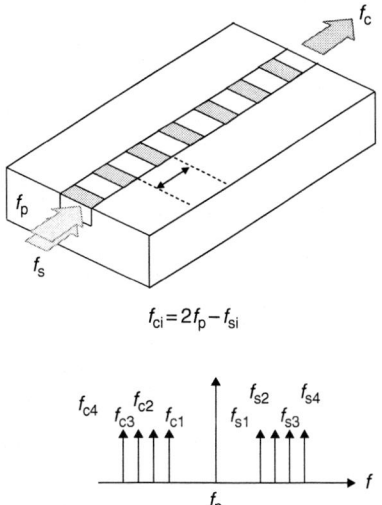

$$f_{ci} = 2f_p - f_{si}$$

Figure 12 Structure of a PPLN wavelength converter. The inset is the optical spectrum alignment for signal, pump, and wavelength conversion light

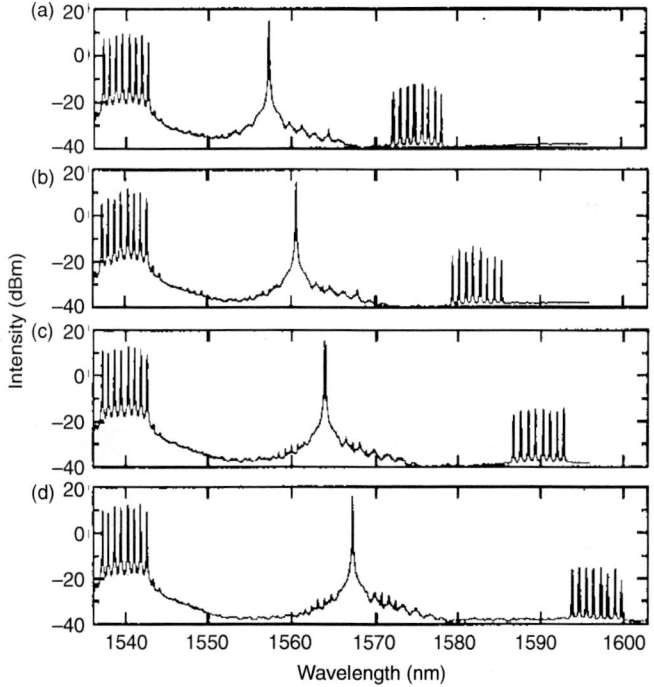

Figure 13 Optical spectra of signal, pump, and wavelength converted light for a PPLN wavelength converter (After M. Asobe, O. Tadanago, H. Miyazawa, Y. Nishida, and H. Suzuki, *Opt. Lett.* 28, 558 [2003]. With permission.)

OPTICAL FIBER-BASED WAVELENGTH CONVERTER

The wavelength conversion can be performed by using an optical fiber. FWM [40,41], SPM and optical bandpass filtering [42,43], and intrachannel cross phase modulation (IXPM) and intra-channel four wave mixing (IFWM) [44–46] works for wavelength conversion. IXPM and IFWM are the same as XPM and FWM, in principle, but they occur in a single channel (intrachannel), not in interchannel. The schematic configuration for FWM is shown in Figure 14. Signal light and pump light amplified by an EDFA are combined into a single fiber. The wavelength conversion is performed in a dispersion-shifted fiber (DSF) or a high nonlinear optical fiber (HNLF) because the optical phase between signal and pump light should be matched. In the case of SPM- , IXPM- , and IFWM-based wavelength conversion, only signal light is injected into an optical fiber, and no pump light is needed. The wavelength conversion in the optical fiber is not carrier-induced effect, so very fast operation can be expected. So far, 40 Gbps operation has been reported for an all-optical 2R regenerator using IXPM and IFWM [44,45].

COMPARISON OF CHARACTERISTICS BETWEEN EACH WAVELENGTH CONVERTER

The comparison of characteristics between each wavelength converter is summarized in Table 1. In general, a SOA- and an EAM-based wavelength converter has the advantage of small size and small operation power. On the other hand, a PPLN and an optical fiber-based wavelength

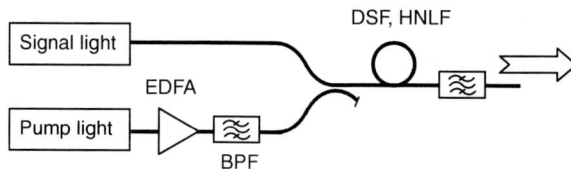

Figure 14 Schematic structure of an optical fiber-based FWM wavelength converter

Table 1
Comparison of characteristics between several wavelength conversion methods

Device	Semiconductor optical amplifier			EA modulator	Electro-optic crystal	Optical fiber
Physical phenomenon	XGM	XPM	FWM	XPM	Second-order optical nonlinearity	FWM SPM
Material	InGaAsP bulk/ MQW	InGaAsP bulk/ MQW	InGaAsP bulk/ MQW/ QD	InGaAsP bulk/MQW	PPLN	DSF HNLF
Speed (Reported)	100 Gbpsec [3]	336 Gbpsec [11]	1 Tbpsec [26]	80 Gbpsec [31]	40 Gbpsec [37]	40 Gbpsec [44]
Extinction ratio (dB)	10	25	25	>10	25	25
Light power (typical) (dBm)	Signal, Probe −10~0	Signal, Probe −10~0	Signal: −20~0 Probe:20	Signal: 15 Probe: 10	Signal: −10 Probe: 20	Signal: 20 (SPM case)
Size (typical)	1 mm	1 mm	1 mm	100 μm	50 mm	>10 km (DSF) 2 km (HNLF)

converter is superior in terms of ultrafast operation speed in principle. In a SMZ switch, however, the relatively high speed operation is possible.

REFERENCES

1. C. Joergensen, S. L. Danielsen, K. E. Stubkjaer, M. Shilling, K. Daub, O. Doussiere, F. Pommerau, P. B. Hansen, H. N. Poulsen, A. Kloch, M. Vaa, B. Mikkelsen, E. Lach, G. Laube, W. Idler, and K. Wunstel, "All-optical wavelength conversion at bit rates above 10 Gb/s using semiconductor optical amplifiers," *IEEE J. Select. Top. Quantum Electron.*, 3, 1168–1180 (1997).
2. T. Durhuus, B. Mikkelsen, C. Joergensen, S. L. Danielsen, and K. E. Stubkjaer, "All-optical wavelength conversion by semiconductor optical amplifier," *J. Lightwave Technol.*, 14, 942–954 (1996).
3. A. D. Ellis, A. E. Kelly, D. Nesset, D. Pitcher, D. G. Moodie, and R. Kashyap, "Error free 100 Gbit/s wavelength conversion using grating assisted cross-gain modulation in 2 mm long semiconductor amplifier," *Electron. Lett.*, 34, 1958–1959 (1998).
4. T. Durhuus, B. Fernier, P. Garabedian, F. Leblond, J. L. Lafragette, B. Mikkelsen, C. G. Joergensen, and K. E. Stubkjaer, "High-speed all-optical gating using a two-section semiconductor optical amplifier structure," in *Proceedings of the Conference on Lasers and Electrooptics* (CLEO1992), CThS4 (1992).
5. C. Joergensen, T. Durhuus, C. Braagaard, B. Mikkelsen, and K. E. Stubkjaer, "4 Gb/s optical wavelength conversion using semiconductor optical amplifiers," *IEEE Photonics Technol. Lett.*, 5, 657–659 (1993).
6. B. Glance, J. M. Wiesenfeld, U. Koren, A. H. Gnauck, H. M. Presby, and A. Jourdan, "High performance optical wavelength shifter," *Electron. Lett.*, 28, 1714–1715 (1992).
7. T. Durhuus, C. Joergensen, B. Mikkelsen, R. J. S. Pedersen, and K. E. Stubkjaer, "All optical wavelength conversion by SOA's in a Mach–Zehnder configuration," *IEEE Photonics Technol. Lett.*, 6, 53–55 (1994).
8. D. M. Patrick and R. J. Manning, "20 Gbit/s wavelength conversion using semiconductor nonlinearity," *Electron. Lett.*, 30, 252–253 (1994).
9. L. H. Spiekman, U. Koren, M. D. Chien, B. I. Miller, J. M. Wiesenfeld, and J. S. Perino, "All-optical Mach–Zehnder wavelength converter with monolithically integrated DFB probe source," *IEEE Photonics Technol. Lett.*, 9, 1349–1351 (1997).
10. S. Nakamura, Y. Ueno, and K. Tajima, "168-Gb/s all-optical wavelength conversion with a symmetric-Mach-Zehnder-type switch," *IEEE Photonics Technol. Lett.*, 13, 1091–1093 (2001).
11. S. Nakamura, Y. Ueno, and K. Tajima, "Error-free all-optical demultiplexing at 336 Gb/s with a hybrid-integrated symmetric Mach–Zehnder all-optical switch," in *Proceedings of the 2002 Optical Fiber Communication Conference* (OFC2002), FD3-1-FD3-3 (2002).
12. B. Lavigne, P. Guerber, P. Brindel, E. Balmefrezol, and B. Dagens, "Cascade of 100 optical 3R regenerators at 40 Gbit/s based on all-optical Mach–Zehnder interferometers," in *Proceedings of the 27th European Conference on Optical Communication* (ECOC2001), Amsterdam, (2001), pp. 290–291.
13. J. Leuthold, G. Raybon, Y. Su, R. Essiambre, S. Cabot, J. Jaques, and M. Kauer, "40 Gbit/s transmission and cascaded all-optical wavelength conversion over 1000000 km," *Electron. Lett.*, 38, 890–892 (2002).
14. A. E. Kelly, I. D. Phillips, R. J. Manning, A. D. Ellis, D. Nesset, D. G. Moodie, and R. Kashyap, "80 Gbit/s all-optical regeneration wavelength conversion using semiconductor optical amplifier based interferometer," *Electron. Lett.*, 35, 1477–1478 (1999).
15. W. Idler, K. Daub, G. Laube, M. Schilling, P. Wiedemann, K. Dütting, M. Klenk, E. Lach, and K. Wünstel, "10 Gb/s wavelength conversion with integrated multiquantum-well-based 3-port Mach–Zehnder interferometer," *IEEE Photonics Technol. Lett.*, 8, 1163–1165 (1996).
16. Y. Hashimoto, R. Kuribayashi, S. Nakamura, K. Tajima, and I. Ogura, "Transmission at 40 Gbps with a semiconductor-based optical 3R regenerator," in *Proceedings of the 29th European Conference on Optical Communication* (ECOC2003), Rimini, (2003), pp. 90–91.
17. Y. Shibata, N. Kikuchi, S. Oku, T. Ito, H. Okamoto, Y. Kawaguchi, Y. Kondo, Y. Suzuki, and Y. Tohmori, "Filter-free all-optical wavelength conversion using Sagnac interferometer integrated with parallel-ampifier structure (SIPAS)," *Electron. Lett.*, 38 (2002).
18. V. M. Menon, W. Tong, C. Li, F. Xia, I. Glesk, P. R. Prucnal, and S. R. Forrest, "All-optical wavelength conversion using a regrowth-free monolithically integrated Sagnac interferometer," *IEEE Photonics Technol. Lett.*, 15, 254–256 (2003).

19. K. Tajima, S. Nakamura, and Y. Sugimoto, "Ultrafast polarization-discriminating Mach–Zehnder all-optical switch," *Appl. Phys. Lett.,* 67, 3709–3711 (1995).

20. Y. Ueno, S. Nakamura, K. Tajima, and S. Kitamura, "3.8-THz wavelength conversion of picosecond pulses using a semiconductor delayed-interference signal-wavelength converter (DISC)," *IEEE Photonics Technol. Lett.,* 10, 346–348 (1998).

21. R. Inohara, M. tsurusawa, K. Nishimura, and M. Usami, "Experimental verification for cascadability of all-optical 3R regenerator utilizing two-stage SOA-based polarization discriminated switches with estimated Q-factor over 20 dB at 40 Gbit/s transmission," in *Proceedings of the 29th European Conference on Optical Communication* (ECOC2003), Rimini, pp. 88–89 (2003).

22. H. Nakajima, "Intracavity nearly degenerate four-wave mixing in a (GaAl)As semiconductor laser," *Appl. Phys. Lett.,* 47, 769–771 (1985).

23. M. C. Tatham, G. Sherlock, and L. D. Westbrook, "20-nm optical wavelength conversion using nondegenerate four-wave mixing," *IEEE Photonics Technol. Lett.,* 5, 1303–1306 (1993).

24. H. Kuwatsuka, H. Shoji, M. Matsuda, and H. Ishikawa, "Nondegenerate four-wave mixing in a long-cavity $\lambda/4$-shifted DFB laser using its lasing beam as pump beams," *J. Quantum Electron.,* 33, 2002–2010 (1997).

25. H. Kuwatsuka, T. Simoyama, and H. Ishikawa, "Enhancement of third-order nonlinear optical susceptibilities in compressively strained quantum wells under the population inversion condition," *J. Quantum Electron.,* 35, 1817–1825 (1999).

26. H. Kuwatsuka, T. Akiyama, B. Little, T. Shimoyama, and H. Ishikawa, "Wavelength conversion of picosecond optical pulses using four-wave mixing in a DFB Laser," in *Proceedings of the 26th European Conference on Optical Communication* (ECOC2000), Vol. 3, (2000), pp. 65–66.

27. T. Akiyama,. H. Kuwatsuka, N. Hatori, Y. Nakata, H. Ebe, and M. Sugawara, "Symmetric highly efficient (~0 dB) wavelength conversion based on four-wave mixing in quantum dot optical amplifiers," *IEEE Photonics Technol. Lett.,* 14, 1139–1141 (2002).

28. N. Edagawa, M. Suzuki, and S. Yamamoto, "Novel wavelength converter using electroabsorption modulator," *IEICE Trans. Electron.,* E81-C, 1251–1257 (1998).

29. T. Otani, T. Miyazaki, and S. Yamamoto, "Optical 3R regenerator using wavelength converters based on electroabsorption modulator for all-optical network applications," *IEEE Photonics Technol. Lett.,* 12, 431–433 (2000).

30. K. Nishimura, M. Tsurusawa, and M. Usami, "First demonstration of 40 Gbps wavelength conversion with no pattern effect utilizing cross-phase modulation in an electroabsorption waveguide", in Proceedings *of the 2001 IEEE International Conference on Indium Phosphide and Related Materials* (IPRM2001), (2001), pp. 444–447.

31. K. Nishimura, R. Inohara, M. Tsurusawa, and and M. Usami, "80 Gbit/s wavelength conversion using MQW electroabsorption modulator in delayed-interferometric configuration," *Electron. Lett.,* 39, 792–794 (2003).

32. T. Suhara and H. Nishihara, "Theoretical analysis of waveguide second-harmonic generation phase matched with unioprm and chirped grating," *J. Quantum Electron.,* 26, 1265–1276 (1990).

33. M. Fujimura, A. Shiratsuki, T. Suhara, and H. Nishihara, "Wavelength conversion in LiNbO$_3$ waveguide difference-frequency generation devices with domain-inverted gratings fabricated by voltage application," *Jpn. J. Appl. Phys.,* 37, L659–L662 (1998).

34. C. Q. Xu, H. Okayama, and M. Kawahara, "1.5 mm band efficient broadband wavelength conversion by difference frequency generation in a periodically domain-inverted LiNbO$_3$ channel waveguide," *Appl. Phys. Lett.,* 63, 3559–3561 (1993).

35. K. R. Parameswaran, R. K. Route, J. R. Kurz, R. V. Roussev, M. M. Fejer, and M. Fujimura, "Highly efficient second-harmonic generation in buried waveguides formed by annealed and reverse proton exchange in periodically poled lithium niobate," *Opt. Lett.,* 27, 179–181 (2002).

36. M. Asobe, O. Tadanaga, H. Miyazawa, and H. Suzuki, "Parametric wavelength conversion and amplification using damage-resistant Zn:LiNbO$_3$ waveguide," *Electron. Lett.,* 37, (2001).

37. M. Asobe, O. Tadanaga, H. Miyazawa, Y. Nishida, and H. Suzuki, "Multiple quasi-phase-matched LiNbO$_3$ wavelength converter with a continuously phase-modulated domain structure," *Opt. Lett.,* 28, 558 (2003).

38. M. Asobe, H. MIyazawa, O. Tadanaga, Y. Nishida, and H. Suzuki, "40 Gbit/s×6 channel wavelength conversion using a quasi-phase matched LiNbO3 waveguide module," in *Proceeding of the 7th OptoElectronics and Communication Conference* (OECC2002), PD2–8 (2002).

39. J. Yamawaku, H. Takara, T. Ohara, K. Sato, A. Takada, T. Morioka, O. Tadanaga, H. Miyazawa, and M. Asobe, "Simultaneous 25 GHz-spaced DWDM wavelength conversion of 1.03 Tbit/s (103×10 Gbit/s) signals in PPLN waveguide," *Electron. Lett.*, 39, 1144–1145 (2003).

40. K. Inoue, "Four-wave mixing in an optical fiber in the zero-dispersion wavelength region," *J. Lightwave Techol.*, 10, 1553–1561 (1992).

41. S. Watanabe and T. Chikama, "Highly efficient conversion and parametric gain of nondegenerate forward four-wave mixing in a singlemode fibre," *Electron. Lett.*, 30, 163–164 (1994).

42. P. V. Mamyshev, "All-optical data regeneration based on self-phase modulation effect," *in Proceedings of the 24th European Conference on Optical Communication* (ECOC1998), Vol. 1 (1998), pp. 475–476.

43. M. Matsumoto, "Analysis of optical regeneration utilizing self-phase modulation in a highly nonlinear fiber," *IEEE Photonics Technol. Lett.*, 14, 319–321 (2002).

44. G. Raybon, Y. Su, J. Leuthold, R.-J. Essiambre, T. Her, C. Joergensen, P. Steinvurzel, K. Dreyer, and K. Feder, "40 Gbit/s pseudo-linear transmission over one million kilometerrs," in *Proceeding of the 2002 Optical Fiber Coomunication Conference* (OFC2002), FD10-1–FD10-3 (2002).

45. Y. Su, G. Raybon, R. -J. Essiambre, and T. -H. Her, "All-optical 2R regeneration of 40-Gb/s signal impaired by intrachannel four-wave mixing," *IEEE Photonics Technol. Lett.*, 15, 350–352 (2003).

46. R. -J. Essiambre, B. Mikkelsen, and G. Raybon, "Intra-channelcross-phase modulation and four-wave mixing in high-speed TDM systems," *Electron. Lett.*, 35, 1576–1578 (1999).

47. K. Morito, M. Ekawa, T. Watanabe, and Y. Kotaki, "High-output-power polarization-insensitive semiconductor optical amplifier," *J. Lightwave Technol.*, 21, 176–181 (2003).

48. K. Magari, M. Okamoto, and Y. Noguchi, "1.55 μm polarization-insensitive high-gain tensile-strained-barrier MQW optical amplifier," *IEEE Photonics Technol. Lett.*, 3, 998–1000 (1991).

49. L. F. Tiemeijer, P. J. A. Thijs, T. van Dongen, R. W. M. Slootweg, J. M. M. van der Heijden, J. J. M. Binsma, and M. P. C. M. Krijn, "Polarization insensitive multiple quantum well laser amplifiers for the 1300 nm window," *Appl. Phys. Lett.*, 62, 826–828 (1993).

50. T. Ito, N. Yoshimoto, K. Magari, and H. Sugiura, "Wode-band polarization-independent tensile-strained InGaAs MQW-SOA gate," *IEEE Photonics Technol. Lett.*, 10, 657–659 (1998).

Wavelength Multiplexer/Demultiplexer (MUX/DEMUX in WDM)

Hiroshi Takahashi

INTRODUCTION

Since the mid-1990s, wavelength division multiplexing (WDM) transmission systems have been needed to meet the huge demand for data communication resulting from the worldwide spread of the Internet. With these systems, signals at different wavelengths are mixed and transmitted through a single optical fiber and this technology provides us with a high per-fiber transmission capacity and low communication costs. A *wavelength multiplexer* is a device that can combine different wavelength signals from plural light sources and output them to a single optical fiber transmission line in a WDM system with no excess loss. It should be noted that a simple device such as a star coupler with N input ports theoretically has an excess loss of 10 log N dB. A *wavelength demultiplexer* is a device that can divide a multiplexed signal incorporating different wavelengths, and output the result to plural optical receivers. Hereafter, the wavelength multi/demultiplexer is referred to simply as a multi/demultiplexer.

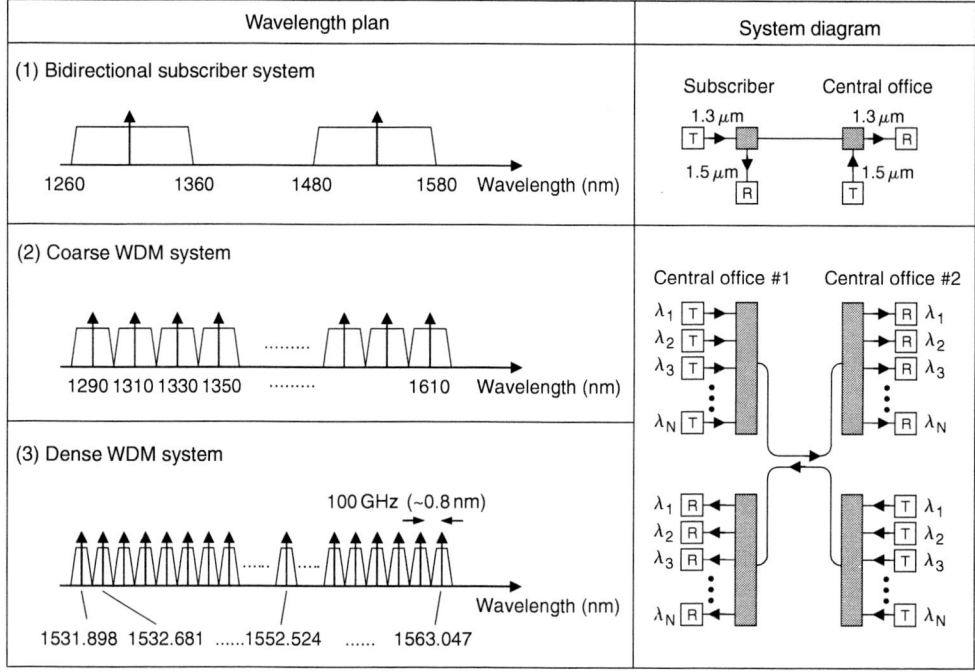

Figure 1 Wavelength plan and system diagram of WDM

Wavelength division multiplexing systems are categorized into three types based on their wavelength channel spacing as shown in Figure 1. The simplest is the bi-directional type. Lasers operating at two different wavelengths (usually 1.3 and 1.55 μm) are used for the two signal transmission directions. The optical filter used in this case does not need a sharp wavelength response and the well-known dielectric high/low pass filters fabricated by established techniques are sufficient. The second type is the coarse WDM (CWDM) in which the channel spacing is 20 nm. This spacing is decided so that the wavelength variation of the light source (usually a DFB laser diode) is about 10 nm. The laser diode requires neither temperature nor wavelength control and this leads to a reduction in system cost. A CWDM system typically has eight (1470, 1490, ..., and 1610 nm) or sixteen (1290, 1310, ..., and 1610 nm) channels. Since conventional optical amplifiers, mainly erbium-doped fiber amplifiers (EDFA), do not còver such a wide wavelength range, the CWDM is well suited to short-haul, low-cost systems such as metropolitan area networks (MAN) and local area networks (LAN). The channel spacing is wide enough for dielectric thin film filter technology to be used. The third category is dense WDM (DWDM) where the channel spacing is about 0.8 nm. This corresponds to a 100 GHz frequency spacing in the 1.5 μm band (as standardized by ITU-T). With this spacing, it is possible to multiplex forty channels into the 35 nm operation bandwidth of the EDFA. Since dielectric thin film filters select only one wavelength and therefore we need the same number filters as wavelength channels to build a multi/demultiplexer, a diffraction grating-based spectrometer type device such as an arrayed-waveguide grating (AWG) is more suitable for DWDM because a single grating can multi/demultiplex plural wavelength channels.

DEFINITIONS OF CHARACTERIZING PARAMETERS FOR MULTI/DEMULTIPLEXERS

The characteristics of the multi/demultiplexers are very important when designing WDM systems. Here, we discuss the definitions of parameters used to characterize multi/demultiplexers with a typical transmission spectrum from a port of a multi/demultiplexer as shown in Figure 2.

Center wavelength: The center of a wavelength range where the relative transmittance to its peak is more than −3 dB (50%). This is often called a 3 dB center. In some cases, a 1 dB center is used instead of a 3 dB one.

Wavelength accuracy/error: The wavelength shift between the measured and designed center wavelengths. In DWDM systems based on the ITU-T standard, the designed center wavelengths are on a 100 GHz spaced grid/comb anchored at 193,100 GHz (1552.524 nm). Typically, the error is within ±0.05 nm in DWDM.

Insertion loss: A decrease in transmittance at a center wavelength, or a given wavelength at which the light source is in the system design.

Bandwidth: The wavelength range where the relative transmittance is more than −3 dB from its peak. This is often called the 3 dB bandwidth. Other definitions such as 0.5 or 1 dB bandwidths are also frequently used. The bandwidth is designed not by the signal bandwidth but by the wavelength accuracy of light source because the frequency efficiency (ratio of signal bandwidth to channel spacing) is as low as 0.1 bit/sec/Hz in conventional 10 Gbps DWDM systems with a 100 GHz spacing.

Clear bandwidth: The wavelength range where the relative transmittance is greater than a predetermined value, for example, −0.5, −1, or −3 dB around a designed center wavelength. If the wavelength error has the ideal value of zero, the clear bandwidth coincides with the aforementioned bandwidth. If not, the clear bandwidth is narrower than the conventional bandwidth by a value of (2 × wavelength error). This means that the clear

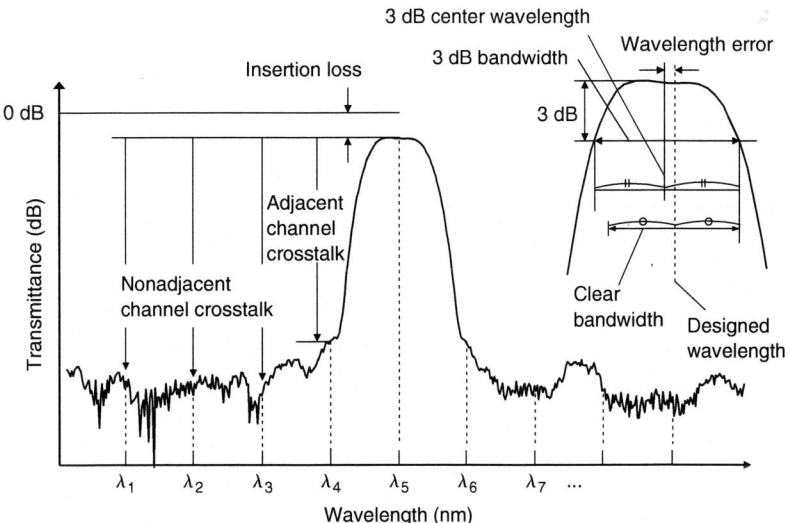

Figure 2   Definitions in multi/demultiplexer spectral response

bandwidth is a good index for presenting an effective range in which the light source wavelength can be when we take account of the wavelength error.

Ripple: Transmittance variation (difference between minimum and maximum transmittance values) in the vicinity of the center wavelength.

Adjacent channel crosstalk: An index of the degree of suppression of the adjacent wavelength channel signal. It is defined as the ratio of the transmittance values of a given channel and the adjacent channel. It is usually a negative dB value.

Non-adjacent channel crosstalk: An index of the degree of suppression of wavelength channel signals other than that of the adjacent channel. This is often defined as the worst ratio of the transmittance values of a given channel and the nonadjacent channels. It is usually a negative dB value.

Total crosstalk: The sum of the relative optical crosstalk power over all the WDM channels except for a given channel (totally 2 adjacent channels and N-3 nonadjacent channels in an N-channel WDM). To prevent signal-to-noise ratio degradation due to crosstalk light, the total crosstalk is typically required to be less than −15 dB.

PDL: Polarization dependent loss. This is defined as the insertion loss fluctuation when the state of polarization of the incident light changes.

PDλ: Polarization dependent wavelength. This is defined as the measured center wavelength difference between orthogonal two polarization states.

ACTUAL MULTI/DEMULTIPLEXERS

Dielectric Thin Film Filter

A thin film filter has a multilayered structure composed of alternately deposited high (H) and low (L) refractive index films as shown in Figure 3. Ta_2O_5 ($n = 2.08$) and SiO_2 ($n = 1.45$) are commonly used as the H and L film materials, respectively. The structure is a cascaded, multi-cavity Fabry–Perot resonator consisting of half-mirrors and spacers. Only the resonant wavelength passes through the filter and the other wavelengths are reflected by it. The multiresonator structure is designed to widen the pass wavelength bandwidth. There are typically five cavities for narrow band filters in DWDM systems, and nine or more for wide band filters. The basic fabrication technology of this filter was developed for color filters, reflection and antireflection coatings on mirrors, lenses and eyeglasses in classic optics. After the late 1990s, with a view to using this filter in DWDM systems, research focused on improving performance by obtaining high wavelength accuracy, which had been degraded by initial wavelength errors during

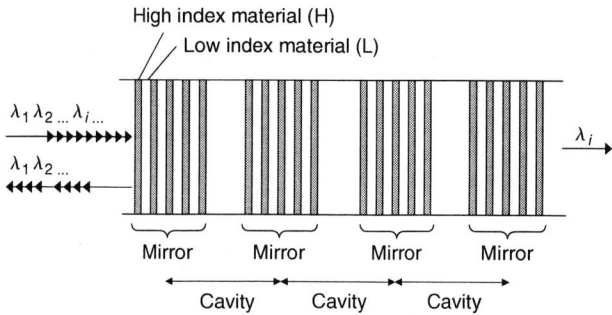

Figure 3   Basic structure of dielectric thin film filter

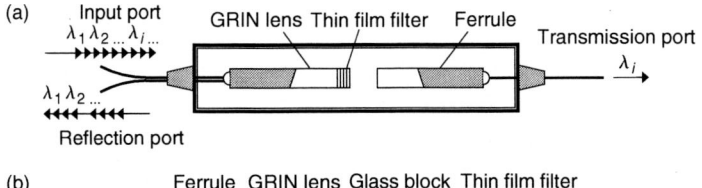

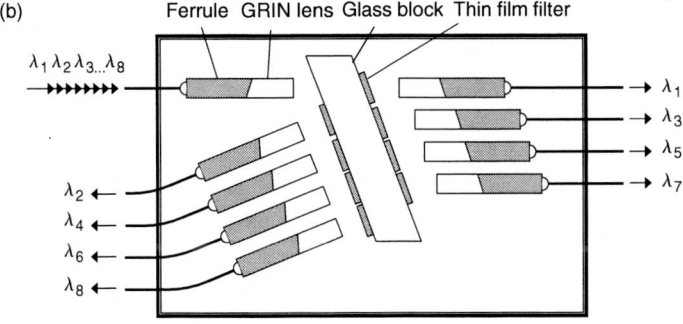

Figure 4 Basic packaging structures of dielectric thin film filter ([a] one channel selection, [b] multi channel selection)

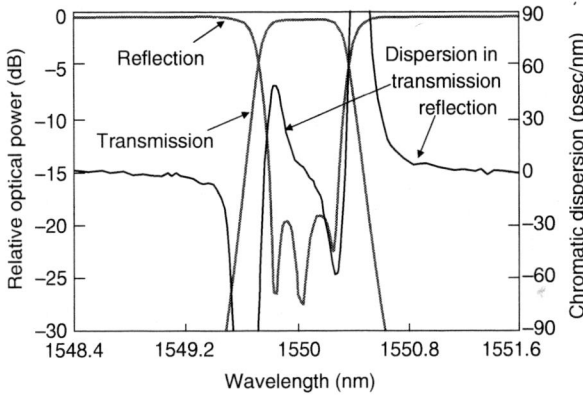

Figure 5 Typical spectral response of thin film filter for 100 GHz spacing WDM

fabrication and the temperature dependent wavelength shift. Narrow bandwidth filters with a high wavelength accuracy of 0.1 nm or better are now available.

Compared with other technologies described later, thin film filters have advantages such as a flat spectral response in the vicinity of the pass wavelength, a low insertion loss, and a low PDL. Figure 4 shows typical packaging of a three-port (input, reflection, and transmission) device using a thin film filter. Light from the input fiber is collimated with a gradient index (GRIN) lens. A resonant, selected wavelength passes the filter attached to the lens endface and reaches the transmission port. Other wavelengths are reflected back. To construct a multi-channel multi/demultiplexer, the input fiber of the next channel package is connected to the reflection port. The alternate structure for compact packaging shown in Figure 4 is also possible. Even if we use this alternate packaging, thin film filters are not suitable for high channel count DWDM systems because of their fabrication cost and device size. They are commonly used in bidirectional units in fiber-to-the-home (FTTH) systems, and in multi/demultiplexer units in low channel count coarse/dense WDM systems.

Figure 5 shows typical spectral responses of a thin film filter designed for 100 GHz spacing DWDM. The transmission spectrum is excellent with a low insertion loss, high flatness and a low crosstalk level. As the channel spacing increases, the flatness becomes more obvious. For example, it has a box-like spectral curve when designed for 20 nm spacing CWDM. One drawback is the chromatic dispersion characteristic because the thin film filter is based on the Fabry-Perot resonator. The chromatic dispersion changes greatly within −50 to 50 psec/nm in the transmission band in Figure 5. With this value, the optical signal deterioration is obvious in high bit rate systems of more than 10 Gb/sec. Since the chromatic dispersion is theoretically in inverse proportion to the square of the bandwidth, it is negligible in CWDM applications. The basic technology and recent progress on this technology are summarized in references [1,2]

Arrayed-Waveguide Grating

An AWG is composed of hundreds of channel waveguides of different lengths and it acts as a transmissive diffraction grating [3–5]. An AWG is usually integrated with input and output waveguides and two slab waveguides as shown in Figure 6 by using planar lightwave circuit technology. Silica glass is commonly used as the waveguide material because a silica waveguide has a very low propagation loss, can be efficiently coupled to a single-mode fiber, and is extremely reliable from both the physical and chemical points of view. The optical circuit shown in Figure 6 is the same as that of a conventional spectrometer. The concave alignment of the waveguide ends of the AWG act as a lens and the AWG acts as a diffraction grating. Accordingly, different wavelength lights are separated to the respective ports when the wavelength multiplexed light is launched into the input port. The polarization dependent wavelength can be eliminated with a half waveplate inserted in the middle of the AWG [6]. The AWG is superior to the thin film filter in high channel count WDM systems because it can simultaneously multi/demultiplex more than two wavelength channels. 40-channel devices are available for use in practical systems. For details, see *Arrayed-waveguide grating*.

The transmission spectrum in the vicinity of the center wavelength is normally Gaussian because the mode field pattern at the junction of the input waveguide and the input slab determines the spectrum and is approximately Gaussian. It is possible to obtain other spectra by changing the mode field. With a wide input waveguide aperture at the junction, the spectrum is a wide Gaussian spectrum. With a parabolic input waveguide, it becomes a flat-top curve [7]. Figure 7 shows three typical spectra of an AWG with a channel spacing of 0.8 nm. The 3 dB bandwidths for Gaussian, flat-top, and wide Gaussian devices are 0.4, 0.6, and 0.7 nm, respectively.

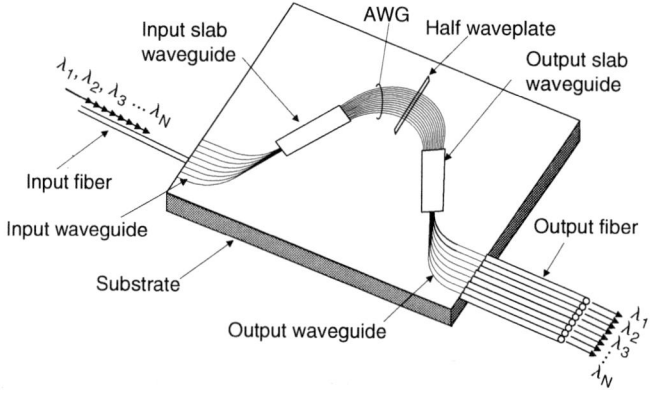

Figure 6   Arrayed waveguide grating multi/demultiplexer

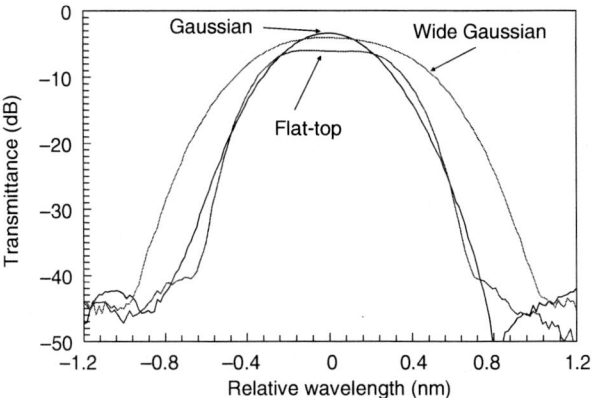

Figure 7   Typical spectral responses of three types of AWG

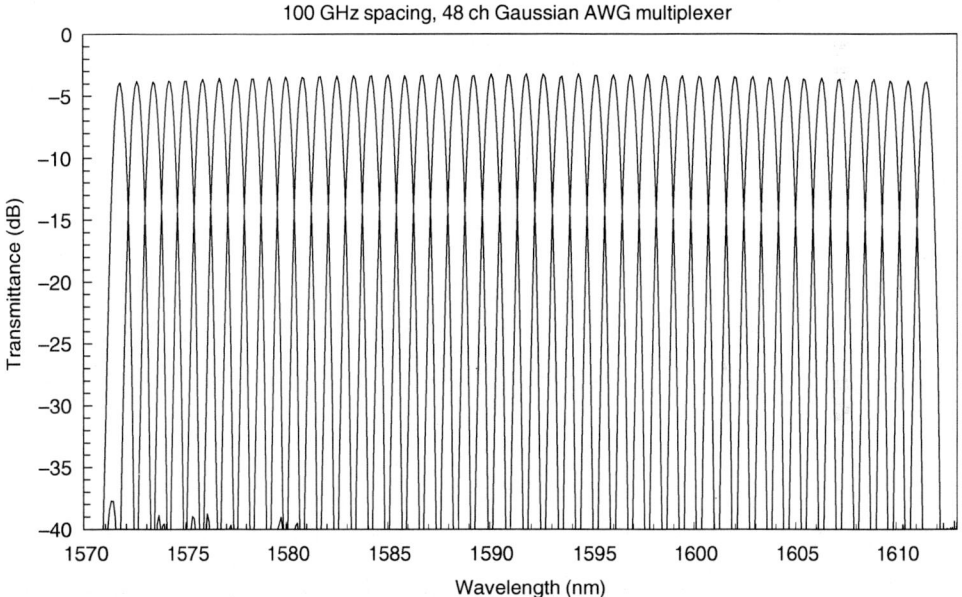

Figure 8   Typical transmission spectrum of AWG

The transmittance, bandwidth, and adjacent channel suppression have a trade-off relationship, and one of the three types or an intermediate design is chosen according to the system design. Figure 8 is a typical transmission spectrum for a 100 GHz spacing, 48-channel, Gaussian AWG. The insertion loss variation is very small (<1 dB) and all 48 channels have the same spectrum. In addition, the chromatic dispersion of the AWG is theoretically zero. These are the reasons why the AWG is used in ordinary DWDM systems that have 16, 32, or 40 wavelength channels with a 100 GHz spacing and a 10 Gb/sec bit rate.

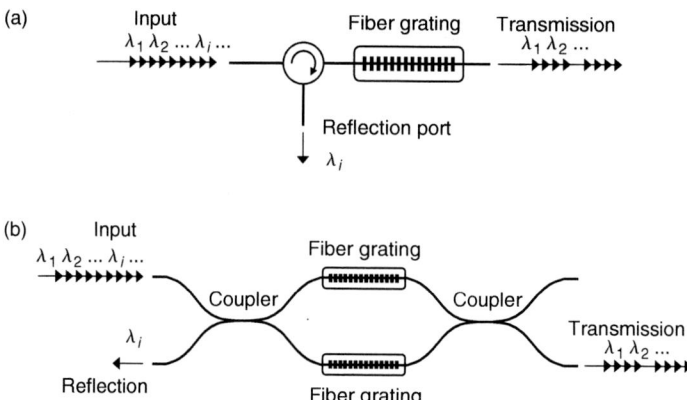

Figure 9   Two ways for separating reflection from fiber grating ([a] circulator type, [b] Mach–Zehnder type)

Other Technologies

A *fiber Bragg grating* that reflects its Bragg wavelength is also used as a multi/demultiplexer. The fabrication technique and characteristics are described in *Fiber Bragg Grating (FBG)* by K. Kikuchi. The chief characteristic is that the insertion loss is extremely low because the grating is formed in a conventional single-mode fiber and there is no excess loss such as the collimator loss in a thin film filter module and the fiber/waveguide coupling loss in an AWG module. By chirping the grating pitch, it is possible to obtain a very flat, wide reflection spectrum. One drawback is that it requires a circulator or a Mach-Zehnder structure to separate the reflected light from the input light as shown in Figure 9.

  Miniature spectrometers based on a diffraction grating are also candidates for multi/demultiplexers. They are divided into two types; one based on a planar waveguide and the other on bulk optics. With the first type, a reflection grating is formed in the slab waveguide by using photolithography, dry etching, and metal coating. The reflectivity and wavelength resolution strongly depend on the verticalness of the etched wall and the accuracy of the grating pitch, which are currently insufficient for realizing the low insertion loss and high wavelength resolution required for practical multi/demultiplexers. With the bulk optics type, a diffraction grating is available that has excellent characteristics, which make it suitable for multi/demultiplexers. However, in principle, this type needs a special optical output structure in which the spacing of the demultiplex-side fibers is narrow since the ratio between spacing and bandwidth in the wavelength channels is approximately the same as the ratio between the fiber spacing and the fiber core diameter. It is possible to realize this structure by arraying specially etched fibers with a small-outer diameter or by using a pitch converter made of a planar lightwave circuit. Unfortunately, these approaches increase both size and assembly cost.

  For the reasons mentioned above, thin film filters and AWGs are now commonly used in practical WDM systems and the application of the other technologies is limited to special cases.

REFERENCES

1. H. A. Macleod, *Thin Film Optical Filters*, 3rd ed. Institute of Physics Publishing, Philadelphia, PA, 2001.
2. R. B. Sargent, "Recent Advances in thin Film Filters," Technical Digest of Optical Fiber Communication, TuD6, 2004.

3. M. K. Smit, "New focusing and dispersive planar component based on an optical phased array," *Electron. Lett.* 24, 385–385, 1988.

4. H. Takahashi, S. Suzuki, K. Katoh, and I. Nishi, "Arrayed-waveguide grating for wavelength division multi/demultiplexer with nanometer resolution," *Electron. Lett.*, 26, 87–88, 1990.

5. C. Dragone, C. A. Edwards, and R. C. Kistler, "Integrated optics $N \times N$ multiplexer on silicon, *IEEE Photonics Technol. Lett.* 3, 896–899, 1991.

6. Y. Inoue, Y. Ohmori, M. Kawachi, S. Ando, T. Sawada, and H.Takahashi, "Polarization mode converter with polyimide half waveplate in silica-based planar lightwave circuits," *IEEE Photonics Technol. Lett.* 6, 626–628, 1994.

7. K. Okamoto and A. Sugita, "Flat spectral response arrayed-waveguide grating multiplexer with parabolic waveguide horns," *Electron. Lett.* 32, 1661–1662, 1996.

Y

Y-BRANCH

Yasuo Kokubun

A waveguide Y-Branch using single-mode waveguide can easily be designed and fabricated by a proper waveguide pattern with small branching angle. The branching angle must be smaller than the propagation angle of the fundamental mode. If the branching angle is small enough, the branching occurs adiabatically and the branching loss can be reduced to almost zero.

However, in the case of the combiner that can be obtained by reversing the input and output ends, the situation is different. A principle 3 dB loss occurs, if the light is incident on one of input ports of the combiner. This is understood by the superposition of in-phase and out-of-phase incidence as shown in Figure 1. When the light is incident on the upper port of the combiner, this is expressed by the superposition in-phase and out-of-phase incidence as shown in Figure 1 (a) and 1 (b). Since the power coupled to the out-of-phase component is radiated in the tapered portion of the combiner and the remaining in-phase component is guided into the single mode output port. Therefore, 3 dB loss always occurs. When the light is incident on both input ports, the loss depends on the phase difference between two incident light waves.

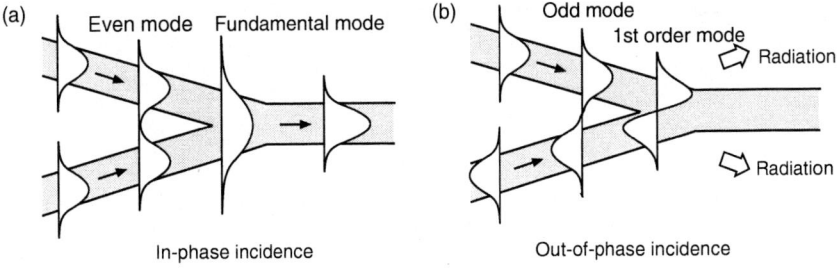

Figure 1 Waveguide combiner in which input and output ends are reversed from Y-branch. (a) In-phase incidence and (b) out-of-phase incidence.

REFERENCES

1. H. Yajima, "Coupled mode analysis of dielectric planar branching waveguides," *IEEE J. Quantum Electron.*, QE-14, 749–755, 1978.
2. S. K. Burns and A. F. Milton, "An analytic solution for mode coupling in optical waveguide branches," *IEEE J. Quantum Electron.*, QE-16, 446–454, 1980.

Contributors/Initial of Chapters

Shigeyuki Akiba / D
International Network Department
Network Engineering Division
KDDI Corporation
Tokyo
Japan

Y. Akulova / T
Agility Communications, Inc.
Santa Barbara
California
U.S.A.

Shigehisa Arai / D
Quantum Nanoelectronics Research Center
Tokyo Inst of Technology
Tokyo
Japan

Masahiro Asada / Q
Department of Electrical and
Electronic Engineering
Tokyo Inst of Technology
Tokyo
Japan

Toshihiko Baba / P
Department of Electrical and
Computer Engineering
Yokohama National University
Yokohama
Japan

J. S. Barton / T
Department of Electrical and
Computer Engineering
University of California
Santa Barbara
California
U.S.A.

Antao Chen / P
Applied Physics Laboratory
University of Washington
Seattle
Washington
U.S.A.

C. W. Coldren / T
Agility Communications, Inc.
Santa Barbara
California
U.S.A.

Larry A. Coldren / T
Department of Electrical and
Computer Engineering
University of California
Santa Barbara
California
U.S.A.

Urban Eriksson / T
Department of Microelectronics and
Information Technology
Royal Inst. of Technology (KTH)
Kista
Sweden

G. A. Fish / T
Agility Communications, Inc.
Santa Barbara
California
U.S.A.

Hiroyuki Fujita / M
Center for International Research on
Micromechatronics
Institute of Industrial Science (CIRMM)
University of Tokyo
Tokyo
Japan

Kazuhito Furuya / O, W
Quantum Nanoelectronics Research
Center
Tokyo Inst of Technology
Tokyo
Japan

Kenya Goto / O
Department of Information and
Communication Technology
School of High-Technology for
Human Welfare
Tokai University
Shizuoka
Japan

Harald Herrmann / A
Angewardte Physik
Universitat-GH Paderborn
Paderborn
Germany

Fred Hesimann / P
Consultant
Colts Neck
New Jersey
U.S.A.

Yoshiaki Hibino / T
NTT Telecommunications Energy
Laboratories
Atsugi
Japan

Kenichi Iga / L, M, O, S, V
Japan Society for the Promotion of
Science (JSPS)
Tokyo
Japan

Kyo Inoue / O
Department of Electronics
Osaka University
Osaka
Japan

Yasushi Inouye / N
Graduate School of Frontier Biosciences
Osaka University
Osaka
Japan

Stefan Irmscher / T
Department of Microelectronics and
Information Technology
Royal Inst. of Technology (KTH)
Kista
Sweden

Masayuki Izutsu / L
The National Institute of Information and
Communications Technology (NICT)
Tokyo
Japan

L. Johansson / T
Department of Electrical and
Computer Engineering
University of California
Santa Barbara
California
U.S.A.

Satoshi Kawata / N
Department of Information and
Physical Sciences
Osaka University
Osaka
Japan

Kazuro Kikuchi / F, T
Research Center for Advanced Science and
Technology (RCAST)
University of Tokyo
Bunkyo-ku
Tokyo
Japan

Katsumi Kishino / I
Sophia University
Tokyo
Japan

Kohroh Kobayashi / T
P&I Laboratory
Tokyo Inst of Technology
Yokohama
Japan

Yasuo Kokubun / A, D, L, M, O, Y
Department of Electrical and
Computer Engineering
Yokohama National University
Yokohama
Japan

Masnori Koshiba / W
Division of Electronics and
Information Engineering
Hokkaido University
Sapporo
Japan

Fumio Koyama / F, M
P&I Laboratory
Tokyo Inst of Technology
Yokohama
Japan

Kazuo Kuroda / F
Institute of Industrial Science
Department of Material and Life
University of Tokyo
Tokyo
Japan

Robert Lewén / T
Department of Microelectronics and
Information Technology
Royal Inst of Technology (KTH)
Kista
Sweden

Christi K. Madsen / O
Department of Electrical Engineering
Texas A&M University
Zachry Engineering Center
College Station
Texas
U.S.A.

Tohru Maruno / P
R&D Vision Department III
NTT Corporation
Tokyo
Japan

Tetsuya Mizumoto
Department of Electrical and
Electronic Engineering
Tokyo Inst of Technology
Tokyo
Japan

Ken Morito / S
Optical Semiconductor Devices
Laboratory
Fujitsu Laboratories Ltd.
Atsugi
Japan

Masataka Nakazawa / E
Coherent-Wave Department
Research Institute of Eectrical
Communication
Tohoku University
Sendai
Japan

Shu Namiki / R
National Institute of Advanced Industrial
Science and Technology (AIST)
Tsukuba
Japan

Renshi Sawada / O
Department of Intelligent Machinery
and Systems
Kyushu University
Fukuoka
Japan

Yasuharu Suematsu / I
National Institute of Informatics
Tokyo
Japan

Toshiaki Suhara / P
Department of Electronics
Osaka University
Osaka
Japan

Hong-Bo Sun / N
RIKEN
Saitama
Japan

Hiroshi Takahashi / A, W
Hyper-photonic Component Laboratory
NTT Photonics Laboratories
Atsugi
Japan

Lars Thylén / T
Department of Microelectronics and
Information Technology
Royal Inst. of Technology (KTH)
Kista
Sweden

Chen Tsai / R
Electrical Engineering and
Computer Science
School of Engineering
University of California
Irvine
California
U.S.A.

Hiroyuki Uenohara / W
P&I Laboratory
Tokyo Inst of Technology
Yokohama
Japan

Katsuyuki Utaka / M
Department of Electrical Engineering and
Bioscience
Waseda University
Tokyo
Japan

Osamu Wada / O
Department of Electrical and
Electronics Engineering
Kobe University
Kobe
Japan

Urban Westergren / T
Department of Microelectronics and
Information Technology
Royal Inst. of Technology (KTH)
Kista
Sweden

Kazuhisa Yamamoto / S
Storage Media Systems Development Center
Matsushita Electric Industrial Co., Ltd.
Osaka
Japan

Yoshihisa Yamamoto / S
Edward L. Ginzton Laboratory
Stanford University
Stanford
California
U.S.A.

Index